International Conference on Civil and Environmental Engineering Technologies (ICCEET 2019)

IOP Conference Series: Materials Science and Engineering
Volume 584

Najaf, Iraq
23 – 24 April 2019

ISBN: 978-1-5108-9457-0
ISSN: 1757-8981

Printed from e-media with permission by:

Curran Associates, Inc.
57 Morehouse Lane
Red Hook, NY 12571

Some format issues inherent in the e-media version may also appear in this print version.

Institute of Physics
Dirac House, Temple Back
Bristol BS1 6BE UK

Phone: 44 1 17 929 7481
Fax: 44 1 17 920 0979

techtracking@iop.org

Additional copies of this publication are available from:

Curran Associates, Inc.
57 Morehouse Lane
Red Hook, NY 12571 USA
Phone: 845-758-0400
Fax: 845-758-2633
Email: curran@proceedings.com
Web: www.proceedings.com

Table of contents

Volume 584

International Conference on Civil and Environmental Engineering Technologies

23–24 April 2019, Najaf, Iraq

Accepted papers received: 3 July 2019
Published online: 15 August 2019

Preface

Editorial material

Peer review statement

Papers

Bond the Gap between Academic and Industry Requirements for Undergraduate Civil Engineering Students in Iraq

Zaher Mohammed Naji, Ammar Dakhil and Samer Al Jasim.....1

Factors effecting on the compressive and tensile strength of reactive powder concrete made with local materials

A H Chkheiwer and J A Kadim.....12

The Combined Effect of Ferric Chloride and Cationic Poly-Electrolyte CPE on Coagulation of Grey Wastewater

Huda Rida Hashim and Ali Jwied Jaeel.....23

The Evaluation of Khorasan Razavi Range for the construction of wind power plants through multi-criteria decision making in GIS environment

A Sh Alhameedah, R Shad and M Ghaemi.....33

A vision of sustainable & smart knowledge based urban development in Iraqi cities: adopting GIS to determine knowledge capital elements and strategies in the city of Hila

Rashaa M Al Khafaji.....46

Influence of Curing Conditions and Filler Type on Mechanical Properties of Cold Asphalt Emulsion Mixtures

Mohammed Abbas Al-Jumaili and Ahmad Saleem Abdulateef Shakoree.....56

Numerical analyses on seismic behavior of helical concrete columns

Alaa M Al-habbobi.....65

Experimental Investigation of High Strength Precast Reinforced Concrete Walls used (Vierendeel Truss Form)

Mustafa AbdulHussein Hassan, Hadi Nasir and Jamal Saeed.....75

The Role of Cooling Degree Days on determining the insulation of building Envelop in a hot climate (Iraqi cities as an example)

Susan Abed Hassan.....86

Effect of Nano-Silica and Micro Steel Fiber on Compressive Strength Development of Fly Ash Geopolymer Paste Cured Under Ambient Temperature

Noor S. Hadi, Wissam D. Salman and S Safie Mahdi Oleiwi.....95

Evaluation of vertical drain with different materials

Ali Akhtarpour, Mohammed Sh Mahmood and Ameer Alali.....102

Improvement of Multi-Lane Airport Highway in Baghdad-Iraq

Basim Jrew and Ahmed Hadi.....112

Punching shear behaviour of solid and bubble reinforced light Weight aggregate concrete two-way slabs

Maha habeeb, Adel A Al-Azzawi and faiq Al- Zwainy.....124

Investigation of Binder Course Cold Asphalt Emulsion Mixture Properties Containing Cement and GGBS

Ammar A Shalan, Nibras A Al-Sahaf and Abbas Al-Hdabi.....137

Structural Behavior of Eccentrically Loaded R.C. Columns under Fire Temperature Loading

A H Kadhim, M M Rasheed and H K Jarallah.....149

Turbidity, Color and Chemical Oxygen Demand Removals from Synthetic Textile Wastewater Using Chitosan as a Coagulant

Hala A Abdullah and Ali J Jaeel.....164

Effect of Embedment Depth on Raft Foundation Settlement Under Seismic Load

Abeer A A Hanash, Mahmoud D Ahmed and Abdulmotalib I Said.....171

A Conflict Index to Assess Traffic Safety at Intersections

Raghad Zeki Abdul Majeed and Hussein A. Ewadh.....181

Asphalt damage reservation due to self-repair: A literature review

T Al-Mansoori, J Hussein and H Aziz.....189

Effect of Hybrid Polypropylene - Carbon Fiber on Self-Compacting Mortar

Muataz Ibrahim Ali, Asmaa Abdul Jabbar Jamel, Ahmed Mohammed Ahmad and Anwer Muneer Ali.....196

Swelling Control of Expansive Soils Using Cement Dust

Alaa D. Almurshedi, Jawad K. Thajeel and Hadeel Ch. Dekhn.....207

Investigating the influence of cement replacement by high volume of GGBS and PFA on the mechanical performance of cement mortar

Ali Abdulhussein Shubbar, Ali Al-Shaer, Rasha Salah AlKizwini, Khalid Hashim, Hayder Al Hawesah and Monower Sadique.....216

Solid Waste Situation in Thi-Qar Governorate

Saleem Ethaib.....223

Decolourization of dye solutions by electrocoagulation: an investigation of the effect of operational parameters

Khalid S. Hashim, N H Al-Saati, Saif Salah Alquzweeni, Salah L. Zubaidi, Patryk Kot, L. Kraidi, Ameer H. Hussein, Rafid Alkhaddar, Andy Shaw and Reham Alwash.....232

Development of Risk Optimisation Model for Oil and Gas Transportation Pipeline Routes

L Kraidi, R Shah, W Matipa, F Borthwick and Khalid Hashim.....240

Using LARS –WG model for prediction of temperature in Columbia City, USA

Salah L. Zubaidi, Patryk Kot, Khalid Hashim, Rafid Alkhaddar, Mawada Abdellatif and Yousif Raad Muhsin.....248

Effect of void Ratio of Inner Steel Tube on Compression Behavior of Double Skin Tubular Column

Hanaa A. Ali and Wissam D. Salman.....257

Effect of Wasit Thermal Power Plant on Water Quality of Tigris River Downstream to Al Zubaidiayh City

Hiba Alaa and Ali Jwied Jaeel.....266

Water quality index of Tigris River on Waist Governorate for aquatic life

Huda Yousif Enad and Ali Jwied Jaeel.....274

Effect of increasing floors number on the natural frequency compared with further masses added to steel building models

Ahmed A. Alalikhan, Mohsen A. Habelalmateen and Haider A. Al-Tameemi.....281

Recycling of disposal Polypropylene Blister Tablets and Strapping ties as fiber reinforcement for Pervious concrete

Ahmed M Ahmed, Ahmed H Hussein and Mohammed T Hammood.....289

Influence of using alum sludge as an aggregate on hot mix Asphalt

Sahar A. Ali Dawood, Hayder Abbas Obaid, Tameem Mohammed Hashim and Hadeel Kareem Jasim.....298

Enhancement of Rigid Pavement Capacity Using Synthetic Discrete Fibers

Ahmed Alsabbagh, Salam Wtaife, Alaa Shaban, Nakin Suksawang and Emad Alshammari.....307

Improvement Properties of Self -Healing Concrete by Using Bacteria

Z M Hussein, A H Abedali and A S Ahmead.....315

Influence of current density and electrodes spacing on reactive red 120 dye removal from dyed water using electrocoagulation/electroflotation (EC/EF) process

B A Abdulhadi, P Kot, K S Hashim, A Shaw and R Al Khaddar.....325

Chitosan as a Widely Used Coagulant to Reduce Turbidity and Color of Model Textile Wastewater Containing an Anionic Dye (Acid Blue)

Hala A Abdullah and Ali J Jaeel.....332

Comparison of Compressibility Behaviour of Lime-Cement Stabilized Lateritic Clay Soil Contaminated by Heavy Metals

K A Saeed, K A Kassim, H Nur and N Z M Yunus.....339

Dissipation of Energy of Flow by Conventional Type of Gabion Weir

A M Al- Fawzy and F M Al- Mohammed.....349

Experimental analysis of static capacity of fibrous concrete cambered slabs after impact loading effect

Al Kulabi Ahmed Kamil and Al Zahid Ali Adnan.....358

The behavior of Lightweight Aggregate Concrete Made with Different Types of Crushed Bricks

H Adem, E Athab, S Thamer and AT Jasim.....370

Hot Asphalt Mixture Characteristics with Nano-Metakaolin Materials

Mustafa A Kareem, Abbas Al-Hdabi, Nibras A Al-Sahaf and Laith F Hasan.....379

The Effect of the Shear Connectors Layout on the Sustainable Lightweight Concrete Sandwich

Laith Sh. Rasheed and Suad Abbas Ali.....390

Tire Rubber Waste for Improving Gypseous Soil

L N Snodi and I S Hussein.....400

The Behaviour of Reinforced Concrete Bridge Piers Using Different Types of Concrete Mix

Ali H Al-Mamoori, Wajde S Alyhya and Hajir A Al-Hussainy.....406

Performance of piled raft foundation supported by either connected or unconnected piles

Athraa M J Alhassani and Ala Nasir Aljorany.....415

Intelligent non-destructive technique for crack existence inspection in a structure using Particle Swarm Optimization (PSO)

Saad Al-Wazni and Ahmed Y Azainul-abideen.....423

Experimental Investigation of Concrete Columns Enhanced by PVC Tubes

Murtaza Allawi Raheemah and Sa'ad Fahad Resan.....432

Mechanical Properties of Lightweight Aggregate Moderate Strength Concrete reinforcement with Hybrid Fibers

Haider Araby Ibrahim, Mohammed Bally Mahdi and Basim Jabbar Abbas.....442

Evaluation of the Performance of MBR-RO Technology for Treatment of Textile Wastewater and Reuse

S H Zahraa and H A Gzar.....453

Numerical analysis of thin ring foundations under different loading conditions

Adel A Al-Azzawi and Khalida A Daud.....462

Properties of different types of concrete containing waste tires rubber- a review

A S Ali and T M Hasan.....470

Experimental study on torsional behavior of fiberious reifoced concrete beams with different concrete strength

Waleed A. Waryosh, Saad Khalaf mohaisen and Raed Hamed Dkhel.....479

Evaluation of the Dynamic Behavior of Green Concrete under Impact Load by Using CEB-FIP Code

Zainab H. Abdulabbas, Jumana H. Sahib and Nabaa N. Bashboosh.....488

Hydraulic conductivity estimation by using groundwater modelling system program for upper zone of Iraqi aquifers

T S Khayyun and H H Mahdi.....496

Experimental investigation for the behavior of concrete slab rests on sandy soil contains cavities

Muhammad Abed Attiya, Laith Jawad Aziz and Ola Mazen Makki.....507

Bearing capacity and settlement of shallow footings on liquefiable soil subjected to seismic loading

Ahmed M. Sadeq Al-Salakh and Bushra Suhale Albusoda.....517

Assessment of Soil Tunnel Interaction in Sand Soil

Omar Yaseen Almashhadany and Bushra Suhale Albusoda.....527

Seismic Behavior of Composite Simply Supported Bridge Decks supported on Elastomeric Bearings

Sara k AL Rubaee and Ammar A Abdul Rahman.....537

Effect of Partial Replacement with Crushed Concrete on Settlement and Bearing Capacity of Soft clay

Maki Jafar Mohammed Al-Waily.....546

An experimental study of flow over V- shape crump weir crest

Hasan Mahdi Mohammed Al-Khateeb, Jumana Hadi Sahib and Huda Hadi Hasan Al-Yasisri.....555

The effect on the interfacial strength of the length/width ratio of FRP sheet bonded to concrete

W. Al-Juboori and L. Weekesb.....566

Evaluation Al-Safeena and Sayed Jawda Intersections in Kerbala City

Huda M. Al-Murshidy, Hussein A. Ewadh and Raid R. A. Almuhanna.....580

Experimental study for strengthening octagonal castellated steel beams using circular and octagonal ring stiffeners

H W Al-Thabhawee and A Mohammed.....594

IOP Conf. Series: Materials Science and Engineering **584** (2019) 011001 doi:10.1088/1757-899X/584/1/011001

Editorial material

1. Introduction

The title of the conference: International Conference on Civil and Environmental Engineering Technologies

The date of the conference: 23-24th April 2019

The location of the conference: University of Kufa, Al-Najaf, Iraq

Abbreviation of title of the conference: ICCEET 2019

ICCEET was established at the University of Kufa with the aim of promoting the acquisition and transfer of civil and environmental engineering technologies for advancing research, design, and construction practices that can be utilised for improving the quality of life for all people in our territory. ICCEET is going to bring people from different marbles in civil & environmental engineering and merge them in one forum. In ICCEET, you'll find a wide range of disciplines to include all fields of civil and environmental engineering including construction, modern GIS, geotechnical, structural, environmental, water resources, and transportation where it hoped that useful discussions and collaborations can be instigated.

2. Scientific committee members

1. Prof. Rafid Al Khaddar, Liverpool john Moores University, UK, r.m.alkhaddar@ljmu.ac.uk
2. Prof. Louay N. Mohammad, Louisiana State University, USA, louaym@lsu.edu
3. Prof. Mohammed Al-Jumaili, University of Kufa, Iraq, mohammedah.aljumaili@uokufa.edu.iq
4. Prof Mohammad Kabir, Amirkabir University of Technology, Iran, mzkabir@aut.ac.ir
5. Prof Ali Khodaii, Amirkabir University of Technology, Iran, akhodaii@gmail.com
6. Prof. Basim Jrew, Isra University, Jordan, basim.jrew@iu.edu.jo

7. Prof. Mohammed Yousif Fattah, University of Technology, Iraq, myf_1968@yahoo.com
8. Prof. Zainab Ziad Ismail, University of Baghdad, Iraq, zismail9@gmail.com
9. Prof. Haitham H. Muteb, University of Babylon, Iraq, dr_haitham_67@yahoo.com
10. Dr. Monower Sadique, Liverpool John Moores University, UK, M.M.Sadique@ljmu.ac.uk
11. Prof. Tareq Salih Al-Attar, University of Technology, Iraq, thshkhma@yahoo.com
12. Prof. Saleh Issa Khassaf, Basrah University, Iraq, Salehissakh@gmail.com
13. Asst. Prof. Ali Talib Albozwain, University of Kufa, Iraq, alit.albozwaida@uokufa.edu.iq
14. Asst. Prof. Ali Naji Diebil, University of Kufa, Iraq, alin.diebil@uokufa.edu.iq
15. Dr. Khalid Hashim, Liverpool John Moores University, UK, K.S.Hashim@ljmu.ac.uk
16. Asst. Prof. Hasan Mahdi Al-Khateeb, University of Kufa, Iraq, hasan.alkhateeb@uokufa.edu.iq
17. Asst. Prof. Ali Akhtarpour, Ferdowsi University of Mashhad, Iran, akhtarpour@um.ac.ir
18. Asst. Prof. Husain Ali Al-Aboodi, University of Kufa, Iraq, hussein.alabboodi@uokufa.edu.iq
19. Dr. Alaa Abd Ali, University of Kufa, Iraq, alaa.abdali@rutgers.edu

IOP Conf. Series: Materials Science and Engineering **584** (2019) 011001 doi:10.1088/1757-899X/584/1/011001

2019

ICCEET

INTERNATIONAL CONFERENCE ON CIVIL AND ENVIRONMENTAL ENGINEERING TECHNOLOGIES

Peer review statement

All papers published in this volume of *IOP Conference Series: Materials Science and Engineering* have been peer reviewed through processes administered by the proceedings Editors. Reviews were conducted by expert referees to the professional and scientific standards expected of a proceedings journal published by IOP Publishing.

IOP Conf. Series: Materials Science and Engineering **584** (2019) 012001 doi:10.1088/1757-899X/584/1/012001

Bond the Gap between Academic and Industry Requirements for Undergraduate Civil Engineering Students in Iraq

Zaher Mohammed Naji[1], **Ammar Dakhil**[1*], **Samer Al Jasim**[1]
[1] Department of Civil Engineering, College of Engineering, University of Basrah, Iraq

*Corresponding Author email: ammarasha@yahoo.com

Abstract. In the last twenty years, civil engineering industry has been improved significantly due to the expanded use of technology and innovations in project management. This progress represented as a challenge to universities to provide competence engineers with the right capabilities to be fit in the new job requirements and able to lead this industry to the new era. As this problem considered a global issue, it has a particular situation in Iraq. This resulted from the sharp change in the Iraqi construction industry after recent political change in 2003. This study aims to investigate the current industry expectation from graduated civil engineers and compare it with actual competences that engineers learned from their study in Iraqi universities. This research implements both a qualitative and quantitative approach to collect valuable and reliable data. This has been achieved by sending an online questionnaire to senior engineers working for both public and private organizations. In addition, another set of questions send to students to evaluate their knowledge to identify the gap. This study identified a considerable gap between current civil engineering education and industry expectation, which mainly resulted in project management skills. This gap needs to be addressed in the near future to make sure that Iraqi civil engineers maintain their chances to get hired from foreign companies and be able to lead the Iraqi construction industry to the global market.
Keywords: Civil, engineering, education, industry, knowledge

1. Introduction

Progressively, Iraqi engineers are facing several kinds of challenges including speedy technological innovation, foreign companies with different responsibilities and the needs to compete globally [1-2]. Therefore, industry, Local authorities and even the central government announced to universities lecturers to review their traditional way of teaching and start to prepare civil engineering students by improving not only technical competencies but also include professional capabilities [3-5]. This trend is a global issue, where most of the recent researches try to identify those competencies that may include: communication skills, business skills, teamwork skills, creativity, lifelong learning skills, and problem-solving skills [3].

Despite the importance of this topic, there are few research and case studies which try to highlight this problem [6-8]. To emphasize this current situation, [9] stated that in 2004 almost a quarter of employers in the USA informed that engineering alumni were less capable in problem-solving and fewer responsive of organizational contexts and restrictions compare to former students ten years earlier. A third of the university lecturers who educated these students correspondingly described that they were not have sufficient capability in math and science and had not as good as technical skills than their indications [10]. In addition, other research also states that numerous engineering students keep on unreliable about what engineering is and what engineers do, even by the time they graduate [11].

Finally, it very important to explain to students what they may expect in their future career otherwise they will feel that they not prepared to be a civil engineer, which may influence some graduates to abandon civil engineering altogether. The lack of research in this particle field can result in a huge gap between academic requirements that every civil engineering student need to meet to graduate and the actual industry skill requirements. To illustrate this problem, in the USA where the education system has improved significantly, National Science Foundation stated that two years after graduation, 28% of all engineering graduates from the classes of 2008 and 2009 were working in fields not related to engineering, including mathematics and computer science. Therefore, this study aims to identify the current gap between academic and industry requirement for civil engineering students in Iraq. This will start by identifying this requirement first and then evaluate final year civil engineering students just before their graduation ceremony.

2. Academic Requirements

According to the Ministry of higher education and research in Iraq, the civil engineering students must study for four years and at least get 50 of 100 to pass each year and graduate. During these four years, different kind of topics will demonstrate starting from basics in the first two years to the most advanced ones in the fourth year. By investigating the available sources on the internet, which include Iraqi Universities websites, it can be concluded that to get civil engineering bachelor in Iraq you must have at least basic knowledge regarding the topics that listed in the Table (1).

Table 1. Fundamental Academic skills and knowledge for civil engineering students in Iraq.

No	Knowledge and Skills	No	Knowledge and Skills
1	Construction material	9	Math
2	Soil mechanics	10	Steel structure
3	Concrete mixture	11	Estimation
4	Structural analysis and design	12	Surveying
5	Foundation design	13	Traffic management
6	Method of construction	14	Project management
7	Fluid mechanics	15	Water and sewage system
8	Irrigation system	16	Hydraulics

3. Industry Skills

The recent significant developments and innovations in the construction industry produce huge pressure on graduate engineers to be qualified for the new job. In other word, companies start to enhance the requirements list for their new civil engineer vacancies with skills and knowledge beyond just academically needs. Therefore, fresh civil engineers are currently facing challenges, these challenges can be summarised in the list below [12-13]:

1- They never heard about these requirements in their academic journey, which can produce frustrated feelings.
2- There is a lack of professional bodies that can support them to learn such skills and knowledge inside Iraq.
3- This situation can create reluctance of engineers for foreign companies that can increase the number of unemployed numbers of engineers.

For the aforementioned reasons, it is very important to find out what these requirements are and how the university as significant knowledge provider can help to solve the problems and reduce the current gap between the industry and academia.

Profound investigation through literature has been implemented to identify the current industry skills and knowledge that a new civil engineer must have to be qualified for new jobs in an international company. Twenty-two areas of knowledge and skills have been recognized, each area have described briefly in Table (2) [14-20]. Most of aforementioned references' data published by The National Bureau of Economic Research.

Table 2. List of new skills and knowledge required by industry for civil engineers.

No	Knowledge and Skills	Description
1	Business knowledge	Awareness of the organization's business needs
2	Communication skills	Ability to express oneself effectively, in written/oral reports and when working with different groups of people
3	Content knowledge	Understanding of technical content, including engineering, science, and math
4	Context knowledge	Awareness of contextual issues affecting engineering solutions, e.g., safety, finances
5	Creativity skills	Ability to develop original ideas or solutions
6	Documentation skills	Ability to take good notes and keep track of records
7	Equipment/processes knowledge	Understanding of equipment and processes needed to do a job
8	Hands-on skills	Ability to make or build an object using tools/processes
9	Information finding skills	Ability to locate resources and information
10	Leadership skills	Ability to provide direction for a team or project
11	Logic skills	Ability to draw conclusions using reasoning
12	Organizational culture/ background knowledge	Understanding the organization's culture and background
13	Organizational hierarchy knowledge	Understanding the organization's labor hierarchy
14	Organizational policies/ procedures knowledge	Understanding the organization's policies and procedures
15	Problem-solving skills	Ability to define and solve engineering problems
16	Programming skills	Ability to using analysis and design programme
17	Project management skills	Ability to plan, organize, and manage project resources
18	Self-directed learning skills	Ability to acquire new knowledge and skills on one's own
19	Self-motivation skills	Ability to focus one's energy and effort towards doing one's work, without influence from other people
20	Teamwork skills	Ability to function on teams
21	Time management skills	Ability to manage one's time and meet deliverables in a timely manner
22	Work ethic	Willingness to keep working until the task is finished

4. Methodology

Questionnaires are frequently used techniques in collecting survey data, often numerical, which tends to be easy to observe [21]. Structured questionnaires have been used in this research to collect the data, which is basically to evaluate the current nearly graduate civil engineering students against industry professional expectations regarding different academic and industry skills and knowledge. Each student has been asked to evaluate his knowledge in term of six levels. Level 0 means not exist, level 1 basic knowledge, level 3 intermediate knowledge, level 5 advance knowledge. Two additional levels have been added between 1, 3, and 5 to give the responder more flexibility to evaluate himself. Civil engineering students at the University of Basra considered as the main sample for this research. In addition, these questionnaires sent to a number of professionals in both public and private organizations using Google Survey Tools. These professionals currently in a position to form the requirements for new civil engineering vacancies. SPSS software has been used to evaluate the responses which not fully complete answers will be neglected.

5. Results

As mentioned previously, our sample fundamentally divided into two main parts. Firstly, the professionals which will be asked to express their expectations from new civil engineers in term of the academic and industrial area of knowledge and skills. Secondly, civil engineering students that in their final year and they will graduate in a few days. They will be requested to assess themselves according to the same criteria that sent previously to the professionals. 90% of the responders did answer all the questions. However, the remaining 10% was neglected where it is not practical to involve them in the analysis.

Before starting to examine the results, it is important to identify their workplace type and its implication on the requirements. From Figure (1), it can be seen that 58% of the responders are working for private companies and the rest work for public companies. This will help to examine the company type on their requirements. To illustrate the gap between the industry expectations and academic current knowledge, the resulted discussed in three subsections as follows:

Figure 1. Professional's workplace type.

5.1. Industry knowledge

Before discussing the results, it will be important to identify the differences between skills and knowledge. Knowledge states the level of learning and understanding concepts, principles and information regarding a particular subject(s) by a person through different sources books, media, encyclopedias, academic institutions, and other sources. Skill represents the ability to use that information and applying it in a context. Therefore, a different area of knowledge required by the industry will be discussed first.

Figure (2) represents that generally new civil engineers are falling behind in term of required industry knowledge. This phenomenon can be easy identified in context, content, and business knowledge. The situation is less critical with organizational hierarchy and culture where civil engineers have quite a good level of knowledge regarding these areas. Industry professional with more 60% state that equipment and process knowledge are important and civil engineers must have at least level 2 of proficiency or more, the current level of this area shows that 50% of the students have no idea about this area or just basic concepts. Only work ethics knowledge area shows a quite good balance between industry requirement and current civil engineers level.

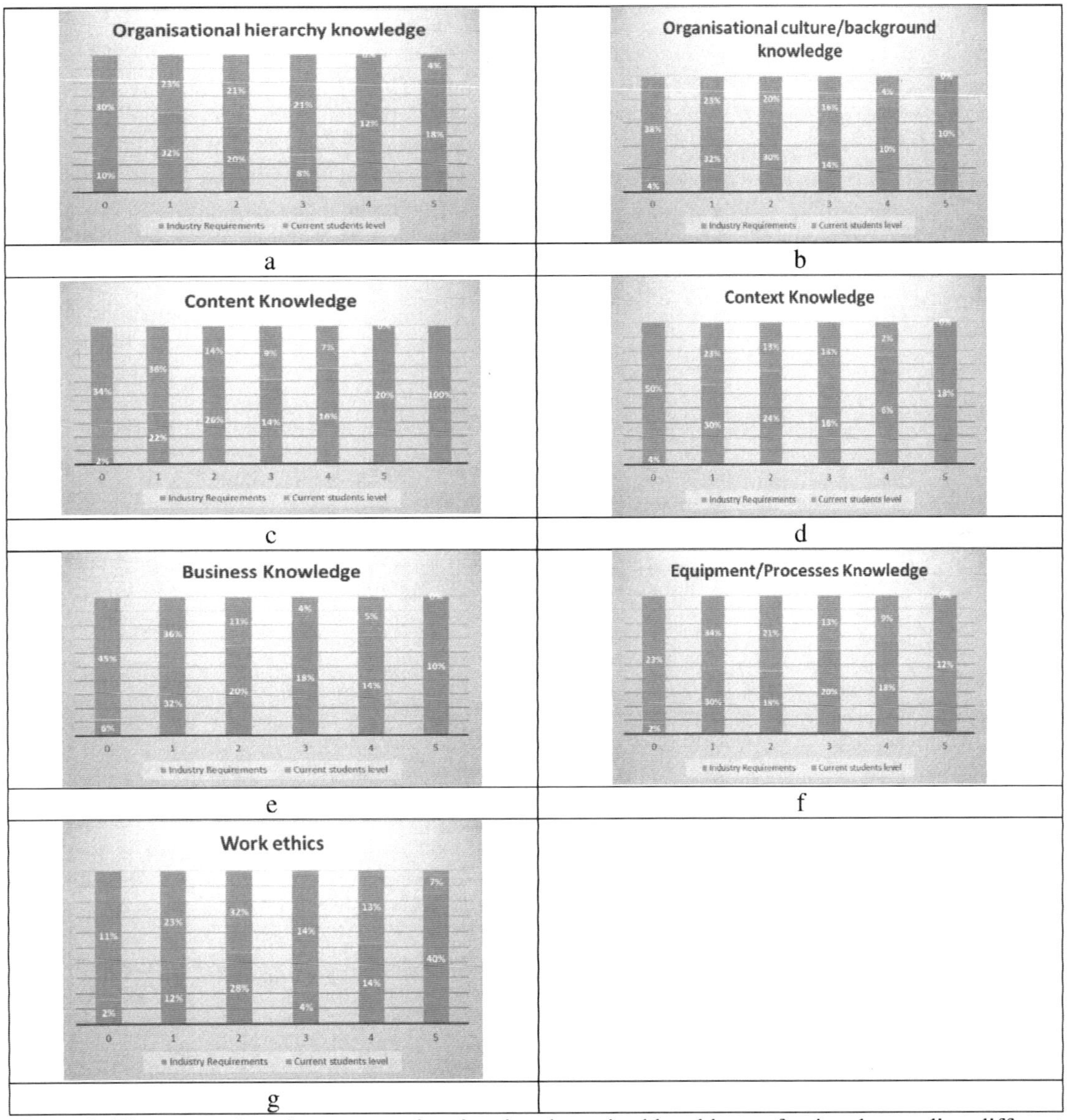

Figure 2. The comparison between student level and required level by professional regarding different areas of knowledge.

5.2. Industry skills

As stated before, skills mean the ability to use knowledge. From Figure (3), it can be concluded that commonly student level is also less than the industry expectation, which requires improve the current curriculums to overcome these gaps.

It can be seen that the current situation of time management, teamwork, documentation, and self-motivation skills, current civil engineering students are meeting the industry opportunities. The main

IOP Conf. Series: Materials Science and Engineering **584** (2019) 012001 doi:10.1088/1757-899X/584/1/012001

reason beyond that is the way of engineering currently studying that required from them to manage their time and work with groups to perform some task such as lab experiments.

The situation of the other skills different where the student currently is not good enough to perform industry necessities such as Information management, logic, project management, and self-directed learning, hand on, communication, and creativity skills. Despite project management is currently part of academic study, the students' level of understanding still not mature enough which require to improve the practical implication of PM knowledge and implemented as apart of study. The way of teaching and evaluating students currently in Iraqi universities prevent skills to be developed such as logic, creativity, and self-directed learning by not giving students any chance to create new ideas or change the way of giving the information. In addition, even with students have been taught programming and information management skills practically in the academic period however they still beyond industry potentials. The main reason for this gap may occur because of unqualified lecturers that are teaching these skills are not a specialist in programming or them just civil engineers, which prevent any advance development in these skills. Communication and hands-on skills are facing the same situation, which really they need to be addressed, or at least introduced to students through their academic study.

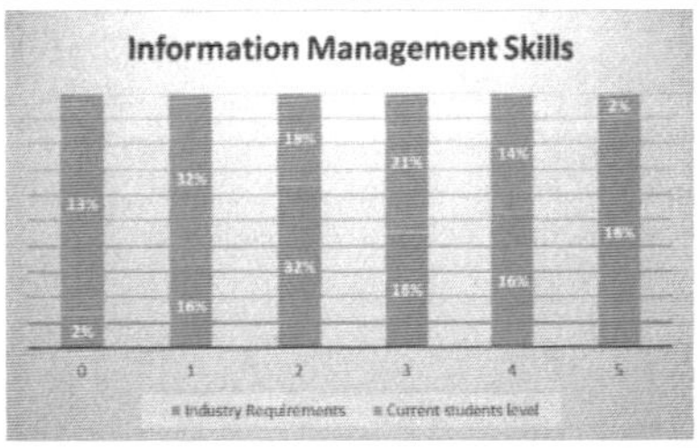

a

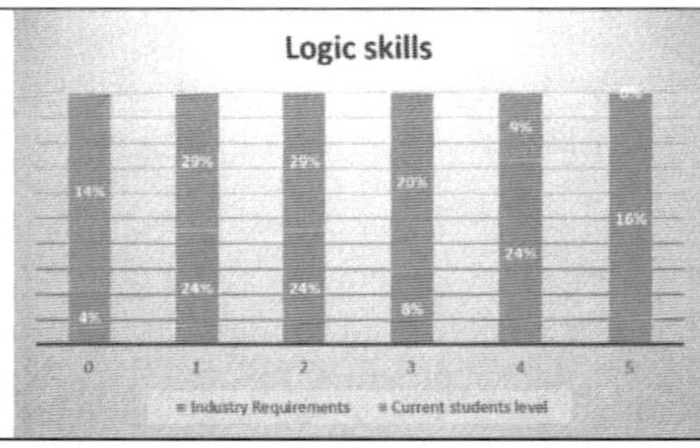

b

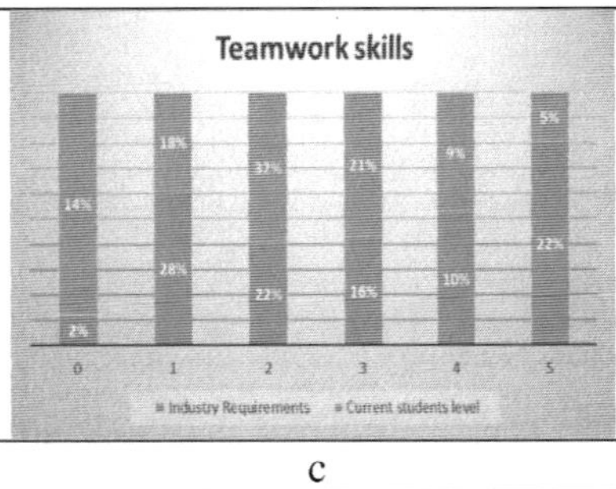

c

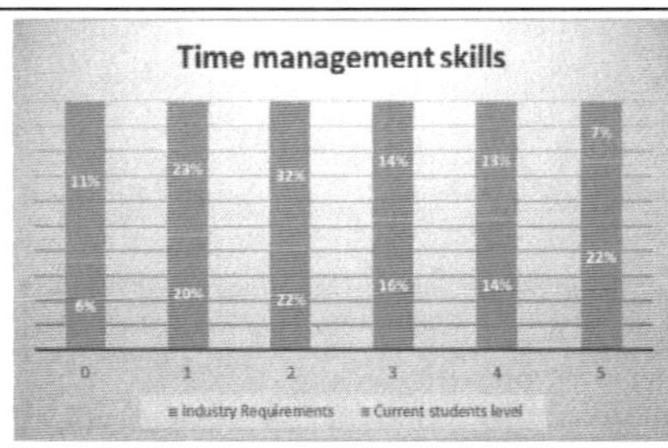

d

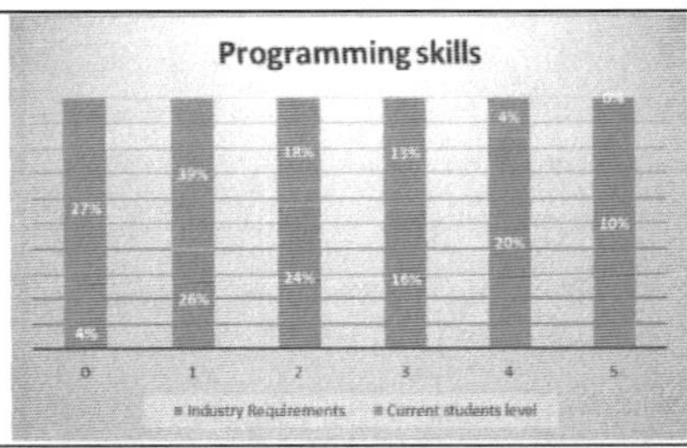

e

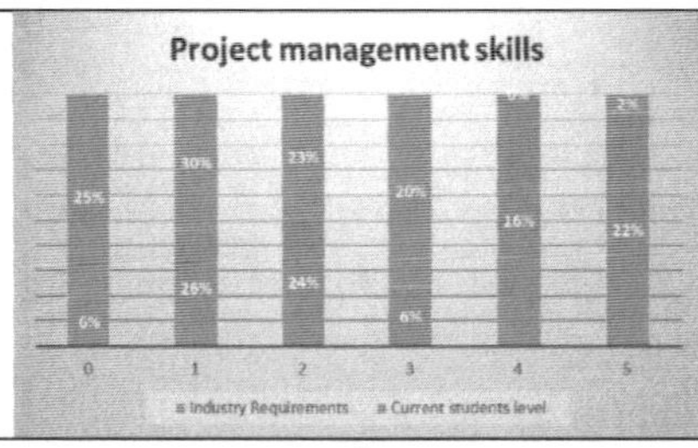

f

g

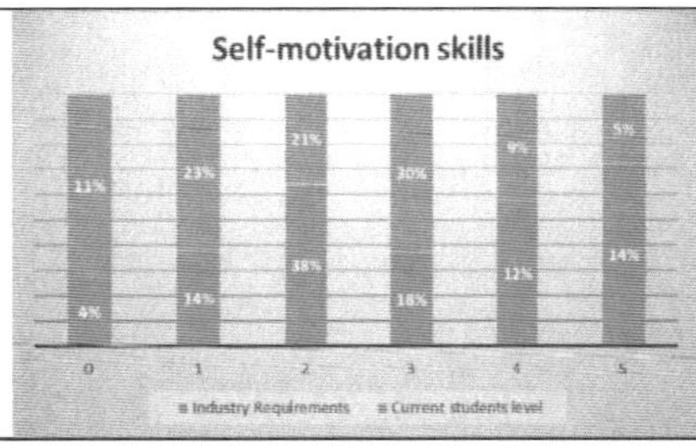

h

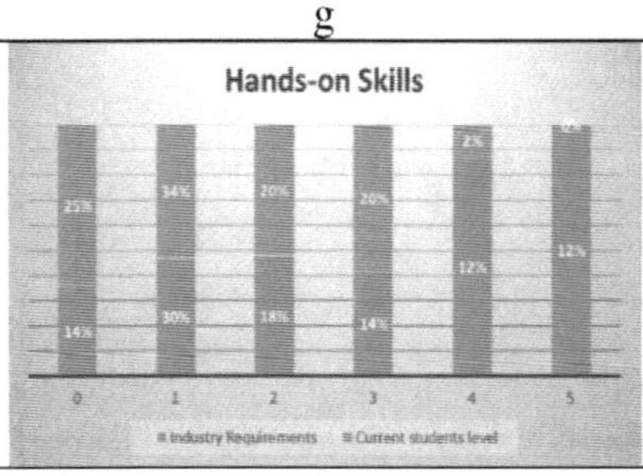

i

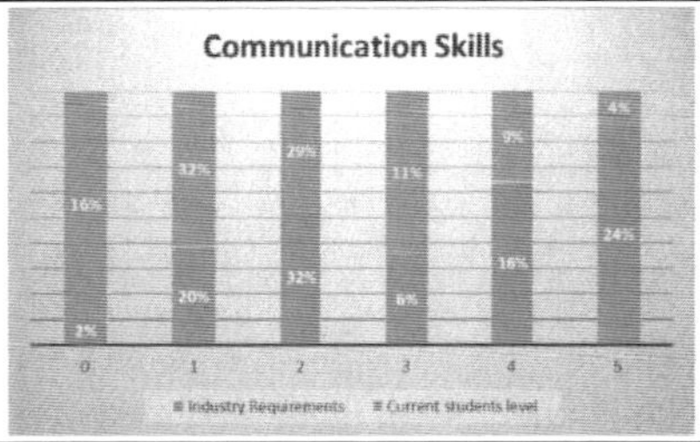

j

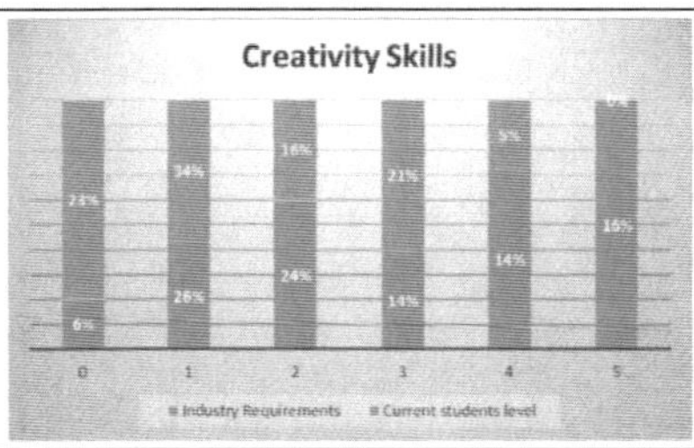

k

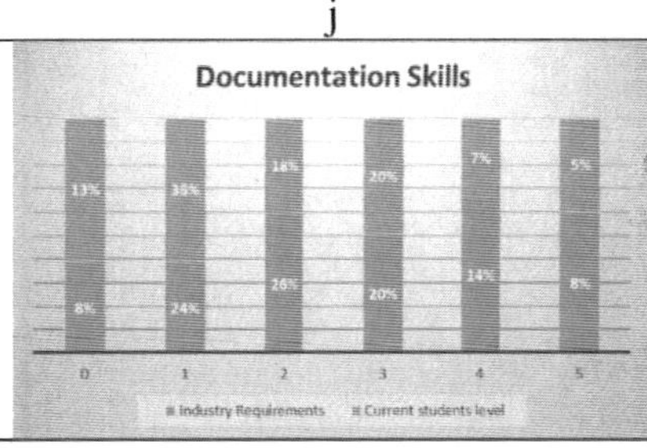

m

Figure 3. The comparison between student level and required level by professional regarding different areas of skills.

5.3. Academic Knowledge skills

The situation is completely different in academic knowledge and skills, where student level of understating is much more advanced that is really required by industry. This means that there are huge efforts put in place to improve student knowledge in a certain area where they may need it or will never use it in their career life.

For example, it can be seen from Figures 4 and 5 that students are at present more advance in hydraulics, math, soil, and traffic management while the industry state that medium level of knowledge will be sufficient. This means that the time and effort have been used for these areas can be used to improve other gaps identified in industry sections.

The other area has quite a stability between academic and industry, which can be considered as an achievement for Iraqi universities. This may result from lecturers' expertise in the construction industry field as consultants, which qualify them to amend their study material to match the industry anticipations.

The only areas of academic knowledge that show different manner are concrete and construction material. Where the student is facing a lack of advance understanding. The main reason is the huge development in construction material and the ways to improve concrete properties by additives. These improvements accelerate much faster than the enhancements in the curriculums, which produce a growing gap that can only be addressed by keeping such topic alive and can be improved regularly.

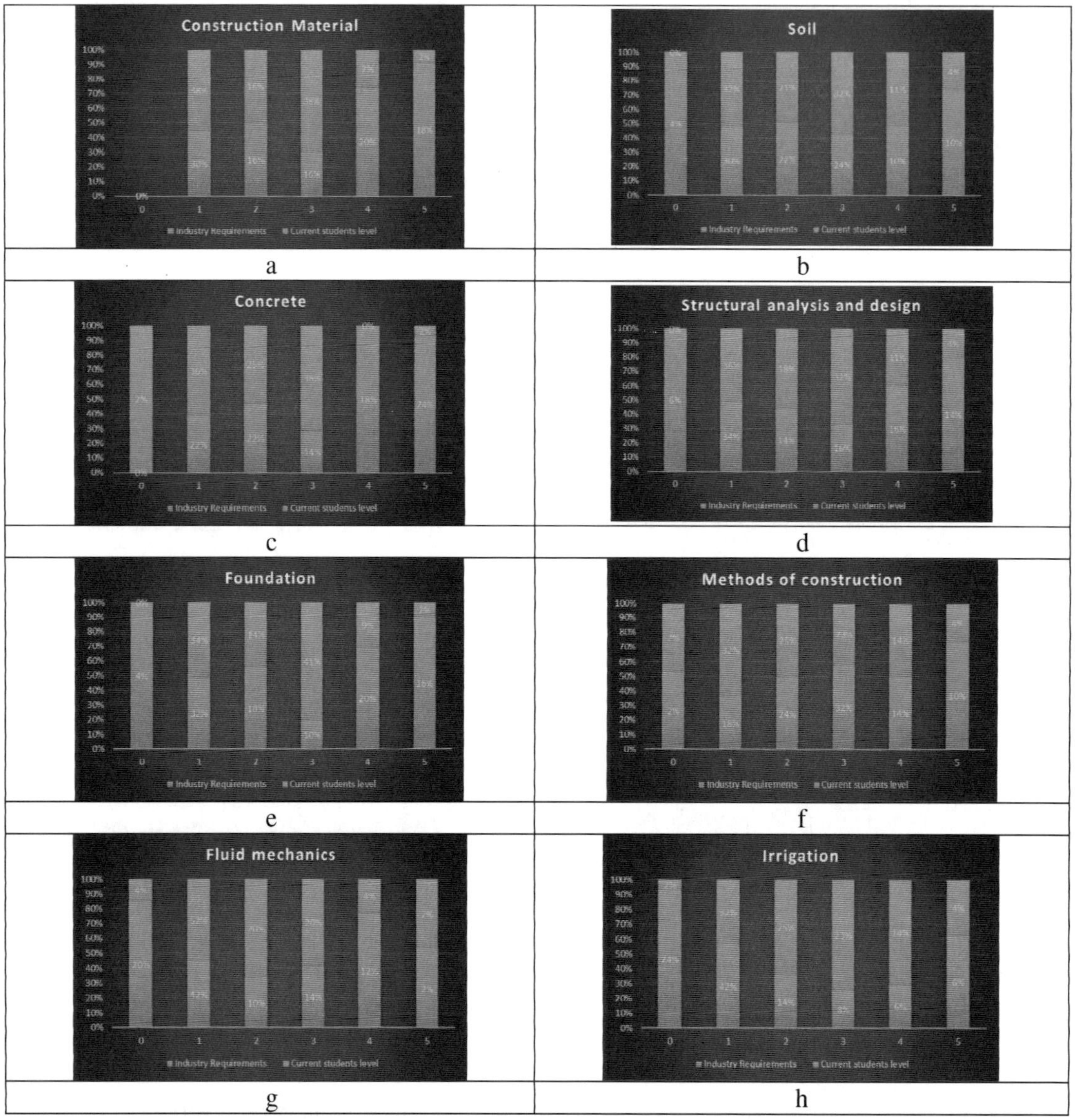

Figure (4): The comparison between student level and required level by professional regarding different areas of academic knowledge.

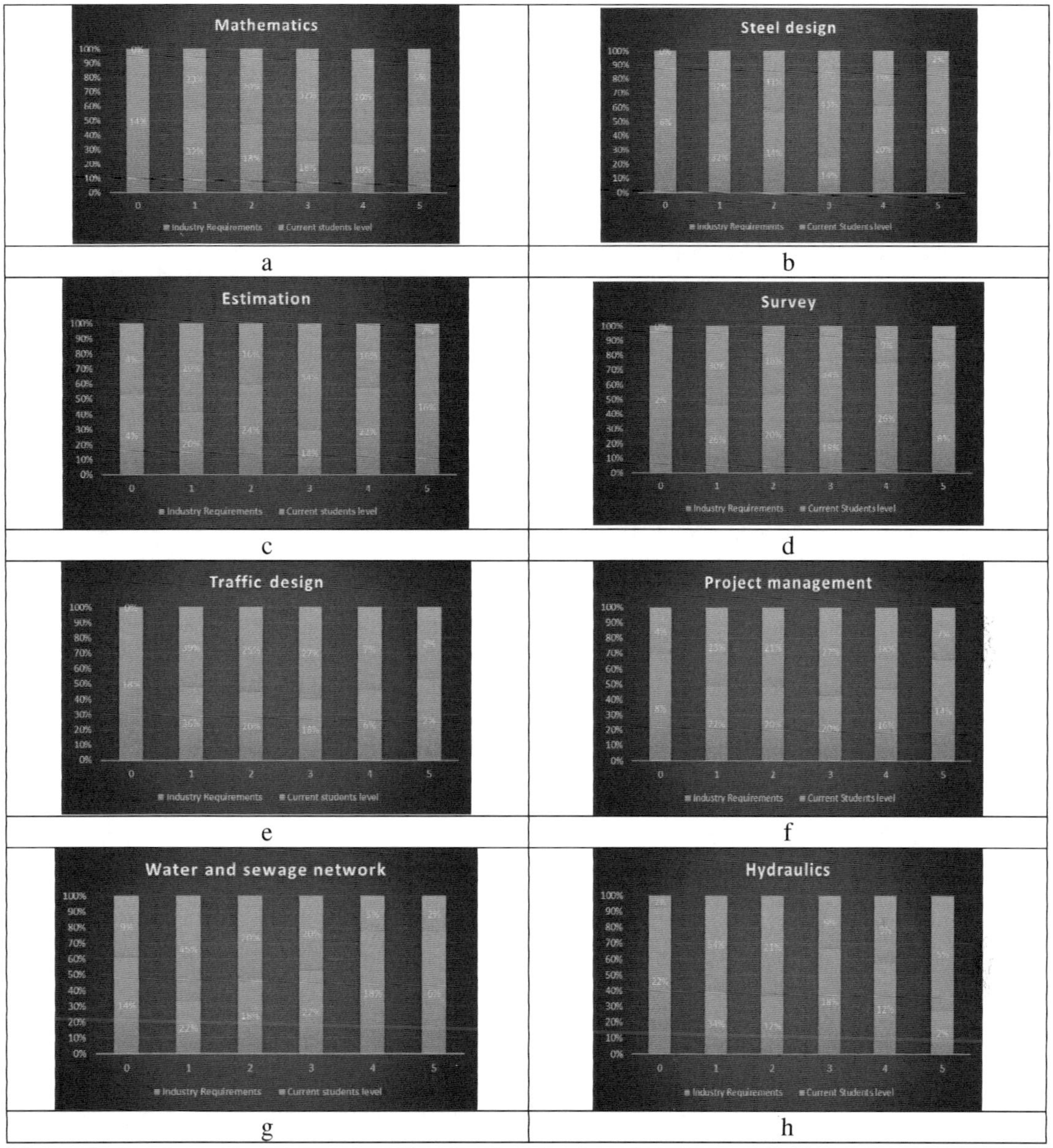

Figure 5. The comparison between student level and required level by professional regarding different areas of academic knowledge.

6. Conclusion

To conclude, it can be seen that there are huge efforts have been deployed to improve student knowledge in certain academic areas where they may not need it in their industry career as professional stated in their answers. This will lead that the universities and the ministry of Higher Education need to put in place such a framework to evaluate current engineering curriculums and make sure that they will help fresh engineering to secure their place in the highly competitive industry.

References

[1] Al-Zwainy, F. M. S., Abdulmajeed, M. H., & Aljumaily, H. S. M. (2013). Using multivariable linear regression technique for modeling productivity construction in Iraq. *Open Journal of Civil Engineering*, *3*(3), 127-135.

[2] Bekr, G. A. (2015). Causes of delay in public construction projects in Iraq. *Jordan Journal of Civil Engineering*, *159*(3091), 1-14.

[3] Al-Janabi, S., & Urban, J. E. (2011, April). Strategic plan for enhancing higher education computing degree programs in Iraq. In *2011 IEEE Global Engineering Education Conference (EDUCON)* (pp. 94-100). IEEE.

[4] Maranzana, N., Segonds, F., Lesage, F., & Nelson, J. (2012, July). Collaborative design tools: a comparison between free software and PLM solutions in engineering education. In *IFIP International Conference on Product Lifecycle Management* (pp. 547-558). Springer, Berlin, Heidelberg.

[5] Al-husseini, S., & Elbeltagi, I. (2018). Evaluating the effect of transformational leadership on knowledge sharing using structural equation modeling: the case of Iraqi higher education. *International Journal of Leadership in Education*, *21*(4), 506-517.

[6] Diwakar, V. (2015). The effect of armed conflict on education: evidence from Iraq. *The Journal of Development Studies*, *51*(12), 1702-1718.

[7] Kaghed, N., & Dezaye, A. (2009). Quality assurance strategies of higher education in Iraq and Kurdistan: A case study. *Quality in Higher education*, *15*(1), 71-77.

[8] Mazawi, A. (Ed.). (2017). *World yearbook of education 2010: education and the Arab'world': political projects, struggles, and geometries of power*. Routledge.

[9] Brunhaver, S. R., Korte, R. F., Barley, S. R., Sheppard, S. D., Freeman, R., & Salzman, H. (2017). *Bridging the gaps between engineering education and practice*. Forthcoming, National Bureau of Economics Research.

[10] Lattuca, L. R., Terenzini, P. T., Volkwein, J. F., & Peterson, G. D. (2006). The changing face of engineering education. *BRIDGE-WASHINGTON-NATIONAL ACADEMY OF ENGINEERING-*, 36(2), 5.

[11] Matusovich, H., Streveler, R., Miller, R., & Olds, B. (2009). Competence in engineering: A tale of two women. In *American Society for Engineering Education*.

[12] Cheah, C. Y., Chen, P. H., & Ting, S. K. (2005). Globalization challenges, legacies, and civil engineering curriculum reform. *Journal of Professional Issues in Engineering Education and Practice*, *131*(2), 105-110.

[13] Kevern, J. T. (2010). Green building and sustainable infrastructure: Sustainability education for civil engineers. *Journal of Professional issues in engineering education and practice*, *137*(2), 107-112.

[14] Clevenger, C. M., Brothers, H., Abdallah, M., & Wolf, K. (2017). Initial assessment of a newly launched interdisciplinary construction engineering management graduate program. *Journal of Professional Issues in Engineering Education and Practice*, *143*(3), 04017001.

[15] Dixit, S., Mandal, S. N., Sawhney, A., & Singh, S. (2017). Relationship between skill development and productivity in construction sector: A literature review. *Int. J. Civ. Eng. Technol*, *8*(8).

[16] Ozbek, M. E., & Clevenger, C. M. (2017). Collaboration in construction academia. *Journal of Construction Engineering and Management*, 143(9), 02517002.

[17] Vemury, C. M., Heidrich, O., Thorpe, N., & Crosbie, T. (2018). A holistic approach to delivering sustainable design education in civil engineering. *International Journal of Sustainability in Higher Education*, *19*(1), 197-216.

[18] Leite, F., Cho, Y., Behzadan, A. H., Lee, S., Choe, S., Fang, Y., & Hwang, S. (2016). Visualization, information modeling, and simulation: Grand challenges in the construction industry. *Journal of Computing in Civil Engineering*, *30*(6), 04016035.

[19] Carter, D. F., Ro, H. K., Alcott, B., & Lattuca, L. R. (2016). Co-curricular connections: The role of undergraduate research experiences in promoting engineering students' communication, teamwork, and leadership skills. *Research in Higher Education*, *57*(3), 363-393.

[20] Wu, P., Feng, Y., Pienaar, J., & Zhong, Y. (2015). Educational attainment and job requirements: Exploring the gaps for construction graduates in Australia from an industry point of view. *Journal of Professional Issues in Engineering Education and Practice*, *141*(4), 06015001.

[21] Dörnyei, Z., & Taguchi, T. (2009). Questionnaires in second language research*: Construction, administration, and processing*. Routledge.

Factors effecting on the compressive and tensile strength of reactive powder concrete made with local materials

A H Chkheiwer[1] and J A Kadim[2]

[1] Civil Eng. Department, Engineering College, University of Basrah, Basra, Iraq.
Email: aqeelcivil@yahoo.com
[2] Civil Eng. Department, Engineering College, University of Basrah, Basra, Iraq.
Email: jafaarahmed@yahoo.com

Abstract. Ultra High Performance Concrete (UHPC) represent as one of the most concrete progressive development so that it can be improved the strength, durability, reliability, and resilience of the concrete structures. The availability of local materials play the main role in the major step in making the economic mix jobs by saving their materials and effort and decrease the concrete cost. So in this paper, the parameters including the curing regimes, of sand gradation, binder type, and content, and steel fiber content were investigated to show their effect on concrete's compressive strength. The obtained results showed that the curing regime that consists of 5 days (2 days at 60°, 3 days at 80°) given the highest strength than other curing regimes. Finer sand, Silica Fume, steel fibers cause a considerable increasing in compression strength. Various relationship were obtained illustrating the effect of studied parameters with concrete strength.

Keywords: Reactive powder concrete; Ultra-high strength concrete; High range water reducers; Hot water curing; Silica fume; Steel fiber.

1. Introduction

Reactive Powder Concrete (RPC) was developed in 1990's by a French company was known as Bouygues so that the new concrete technology gives an ultra-high strength and more ductility composite material with enhancing of mechanical properties. The basic idea of RPC type is the behavior optimization of microstructure by making a precise mix gradation for most particles in the concrete skeleton to achieving the maximum concrete density. RPC consists of very fine sand, super plasticizers, silica fume, steel (reinforcement) fiber, and cement that are resulting in a cementation mixture with the very low water-cement ratio (w/c). Briefly, it is not considered as concrete materials because of the nonexistence of the coarse aggregate in the cement mixture.

The non existence coarse aggregate can be represented as a big feature relating the microstructure system of RPC so that enhancement the performance behavior of RPC in order to trim down the heterogeneity between the cement base medium and the aggregate. Today, RPC is represented as an advanced technological progressive for the buildings construction industry. RPC is already prefer used for a special applications such as bearing units of a bridge girder, end block of pre-stress units, finder systems in port engineering, base units in power plants, medium to long span bridge project, highway structures subjected to a harmful environmental condition, etc…[1].

Since the task of RPC function is the removal of coarse aggregates causing an optimization of the granular framework by increasing the ratio of surface area over particles volume, leading to increasing the cementation matrix density by away of a post set heat treatment. The using of UHPC available local materials can make further choices and opportunities for the applications of UHPC in the buildings, transportations, in addition to the underground structures. The investigation of UHPC constituent materials will contribute to improving the concrete mechanical properties as discussed in this work so that this will easily lead to more practical development and to simplify UHPC mixture proportions as specified by the experimental new program. According to the composition of UHPC mixture, the mode of failure can be explosive, and this result can be recognized to the higher compressive strength resulting from the brittle nature of UHPC. The additional steel fibers will be used to eliminate this mode of failure in which the steel fibers enhance and increase the flexural capacity in addition to the performance of UHPC [1,2].

The estimated stress-strain curves of UHPC display a linear elastic response to 90% of the ultimate concrete strength compared to 45% of the ordinary concrete so this property is essentially so useful in practice for reducing deformation, including both deflection and cracking, as the concrete structure can be better to resist the applied external or internal loads. In the manufacturing of UHPC, the temperature is considered as one of the most important factors during the curing process which means the curing system has an influence effect on the mechanical properties such as a heat-curing system can largely raise the in the early days the age compressive strength and improve the ultimate compressive strength. A usual system of UHPC curing consists of two stages so that in the first stage the concrete is treated by placing it in an appropriate temperature for avoiding moisture losing until researching the required situate while in the second stage, the temperature of curing could be increased for accelerating the compressive strength.

Hence, different UHPC mixes have been employed with several ratios and various filler material, high range water reducers (HRWR), steel fibers, and accelerator means. The UHPC filler material is considered as an important component that represents closely about 50% by weight, therefore, the filler material type is straightforwardly affect the cost of UHPC as its compressive strength. Consequently, using the available local materials leads to decrease in the cost and give confidence in using UHPC in the practice. But in sometimes, according to the nature of the material location, using of these local materials may be lower the mix strength. For present work, the researchers employed UHPC mixtures using available local fine sand as a filler material. Thus, the mix quantities of the UHPC constituent materials are based on the recommended values associated to the subject which make the resulting UHPC mixtures have a compressive strength at least equal to 150 MPa [3,4].

2. Literature review

During a few years since the 2000s, a sensible number of studies had been conducted to deal with the mixture properties and proportions to increase the mechanical properties of UHPC. As an example in the United States, the Federal Highway Administration [5] which is one of many organizations that interesting to investigate UHPC development applications. In the literature of this subject area, there are two main progresses in the UHPC field. The first one interest on the improved UHPC mechanical properties such as the resulting mixture strength (compressive, tensile, shear, and flexure), and the durability correlated properties so these developed properties are achieving using enhancing the UHPC mixture proportion. The second development focuses on the UHPC applications to raise its use in many types of both design and construction stages related to different concrete structures. Below some researchers work relate to high performance concrete.

Richard [6] found 35 kg/m³ of super plasticizer was suitable to give a slump flow of 360 mm in which this amount of HRWR was closely larger than the allowance amount used in the Ductal which equal to 30.7 kg/m³. However, using large dosages of super plasticizer may be producing a large setting time value of UHPC, hence for solving this potential undesired effect, an accelerator agent may be applied to reduce the setting time. The accelerator disbands the cement particles in water which make accelerating the reaction processes leading to reducing the concrete setting time. As mentioned to Lafarge, an accelerator dosage of 30 kg/m3 is recommended for UHPC which used as a considerable amount of super plasticizer.

Tayfun [7] work with the microstructure and flexural behavior of reinforced (steel-fiber) concrete formed with the different volume fraction of steel fibers and specimens aspect ratio. Concrete specimens of the prismatic shape of dimensions 100×100×350 mm were employed with and without steel fiber (two different steel fiber types each one was hooked-end). The flexural strength had been distinct for each age while the crack widths had been calculated after the greatest bearing loads. The microstructure of steel fiber reinforced concrete (SFRC) had been considered by two means (scanning electron microscopy and optical microscopy). The results indicated that the microcopy images could be used for demonstrated the characteristic of SFRC bond where a good bond was illustrated between the concrete matrix interface zone and steel fiber by using polarizing microscopy. SFRC flexural strength became enlarger with increasing both the concrete age and fiber volume fraction.

Kumar et. al. [8] work with a very dense matrix mix of Reactive Powder Concrete (RPC) that composed of very fine powders (cement, sand, quartz powder and silica fume), steel fibers (optional), and a super plasticizer and using these components to give the ultra-high strength and durability properties. The performance of reactive powder concrete had been investigated in which RPC was made containing fly ash Ground Granulated Blast Furnace Slag (GGBS) as a replacement for cement at three percentages of 5%, 10%, and 15% and in order to compare these result of cement replaced mixture, a specimen without cement replacement (RPC) were also cast. The obtained results showed that a development related to the strength (compressive and flexural) in addition to the elastic modulus for cement replaced mixes.

3. Paper Plan

This paper consists of an experimental program which carried out to investigate the suitability amount of silica fume, and steel fiber in the production of high strength concrete with using of available local materials to reach a new approach in producing UHPC from available local materials. Hence, the first step focus on the investigating of the influence of the curing regimes on the concrete behavior (strength types) measured by different ages so that the curing regimes are conducted that including 6 curing types which given as follows: 2 days at 80°C, 3 days at 80°C, 4 days at 80°C, 7 days at 20°C, 3days at 80°C, and 2 days at 60°C followed by 3 days at 80°C. Then, the experimental procedures with their results are presented to UHPC mixtures. Moreover, the obtained results are summarized in tables and figures given by the different relationship with the deep discussion.

The text of your paper should be formatted as follows:

4. Experimental program

The experimental program consists of many different stages beginning from the selection of the job materials followed by mixing them and casting them into their cubes and prisms molds and finally subjected to the curing system.

4.1 Materials

RPC compose of extremely fine powder materials such as cement, sand, and silica fume in addition to steel fibers (added as mix specified), and super plasticizer material that used at its best possible dosage to decrease the water cement ratio (W/C) and getting better the concrete workability.

4.1.1 Cement

In this research, Portland cement Type V (Aljisir sulfate resisting) which produced in Iraq was used in which its chemical and physical properties satisfy the requirements of Iraqi specification No.5.

4.1.2 Fine aggregate

Natural sand with four maximum sizes of 2.63, 1.18, 0.6 and 0.3 mm was used. The content of Sulfate of fine aggregate was found within the permissible limits of the Iraqi specification No. 45.

4.1.3 Silica fume

Silica fume is consequential from the reduction of high purity quartz with coal or coke, and wood chips in an electric arc furnace during the silicon metal or silicon alloys production. Using silica fume in concrete has engineering potential and economic. It added to the UHPC mixes of this work. The percentages used were 10%, 15%, 20%, 25%, and 30% as partial replacement of cement weight. The tested silica fume is conformed and satisfy the physical and chemical requirements of ASTM C1240 specifications.

4.1.4 Steel fibers

In general, the using of this specific steel fiber enhances the initial values of the compressive and tensile strength of RPC. Hooked steel fibers used throughout the experimental program as shown in figure 1 in which hooked type has enlarged the connection of hydrated cement paste around steel fibers surfaces. Here, the steel fiber has the following properties diameter 0.5mm, length 30mm, density 7800 kg/m³, and the ultimate tensile strength not less than 850 MPa.

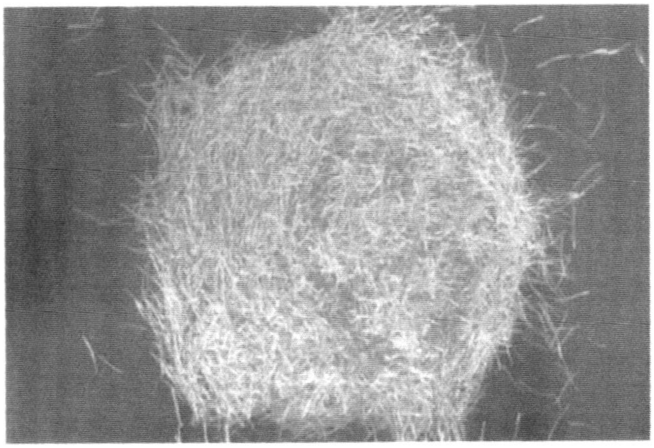

Figure 1. Hooked steel fibers.

4.2 Testing procedure

Test procedure is divided into two categories such that the first one relates to a suitable selection of mix properties and proportion while the second deals with the mixing operation in the laboratory.

4.2.1 Mix proportions

Table 1 shows the UHPC mixes proportion which is essential for getting the minimum compressive strength at least 150 MPa. In this table, silica fume accounts 10% to 30 % of the binder for mix non contain steel fiber and 25% for mixes without steel fiber. Using silica fume is essential to reach an excessive compressive strength in addition to the high durability because silica fume makes accelerating to the pozzolanic reactions which produce additional calcium silicate hydrates (C-S-H) that consequently fill the voids exist in the paste matrix. Since the present work focuses on finding the most effectual silica fume content for developing UHPC by using the locally available materials achieve two main goals, the first goal is to obtain a mix satisfy the adequate compressive strength and the second goal is to minimize the cost of UHPC.

4.2.2 Mixing procedure

The following mixing procedure was employed for each patch of (HPC) under specified laboratory circumstances: The quantity of silica fume in dry condition state was mixed with the necessary cement for period 5 minutes to make sure that silica fume was carefully dispersed through cement particles. Then, the fine sand was added to the mixer and mixed for 5 minutes. Later, the super plasticizer was dissolved in pure water and the liquid of super plasticizer and water was added to the rotary machine mixer so that the whole mix ingredients by mixing for a sufficient time. After that, the mixer was stopped, and the mixing was continued by hand particularly for parts which not reached by the mixer blades. Finally, the mixer was operated for 5 minutes to attain sensible fluidity. Steel fibers were slowly distributed in the uniform state through the for 5 minutes for the period of the mixing process, and later

IOP Conf. Series: Materials Science and Engineering **584** (2019) 012002 doi:10.1088/1757-899X/584/1/012002

the mixing process was continued for an additional 3 minutes. Finally, the total times requires for the mixing one batch is approximately 30 minutes. Two types of test samples are required in which the first one is cubed specimens to estimate the concrete compressive strength and the second is the prisms specimens to determine the flexural (tensile)concrete strength.

Table 1. Mix materials proportions

Sample	Cement [a]	Fine Sand [a]	Silica Fume [b]	Super Plasticizer [b]	Water-Cement ratio (w/c) [c]	Steel Fiber [d]
1	1	1	10	6	18	---
2	1	1	15	6	18	---
3	1	1	20	6	18	---
4	1	1	25	6	18	---
5	1	1	30	6	18	---
6	1	1	25	6	18	0.5
7	1	1	25	6	18	1.0
8	1	1	25	6	18	1.5
9	1	1	25	6	18	2.0
10	1	1	25	6	18	2.5

[a] Cement an fine sand are expressed by fraction weight proportion of total mix such as 1:1.
[b] Silica fume and superplastisizer are expressed as percent weight of cement.
[c] Water-cement ratio was expressed as percentage ratio.
[d] Steel fiber is expressed as percent volume of total mix.

5. Results and discussion

The study result is divided into four sections, so each section focuses on the required investigated parameter in order to illustrative its effect on the strength gained.

5.1. Curing regime

Table 2 shows the effect of different methods of hot curing on the compressive strength. Herein results indicate that the highest compression strength that can be obtained by increasing the curing period which reaches to 90 days and this value of strength can be obtained in hot curing with 2 days at 60°C + 3 days at 80°C.

Table 2. Compression of UHPC strengths for different curing conditions.

Curing regime	Compression Strength (MPa)
2 days at 80°C	56.3
3 days at 80°C	63.5
4 days at 80°C	66.2
7 days at 20°C + 3 days at 80°C	50.1
2 days at 60°C + 3 days at 80°C	89.0

5.2. Sand gradation

Table 3 describes the results of four graded of sand with corresponding compressive strength in which these results were shown in figure 2. It can be observed that the mixture compressive strength become lesser with the increase of the maximum size of fine aggregate used that given as logarithm relationship. Also, it is observed that 0.3 mm maximum size exhibited 39% increase in compressive strength compared with the maximum size of fine aggregate equal to 2.36 mm. According to the obtained results, the following expression was found between the concrete compressive strength (f_{cu}), here known as grade strength, with the size of fine sand (S_{pass}) which given as:

$$f_{cu} = 16.48\log_e(S_{pass}) + 100.8 \quad (1)$$

with a modulus of correlation equals to 0.9305

Table 3. Compression strength of concrete mixture with a gradation of fine sand.

Sieve size passing (mm)	Compression Strength (MPa)
2.36	89.00
1.18	97.17
0.60	104.00
0.30	124.41

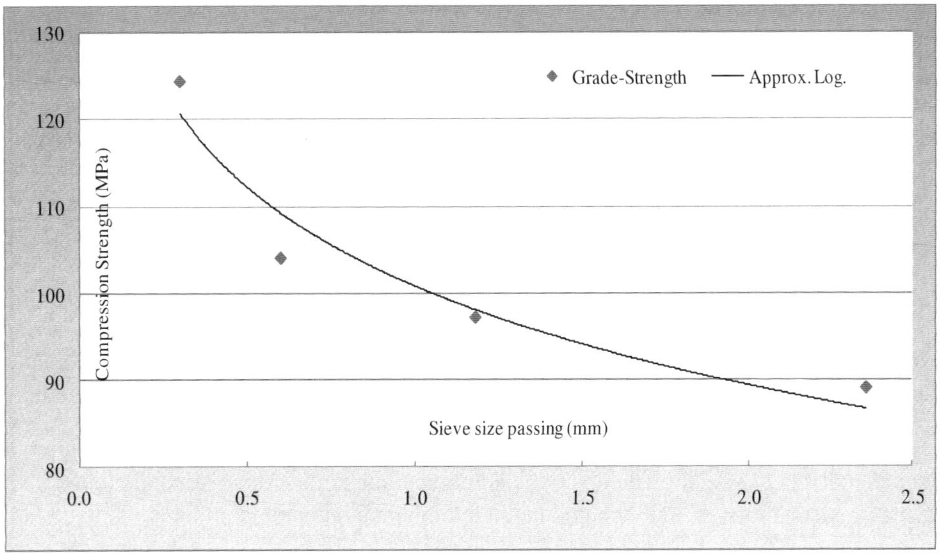

Figure 2. The relationship between compression strength of concrete mixture with a grade of fine sand.

5.3. Silica fume

Different proportions of silica fume were selected, and their result is given in Table 4, and their effect is shown in figure 3. It is noticed from Table 4 that the increasing of silica fume ratios increases the compressive strength but with less amount in which a small compressive strength had been gained after 25% ratio of Silica Fume. The effect of Silica fume is well recognized in figure 3 that display a nonlinear relationship between the concrete compressive strength (f_{cu}) and the silica fume contents (S_i) which given as:

$$f_{cu} = 0.008S_i^3 - 0.3311S_i^2 + 9.9607S_i + 23.88 \quad (2)$$

so that this relationship gives value for the modulus of correlation equals to 0.9972.

Table 4. Compression strength of concrete mixture with different silica fumes content.

Silica Fume (%)	Compression Strength (MPa)
10	94.0
15	112.3
20	120.0
25	126.0
30	127.1

It is so clear that the effect of silica fume is enhancing the mechanical characteristics of the hardened concrete by to make these properties better than the case without using it noting that the basic physical principle of its action in the concrete mix is a closely similar to that of the filler action. Due to its fineness, silica fume can well be placed into the interval between the cement particles as the similar pattern of sand that filling the voids between coarse aggregate and cement particles to enclose these voids, in addition, the space between sand grains.

Generally, the silica fume used to increase the concrete strength of considerable amount. In addition, since silica fume has high surface area and high content of amorphous silica , therefore the chemical reaction of silica fume is huge active pozzolan reacts more rapidly than usual pozzolans. Consequently, from Table 4, it can be found that the optimum dose of silica fume is 25% (by weight) when used as partial replacement of ordinary Portland cement (OPC).

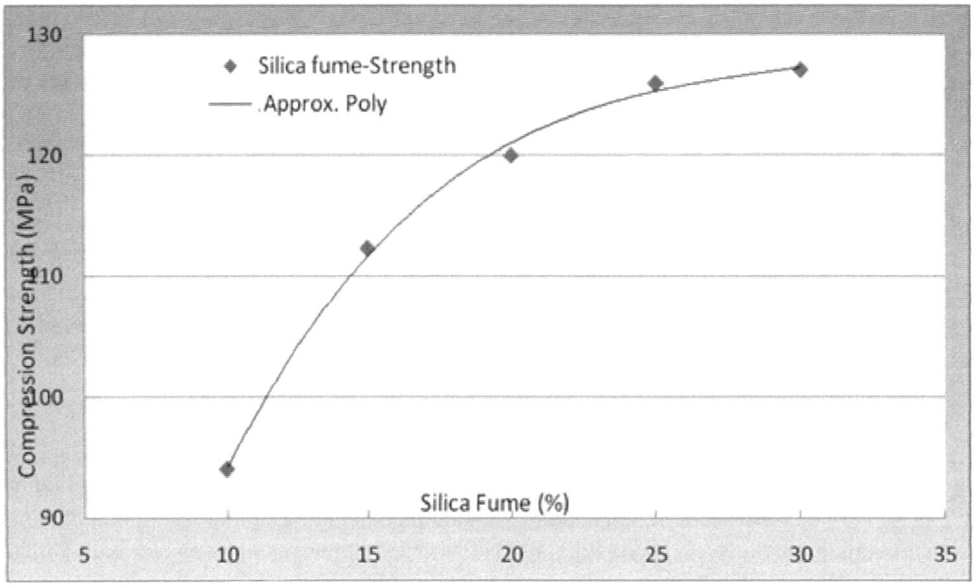

Figure 3. Relationship between compression strength of concrete mixture with silica fume ratio.

5.4 Steel Fiber

Six steel fiber ratios were used and their results with high performance concrete (HPC) and UHPC strength (more than 150 MPa) are given in Table 5. These results indicate a large benefit of the steel fiber in the high performance concrete (both HPC and UHPC) strength types and it is important to conclusion that steel fiber more than 1.6% change the concrete type to UHPC which has a high effect on both compression and tension strength The microstructures of RPC at early ages in the young and mature pastes state can be shown in various figures that given by Tayfun [8] and ACI 234 [9].

Figure 4 shows the trend of steel fiber of high performance concrete (UHPC-HPC) strength relationship which takes a linear nature regardless strength type with the modulus of regression equals 0.99. Two relationships are obtained here in which the first between the compressive concrete strength and the steel fiber content (S_f) while the second between the flexural concrete strength (f_{ct}), sometimes known as the tensile concrete strength, with the steel fiber content and these two relationship are given below as:

$$f_{cu} = 14.823S_f + 126.3 \quad (3)$$

$$f_{ct} = 8.9714S_f + 7.7524 \quad (4)$$

Table 5. Steel fiber ratio with compression and tensile strength of UHPC

Steel Fiber %	Compression Strength (MPa)	Tensile Strength (MPa)
0.0	124.0	9.0
0.5	135.2	11.8
1.0	143.2	15.5
1.5	148.0	20.4
2.0	156.4	26.5
2.5	162.2	30.6

The effect of the presence of steel fibers on the concrete strength can be illustrated as a very dense matrix is satisfied by optimizing the granular packing of the fine dry powders and this state of microstructure concrete is achieved by increasing the steel fiber content regardless of the economic point of view.

For tensile strength, the UHPC mixture steel fibers of ratio 2.5% increase the strength by 240 % than mixture without steel fibers while this ratio is 31% corresponding to compression strength. These results can be attributed to the action of steel fibers as the interception of steel fibers to the development and growth of cracks formed in the hardened concrete paste, which leads to the suspension of growth of these cracks in addition to not generating additional cracks within the ultra high performance concrete (UHPC) paste. Also, it seen that there is a good quality attachment in interface zones of cement concrete medium and steel fiber which alter in this situation by increasing the bonding between steel and concrete [10].

The above results are so clear as depicted in Figure 5 which displays two samples of crushing concrete cubes with and without steel fibers. These results may comparable with those when using polypropylene fibbers as a solution for sapling resistance despite fiber behavior in high temperature while PP melt and create channels through which the water vapour pressure built-up within HPC as temperature rises is released, which significantly reduces the sapling tendency of HPC under fire conditions [4].

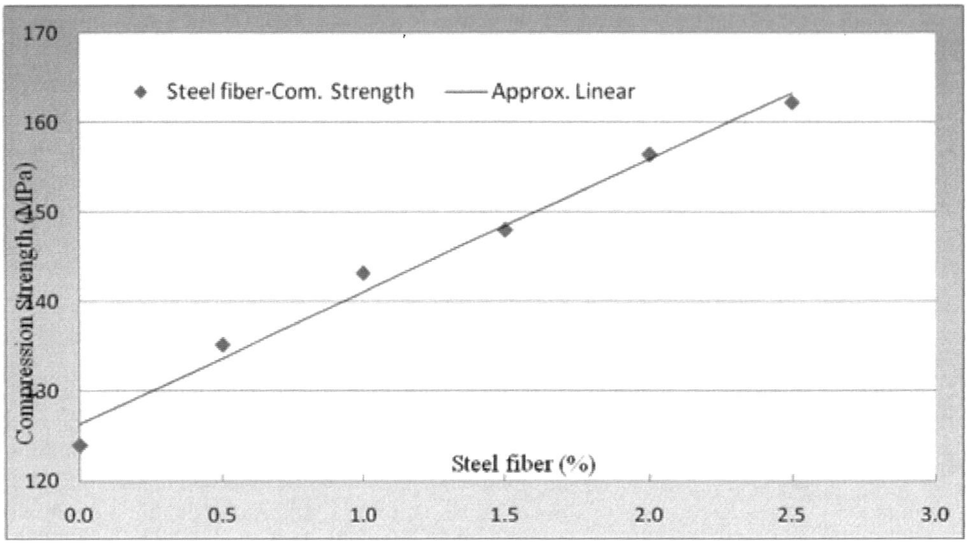

(a)

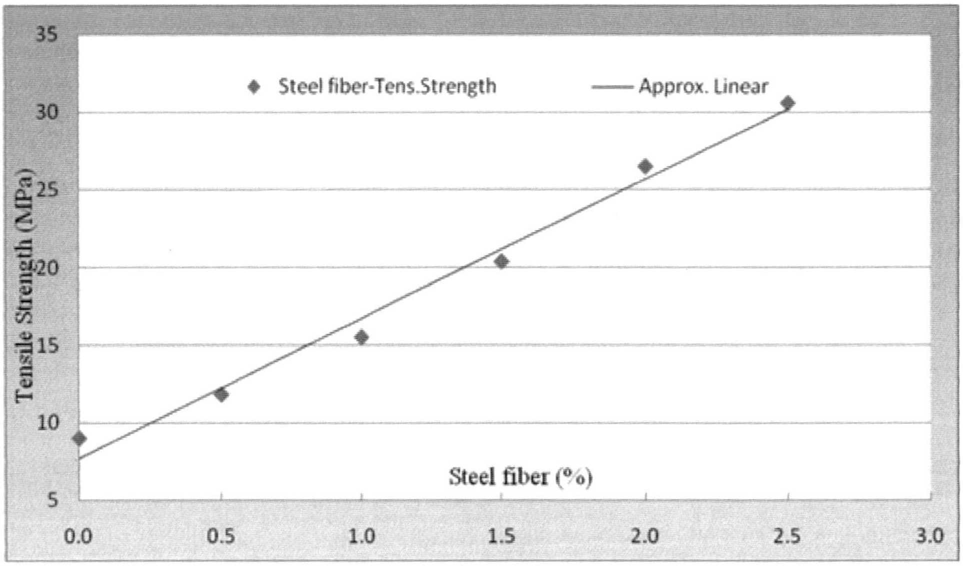

(b)

Figure 4. Relationship between steel fiber ratio with (a) concrete compressive strength; (b) Tensile strength

Figure 5. Concrete cubes crushing samples with and without steel fiber.

6. Conclusions

According to the receiving results, the following conclusions are specified as :

1. The curing regime had an influenced effect on the concrete compressive strength in which curing regime C that consists of 2 days at 60°C followed by 3 days at 80°C, giving the maximum concrete compressive strengths.
2. The added steel fibers cause a considerable increasing in the tensile strength (flexural) than compression strength with a linear relationship.
3. It is achievable for producing UHPC mixtures containing availably local materials to obtain 150MPa compressive strength with the following mixture proportion by weight (cement 1, fine Sand 1, silica fume 25%, water / cementations material ratio 18%, and steel fiber 2.0% by volume).
4. The finer sand causes an increasing in the concrete compressive strength than using the usual gradation sand.
5. Using amount of silica fume larger than 25% had insignificant special effects on the concrete compressive strength so that this ratio of silica can be considered the optimal value.

References

[1] Shetty M S 2005 *Concrete Technology: Theory and Practice* (New Delhi-Ram Nagar) p656.

[2] Kazemi S and Lubell S 2012 A Influence of specimen size and fiber content on mechanical properties of ultra-high- performance fiber-reinforced concrete, *ACI Mater. J.* 109 (6) 675-684, https://doi.org/10.14359/51684165

[3] Schmidt M and Fehling E 2005 Ultra-high-performance concrete: research, development and application in Europe, *7th International Symposium on Utilization of High Strength High Performance Concrete Volume 1*, 51-77

[4] Kadri E H, Duval R, Aggoun S and Kenai S 2009 Silica fume effect on hydration heat and compressive strength of high- performance concrete, *ACI Mater. J.* 106 (2) 107-13, https://doi.org/10.14359/56454

[5] Graybeal B 2014 Development of a Family of Ultra-High Performance Concrete Pi- Girders, FHWA-HRT-14-028

[6] Richard P and Cheyrezy M H 1994 Reactive Powder Concretes With High Ductility and 200 - 800 Mpa Compressive Strength, *ACI Mater. J.* 144 507-18,

[7] Uygunog˘lu T 2008 Investigation of microstructure and flexural behavior of steel-fiber reinforced concrete, *J. Materials and Structures*, 41 (8) 1441–49, https://doi.org/10.1617/s11527-007-9341-y

[8] Kumar A, Udaya A and Sabhahit N 2013 Reactive Powder Concrete Properties with Cement Replacement Using Waste Material, *IJSER* 4 (5) 203-6

[9] ACI Committee ACI 234R-06 2012 Guide for the Use Silica Fume in Concrete (ACI-Farmington Hills, Mich)

[10] Maroliya M K 2012 Micro Structure Analysis of Reactive Powder Concrete, *IJRED* 4 (2) 68-77

IOP Conf. Series: Materials Science and Engineering **584** (2019) 012003 doi:10.1088/1757-899X/584/1/012003

The Combined Effect of Ferric Chloride and Cationic Poly-Electrolyte CPE on Coagulation of Grey Wastewater

Huda Rida Hashim[1] **and Ali Jwied Jaeel**[2]

[1] College of Veterinary Medicine, University of Wasit, Iraq. Email: enghuda80@yahoo.com

[2] Civil Engineering Department, PhD, University of Wasit, Iraq. Email: alijwaid@uowasit.edu.iq

Abstract. The cationic polyelectrolyte is commonly adopted as a coagulant aid to reinforce the function of coagulant agent in industrial and municipal purposes; As a result, ferric chloride could be considered as the major chemical coagulant and Cationic Poly Electrolyte CPE as the coagulant aid. A set of examinations were implemented to minimize the amounts of both ferric chloride and CPE and to increase the removal efficiencies for the contaminants concentrations such as turbidity, total suspended solids (TSS), total dissolved solids (TDS) and chemical oxygen demand (COD) originated in grey wastewater so as to reuse these water for irrigation purposes. The combination of ferric chloride and cationic polyelectrolyte was found to improve the results for pollutants removal efficiency (Turbidity, TSS, TDS, and COD), by (97.5%, 91%, 88% 75%) respectively. The predicted results from artificial neural network (ANN) approach versus that observed from the experimental study for COD, TDS, TSS and turbidity removal showed an acceptable agreement for all models. All the parameters had a value of R^2 of (0.9972, 0.9993, 0.9992, 0.9985) respectively

Keyword: Coagulation, ferric chloride, coagulant aid, wastewater treatment

1. Introduction

Gray water, wastewater produced by house actions such as cleansing, rinsing, bathing and laundering expelling toilet waste [1]. Grey water is commonly less polluted than toilet water; it has less parasites and about 90% lesser nitrogen than toilet water [2]. The greywater reclaiming increased in the recent years, principally as available clarification to water scarcity in developing countries, and it is fulfilling a fast growth on general level nowadays [3]. By recycling greywaters the drinking water demand per capita and the amount of the generated and drained wastewater can be remarkably decreased. A critical issue in greywater reusing is the water quality standardizations. Water analysis offers several of appropriate and effective estimating instruments, which assist the specific planning of the pollutants. The seniority of the applicable studies examines the organic, nutriment and biological pollution in greywaters. [4]. The activity of coagulation has been known as a pre-process that achieves filtration. In this process, coagulants cause rawer particles in water and these particles are separated from the water by sedimentation or filtration processes. Primarily, metal salts like alum, ferric sulphate, ferrous sulphate, ferric chloride, and anionic cationic and non-ion organic polymers are called coagulants. Coagulants and/or flocculants are necessary to reduce these particles, by discontinuing their stability. Agglomeration, generated in this manner, can be removed from the polluted waters [5]. In most conditions aluminium and ferric compounds are adopted to neutralize charge. Aluminium sulphate has been adopted as a major chemical usage for water treatment operation for more than 300 years. Because of its cheapest price as compared with other materials. However, Ferric chloride considers superior appropriate for its usage on water treatment operation in many countries such as Australia, United States of America and Canada. Because it displays an efficient solution for particles destabilization, reduction of total suspended solids and total dissolved solids, induction of flocculation and also improvement of filtration. In many cases the process cationic polyelectrolytes are more promising flocculants to undermine the polluting particles that are negatively charged. The electrostatic reaction provides strong adsorption that neutralizes the surface of the particles leading to the reflection of the charge. Furthermore, the flocculation may occur due to the reduction of the particle surface charge and the low electrical repulsion between them. Many studies conducted on polyelectrolyte of maxim charge density are simply effective as, in a given dose; they have a good ability to make more charge to particle shell. [6]. The purpose of this research is to minimize the contaminants concentrations of the grey wastewater using a combination solution (ferric chloride) and polymers (coagulant aid), so as to get higher removal efficiencies for several pollutants such as turbidity, total suspended solids, total dissolved solids and chemical oxygen demand (COD). So that the treated water can be used for re-using and re-cycling purposes. And also prediction the removal efficiencies of these pollutants by apply ANN technique.

2. Experimental Works

2.1. Formulation of synthetic wastewater

The composition and characteristics of used wastewater are shown in table 1 and table 2.

Table 1. Composition of Synthetic Greywater [7]

Compound	Concentration (mg/l)
Glucose	300
Sodium acetate trihydrate	400
Ammonium chloride	225
Disodium hydrogen phosphate	150
Potassium dihydrogen phosphate	75
Magnesium sulphate	50
Cow dung	225

Table 2. Characteristics of Synthetic Greywater.

Parameter	Unit	*Value*
pH	-	6.5
Turbidity	NTU	60
TSS	mg/l	350
TDS	mg/l	800
COD	mg/l	300

2.2. Chemicals and Coagulants

The coagulants and chemicals used to improvement the removal efficiency, included $FeCl3$, and cationic polymer. Stock solution of ferric chloride was formulated by adding 10gm of $FeCl3$ to 1000 ml of distilled water at 60°c to make a homogenous solution of ferric chloride solution concentration of 1%. Thus, each 1 ml/l of stock solution included 10 mg/l of $FeCl3$. The polyelectrolyte solution was formulated by dissolving 1.0gm of the synthetic polymer into 10 ml of alcohol to be more abandoned into the water, then diluted into tab water at 1000 ml to give solution of concentration 0.1%. Thus, 1 ml/l of solution is corresponding to 1 mg/l polymer. The most preferred dosage of the singular coagulant or combination of more than one chemical was determined by Jar test, the cationic polyelectrolyte used commercially known as (Mangafloc LT 225) (polyalkyenimine) multipurpose polymer with a molecular range (10000-500000).

2.3. Best Dose of Coagulant and Polyelectrolyte
The prime method adopted to calculate the best and optimum coagulant dosage is jar test method. The one litter cylinder jars were filled with wastewater sample. In each jar, a divergent infusion concentration is adjusted with several dosages 10, 20, 30, and 40 mg / l. During the experiment, the jars submitted to a fast mixing velocity 120 revolutions per minute for 2 min so as to disperse the coagulant in all solution and then a gentle mixing velocity 2 revolutions per minute for 30 min for so as to get flocs generation. Then after that, the mixers are switched off and taking a period of 30 minutes for flocs sedimentation process. After the sedimentation period has finished, samples were taken from a distance 3-4 cm under the water surface of each beaker by a medical syringe. The samples were tested for determination the removal of turbidity, TDS, TSS and COD. The best dose was determined for ferric chloride, polyelectrolyte, and them combination.

2.4. Artificial Neural Network Approach (ANN)
The artificial neural network (ANN) is computing system that considered nonlinear statistical data modeling tools. This system designed to solve the most complex problems and simulate the way as the human brain processes and analyzes the information. ANN is used to predict the physical, chemical, and biological parameters in this study by interpolating and extrapolating the data. ANN is a mathematical model can be described by neurons. Each neuron acquires an input information, which is the biased sum total of the outputs of previous neurons. The neuron behaves on this input by mechanism of an activation function and a bias utility. ANN structure contains at least three components, the first one called as input layer (ferric chloride dosage, CPE dosage, initial COD, initial TSS, initial TDS and initial turbidity), the second component is the hidden layer and the third part noun as an output layer (COD, TSS, TDS and turbidity removals) as shown in figure 1. ANN structure may have two or more hidden layers; each hidden layer has a learning value to maximum the transforming of input layer. In this study, the sigmoid activation function was adopted to model the experimental data, weights were measured using Quick Propagation Algorithm (QP) and the connection between neurons of each layer to solve and transfer the experimental data using Multilayer Normal Feed Forward (MNFF).

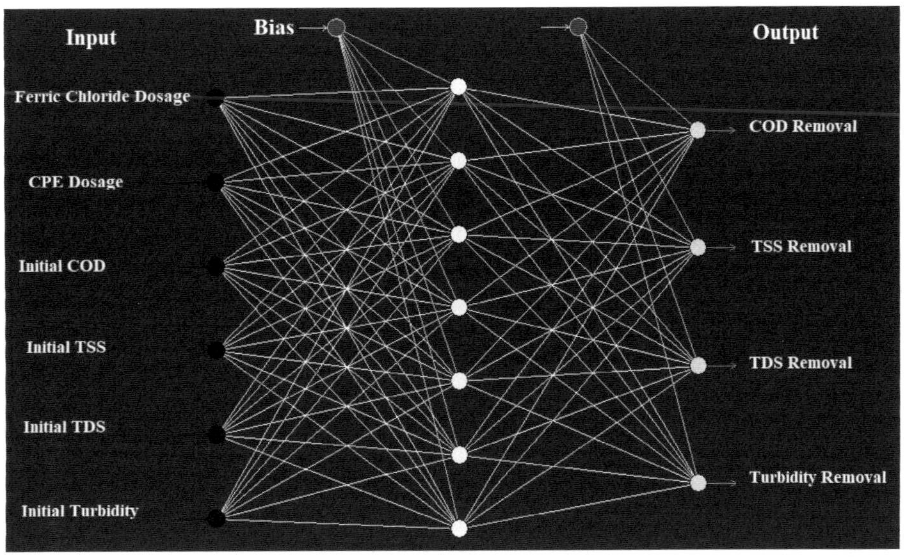

Figure 1. ANN structure for the model.

3. Results and Discussion

3.1. Using Ferric Chloride only

Figure 2 to figure 5 show the effect of various ferric chloride doses, on wastewater quality parameters like (Turbidity, TSS, TDS and COD) with initial values of (60 NTU, 350 mg/l, 800 mg/l and 300 mg/l) respectively. From figure (2) it appears that ferric chloride dose of 25 – 35 mg/l gave the maximum turbidity removal efficiency (92%) while the maximum TSS reduction was 88% for the same range of dosage (25 – 35 mg/l) as shown in figure 3. The effect of ferric chloride effect on TDS removal efficiency could be shown from figure 4), the maximum removal was 87% was done using coagulant dose of 20 mg/l. Figure (5) illustrated the COD removal efficiency of using ferric chloride, the removal was 66% using a dosage of 30 mg/l.

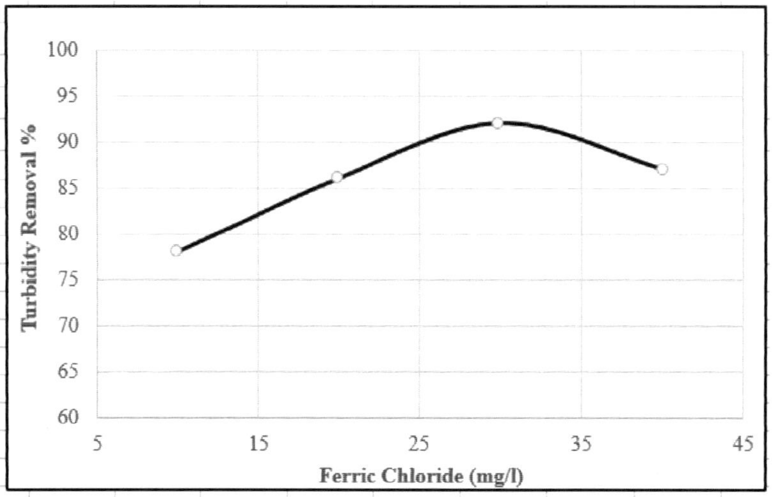

Figure 2. Effect of ferric chloride only on turbidity removal.

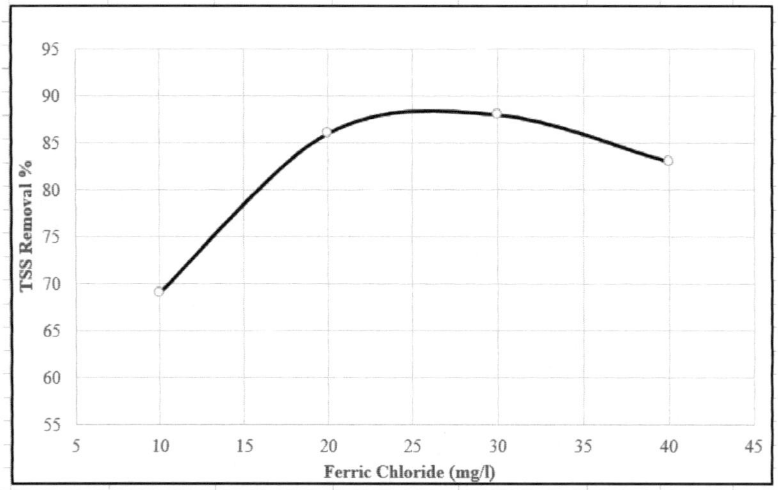

Figure 3. Effect of ferric chloride only on TSS removal.

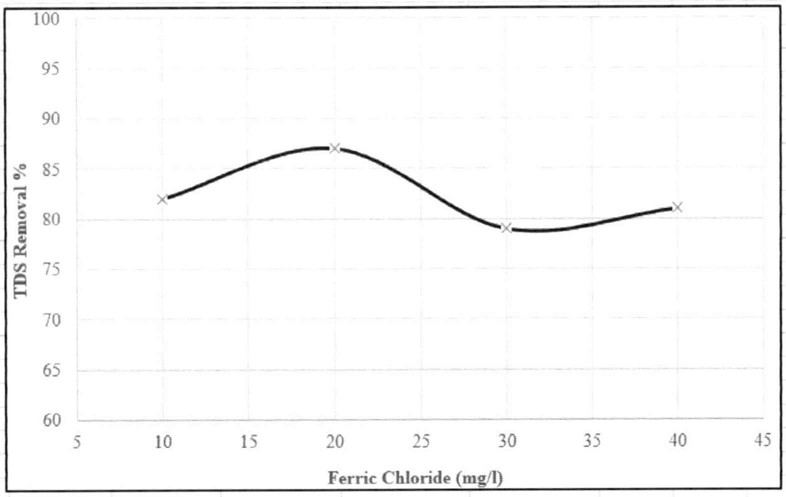

Figure 4. Effect of ferric chloride only on TDS removal.

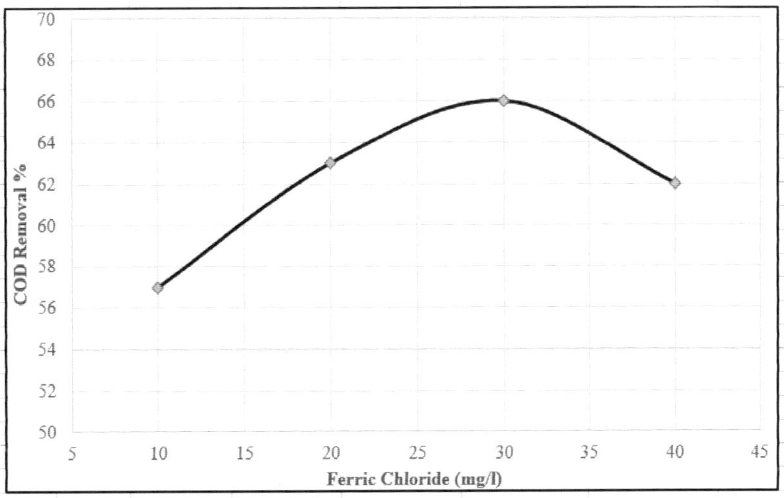

Figure 5 Effect of ferric chloride only on COD removal.

3.2. Combination Effect of Ferric Chloride and CPE

The cationic polyelectrolyte is commonly adopted as a coagulant aid to reinforce the function of coagulant agent in industrial and municipal purposes; As a result, ferric chloride will be considered as the major chemical coagulant and CPE as the coagulant aid. In this manner, a set of experiments were carried out on the treatment of grey wastewater of relatively low turbidity by ferric chloride at dose rang of 10 to 40 mg/l and adding CPE in the range of 1 mg/l to 3 mg/l by using an increment of 0.5 mg/l) to support the flocs formation as well as the sedimentation behavior. The adoption of the little turbidity scale for the synthetic wastewater is depended on the fact that reutilizing practices concern treated wastewater close to hand for discharging. The effect of different ferric chloride dosages on the removal efficiency of turbidity, total suspended solids, total dissolved solids, electrical conductivity and chemical oxygen demand of the grey wastewater at many levels of CPE (1, 1.5, 2.0, 2.5, and 3 mg/l). The initial pollutants concentrations for (Turbidity 60 NTU), (TSS 350 mg/l), (TDS 800 mg/l) and (COD, 300 mg/l). These results are typically plotted in figures 6 to 10.

3.2.1 Turbidity

As shown previously in figure 2 the total removal efficiency was 92% and achieved by using 30 mg/l as ferric chloride coagulant without any addition of coagulant aid. Figure 6 shows that the total removal efficiency increased to 97.5% and decreasing ferric chloride dosage to 20 mg/l when a combination of 2 mg/l of CPE is used

3.2.2 Total Suspended Solids (TSS)

As shown previously in figure. 3 the total removal efficiency was 88% and achieved by using 30 mg/l as ferric chloride coagulant without any addition of coagulant aid. Figure 7 shows that the total removal efficiency increased to 91.14% and decreasing ferric chloride dosage to 20 mg/l when a combination of 2 mg/l of CPE is used

3.2.3 Total Dissolved Solids (TDS)

As shown previously in figure. 4 the total removal efficiency was 87% and achieved by using 20 mg/l as ferric chloride coagulant without any addition of coagulant aid. Figure 8 shows that the total removal efficiency increased to 88% when a combination of 1 mg/l of CPE is used.

3.2.4 Chemical Oxygen Demand (COD)

The combinations of coagulant and coagulant aid have an important impact on the COD removal efficiency as illustrated in figure 10 when using CPE by dose of 2.5 mg/l, the COD removal efficiency increased to 75% using ferric chloride of 20 mg/l while the COD removal was 65% using 30mg/l ferric chloride without additional of CPE. This combination of CPE had a clear effect in decreasing the dosage of coagulant and increasing removal efficiency. Similar results had been got by [8] at their study of the effect of several coagulant aids (polymers) on the COD removal efficiency. From the overall results it can be seen that the arrangement of CPE with ferric chloride leads to minimize the amounts of both ferric and CPE and increasing the removal efficiencies for the pollutants in grey wastewater. The reduction in quantities of ferric chloride and CPE is essential for economical purposes and operation conditions, addition to reducing the sludge, the disposal complications and diminishes filter blocking problems during filtration process; reduction of ferric chloride dosages means constrained increment of pH values.

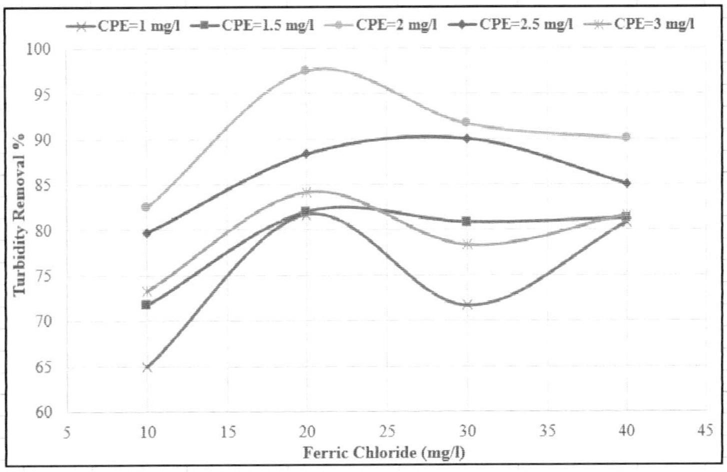

Figure 6. Effect of combination addition of ferric chloride and CPE on turbidity removal.

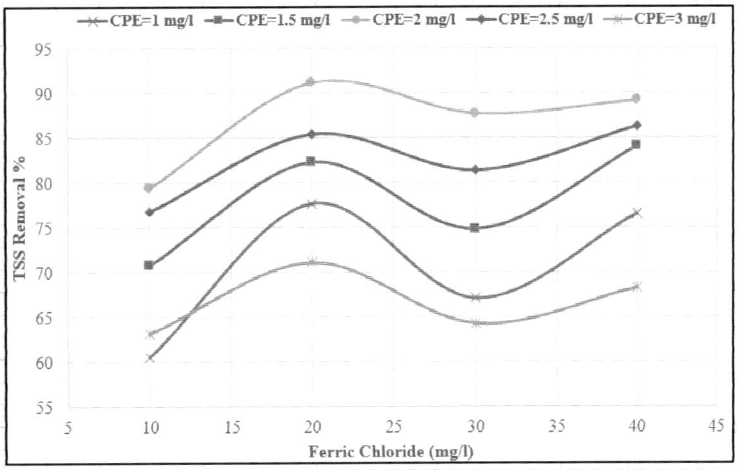

Figure 7. Effect of combination addition of ferric chloride and CPE on the TSS removal.

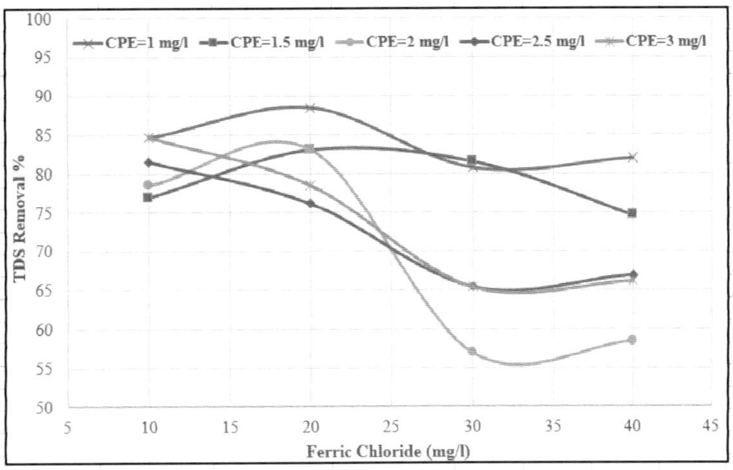

Figure 8. Effect of combination addition of ferric chloride and CPE on the TDS removal.

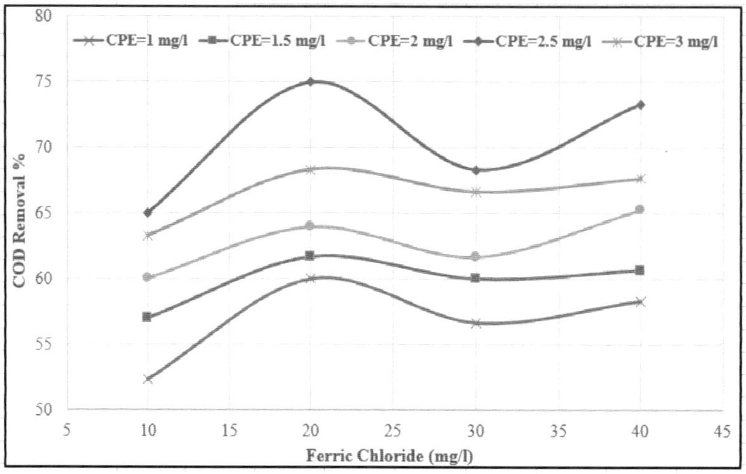

Figure 9. Effect of combination addition of ferric chloride and CPE on the COD removal.

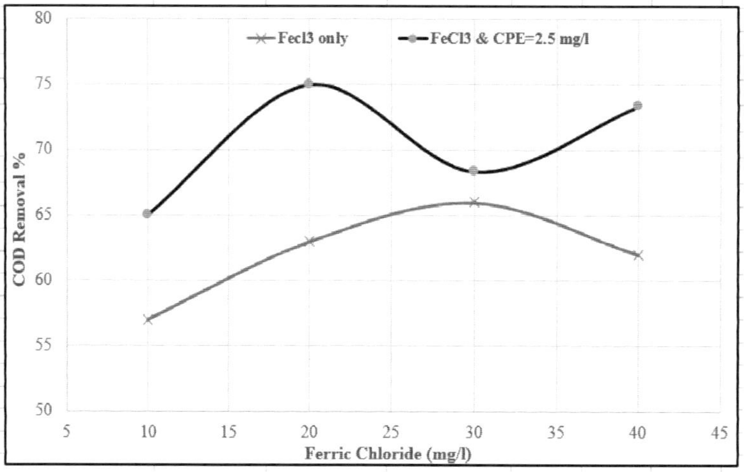

Figure 10. Effect of combination addition of ferric chloride and CPE on COD removal efficiency with initial (COD 300 mg/l).

3.2.5 ANN Model Output

The ANN model accuracy and also time spend to reach the best solution, that gives the best coincidence with the observed data for the studied phenomenon, depending on number of neurons in the hidden layer. Hence, many numbers of hidden layers were examined and the number of layers that satisfied RMSE less than 0.05 were adopted. Figures 11 to 14 show the predicted results from ANN approach versus that observed from the experimental study for COD, TDS, TSS and turbidity removal respectively. The results showed an acceptable agreement for all models. All the parameters had a value of R^2 of (0.9972, 0.9993, 0.9992, 0.9985) respectively.

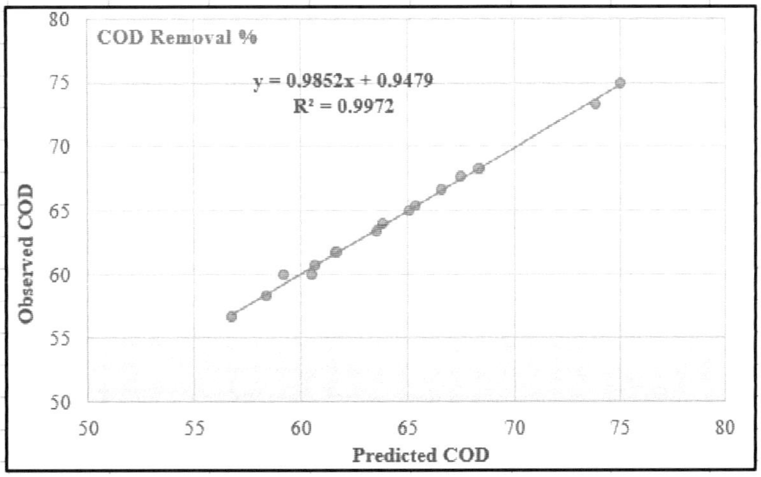

Figure 11. Predicted- Observed fitted line of ANN Results for COD.

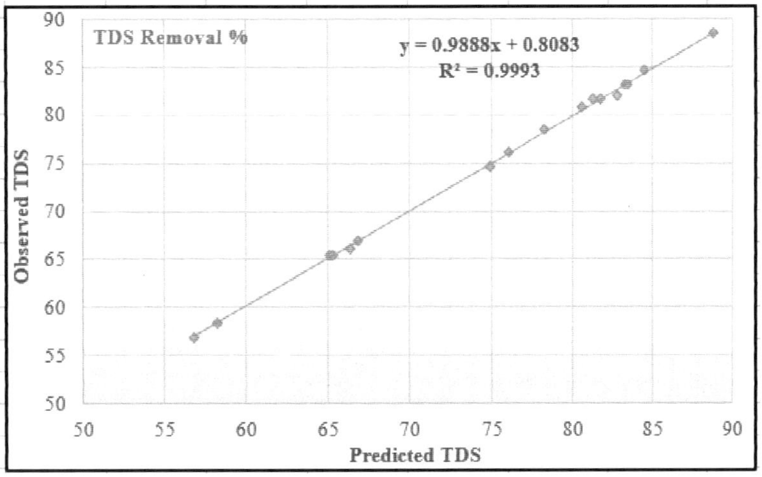

Figure 12. Predicted- Observed fitted line of ANN Results for TDS.

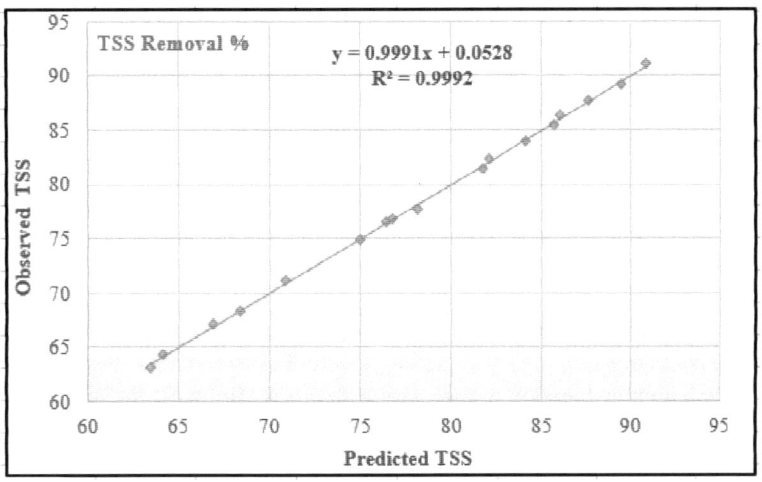

Figure 13. Predicted- Observed fitted line of ANN Results for TSS.

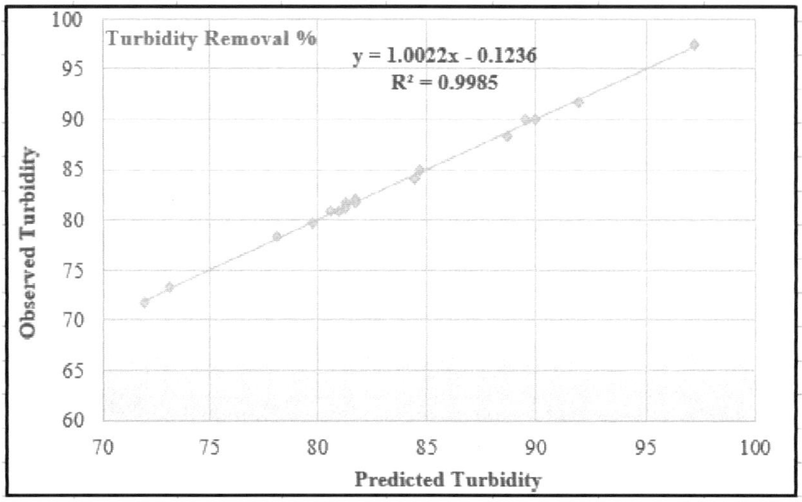

Figure 14. Predicted- Observed fitted line of ANN Results for turbidity.

4. Conclusion

Pollutants removal efficiency (Turbidity, TSS, TDS, and COD), by (92%, 88%, 87% 66%) respectively could be obtained by using ferric chloride solution of concentration 20-35 mg/L only without addition of cationic polyelectrolyte. The combination of ferric chloride and cationic polyelectrolyte was found to improve the results for pollutants removal efficiency (Turbidity, TSS, TDS, and COD), by (97.5%, 91%, 88% 75%) respectively addition to decrease the dosage of the used coagulant. The reduction in quantities of ferric chloride and CPE is essential for economical purposes and operation conditions, addition to reducing the sludge, the disposal complications and diminishes filter blocking problems during filtration process; reduction of ferric chloride dosages means constrained increment of pH values. Finally, the predicted results from artificial neural network (ANN) approach versus that observed from the experimental study for COD, TDS, TSS and turbidity removal showed an acceptable agreement for all models. All the parameters had a value of R2 of (0.9972, 0.9993, 0.9992, 0.9985) respectively.

References

[1] Morel, A. Diner, S., 2006 Greywater Management in Low and Middle-Income Countries, Review of different treatment systems for households or neighborhoods. Swiss Federal. Institute of Aquatic Science and Technology (Eawag). Dübendorf, Switzerland.

[2] Macungie, E. Sakuringwa, S., 2007. Greywater reuse: A strategy for water demand management in Harare. Physics and Chemistry of the Earth 32, 1231-1236.

[3] Al-Jayyousi OR, Greywater reuse: towards sustainable water management, Desalination, 156, 2003, 181-192.

[4] Lu W, Leung AYT, A preliminary study on potential of developing shower/laudry wastewater reclamation and reuse system. Chemosphere 52, 2003, 1451-1459.

[5] Rahbar, M.S., Alipour, E. and Sedighi, R.E. 2006 Color Removal from Industrial Wastewater with a novel Coagulant Flocculation Int. J. Environ. Sci Tech. vol.3, No.1, pp : 79-88

[6] Sostar-Turk, S, Petrinic, I, Simonic M, Laundry wastewater treatment using coagulation and membrane filtration. Resour. Conserv. Recy. 44, 2005, 185–196.

[7] G.Ramon, M.Green, R. Semiat and, D. Dosoretz, Low strength greywater characterization and treatment by direct membrane filtration Desalination, vol 170, pp. 241-250, 2004.

[8] Pia A.B., J.A. Mendoza J.A, Alcover L.R. and Iborr A. 2003 Comparison between Nano Filtration and Ozonation of Biologically Treated Textile Wastewater for its Reuse in the Industry Desalination 157 pp: 81-86.

The Evaluation of Khorasan Razavi Range for the construction of wind power plants through multi-criteria decision making in GIS environment

A Sh Alhameedah[1], R Shad[2], M Ghaemi[3]

[1] MSc graduated of Civil Engineering-GIS, Dhi Qar, Iraq. Email: ah31101988sh@gmail.com
[2] Assistant Professor of Civil department, Engineering Faculty, Ferdowsi University of Mashhad, Mashhad, Iran. Email: r.shad@um.ac.ir
[3] Assistant Professor of Civil department, Engineering Faculty, Ferdowsi University of Mashhad, Mashhad, Iran. Email: mghaemi270@gmail.com

Abstract. The purpose of this research was to determine the suitable location of wind power plants in Khorasan Razavi province according to the climate criteria and conditions (wind speed, prevailing wind speed, pressure and temperature), geography (elevation, slope), socioeconomic factors (distance, distance from the cities and distance from the villages), environmental factors (distance from protected areas, land use and distance from the river) and geology (distance from faults). The process has been carried out in the geographical information system environment using the analytical hierarchical process. Based on the role and effect of these factors, a map of the Effective Factors was prepared in ArcGIS10.6 environments. The weight of the criteria was determined by the Analytical Hierarchy Process Analysis (AHP) method in the EC2000 software environment. The ArcGIS10.6 software was used for spatial modelling, analysis and overlaying the different layers. The final map of suitable wind power plants was carried out in four different classes (very suitable, suitable, somewhat suitable and unsuitable). The results showed that 10.37% (1239871 ha) of the area was identified as very suitable for construction, including areas in the northwest of the province (around Sabzevar) and relatively slightly in the southeast (around Taibad) and eastern area (around Sarakhs). The results also showed that the GIS, as a decision support system in building a power plant, can be used to prepare data and prioritize experts' models and opinions regarding various factors in choosing the appropriate location for an effective wind power plant.

Keywords: Wind power site location, Analytical Hierarchy Process (AHP), Geographic Information System (GIS), Razavi Khorasan Province.

1. Introduction

One of the best ways to supply energy is the wind energy that is produced by installing wind turbines [1]. The rapid development of wind energy extraction technologies has made it a safer alternative to today's energy systems [2]. Many countries have started to extract this kind of energy because of the easy installation and rapid commissioning of wind turbines and low costs [3]. Among the studies that carried out in this regard is Ebrahimi and Kamalimoghadam [3] who used the GIS to allow the construction of wind farms in the southern Khorasan province. Bennui et al. used the GIS, multi-criteria decision making, and hierarchical analysis process (AHP) to locate wind turbines in Thailand.

The study found that the East Coast of Thailand was one of the most prone areas for wind turbines [5]. Soltani et al. examined the potential of wind energy for determining the feasibility of establishing wind power plants in Amir Abad, Iran [6]. Nooralahi et al. evaluated the potential of the western region in Iran to construct a wind power plant, taking into account technical, environmental, and economic criteria in the GIS environment [7]. Based on the wind direction and velocity data for three hours wind stations in Kermanshah, Islamabad, West, Ravansar, Kangavar and Sarpol Zahab, over the course of 9 years (2009-1997), Mohammadi et al. examined the potential of these areas for using wind energy [8]. The results showed that the three stations of Ravansar, Sarpol Zahab and Kangavar had the potential to produce wind energy. The Islamabad area of the West was suitable for using wind energy if using high wind turbines, and Kermanshah does not have the potential to use wind power [8]. Nasrollahi, in a study using the GIS, prioritized the use of wind and power plants in different parts of Iran [9]. The results indicated high potential of the cities of Manjil, Khoreh, Birjand, Zabol and Sabzevar for the construction of wind farms. Nguyen (2007) [10], considering the suitable criteria for building wind farms in Vietnam, considered the potential of this country to be the use of wind energy. Prabamroong et al. evaluated the development of wind farms in Thailand using GIS, taking into account 20 criteria such as distance from city center, distance from roads, distance from fault and etc. [11]. The results showed that about 30% of the total area was suitable for installing a wind power plant and most of the places were located in rural areas. In Bazzi, (2008) [12], wind turbines were used to design wind farms in Lebanon using the GIS to meet the data requirements of wind, urban areas, turbine types, and the distance between wind turbines. Using GIS and remote sensing, identified suitable sites for the construction of windmills in the Andaman and Nicobar Islands, India [13]. The results showed that Andaman and Nicobar Islands had a high potential for wind energy and had good conditions for the construction of windmills. Due to the windy situation of the Khorasan Razavi province climate, there is a lack of related studies in the use of wind power capacity and the failure to complete a study in this regard. Therefore, the purpose of this study was to obtain the necessary factors for selecting suitable sites for the construction of Wind power plants. These factors are our Major Criteria such as climate, geography, socio-economic, environmental and geology.to this end, according to selected criteria, Geospatial information systems (GIS) have been used to evaluate the appropriate locations of wind power plants in the study area.

The climate criterion in the weighting process was more important than the other criteria and thus had a higher weight. Climate parameters are wind speed, prevailing wind speed, pressure and temperature from climatic options [8]. Geographic criteria are height from sea level and slope. In areas with high altitudes and roughnesses, it is difficult to move equipment. Moreover, an increase in altitude is effective in wind speeds, and as altitude increases, wind speed increases [14]. The socio-economic criterion includes options for minimum distance from communication paths, at least distance from cities and villages (population centers). The more distance the communication paths area, the easier it will be to access the place. Availability of access roads, as well as being close to areas where access to the national electricity grid and electricity consumers can increase the district's potential for the construction of a wind power plant [15]. Cities and populations with a large population may be affected by wind farms in terms of safety, noise and landscape. Environmental criteria include distance from protected areas, land use and distance from rivers. Protected areas have been protected because of the presence of valuable forests or valuable meadows and beautiful landscapes to prevent the degradation of vegetation and animal life. Wind farms have a negative impact on the inherent nature of these areas due to the addition of a natural technology factor [16]. The only geological criterion option in this study is the distance from the faults. the mild fault activity causes major changes in the earth's crust and the stability of the regions. Therefore, the selected location for the power plant should be located at an appropriate distance from the fault center [17-18].

1.1. Area of study

Khorasan Razavi province is one of the provinces of Khorasan located in northeastern Iran and its capital is Mashhad. The area of Khorasan Razavi province is 11544664.7744 ha, which covers 7% of the total area of Iran. Forty-nine point two % of the province is mountainous and 50.8% is plain. This province consists of 4 watersheds including Atrak, Qarahagh, Central Desert and East of Iran. Figure 1 shows the geographic location of the Khorasan Razavi province and its counties. Khorasan Razavi province is very diverse due to the vast variety of natural conditions, and each of its different areas has certain characteristics. The province has a total length of 531.6 km from the north and northeast, with a common border with Turkmenistan from the east, with a total length of about 302 km, and a border with Afghanistan. The highlands can be divided into northern and southern highlands. The northern heights of Khorasan are generally eastern-western, while the south-east highlands are north-south. The highest point of the province is Binalood Peak at 3615 meters and the lowest point of the province in the Sarakhs plain with a height of 299 meters above sea level.

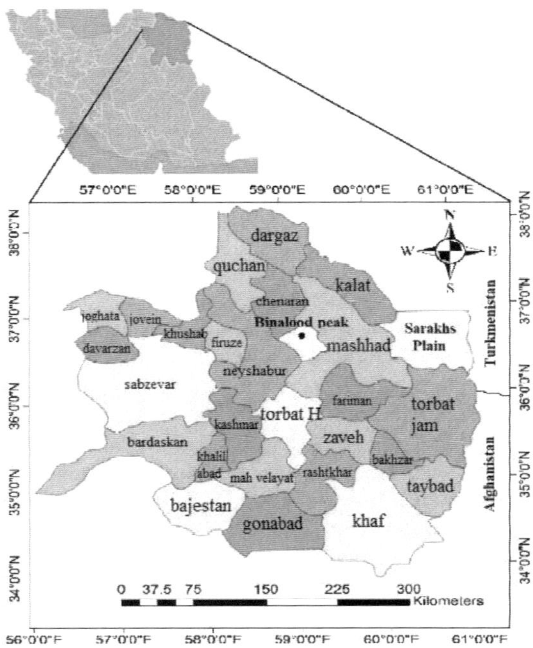

Figure 1. Location of study area (Khorasan Razavi province).

2. Materials and methods

2.1. Research Methodology

In order to locate the appropriate areas for wind power plant construction, the first layer of information is 5 main criteria and 13 sub-criteria (1) climate (wind speed, wind speed prevailing, pressure and temperature), (2) socio-economic factors (distance from towns and villages and distance from roads communication), (3) environmental factors (distance from the waterways, land use and distance from protected areas), (4) geographic (height from sea level and slope), and (5) geology (distance from faults); and from the integration of the criteria and sub-criteria maps, a final map of the construction sites of the wind farm was prepared (Figure 2).

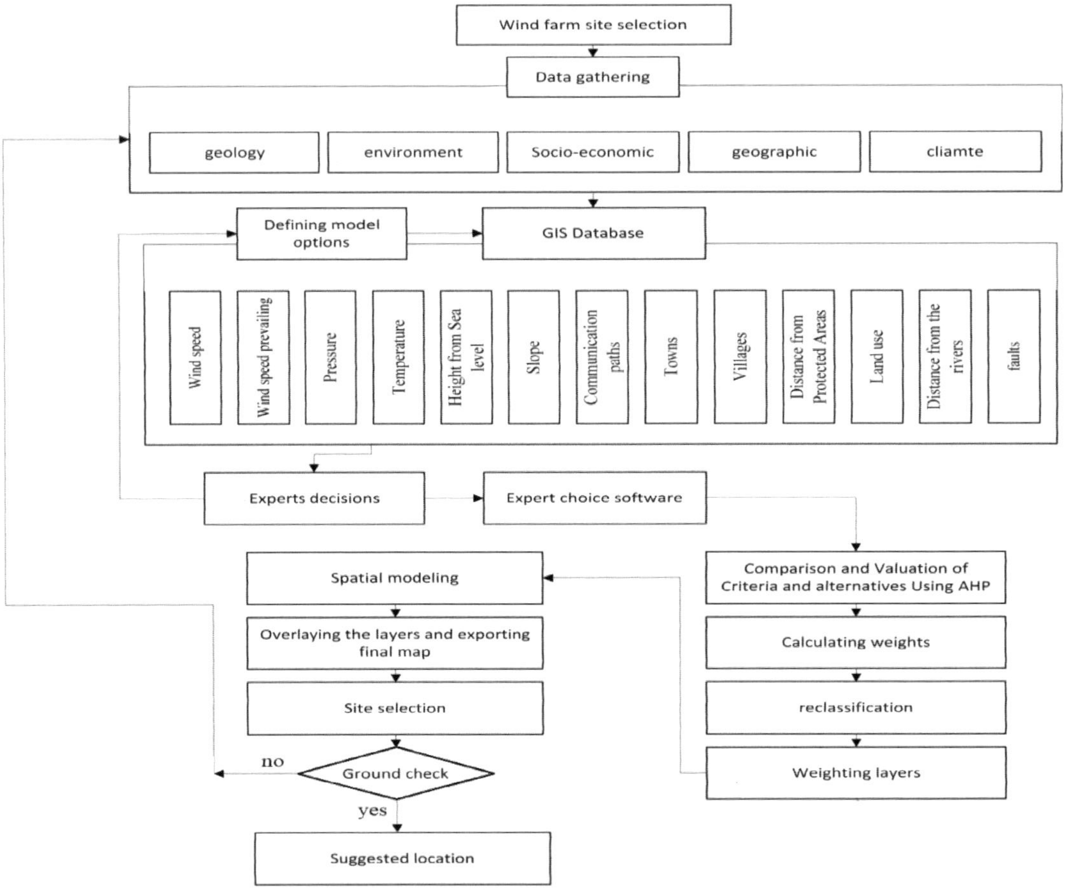

Figure 2. Stages of preparation and integration of information layers and preparation of final maps of suitable sites for wind power plant construction.

Maps of distances from cities, villages, protected areas, communication paths, streams and faults, after providing data related to each of the mentioned criteria from relevant organizations (Table 1) by using the Spatial Analysis Tools extension in Arc Gis toolbax and the Distance function in the Arc GIS 10.6 software environment. Wind velocity map, prevailing wind speed, pressure and temperature. After creating the dotted point 13 of the synoptic station in the Arc GIS 10.6 environment, we use the Inverse Distance Weighting (IDW) method and according to the preferred classes (Table 2), it turned out that the map of the classes of effective options for the construction of a wind power plant is shown in Figures 3-7.

Table 1. Research data

Data name	Type	Aim	source
wind speed	Excel	Preparation of zoning map	**Meteorological Organization**
Wind speed prevailing	Excel	Preparation of zoning map	**Meteorological Organization**
Pressure	Excel	Preparation of zoning map	**Meteorological Organization**
Temperature	Excel	Preparation of zoning map	**Meteorological Organization**
DEM	raster	Preparing a gradient map	**Prepared from digital cameras**
Land use	vector	Reference data for location analysis	**Produced by the researcher**
Network of watercourses	vector	For location analysis	**Regional Water Organization**
protected areas	vector	For location analysis	**Department of Environment**
City Network	vector	For location analysis	**Agricultural and Natural Resources Research Center**
Village network	vector	For location analysis	**Agricultural and Natural Resources Research Center**
Communication path network	vector	For location analysis	**Agricultural and Natural Resources Research Center**
Faults	vector	For location analysis	**Geology organization**

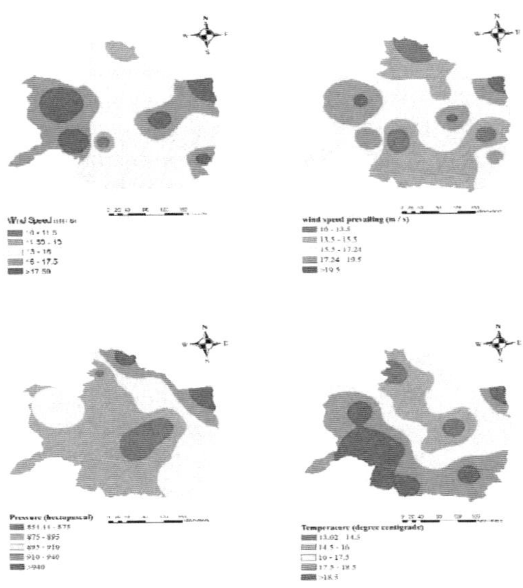

Figure 3. Map of the subclasses of climatic classes.

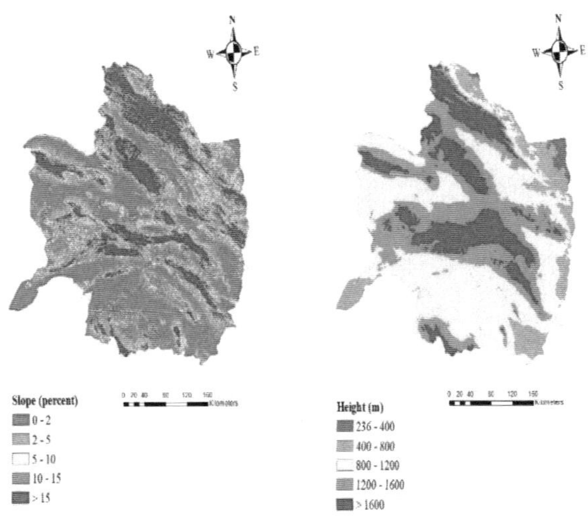

Figure 4. Map of the subclasses of geographic.

IOP Conf. Series: Materials Science and Engineering **584** (2019) 012004 doi:10.1088/1757-899X/584/1/012004

Table 2. Land use classes for the construction of wind power plants and their desirability

Subcategories	**Inappropriate**	**fairly inappropriate**	**Somewhat appropriate**	**Proper**	**Very convenient**
wind speed (m/s)	10-11.5	11.5-13	13-16	16-17.5	**>17.5**
Wind speed prevailing (m/s)	10-13.5	13.5-15.5	15.5-17.24	17.24-19.5	**>19.5**
Pressure (hectopascal)	875-854	875-895	895-910	910-940	**>940**
Temperature (°C)	13-14.5	14.5-16	16-17.5	17.5-18.5	**>18.5**
Height from sea level (m)	>1600	0-400	400-800	800-1200	**1200-1600**
Slope (percent)	>15	10-15	5-10	2-5	**0-2**
protected areas (km)	0-2	2-4	4-6	6-8	**>8**
Land use	Agriculture	Forest	lake meadows and landscapes	wetlands and rocky marshland	**sand**
The watercourses (km)	0-1	1-2	2-3	3-4	**>4**
Cities (km)	0-3	>12	9-12	6-9	**3-6**
Villages (km)	0-2	>11	8-11	5-8	**2-5**
communication paths (km)	>8	6-8	4-6	2-4	**0-2**
Faults (km)	0-1	1-6	6-10	10-15	**>15**

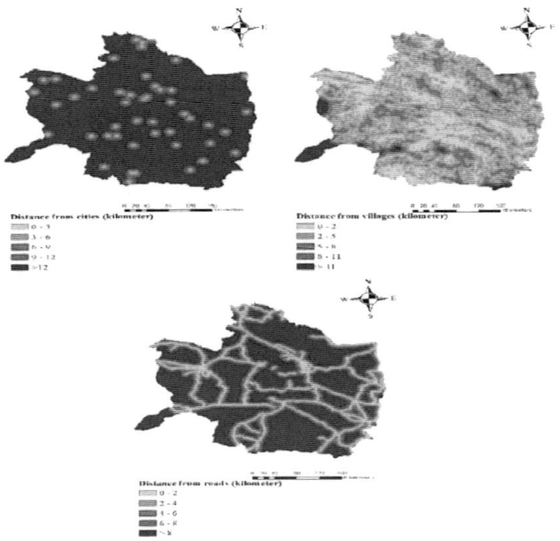

Figure 5. Map of the subclasses of geology classes.

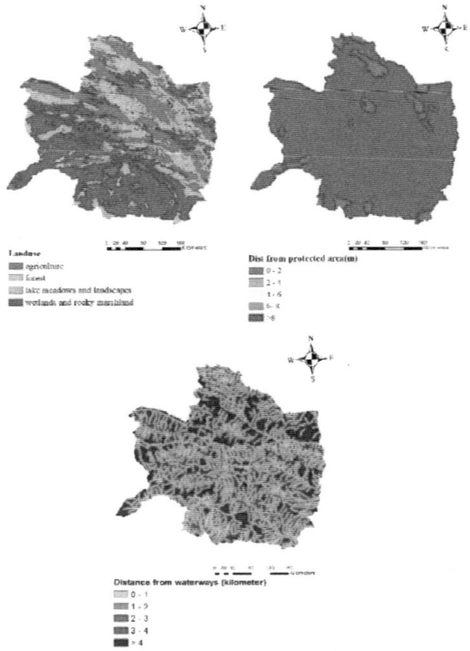

Figure 6. Map of the subclasses of socio-Economic classes.

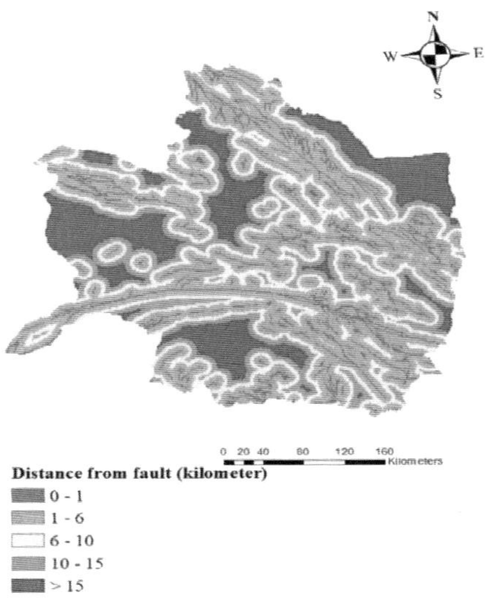

Figure 7. Map of the subclasses of the ecological class.

The weight of the parameters is obtained by pairwise comparing between them and using the analytical hierarchical process [19]. A fair comparison between the parameters was made according to experts' judgment and based on that, the importance and weight of each criterion were determined. Then, by multiplying the map of each criterion in its weight and aggregating the weighted layers, the final map of land suitability was calculated with equation 1.

The final weight of the option j = $\sum_{k=1}^{n}\sum_{i=1}^{m} W_k W_i(g_{ij})$ (1)

In this regard, W_k is the significance coefficient of k,, W_i the significance coefficient of sub-criterion i, g_{ij} is the value of j in relation to the sub-criterion i [20]. The map was categorized based on four equal parts in four classes, i.e. inappropriate, relatively appropriate, appropriate and very appropriate.

In this research, weight of the criteria and sub-criteria in the Expert Choice2000 software environment was determined by collecting expert opinions and identifying the value and importance of the criteria against each other and finally creating Paired comparison matrix of criteria. According to Table 3, the results showed that among the main criteria, climatic criteria, with 0.515, had highest degree of importance and geological criterion, with 0.08, had the least importance in the construction of wind power plants. Also, among the sub-criteria of climate, the wind speed with 0.549 and the temperature with 0.074 had the highest and lowest values respectively (Table 4). Among the geographic criteria, height from the sea level with value of 0.8 was the most important (Table 5) because it plays a big role in the location of the wind power plant and, in geological field, the distance from the faults as the only importance factor in locating the wind power plant. In the socioeconomic factors, the distance from communication paths with the importance of 0.655 was the most important (Table 6). Among the environmental criteria, the distance from protected areas with a significance of 0.54 was the highest importance (Table 7). Pairwise comparison of the criteria showed that the climate and geology, respectively, had the most and the least importance among the effective criteria in locating the wind power plant in the study area.

Table 3. Comparison of the main layer's criteria in the location of wind power plants

criteria	The climate	Geographical	Economic-social	environmental	Geology	weight
The climate	1	3	6	6	6	0.515
Geographical	-	1	3	3	3	0.221
Economic-social	-	-	1	3	3	0.129
environmental	-	-	-	1	3	0.082
Geology	-	-	-	-	1	0.053
Compatibility factor						0.08

Table 4. Comparison of climatic sub-criteria in locating wind power plants

Climate sub criteria	wind speed	Wind speed prevailing	Pressure	Temperature	weight
wind speed	1	3	5	5	0.549
Wind speed prevailing	-	1	3	3	0.248
Pressure	-	-	1	3	0.129
Temperature	-	-	-	1	0.074
Compatibility factor					0.07

Table 5. Comparison of geographic subcategories in location of wind power plants

Geographical sub criteria	Height from Sea level	Slope	weight
Height from Sea level	1	4	**0.8**
Slope	-	1	**0.2**
Compatibility factor			**0**

Table 6. Comparison of socio-economic sub-criteria in locating wind power plants

socio-economic sub-criteria	Communication paths	Towns	Villages	weight
Communication paths	1	3	6	0.655
Towns	-	1	3	0.25
Villages	-	-	1	0.095
Compatibility factor				0.02

Table 7. Comparison of environmental sub-criteria in the locating of wind power plants

environmental sub-criteria	Distance from Protected Areas	Land use	Distance from the rivers	weight
Distance from Protected Areas	1	2	3	0.54
Land use	-	1	2	0.297
Distance from the rivers	-	-	1	0.163
Compatibility factor				0.008

2.2. Final map and integration of criteria

After weighing the effective layers in the location of wind power plants using the Analytical Hierarchy Process (AHP) process, Arc GIS 10.6 was used to combine and overlap the maps. Finally, the map of suitable locations for construction of wind power plants in the final deck was categorized in four classes including, inappropriate, relatively suitable, appropriate and very appropriate (Figure 8). Areas suitable for the construction of wind power plants were in the appropriate classroom, in the western area and a small area in the east and southeast with an area of 1239871 hectares. Restricted areas of the province with inappropriate potential are located in the north and center of the province with an area of 3477116 hectares (Table 8).

Table 8. Specifications of the location of wind power plants by AHP method

Row	class	area (ha)	Portion of province
1	Inappropriate	3477116.8062	30.11%
2	Somewhat appropriate	4369444.5484	37.84%
3	Suitable	2458231.6228	21.29%
4	Very convenient	1239871.7970	10.73%

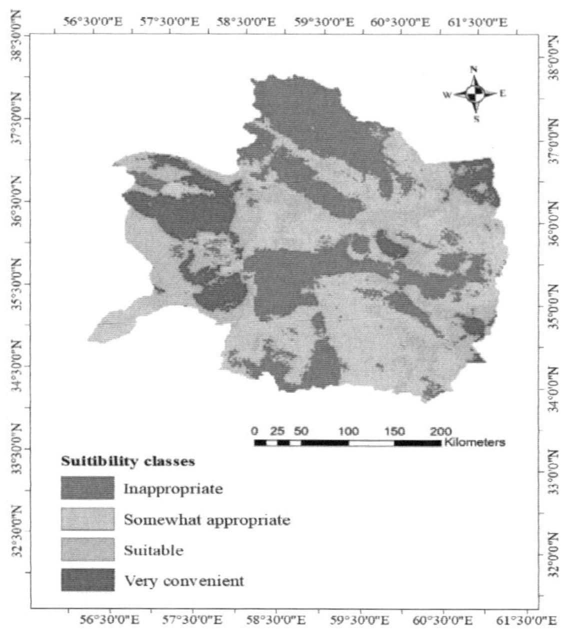

Figure 8. Mapping the location of wind power plants by AHP method.

3. Discussion

In this study, the criteria for wind farm site selection were categorized into 5 classes including climatic, socioeconomic, environmental, geographic and geological groups. The climate criterion in the weighting process was more important than the other criteria and thus had a higher weight. Climate parameters are wind speed, prevailing wind speed, pressure and temperature from climatic options. Wind speed is the most important factor in installing wind power plants. The higher wind speed, the greater the wind turbine power produced. Because the wind energy rotates two or three feathers that are placed behind the wind turbine rotor and As the wind speed increases, the turbine output will increase. Usually, the minimum wind speed for starting wind turbines is between 3 and 4 m/s depending on the type and design turbine [21]. According to the wind speed classification map of the province, more than 95 per cent of the province area has a wind speed of more than 4 m/s, which indicates the potential of the region to build a wind power plant in the region. The higher the wind speed, the greater the wind power produced by wind turbines. The dominant wind direction is very important when using wind energy, and wind turbines must be installed in places that have the least obstacles in the dominant wind directions [8-9]. Increasing the height leads to lower air pressure. This reduction in pressure will be effective on the output power from turbines, because the pressure affects the performance of turbine components [22]. Hence, the pressure is also one of the most important parameters in the construction of wind power plants. The other parameter of the climate that contribute to the location of the wind power plant is temperature. Increasing the height reduces the air temperature. Reducing the temperature will be effective in the amount of power to be extracted from the turbine [15].

Geographic criteria are height from sea level and slope. In areas with high altitudes and roughnesses, it is difficult to move equipment. Moreover, an increase in altitude is effective in wind speeds, and as altitude increases, wind speed increases. So, the height should not be too high to make any significant changes in the amount of turbins power [14]. The slope of the area is also an important factor in the proper operation of turbines and their installation. The slope of the earth will increase turbine maintenance costs, shortened life and reduced energy production. The slope should not be too high to prevent the installation of a turbine [10].

The socio-economic criterion includes options for minimum distance from communication paths, at least distance from cities and villages (population centers). The more distance the communication paths area, the easier it will be to access the place. Availability of access roads, as well as being close to areas where access to the national electricity grid and electricity consumers can increase the district's potential for the construction of a wind power plant [15]. Cities and populations with a large population may be affected by wind farms in terms of safety, noise and landscape. On the other hand, these population centers are considered as major consumers of power, and the proximity of the chosen location makes them closer to the centers of production and power consumption [10]. Due to the nature of wind energy that focuses on production, often in remote and rural areas, the development of this industry, both in developed and Developing countries, has led to clear developments in rural areas.

One of the important factors in locating wind power plants is considering environmental issues. Environmental criteria include distance from protected areas, land use and distance from rivers. Protected areas have been protected because of the presence of valuable forests or valuable meadows and beautiful landscapes to prevent the degradation of vegetation and animal life. Wind farms have a negative impact on the inherent nature of these areas due to the addition of a natural technology factor [16]. The purpose of land use survey is to determine the value of land for locating wind power plants. This classification is based on the types of agricultural land use, pastures and forest lands, lakes, wetlands and lands with intersection, rocky, Non-fertile and smooth sand. The wind turbines located on the beaches and close to them and adjacent to the rivers, will have adverse effects on these areas. These areas are also ecologically the habitat of many specific birds, which have always been important for their care [16].

The only geological criterion option in this study is the distance from the faults. the mild fault activity causes major changes in the earth's crust and the stability of the regions. Therefore, the selected location for the power plant should be located at an appropriate distance from the fault center [17-18]. According to the results, the zones identified in the region with the most suitable location accounted for 10.73% of the total area with an area of 1239871 ha. High-power areas have been designated in the selected regions in order to locate wind power plants in the area. These areas are in the western region and small areas in the east and southeast of Khorasan Razavi province.

4. Conclusion

In this research, after determining the effective factors in locating wind power plants and their role, spatial information layers were prepared and by combining these layers, suitable sites for the construction of wind power plants were determined. Deciding on the right location for a power plant requires simultaneous attention to a number of factors that the GIS allows for the integration of spatial layers. Selected sites are fully influenced by the parameters involved in the analysis and the relevant weights. Analytical Hierarchical process allows for proper weighting of options and criteria by creating a hierarchical structure to find suitable weight and sites for the construction of wind power plants.

One of the challenges in this study was to convert the daily statistical data of wind speed, prevailing wind speed, temperature and pressure into information layers, which were provided by 13 weather stations throughout the province of Khorasan Razavi. Hence, The survey data was compiled from the beginning of 2014 to the end of 2016, and average of this data was taken. Then, using interpolation in Arc Gis10.6 software, these values were calculated for the whole Khorasan Razavi province.
Another problem in this study was the accuracy of the layers of information provided, including the streets and cities, villages, protected areas, waterways and faults, because the higher the accuracy of the input layers, the higher the precision is.
We can use other multi-criteria weighing methods to locate wind power plants and compare the results with the results of this study.

References

[1] Rahimzade F and Jahangiri Z 2002 A suitable location for using wind energy in the country. *Proceedings of the 3rd Fuel Conservation Optimization Conference in Tehran Building.*

[2] Zaim S, Sevkli M, Camgöz-Akdağ H, Demirel OF, Yayla AY and Delen D 2014 Use of ANP weighted crisp and fuzzy QFD for product development .*Expert Systems with Applications*, 41(9): 4464 -4474. (DOI: 10.1016/j.eswa.2014.01.008).

[3] Yang YQ, Wang SQ, Dulaimi M, Low SP 2003 A fuzzy quality function deployment system for buildable design decision-makings. *Automation in construction*, 12(4): 381-93 .(DOI: 10.1016/S0926-5805(03)00002-5).

[4] Ebrahimi M and Kamalimoghadam M 1388 Application of GIS in the feasibility study of wind power plants. Case Study: South Khorasan. *First Con. Renewable Energy and Dispersal Production of Iran*. University of Birjand, 7.

[5] Bennui A, Rattanamanee P, Puetpaiboon U, Phukpattaranont P and Chetpattananondh K 2007 Site selection for large wind turbine using GIS *PSU-UNS. Int. Con. Engineering and Environment*, Thailand. 561-566.

[6] Soltani B, Gholamian A and Farahani K 2010 Investigation of wind energy potential in Amirabad for the purpose of feasibility study of the establishment of a wind power plant. *Iranian Journal of Energy*, 16.

[7] Nooralahi Y, Ashraf S and Zamani M 2011 West Wind Power Wind Power Using Geographic Information System (GIS*). Iranian Journal of Energy*, 21.

[8] Mohamadi H, Rostami S and Taghavi F 2012 Potentiometer of wind energy in Kermanshah province. *Natural Geography Research*, 13.

[9] Nasrolahi R 2007 site Location of wind power plants using multivariate analysis methods. 120.

[10] Nguyen KQ 2007 Wind energy in Vietnam :Resource assessment, development status and future implications. *Energy Policy*, 35(2): 1405 -13 .(DOI: 10.1016/j.enpol.2006.04.011).

[11] Prabamroong A, Manomaiphiboon K ,Chanaprasert W, Rajpreeja N 2009 Development of a GIS decision support system for wind farm installations in Thailand: current state and results, *SEE*, 690-95.

[12] Bazzi AM 2008 GIS-based wind farm site selection in Lebanon Electro/Information Technology. IEEE. ICEE. xplore, 201-204.

[13] Iyappan L, Kasinatha Pandian P, 2012 Identification of potential wind farm locations in Tirumangalam Taluk using geospatial information technology. *IJERA*, 2(3): 1578 -83.

[14] Banos R, Manzano-Agugliaro F, Montoya F, Gil C, Alcayde A and Gómez J 2011 Optimization methods applied to renewable and sustainable energy A review, *Ren and Sust. Energy. Rev.* 1753-66

[15] Asgharipoor J, Morshedi and Borna 2012 Application of Geographic Information System in Locating Wind Power Plants (Case Study: Khuzestan Province) *J. RSGIS. Natural Resources Science.*

[16] Baban SM and Parry T 2001 Developing and applying a GIS-assisted approach to locating wind farms in the UK. Renewable Energy, 59-71.

[17] Beheshtifar M, Sadimesgari M and Karimi M 2006 Location of gas stations using GIS. *Twenty-first Int Power Con*, 7.

[18] Sadeghi Z, Dalalbashi V and Hori H 2013 Prioritizing factors affecting the location of renewable energy power plants (solar energy and wind power) in Kerman province using GIS and multi-criteria decision-making techniques. *Journal of Planning and Energy Policy Research*, 17.

[19] Saaty TL 1990 How to make a decision: The analytic hierarchy process. *European Journal of Operational Research*, 48(1): 9-26 .(DOI: 10.1016/0377-2217(90)90057-I).

[20] Malczewski J 2000 On the use of weighted linear combination method in GIS: common and best practice approaches, *Transactions in GIS*, 4(1): 5 -22. (DOI: 10.1111/1467-9671.00035).

[21] Yue C-D and Wang S-S 2006 GIS-based evaluation of multifarious local renewable energy sources: a case study of the Chigu area of southwestern Taiwan. *Energy Policy*, 34(6): 730-42 . (DOI: 10.1016/j.enpol.2004.07.003).

[22] Li D S, Li R N, Wei L J, Wang X Y, Qiang Y and Li Y R 2013 Comparison of the pressure distribution of a wind turbine blade based on field experiment and CFD. *Materials Science and Engineering*, 52(052004). (DOI:10.1088/1757-899X/52/5/052004).

IOP Conf. Series: Materials Science and Engineering **584** (2019) 012005 doi:10.1088/1757-899X/584/1/012005

A vision of sustainable & smart knowledge based urban development in Iraqi cities: adopting GIS to determine knowledge capital elements and strategies in the city of Hila

Rashaa M Al Khafaji[1]

1Department of Architectural Engineering, Collage of Engineering, Babylon University, Hila, Iraq. Email: rashaa.malik@gmail.com

Abstract. The research seeks to explain the dimensions of development in cities according to the foundations and mechanisms of knowledge capital, especially urban capital, which contributes to enhancing the characteristics that create successful knowledge cities and support their strategies. A local case study was conducted to determine the potential of applying knowledge-based urban development strategies, and in accordance with the specific analytical framework for the characteristics and indicators of knowledge cities. The analytical framework for knowledge cities indicators will be adopted as a baseline for the analysis of the characteristics of the cluster and diversity. Two basic studies included: the structural plan for Babil province (five phases prepared between 2009 and 2010), and the master plan for Hila 2006. A survey method applied to collect the required, and use the latest available statistics about the city. The master plan of the city was analyzed, updated for specific indicators through GIS method. Opportunities for development of two axes were developed in the center of the northern and southern university. Implementation of the cluster strategy. In the center of the ancient city of Babylon, with knowledge gathering support strategy, and knowledge capital indicators are monitored according to the framework tested to determine the course of the three strategies.

Key words: urban development, knowledge cities planning, GIS.

1. Introduction

Cognitive transformation in cities is generally defined and then the identification of basic levels of transformation occurs through which the characteristics of each wave of transformation in the human civilization are determined and the latest wave is based on high-value-added knowledge activities. The international perspective is the main framework of KC strategies, showed in 'Figure 1'.

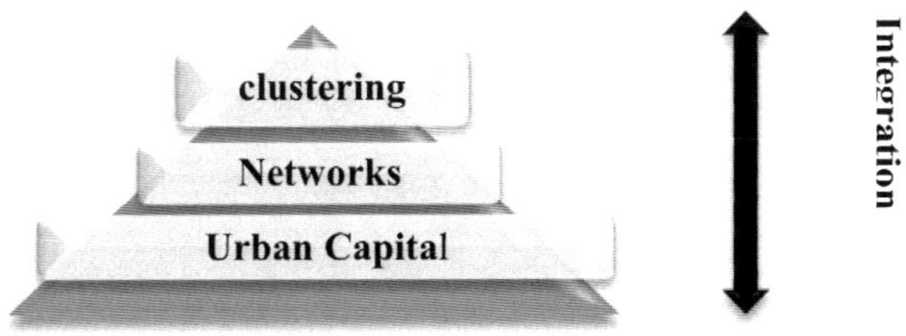

Figure 1. Integrated Perspective of KC Development Strategies, (by researcher)

2. Knowledge clustering Strategy and Knowledge Networking Strategy indicators

"The concept of Knowledge City(KC) is not confined to the physical planning of the sites of knowledge groups" [1] The concept is about strengthening communication and enhance full control (by legislation) for the design and management of private and public projects and investments" [2]. "Networks mean enhancing functional relationships affecting products and services. The value of the connection may be greater than the actual value of the product" [3]. These goals are summarized into two strategies that are showed in 'Figure 2', and leaded to knowledge agglomeration as an urban characteristic of KC.

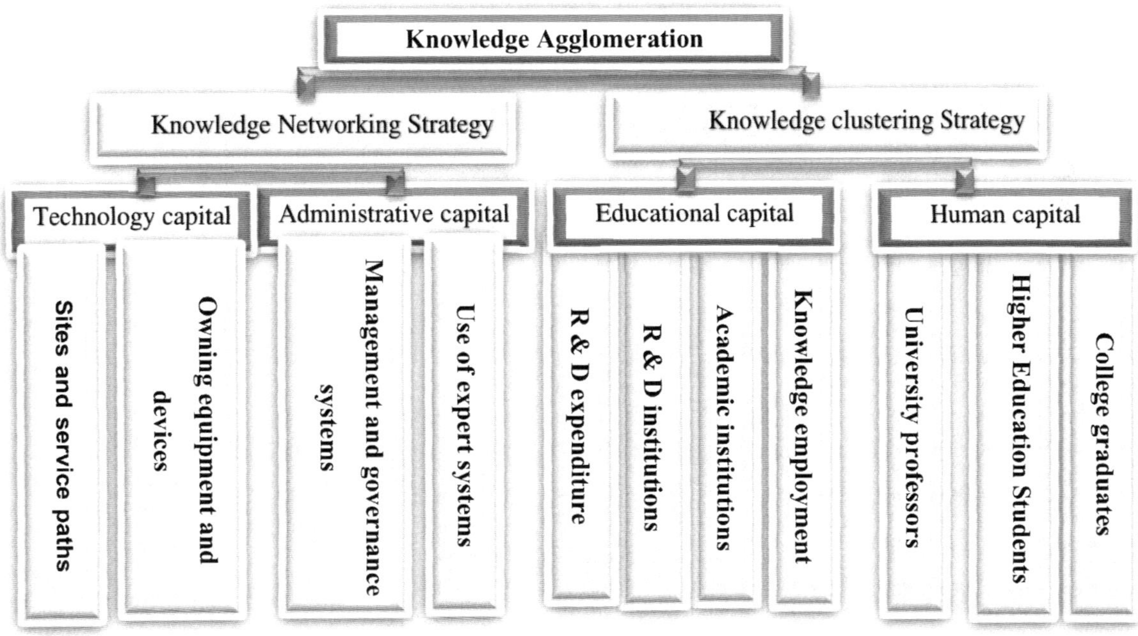

Figure 2. Vocabulary of the interpretation of the knowledge agglomeration characteristic, (by researcher).

3. Urban Capital Support Strategy indicators:

"Knowledge-City strategies refer to the process of interlinking these elements according to the term" City branding" [4], for example, marketing the city with a new attractive identity "based on several elements such as; policy, city management and external relations, tourism and private sector institutions and agreements, investments, events and culture, heritage and urban landmarks, and the community" [5]. With the aim of finding new dimensions of existing assets of value, or reading a new value of existing assets in the city, urban capital support strategy leads to urban diversity as an important character of KC that showed in 'Figure 3'.

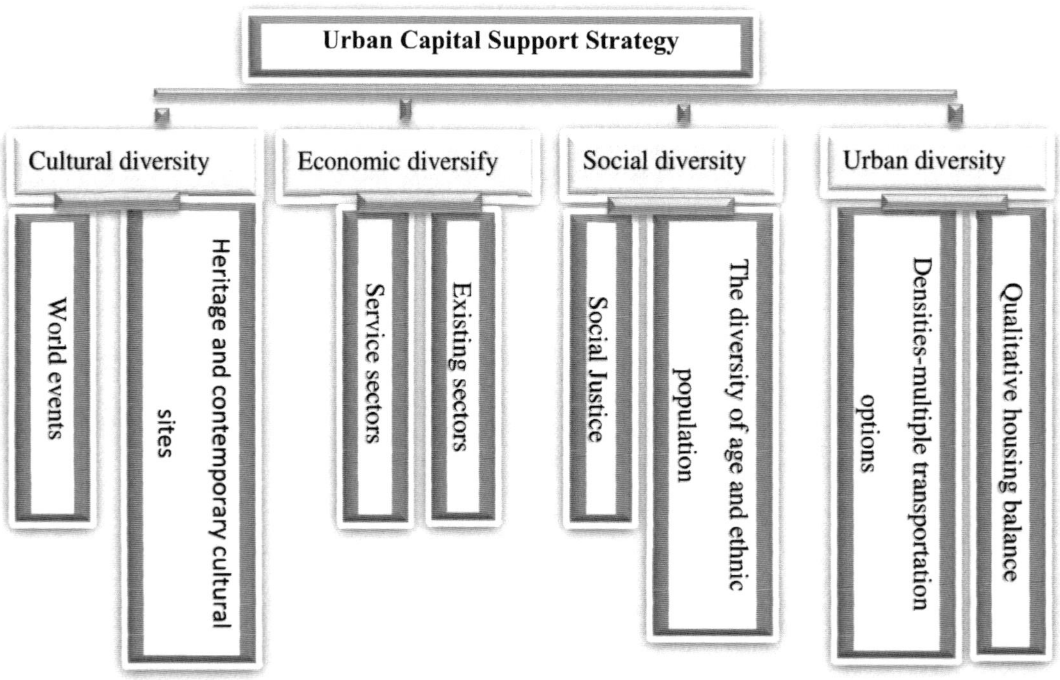

Figure 3. Vocabulary of the Interpretation of Cognitive Diversification, (by researcher)

4. Adopting GIS method to investigate KC characteristics in the study area:

The method depends on many global studies on KC strategies in cities like; Melbourne, Shanghai [6], Munich [7], and Amsterdam [8] & [9]. The application depends on using GIS method in order to prepare a vision of Figure 2 & Figure 3. Each one of these indicators investigated and represented in separated maps. Interventions to support the three strategies in the study area are developed according to the nature and opportunities of application. The composition of the activities, population and employment were organized according to the knowledge capital indicators, in order to produce results on the possibilities and opportunities of realizing the concept by collecting information from the relevant state departments and the private survey see; (Strategies development studies for Hila, 2009-2010) [10], & (Hila master plan development 2007) [11-13].

5. Discussion and analysis of application results:

Analysis of the potential of the city of Hila according to the framework of the three strategies:

5.1 Knowledge clustering;

There are two main sites identified as core of nucleus to encourage knowledge groups in the city of Hila:

First- Universities site (Babil University Center, Future College, and The Islamic University) and the existing and specialized industrial zones in the south of Hila city.

Second- Universities site (University of Babylon at the site of Morjan) and the existing and allocated industrial areas located north of the city of Hila, towards the province of Baghdad.

The importance of identifying these two sites for the implementation of the knowledge gathering strategy is focused on achieving the following:

a- Strengthening educational and human capital: These sites represent the main entrance to the city of Hila. They linked with the movement's axes (Street 40 and Street 60), which include several governmental and private colleges such as the Faculty of Medicine, Pharmacy, Dentistry and Science collages of the University of Babylon and the University of Hila and the Colleges of Future private University.

b- The city administration can take the initiative, as in the Shanghai experience [6] to support the change of land use in these two sites (which have been designated within the design as industrial zones) for technology-intensive activities and services in support of knowledge activity.

5.2 Knowledge Networks;

The network building strategy is the vision of the city from the perspective of maximizing access to intensifying the active exchange of information, collective learning and innovation, intelligent deployment of residential and commercial uses. Networks rely on interconnections to pass and improve horizontal links that extend across the boundaries of individual institutional organizations or vertical links within the organizational structure of the institution arises because of the response to the administrative and technological capital mechanisms by linking the (knowledge-intensive) clusters to the supply and exchange networks of knowledge and services. The most important obstacles to the construction of such networks in the city of Hila lack of administrative vision and imbalance in the relations between the parties to the economic process (planning, service and legal). The other is the response to the technological capital mechanisms of spreading technologies within the city's society and employing techniques to improve the productive performance of the working groups. A high percentage of technological capital indicators have emerged for the availability of equipment and service networks, with a lack of employment in the field of promoting innovative work (the use of technologies in higher education institutions is limited to scientific research and education).

5.3 support urban capital diversity;

The indicators of strategy of the urban capital diversity supporting in Hila city found in two locations:

- The location of the northern gate of the city, linking the city of Hila to the capital Baghdad and the location of the old city center. The indicators of diversity in this site are the following dimensions:

Urban dimension: includes broad and diverse requirements for urbanization patterns, urban form, scale, coverage, variables related to quality of life, open space patterns, etc. These variables have been neutralized (in the present research as it focuses on defining a general framework for the study of urban transformation and the concept of cities of knowledge) following:

- Multiple housing options: Includes the availability of land suitable for the expansion of housing and the diversity of housing patterns in terms of size and nature of occupancy of housing units and housing vertical and horizontal.

The proposed master plan of the city proposed to empty the old area of the city from residential use and allocate it for commercial use, while providing new residential expansions within the boundaries of the old center and the south-east side of the city (the new design area 470).

The urban housing strategy is based not only on the type of housing, but also on the importance of providing a variety of housing styles to meet the needs of the groups that attract the knowledge-intensive communities (universities and productive institutions) Technology and research and development activities), which are the youngest groups of researchers, students and smaller families. The types of housing available near the proposed knowledge clusters include the following:

- Separate or semi-detached housing type: This is the most common practice in Hila currently consisting of separate or semi-detached residential buildings consisting of one or two floors

- Multi-story housing: located in several clusters of apartment buildings consisting of 3 to 5 floors. This pattern will continue to form the main component of the housing block in the future. Also, a new set of patterns can be included, and coordinate with the locations of clusters identified as part of the strategy of knowledge groupings. This can illustrated especially in the presence of large tracts of unused land near the proposed sites, for the implementation of the cluster strategy and the existence of multi-story residential complexes on the road Hila - Najaf and near the southern location of the University of Babylon).

- Other styles: Some of the multi-story buildings and uses began to appear along the arterial streets in the city, which represents a balance of residential styles more diverse and appropriate to the requirements of student groups and small families.
- The locations of the proposed knowledge groups represent a variety of opportunities to support the housing balance, but unevenly. The proposed southern location offers larger areas of expansion. This feature is not available for the purpose of residential expansion in the northern location of the proposed bloc due to the density of housing in this area. However, this area is characterized by the existence of the river and large areas near the university site itself. These characters give a strong chance to reuse cultural sites that accommodate the events of the university and the city.
- *Multiple options of transportation:*

The path of the river connects the archaeological site in Babylon with the old Hila Center (the site of many heritage sites and shopping centers). This provides the possibility of connecting both heritage and archaeological centers with river paths and organizing land uses to strengthen this axis (heritage sites - open spaces and archaeological city). The lack of efficiency and inadequacy of existing roads is noted to accommodate the increasing density of vehicular traffic, while there is potential for developing river transport between these two sites, thus expanding mobility options for the population. The presence of the bridge project (the incomplete bridge of the mother) can be the focal point between the area, the city center and the southern university site (the proposed nucleus in the knowledge gathering strategy). This enhances the densities and expands transportation options.) To build bridges to connect the two banks of the river in that area.

-*The social dimension*: The city of Hila generally has strong foundations for social diversity. The role of the proposed two elements in the strategy of supporting urban capital in promoting social diversity is accompanied by its role in achieving cultural diversity. It is attractive to visitors and strengthens the axis between the archaeological site and heritage center of Hila. Proposed to revive the role of the city's urban center in social and economic life.

- *The economic dimension*: The axes between the city and the proposed nucleus are characterized by the following:
- The development of traditional sectors, which include the industrial zones (the industrial neighborhood in the southeast of the city and the industrial facilities located on the main axes and their branches between the north and south of the city, and these areas include most of the areas defined by the basic design for the expansion of industrial activity.

The proximity of these proposed sites to the knowledge groups (universities) is preparing to form an integrated nucleus in the future to be a gathering of knowledge-intensive activities.

- The possibility of developing services that supporting the activity along Street 60. This street is characterized by administrative and government use, hotel services and restaurants while the Street 40 is witnessing large changes driven mostly by the private sector where increased areas for commercial use (retail and hotel services are available), which focuses most of the private companies in the city, (Travel and tourism services, Internet companies, contracting, public investments, etc.)

-*The Cultural dimension*: The importance of the proposed knowledge-gathering site in the north of the city is to enhance the cultural dimension due to its proximity to the archaeological site of the ancient city of Babylon and near the Shatt al-Hila. This site extends to the center of the new city through the hub of the road Hila - Baghdad and the intersection of Street 40 to link between the old sites in the Babylonian archaeological area and religious sites as the site of the response of the sun). The tourism sector is a great opportunity to attract tourists and employment of labor because of the presence of elements of this sector in the city.

The river-link hub also provides the opportunity to extend the activities of the annual Babel Festival to the old city center and to create new events and expand them to include activities related to the heritage character of the city center.

6. Conclusions

Urban capital represents an important catalyst in organizing the strategic vision of transforming a city into a knowledge-based urban landscape.

- Urban capital enhances the indicators of other components of knowledge capital according to their priorities. Accordingly, knowledge groups are formed. Knowledge networks are established and the pattern of their spread and expansion is determined.

The city of Hila has been evaluated according to the theoretical framework indicators; possess particular hierarchical arrangements for the three knowledge cities' mechanisms and strategies (group support, support for building networks and support for urban capital diversity)

- The urban form and the pattern of expansion of the case study as a special and distinct case and this confirms the key role of urban capital indicators in activating the spread of clusters based on knowledge activities and activating the construction of knowledge networks locally and regionally. The result of these conclusions arranged as visions maps for Hila by using GIS method, and the maps showed in Figures 4, 5 and 6.

IOP Conf. Series: Materials Science and Engineering **584** (2019) 012005 doi:10.1088/1757-899X/584/1/012005

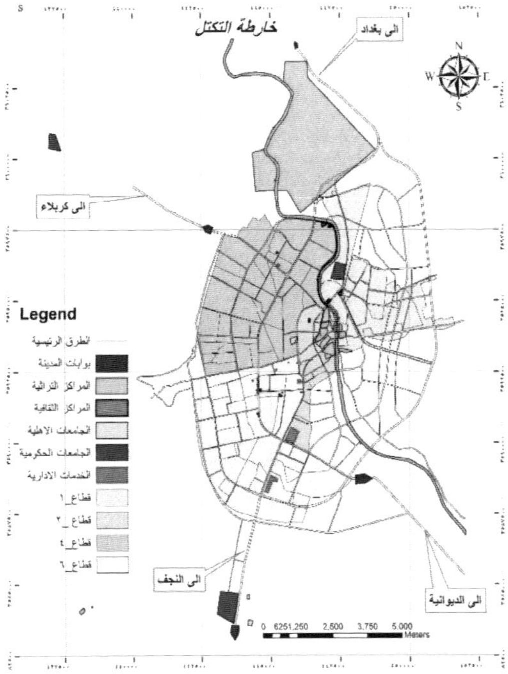

Figure 4. Knowledge agglomeration indicators in Hila (by researcher).

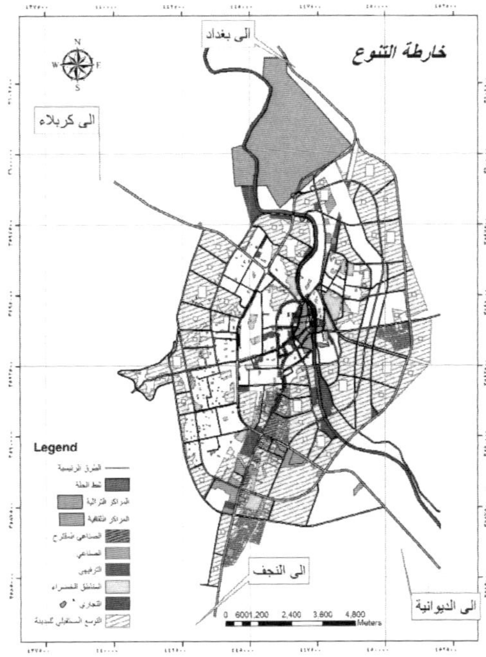

Figure 5. Urban capital diversity indicators in Hila (by researcher).

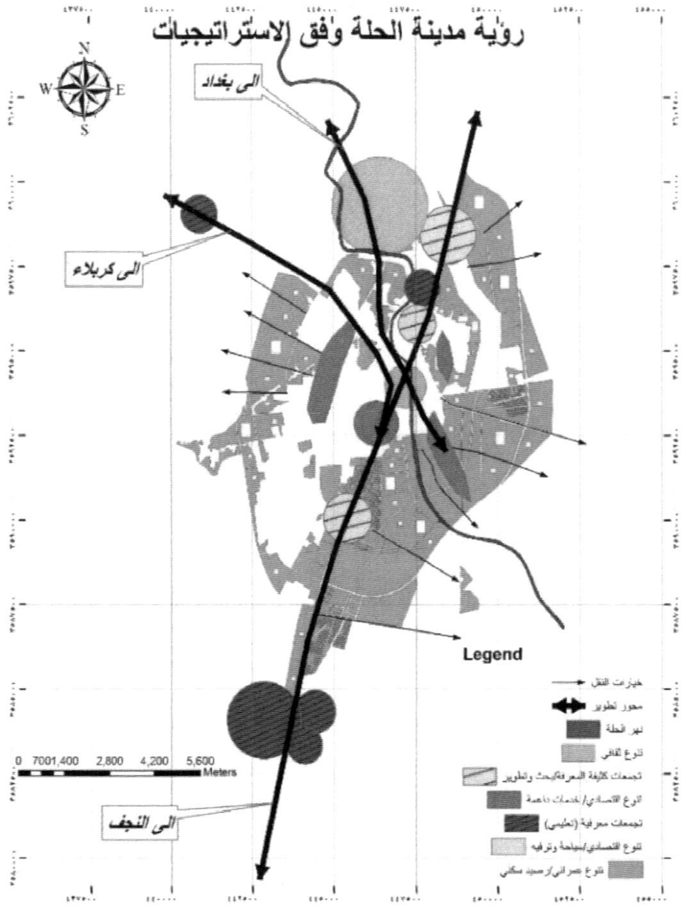

Figure 6. Future vision of the urban structure of Hila city according to strategies to support the urban capital and knowledge networks and networks (by researcher).

7. Recommendations

Adoption of new trends (imposed by the mechanisms of knowledge capital) as basic principles of urbanization. In this regard, the application of knowledge city strategies provides an integrated perspective of the city as a group of potential opportunities for interaction and not as isolated elements. Also, using GIS as an adopting method for explaining cities future visions for smart development which direct the resources towards the most promising urban spots.

8. References

[1] Ruiz, Víctor-Raúl López,*et al* 2014 *Knowledge-city index construction: An intellectual capital perspective, Expert Systems with Applications* (Elsevier Ltd vol.41) pp5560-5572.

[2] SGS Economics & Planning Submission2011 *Inquiry into a State-Based Reform Agenda Victorian* (Melbourne: SGS Economics and Planning Pty Ltd) p7.

[3] Yigitcanlar, Tan, O'Connor, Kevin & Westerman, Cara 2008 *Melbourne's knowledge-based urban development experience* (Elsevier vol.25) pp 63-72.

[4] Van Winden ,Willem 2010 *Outlook on Europe knowledge and the European City* (New York: Blackwell Publishing Ltd vol 101) pp 100–106.

[5] Nielsen, Andreas & Mathias Dackheden 2005 *The Knowledge City Concept The :Ideal Conditions of a Knowledge City* (Lund :School of Economics and Management, Lund University) pp 34-35.

[6] Chen, Y 2014_*Shanghai City Strategy 2050: Road Map to Knowledge City* Conf. Paper at: https://www.researchgate.net, 29/ October/ 2016.

[7] Musterd, Sako & Gritsai 2013 Conditions for 'Creative Knowledge Cities Findings from a comparison between 13 European metropolises,Going creative (Amsterdam: AISSR).

[8] Krätke, Stefan 2011 *The Creative Capital of Cities* (Oxford: Blackwell Publishing) p25.

[9] Elena, Cigu 2015 *The making of knowledge cities in Romania* (Romania: Elsevier Proc. Economics and Finance no.32) pp 534 – 541.

[10] Ministry of municipalities and public works, Directorate General of urban planning 2010 Structural Plan of the Province of Babylon: phase three report-planning alternatives.

[11] Ministry of municipalities and public works, Directorate General of urban planning 2007 Updating the Master Plan for the City Hill; Strategies to Develop Hila, stage 6.

[12] Ministry of municipalities and public works Directorate General of urban planning 2009 Structural Plan of the Province of Babylon.

[13] Ministry of planning, Central Bureau of statistics 2014 Economic and social survey report for the family in Iraq.

Influence of Curing Conditions and Filler Type on Mechanical Properties of Cold Asphalt Emulsion Mixtures

Mohammed Abbas Al-Jumaili [1], Ahmad Saleem Abdulateef Shakoree[2]

[1] Professor, Head of Civil Engineering Department at University of Kufa, Faculty of Engineering, Najaf, Iraq. Email: mohammedah.aljumaili@uokufa.edu.iq
[2] Ph.D. Candidate, Al-Mustansiriah University, Faculty of Engineering, Highways and Transportation Department. Iraq. Email: ahmadshakoree@yahoo.com

Abstract. Strength enhancement in cold asphalt emulsion mixtures (CAEMs) is continuing and reliant on the rate at which curing proceeds. The initial life strength is low and exhibits a challenge in asphalt pavement design. The solution can be made through subjecting the specimens to accelerate laboratory curing for achieving practical conditions. Kind of mineral filler, asphalt emulsion type, curing time and temperature, and moisture content is five major factors in early age advancement of designing properties. By studying these factors, there is a requisite to build up a curing system that simulates as reasonably as actual conceivable conditions of pavement during early life. This study examines the requirement for speeded curing strength that is reasonable by concentrate distinctive curing protocols and the impact on mechanical properties of cold asphalt emulsion mixes. The experimental configuration is cautiously chosen so as to the curing conditions incorporate protocols which are trailed by various literary works and furthermore circumstances that are required for simulating different times and temperatures. The examination incorporates both cement kiln dust (CKD) and limestone mineral filler (LMF) as a filler to be utilized in the blends. One of the findings of this study is, the inclusion of (CKD) leads to strong results better than those of lime stone mineral filler mixes.

Key words: cement kiln dust, cold asphalt emulsion mixtures, curing, and mechanical properties

1. Introduction

Completely, the emulsion of asphalt cured when the water beside any volatile oils have evaporated or expulsed then adhesive bond strength is built up between the aggregate and binder and aggregate. The water can be eliminated by using pressure (compacting and rolling), evaporation, and by aggregate absorption [1]. Curing, through setting and breaking stages, encourages the improvement of mechanical properties of the blend [2]. In 2009, (AEMA) clarifies that the main considerations which influence curing and breaking of asphalt emulsions incorporate: emulsifier kind and quantity, plus climate environments as hotter and elevated temperatures are positive to the setting of asphalt emulsions in order to facilitate curing [3]. Table(1) outlines key past investigations on curing regimes for (CAEMs).

Table 1. Summary of Previous Investigations of (CAEMs) Curing Regimes [4] [5].

	Reference	Curing Techniques
1	Terrel and Wang (1971).	1- One, three, and seven days curing at 24 C. 2- Three days curing in the kiln 49 C.
2	Mamlouck et al. (1981).	1- One and three days at 24 C. 2- Three days in the kiln.
3	(Ruckel et al , 1982), (Mauquisc, 2003), and(Jenkins et al, 2004).	1- Simulating intermediate curing of 24 hrs at 40 C° . 2- Simulating long term curing of 72 hrs at 40 C°.
4	Anderson and Thompson (1995).	Curing until reaching a constant weight at 38 C.
5	Serfass et al. ,2004.	14 days at 35 C° for accelerated curing .
6	UK-MCHW .	(Either 40 ± 2 C or 20 ± 2) for 28 days to simulate locally the first year of the road conditions.
7	Quick and Guthrie (2011).	One year fileld curing.

In 2002 (Montepara and Giuliani), expressed that treated asphalt layers displayed untimely distress rapidly next to construction which unfavorably influenced pavement performance. The exploration demonstrated that cold asphalt emulsion blends showed moderate initial life strength and construction is better in dry conditions [6].

Serfass et al., (2004) submitted a few rules regarding accelerated curing that consist of: the curing technique ought to create components as near their in-place maturity state and should not fundamentally age the asphalt binder. The curing protocol ought to be as short as conceivable to avoid aging of the bitumen and the lab apparatus hardware ought not to be excessively complex. In spite of the fact that curing methodology has been standard in numerous countries, the regimes are changed and a reliable system is as of now not accessible [7].

Additives are utilized to accelerate the curing rate that gives initial strength to blend in beginning early life time. In view of cement is an expensive material in this way, it tends to be utilized the waste materials rather than cement, since they are accessible, cheap, and could be recycled, for example, cement kiln by product dust (CKD) which is collected during Portland cement production, and limestone mineral filler dust (LMF).

This work investigates, during (CAEMs) early life, the requirement for speeding up and that is sensible by evaluating and studying different curing protocols and their impact on mechanical attitude by utilizing (CKD) and (LMF) as a filler material .

2. Material Characterization

Four material types were evaluated in order to accomplish the objectives of this study. These materials are: natural sand, natural aggregate, cement kiln dust, and limestone mineral as filler type. All studied materials were isolated, graded, sieved, and mixed in order to make a job mix formula (JMF) that meet the gradation required for coarse binder blends according to Iraqi Standard Specification for Roads and Bridges [8]. This gradation accommodated hot mix asphalt is applied since no native standard gradation for cold asphalt blend is accessible yet. The gradation of the considered materials is detailed in Figure (1). The midpoint limits gradation according to [8] requirements are utilized in setting up these blends, where the upper limit characterize the fine gradation, and the lower limit represents the coarse gradation as shown in Figure (2).

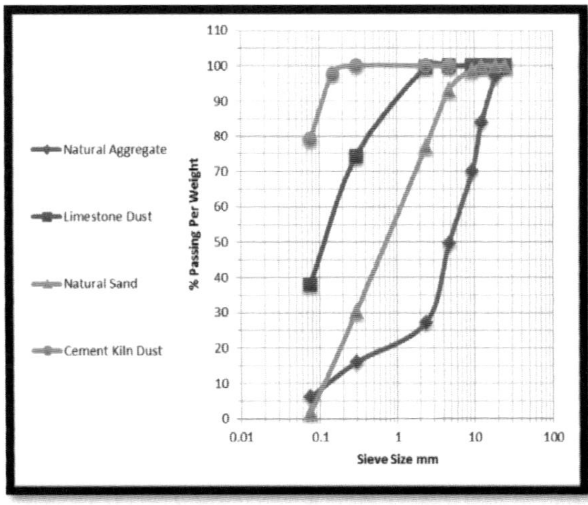

Figure 1. Aggregate and Filler Gradation.

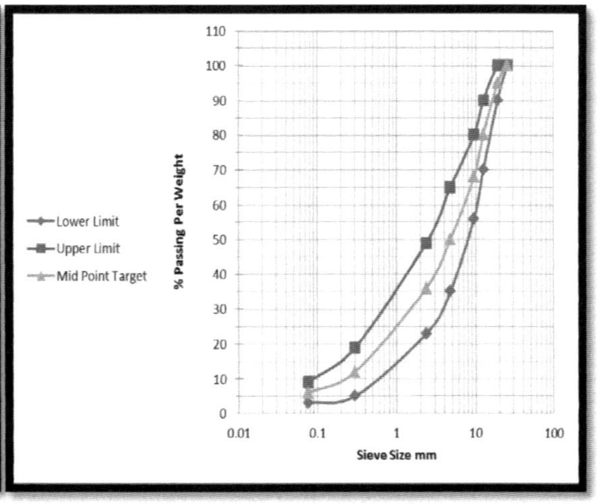

Figure 2. Adopted (JMF) Gradation.

The cationic slow breaking dark brown emulsified bitumen (CSS-1) was used in this investigation because it is more prominent than other emulsion kinds and it is entirely appropriate with the adopted aggregate gradation requirements [9]. The test results demonstrated that according to (ASTM D6997) [10] the asphalt residue from distillation was (61 %), and as per [11] the softening point was (51 C°).

3. Mix Design Procedure

As of now, there is no broad mechanical assembly made particularly for the production and design of cold asphalt emulsion mixtures, consequently, those for hot mix asphalt are most commonly utilized. Marshall strategy has been generally implemented for designing cold asphalt mixes in view of Asphalt Institute (Manual Series No. 14), 1989 and this specification was built up in this exploration.

Table (2) state design parameters utilized as acquired from the Asphalt Institute mix design technique for the two blends : (CKD) and (LMF) filler types. It is noticed that the total fluid content (asphalt emulsion content plus pre-wetting water content) for (CKD) blend is higher than that for (LMF) blend. This might be credited to the higher surface area for (CKD) which need and absorb extra liquids comparing to (LMF).

Table 2. The Used Percentages of (CAEMs) Design Parameters.

	% Content			
Category Mix	Initial Residual Bitumen	Initial Emulsion	Pre Wetting Water	Total fluid
(CAEMs) contains (LMF)	6.75	11	3	14
(CAEMs) contains (CKD)	6.3	10.32	4.5	14.82

The aggregates were washed, oven dried and patched in plastic bags with 1 kg of mass as indicated by the permitted JMF gradation in Figure (2). The pre-wetting water content was added as a thin spray to the dry aggregate mix and blended by a mechanical blender for (60 sec). A spatula was also used for (30 sec) to make sure that the aggregate was cautiously mixed and moistened to be ready for the addition of the bitumen emulsion. The essential bitumen emulsion was then added and blended for another (60 sec). To ensure consistency and homogeneity in the blend, the elements were then blended with hands using a spatula for (30 sec). These methods were blended at ambient temperature and recognized applicable for such mixtures, see [4]. Before the process of compaction, the loose mix would be discovered: if it is too wet, air ventilation may be essential until the mix is adequately loose, neither too wet nor too dry. In order to make cylindrical Marshall samples, each blend was fabricated with (65) mm height and, (101.6) mm diameter.

Curing significantly affects the mechanical properties of (CAEMs). This exploration was planned to give a substantial examination concerning the behavior of (CAEMs) considering:

- The impact of curing duration and temperature,
- The impact of filler type with various curing protocols and
- CAEMs mechanical properties differences according to curing conditioning.

Table (3) describes the strategy for the experimental study.

Table 3. Studied Experimental strategy.

Stage	The temperature of curing °C	Curing Condition	Curing Period (day)	Filler Type
1	25 40	Fully Sealed (FS)	3	CKD
2	25 40	Partially Sealed (PS)	3	CKD
3	25 40	Fully Sealed (FS)	3	LMF
4	25 40	Partially Sealed (PS)	3	LMF
5	25 40	Fully Sealed (FS)	7	CKD
6	25 40	Partially Sealed (PS)	7	CKD
7	25 40	Fully Sealed (FS)	7	LMF
8	25 40	Partially Sealed (PS)	7	LMF

The suggestion of partially sealed specimens conditions was presented to simulate a typical site job conditions with a coarse binder layer acting as a surface layer in order to sustain the traffic loading so that the samples will be impenetrable from bottom and sides (partially sealed). While Fully sealed specimens represent asphalt binder coarse layer that had covered by wearing surface layer. The selection of curing temperature was as the most conservative estimation to simulate the real performance of (CAEMs) on the field in addition to avoid any primary aging of the bitumen [4] and [7]. The (25 C°) represents the ambient temperature and the (40 C°) which less than asphalt residue softening point, was chosen to simulate moderate hot weather and to accelerate curing conditions.

The first attempt to accelerate the curing time of (CAEMs) was made by compacting the specimens in two layers (two lifts) in this manner enabling the moisture to escape quicker from each layer and as a result of that, diminishing the general curing time. Impact Marshall Hammer, which can be considered a fast loading rate, was used for compacting the specimens; (75) blows to each face. When compaction was finished, the samples were left in the mold for 24 hours and this was followed by curing as per the strategy as expressed in Table (2). Plate (1) displays the samples in their particular lab curing conditions.

Plate 1. Left to Right: Unsealed, Partially Sealed, and Fully Sealed Samples.

4. Testing Methods and Results

In order to describe the mechanical characteristics of CAEMs, there are several tests that could be implemented in the laboratory. In this work, Marshall Stability test and Indirect Tensile Strength test (ITS) were adopted. In the Marshall test, it is important to measure the resistance to plastic flow of cylinder-shaped samples for asphalt paving mixture. Whereas, the load is applied in a direction vertical to the cylindrical axis by means of the Marshall machine. While Indirect tensile strength test involved Marshall samples loading with uniform compressive load at 2 inches per minute rate allocating parallel to and along the perpendicular diametric level over (0.5) inch wide steel strips which have a concave curve at the interface with the samples. For these specimens, splitting fail along the vertical diameter plane will be occurring. The process of the ITS test is followed by ASTM D6931.

4.1 Impact of Curing Conditioning

Figures 3 and 4 present indirect tensile strength test results as achieved for (CAEMs) produced by cationic slow setting bitumen emulsion conditioned for 3 and 7 days at two different temperatures for

both filler types : (CKD) & (LMF). The results as seen, show that the impenetrable fully sealed (FS) samples has the lowermost strength mainly those cured at (25°C) with values of (125 KPa) and (162 KPa) respectively for both (CKD) and (LMF) mixes for three days curing. The uppermost strength values for the (FS) mixes were detected on samples cured at (40°C) with values of (304 KPa) and (249 KPa) for both (CKD) and (LMF) mixtures respectively at (7) days curing. At all curing temperatures, partially sealed (PS) specimens with (CKD) reached the highest strength.
Marshall stability test results are displayed in Figures(5 & 6). The tendency detected for the test results is similar to that for(ITS). The lowermost strength values were also identified for the (FS) samples cured at (25°C) for both studied mixes. Also, at all curing temperatures, partially sealed (PS) specimens with (CKD) reached the highest strength. The added emulsion contributes to the cohesion properties of the aggregate and causes shear improvement inside the (CAEMS) using (CKD) filler as can be seen in Table (4).

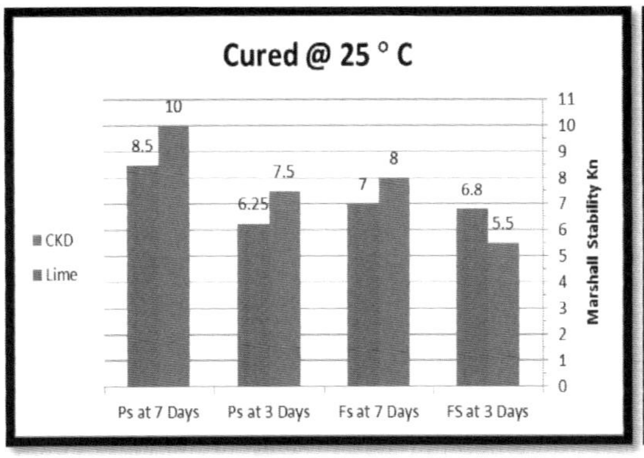

Figure 3. ITS of Specimens Cured at 25°C.

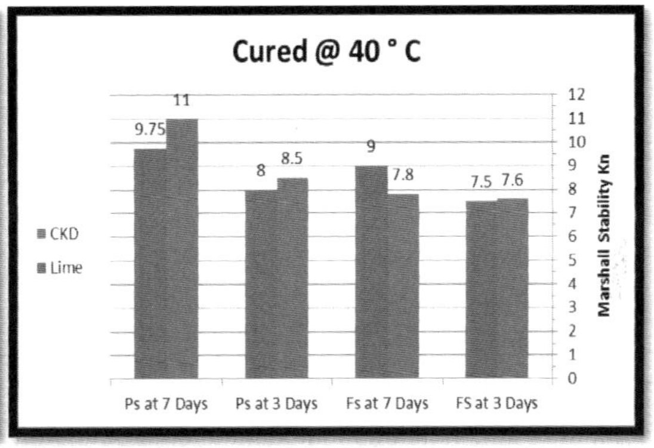

Figure 4. ITS of Specimens Cured at 25°C & 40 °C.

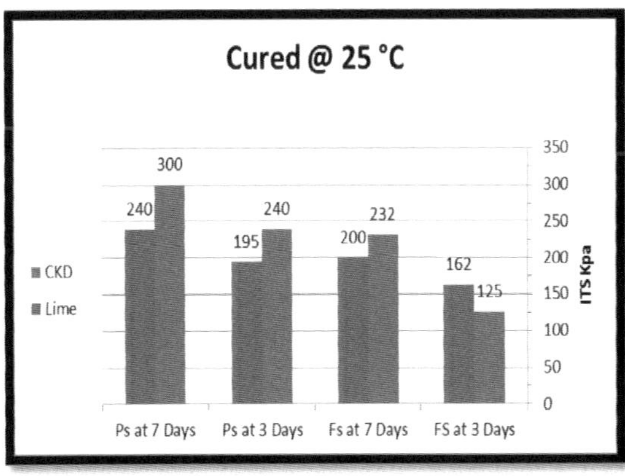

Figure 5. Marshall Stability of Specimens Cured at 25°C.

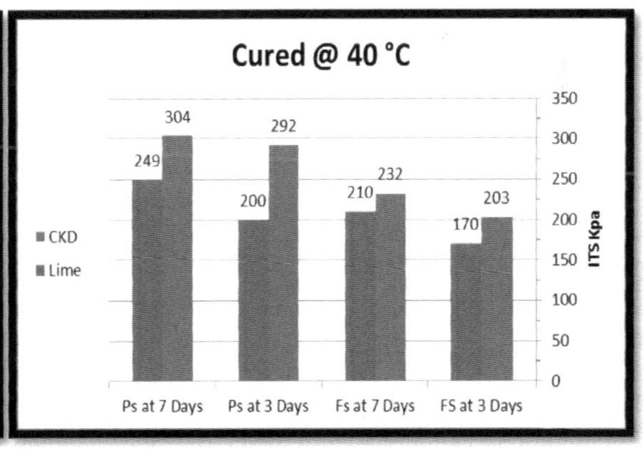

Figure 6. Marshall Stability of Specimens Cured at40 °C.

4.2 Influence of Temperature on Curing

The increase of the curing temperature from (25 °C to 40 °C) accelerated water evaporation and asphalt droplets breaking which increases cohesion and friction angle of the mix. Also, moderate temperature, i.e., 40 °C, will motivate the hydration development process of the cementations filler with water. Increasing curing temperatures lead to increase in the (ITS) and Marshall stability development. This was obvious irrespective of the curing conditioning implemented. Higher rates of strength increase which led to higher maximum values. Is a result of higher curing temperatures.

During the curing process and increasing of temperature, a major phenomenon occurring is moisture loss. As presented in Figures 7 & 8, there is a substantial difference between (FS) and (PS) samples and between 3 and 7 days curing, particularly with (CKD) mixtures. However, it is obvious from the final water content differences, that in curing, moisture loss is not the only mechanism involved. Elevated temperature is undoubtedly responsible for extra strength improvement, probably by enabling aggregate- bitumen adhesion.

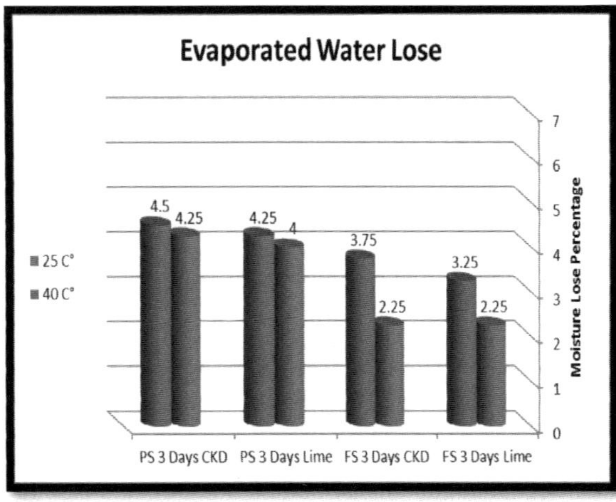

Figure 7. Mixture Moisture Evolution at 3 days.

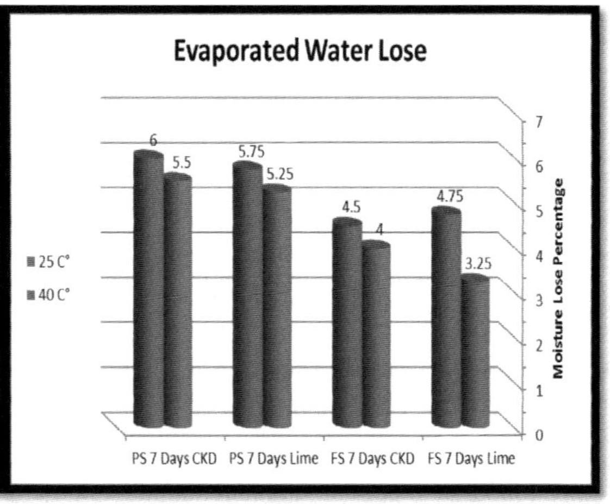

Figure 8. Mixture Moisture Evolution at and 7 days.

4.3 Influence of (CKD) and (LMF)

In all curing temperature, the addition of (CKD) increased the strength for both FS and PS conditions with a small variation comparing with (LMF) mixes. In terms of strength, the effect was somewhat less but still important. Partly, this increase is due to the elimination of water from the structure during hydration and partly due to the creation of cementitious cement.

The same behavior was conducted for (LMF) mixes in which, by mixing water with lime, certainly, one gets slaked lime. The chemical formula $Ca(OH)_2$ characterizes slaked lime, and it is written, in shorthand, [$CaO + H_2O \rightarrow Ca(OH)_2 + \Delta$].The delta or triangle sign shows heat. The extent of energy formed is an indicator of the degree lime tends to reduce water from its surroundings [12]. Because of that early heating and, as can be shown in Figures(3 to 6), for (3) days curing at (25 C°) and fully sealed specimens, the strength of (LMF) mixtures is higher than (CKD) mixes.

4.4 Influence of Curing Time

The selection of three and seven days curing time may be attributed to the following points:

- It is essential to accelerate curing route to achieve an early sign of short term strength (less than one week).

- The slower time for curing regime for mixtures contains cementitious materials (e.g. (CKD) and lime) is necessary as a hydraulic binder can hydrate more quickly at elevated temperatures causing in sample cracking and untrustworthy evaluation.

As can be seen from previous figures, strength values for the (FS) samples, cured at (40°C) for (3) days were about (60 – 70 %) of those attained at (7) days and with respect to the Marshall stability marks, the ratio is around (70%).

5. Conclusions

The findings of this study are summarized as follow:

1. The temperature of curing slightly affects the strength values rate, and partially sealed specimens were cured faster than fully sealed ones.
2. Overall, the inclusion of cement kiln dust in the (FS) and (PS) cases leads to ~~strength~~ better results than those of lime stone mineral filler mixes. However, for the fully sealed case and at three days curing time, (LMF) mixture is the best choice this means avoiding the effects of damage during early life.
3. In order to accelerate curing time for (CAEMs) so as to estimate its early life mechanical properties, the following regimes are adopted :

A. For mixes containing CKD mineral filler: curing of the partially sealed specimen at 40 °C for 7 days.
B. For mixes containing (LMF) : curing of the partially sealed specimen at 40 °C for 7 days.

4- Strengths results at 3 days are nearly (60-70%) of those at 7 days, showing the hazard of damage due to early trafficking.

References

[1] Austroads, 2008. *Bitumen emulsions*. AP/T107/08, Sydney, Australia.
[2] Thom, N. H., 2014. *Principles of Pavement Engineering*. Thomas Telford Publishing Limited, London,
[3] Asphalt Emulsion Manufacturers Association (AEMA), 2009. *Basic Asphalt Emulsion Manual*, Manual Series No. 19, Third Edition, USA.
[4] Ojum, C.K., 2015. *The Design and Optimization of Cold Asphalt Emulsion Mixtures*. Ph. D Dissertation, University of Nottingham.
[5] Barbod, B , and Shalaby, A. 2014, *Laboratory Performance of Asphalt Emulsion Treated Base for Cold Regions Applications,* Paper prepared for presentation at the Characterization of Soils and Pavement Materials for Mechanistic Empirical Pavement Design Session of the Conference of the Transportation Association of Canada Montreal, Quebec.
[6] Montepara, A. and Giuliani, F. 2002, *A Study on Design and performance of Recycled Pavement Cold Stabilized with Cement and Bituminous Emulsion*. Proceeding of 4th European Symposium on Performance of Bituminous and Hydraulic Materials in Pavement, BITMAT 4, University of Nottingham , A.A. Balkema Publishers, Netherlands, pp 213-217.
[7] Serfass, J-P., Poirier, J-E., Henrat, J-P and Carbonneau, X. 2004,. *Influence of Curing on Cold Mix Mechanical Performance*. Proceedings of the 6th International RILEM Symposium on Performance testing and Evaluation of Bituminous Materials, Zurich, (ed. Partl, M.N.), RILEM Publications, Bagneux, pp 81-87.

[8] SSRB,2003, "*Standard Specifications for Roads and Bridges*", Section R9 Hot Mix Asphaltic Concrete Pavement, State Organization for Roads and Bridges, Design and Studies Department, Ministry of Housing and Construction, Republic of Iraq.

[9] Asphalt Institute. 1989, *Asphalt Cold Mix Manual*, manual series No.14 , Maryland,USA.

[10] ASTM D6997, 2012 *Standard Test Method for Distillation of Emulsified Asphalt*, American Society for Testing Material, West Conshohocken, United States.

[11] ASTM". D6913. 2012, "*Standard test method for indirect tensile (IDT) strength of bituminous mixtures*, American Society for Testing Material, West Conshohocken, PA 19428-2959, United States.

[12] Al-Jumaili, M. A. and Ahmad S. A. *2018 " Impact of aggregate and filler type on cold asphalt mixture attitude*", International Conference on Advance of Sustainable Engineering and its Application (ICASEA).

IOP Conf. Series: Materials Science and Engineering **584** (2019) 012007 doi:10.1088/1757-899X/584/1/012007

Numerical analyses on seismic behavior of helical concrete columns

Alaa M Al-habbobi

Civil Engineering Department, Faculty of Engineering, University of Kufa, Najaf, Iraq. Email: alaa.alhabbobi@uokufa.edu.iq

Abstract. This paper presents the behavior of the helical reinforced concrete columns (HCC) subjected to horizontal 2D seismic motion excitation combined with the static axial load. Helical concrete columns are new types of elements which arose in the new twisted buildings, providing a special geometry by twisting the top section by an angle (ϕ), furthermore, offsetting the top section by a considerable distance respecting to the bottom section. In order to study seismic behavior of the HCC under a seismic load of the short period using finite element software, numerical simulation is carried out in this paper using ANSYS V 18.1 software. A brief comparison of the HCC behavior was made between different cases of conventional reinforcement in the longitudinal and transverse directions, as well as the different twisting angles (ϕ) of column sections. The comparison illustrated that all columns response with giving the same mode shapes of vibration in the modal analysis with different values of natural frequencies and relative amplitude deformations. However, the difference be very noticeable when changing the section size of the columns comparably with changing the twisting angles of the section, where increasing the section size of HCC from 250 mm to 400 mm cause increase the natural frequency by 55.4% for second mode shape, while it cause decreasing the relative amplitude response of HCC by 52.5% for sixth mode shape. The transient analysis of the six types of HCC excited by horizontal 2D seismic motion indicates the tendency of HCC to increase its responses, like twisting and buckling deformations. The deform shape-time history of 6 HCC demonstrated that with increasing (shear stress path section distance) S of HCC column which depends up on (ϕ value), the top section would have less stresses.

1. Introduction

Recently, designers of building face challenging models in designing structures. An earthquake brings huge losses in the development of human society repeatedly, which is one of the most serious natural disasters facing the human. After the damage caused by the 1971 San Fernando earthquake, the design requirements have been modified to accommodate the need for columns to achieve sufficient strength and ductility during seismic activity [1]. Traditional seismic design methods focus on increasing column shear reinforcements, increasing the strength of concrete, and increase the sizes of the column section to reduce the earthquake disasters, which is a negative and passive seismic counter measure. The technical aspect of the model-making industry component is a complex one where it involves the modeling of very subtle structures such as twisted towers. Several applications include Infinity (Cayan) tower in Dubai Marina, Moscow evolution tower and Majdoul tower 'Figure 1'. These high rise concrete buildings are new architectural models represents a challenge to the structural designer engineers especially when including the seismic hazard. However, it is useful to study the behavior of seismically excited helical concrete column (HCC) of the twisting building. HCC is a preeminent member in the twisted concrete towers, which has two fundamental and functional variables delimit the column geometry. The first variable is the twisting angle of the top section (ϕ in degrees) with respect to its bottom section, while the second is the offsetting of the top section origin coordinate with respect to its bottom section coordinate. See 'Figure 2'.

Figure 1. (a) Cayan tower, (b) Evolution tower and (c) Majdoul tower [2].

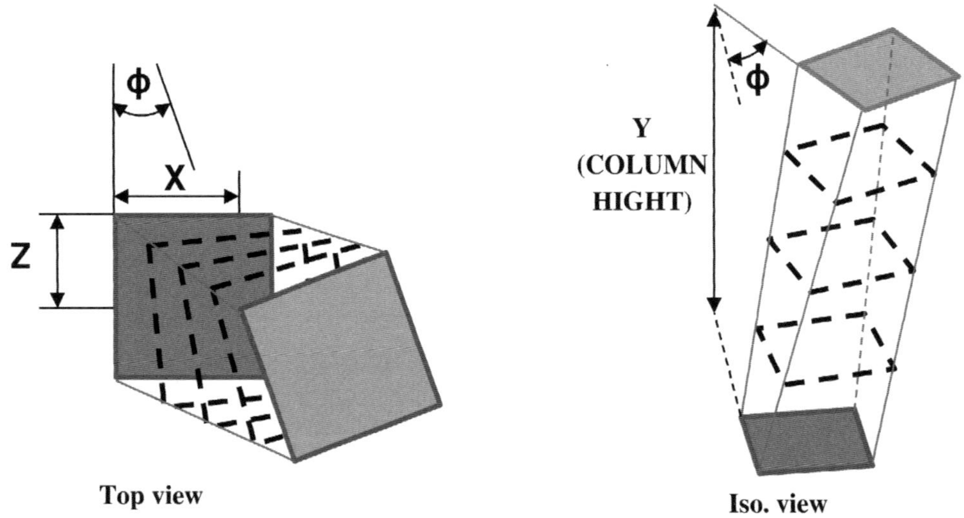

Figure 2. Geometry and Orientation of the HCC

The drifts of floors in the structure affect the damages locations during an earthquake occurrence. The damage will happen in the less stiffness elements of the structure, and if columns being those elements, the drift locates in few first floors and may overtake the columns capacity. Otherwise, increasing stiffness of columns represents the increase better distribution of damages location over the entire height of the structure. It is preferred that the damages occur in the beams than columns. Because beams carry local gravity loads of the floor while columns carry all loads from local and up floors. So, the failure in columns is more serious than failure in beams. This method of failure distribution (beam /column) is the basic principle in seismic behavior of the structures when an earthquake occurs [3]. In the columns, the shear failure is more significant particularly in the reinforced concrete structure during an earthquake, this type of failure can cause a high reduction in axial load-carrying capacity and lateral strength. In such a case of columns design, it is preferable to neglect the shear resistance of the concrete part, and the shear will be resisted by reinforcing steel only [3]. In the last decades, the prescriptions for RC structural seismic design have produced minor improvements in the steel reinforcement technology. In the fabrication and handling of reinforcement, an important amount (30 to 65 percent) of the resources is dedicated to transversal reinforcement stirrups [4]. However, according to the author's knowledge, there are no researches to study the seismic behavior of helical columns, taking into account the effects of changing the twisting angle of the column section (ϕ) and the various amounts of reinforcement. The information above will illustrate the prominence of studying the efficiency of the reinforcement of the HCC columns of the twisting building under the horizontal 2D seismic loading.

2. COMPUTATIONAL MODELING

In the present study, a dynamic nonlinear three-dimensional finite element analysis software ANSYS Release 18.1 is used. The earthquake intensity of two dimensions ground motion displacements-time data, material properties and its representation in ANSYS have been dealt with. As well as, procedures of modeling and solution techniques are presented in this research. ANSYS V.18.1 program was used in this research to simulate the vibration of 22 HCC. Table 1 illustrates the dimensions and reinforcing details of the columns modeled.

Table 1. details of helical concrete columns modeled

No.	Column Name	Section Dimensions (mm)	Height of column (mm)	Angle of twisting section	Reinforcement Diameter size (mm)	
					Longitudinal	Stirrups
1	$C_{15}L_{12}S_{12}$				8ϕ12 mm	ϕ12@200 mm
2	$C_{15}L_{16}S_{10}$				8ϕ16 mm	ϕ10@200 mm
3	$C_{15}L_{16}S_{12}$	300X300	3000	15 deg.	8ϕ16 mm	ϕ12@200 mm
4	$C_{15}L_{16}S_{16}$				8ϕ16 mm	ϕ16@200 mm
5	$C_{15}L_{25}S_{12}$				8ϕ25 mm	ϕ12@200 mm
6	$C_{30}L_{12}S_{12}$				8ϕ12 mm	ϕ12@200 mm
7	$C_{30}L_{16}S_{10}$				8ϕ16 mm	ϕ10@200 mm
8	$C_{30}L_{16}S_{12}$	300X300	3000	30 deg.	8ϕ16 mm	ϕ12@200 mm
9	$C_{30}L_{16}S_{16}$				8ϕ16 mm	ϕ16@200 mm
10	$C_{30}L_{25}S_{12}$				8ϕ25 mm	ϕ12@200 mm
11	$C_{45}L_{12}S_{12}$				8ϕ12 mm	ϕ12@200 mm
12	$C_{45}L_{16}S_{10}$				8ϕ16 mm	ϕ10@200 mm
13	$C_{45}L_{16}S_{12}$	300X300	3000	45 deg.	8ϕ16 mm	ϕ12@200 mm
14	$C_{45}L_{16}S_{16}$				8ϕ16 mm	ϕ16@200 mm
15	$C_{45}L_{25}S_{12}$				8ϕ25 mm	ϕ12@200 mm
16	$C_{60}L_{12}S_{12}$				8ϕ12 mm	ϕ12@200 mm
17	$C_{60}L_{16}S_{10}$				8ϕ16 mm	ϕ10@200 mm
18	$C_{60}L_{16}S_{12}$	300X300	3000	60 deg.	8ϕ16 mm	ϕ12@200 mm
19	$C_{60}L_{16}S_{16}$				8ϕ16 mm	ϕ16@200 mm
20	$C_{60}L_{25}S_{12}$				8ϕ25 mm	ϕ12@200 mm
21	$C_{250X250}L_{16}S_{12}$	250X250	3000	30 deg.	8ϕ16 mm	ϕ12@200 mm
22	$C_{400X400}L_{16}S_{12}$	400X400	3000	30 deg.	8ϕ16 mm	ϕ12@200 mm

However, three dimensional 8-nodes brick element (SOLID 65) is used in this study to simulate the behavior of concrete for the 3D modeling of solids. The elements are defined by eight nodes having three degrees of freedom at each node; translations of the nodes are in x, y, and Z-directions. The most important aspect of this element is the treatment of nonlinear material properties. The element is capable of cracking in tension and crushing in compression [5]. Reinforcing bars simulated using (LINK 180) with different sections of bars. The bars are capable of tension and compression, but not shear. They are also capable of plastic deformation and creep [5]. This research aims to study numerically the modal and transient analysis and the effect of two dimensional seismic loading up on 6 types of HCC (Helical concrete columns) with different angles of twisting section and the size of the cross-section of columns. Modal analysis of 22 types of HCC is also conducted for different parameters, such as longitudinal reinforcement amount, diameter of the stirrups, Angle of twisting the column section (ϕ) as well as the size of the cross-section of columns. The numerical study is performed using ANSYS V 18.1. 'Figure 3' illustrates HCC modeling.

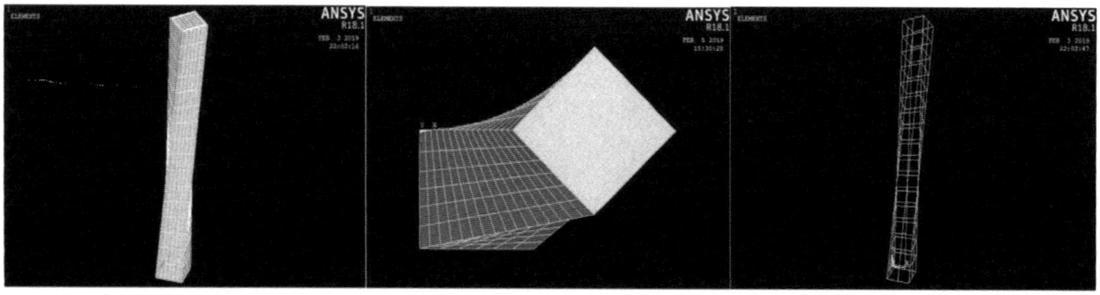

Figure 3. geometry and modeling: (a) Isometric view, (b) Top view and (c) skeleton of reinforcement

3. Modal Analysis

Modal analysis is a technique used to determine the structure's vibration characteristics, such as the natural frequencies, the mode shapes, mode participation factors (how much a given mode participates in a given direction) and most fundamental of all the dynamic analysis types. The deformed shape of the structure at a specific natural frequency of vibration is termed its normal mode of vibration. The natural frequencies of a structure are the frequencies at which the structure naturally tends to vibrate if it is subjected to a disturbance. Natural frequencies and mode shapes are functions of the structural properties and boundary conditions, and the main reasons to compute the natural frequencies and mode shapes of a structure is to assess the dynamic interaction between a component and its supporting structure. If the structural properties change, the natural frequencies change, but the mode shapes may not necessarily change. For example, if the elastic modulus of the member is changed, the natural frequencies change, but the mode shapes remain the same. If the boundary conditions change, then the natural frequencies and mode shape both changes [6]. ANSYS software calculates the mode, and according to the mechanical model, the basic characteristic equation is:

$$[K]\{\ddot{\Phi}_\mathrm{i}\} = \omega_\mathrm{i}^2[M]\{\Phi_\mathrm{i}\} \tag{1}$$

Where: [K] is stiffness matrix, [M] is a mass matrix, $\{\Phi_\mathrm{i}\}$ is the eigenvector mode shape for the ith vibration mode ω is natural frequencies of the structure and ω_i is natthe ural frequency for the ith vibration mode. By solving the characteristic equation (1) we get:

$$\omega_i = \frac{\{\Phi_\mathrm{i}\}^\mathrm{T}[K]\{\Phi_\mathrm{j}\}}{\{\Phi_\mathrm{i}\}^\mathrm{T}[M]\{\Phi_\mathrm{j}\}} \tag{2}$$

Now, it can obtain the mode $\{\Phi_\mathrm{i}\}$ by introducing equation (2) into equation (1).

Based on the fact that real eigenvalue analysis is the basis for many types of dynamic response analyses. Therefore, an overall understanding of normal modes analysis as well as knowledge of the natural frequencies and mode shapes of the particular structure is important for all types of dynamic analysis. Modal analysis of 22 HCC is studied in this research for different parameters such as longitudinal reinforcement amount, the diameter of the stirrups, the angle of twisting the column section (ϕ) as well as the size of the cross-section of columns. The numerical study performed using ANSYS V.18.1. Table

2 illustrates the natural frequencies and the maximum deformation of the 22 HCC columns. It can be seen prominently that for the same mode of all HCC columns have the same size are approximately have close values of natural frequency, despite the difference in the other parameters. Furthermore, all the studied HCC showed the same modal shape despite having a different relative amplitude of deformations. See 'figure 4'. This is attributable that all the 22 HCC have the same modulus of elasticity and the same way of mass distribution per unit length, see equations. (1) & (2), as well as, not forget that all HCC analyzed with a same end supporting (method of supporting) which is constraining the bottom end and keeping the top end free.

Table 2. modal analysis results of 22 helical concrete columns

Column Name	Natural Frequency / relative amplitude of deformations (mm)								
	Mode1	Mode2	Mode3	Mode4	Mode5	Mode6	Mode7	Mode8	Mode9
$C_{15}L_{12}S_{12}$	0.15644	0.15651	0.94291	0.94331	1.34350	2.40252	2.5136	2.51446	4.043
	0.02373	0.02372	0.02384	0.02381	0.02888	0.01704	0.02444	0.02419	0.02875
$C_{15}L_{16}S_{10}$	0.16123	0.16131	0.96835	0.96876	1.31113	2.44493	2.56959	2.57048	3.94584
	0.02334	0.02332	0.02343	0.02341	0.02819	0.01672	0.02396	0.02374	0.02805
$C_{15}L_{16}S_{12}$	0.16027	0.16035	0.96249	0.96291	1.30176	2.4312	2.55363	2.55449	3.91736
	0.02319	0.02318	0.02326	0.02324	0.02799	0.01663	0.02377	0.02355	0.02785
$C_{15}L_{16}S_{16}$	0.15790	0.15798	0.94800	0.94840	1.27875	2.39716	2.51415	2.5149	3.8474
	0.02283	0.02281	0.02284	0.02282	0.02751	0.0164	0.02329	0.02307	0.02736
$C_{15}L_{25}S_{12}$	0.17022	0.17030	1.00974	1.01016	1.18788	2.50809	2.63863	2.63948	3.57442
	0.02164	0.02163	0.02158	0.02161	0.02559	0.01551	0.02188	0.02171	0.02544
$C_{30}L_{12}S_{12}$	0.15381	0.15397	0.92515	0.92604	1.34803	2.39906	2.47394	2.47583	4.06351
	0.02412	0.02412	0.02384	0.02388	0.03006	0.01759	0.02429	0.02394	0.02951
$C_{30}L_{16}S_{10}$	0.15857	0.15873	0.95064	0.95155	1.31500	2.44278	2.53054	2.53249	3.96387
	0.02371	0.02370	0.02343	0.02348	0.02933	0.01721	0.02380	0.02351	0.02877
$C_{30}L_{16}S_{12}$	0.15760	0.15777	0.94476	0.94566	1.30547	2.42872	2.51450	2.51641	3.93495
	0.02355	0.02356	0.02326	0.02331	0.02912	0.01712	0.02362	0.02332	0.02856
$C_{30}L_{16}S_{16}$	0.15522	0.15538	0.93022	0.93109	1.28205	2.39388	2.47488	2.47672	3.86393
	0.02319	0.02319	0.02283	0.02289	0.02862	0.01689	0.02316	0.02286	0.02805
$C_{30}L_{25}S_{12}$	0.16740	0.16758	0.99182	0.99275	1.18965	2.50771	2.60031	2.60222	3.58509
	0.02194	0.02194	0.02158	0.02170	0.02659	0.01591	0.02176	0.02154	0.02601
$C_{45}L_{12}S_{12}$	0.14971	0.15001	0.89803	0.89966	1.33617	2.39344	2.41179	2.41428	4.05640
	0.02482	0.02481	0.02418	0.02421	0.03206	0.02045	0.02568	0.0242	0.03078
$C_{45}L_{16}S_{10}$	0.15454	0.15485	0.92402	0.9257	1.30170	2.4397	2.46939	2.4725	3.95062
	0.02436	0.02435	0.02376	0.02381	0.03123	0.01898	0.02482	0.02377	0.02995
$C_{45}L_{16}S_{12}$	0.15358	0.15389	0.91817	0.91984	1.29207	2.42528	2.45347	2.4565	3.92137
	0.02420	0.0242	0.02358	0.02364	0.03101	0.01895	0.02466	0.02358	0.02973
$C_{45}L_{16}S_{16}$	0.15121	0.15152	0.90371	0.90534	1.26843	2.38954	2.41417	2.41698	3.84956
	0.02382	0.02381	0.02315	0.02320	0.03045	0.0189	0.02428	0.02311	0.02920
$C_{45}L_{25}S_{12}$	0.16347	0.16380	0.96591	0.96764	1.17301	2.50793	2.54469	2.54469	3.55728
	0.02247	0.02247	0.02186	0.02201	0.02821	0.01730	0.02179	0.02179	0.02691
$C_{60}L_{12}S_{12}$	0.14367	0.14415	0.86156	0.86418	1.31050	2.3235	2.33033	2.38616	4.05845
	0.02577	0.02577	0.02443	0.02452	0.03490	0.02365	0.02422	0.01939	0.03300
$C_{60}L_{16}S_{10}$	0.14862	0.14912	0.88817	0.89087	1.27438	2.38356	2.39065	2.43539	3.94247
	0.02526	0.02526	0.02401	0.24119	0.33933	0.02322	0.02378	0.0192	0.03201
$C_{60}L_{16}S_{12}$	0.14767	0.14817	0.88236	0.88504	1.26469	2.36772	2.37473	2.42055	3.91267
	0.02509	0.02509	0.02382	0.02394	0.03368	0.02301	0.02358	0.01905	0.03177
$C_{60}L_{16}S_{16}$	0.14533	0.14582	0.86802	0.87065	1.24093	2.32866	2.33547	2.38383	3.83955
	0.02468	0.02469	0.02336	0.02349	0.03306	0.02249	0.02308	0.0187	0.03118
$C_{60}L_{25}S_{12}$	0.15775	0.15828	0.93072	0.93350	1.14233	2.45861	2.46573	2.50924	3.52535
	0.02319	0.0232	0.02206	0.02231	0.03048	0.02123	0.02172	0.01764	0.02851
$C_{250X250}L_{16}S_{12}$	0.13213	0.13225	0.79179	0.79243	1.26938	2.10501	2.10664	2.45004	3.82515
	0.02753	0.02753	0.02706	0.02702	0.03412	0.02733	0.02704	0.02967	0.03363
$C_{400X400}L_{16}S_{12}$	0.20523	0.20554	1.17083	1.17234	1.32054	2.40745	2.91635	2.91975	3.97214
	0.01817	0.01814	0.01866	0.01842	0.02272	0.01298	0.02323	0.02246	0.02182

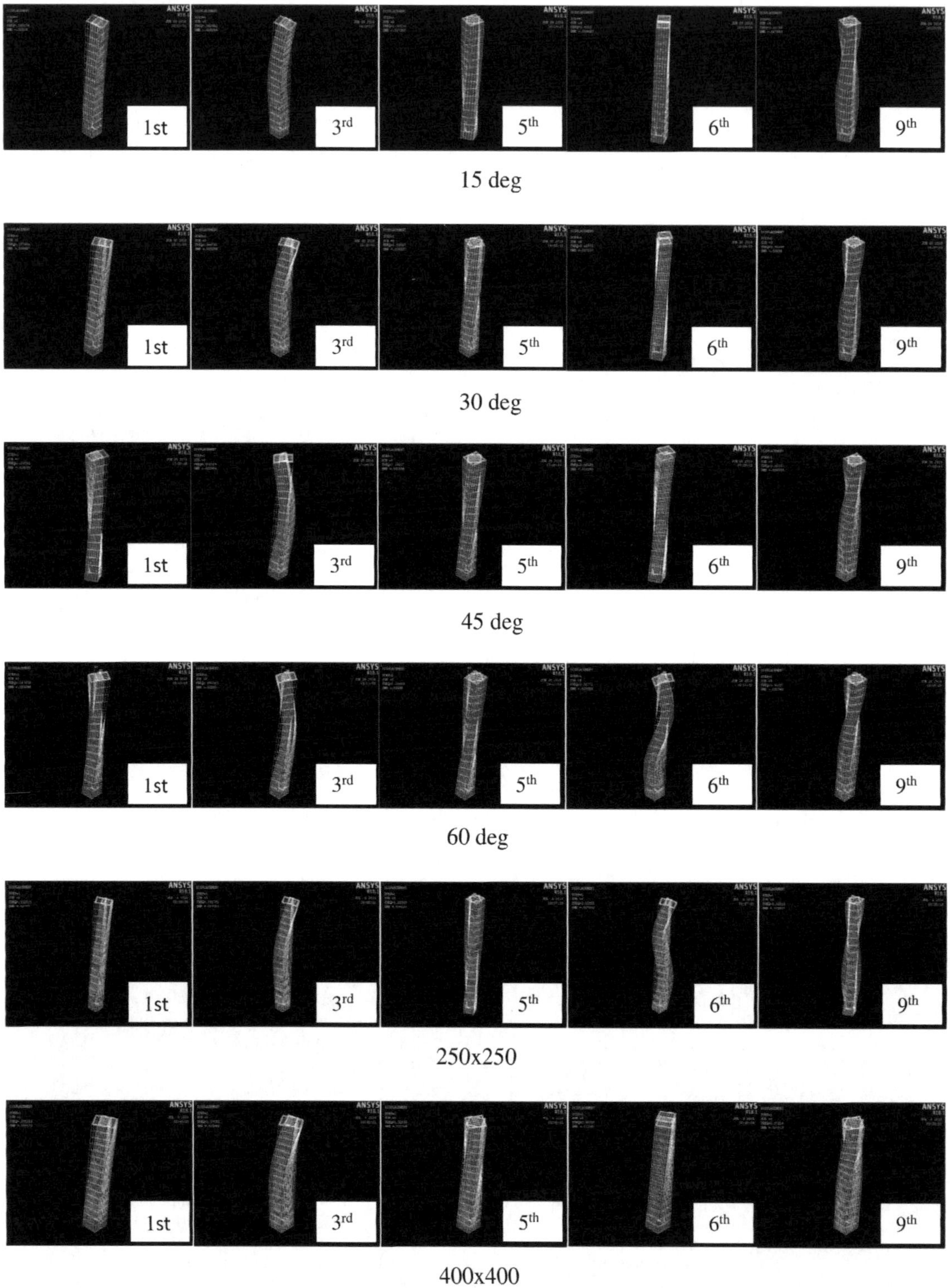

Figure 4. mode shapes of different types of helical concrete columns.

However, the simple difference in the values of natural frequency/amplitude of deformations that is illustrated in table 2 May ascribable to the difference in the moment of inertia values of HCC.
From the modal shapes figures of the six types of HCC that all reinforced with 16 mm longitudinal bars and 12 mm stirrups, It can preeminently see that with increasing twisting angle of the top section (ϕ) and keeping the size of the concrete section constant, there is an increase of the twisting deformation and the buckling shapes; see 'figure 4'. This means that an increase in the flexural, torsional, and shear stresses under the seismic loading which leads to the needing for the increase the overall reinforcement of the HCC with the increasing (ϕ).
Despite of the increase in the section size of column lead to increase the mass of the column, but the modal analysis revealed that increasing section size of HCC from 250 mm to 400 mm the natural frequency of the column might increase by 55.4% for the second mode shape; this is due to the increase of moment of inertia value. This increasing of column section size is causing a decrease in the relative amplitude response of HCC by 52.5% for the sixth mode shape. However, increasing the longitudinal reinforcement diameter bars from 12 mm to 25 mm with keeping stirrups diameter 12 mm constant, the natural frequency decreased by 13.1% at the 9th mode shape of HCC with (ϕ=60 deg.) and increase the relative deformation by 15.4% at the 6th mode shape of HCC with (ϕ=45 deg.). This is normal due to the increase of the total mass of the column. While changing the stirrups diameter reveals a trivial change in the values of natural frequency and the relative amplitude deformation not reach 2.6 % and 3.5 %, respectively.

4. Transient Analysis

Transient Dynamic Analysis is a technique used in determining the response of a structure to arbitrary time-varying loads such as an explosion. The input loads data is a function of time, and the output data are time-varying displacements and other derived quantities such as stresses and strains. The earthquake effects on structures are always in three directions longitudinal, transverse, and upward direction. The upward seismic effect usually is resisted by the structure weight, while the other effects (longitudinal and transverse) are the principal effects [7]. Generally, horizontal inertia forces cause the most damage to a structure during an earthquake. However, recent observations and analysis of earthquake ground motion have shown that the vertical motion in the structure should be completely ignored [8]. In reality, however, seismic excitations may occur horizontally in the latitude or transverse directions or include components of both directions (bi-directional loading) in the case of ground excitation at an angle to the column support. However, the response of multi degrees of freedom (MDOF) systems to a transient signal may be calculated by decomposing the system into series of single degree of freedom (SDOF) systems, calculating the response of each in the time domain and then algebraically combining the response history to obtain the response of the MDOF system [9].
ANSYS V 18.1 offers two methods of transient dynamic analysis: Full transient and modal superposition methods. The full transient method is the most general of the two methods because it allows all types of nonlinearities to be included. The full method uses the full system matrices to calculate the transient response (no matrix reduction). However, the full method accepts all types of loads: nodal forces, imposed (nonzero) displacements, element loads (pressures and temperatures) and allows tabular boundary condition specification via table type array parameters. Transient dynamic analysis cannot accept the acceleration wave, so that must be integrated to convert to another loads-time type, such as displacement-time excitation [5].
In this study the full transient analysis was used to investigate the response of 6 HCC subjected to horizontal 2D seismic excitation (displacement-time excitation), four of them having different ϕ values (15, 30, 45 and 60 deg.) and constant section size 300*300mm, while the other two having section sizes (250*250 mm) and (400*400 mm) with a constant ϕ = 30 degree. This finite element analysis will involve modeling HCC. Each column will have an axial load package equal to 140 kN applied at the top of the columns to simulate the dead loads of the superstructure, and this axial load represents the highest static load value that could be carried by the weakest columns. HCC will also be subjected at its supports

in the bottom end to a seismic loading horizontally in x and z directions (the y-direction had been kept constrained). The seismic loading is applied at the base of the HCC. The seismic data used in this study is (displacement vs. time) in two directions. The horizontal (x-direction) seismic signal data were simulated according to a part of Loma-Preita earthquake displacement-time history, see 'figure (5-a)'[2]. While, the horizontal (z-direction) seismic signal data was simulated according to an arbitrary small sign data, see 'figure (5-b)'. The current transient dynamic analysis of the HCC columns takes into account inertia-gravity effect.

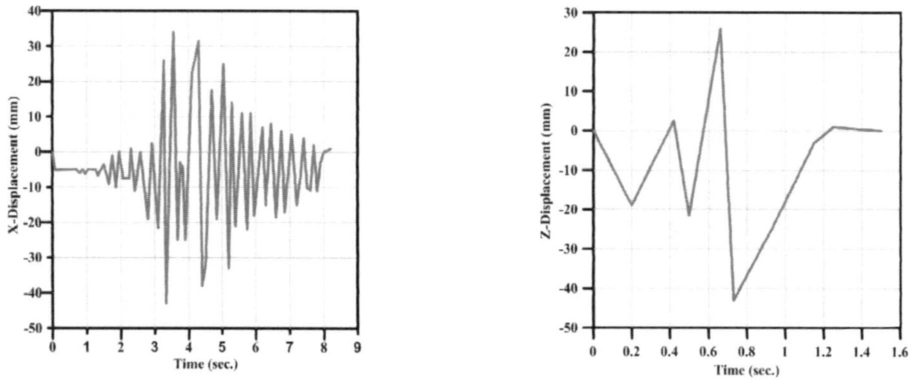

Figure 5. dimensional seismic motions: (a) Latitude displacement (b) transverse displacement.

HCC behave differently than the normal geometry columns under the seismic excitation. However, the numerical 2D transient analysis shows preeminently that HCC tends to boost its responses, like twisting and buckling deformations, especially with existing of axial load. The boosting in the responses of HCC necessitate more attentions for the reinforcement details. The deform shape-time history of 6 HCC demonstrated that with increasing stress path section distance (S) which is the distance of the top section of HCC column that will transfer shear stresses along with seismic motion path, 'figure 6', the top section would have less shear stresses.

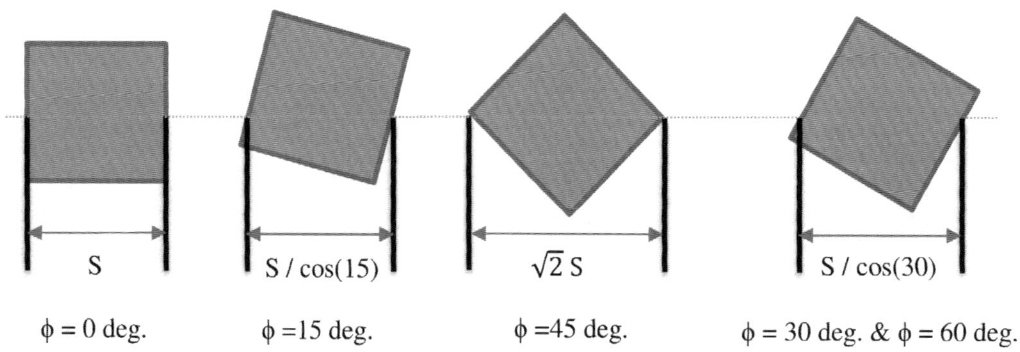

Figure 6. effect of section twisting angle up on the stress path section distance.

This is conducted from the comparison of four 300x300 mm HCC having different (ϕ) at the same moment during the time of the seismic motion, see 'figure 7' and 'figure 8'. This is logical due to the increase in the resistance path of the shear stresses crossing section. However, increasing the size of the concrete section of the HCC affect as increasing the strengthening energy against the shear stresses resulted by the torsional stirrings and the lateral movements. Furthermore, increasing the twisting angle lead to increase in the overall deformations of HCC. Figure 9 shows the differences in the time-displacement of the HCC columns during the seismic excitation duration according to varying the twisting angle (ϕ) and increasing section size.

IOP Conf. Series: Materials Science and Engineering **584** (2019) 012007 doi:10.1088/1757-899X/584/1/012007

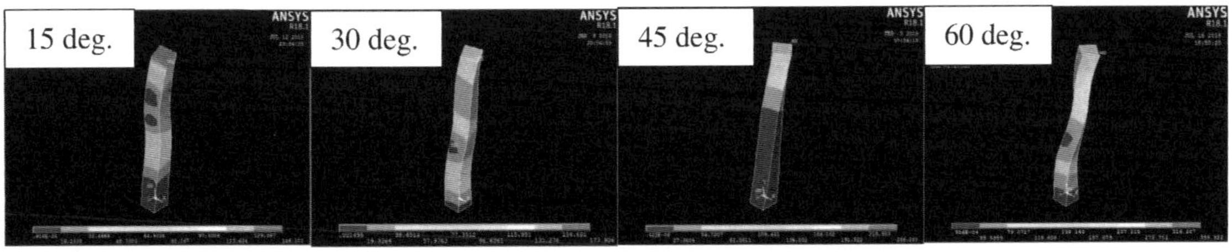

Figure 7. effect of stress path section S distance up on the stress of top section of 300x300 mm columns after 6 sec of beginning the seismic excitation.

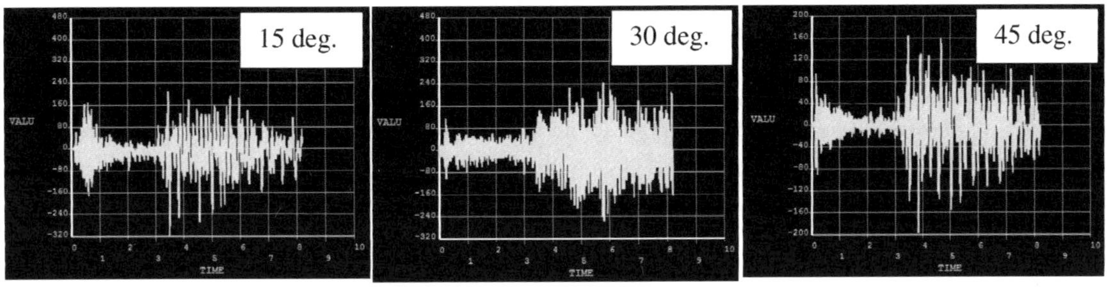

Figure 8. Time-stress results of 300x300 mm HCC with different (ϕ) during the seismic excitation duration.

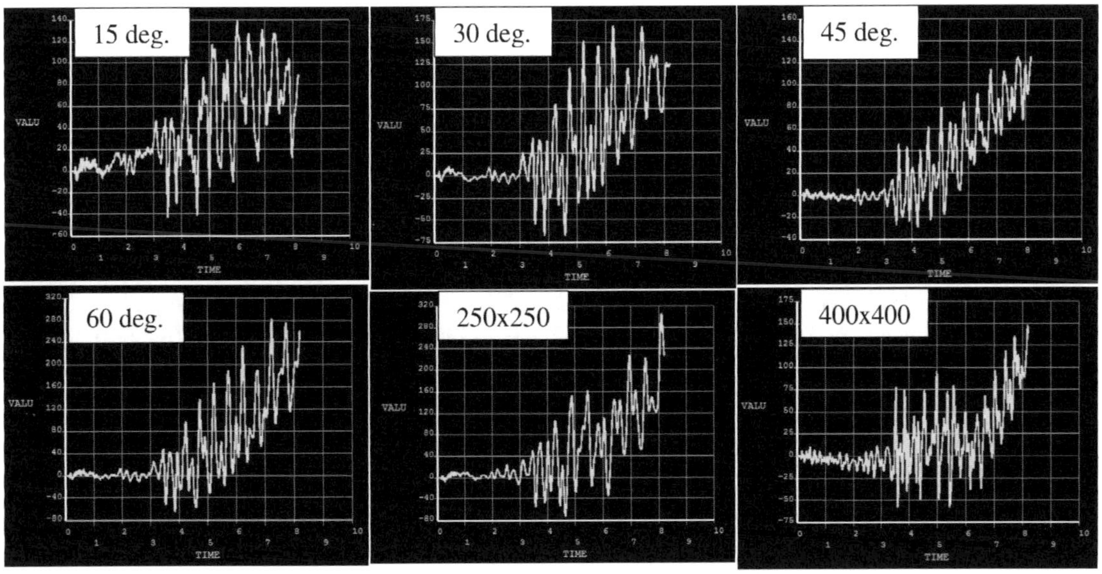

Figure 9. Time-displacement results of HCC with different (ϕ) and section size during the seismic excitation duration.

The displacement value at a specific moment of seismic excitation period depends on the direction of the seismic movement at that time relative to the direction of (S). If the direction of the excitation movement is in a parallel direction to (S), it will reduce the displacement response of the column even though the twisting angle (ϕ) of the section is large. This behavior attributed to the increasing the moment of inertia of the section in that direction. This explains why the column response displacement values at 45 angle are less than displacement values in cases of 15 and 30.

5. Conclusion

The main conclusions of the present study can be summarized as the following:

- The helical reinforced concrete columns that have the same method of loading and supporting will have the same mode shapes, despite them natural frequency values.
- Increasing the HCC section size will increase the natural frequency and decreasing the relative responses under the seismic excitation.
- The reinforcement amount of the HCC will effect up on the natural frequency as well as the deformations values and strength of the columns to the seismic loads despite it did not affect the mode shapes.
- From the shapes of disturbance of the HCC response excited by 2D horizontal seismic motion, it was found that with changing the twisting angle of the column section there is a changing in the shear stress subjected at the top section of HCC, due to change the stress path section distance.
- A helical concrete column, exited by the 2D seismic load which provides a mechanism to accommodate moment in all directions, should have a considerable design for supporting vertical loads, twisting and shear stresses.

References

[1] Schlick B (2002) Finite Element Analysis of Reinforced Concrete Bridge Pier Columns Subjected to Seismic Loading (University of Massachusetts) Department of Civil and Environmental Engineering 236 Marston Hall, Amherst. ASME.
[2] https://www.pinterest.com/
[3] Moehle J, Hooper J and Lubke C 2008 Seismic Design of Reinforced Concrete Special Moment Frames: A Guide for Practicing Engineers. NEHRP. (National Institute of Standards and Technology). GCR 8-917-1.
[4] Doniga, C, Atanasiu G 2012 Improving Seismic Safety of RC Elements by Using Advantages of Rectangular Spiral Stirrups. 15 WCEE.(Technical University). Romania. http://www.iitk.ac.in/
[5] ANSYS Help. ANSYS Release.18.1, Inc, 2017. https://www.ansys.com/
[6] MSC Nastran 2012 Dynamic Analysis User's Guide. https://www.mscsoftware.com/
[7] Merritt, F and Ricketts, J 2000 Building Design and Construction Handbook, 6th Edition, McGRAW-HILL,.
[8] Aung T, Win T and Kyaw N. Analysis and Design of Reinforced Concrete Structures with Spring Base Isolation, (Mandalay Technological University).
[9] Elnashai A and Di Sarno L 2008 Fundamentals of Earthquake Engineering, (Jon Wiley and Sons Inc.), First published. www.Architects.Ir

IOP Conf. Series: Materials Science and Engineering **584** (2019) 012008 doi:10.1088/1757-899X/584/1/012008

Experimental Investigation of High Strength Precast Reinforced Concrete Walls used (Vierendeel Truss Form)

Mustafa AbdulHussein Hassan[1], Hadi Nasir[2*], and Jamal Saeed[3]

[1] Graduate student, Civil Engineering Department, Al-MustansiryahUniversity, Baghdad, Iraq.
[2] Assistant Professor, Civil Engineering Department, Al-Mustansiryah University, Baghdad, Iraq.
[3] Assistant Professor, Civil Engineering Department, Al-Mustansiryah University, Baghdad, Iraq.

*Corresponding author email: hadinasir@gmail.com

Abstract. This paper presents an experimental investigation to study the behavior of Vierendeel Truss used instead of bearing wall in precast construction. The experimental program contains casting and testing (8) R.C wall panels specimens with opening with right angle corner, the dimension of the opening (300*350*75) mm which it is length, width, and thickness. The dimension of the R.C wall panel specimens was (1000*750*75) mm which they are length, width and thickness, respectively and all the specimens contain cantilever portion (100*60*75) mm length width and thickness respectively. In this study the main variable is the type of concrete and the compressive strength. The wall panel specimens were tested with uniformly distributed load with an eccentricity of (67.5) mm. The specimens were simply supported to represent the supports of the truss. The wall panels were divided into (2) groups, the first group is reactive powder concrete with (4) specimens and the second group is high strength self-compacted concrete with (4) specimens. The failure load, load – deflection curves crack pattern was studied. All the panels were deflected in single curvature in the vertical direction of loading. Opening causes concentrated stress at the corners of opening and initiated the cracks and failure.

1. Introduction

1.1 Vienedeel Truss

In 1896 professor Arthur Vierendeel developed a rigid frame having rigid joints, this frame comprising from upper and lower chord with vertical members between the top and bottom chords. A Vierendeel truss is hyper static frame composed of a series of rectangular or trapezoidal panels without diagonal members. These types of frames depending on the rigidity of joints for stability. This type of trusses does not have the usual triangular voids which used in a pin joint, rather employing rectangular opening with a rigid connection in the elements, which unlike conventional truss must also substantial bending forces. Vierendeel frames are usually subjected to bending while the individual members in it are subjected to bending moments and shear force in addition to direct tension of compression. Viernedeel frames used in structural when the free space between the top and bottom chord is required. [1]
Vierendeel truss offers some esthetic qualities, has simple details because of the limited number of members at a joint, is easier to form and place and can be pre-cast or cast in place. Figure (2) shows different types of Vierendeel truss and applications.

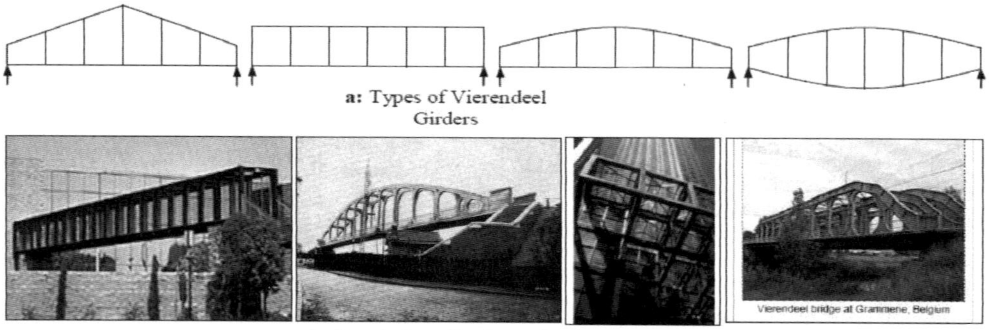

b: Practical applications of Vierendeel

Figure 1. Different Types of Vierendeel Truss and its application.

1.2 Self-Compacted Concrete

Self-compacting concrete (SCC) is a highly flow- able, non-segregating concrete that can spread into place, fill every corner of the formwork, and encapsulate the reinforcement without any mechanical consolidation or compaction and characterized by high resistance to segregation. SCC has also been described as self-placing concrete, and self-leveling concrete, which are all subsets of SCC. [2, 3]
Self-compacting concrete consists of the same components as conventionally vibrated normal concrete, which are cement, aggregate, water, and admixtures. However, the high amount of super-plasticizer for better workability, the high powder content as "lubricant" for the coarse aggregates [4].

1.3 Reactive Powder Concrete

Reactive powder concrete (RPC) is one of the latest and most important developments in concrete technology, it has superior mechanical properties such as; high strength, high ductility, high durability, limited shrinkage, high resistance to corrosion and abrasion and improvement in tensile cracking resistance, post cracking strength, ductility, and energy absorption capacity [5].RPC is a mixture of cement, silica fume, fine sand, high range water reducer, water and steel fibers without coarse aggregate to enhance the homogeneity [6].Owing to the fineness of silica fume and the increased quantity of hydraulically active components, it has been called reactive powder concrete [7].

2. Materials

RPC is a mixture of cement, silica fume, fine sand, high range water reducer, water and steel fibers without coarse aggregate. Self-compacted concrete contains the main compound of the mixture which it is cement, fine aggregate, and coarse aggregate and water but an amount of super-plasticizer which named commercially as Glenium 51 to enhance the workability. The materials were according to Iraqi specifications.

3. Mix Proportion

All the mix proportion was selected according to the previous researchers.

3.1. Reactive Powder Concrete

Table 1. Shows the proportion of reactive powder concrete

Group No.	Cement Kg / m3	Sand Kg/ m3	Silica Fume %	Super-plasticizers %	Steel Fibers %	W/C ratio
3	1000	1000	15	6	1	0.25

Silica fume by wt. = 150 kg/m^3
Super-plasticizer by wt. = 69 lit/m^3

IOP Conf. Series: Materials Science and Engineering **584** (2019) 012008 doi:10.1088/1757-899X/584/1/012008

3.2 High Strength Self-Compacted Concrete Proportion

Table 2. Shows the proportion of Self-Compacted concrete.

Group No.	Cement Kg/m^3	Sand Kg/m^3	Gravel Kg/m^3	Silica Fume Kg/m^3	Super-plasticizer Lit/m^3	Water Lit /m^3
2	450	860	790	50	20	150

4. Form Work

The form work used for casting the concrete of the wall panels was fabricated from rectangular timber planks named commercially as ply-wood with (20) mm thickness, this used for the bottom and opening. For the side of the form was made from the type of timber named commercially as jam. The sides of the form work were connected by screws, and these were connected with the bottom face by screw also, and the openings were connected by screws to the bottom face. Figure (2) shows Form work of the R.C Wall Panel Specimens

Figure 2. Form work of the R.C Wall Panel Specimens.

5. Panel Designation and Dimensions

Panels are designated as (Wx1-x2); the number (x1) refers to the number of the group. Group 1 was for the normal strength concrete; group 2 was the high strength self-compacted concrete; group 3 was for the reactive powder concrete. The number (x2) refers to the number of the specimen within the group. Table (3) shows the group of panels. The tested wall panels were designed to have appropriate dimensions that can be manufactured, handled and tested as easy as possible. The panel dimensions were (1000 *750 * 75) mm as length, width, and thickness respectively, and contain a cantilever portion on edge in one direction which dimension was (1000 * 100 * 60) mm and these were inclined inside the panel. The wall panel also contains two opening with a dimension of (350 * 300 * 75) mm with right angles. Figure (3) shows the dimensions of the wall panel's specimens.

IOP Conf. Series: Materials Science and Engineering **584** (2019) 012008 doi:10.1088/1757-899X/584/1/012008

Table 3. Wall panels Designation and Dimensions.

Group No.	Wall Panels	Length mm	Height mm	Thickness mm
Group 1	W2-1	1000	750	75
	W2-2	1000	750	75
	W2-3	1000	750	75
	W2-4	1000	750	75
Group 2	W3-1	1000	750	75
	W3-2	1000	750	75
	W3-3	1000	750	75
	W3-4	1000	750	75

6. Steel Reinforcement

All the (8) wall panels specimen were reinforced with (8mm) as main reinforcement for the beam and columns and the cantilever portion, and (6 mm) for both ties and stirrups. The reinforcement used in these wall panels was satisfying the minimum required ratio of ACI- code [8]. Figure (4) shows the reinforcement of the specimens.

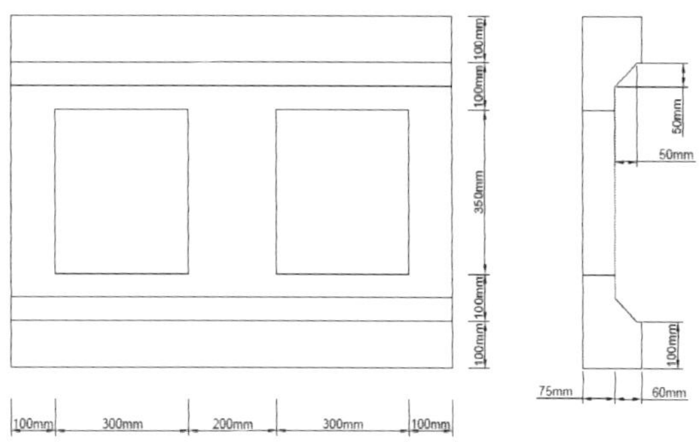

Figure 3. Wall panel dimensions.

Figure 4. Reinforcement of the R.C. wall specimens.

7. Testing

The preparation before testing includes painting all the specimens with white paint in order to make the cracks clear and apparent. The main testing machine is a universal testing machine (8551 M.F.L system) available in the Structural Lab.in Civil Eng. College of Al-Mustansiriyah University as shown in figure (5). The panels are tested by this machine after making some arrangement to the supports and measure the span between supports and marking the required eccentricity and put the frame which is used to apply uniform line load and attach all the dial-gauges. The compressive strength was test by using this device, the average of (3) cylinders (150*150*300) mm was recorded in Table (4).

Figure 5. Testing of R.C wall panel specimens.

Table 4. Compressive Strength.

Type of concrete	f'c MPa
Reactive Powder Concrete	112
High Strength Self-Compacted Concrete	70.2

8. Cracking load

Any prominent crack in R.C units greatly detracts from the appearance. Excessive types of cracking affect durability and can lead to corrosion of reinforcement although strength may not be affected. [9].Cracking load (first crack load) is that load at which the first visible surface crack is seen by the naked eye on the surface of the wall, ultimate load is the maximum load capacity of the specimen can reach before its failure Table (5) shows the first crack load and ultimate load and the ratio between them.

Table 5. First Ultimate load and ratio between them.

Group	Wall panel	First crack load (Pcr)	Ultimate load (Pu)	Pcr/Pu*100%
Group 1	W2-1	92.5	267.5	34.57
	W2-2	92.5	262.5	35.23
	W2-3	95	202.5	46.91
	W2-4	72.5	157.5	46.03
Group 2	W3-1	210	395	53.16
	W3-2	165	342.5	48.17
	W3-3	100	412.5	24.24
	W3-4	92.5	402.5	22.98

9. Cracking Pattern

The crack patterns for typical specimens observed on the tension and compression face of the wall panels. All these photographs are taken for the panels after the failure of these panels and marking the visible crack with colored lines as can as possible. The cracks were observed in tension and compression faces of the specimens.

(Compression face) (Tension face)

Figure 6. Crack pattern for W2-1

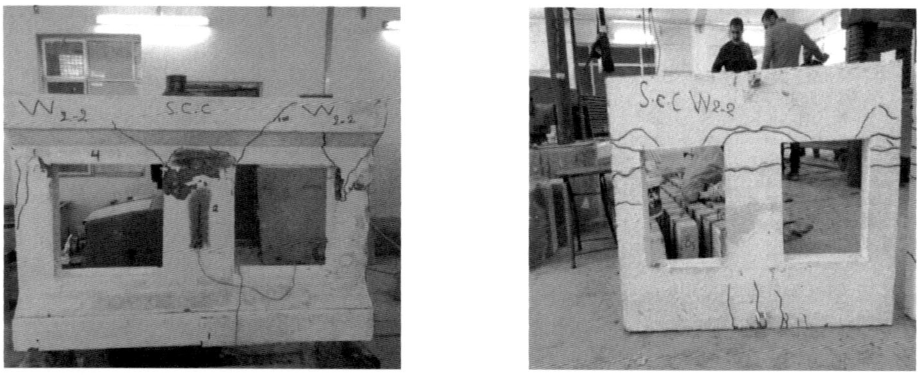

(Compression face) (Tension face)

Figure 7. Crack pattern for W2-2.

10. Deflection Characteristics

Graphically, shows the load versus deflection profiles for R.P.C, S.C.C. During the test, the applied load and the corresponding deflections, at the mid center of the lower chord, the mid center of the middle column and the bottom edge of the specimen and mid of the opening were recorded using dial gauges (accuracy 0.01mm). All the tests are carried out under the condition of load control of 10kN increments. At the beginning of each test, a small load is applied (about 2kN) to seat the supports and loading system, then the load is released, and reloading is carried out in a rate with an increment of 10kN. Figure (15) shows the positions of dial gauges.

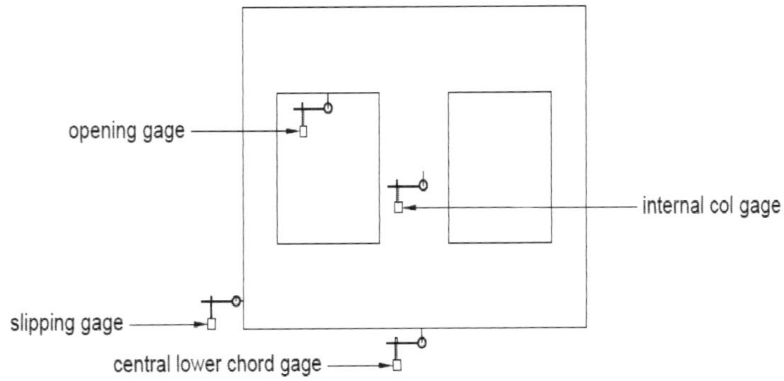

Figure 8. The position of dial gauges.

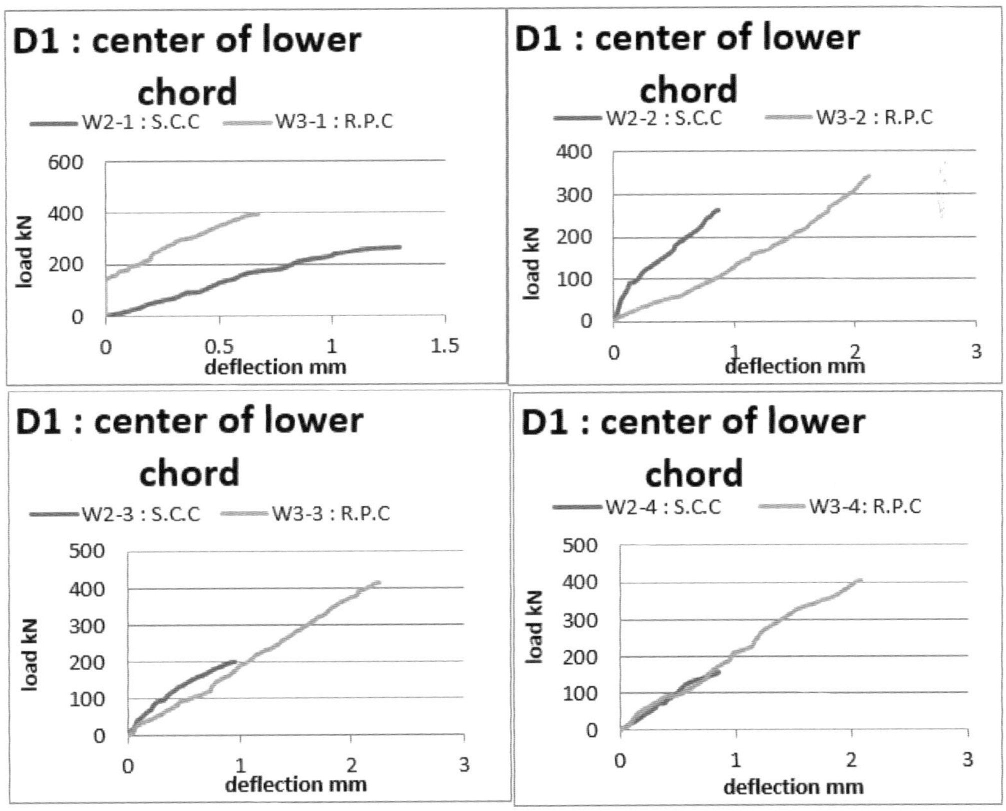

Figure 9. D1: centre of the lower chord.

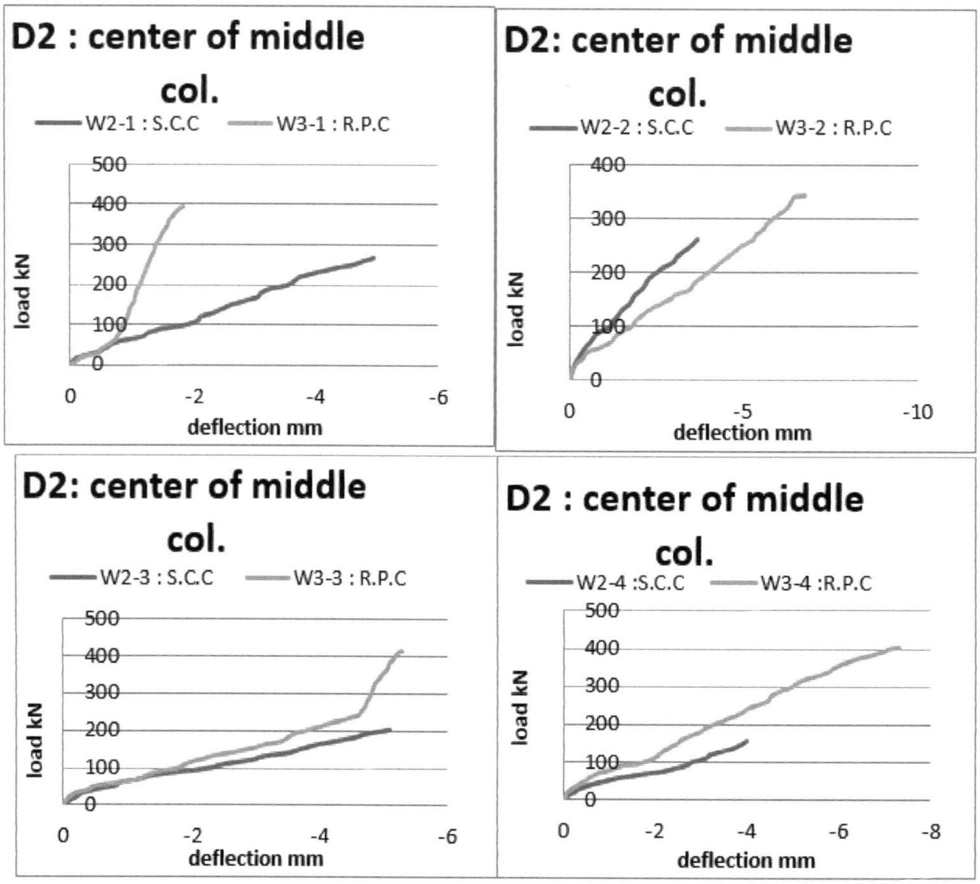

Figure 10. D2: centre of the lower chord.

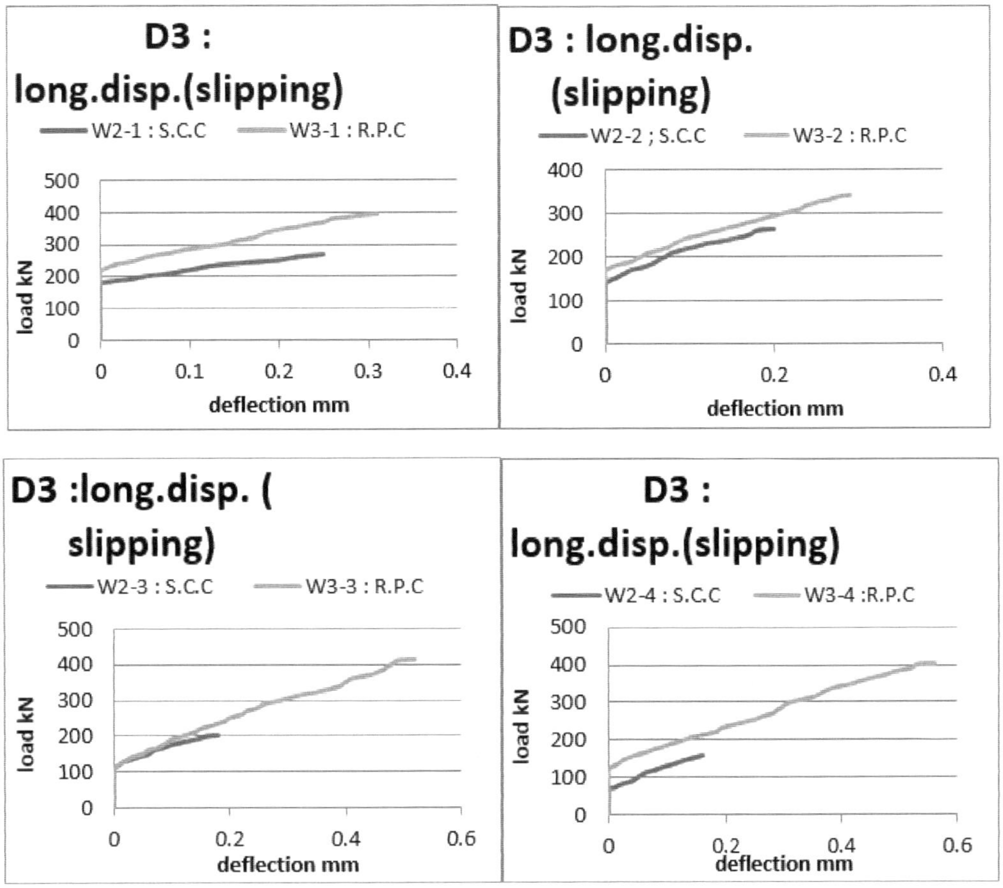

Figure 11. D3: long. Disp. (slipping).

11. Mode of Failure

The modes of failure of R.C concrete wall specimens tested under axial eccentric distributed line load are:

1. Cracking and crushing and bending in upper chord
   W1-1, W1-2, W1-3and W1-4 .Cracking and bending in upper chord
   W3-1, W3-2, W3-3and W3-4.
2. The panels were deflected in single curvature in the vertical direction until the failure of specimens due to the eccentric load applied.
3. Bending in the upper chord due to eccentric load and thus the columns were subjected to uniaxial action and this led to the curvature of columns (buckling).

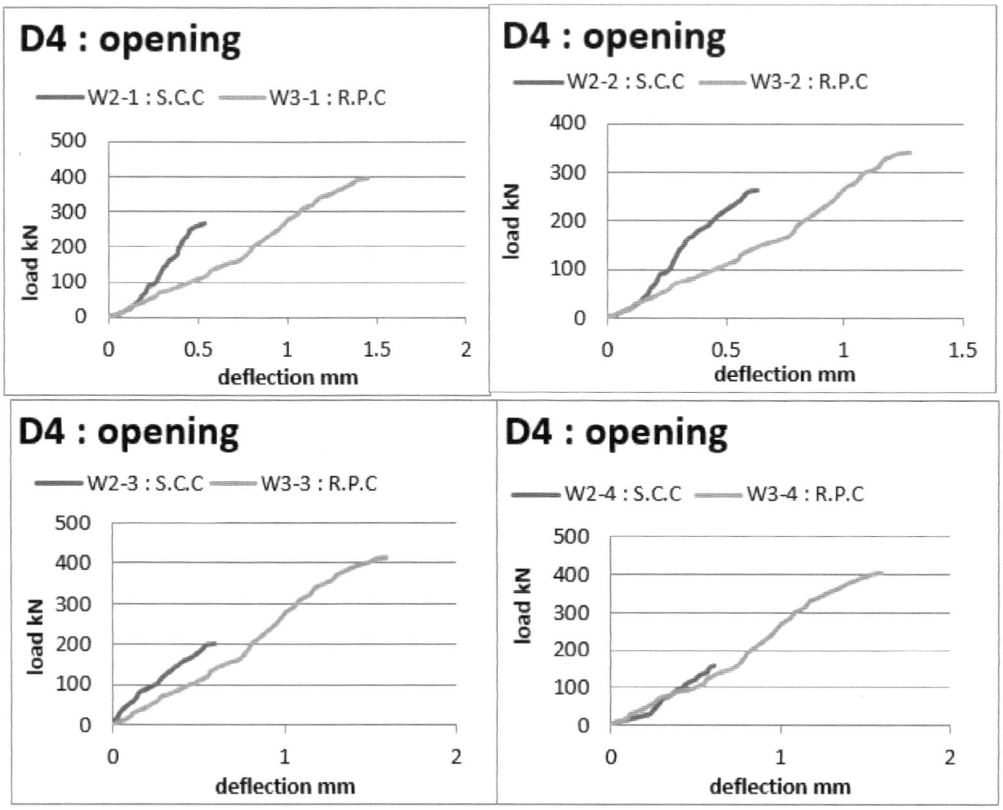

Figure 12. D4: opening.

12. Conclusion

- The cracking load of S.C.C is about (35-47) % of the failure load. The cracking load of R.P.C is about (23-54) % of the failure load.
- R.P.C show the high value of the failure load (ultimate load) than S.C.C; this is because the existences of silica fume and a steel fiber in the mixture and that shows ductile failure behavior.
- The addition of silica fume and very fine sand in R.P.C led to make the mixture dense and improve it is a behavior to load.
- For S.C.C specimens, the main failure of the specimens in this group is at the joint between the upper chord and columns. The concrete is cracked and crushed in the compression face. For R.P.C specimens, the main failure of the specimens in this group is at the joint between the upper chord and columns. The concrete in this point shows different behavior that in the S.C.C which the concrete in these specimens is cracked with little crushing of concrete because of the existence of steel fibers in the mixture. The steel fibers arrest the crack and enhance the ductility.
- Cracks in tension face of columns of the specimens and the cracks are non-straight. The crack in the columns is more than the upper and lower chords.
- In the lower chord, flexural cracks are appeared during the test and before the shear cracks appeared.
- The lower chord work as tension member and this lead to that the specimen work as truss and this is why the dial gauge which used in the bottom edge of the specimen (long. Displacement) recorded value of deflection.
- The reason for cracks in tension face of columns is the bending moment formed due to eccentric axial load and the cracks increases with the increasing of the load.

IOP Conf. Series: Materials Science and Engineering **584** (2019) 012008 doi:10.1088/1757-899X/584/1/012008

- Cracks in the compression face of specimens appeared in the upper chord and started from the corner between the columns and upper chord and with 45' this is due to concentrating the stress at corners of the opening.
- It should know that the panel W2-4, the frame during loading failed before it reaches its ultimate load.

References

[1] Raju, N. K., "Advanced Reinforced Concrete Design", CBS Publishers &distributers, First Edition, PP. 360, 1986.

[2] ACI Committee 237R-07, "Self-Consolidating Concrete", Copyright by the American Concrete Institute, Farmington Hills, MI., First Printing, PP. 30, April, 2007.

[3] Neville, A.M., "Properties of Concrete", Fifth and Final Edition, Wiley, New York and Longmont, London, PP.844, 2010.

[4] Dehn , F., Holschemacher , K. and Weibe D., "Self-compacting Concrete (SCC) – Time Development of the Material Properties and the Bond Behavior", The Laboratory for Combustion and Energy Research (LACER) No. 5, pp.115-123, 2000.

[5] Richard, P.and Cheyrezy, M., "Reactive Powder Concrete with High Ductility and 200-800 MPa Compressive Strength", ACI, SP144-24, pp. 507-518, 1994.

[6] Raj , J. and Jeenu, G., "Flexural Behavior of UHPC-RC Composite Beams", Proceedings of International Conference on Technological Trends (ICTT), College of Engineering, Trivandrum, India, 5 pp, 2010.

[7] Sadrekarimi, A., "Development of a Light Weight Reactive Powder Concrete", Journal of Advanced Concrete Technology, Japan Concrete Institute, Vol.2, No.3, pp.409-417, October 2004.

[8] ACI 318-08, "Building code requirement for reinforced concrete", American Concrete Institute, USA, 2008

[9] Macginley, T.J. and Choo, B.S., "Reinforced Concrete Design Theory and Examples ", Taylor & Francis e-Library, Second Edition, 2003.

IOP Conf. Series: Materials Science and Engineering **584** (2019) 012009 doi:10.1088/1757-899X/584/1/012009

The Role of Cooling Degree Days on determining the insulation of building Envelop in a hot climate (Iraqi cities as an example)

Susan Abed Hassan[1]

[1] Department of Architectural engineering, College of Engineering, Al- Nahrain University-Iraq.
Email: dr.susanabedhassan@eng.nahrainuniv.edu.iq , suzana302002@yahoo.com

Abstract. Last decades witnessed rise in global average temperature for cities, consequently with the problem of rising needs for energy results from the growth of populations in cities. Recent studies concentrated on degree days method to calculate energy consumption for buildings, the main problem of this research is the role of cooling degree days on determining insulation of buildings envelope in a hot climate and for Iraqi cities as an example. The research aims to determine insulation of buildings envelope according to cooling degree days which reduce energy consumption; the results showed that there were differences in cooling degree days for main Iraqis cities, and there was an effect of buildings envelope insulation on cooling energy lost, the insulation for walls reduced consumption energy by 70-80%. Also, the insulation for ceiling reduced consumption energy by 65%, as compared to the same building materials without insulation.

1. Introduction

These Global temperatures had raised at the end of the 20th century and the beginning of the 21st century, the high record temperatures was one of the most prominent indicators of global warming. That affected buildings cooling energy loads especially in a hot arid climate [1]. There were many ways to calculate the energy usage for cooling and heating buildings, but the method of calculating the degree of days was one of the main types. Degree days depended on calculation temperature of the external environment, and its association with the internal temperature of buildings. The previous studies did not adequately address this indicator in Iraq, but the international studies highlighted many aspects of research and the latest of these studies. Indraganti et al. (2016) [2] examined the effect of increasing the degree days whether in cooling or heating in five different climatic zones in Saudi Arabia and for nine years with difference of the basic comparison class in the case of heating between (14, 16, 18 °C) in the case of cooling between (18, 20, 22, 24, 28 °C) and control of energy consumption (for cooling and heating) of buildings for the same cities. The study showed increase energy consumption in recent years with rising temperatures degree days. Mishra, et al., (2012) [3] studied the effect of degree days on determined the thickness of thermal insulation for walls and ceiling for buildings that minimize loss of heat through external walls, window, and ceiling, for four different climatic zones in India. The research found that optimum insulation thickness varies between 0.1446m and 0.2077 m, and energy reduction between 500.03 Rs/m^2 and 1014.27 Rs/m^2, and Payback period varies between 0.87 to 1.2374 years for External walls. Yuan, et al., (2017) [4] studied the relationship between insulation for the exterior walls of buildings and energy and CO_2 emissions in different places in China. Using life-cycle cost analysis (LCCA) and degree-days (DD) method. . The research found that thermal insulation of exterior walls is more effective in Severe Cold and Hot Summer & Cold Winter climatic zones of China for both total energy cost saving and CO_2 emissions reduction per unit area of the exterior building walls. Melo, et al., (2015) [5] Studied the effects of insulation materials on energy performance for Brazil commercial

IOP Conf. Series: Materials Science and Engineering **584** (2019) 012009 doi:10.1088/1757-899X/584/1/012009

buildings. Depend on the climate of the building site. The research found that contradict the maximum limits of overall heat transfer coefficient for walls given in the ASHRAE Standard 90.1, at least for hot climates dominated by cooling energy use.

2. Degree days

Building During occupancy consumes more energy than energy for producing it. Recent studies had shown that energy consumption for buildings reached 40% of total energy consumption in the cities [6]. The consumption of buildings for energy is affected by several external environmental factors (external temperature, solar radiation intensity, relative humidity, wind, etc.), while buildings affected by its design, envelop materials and components in the calculation of consumed energy. The index of degree days is the simplest method to calculate the energy needs for buildings. It adopted by Lt-Gen. Sir Richard Strachey, during his study of the effect of daily temperature diversity across seasons on plants and seasons of growth according to their climatic zones. It defined as the sum of the differences between the hourly temperature and the base degree [7]. The degree of the base represents the degree to which the balance between the external air temperature and the temperature of the building, which does not require the building to energy for cooling or heating to achieve thermal comfort [8]. The degree days index depends on the range of temperature variations over time, thus determine both the length and duration of outdoor temperatures with the calculation of the difference between the temperature that achieves thermal comfort in the building and temperature of the outdoor air. For example, if the temperature outside in the winter 10 degrees Celsius and the temperature required to be provided in the building to achieve the human comfort of 22 degrees Celsius, this means that heating system should provide heating equal to raise the temperature inside 12 degrees Celsius. This contributes to knowledge of the energy needed to heat or cool buildings in summer. The importance of the degree of days [9] :-
- Determination of energy required for cooling and heating of buildings.
- Evaluation of thermal loss of buildings to determine the required thermal insulation and other energy saving methods.
- Control of energy consumption in buildings [10].

The degree days can be divided into two main types, the degree days of heating (HDD) and the degree days of cooling (CDD), which are two indices to calculate the energy consumed for cooling and heating in buildings of different types. The degree days of heating (HDD) measure the number of degrees and periods on days when the temperature of the air outside the base degree, which can calculate the amount of energy, spent for heating. Degree days of cooling (CDD) also measure the day periods when the external air temperature is above the base level, which enables the calculation of the amount of energy spent for cooling [11].

3. Degree Days calculation

There are many types to calculate degree days, depend on the factors influence in the calculation of it, as fallowing:

- (ASHRAE) method: American Society of Heating, Refrigeration and Air-conditioning Engineers
- (UKMO) method: United Kingdom Meteorological Office
- (S-K) method: Schoenau and Kehrig's
- Hitchin's method
- Mean degree-hours [2]

Mean degree hours were the most appropriate type to experience the daily changes in temperatures. This depends on the highest and lowest outside temperatures [12]. To calculate the Mean degree- hours three basic factors are included, (T_{max}) Daily maximum temperature, (T_{min}) Daily minimum temperature, and a (T_{base}) base temperature, nominated by the user as an estimate of the outside air temperature at which no artificial heating (or cooling) is required. Base temperature is related to the degree to which thermal comfort is achieved within buildings. Previous studies have relied on the determination of the base score for the cooling and heating calculations. For example, the 18.3 ℃ was adopted by the United States Heating and Ventilation Organization (ASHRAE). In the United Kingdom, 15.5 ° C was adopted and

15 ° C was adopted in Germany are associated with varying temperatures and annual and cumulative rates in each of these states [10].

4. The relationship between the degree of days and heat transfer coefficient

The calculation of energy consumption in buildings is based on the value of thermal insulation for the buildings envelop which include (walls, ceiling, and floor). The most important indicator of insulation calculation is the total heat transfer coefficient U-Value, which represents the ratio of heat transfer across a given area with different temperatures. It is based on the properties of building materials and is measured in watts per square meter in a single Celsius degree ($W / m^2 / {}^0C$). The relation between heat transfer coefficient, thermal loss, and the base temperature, according to the equation (1) [3]:

$$Q = U(T_b - T_a) \tag{1}$$

Where
Q= thermal loss
U= heat transfer coefficient
T_b = Base Temperature (°C)
T_a = average Temperature (°C)
The relationship between degree days and the thermal loss for building envelop is expressed in the equation (2):

$$QA = 86400DD \times U \tag{2}$$

Where
DD= Degree days

5. Research objectives

The main objective of this research is to contribute to the understanding the role of cooling degree days on determining the building envelope insulation in a hot dry climate. This goal can be achieved through:

1- Investigating the degree days for cooling for the main cities in Iraq for the periods of time (2010-2017).
2- Identifying the building envelope insulation according to the degree of the day of cooling which contributes to reducing the energy consumption during summer in the hot, dry climatic characteristics of Iraq.

6. Methodology

The research studied the degree days for the main cities in Iraq (Baghdad, Basrah, Khanaqen, Erbil, and Haditha) for the periods of time (2012-2017). The calculation requires measurements of maximum and minimum outside air temperatures (T_{max} and T_{min}). According to Iraq weather history and a 'base temperature' T_{base}, for all the months of the recent years were used, to estimate cooling degree days from the following formulae:

If the minimum temperature is larger than base temperature $T_{min} > T_{base}$ the equation (3) is applicable:

$$CDD = \frac{(T_{max}+T_{min})}{2} - T_{base} \tag{3}$$

If the minimum temperature is equal or smaller than base temperature $T_{min} <= T_{base}$ the equation (4) is applicable:

$$CDD = \frac{(T_{max}+T_{min})}{2} - \frac{T_{base}-T_{min}}{4} \tag{4}$$

And the relationship between Cooling Energy Lost and Cooling degree days and the heat transfer coefficient is expressed in the equation (5) [13]:

$$Cooling\ Energy\ Lost = (24 \times CDD\ Annual \times U)/1000 \text{ kWh/year/m}^2 \tag{5}$$

7. Results

7.1. Comparison of Cooling Degree Days in Iraq main cities

A comparison was made between Iraq main cities average cooling degree days, for different base temperatures 16.5, 17, 18, 19, 20, 21, 22, 23, 24, 25, 26 °C for the periods of time (2012-2017) are shown in Figure (1). It is obvious that cooling degree days are higher in 16.5 °C base temperature in all Iraqi cities. Also there were differences in cooling degree days for main Iraqis cities.

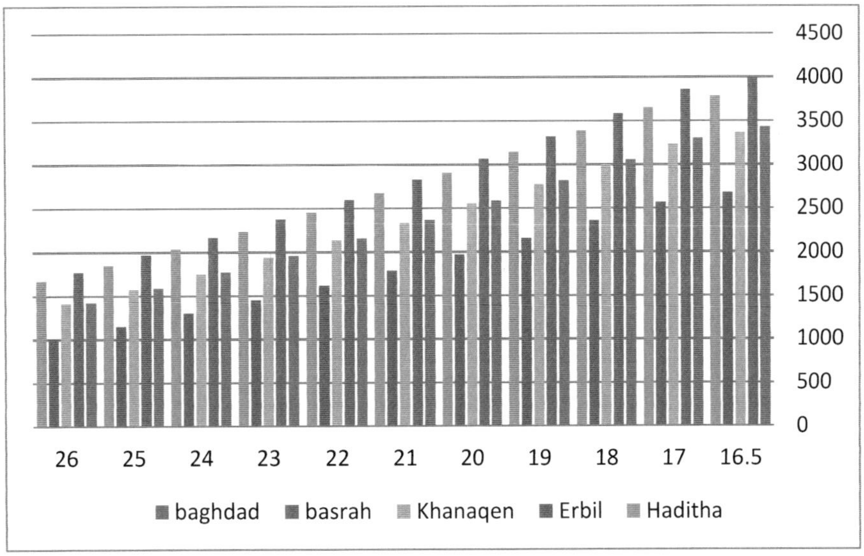

Figure 1. Average cooling degree days (2012-2017) in main Iraq cities (Baghdad, Basrah, Khanaqen, Erbil, and Haditha).

7.2. Comparison of the effect of the building envelope insulation on cooling energy lost

The common building materials in Iraqi cities which used in building envelope have different U-Value, as shown in Table 1.

Table 1. The common building materials in Iraqi cities and its U-Value

	Symbol	Building envelop components	U-Value
Wall	W1	20 mm plaster+240 mm clay brick +20 mm cement finish	1.429
	W2	20 mm plaster+240 mm Thermostone +20 mm cement finish	0.723
	W3	20 mm plaster+240 mm Cement Block +20 mm cement finish	1.515
	W4	20 mm plaster+240 mm clay brick +RSI 1.2 board insulation +20 mm cement finish	0.516
	W5	20 mm plaster+240 mm Thermostone +RSI 1.2 board insulation +20 mm cement finish	0.396
	W6	20 mm plaster+240 mm Cement Block +RSI 1.2 board insulation +20 mm cement finish	0.370
	W7	20 mm plaster+240 mm clay brick +RSI 2.5 board insulation +20 mm cement finish	0.316
	W8	20 mm plaster+240 mm Thermostone +RSI 2.5 board insulation +20 mm cement finish	0.267
	W9	20 mm plaster+240 mm Cement Block +RSI 2.5 board insulation +20 mm cement finish	0.253
Ceiling	C1	40mm precast concrete tile+ 10mm river sands+ 20 mm tar+150 mm concrete slap+20 mm plaster	1.1696
	C2	40mm precast concrete tile+ 10mm river sands+ 20 mm tar+25 mm polystyrene boards+ 150 mm concrete slap+20 mm plaster	0.5931
	C3	40mm precast concrete tile+ 10mm river sands+ 20 mm tar+50 mm polystyrene boards+ 150 mm concrete slap+20 mm plaster	0.3975

The effect of different building envelop materials on consume cooling energy for a main Iraq cities according to an average cooling degree are shown in Tables 2-11 according to equation (5).

Table 2. Cooling energy lost for cooling degree days 17°C

	Symbol	U-Value	Cooling energy lost in kWh/year/m^2				
			Basrah	**Baghdad**	**Erbil**	**Khanaqen**	**Haditha**
Wall	W1	1.429	132.38	113.24	88.07	111.01	125.14
	W2	0.723	66.97	57.29	44.55	56.16	63.31
	W3	1.515	140.34	120	93.37	117.69	132.67
	W4	0.516	47.80	40.89	31.80	40.08	45.18
	W5	0.396	38.04	31.38	24.40	30.76	34.68
	W6	0.370	35.54	29.32	22.80	28.74	32.40
	W7	0.316	30.35	25	19.47	24.54	27.67
	W8	0.267	25.65	21.15	16.45	20.74	23.38
	W9	0.253	24.06	20	15.59	19.65	22.15
Ceiling	C1	1.1696	108.35	92.68	72.08	90.86	102.42
	C2	0.5931	54.94	47	36.55	46.07	51.94
	C3	0.3975	36.82	31.5	24.49	30.88	34.81

Table 3. Cooling energy lost for cooling degree days 18°C.

	Symbol	U-Value	Cooling energy lost in kWh/year/m^2				
			Basrah	**Baghdad**	**Erbil**	**Khanaqen**	**Haditha**
Wall	W1	1.429	122.95	104.7	80.90	102.81	116.36
	W2	0.723	62.2	52.97	40.93	52.02	58.87
	W3	1.515	130.35	111	85.77	109	123.36
	W4	0.516	44.39	37.8	29.21	37.12	42.01
	W5	0.396	34.07	29	22.41	28.49	32.24
	W6	0.370	31.83	27.11	20.94	26.62	30.12
	W7	0.316	27.18	23.15	17.89	22.73	25.73
	W8	0.267	22.97	19.56	15.11	19.21	21.74
	W9	0.253	21.76	18.53	14.32	18.20	20.60
Ceiling	C1	1.1696	100.63	85.69	66.21	84.15	95.24
	C2	0.5931	51.03	43.45	33.57	42.67	48.29
	C3	0.3975	34.20	29.12	22.50	28.60	32.36

Table 4. Cooling energy lost for cooling degree days 19°C.

	Symbol	U-Value	Cooling energy lost in kWh/year/m^2				
			Basrah	**Baghdad**	**Erbil**	**Khanaqen**	**Haditha**
Wall	W1	1.429	113.86	96.47	74.04	94.89	107.82
	W2	0.723	57.6	48.81	37.46	48.01	54.55
	W3	1.515	120.715	102.28	78.50	100.60	114.31
	W4	0.516	41.114	34.83	26.73	34.26	38.93
	W5	0.396	31.55	26.73	20.51	26.29	29.88
	W6	0.370	29.48	24.97	19.17	24.57	27.91
	W7	0.316	25.17	21.33	16.37	20.98	23.84
	W8	0.267	21.27	18	13.83	17.73	20.14
	W9	0.253	20.15	17	13.10	16.80	19.09
Ceiling	C1	1.1696	93.19	78.96	60.60	77.67	88.25
	C2	0.5931	47.25	40	30.73	39.38	44.75
	C3	0.3975	31.67	26.83	20.59	26.39	29.99

Table 5. Cooling energy lost for cooling degree days 20°C.

	Symbol	U-Value	Cooling energy lost in kWh/year/m^2				
			Basrah	**Baghdad**	**Erbil**	**Khanaqen**	**Haditha**
Wall	W1	1.429	105.25	88.65	67.76	87.31	99.59
	W2	0.723	53.25	44.85	34.28	44.17	50.39
	W3	1.515	111.588	93.99	71.84	92.57	22.02
	W4	0.516	38.006	32	24.47	31.52	38.93
	W5	0.396	29.16	24.56	18.77	24.19	27.59
	W6	0.370	27.25	22.95	17.54	22.60	25.78
	W7	0.316	23.27	19.6	14.98	19.30	22.02
	W8	0.267	19.66	16.56	12.66	16.31	18.60
	W9	0.253	18.63	15.69	11.99	15.45	17.63
Ceiling	C1	1.1696	86.14	72.56	55.46	71.46	81.51
	C2	0.5931	43.68	36.79	28.12	36.24	41.33
	C3	0.3975	29.27	24.66	18.85	24.28	27.70

Table 6. Cooling energy lost for cooling degree days 21°C.

	Symbol	U-Value	Cooling energy lost in kWh/year/m^2				
			Basrah	**Baghdad**	**Erbil**	**Khanaqen**	**Haditha**
Wall	W1	1.429	96.98	81.11	61.25	80.11	91.70
	W2	0.723	49.07	41.03	30.99	40.53	46.39
	W3	1.515	102.82	85.99	64.93	84.93	97.22
	W4	0.516	35.02	29.28	22.11	28.92	33.11
	W5	0.396	26.877	22.47	16.97	22.2	25.41
	W6	0.370	25.11	21	15.85	20.74	23.74
	W7	0.316	21.44	17.93	13.54	17.71	20.27
	W8	0.267	18.12	15.15	11.44	14.96	17.13
	W9	0.253	17.17	14.36	10.84	14.18	16.23
Ceiling	C1	1.1696	79.38	66.38	50.13	65.57	75.06
	C2	0.5931	40.25	33.66	25.42	33.25	38.06
	C3	0.3975	26.97	22.56	17.03	22.28	25.50

Table 7. Cooling energy lost for cooling degree days 22°C.

	Symbol	U-Value	Cooling energy lost in kWh/year/m^2				
			Basrah	**Baghdad**	**Erbil**	**Khanaqen**	**Haditha**
Wall	W1	1.429	89.06	73.90	55.35	73.15	84.09
	W2	0.723	45.063	37.39	28.0	37.01	42.54
	W3	1.515	94.42	78.35	58.68	77.55	89.15
	W4	0.516	32.161	26.68	19.98	26.41	30.36
	W5	0.396	24.68	20.48	15.33	20.27	23.30
	W6	0.370	23.06	19.13	14.33	18.94	21.77
	W7	0.316	19.69	16.34	12.24	16.17	18.59
	W8	0.267	16.64	13.80	10.34	13.66	15.71
	W9	0.253	15.76	13.85	9.8	12.95	14.88
Ceiling	C1	1.1696	72.89	60.49	45.30	59.87	68.82
	C2	0.5931	36.96	30.67	22.97	30.36	34.90
	C3	0.3975	24.77	20.55	15.39	20.34	23.39

IOP Conf. Series: Materials Science and Engineering **584** (2019) 012009 doi:10.1088/1757-899X/584/1/012009

Table 8. Cooling energy lost for cooling degree days 23°C.

	Symbol	U-Value	Cooling energy lost in kWh/year/m2				
			Basrah	Baghdad	Erbil	Khanaqen	Haditha
Wall	W1	1.429	81.45	67.18	55.35	66.46	76.61
	W2	0.723	41.211	33.99	25.16	33.62	38.76
	W3	1.515	86.355	71.22	52.72	70.46	81.22
	W4	0.516	92.412	24.26	17.95	24	27.66
	W5	0.396	22.57	18.61	15.33	18.41	21.23
	W6	0.370	21.09	17.39	14.33	17.20	19.83
	W7	0.316	18.01	14.85	10.99	14.69	16.94
	W8	0.267	15.21	12.55	9.29	12.41	14.31
	W9	0.253	14.42	11.89	8.80	11.76	13.56
Ceiling	C1	1.1696	66.66	54.89	40.7	54.40	62.70
	C2	0.5931	33.80	27.88	20.63	27.58	31.79
	C3	0.3975	22.65	18.68	13.83	18.48	21.31

Table 9. Cooling energy lost for cooling degree days 24°C.

	Symbol	**U-Value**	**Cooling energy lost in kWh/year/m^2**				
			Basrah	**Baghdad**	**Erbil**	**Khanaqen**	**Haditha**
Wall	W1	1.429	74.31	60.7	44.41	60.18	70.03
	W2	0.723	37.601	30.71	22.47	30.45	35.43
	W3	1.515	78.79	64.35	47.08	63.81	74.24
	W4	0.516	26.83	21.91	16.03	21.73	25.28
	W5	0.396	20.59	16.82	12.30	16.67	19.40
	W6	0.370	19.24	15.71	11.49	15.58	18.13
	W7	0.316	16.43	13.42	9.82	13.30	15.48
	W8	0.267	13.88	11.34	8.29	11.24	13.08
	W9	0.253	13.15	10.74	7.86	10.65	12.39
Ceiling	C1	1.1696	60.82	49.68	36.35	49.26	57.31
	C2	0.5931	30.84	25.19	18.43	24.98	29.06
	C3	0.3975	20.67	16.88	12.35	16.74	19.48

Table 10. Cooling energy lost for cooling degree days 25°C.

	Symbol	**U-Value**	**Cooling energy lost in kWh/year/m^2**				
			Basrah	**Baghdad**	**Erbil**	**Khanaqen**	**Haditha**
Wall	W1	1.429	67.35	54.53	39.40	54.08	63.44
	W2	0.723	34.7	27.58	19.93	27.36	32.10
	W3	1.515	71.411	57.81	41.77	57.33	67.26
	W4	0.516	24.322	19.69	14.22	19.52	22.91
	W5	0.396	18.66	15.11	10.92	14.98	17.58
	W6	0.370	17.74	14.11	10.20	14	16.42
	W7	0.316	14.89	12.05	8.71	11.95	14.03
	W8	0.267	12.58	10.18	7.36	10.10	11.85
	W9	0.253	11.92	9.65	6.97	9.57	11.23
Ceiling	C1	1.1696	55.13	44.63	32.25	44.26	51.93
	C2	0.5931	27.95	22.63	16.35	22.44	26.33
	C3	0.3975	18.73	15.16	10.96	15.04	17.64

Table 11. Cooling energy lost for cooling degree days 26°C.

	Symbol	U-Value	Cooling energy lost in kWh/year/m^2				
			Basrah	**Baghdad**	**Erbil**	**Khanaqen**	**Haditha**
Wall	W1	1.429	60.77	48.7	34.70	48.39	57.27
	W2	0.723	30.74	24.63	17.56	24.48	28.97
	W3	1.515	64.42	51.63	36.79	51.30	60.72
	W4	0.516	21.944	17.58	12.53	17.47	20.68
	W5	0.396	16.841	13.49	9.61	13.41	15.87
	W6	0.370	15.73	12.60	8.98	12.52	14.82
	W7	0.316	13.43	10.76	7.67	10.70	12.66
	W8	0.267	11.35	9.09	6.48	9.04	10.70
	W9	0.253	10.75	8.62	6.14	8.56	10.14
Ceiling	C1	1.1696	49.74	39.85	28.40	39.60	46.87
	C2	0.5931	25.22	20.21	14.40	20.08	23.77
	C3	0.3975	16.90	13.54	9.65	13.46	15.93

The results showed that different wall materials (clay brick, Thermostone, and cement block) with RSI 2.5 board insulation and 26 °C base temperature were more efficient in energy consumption for cooling for all tested Iraqi cities, and concrete ceilings with 50 mm polystyrene boards insulation was and 26 °C base temperature were more efficient in energy consumption for cooling for all tested Iraqi cities.

8. Conclusion

The cooling degree days can be used by the designers to predicate cooling energy consumption for building envelop in a hot climate. In this study average cooling degree days was investigated for years (2012-2017), in main Iraq cities (Baghdad, Basrah, Khanaqen, Erbil, and Haditha) with base temperatures of 16.5,17,18,19,20,21,22,23,24,25, and 26 °C were chosen to calculate cooling degree-days. The higher base temperatures 26 °C related to lower cooling degree days in all Iraqi cities and better cooling energy consumption.

Also, calculations were made to show the effect of building envelope insulation on energy consumption according to cooling degree days for different base temperatures to determine the most efficient in energy consumption. The most important factor affected by energy consumption was the insulation on walls that reduced energy consumption to 70-80% as compared with the same wall materials without insulation. Also, the insulation on ceiling reduced energy consumption to 65% as compared with the same ceiling materials without insulation.

References

[1] Hassan Susan 2018 *The environmental criteria as a strategy to adapt climate changes in Iraq.* International Journal of Engineering and Technology 7 (4.20) 496-501

[2] Indraganti Madhavi Boussaa Djamel 2016 *A method to estimate the heating and cooling degree-days for different climatic zones of Saudi Arabia* Journal of Building Services Engineering Research & Technology The Chartered Institution of Building Services Engineers DOI: 10.1177/0143624416681383 bse.sagepub.com

[3] Subhash Mishra Dr J A Usmani Sanjeev Varshney 2012 *Optimum Insulation Thickness Of The External Walls And Roof For Different Degree-Days Region* International Journal of Engineering Research & Technology (IJERT) Vol 1 Issue 7

[4] Jihui Yuan Craig Farnham and Kazuo Emura 2017 *Optimum insulation thickness for building exterior walls in 32 regions of China to save energy and reduce CO_2 emissions* Sustainability 9, 1711; doi:10.3390/su9101711

[5] Melo Ana Paula Lamberts Roberto Versage Rogério de Souza Yi Zhang 2015 *Is thermal insulation always beneficial in hot climate?* 14th Conference of International Building Performance Simulation Association Hyderabad India Dec. 7-9.

[6] Alsammarae Aiham 2005 *Iraq's electrical system status and future plan* Electric power in Chicago.

IOP Conf. Series: Materials Science and Engineering **584** (2019) 012009 doi:10.1088/1757-899X/584/1/012009

[7] D Thevenard 2011 *Methods for estimating heating and cooling degree-days to any base temperature* ASHRAE Transactions vol. 117,pp. 1, pp. 884-891,

[8] Mc Gilligan C Natarajan S Nikolopoulou M 2011 *Adaptive comfort degree-days: a metric to compare adaptive comfort standards and estimate changes in energy consumption for future UK climates* Energy Build 43,: 2767–2778 (ASHRAE 2001).

[9] American Society of heating refrigerating and air-conditioning engineers 2001 *Fundamentals Handbook* Atlanta, GA.

[10] Moustris K P P T Zacharia I K Larissi P T Nastos and A G Paliatsos 2011 *Cooling and heating degree-days calculation for representative locations within the greater Athens area*, Greece.Proc 12th Conf on Environmental Science and Technology Rhodes Greece Global NEST A1279–A1286.

[11] Day Tony Franklin Bryan Fry Martin Holme Michael 2006 *Degree-days theory and application* The Chartered Institution of Building Services Engineers 222 Balham High Road, London SW12 9BS http://trade.gov/static/iraq_electricalppt.pdf

[12] Carbon Trust 2007 *Degree days for energy management A practical introduction* Carbon Trust pp.20.

[13] Boujnah Mohamed Jraida Kaoutar Farchi Abdelmajid Mounir Ilham 2016 *Comparison of the Calculation Methods of Heating and Cooling Degree-Days in Three Different Cities in Morocco* International Journal of Current Trends in Engineering & Technology ISSN: 2395-3152 Volume: 02, Issue: 06

Effect of Nano-Silica and Micro Steel Fiber on Compressive Strength Development of Fly Ash Geopolymer Paste Cured Under Ambient Temperature

Noor S. Hadi[1], Wissam D. Salman[2*] and S Safie Mahdi Oleiwi[3]

[1]MSc. Student Department of Civil Engineering,College of Engineering, University of Diyala, Iraq.
[2]Asst. Prof.Dr. Department of Civil Engineering,College of Engineering, University of Diyala, Iraq.
[3]Lec. Dr. Department of Civil Engineering,College of Engineering, University of Diyala, Iraq.

*Corrisponding author: dr_wissam80@yahoo.com

Abstract. This paper present the development on compressive strength of fly ash geopolymer paste under effectiveness of nano-silica and micro steel fiber. Twelve geopolymer paste mixes under two series, the first series with replacement (0% to 1.4%) by increment (0.2%) of binder by nano-silica and using the optimum concentration of nano silica in second series with addition (0.5% to 2%) by increment (0.5%) of micro steel fiber. The mixes were using fly ash as binder with Sodium silicate (Na2SiO3) and sodium hydroxide (NaOH) solutions as the alkali activators. (Na2SiO3 to NaOH) ratio was (2.0) with molarity of sodium hydroxide (10 M), and the specimens cured under room temperature $(35\pm2)^0$c. Results Show that the compressive strength of fly ash geopolymer paste can be improved as high as 76.5MPa at 28 days by add 0.8% nano silica and 1% micro steel fiber which were the optimum proportion that caused noteworthy improvement of compressive strength.

Keywords: nano silica, micro steel fiber, geopolymer paste and compressive strength

1. Introduction

Production of Portland cement is known to expend considerable energy and product large amounts of carbon dioxide which contributes in global warming [1, 2]. To produce ton of Portland cement produces ton of carbon dioxide in the atmosphere, so increasing environmental worry with the using of concrete in construction have led to search for materials more sustainable and environmentally friendly [3]. Fly ash (FA) is one of the most common industrial products may be used as bonding materials. It has been used as alterative of cement and enhance the physical, chemical and mechanical properties of concrete, CO2 emissions may be reduce by 80% when using geopolymer with compared with the use of ordinary Portland cement [4]. Geopolymers are a three-dimensional structure materials, synthesized by alkaline activation consist of an aluminosilicate. Geopolymerization process depend on the reaction of silica and alumina with alkali solution. The constituents may be any industrial or agricultural by-products having silica and alumina. These aluminosilicate reactive materials are rapidly dissolved in alkaline solution resulting in formation of free SiO4 and AlO4 tetrahedral units [5]. In case of using geopolymer, high temperature in early curing stage is essential to provide enough strength increases to access high mechanical properties [6]. However, there are several practical issues with the application of heat curing in large scale structures; so nano-silica (NS) is one of the common mineral admixtures to accelerate, where use of nano-silica is gaining wider attention due to its significant effect on the microstructural and mechanical properties of Portland cement based binders [7]. The aim of this paper is to show the optimum quantity of nano-silica and its effect on

IOP Conf. Series: Materials Science and Engineering **584** (2019) 012010 doi:10.1088/1757-899X/584/1/012010

compressive strength of fly ash geopolymer paste, then improve the compressive strength by addition different dosages of micro steel fiber for fly ash geopolymer paste.

2. Experimental Context

2.1 Material

Fly ash class F (FA) was using in this study and nano-silica (NS) were obtained from a commercial provider in Turkey. The average particle diameter of nano-silica was 15-20 nm. The chemical composition of the fly-ash and nano-silica shown in Table 1.

Micro steel fiber with length 6mm and aspect ratio 30, yield strength equal to 1500N/mm, young modulus 200GPa and density 7850kg/m3. The alkaline activator was compound of sodium silicate solutions (UAE origin) and sodium hydroxide (Kuwait origin). The molarity of sodium hydroxide solution 10M was prepared by mixing with pure water, and the chemical composition of sodium silicate solution was Na_2O 13.1-13.7%, SiO 32-33%, specific gravity 1.534-1.551 and viscosity 600-1200. The using material shown in Figure 1.

Table 1: Chemical composition of fly ash and Nano-silica

Chemical analysis	Fly ash (*wt. %*)	Nano silica (*wt. %)*
SiO_2	57.2	>99.8
Al_2O_3	24.4	-
CaO	4.24	-
Fe_2O_3	6.1	-
MgO	2.4	-
SO_3	0.29	-
K_2O	2.37	-
Na_2O	0.38	-
LOI	1.58	<1.5
Specific gravity	2.25	2.2
Surface volume ratio (*m^2/g*)	-	170-230

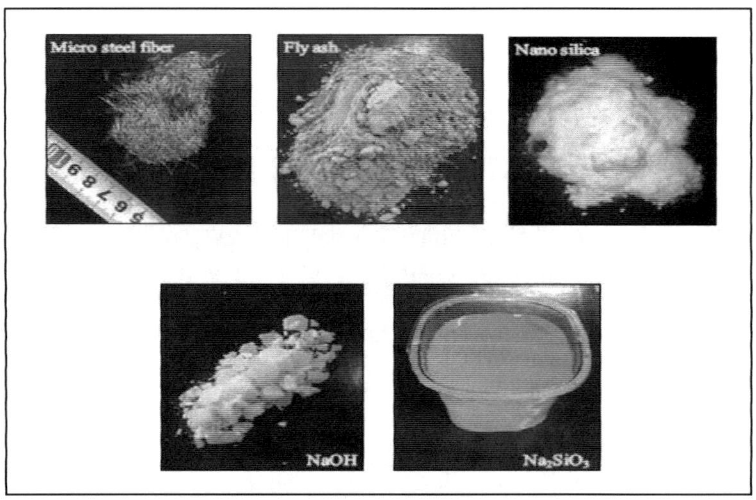

Figure 1. Used materials.

2.2 Geopolymer Mixes

Twelve geopolymer paste mixes were prepared in two series, the mix proportions are given in Table 2. The main binder has been fly ash with replacement (%0 to 1.4% increment 0.2%) by nano-silica in first series, in second series using the optimum concentration of nano silica with addition (0.5% to 2% increment 0.5%) of micro steel fiber. The alkaline liquid was prepared before 24 hour of mixing with ratio of sodium silicate to sodium hydroxide 2.0 at each mix. The samples were named according to the proportion of nano silica (NS) and micro steel fiber (SF) used in the mix.

Table 2. Mix proportions of fly ash geopolymer pastes.

Mix	Binder (*wt%*)		F/B	Solution (*kg/m³*)		SF (wt%) of volume
	FA	NS		NaOH (*10M*)	Na_2SiO_3	
FA-NS0.0-SF0.0[a]	100	0	0.42	236	474	0
FA-NS0.2-SF0.0	99.8	0.2	0.42	236	474	0
FA-NS0.4-SF0.0	99.6	0.4	0.42	236	474	0
FA-NS0.6-SF0.0	99.4	0.6	0.42	236	474	0
FA-NS0.8-SF0.0	99.2	0.8	0.42	236	474	0
FA-NS1.0-SF0.0	99	1.0	0.42	236	474	0
FA-NS1.2-SF0.0	98.8	1.2	0.42	236	474	0
FA-NS1.4-SF0.0	98.6	1.4	0.42	236	474	0
FA-NS0.8-SF0.5	99.2	0.8	0.42	236	474	0.5
FA-NS0.8-SF1.0	99.2	0.8	0.42	236	474	1.0
FA-NS0.8-SF1.5	99.2	0.8	0.42	236	474	1.5
FA-NS0.8-SF2.0	99.2	0.8	0.42	236	474	2

[a] FA-NS0-SF0 represents the mix having: fly ash, 0% nano-silica (NS) and 0% steel fiber (SF) in the fly ash geopolymer paste.

2.3 Mixing Procedure and Curing

The alkaline solution was production before 24 hour of working [8], NaOH flakes were dissolved in pure water to obtain the desired molarity and mixing sodium hydroxide solutions with sodium silicate at the required proportion. Firstly, mixing nano-silica particles with the alkaline solution by using electric-mixer because nano-silica may not disbanded well through moistening and caused agglomerations in the mixture. So, a disband of nanoparticles is very important to obviate agglomeration which may be have negative effect on the reaction. The distributed of nano-silica in the alkaline liquid before and after mixing, shown in Figure 2.

For prepare each mix of the first series, putting weighted fly ash in the mixer and addition of the solution with nano silica and mixing it for five minutes, and for prepare each mix of the other series putting weighted fly ash with required dosage of micro steel fiber in the mixer and mixing it for two minutes then following by addition of the solution with nano silica and mixing it for five minutes [9, 10]. The pastes were cast in 50×50×50 mm³ cubes accordance with the ASTM C109 standard [11].Then using a table vibrator for compacting the specimens for 30 second in two layers and demolded after 48 hour of casting. The specimens were putting in isolated bags during curing period and cured in room temperature (35 ± 2)^{0}C. Compressive strength was tested at the age 7, 14 and 28 days. The stages of the work shows in Figure 3.

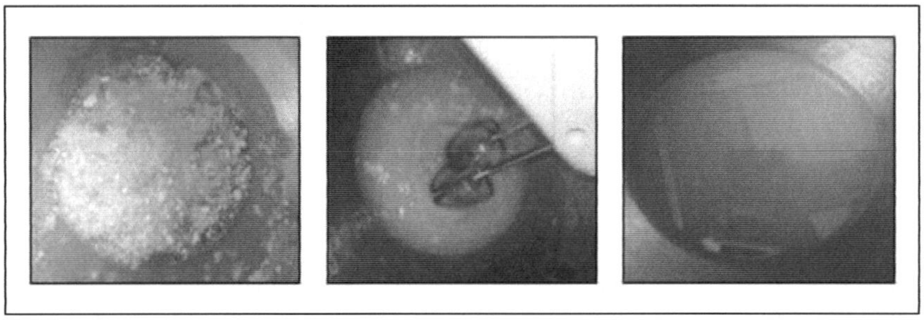

Figure 2. Alkaline solution with nano-silica.

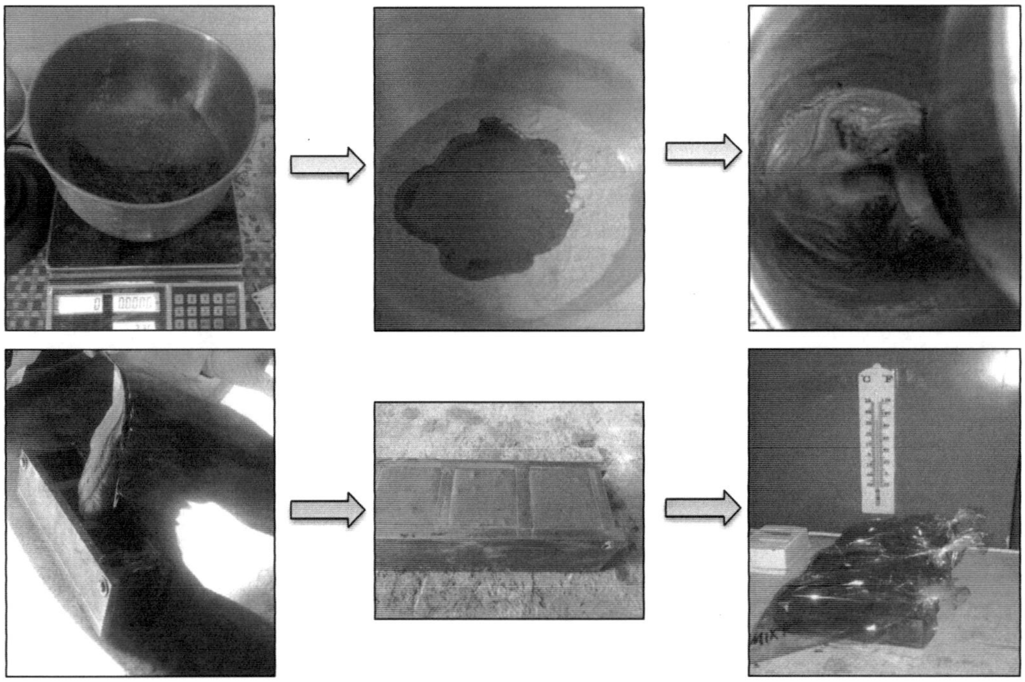

Figure 3. Stages of casting and curing.

3. Results and discussion

3.1. Compressive strength development under effectiveness of nano-silica

Effect of the different concentrations of nano-silica on the compressive strength shows in Figure 4. In 28-days the compressive strength of the reference (with zero nano-silica) was 34.5MPa which increased to 71.8MPa by the additive nano-silica. Figure 4, shows that the mixes containing 0.8% nano-silica have higher compressive strengths than the other mixes, its higher than the reference mix by (41%, 63% and 108%) at (7, 14 and 28) days respectively. And in general, all the mixes which contain nano-silica were higher than that of the reference mix, its higher than the reference mix by (15%, 8% and 33%) at (7, 14 and 28) days respectively for mixes containing 0.2% nano-silica, (24%, 21% and 58%) at (7, 14 and 28) days respectively for mixes containing 0.4% nano-silica, (28%, 43% and 83%) at (7, 14 and 28) days respectively for mixes containing 0.6% nano-silica, the mixes containing 0.8% nano-silica have higher compressive strengths but when addition more contain of nano-silica cause drop in result but it's remaining higher than that of the reference mix, whereas, at mixes containing 1% nano-silica compressive strength was higher than the reference mix by (41%, 63% and 101%) at (7, 14 and 28) days respectively, (35%, 54% and 97%) at (7, 14 and 28) days respectively for mixes containing 1.2% nano-silica and (24%,43% and 86%) at (7, 14 and 28) days respectively for mixes containing 1.4% nano-silica. This means that the geopolymer matrix becomes stronger by replace proportion of binder by nano-silica. It was also observed similar behavior by other researchers [12]. In fact, the nano-silica was controlling on geopolymerization process, it's made as acceleration factor; because of its high surface area then its reacts faster, but increasing the nano-silica on the desired limit causes declined in the strength. This is mean the replacement of binder by 0.8% of nano-silica was suitable for reaction and any addition was more than 0.8% did not partake in the reaction and formed as weakness areas.

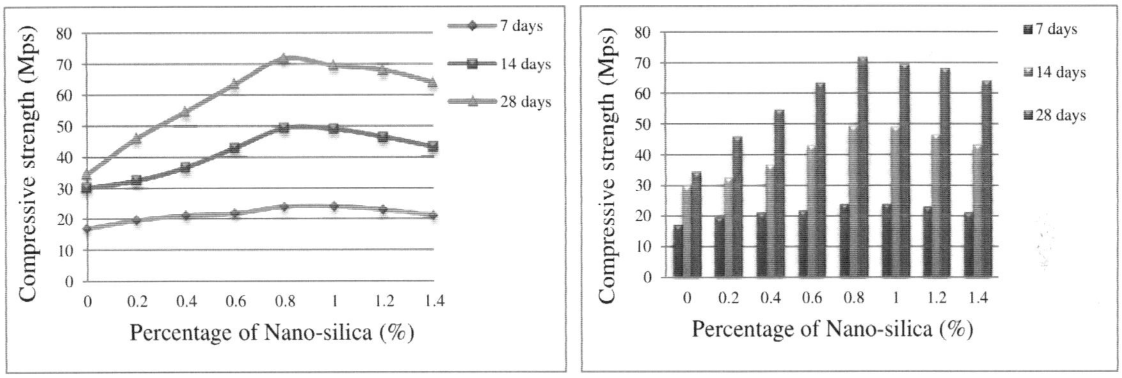

Figure 4. Compressive strength development under effectiveness of nano-silica.

3.2. Compressive strength development under effectiveness of micro steel fiber

The results presented a slight develop in compressive strength when adding different dosages of micro-steel fiber. The highest increase was show in dosage 1% of micro steel fiber, its higher than the reference mix by (4%, 7.5% and 6.5%) at (7, 14 and 28) days respectively, after which there was causes declined in the strength. At other dosages the develop in compressive strength was very simple not exceed 3%.

In fact, the micro steel fiber has a minor effect on compressive strength development, but it may be improve other properties of the geopolymer paste such as bonding, flexural and tensile strength. The effect of micro steel fiber on the compressive strength shown in Figure 5.

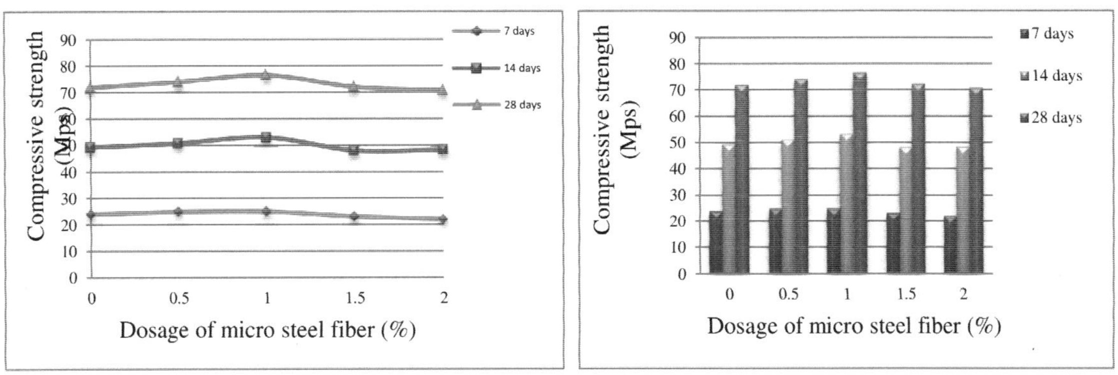

Figure 5. Compressive strength development under effectiveness of micro steel fiber.

IOP Conf. Series: Materials Science and Engineering **584** (2019) 012010 doi:10.1088/1757-899X/584/1/012010

4. Conclusions

The compressive strength development of fly ash geopolymer pastes under effectiveness of different concentrations of nano-silica and different dosages of micro steel fiber was studied under the ambient curing. The following conclusions can be found:

- Compressive strength of fly-ash geopolymer paste was increase with the addition of nano-silica when cured under ambient temperature. The proportion of replace 0.8% of binder by nano silica was the optimum which showed best increase in compressive strength, then when increase the proportion of nano-silica caused decline in compressive strength. Compressive strength at the optimum proportion of nano-silica was higher than the reference by 41%, 63% and 108% at 7, 14 and 28 days respectively.
- Micro steel fiber was showed a simple increase in compressive strength and the optimum proportion of micro steel fiber was 1% additive but it's may be more developed other properties of geopolymer paste. Compressive strength at the optimum proportion of micro steel fiber was higher than the reference mix by 4%, 7.5% and 6.5% at 7, 14 and 28 days respectively.

References

[1] Turner LK, Collins FG. Carbon dioxide equivalent (CO2-e) emissions: A comparison between geopolymer and OPC cement concrete. *Construction and Building Materials*. 2013 Jun 1;43:125-30.

[2] Ibrahim AM, Oukaili NK, Salman WD. Flexural behavior and sustainable analysis of polymer bubbuled reinforced concrete slabs. *Fourth Asia Pacific Conference on FRP in Structures* (APFIS 2013) 11-13 December, Melbourne, Australia © 2013 International Institute for FRP in Construction.

[3] Lawrence CD. The production of low-energy cements. *Lea's chemistry of cement and concrete*. 2003 Nov 12:421-70.

[4] Li Z, Ding Z, Zhang Y. Development of sustainable cementitious materials. *InProceedings of the international workshop on sustainable development and concrete technology*. 2004 May 20 (Vol. 1, No. 1, pp. 55-76). Iowa State University.

[5] Yunsheng Z, Wei S, Qianli C, Lin C. Synthesis and heavy metal immobilization behaviors of slag based geopolymer. *Journal of hazardous materials*. 2007 May 8; 143(1-2):206-13.

[6] Mermerdaş K, Algın Z, Oleiwi SM, Nassani DE. Optimization of lightweight GGBFS and FA geopolymer mortars by response surface method. *Construction and Building Materials*. 2017 May 15;139:159-71.

[7] Aggarwal P, Singh RP, Aggarwal Y. Use of nano-silica in cement based materials—A review. *Cogent Engineering*. 2015 Dec 31; 2 (1):1078018.

[8] Oleiwi SM, Algın Z, Nassani DE, Mermerdaş K. Multi-Objective Optimization of Alkali Activator Agents for FA-and GGBFS-Based Geopolymer Lightweight Mortars. *Arabian Journal for Science and Engineering*. 2018 Oct 1:1-5.

[9] Mermerdaş K, Manguri S, Nassani DE, Oleiwi SM. Effect of aggregate properties on the mechanical and absorption characteristics of geopolymer mortar. *Engineering Science and Technology, an International Journal*. 2017 Dec 1; 20(6):1642-52.

[10] Salman WD, Mubarak HM and Mahmood MS. Structural behavior and mechanical properties of ferrocement slab panels incorporating fibers. *International Journal of Civil Engineering and Technology* (IJCIET), 9(11),.2018, pp. 2289–98.

[11] American Society of testing materials, Standard Test Method for Compressive Strength of Hydraulic Cement Mortars, (Using 2-in. or [50-mm] Cube Specimens). 2002 ASTM C109/C109M.

[12] Rashad AM, Ouda AS. Thermal resistance of alkali-activated metakaolin pastes containing nano-silica particles. *Journal of Thermal Analysis and Calorimetry*.:1-2.

IOP Conf. Series: Materials Science and Engineering **584** (2019) 012010 doi:10.1088/1757-899X/584/1/012010

Acknowledgments
I would like to express my thanks and appreciation to Diyala University College of Engineering laboratory management for founding this research program.

Evaluation of vertical drain with different materials

Ali Akhtarpour[1], Mohammed Sh Mahmood[2], and Ameer Alali[3]

[1] Ph.D., Asst. Prof., Civil Eng. Dept., Ferdowsi University of Mashhad, I.R. Iran. Email: akhtarpour@um.ac.ir
[2] Ph.D., Asst. Prof., Civil Eng. Dept., University of Kufa, Iraq. Email: mohammedsh.alshakarchi@uokufa.edu.iq
[3] Formerly M.Sc. Student, Civil Eng. Dept., Ferdowsi University of Mashhad, I.R. Iran. Email: ameeraljabery1987@gmail.com

Abstract. Accelerate the drainage process and consolidate saturated clay soils is important to achieve stability of the construction. Vertical drains, with a suitable grid and depth, the result of shortening the path of water flow inside the soil and soil consolidation rate increases from several years to several months. Sombar earth dam, a case study, in Gholaman village in North Khorasan province, Iran, is used to provide agricultural and drinking water in the area. The recent paper shows a comparison between the results of Sombar dam stability with vertical drains of two different materials using the Mohr-Coulomb model with the aid of GeoStudio 2012 computer software. The drains materials are sand and geotextile and the investigated parameters are permeability ratio (ky/kx), drains geometry and time of the dam construction. The results are presented as ratios, settlement ratio (SSAND/SGEOTEX) and safety factor ratio (FSSAND/FSGEOTEXT). The results revealed that there are clear increases in the consolidation settlement ratio, but, slightly increase in safety factors ratio using sand drains comparing to the geotextile. The ky/kx has a clear effect on settlement and factor of safety in sand drains.

1. Introduction

Soft clay has a low bearing capacity and high compressibility features. Thus, it is necessary to apply improvement techniques prior to construction to avoid unacceptable differential settlement [1]. To increase the soft soil shear strength, using vertical drains for preloading of soft clay is one of the common methods, in addition, to control its post-construction settlement. Various types of vertical drains were adopted and have been studied, including gravel piles, geosynthetic vertical drains and sand drains [2, 3]. Consolidation time can be reduced from about 1 to 5 years without a vertical drain to about 50 days with prefabricated vertical drains (PVDs) [4]. The efficiency of the vertical drainage method is a function of their distances, depth and diameter of the boreholes and more efficiency in alloys with horizontal permeability exceeding vertical permeability [5]. Depending on scaled model tests, the using of PVDs is successful for the improvement of ultra-soft with increasing in the undrained shear strength and a substantial amount of pore pressure dissipation [6].

The Mohr-Coulomb models (MC) are simple and easily determinable model parameters in numerous engineering applications and the stress-strain response depends on the liquid limit value [7]. The linear elastic-perfectly plastic model (or MC) is one of the most commonly used pressure-sensitive constitutive models [8]. It is necessary for a two-dimensional (2D) finite-element analysis with the plane-strain model to convert the system of vertical drains into an equivalent drain wall [9].

The variation of a factor of safety (increase) with time is primarily influenced by the rate of dissipation of excess pore pressure within the depth of soil up to which the critical slip surface passes. Installation of vertical drains also decreases the post-construction settlement of the embankment and helps to settle the embankment more uniformly [10].

In the recent paper, the authors investigate the effect of vertical drains (geotextile and sand) in the foundation soil on the stability of the Sombar dam, using Mohr-Coulomb model with aid of GeoStudio 2012 [11] computer software. The authors develop a reliability analysis of the stability due to the vertical drains for the different studied parameters.

2. Theoretical Background

Baron [12] provided a method for analyzing a cylindrical element of soil with a vertical drain in its center. The theory was based on the simplification of the hypothesis of single-dimensional consolidation of the hypocrisy. The governing differential equation as in Eq. 1:

$$\frac{\partial u}{\partial t} = C_h\left(\frac{1}{r}\frac{\delta u}{\delta r} + \frac{\delta_u^2}{\partial_r^2}\right) \tag{1}$$

Where u: excess water pressure in the hole due to the increase of tension, t: time, Ch: the horizontal coefficient of consolidation and r: the radial distance of the drain. The solution of the above equation is presented in Eq. 2 and Eq. 3 [12]:

$$U_h = 1 - ex\,p\left(\frac{-8T_h}{\mu}\right) \tag{2}$$

$$T_h = \frac{C_h^t}{D_e} \tag{3}$$

Where Uh: the average degree of radial consolidation, Th: dimensionless factor, De: the diameter of soil equivalent and μ is the component that is included in the drain spacing, the drain equivalent diameter, and drain discharge capacity; it is represented as in Eq. 4 and Eq. 5 [12].

$$\mu = \frac{n^2}{n^2-1}\ln(n) - \frac{3n^2-1}{4n^2} \tag{4}$$

$$n = \frac{D_e}{d_w} \tag{5}$$

Where n: the drain spacing, dw: the equivalent diameter of the drain.

An approximate solution for vertical drain is shown based on the hypothesis equal strain for the area affected (disturbed and undisturbed) by the drain. The theory is based on simplifying the assumptions of Darcy law. The governing equations are Eq. 6 to Eq. 9 [13]:

$$U_r = 1 - \exp\left(\frac{-8T_r}{F}\right) \tag{6}$$

$$F = F(n) + F_s + F_r \tag{7}$$

$$F = ln\left(\frac{n}{s}\right) + \left(\frac{k_s}{k_w}\right)ln(s) - 0.75 + \pi z(2l - z^2)\frac{k_h}{q_w} \tag{8}$$

$$s = r_s/r_w \tag{9}$$

Where Fs: the drain area, Fr: resistance, kh: permeability of the undisturbed soil area, kw: permeability in the drain area, ks: permeability of the disturbed soil area l: drainage length and qw: drainage.

The equivalent diameter for tape strips: By using the conventional consolidation theory for vertical drains, we assume that the vertical drainage is in the transverse section of the circle. Most over-drained drains have a trapezoidal shape and this trapezoidal surface can be close to the surface of the circle of the equivalent diameter. Therefore, the diameter dw deformed into a wall drain with a width of "a" and thickness of "b" as in Eq. 10 [14].

$$d_w = \left(\frac{a+b}{2}\right) \tag{10}$$

Various methods have been proposed to consider the coherent effect of vertical and radial consolidation. The simplest version was given by Carroll [15] as in Eq. 11.

$$(1 - U) = (1 - U_v)(1 - U_h) \tag{11}$$

Where U: the total average degree of consolidation, Uv: the average vertical degree of consolidation, and Uh: the average horizontal degree of consolidation.
If the drain capacity is low, a resistance to water movement is created (collection of water near the drain). According to Xie [16], Eq. 12 should be considered to maintain a low level of the wall resistance.

$$\frac{\pi}{4} * \frac{k_h}{q_w} * l_m^2 < 0.1 \tag{12}$$

Where kh: the hydraulic conductivity of the soil in the horizontal direction (m/sec), lm: the length of the drain (m).
The relationship between strain and stresses in Mohr-Coulomb model is completely linear to a point and after this point the stress-strain curve becomes a horizontal line. The potential for failure can be expressed as in Eq. 13 [11].

$$F = \sqrt{J_2}\sin\left(\theta + \frac{\pi}{3}\right) - \sqrt{\frac{J_2}{3}}\, cs(\theta + \frac{\pi}{3})\sin\emptyset - \frac{I_1}{3}\sin\emptyset - c\cos\emptyset \tag{13}$$

When the angle of internal friction is zero, the criterion for collapse in Mohr-Coulomb model becomes as in Eq. 14 [11].

$$F = \sqrt{J_2}\sin\left(\theta + \frac{\pi}{3}\right) - c \tag{14}$$

3. Materials and Method

The earth dam of Sombar has located 2 km from the asphalt road between Gholaman city and the Castle village in northern Khorasan province, Iran. Due to the length of the crown of the dam, the calculations are based on the plane strain state for the maximum barrier section. The dam is a type of earth limestone with a clay core and sandy crust. The site of the dam is mostly fine-grained alluvium with coarse-thick layers up to 30 meters wide across the valley. Figure 1 shows the section in the soil foundation and the dam. The fine-grained alluvium was classified as CL-ML, CL, and sometimes ML, and the coarse aggregates are classified as SC, SC-SM, SM according to the unified soil classification system (USCS). The groundwater level is around 4m from the ground level. The coefficient of consolidation is 2.25*10-3 cm2/sec under vertical stress of 2kg/cm2, 1.1*10-3 cm2/sec under 4kg/cm2 and 1*10-3 cm2/sec under 8kg/cm2 [17, 18]. Table 1 summarizes the soil properties of the foundation and dam.

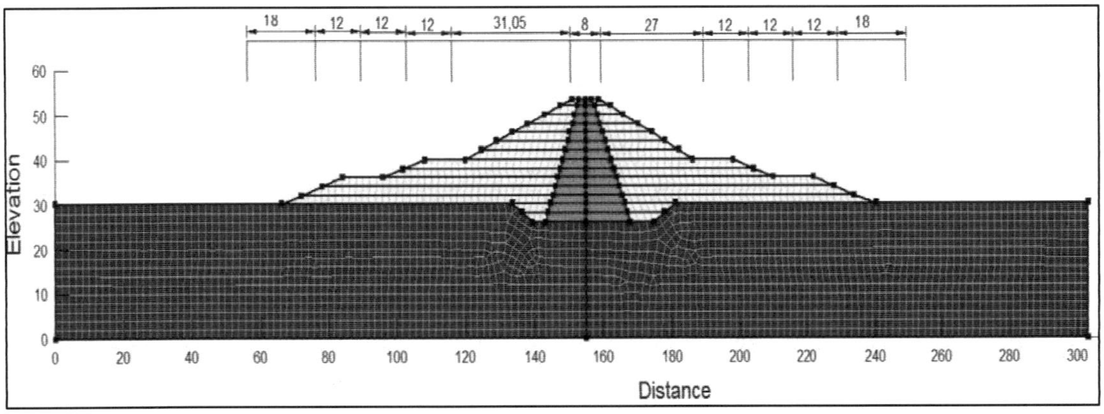

Figure 1. Section in the Sombar Dam and Foundation [18].

Table 1. Summary of Soil Properties the Foundation and Dam.

Soil type	**Test method**	**Unit weight, kN/m³**	**k, m/sec**	**Φ, degree**	**C, kPa**	**E, kPa**	**υ**
	UU			0	80	22900	0.48
Clay core	CU	20	10-9	15	45	22900	0.48
	CD			27	32	19100	0.25
Sandstone	-	22	10-5	42	42	40000	0.2
Foundation Fine-grained	CU	19	10-9	14	14	1416	0.48

Software named "GeoStudio 2012" are used to solve the consolidation equations. Using SIGMA/W and SLOPE/W (from GeoStudio) in sequence facilitates the modelling of staged construction with Multiple analyses method. SIGMA/W together with SEEP/W are performed for A fully-coupled consolidation analysis. Mohr-Coulomb Model (MC) is adopted for the analysis of the dam and the underlined foundation under effective stress. The basic equations of equilibrium of the Hook relationship, Darcy law and equations of contiguity are applied. The investigation program includes the effect of sand and geotextile vertical drains in foundation soil on the dam settlement and safety factor. The drains depth (8, 15, 18 and 24m), the drains spacing (2, 3 and 4m) for each drains depth. These parameters are investigated corresponding to two coefficient of permeability ratios (ky/kx) of 1 and 0.1 and construction stages of 4, 8, 15 and 24 months. Figure 2 shows the distribution of the drains in the foundation of the dam. The presentation of the results in terms of the ratios for comparison purposes, the settlement ratio is SSAND/SGEOTEX and the factor of safety ratio is FSSAND/FSGEOTEX. The lowest values of safety factor are adopted in the comparison.

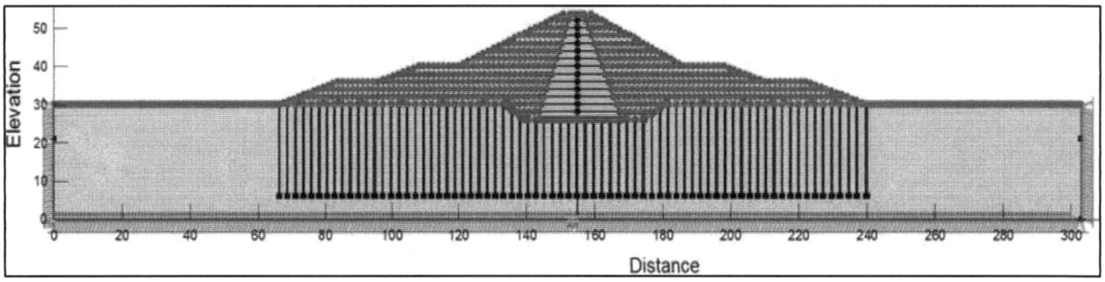

Figure 2. Diagram of the Vertical Drains in the Foundation Soil.

4. Results

Generally, with vertical drains (sand or geotextile) in the foundation soil of the Sombar dam, the rate of consolidation settlement is increased with respect to the no drain condition as probable due to water escape from the soil through the drains. For comparison purposes, this paper investigates the change percentages in settlement and factor of safety using sand or geotextile drains. The settlement ratio (SSAND/SGEOTEXT) and safety ratio (FSSAND/FSGEOTEXT) are discussed.

Figure 3 illustrates the general schematic of dam settlement using vertical drains. The settlement ratios (SSAND/SGEOTEX) show that the settlement rate exhibits increasing with using of sand drains regardless of the drain diameter comparing to the geotextile drains, as shown in Figure 4, but these differences in settlement ratios are not reach +6%. The settlement ratio for the 0.5m sand drain, as be expected, is larger and approximately in constant trend. This condition may be attributed by the high permeability of sand in both directions corresponding to the geotextile material.

Figure 5 illustrates the results of safety factors ratio (FSSAND/FSGEOTEX) in upstream (lowest) for the of 9m depth and 4m spaced drains. One can notice that there is clear difference behavior between FS ratios with respect to 0.3m and 0.5m sand drains. The 0.5m diameter reveals more 100% increase related to the geotextile drain while, unexpectedly, the 0.3m sows less 100% and this may be as a reason of low stiffness of sand drain comparing to geotextile drain.

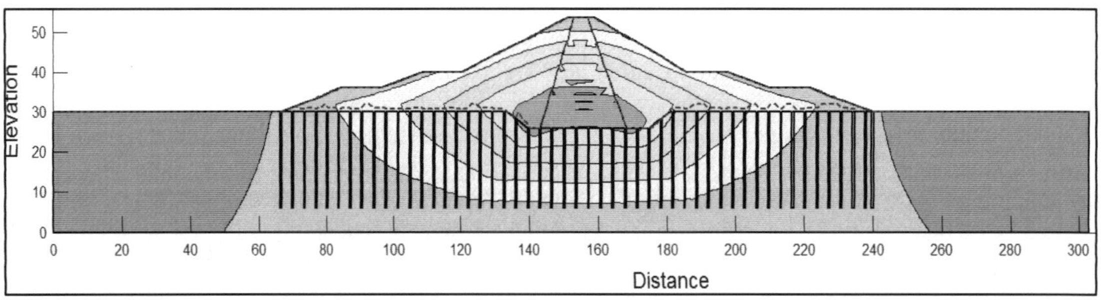

Figure 3. Schematic of the Dam Settlement Using MC for ky/kx=1 after 24 Months.

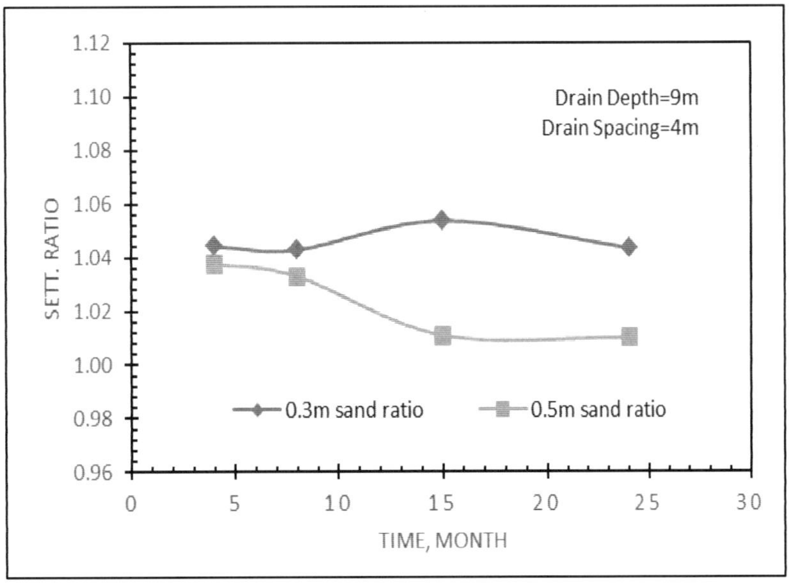

Figure 4. Settlement Ratio versus Time for Selected Drain Geometry.

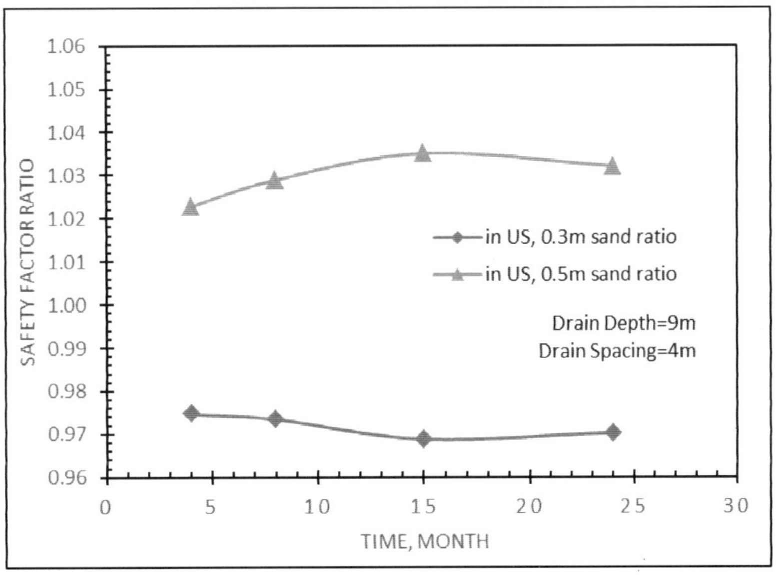

Figure 5. Safety Factor Ratio versus Time for Selected Drain Geometry.

With decreasing the drain spacing, the settlement ratio shows more straight lines with converging trends at the end of the construction time, as shown in Figure 6. There is a high difference between the sand and geotextile drains, sustainably, the sand drains revealed more settlement. One can conclude that the sand drain is more effective for settlement, regardless of the diameter, comparing to the geotextile one. But with decreasing of drain spacing, the factor of safety ratio exhibits a constant trend for the different sand geometry with respect to the geotextile, as shown in Figure 7. The values of safety factor ratios are around the unity especially with advancing in time of the dam construction. This situation, due to a decrease in drain spacings, the stiffness of the drains, with different materials, is more comparable.

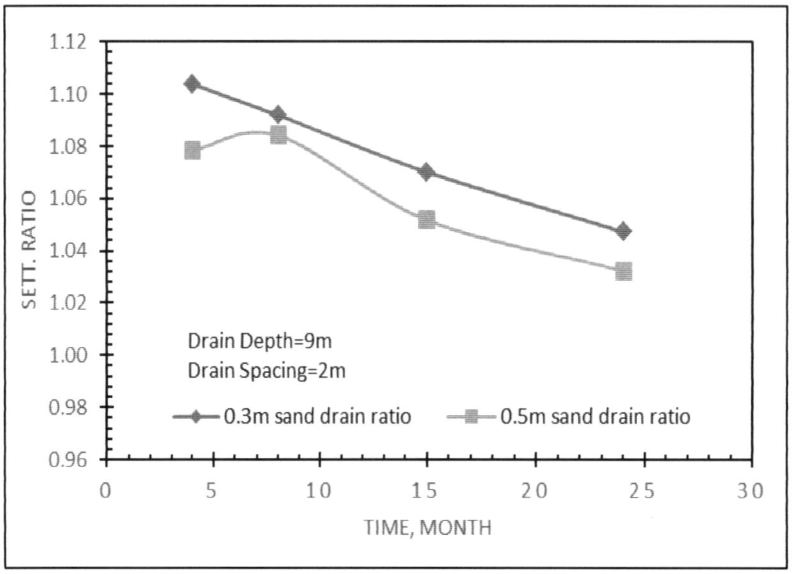

Figure 6. Settlement Ratio versus Time for Selected Drain Geometry.

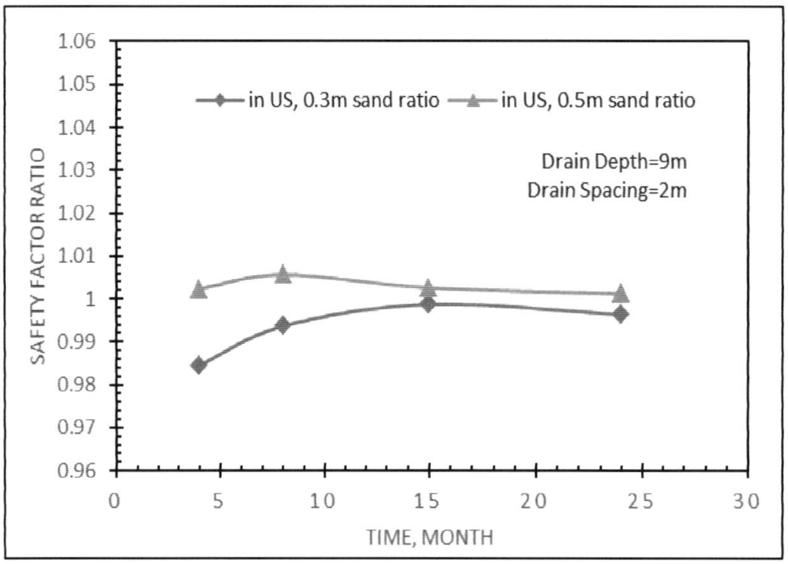

Figure 7. Safety Factor Ratio versus Time for Selected Drain Geometry.

5. Discussion

The recent paper compares the effect of vertical drains in foundation soil with two different materials on the behavior of Sombar earth dam, settlement and factor of safety. This comparison is prepared to make the reliability of the results from the settlement ratio (SSAND/SGEOTEXT) and safety ratio (FSSAND/FSGEOTEXT). The study covered different parameters including drain geometry, drain materials and soil permeability ratio (ky/kx).

There is a clear increasing trend of settlement ratio (SSAND/SGEOTEXT) from different studied drain parameters, i.e., unrelatedly of drain geometry, sand drain contributes a larger settlement rate, as stated by Mahmood et al. [19]. While for safety factor ratio (FSSAND/FSGEOTEXT) from the geotextile is little more than from sand drain, but with decreasing of the drain spacing, the values of FS from both drain types are identical and this may be due to the comparable stiffness in the lowest spacing.

To investigate the effect of ky/kx, Figs. 8 and 9 present a probability density function (pdf) for settlement and factor of safety ratios respectively for different values of ky/kx. The pdf shows the distribution of data (SSAND/SGEOTEXT and FSSAND/FSGEOTEX) around the mean value of the data. These analyses are made using the continuous random variable method with the aid of computer software named "Risk Tools (RT)" after Mahsuli, 2014 [20].
With decreasing in the ky/kx of the soil, the peak "pdf" (mean value) less and the curve is more splayed, as in Figure 8, but, with anisotropic soil condition (kx more than ky), there is a clear increase in the settlement rate from sand drains. An interesting result is in safety factor ratios, with decreasing in ky/kx, the resulted ratios of FS are slightly increased, as shown in Figure 9, however, dissimilarity to settlement ratios, the distribution of the pdf of FS ratios is close to the peak value (mean) in the anisotropic condition. These results indicate that there is a significant effect of permeability ratio (ky/kx) on the stability of the dam with respect to use the sand drains.

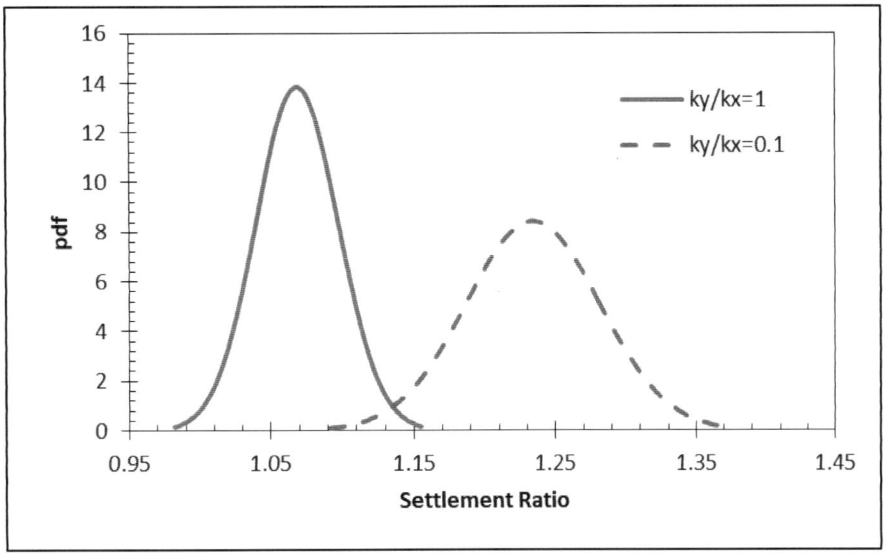

Figure 8. Pdf Curves of Settlement Ratios for Different ky/kx.

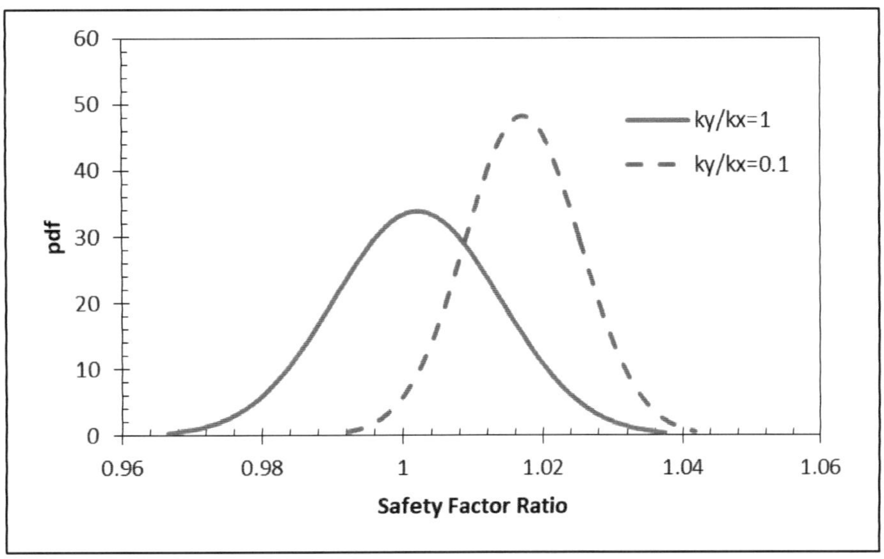

Figure 9. Pdf Curves of Safety Factor Ratios for Different ky/kx.

6. Conclusions

Comparisons are made to investigate the effect of vertical drains in the dam foundation with different materials (sand and geotextile) on the settlement and safety factor as a ratio of the different materials. There are clear increases in the rate of consolidation settlement, but, little in safety factors with using sand drains corresponding to the results of geotextile for different studied parameters. There is a significant effect of ky/kx ratio on the stability of the dam with respect to use the sand drains. With decreasing of the ratio of vertical to horizontal permeability (ky/kx), as in anisotropic soils, the settlement ratios (SSAND/SGEOTEX) remain larger and safety factors ratios (FSSAND/FSGEOTEXT) are slightly above 1 with respect to time and depth change.

References

[1] Indraratna B, Perera D, Rujikiatkamjorn C and Kelly R 2012 Soil Disturbance Analysis Due to Vertical Drain Installation Proc. of the Institution of Civil Engineers: Geotechnical Engineering 168 (3) 236-246. http://dx.doi.org/10.1680/geng.14.00052.

[2] Indraratna B, Rujikiatkamjorn C, Wijeyakulasuriya V, McIntosh G and Kelly R 2010 Soft soils improved by prefabricated vertical drains: performance and prediction, In Almeida M (ed), Symposium on New Techniques for Design and Construction in Soft Clays 227-246, Brazil: Officna de Textos.

[3] Indraratna B, Rujikiatkamjorn C, Balasubramaniam A S, McIntosh G 2012 Soft Ground Improvement Via Vertical Drains and Vacuum Assisted Preloading 30(4) 16-23. DOI: 10.1016/j.geotexmem.2011.01.004.

[4] Dhar A S, Siddique A and Ameen S F 2011 Ground Improvement using Pre-loading with prefabricated Vertical Drains International Journal of Geoengineering Case Histories, geoengineer.org 2(2) 86-104.

[5] Bergado, D T, Anderson L R, Miura N and Balasubramaniam A S 1996 Soft Ground Improvement: In Lowland and Other Environments American Society of Civil Engineers, New York, USA.

[6] Chu J, Bob M W and Choa V 2006 Improvement of ultra-soft soil using prefabricated vertical drain, Geotextiles and Geomembranes 24(6) 339–348.

[7] Sandhya Rani R, Nagendra Prasad K and Sai Krishna T 2014 Applicability of Mohr Coulomb & Drucker Prager Models for Assessment of Undrained Shear Behaviour of Clayey Soils International Journal of Civil Engineering and Technology (IJCIET) 5(10) 104-123.

[8] Coombs W M, Crouch R S and Heaney C E 2013 Observations on Mohr-Coulomb Plasticity under Plane Strain J. Eng. Mech. ASCE 139 1218-1228.

[9] Indraratna B 1997 Plane-Strain Modelling of Smear Effects Associated with Vertical Drains Journal 0f Geotechnical and Geoenvironmental Engineering 123 474-478.

[10] Krishnamoorthy A and Kamal S 2016 Stability of an Embankment on Soft Consolidation Soil with Vertical Drains Journal of Geotechnical and Geological Engineering. DOI: 10.1007/s10706-015-9975-4.

[11] GeoStudio User's Manual (Sigma/w Modelling) 2012 3rd Edition.

[12] Barron R A 1948 Consolidation of Fine-Grained Soils by Drain Wells Trans. ASCE 113 718-742.

[13] Hansbo S 1981Consolidation of Fine - Grained Soils by Prefabricated Drains Proceedings of the 10th International Conference on Soil Mechanics Foundation Engineering Stockholm Balkema Rotterdam The Netherlands 3(4) 677–682.

[14] Indraratna B, Bamunawita C, Redana I W and Mcintosh G 2003 Modelling of prefabricated vertical drains in soft clay and evaluation of their effectiveness in practice Journal of Ground Improvement 7(3) 127-138.

[15] Carillo N 1942 Simple Two-and Three Dimensional Cases in the Theory of Consolidation of Soils Journal of Mathematics and Physics 21(1) 1-5.
[16] Xie K H 1987 Consolidation Theories and Optimization Design for Vertical Drains PhD Thesis Hewing University Chin.
[17] Final Dam Report: Geotechnical studies of the Sombar Dam 2007 (in Persian).
[18] Alali Ameer 2017 The Effect of Vertical Drains On Stability and Settlements During Construction of an Embankment Dam On Saturated Soft Soil (Case Study: Sombar Dam) M.Sc. Thesis Faculty of Engineering, Ferdowsi University of Mashhad, Iran (in Persian)
[19] Mahmood Mohammed Sh, Akhtapour Ali and Alali Ameer A A unpublished Mechanical Behavior of Dam-Vertically Sand Drained Foundation, Case study: Sombar Dam J. Eng. Technol. Sci. Indonesia.
[20] Mahsuli M, Haukaas T 2014 Rt-Risk Tools (software) www.inrisk.ubc.ca.

Improvement of Multi-Lane Airport Highway in Baghdad-Iraq

Basim Jrew[1] and Ahmed Hadi[2*]

[1] Professor, Civil Engineering Department, Isra University, Amman, Jordan. E-mail: basim.jrew@iu.edu.jo

[2] Engineering Project Management, Salah Al Deen, Iraq. E-mail: Ahmed_hadi_92@yahoo.com

Abstract. Transportation industry is basically a service sector component. It is a great importance for traffic engineers and planners to understand and evaluate the quality of service provided by transport facilities. Highway Capacity Manual (HCM), with all its revisions since 1950, is the pioneer document in this area. It quantifies the concept of capacity for a transport facility and lays the foundations for estimating the level of service (LOS) being provided by that facility to its users. This study is an engineering project management. It is about evaluation and improving multi-lane highways in Iraq. The research study was conducted on one main multi-lane airport highway in Baghdad, including: Al- Dora intersection to Al- Rashid interchanges, where this highway is considered as a major highway with high traffic volume in Iraq. HCS-2010 program software was used to evaluate and improve the level of service and traffic conditions for the highway for existing conditions (year of the study-2018) and short-term conditions (2023). The evaluation shows that all segments are operating at LOS E or LOS F for existing conditions, short-term conditions. The segments that operated at LOS E or LOS F for existing, short-term and mid-term conditions improvements are suggested by modifying the geometric and traffic conditions and changing the highway to a freeway facility; in order to operate at better LOS. A management flow diagram was prepared for decision makers to be a guide for future planning programs.

Keywords: Airport highway in Iraq, Highway capacity manual (HCM-2010), Highway capacity software (HCS- 2010), Level of service (LOS), Multi-lane highway.

1. Introduction

Roads and highways are a major part of the transportation infrastructure in Iraq and play a substantial role in the local economy and community development. High quality of service of these facilities is essential to ensure safe, cost-effective and daily traffic operations. There has been a rapid growth in the population of Iraq, which led to an increase of demand in each and every aspect of our lives. Therefore, the increase of demand is directly proportional to the increase of the usage of vehicles and highway capacity.

Highways in Iraq need to be reconsidered and future plans should be prepared to improve the quality of service of these facilities. With the growing number of vehicles, the quality of service decreased, so that it became necessary to study the causes of the problem and find the appropriate solutions, either for the current or future conditions.

Transportation industry is basically a service sector component. It is of great concern for traffic engineers and planners to understand and evaluate the quality of service provided by transport facilities. Highway Capacity Manual (HCM) [1], with all its revisions since 1950, is the pioneer document in this area. It quantifies the concept of capacity for a transport facility and lays the foundations for estimating the level of service (LOS) being provided by that facility to its users.

Multi-lane highways are an important type of facilities, in which there should be no obstructions to the movement of vehicles along the road. Such facilities represent the majority of the highway system in most countries. Highway Capacity Manual (HCM 2010) [2] defined traffic density in terms of passenger cars per mile per lane as the primary level of service (LOS) measure for multi-lane highways. It also defined the free flow speed (FFS) as the primary capacity measure for the same type of highway facility and HCM is the pioneer document in management and evaluation of capacity and quality of service of various highway facilities, including freeways highways, arterial roads, roundabouts, signalized and un-signalized intersection, urban highways, as well as the effects of mass transit, pedestrians and bicycles on the performance of such systems.

2. Previous Researches

Jrew et al. (2018) [3], in their study "Management and Evaluation of Multi-lane Highway (case study)/Jordan" were used HCS-2000 and HCS-2010 software for the existing condition (year 2014), short-term condition (year 2019) and mid-term condition (year 2024). The evaluation shows that some of the highway segments are operating at LOS-E or LOS-F for short-term and mid-term conditions.
Jrew et al. (2017) [4], in their study "Management and Evaluation of Two Lane Two way Highways in Jordan (Case Study)". The updated HCS 2010 software were used to evaluate and improve the level of service and traffic conditions at each segment for existing (year of the study-2016), short-term (2021) and mid-term (2026). The evaluation show that some segments are operating at LOS E or LOS F for existing and short-term condition.
Yadav et al. (2014) [5], in their study "Roadway Capacity Estimation for Multi-lane Interurban Highways in India", focused on the estimation of roadway capacity for multilane interurban highways in India. Two sections of 60-meter length were identified as a study case on NH-2 (Delhi-Mathura corridor), where the studied highway is a four-lane divided national highway. The sections were so chosen in order to determine the effects of varying road geometry, pavement service conditions and possible variations in traffic conditions.
Semeida, A. M. (2013) [6], studied the relationships between the road geometric characteristics, heave vehicles (HV), LOS and capacity for rural multi-lane highways in Egypt using 45 different sites located in desert and agricultural highways. The result of the study found that the most influential variables on LOS and capacity in desert sites are HV then lane width (LW), while in agricultural sites are LW then existence of number of access points.
Robertson et al. (2013) [7], developed procedures for calculating free flow speed (FFS) on highway facilities with higher posted speed limits and methods for estimating LOS on multilane highways with FFS up to 80 mph .The study recommended to use Texas A&M Transportation Institute (TTI) speed prediction equations for multilane highways with speeds greater than 60 mph.

3. Study Area

The selected highway is of economic and cultural importance. It is an international highway that connects the major recreational and religious places in the southern governorates to the capital city of Baghdad and connects the southern governorates to the northern governorates. Also, it connects the capital city of Baghdad with the province of Basra in the south and the province Al-Anbar in the west.
The selected section of the multi-lane highway starts from Al- Dora intersection in Baghdad to Al-Rashid interchanges which are two branches with a length of 15 km. The multi-lane highway connects the freeway system in Iraq to the expressway system in Baghdad city, including International Baghdad Airport, as shows in Figure (1).
The street suffers from serious congestions at different times of the day and has a high percent of heavy trucks which transfer goods from and to the capital city of Baghdad. The highway was divided to six segments, the segment boundary was determined based on traffic volume, number of access points, PHF and % of HV as show in Figure (1).

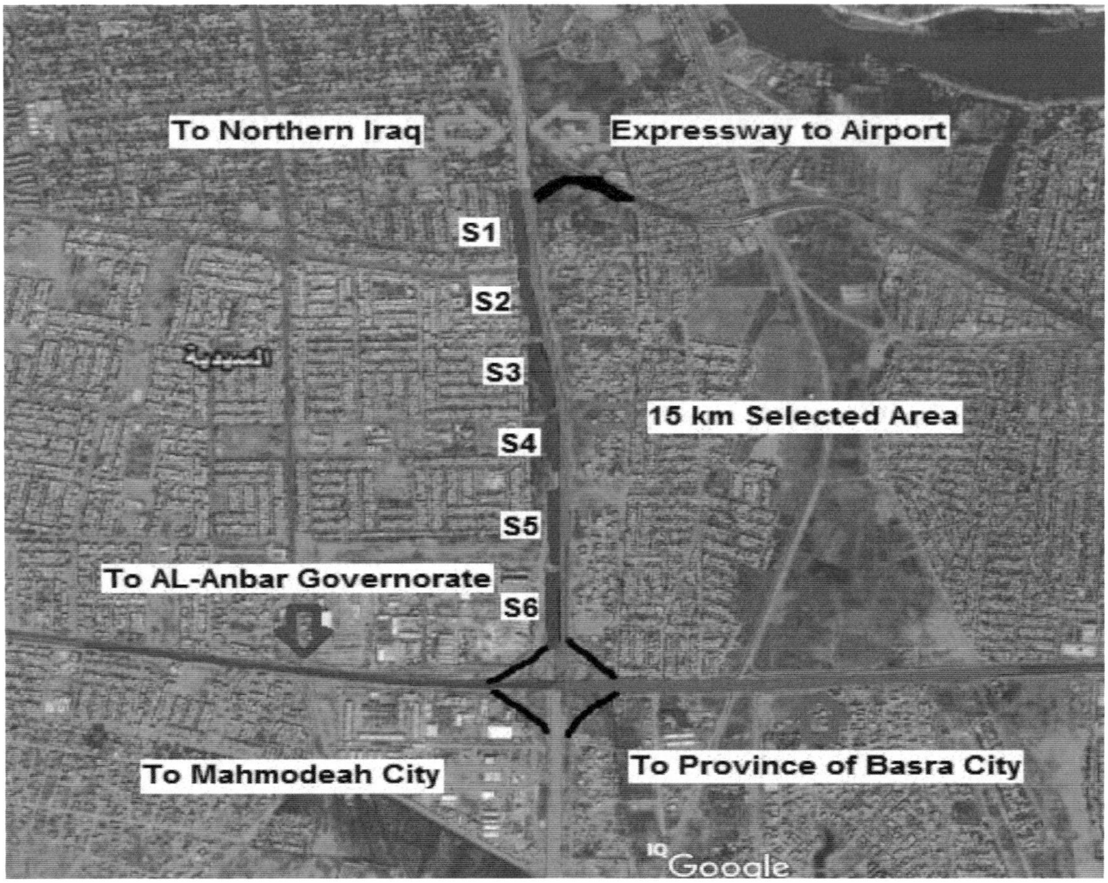

Figure 1. Aerial photograph (screen shot) from Al- Dora intersection to Al- Rashid interchanges (Google earth program).

4. Data Collection

The data collected for the highway was divided into three types: road geometric data, vehicle speed data and traffic volume data. All the traffic data was collected by observation, which is the main source of the data needed for this study. Through positioning of surveillance cameras at each section of the highway, necessary input data was obtained for evaluation and improvement.

The Traffic Department at the Ministry of Interior was contacted to get the registered vehicles by state institutions in governmental, public and mixed sectors, in order to obtain the yearly growth of traffic volume on Iraqi highways. Traffic volume collected in the past years shows that the number of vehicles increases with an annual growth rate (g) of 6.2%.

Field investigation of the study area was conducted several times in order to find out the geometric and traffic data for the highway, define the highway segments and collect the data needed for the analysis. For the missing data, the ideal conditions and default values were used as recommended by HCM.

The posted speed limits were determined by observation for each segment. The Base Free Flow Speed (BFFS) was estimated from the speed limit. According to HCM 2010, the BFFS on multi-lane highways under base conditions is approximately the posted speed plus 5 mph (8 kph) for a speed limit of 50 mph (80 kph) or higher and the posted speed plus 7mph (11 kph) for speed limits less than 50 mph (80 kph).

The demand volume under prevailing conditions (V) for the selected segments must be estimated and converted into the demand flow rate under equivalent base conditions (V_p), in order to define the LOS for each segment. Maximum peak hour volume (PHV) for each segment was obtained from the daily traffic volume.

Based on the Iraqi code of highways, the left side of the highway is for the (NB) traffic direction and the right side of the highway is for the (SB) traffic direction. Therefore, all the access points have a sign

IOP Conf. Series: Materials Science and Engineering **584** (2019) 012012 doi:10.1088/1757-899X/584/1/012012

as (L-exit) at the north direction and a sign as (R-exit) at the south direction.
The geometric and volume input data for each segment in each direction of the selected highway were organized for the existing conditions (year 2018), short-term conditions (year 2023) and mid-term conditions (year 2028) as shown in Table 1, 2 and 3.
The highway from Al- Dora intersection to Al- Rashid interchange is a rural multi-lane highway with two lanes in each direction. 12.95 ft (3.95 m) lane width for each lane, shoulder width 3.9 ft (1.2 m) in each direction and the posted speed limit is 55 mph. (88 kph.).The existing geometric cross-section of the highway is shown in Figure (2). The following descriptions are given for each of the six segments of the highway.
The existing geometric conditions of the multi-lane highway are improved by widening the cross-section 12 ft (3.6 m) in each direction and 6 ft (1.8 m) width of shoulder. The highway is converted from an undivided multi-lane highway into a divided multi-lane highway with concrete barriers. Therefore, the multi-lane highway becomes of three lanes in each direction as shown in Figure (3).
The multi-lane highway is improved to a freeway highway. Therefore, the six segments are combined to two segments with a cloverleaf interchange at the middle. Segment number one in the freeway highway combines segments 1, 2 and 3 in the multi-lane highway and segment number two in the freeway highway combines segments 4, 5 and 6 in the multi-lane highway as shown in Figure (4).

Table 1. Traffic and geometric data of each segment for two directions leading to International Baghdad Airport (existing conditions/ 2018).

Al- Dora intersection to Al- Rashid interchanges								
Right Side (NB)								
Segment Number	Length (mi)	PHV (veh/h)	HV%	PHF	Access Points AP/mi	Posted Speed (mi/h)	Lane * Width (ft)	No. Of Lanes
1	1.2	3186	7.4	0.90	6	55	12	2
2	1.3	2994	6.4	0.89	6	55	12	2
3	2.36	3349	5.4	0.89	4	55	12	2
4	0.98	3362	7.4	0.90	4	55	12	2
5	1.6	2994	7.8	0.90	3	55	12	2
6	1.53	3382	6.5	0.90	5	55	12	2
Left Side (SB)								
Segment Number	Length (mi)	PHV (veh/h)	HV%	PHF	Access Points AP/mi	Posted Speed (mi/h)	Lane Width (ft)	No. Of Lanes
1	1.2	3259	7.6	0.90	5	55	12	2
2	1.3	3011	6.4	0.89	4	55	12	2
3	2.36	3407	6.96	0.89	5	55	12	2
4	0.98	3344	6.7	0.90	4	55	12	2
5	1.6	2237	8.1	0.92	2	55	12	2
6	1.53	3321	6.2	0.90	4	55	12	2

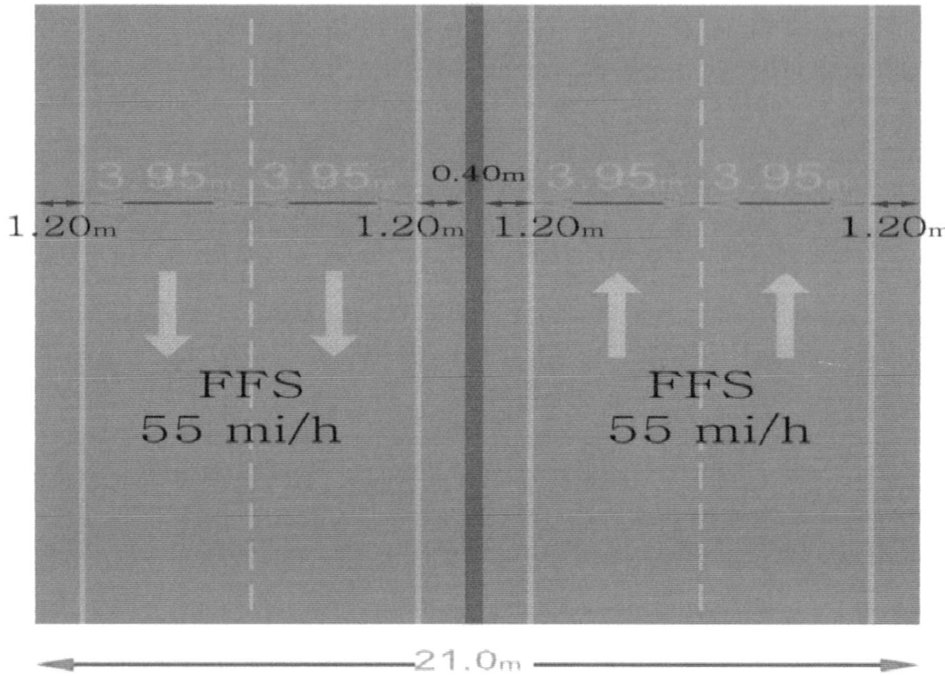

Figure 2. Cross-section of the existing undivided highway.

Table 2. Traffic and geometric data of each segment for north bound direction leading to International Baghdad Airport (after 5 years/ 2023).

Al- Dora intersection to Al- Rashid interchanges								
Right Side (NB)								
Segment Number	Length (mi)	PHV (veh/h)	HV%	PHF	Access Points AP/mi	Posted Speed (mi/h)	Lane * Width (ft)	No. Of Lanes
1	1.2	4303	10	0.90	6	55	12	2
2	1.3	4048	8.6	0.89	6	55	12	2
3	2.36	4524	7.3	0.89	4	55	12	2
4	0.98	4541	10	0.90	4	55	12	2
5	1.6	4045	10.5	0.90	3	55	12	2
6	1.53	4568	8.8	0.90	5	55	12	2
Left Side (SB)								
Segment Number	Length (mi)	PHV (veh/h)	HV%	PHF	Access Points AP/mi	Posted Speed (mi/h)	Lane Width (ft)	No. Of Lanes
1	1.2	4402	10.3	0.90	5	55	12	2
2	1.3	4067	8.6	0.89	4	55	12	2
3	2.36	4602	9.4	0.89	5	55	12	2
4	0.98	4517	9.05	0.90	4	55	12	2
5	1.6	3021	10.9	0.92	2	55	12	2
6	1.53	4486	8.4	0.90	4	55	12	2

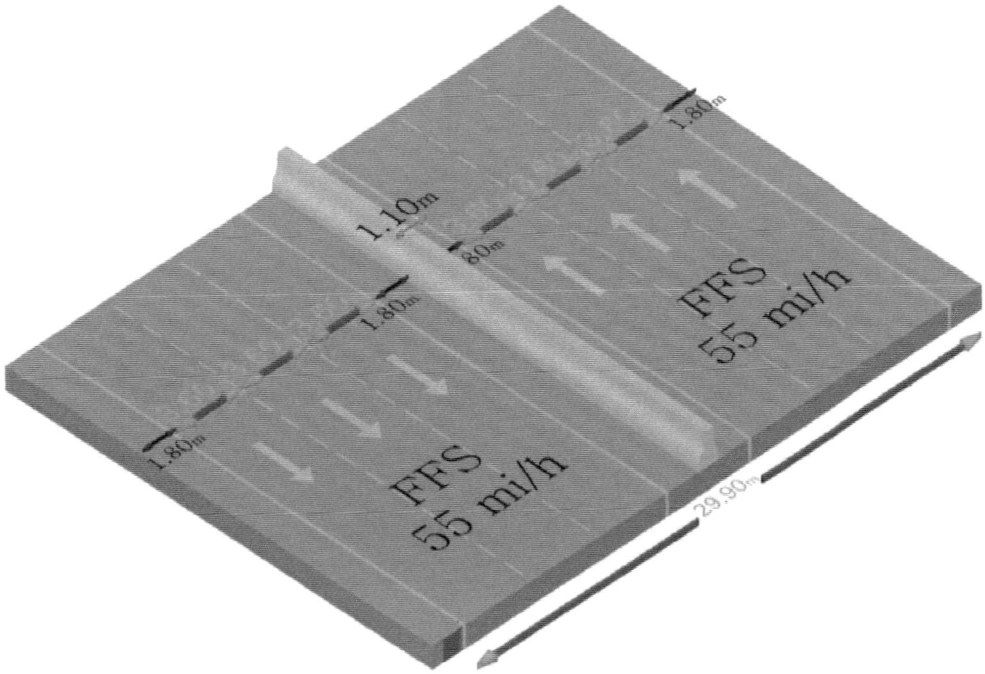

Figure 3. Suggested divided Cross Section of the selected highway.

Table 3. Traffic and geometric data of each segment of the two directions leading to International Baghdad Airport (after 10 years / 2028).

Al- Dora intersection to Al- Rashid interchanges								
Right Side (NB)								
Segment Number	Length (mi)	PHV (veh/h)	HV%	PHF	Ramp Density (R/mi)	Posted Speed (mi/h)	Lane Width (ft)	No. of Lane
1	4.9 (7.9 km)	5455	8.8	0.89	1	70 (112km)	12 (3.6m)	3
2	4.1 (6.6 km)	5509	10.6	0.90	1	70 (112km)	12 (3.6m)	3
Left Side (SB)								
Segment Number	Length (mi)	PHV (veh/h)	HV%	PHF	Ramp Density (R/mi)	Posted Speed (mi/h)	Lane Width (ft)	No. of Lane
1	4.9 (7.9 km)	5549	11	0.89	1	70 (112km)	12 (3.6m)	3
2	4.1 (6.6 km)	5409	10	0.90	1	70 (112km)	12 (3.6m)	3

Figure 4. Proposed cross-section for segment 1 and 2 with One-way service roads for the freeway airport highway

The HCS- 2010 software is in (US units); therefore, the data is provided in both SI and US units for each segment. The data includes the length for each segment, the peak hour volume (PHV), percent of grade (terrain), percent of heavy vehicles, density of access points (AP/mi), posted speed limit, lane width (LW) and number of lanes in each direction (NL).

5. Finding

In this study, the existing traffic conditions at the selected highway (Al- Dora intersection to Al- Rashid interchanges) were evaluated to find out on which Level of Service (LOS) it is expected to be operated in the future situation after five years (as short-term evaluation - 2023) and (mid-term evaluation - 2028). Also, it is important for decision makers to find out the possibility of improvement of each segment based on the updated HCM-2010 methodology. HCS-2010 software computer program were used in this study for compression and validation on each studied segment of the selected highway to develop more accurate management improvement programs.

5.1 Output Results

According to the Policy on Geometric Design of Highways and Streets, 2011, the guideline for level of service selection for rural highways is LOS-C (stable flow). The highest level of service is acceptable for some highway agencies, such as LOS-D (approaching unstable flow). LOS-E (unstable flow or maximum capacity) and LOS-F (forced or breakdown flow) are not acceptable traffic conditions and the highway segments should be improved to better LOS [8].

A lower level of service leads to significant environmental impacts on exposed persons living nearby and may interrupt with different activities in the surrounding area.

Such a condition is a major contributor to the increase in levels of unwanted gases due to increased amount of fuel burned in such area. The increase in number of vehicles due to poor LOS is usually associated with higher levels of pollutions such as gases, noises and psychological discomfort of nearby population. Furthermore, poor LOS results in an increase in traveling time, which translates into an increase in the amount of fuel consumed by vehicles, thus lifting the traveling cost for highway users.

The analysis of each segment of the highway in this study was conducted by using HCS-2010 for the existing, short-term and mid-term conditions and for right direction (SB) and left direction (NB) in each segment. The results HCS-2010 are presented for the selected highway in Tables (4, 5 and 6), as shown below.

Table 4. The output results for the existing conditions of the airport highway using HCS-2010 (year 2018).

Segment Number	Direction	Flow rate Pc/h/ln	FFS mi/h	Average travel speed mi/h	Density Pc/mi/ln	LOS
1	Right	1831	51.9	47.7	38.4	E
	Left	1855	52.2	47.6	39.0	E
2	Right	1740	51.9	48.3	36.0	E
	Left	1742	52.4	48.3	36.0	E
3	Right	1937	52.4	-	-	F
	Left	1981	52.2	-	-	F
4	Right	1933	52.4	-	-	F
	Left	1922	52.4	47.1	40.8	E
5	Right	1729	52.7	53.6	32.3	D
	Left	1264	52.9	55.0	23.0	C
6	Right	1944	52.2	-	-	F
	Left	1900	52.4	47.3	40.2	E

Table 5. The output results for short-term conditions of the airport highway using HCS-2010 (year 2023).

Segment Number	Direction	Flow rate Pc/h/ln	FFS mi/h	Average travel speed mi/h	Density Pc/mi/ln	LOS
1	Right	2510	51.9	-	-	F
	Left	2567	52.2	-	-	F
2	Right	2376	51.9	-	-	F
	Left	2387	52.4	-	-	F
3	Right	2630	52.4	-	-	F
	Left	2701	52.2	-	-	F
4	Right	2648	52.4	-	-	F
	Left	2622	52.4	-	-	F
5	Right	2370	52.7	-	-	F
	Left	1732	52.9	53.6	32.3	D
6	Right	2651	52.2	-	-	F
	Left	2591	52.4	-	-	F

5.2 Suggestions for improvement

The analysis of the selected highway shows that all segments are operated at LOS-F or at LOS-E for the existing, short-term and mid-term conditions. In order to improve the LOS of all segments, the following suggestions should be applied:

1. Increasing the number of lanes from two in each direction to three in each direction for the existing conditions.
2. Improving the geometric condition of the segment to ideal geometric condition in order to increase the Free Flow Speed (FFS).
3. Changing the highway from an undivided multi-lane highway to a divided multi-lane highway for the short term conditions.
4. For the medium term conditions all the access points must be eliminated from all segments and construct one cloverleaf with collector distributer road (CDR) interchange at the middle of the two selected segments. The cloverleaf interchange with CDR is required to eliminate or reduce the traffic weaving in basic freeway segment

IOP Conf. Series: Materials Science and Engineering **584** (2019) 012012 doi:10.1088/1757-899X/584/1/012012

5. The service roads also are suggested on each directions of the freeway (two-lane two-way or One-way direction). The right-of-way for constructing the service roads is available for future short-term and mid-term period in year (2023) and year (2028).
6. The expressway for the selected study area will develop route continuity with the Airport expressway and freeway system.

Table 6. The output results for mid-term conditions of the airport highway using HCS-2010 (year 2028).

Segment Number	Direction	Flow rate Pc/h/ln	FFS mi/h	Average travel speed mi/h	Density Pc/mi/ln	LOS
1	Right	3456	51.9	-	-	F
	Left	3535	52.2	-	-	F
2	Right	3256	51.9	-	-	F
	Left	3271	52.4	-	-	F
3	Right	3604	52.4	-	-	F
	Left	3719	52.2	-	-	F
4	Right	3646	52.4	-	-	F
	Left	3593	52.4	-	-	F
5	Right	3247	52.7	-	-	F
	Left	2384	52.9	-	-	F
6	Right	3634	52.2	-	-	F
	Left	3551	52.4	-	-	F

5.3 Discussion of Results

Based on the suggestions the output results shows improving the LOS from E or F to better LOS. Tables 7, 8 and 9 shows the output results of improvement of all segments for existing, short-term and mid-term periods. Segments will be operated between LOS-C and LOC-D, which are acceptable for both directions.

Table 7. Improvement of existing conditions (year 2018).

Segment Number	Direction	Flow rate Pc/h/ln	FFS mi/h	Average travel speed mi/h	Density Pc/mi/ln	LOS
1	Right	1221	53.5	55	22.2	C
	Left	1255	53.8	55	22.8	C
2	Right	1154	53.5	55	21.0	C
	Left	1161	54.0	55	21.1	C
3	Right	1285	54.0	55	23.4	C
	Left	1320	53.8	55	24.0	C
4	Right	1288	54.0	55	23.4	C
	Left	1281	54.0	55	23.3	C
5	Right	1153	54.3	55	21.0	C
	Left	842	54.5	55	15.3	B
6	Right	1296	53.8	55	23.6	C
	Left	1266	54.0	55	23.0	C

Table 8. Improvement of short-term conditions (year 2023).

Segment Number	Direction	Flow rate Pc/h/ln	FFS mi/h	Average travel speed mi/h	Density Pc/mi/ln	LOS
1	Right	1673	53.5	53.9	31.0	D
	Left	1711	53.8	53.7	31.9	D
2	Right	1584	53.5	54.3	29.1	D
	Left	1591	54.0	54.3	29.3	D
3	Right	1779	54.0	53.3	33.4	D
	Left	1801	53.8	53.2	33.9	D
4	Right	1785	54.0	53.3	33.5	D
	Left	1801	54.0	53.2	33.9	D
5	Right	1580	54.3	54.4	29.1	D
	Left	1154	54.5	55	21.0	C
6	Right	1767	53.8	53.4	33.1	D
	Left	1727	54.0	53.6	32.2	D

Table 9. Improvement of mid-term conditions (year 2028).

Segment Number	Direction	Flow rate Pc/h/ln	FFS mi/h	Average travel speed mi/h	No. of lanes	Density Pc/mi/ln	LOS
1	Right	1286	72.2	69.9	3	18.4	C
	Left	1321	72.2	69.8	3	18.9	C
2	Right	1296	72.2	69.9	3	18.5	C
	Left	1267	72.2	69.9	3	18.1	C

6. Development Management Program

The main objective of this study is to improve the quality of service for the selected segments. This improvement is achieved by increasing the posted speed to the maximum safe speed and increasing the number of lanes in order to decrease density, which in turn improving LOS (LOS F or LOS E) to better LOS (LOS C or LOS D). This improvement will lead to decrease conjunction, delay time, pollution and stopping time and increase safety for drivers with better travel time. The cost estimation for this improvement has to be carried out in order to determine the benefit to cost ratio. The cost estimation is out of the scope of this study and is strongly recommended for any future work in this field.

Management flow diagram is developed for the existing, short-term and mid-term period planning based on the analysis, evaluation and improvement resulting from using HCS-2010 computer programs for the six segments of the selected Al- Dora intersection to Al- Rashid interchange, as shown in Figure (5).

The management flow diagram will assist decision makers to improve the breakdown traffic condition segments that operate at LOS-F or operate at maximum capacity (LOS-E) for existing planning (year 2018), short-term planning (year 2023) and mid-term planning (year 2028) during peak hour periods.

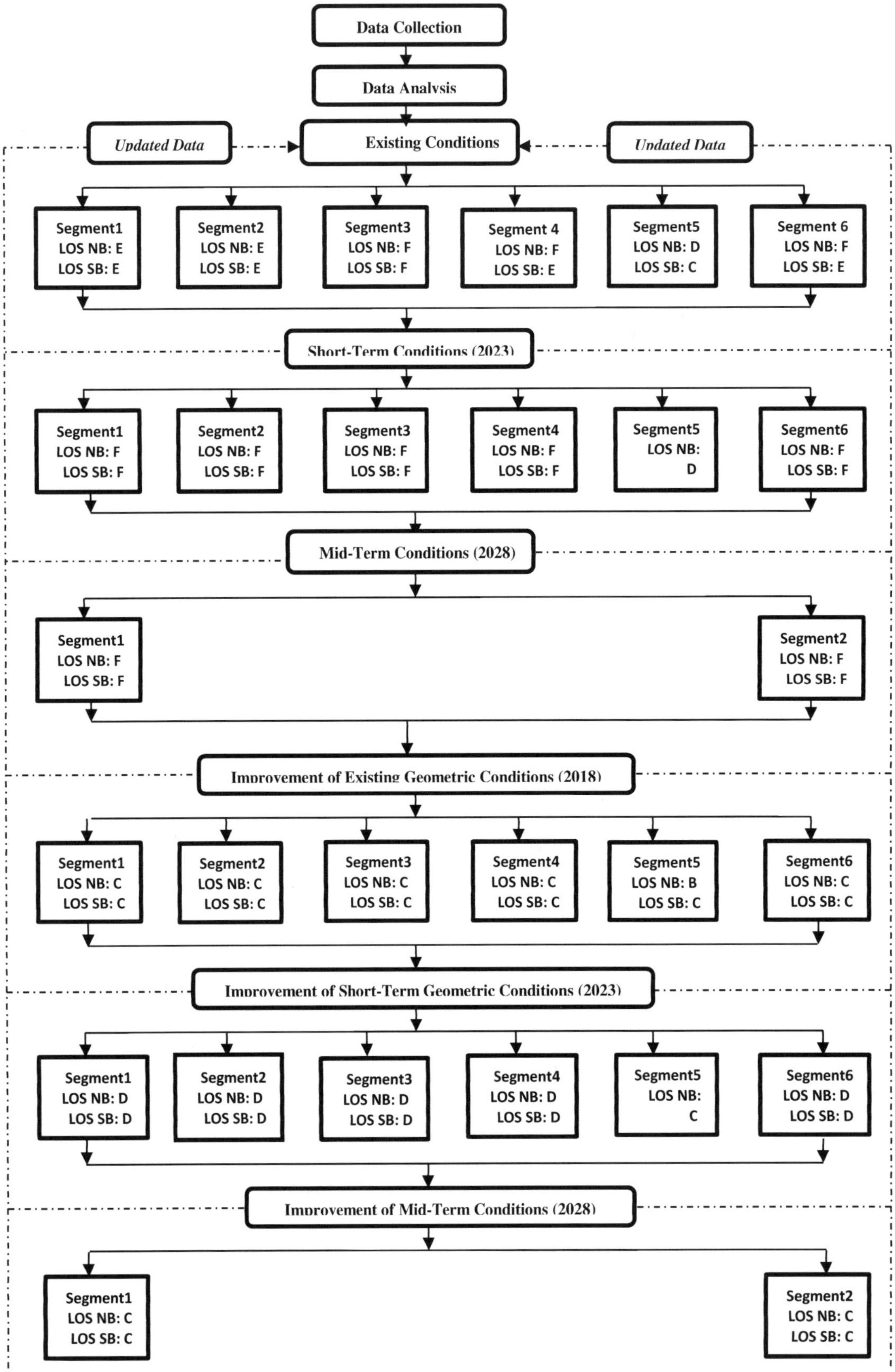

Figure 5. Flow diagram of management components for Al- Dora intersection to Al- Rashid interchange.

7. Conclusions

From the application of updated HCM-2010 on the Al- Dora intersection to Al- Rashid interchange, the following conclusions can be drawn:

1. The existing conditions of the selected highway are operated on LOS-E and LOS-F, except for segment 5 for south bound direction (SB), which is operating at LOS-C and north bound direction (NB), which is operating at LOS-D.
2. Short-term planning shows that segments 1, 2, 3, 4 and 6 for both north and south bound directions (NB and SB) will be operated at LOS-F, except for segment 5 for south bound direction (SB), which will be operated at LOS-D.
3. Mid-term planning shows that segments 1, 2, 3, 4, 5 and 6 for both north and south bound directions (NB and SB) will be operated at LOS-F.
4. Due to the improvement of management program, all segments will be operated at LOS-C and LOS-D, except for segment 5 (SB) for existing condition which will be operated at LOS-B.
5. The improvement in LOS for the previous segments that are operating at LOS-E or LOS-F was due to modified geometric and traffic conditions, such as: increasing the number of lanes from two lanes to three lanes in each direction , changing the highway from an undivided multi-lane highway to a divided multi-lane highway, increasing the posted speed limit in order to decrease the density and upgrading the prevailing geometric conditions to ideal geometric conditions, and changing the highway from multi-lane highway to a freeway.
6. The developed flow diagram for management of Al- Dora intersection to Al- Rashid interchange will assist decision makers in improving the highway to better LOS.

8. Recommendations

The following recommendations are suggested.

1. It is recommended to use the methodology of HCS 2010 for all major rural and suburban highways in Iraq in terms of operational analysis, design and planning.
2. Changing all undivided multilane highways in Iraq to divided multilane highways with ideal geometric conditions for short term period (year 2023).
3. Modifying the airport multilane highway in Iraq to a freeway with full control of access for medium term of period (year 2028).
4. Cost effectiveness is highly recommended for any improvement in a future research.

References

[1] Highway Capacity Manual (HCM). (2010). Transportation Research Board (TRB), National Research Council, Washington, D.C.

[2] Highway Capacity Software (HCS). (2010). McTrans Center, University of Florida. Traffic Software Integrated System (TSIS).

[3] Jrew et al. (2018). "Management of Multi-Lane Highways in Jordan (Case Study)".Jordan Journal of Civil Engineering (JJCE). vol.12, No.1.

[4] Jrew et al. (2017). "Management and Evaluation of Two Lane Two way Highways in Jordan (Case Study)".The Seventh Jordanian International Civil Engineering Conference. 9-11, May 2017.

[5] Yadav et al, (2014). Roadway Capacity Estimation for Multi-lane Interurban Highways in India, Colloquium on Transportation Systems Engineering and Management CTR, CED, NIT Calicut, India.

[6] Semeida, A. M. (2013). "New models to evaluate the level of service and capacity for rural multi-lane highways in Egypt." Alexandria Engineering Journal 52(3): 455-466

[7] Robertson et al. (2013). Determining Level of Service on Freeways and Multi-lane Highways with Higher Speeds.

[8] American Association of State Highway and Transportation Officials, AASHTO. (2011).

IOP Conf. Series: Materials Science and Engineering **584** (2019) 012013 doi:10.1088/1757-899X/584/1/012013

Punching shear behaviour of solid and bubble reinforced light Weight aggregate concrete two-way slabs

Maha habeeb[1], Adel A Al-Azzawi[2], and faiq Al- Zwainy[3]

1 Faculty of Engineering, Al- Nahrain University, Iraq. Email: Eng.maha97@yahoo.com

2 Faculty of Engineering, Al- Nahrain University, Iraq. Email: Dr_adel_azzawi@yahoo.com

3 Faculty of Engineering, Al- Nahrain University, Iraq. Email: Faiq_faiqmohmed@yahoo.com

Abstract. Punching shear is the most important problem in flat slabs, which usually required reducing the self-weight of the slab. A lightweight concrete as well as the bubble deck technology were used to allow for lighter self-weight of the structure. This study aims to investigate the use of lightweight aggregate concrete (LWAC) with the bubble deck technology in flat slab. The behavior of two- way LWAC voided slab (with and without shear reinforcement) under punching shear effect with respect to solid normal weight and solid LWAC specimen were studded. The behavior of six specimens having the same dimensions, slab thickness and flexural steel reinforcement were studded. The main changed variables were concrete type, percent of void and punching shear reinforcement existence. The results show that the load carrying capacity for light weight aggregate concrete voided slab without shear reinforcement decreased by (60%, 53%) with respect to solid normal weight and solid LWAC slabs without shear reinforcement respectively. In addition, the load carrying capacity for LWAC voided slab with shear reinforcement decrease by (67%, 47%) with respect to solid normal weight, solid LWAC slabs with shear reinforcement respectively. Also, for lightweight aggregate concrete, solid and voided slab, shear reinforcement existence did not significantly increase the slab capacity for punching shear. But cause a reduction in maximum deflection with respect to slabs without shear reinforcement.

1. Introduction

In a reinforced concrete structure, the span between columns is the main design limitation of the slab. To design a large slab span between columns, peripheral beams and/or very thick slabs are required. Which leads to increase the weight of the structure because of large amounts of used concrete (1). Another option for reducing weight is the bubble deck technology which uses spheres made of recycled industrial plastic to create air voids while providing strength through arch action. It attempts to utilize the positive aspects of concrete slab construction while minimizing the negative attributes of solid slabs by lightening the self-weight of the structure (2). Many experimental investigations were conducted on voided slab. But, there is no research study on the behavior of lightweight concrete voided slab.
The use of lightweight concrete (LWC) and voids can reduce the self-weight of structures. So, the cross sectional area of structural member will be reduce too.

2. Experimental work

A real flat slab system with dimensions of (7.5*7.5) m supported on columns only, represent the prototype in this research. Its zero-point moment lies approximately at (0.22 L) from the column axis. the punching shear specimen represent the column strip with scale of 1/3 which exposed to punching shear due to column of square cross section, that have the scaled dimension equal to (10*10) cm. The experimental work consists of six two way flat slabs specimens. Simply supported along all edges with dimension equal to (1100*1100*100) mm and designed to fail in punching shear.

A series of tests is carried out, including raw materials, fresh and hardened concrete, as well as tests conducted on specimens under punching shear loads.

2.1. Materials

2.1.1 Lightweight aggregate

Light Expanded Clay Aggregate (LECA) is a porous ceramic product with a uniform pore structure and density ranges from 600 kg/m3 to 1900 kg/m3 see plate (1). It is manufactured in Rotary kilns from raw materials containing clay minerals, between 1100 °C and 1200 °C, resulting in a significant increase in volume due to expansion (3). Structural LECA of gradations 0-8 mm was used to produce the concrete for light weight aggregate specimen. Its bulk dried density equal to 740 kg/m3.

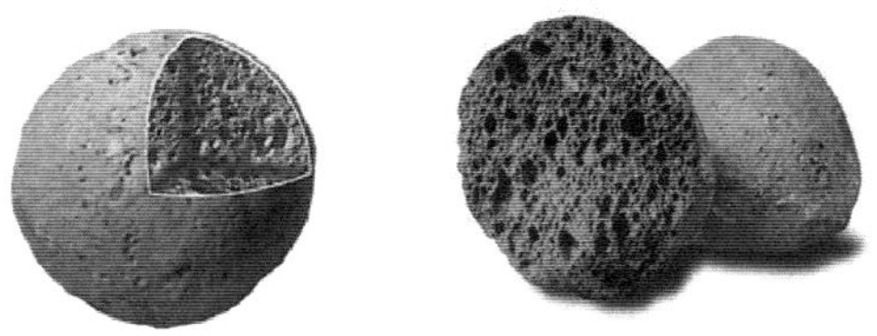

(a) **(b)**

Plate 1. (a) Ordinary leca. (b) Structural leca.

The LECA pebbles internal cellular structure with thousands of air-filled cavities gives thermal and sound insulation properties. Leca aggregate is chemically neutral, with pH of (7-7.2). It does not affect or be affected by other materials, and even preserves materials intact from chemical hazards (4).

2.1.2. Reinforcement

In order to achieve punching failure and to prevent bending failure in the slab specimens. A constant reinforcement ratio, (which was more than what normally used in practice) was selected. Steel bar of 10 mm in diameter, spaced at 65 mm in two directions was used. For punching shear reinforcement, a three vertical bars 6mm in diameter, mechanically anchored at each ends at an angle of 90 degrees. Each three bars were welded on via hooks to steel rail have the dimension (20*1.25) cm as shown in plate (2). Reinforcement arranged at three control perimeter lies at 0.5d, d and 2d from column face as shown in fig (1).

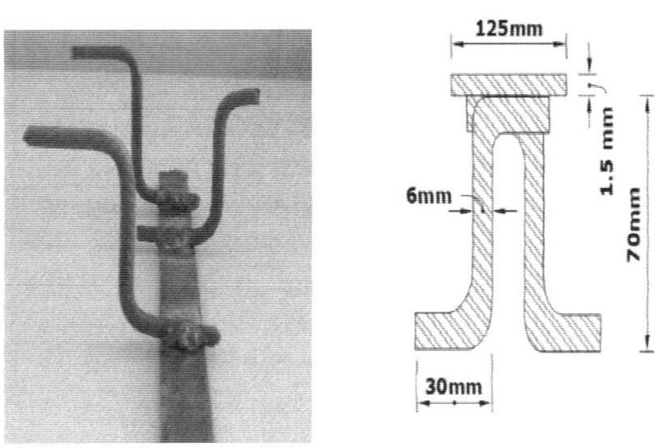

Plate 2. Shear reinforcement

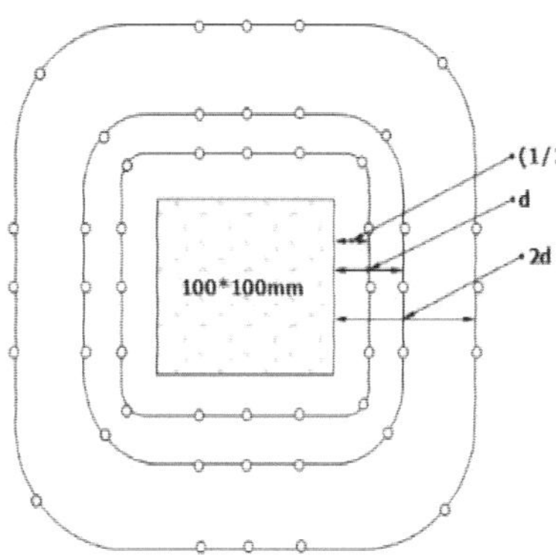

Figure 1. Shear reinforcement Arrangement

2.1.3. Plastic Spheres

Oval with dimensions of (80*80*40) mm made from plastic were manufactured and used. The void should be covered by concrete from all direction with at least one ninth of the ball diameter"(5). Which mean that the maxitmum bubble diameter that could use in a voided slab equal to (0.82 * overal depth of the slab). Figure (2a,b) shows specimens detals.

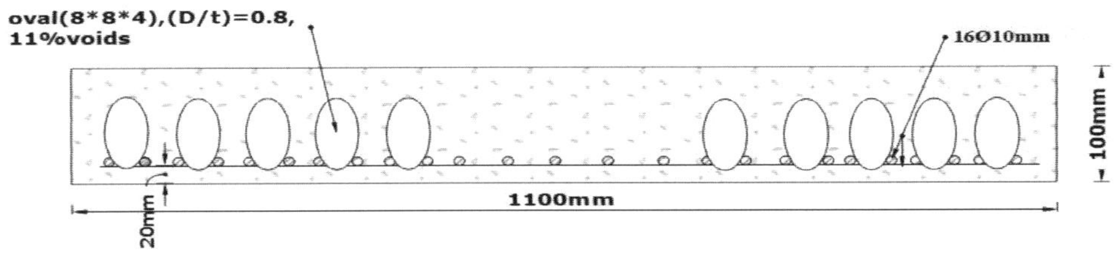

(a) Voided slab without shear reinforcement

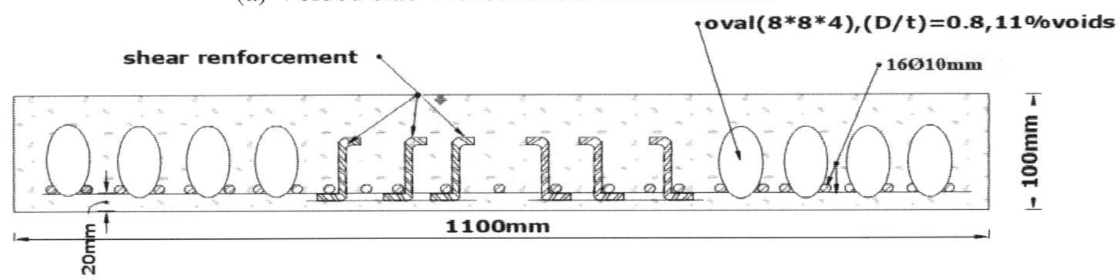

(b) Voided slab with shear reinforcement

Figure 2. Specimens details (a) specimens without shear reinforcement (b) specimen with shear reinforcement

2.2. Mix proportions

Two types of concrete were used in preparing the specimens:

(1)Normal weight concrete was used for casting two solid control specimen. Ordinary Portland cement, Natural sand and normal weight gravel was used as shown in table (1).

(2) Lightweight aggregate concrete was used to cast all other specimens. Ordinary Portland cement, Natural sand and LECA was used (6). As shown in table (1).

Also, to improves the physical and mechanical performance of concrete. High performance concrete hyper plasticizer (HP 580) Epsilone HP 580 was used (7). The Basic mechanical properties for concrete, reinforcement and aggregate were obtained laboratory. To obtained the mechanical properties of the hardened concrete. A (150×150×150 mm) cubic samples and standard concrete cylinders with dimension (150×300 mm) where tested to obtain the compressive strength and Split tensile strength according to ASTM C496 (8). Also, a Prism concrete specimen with dimension (10*10*40) cm where tested to determine the flexural strength capacity, and the same cylindrical samples that used to determine the split tensile strength was tested to find modulus of elasticity first. Specimens details and mechanical properties are shown in table (2).

IOP Conf. Series: Materials Science and Engineering **584** (2019) 012013 doi:10.1088/1757-899X/584/1/012013

Table 1. Mix proportion for normal and light weight concrete

Mix No.	concrete type	Compressive Strength (fc), MPa	Cement Content, Kg/m³	Sand Content, Kg/m³	LECA Content, Kg/m³	Gravel Content, Kg/m³	Water L/ m³	HP580 L/m³
1	normal weight	40	325	550	_	1300	146	**0.9**
2	light weight	40	550	550	500	_	133	**0.57**

Table 2. Specimens' details and mechanical properties

No.	Labeling	slab thickness t(mm)	shear reinforcement	bubble dimension	D/t	voided %	comp. strength cubic	Fr (Mpa)	Ft (Mpa)	Ec (Mpa)	reduction in weight due to [illegible]	reduction in weight due to lightweight	total reduction %
1	SNW 1	100	without	0	0	0	48.7	6.4	3.877	28318.3	0.00	0.00	0.00
2	SNW 2	100	with	0	0	0	49.0	6.8	4.013	28322.5	0.00	0.00	0.00
3	SLW 1	100	without	0	0	0	42.1	3.5	2.222	11292.2	0.00	33.14	33.14
4	SLW 2	100	with	0	0	0	40.7	3.2	2.207	11845.8	0.00	32.25	32.25
5	BLW 1	100	without	80*80*40	0.8	20.9	39.8	3.1	2.033	11671.0	20.95	36.34	57.29
6	BLW 2	100	with	80*80*40	0.8	20.1	39.6	3	2.01	10769.0	20.06	36.37	56.42

2.3. Tests

2.3.1. Test setup and measurements

The tests were done by using a universal testing machine of high load (1000 kN). A steel frame (used to support the slab specimen, act as simply support on all sides.) with dimension of (1000*1000*90) mm welded to a stiff steel beam that fixed to the fitted center of main hydraulic jack and piston of testing machine. The stiff steel beam, the steel frame and the specimen move upward as one unit with the movement of the piston.

Vertical deflection due to specimen loading was measured at three points by using digital dial gauges with Accuracy equal to (0.001 or 0.01) mm. These points were allocated at: center of the tested slab (C) on the tension face while the other two points allocated at distance (0.375L,0.25L from the support center) on the compression face. The readings from these gauges can be recorded for each (0.1 kN). A set of reading ranged from (50-150) were selected depending on the magnitude of ultimate load and severity of changes in load–deflection curve patterns.

To measure the strain in steel reinforcement during the test, two steel strain gages were used on bending reinforcement for (SNW1, SNW2, SLW1, SLW2) specimen. To register the values of stress in reinforcement at failure. To indicate whether the slab failure is due to punching shear or bending.

Before testing, specimens were prepared by painting them to show a clearer picture of crack plate (3a). The position of load, support, dial gauges and concrete strain gages were marked also. After the concrete strain gages fitting, the specimen placed at the exact location on the support. Then the dial gauges centered in their position. As shown in plate (3b)

IOP Conf. Series: Materials Science and Engineering **584** (2019) 012013 doi:10.1088/1757-899X/584/1/012013

(a) **(b)**

Plate 3. (a) Model preparing before test. (b) Loading setup (model after testing before releasing).

Specimens were tested under a static load to study their punching shear behavior as shown in figure (3). The load was applied in load control mode at a load rate of 5 kN/min, and the strain gauges' data were collected by a data acquisition system. Plate (4) represent the specimen after failure.

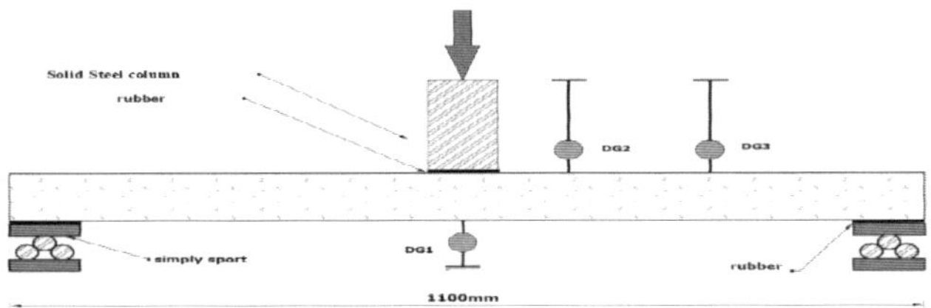

Figure 3. Punching shear setup.

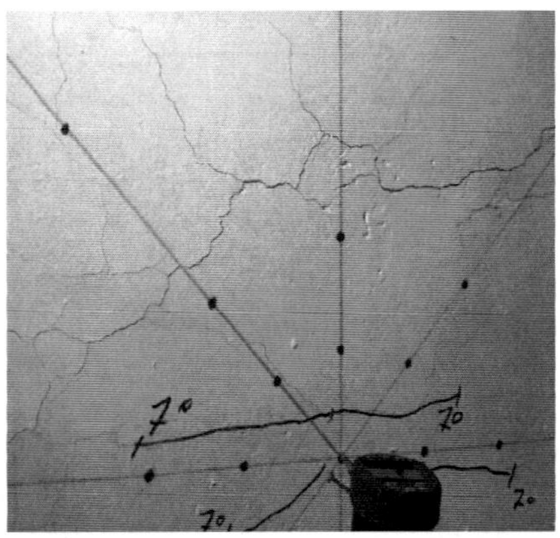

Plate 4. Punching shear specimen after test.

2.3.2. Test results

ACI 318-14 Code and the Euro code 2 were adopted in analyzing and discussing the results of this study. All the factors influencing the section resistance to the punching shear force according to ACI 318-14 Code or Euro code were fixed. The parameters are (concrete compressive strength, the flexural reinforcement ratio ρ, the thickness of the slab, and the ratio of the dimension of supporting column to the effective depth of slab). To give a clear extent to the effect of changing concrete type on the behavior of the section, to be study and know.

In ACI318-14 code, the critical section is parallel to the column face at a distance equal to half the effective depth of the slab (9). While the Euro code take the critical section at 1.5d to 2.0d from the face of the column (10). A critical distance between O.5d- 2d were studded in this research. As shown in figure (4).

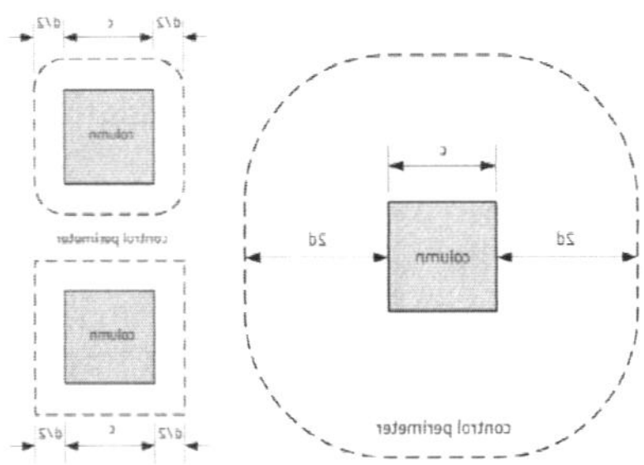

(a) ACI 318 (b) Eurocode 2

Figure 4. Control perimeter specified in ACI 318 and Eurocode2.

3.1. Ultimate Load Capacity and maximum Deflection

Table (3) shows all specimens details, ultimate punching shear load and deflection. The results show that:

Table 3. Specimen's details, ultimate load and deflection.

No.	Labeling	Slab Thickness H (mm)	shear reinforcements	bubble diameter	voided %	d/t	As (mm2)at 1/2 d u=700mm	As (mm2) at d u=1000mm	As (mm2) at 2d u=1600mm	Compression strength	Pu ultimate load (kN)	Max. deflection (mm)
1	SNW 1	100	without	0	0	0	0	0	0	48.7	195.4	10.66
2	SNW 2	100	with	0	0	0	452.57	452.57	452.57	49.0	255.5	12.9
3	SLW 1	100	without	0	0	0	0	0	0	42.1	168	11.81
4	SLW 2	100	with	0	0	0	452.57	452.57	452.57	40.7	159.8	10.2
5	BLW 1	100	without	80*80*40	20.9	0.8	0	0	0	39.8	79.1	12
6	BLW 2	100	with	80*80*40	20.1	0.8	452.57	452.57	452.57	39.6	84.8	8

- The existence of shear reinforcement causing an increase in slab loading capacity for specimen with same concrete type. The increase in the failure load for solid normal weight, bubble light weight concrete equal to (23.5% and 6.7%) respectively. But the solid light weight specimen shows an (4.9%) increase in loading carrying capacity for slab without shear reinforcement. This may be attributed to the slight difference in compressive strength of normal and light weight concrete.
- The failure load of specimens was ranged from (255.5 to 79.1) kN as shown in table (3). So, the reduction percent in load caring capacity between specimens with different concrete type was as shown in table (4).

Table 4. Reduction percent in load caring capacity

No.	The comparison between	Specimen with shear reinforcement	Specimen without shear reinforcement
1	normal and light weight concrete	68.3%.	14%
2	normal and bubble light weight	35.3%.	59.5%
3	solid and voided light weight	51.7 %.	52.9%

- For light weight concrete solid and voided slab, specimen without shear reinforcement show maximum deflection larger than slabs with shear reinforcement. Even though, the punching shear reinforcement existence did not significantly increase the slab capacity for punching shear because of light weight concrete properties control the failure criteria. That's mean, shear reinforcement existence increases the deformation capacity for the specimen. But the effect of reduced mechanical properties of lightweight concrete (relative to normal weight concrete with the same compressive strength) govern the failure criteria.
  While in the normal weight concrete, things seem to be reversed. Punching shear reinforcement existence gave maximum deflection larger by (17.4%). which can be explained by the significantly increase in slab capacity by (23.5%).
- The failure load was ranged from (234 to 79.1) KN as shown in table (2). Which gives 68.3% reduction percent between normal and lightweight concrete. 35.3% between normal and bubble lightweight. And 51.7 % between solid and voided lightweight for specimen with shear reinforcement. While the reduction percent for slabs without shear reinforcement was (14%, 59.5% and 52.9%) respectively.
- For normal weight concrete, punching shear reinforcement existence gave maximum deflection larger by (17.4%). Which can be explained by the significantly increase in slab capacity with (23.5%) increase on the failure load.
- For lightweight concrete solid and voided slab, specimen without shear reinforcement show maximum deflection larger than slabs with shear reinforcement. Even though, the punching shear reinforcement existence did not significantly increase the slab capacity for punching shear.

3.2. Load deflection curves

The central deflection values are plotted against loading for the six models together to explain the differences in behavior more clearly as shown in figure (5).

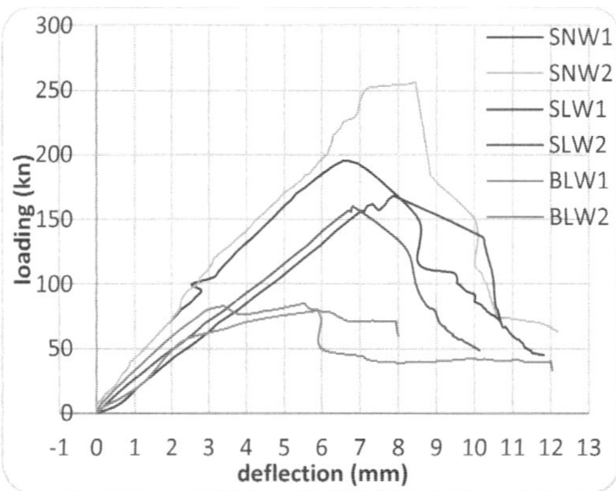

Figure 5. Load-central deflection curve for case one specimen.

1. In general, the load–deflection curve for all specimens shows identical behaviors which seems to be straight with linear relationship until reaching the peak load. Which could be attributed to symmetry for all specimens in geometry of supports and arrangement and applied load. Although, there is a clear difference between the magnitude of maximum load and deflection that occurred before failure because it depends on slab stiffness.
2. All specimens with shear reinforcement before reaching failure load show less deflection at the same load level compare to the specimen have the same geometrical and mechanical properties but without shear reinforcement.

The load deflection curve for all six specimens individual are shown in Figure (6). Three curves represented the three points where deflection was measured (٠.5L, 0.375L and 0. 25L) are shown.

1. The deflation for all solid specimens (normal and lightweight), at point 0.375L shows an asymptotic performance and values close to central deflection. While at 0. 25L the deflection shows less magnitude

than central deflection but with the same behavior. But for lightweight voided specimen the deflection at the three-point show an identical behavior with very close value.
That is means, for lightweight voided specimen. The area around column, which extended a distance equal 2. d from column face in all directions, was moving as a single block and expressing a deflection at the same level approximately. So the punching shear stresses were concentrated at the first row of voids. That located after the solid area around column, and represent the first weakness area.

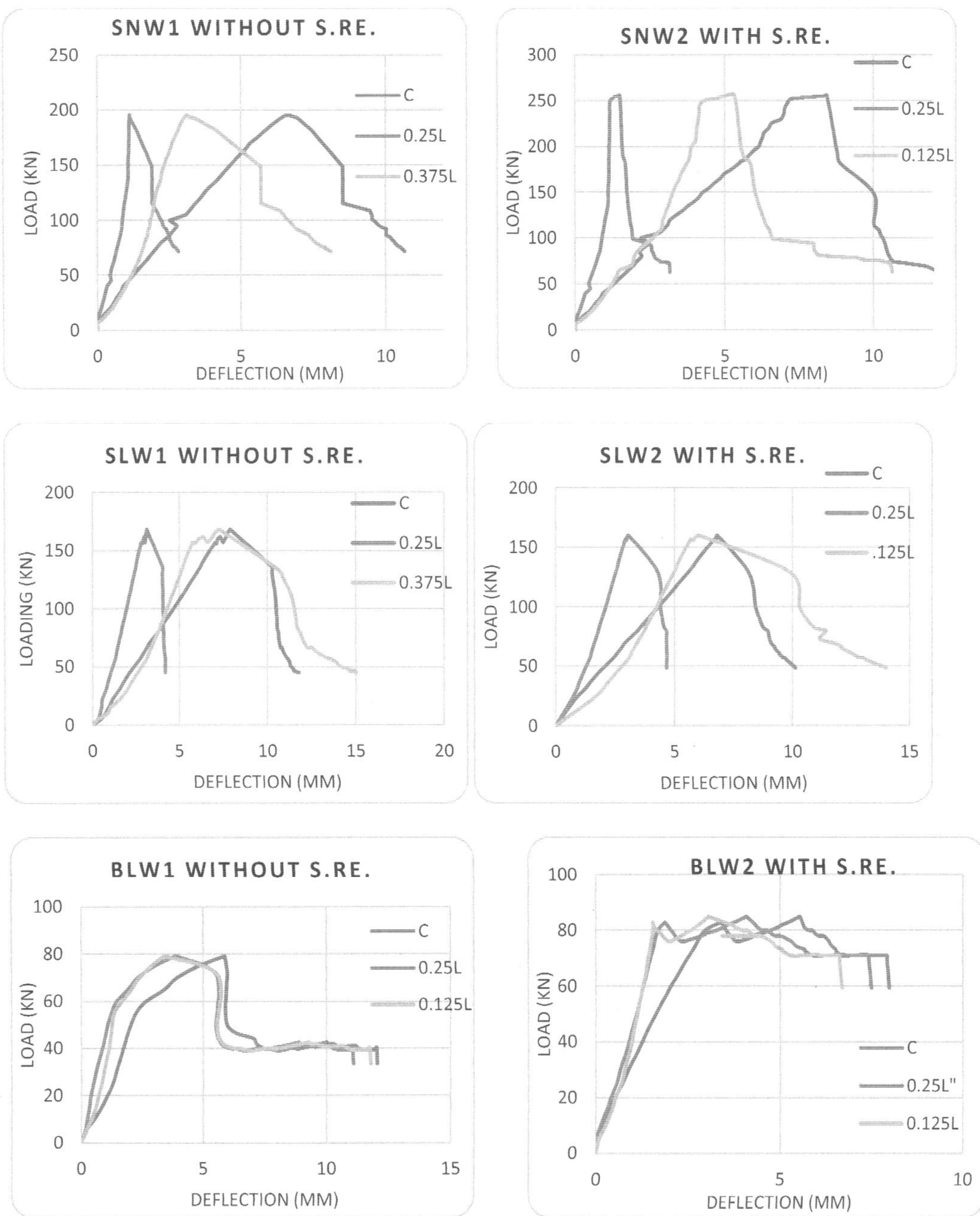

Figure 6. Load deflection curve for individual specimen continue.

3.3.Toughness and deformation factor

The material toughness is the maximum amount of energy it can absorb before fracturing [11]. The term "toughness" represented by the area of the post cracking region under the load deflection curve. [12] Also, the percent of deflection that accord after first crack was calculated to measure specimen's ability to show significant deformation before rupture. Table (5) show toughness value and deformation factor for first case specimen.

$$\text{Deformation factor} = \frac{\Delta u - \Delta cr}{\Delta u} * 100 \qquad (1)$$

The material toughness is the maximum amount of energy it can absorb before fracturing (11). The term "toughness" represented by the area of the post cracking region under the load deflection curve (12) Also, the percent of deflection that accord after first crack was calculated to measure specimen's ability to show significant plastic deformation before rupture. Table (5) show toughness value and plastic deformation percent for first case specimen.

Table 5. Toughness and plastic deformation percent for first case specimen.

Labeling	**concrete density kg/m3**	**toughness (kN*mm)**	**Cracking load (KN)**	**deflection at first crack load (mm)**	**ultimate load (kN)**	**deflection at ultimate load (mm)**	**deformation factor** $\frac{\Delta u - \Delta cr}{\Delta u}$
SNW 1	2281.5	27192.98	57	1.5	195.4	6.6	77
SNW 2	2314.7	42215.14	55	1.5	255.5	8.4	83
SLW 1	1727.5	30371.99	35	1.75	168	7.9	78
SLW 2	1739.1	18019.61	55	2.35	159.8	6.8	66
BLW 1	1686.9	3633.301	57	2.5	79.1	5. 9	58
BLW 2	1686.6	1702.24	54	1.8	84.8	3.4	48

1- For normal weight concrete solid slab, shear reinforcement existence leads to make the slab more tough and ductile by percent equal to (155%, 107%), respectively.
2- For light weight concrete, shear force existence reduces the toughness of slab by (40.7%, 53.2%) for solid and voided slab, respectively.
3- Lightweight concrete solid slab without shear reinforcement (SLW1) absorbed energy larger than solid normal weight (SNW1) by (11.7%). But (SNW2) slab showed higher energy absorption reached to (234%) than (SLW2) because of shear reinforcement effect.
4- Brittle materials have small toughness, because large elastic and plastic deformations lead to absorb large amounts of energy [13]. This means that shear reinforcement existence makes normal weight concrete more ductile while the void and shear reinforcement existence with light weight concrete creating a more brittle specimen.

3.4 Crack pattern

As the specimens start to be loaded, the punching shear force is transferred from column through surrounding slab concrete. Immediately after the stresses with in the concrete reach the ultimate tensile stress, first crack forming. When the loading continuo to increase, the other cracks forming at the slab central region. Cracks start to extended away from the area where higher values of tensile stresses towards the edges of the slab where lower stresses existence until rupture accord.
In general, the behavior of normal weight concrete specimen was different from lightweight concrete. Because of the high percent of small void existence in lightweight aggregate structure (leca). Which leads to different failure crack mode.

IOP Conf. Series: Materials Science and Engineering **584** (2019) 012013 doi:10.1088/1757-899X/584/1/012013

In lightweight concrete, there is a flexural crack noticed on the compression side in addition to column penetration crack. Also, in tension sides there is a clear cracks map with no damaged zone.
While in normal weight concrete, there is no flexural cracks on the compression side. The only crack is the crack formed due to column penetration. But in tension face there is a clear damaged zone. plate (5) show crack patron for all specimen.

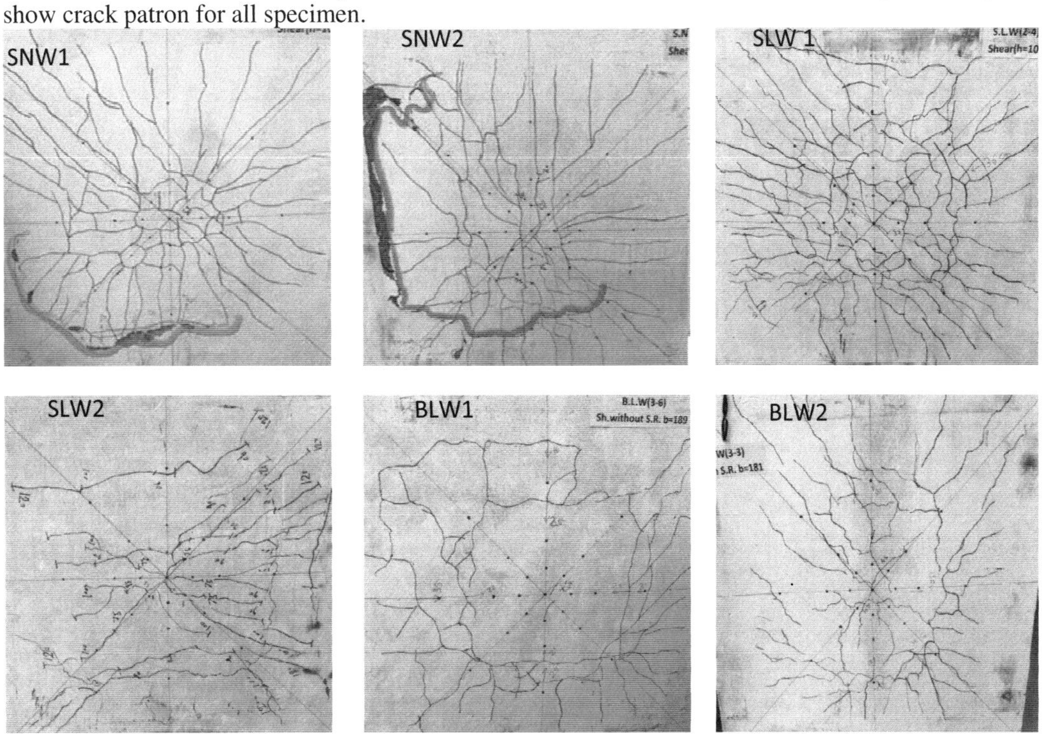

Plate 5. Crack patron for all specimen in case one.

3.5. Failure Zone and Failure Angles
The failure modes for slabs with punching shear reinforcement have several different cases. The position of the effective control perimeter changes depending on concrete mechanical properties, bending reinforcement ratio, shear reinforcement type and ratio (13).
The mode of punching failure was typically pyramid in shape. Which produce an angle (θ) with the tension face of the slab as shown in Fig. (7) (14). Where the distance between columns face to failure perimeter denoted by (Y). And the specimen plan direction named (N, S, E, W) which represent (north, south, east, west) directions.
The angle magnitude ranges between 20° and 45°. Which depends on the type and ratio of flexure and shear reinforcement on the slab (14).

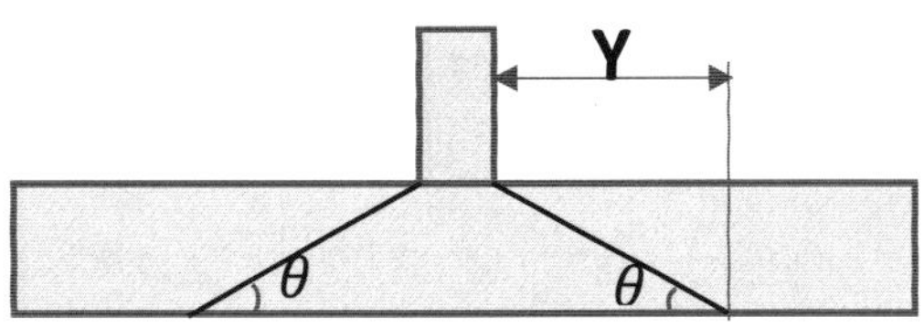

Figure 7. Punching shear failure angle.

Figure (8) shows the failure zone and the effective control perimeter position for all slab specimen in this case. Also table (6) show the magnitude of failure angle and the extend of punching zone in all direction.

IOP Conf. Series: Materials Science and Engineering **584** (2019) 012013 doi:10.1088/1757-899X/584/1/012013

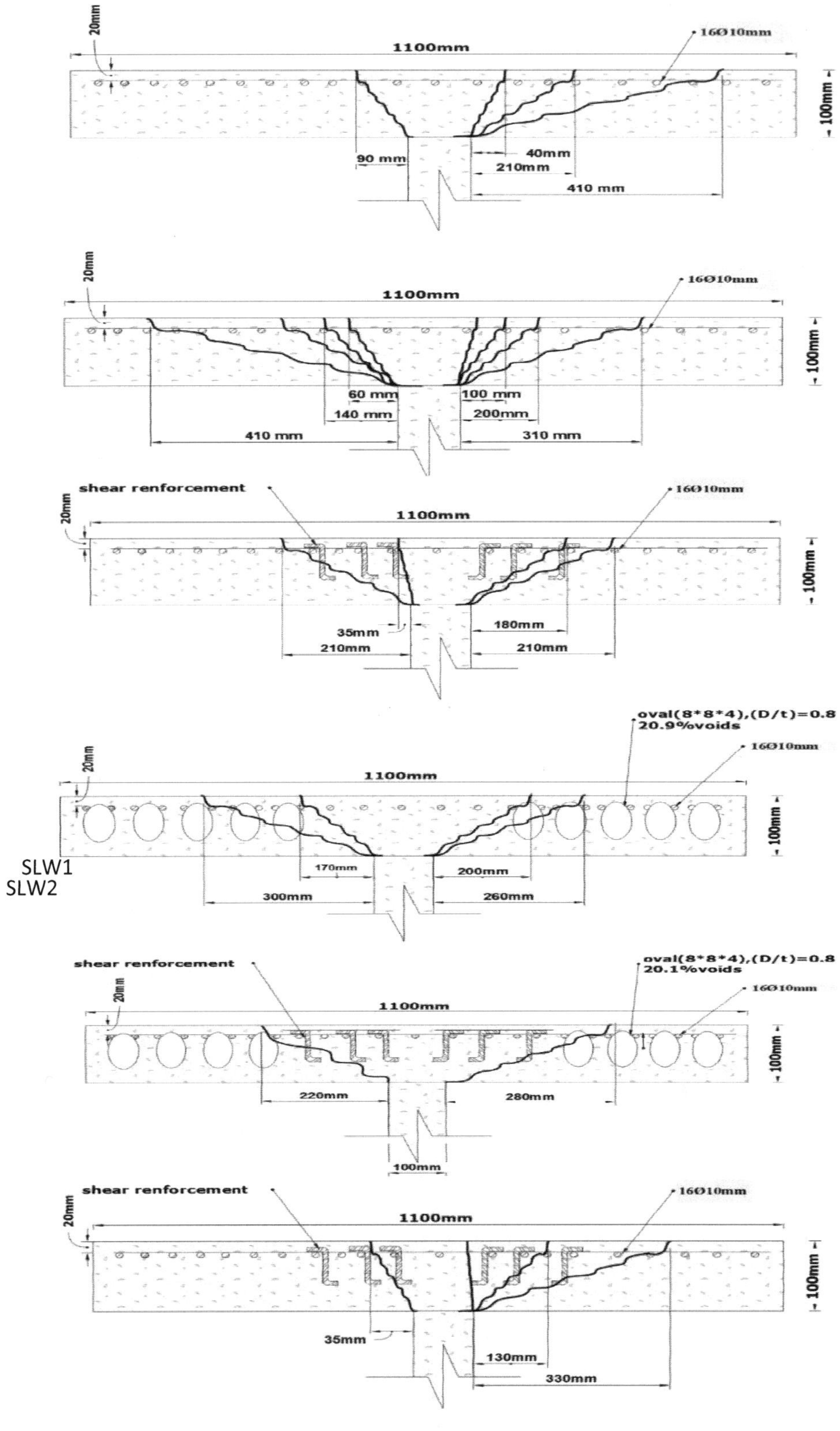

Figure 8. Failure zone and angle.

Table 6. Critical perimeter and failure angles.

Labeling	YN (mm)	YS (mm)	YE (mm)	YW (mm)	Perimeter (mm)	Failure Angles			
						θN	θS	θE	θW
SNW 1	-	290	410	-	760	-	19	13.7	-
SNW 2	370	-	410	310	1410	15.1	-	13.7	17.9
SLW 1	250	150	210	200	2020	21.8	33.7	25.5	26.6
SLW 2	230	220	260	-	2030	23.5	24.5	21	-
BLW 1	210	160	170	200	1880	25.5	32	30.5	26.6
BLW 2	200	230	-	180	1420	26.6	23.5	-	29

4. Conclusion

- The load carrying capacity for lightweight voided slab without shear reinforcement decreased by (60%, 53%) with respect to solid normal weight, solid lightweight slabs without shear reinforcement respectively.
- The load caring capacity for lightweight voided slab with shear reinforcement decrease by (67%, 47%) with respect to solid normal weight, solid lightweight slabs with shear reinforcement respectively.
- For voids ratio equal to (20%), the shear reinforcement existence leads to reduce the deflection magnitudes at ultimate load by (25%, 32.3%) for slabs with hook, shear reinforcement with respect to solid normal weight and solid light weight control specimen without shear reinforcement respectively.
- For normal weight concrete, shear reinforcement existence leads to make the slab more tough and ductile by percent equal to (155%, 107%) respectively. But for lightweight concrete, shear reinforcement existence reduces the toughness of slab by (40.7%, 53.2%) for solid and voided slab respectively.
- In lightweight concrete, there is a flexural crack are noticed on the compression side in addition to column penetration crack. Also, in tension sides there is a clear cracks map with no damaged zone. While in normal weight concrete, there is no flexure cracks on the compression side. The only crack is the crack formed due to column penetration. But in tension face there is a clear damaged zone.

References

[1] Singh M., Saini B 2018, "Analytical and Experimental Study of Voided Slab. In: Singh H., Garg P., Kaur I. (eds) Proceedings of the 1st International Conference on Sustainable Waste Management through Design. ICSWMD 2018. Lecture Notes in Civil Engineering, vol 21. Springer, Cham

[2] Chung, Joo-Hong, et al. "Two-Way Flexural Behavior of Donut-Type Voided Slabs."International Journal of Concrete Structures and Materials 12.1 (2018): 26.

[3] Leca [web source] URL https://leca.ae/

[4] Leca.co.uk [web source] URL https://www.leca.co.uk/ properties-of-leca

[5] BubbleDeck Head Office UK, (2008) " BubbleDeck Voided Flat Slab Solutions /Technical Manual & Documents ", Sole United Kingdom and Channel Islands Distributor.

[6] ACI Committee 211. 2-98 "Standard Practice for Selecting Proportions for Structural Lightweight Concrete (ACI 211.2-91)." American Concrete Institute, 1998.
[7] Weber [web source] URL https://issuu.com/e-weber/docs/product
[8] ASTM C496-86 (2004): "Standard Test Method for Splitting Tensile of Cylindrical Concrete Specimens". Annual Book of ASTM Standards, Vol. 04-02, (2004), PP.259-262.
[9] ACI Committee 318, (2014). Building Code Requirements for Structural Concrete (ACI 318M-14) and Commentary (ACI 318RM-14), American Concrete Institute, Farmington Hills, 503PP.
[10] Eurocode 2 - BS EN 1992-1-1:2004, Design of concrete structures. General rules and rules for buildings published by the British Standards Institution (BSI).
[11] Wikipedia [web source] URL https://en.wikipedia.org/wiki/toughness
[12] Scancemmaterials , [web source] URL: http://www.scancemmaterials.com /pdf/ Scanfibre / - Flexural Toughness Testing.pdf
[13] Oukaili, Nazar K., and luma F. Huseen (2016). "punching shear strength of bubble decks under eccentrical load."
[14] Fernández Ruiz, Miguel, and Aurelio Muttoni. , 2010 "Performance and Design of Punching-Shear Reinforcing Systems." 3rd fib International Congress. No. EPFL-CONF-163107.

IOP Conf. Series: Materials Science and Engineering **584** (2019) 012014 doi:10.1088/1757-899X/584/1/012014

Investigation of Binder Course Cold Asphalt Emulsion Mixture Properties Containing Cement and GGBS

Ammar A Shalan[1], Nibras A AL-Sahaf [2] and Abbas Al-Hdabi [3]

[1] Higher Diploma in civil engineering, Department of civil engineering, Faculty of Engineering, University of Kufa, Kufa , Al-Najaf, Iraq. ammarabd545@gmail.com
[2] Lecturer, Department of civil engineering, Faculty of Engineering, University of Kufa, Al-Najaf, Iraq, nibrasa.alsahhaf@uokufa.edu.iq and nibras.alsahaf@gmail.com
[3] Assist. Professor, Department of civil engineering, Faculty of Engineering, University of Kufa, Al-Najaf, Iraq, abbas.alhadabi@uokufa.edu.iq and alhudabi73@yahoo.com

Abstract. Cold bitumen Emulsion Mixture (CBEM) is a mixing of aggregate with emulsion of asphalt that is blended at room temperature. It has many environmental and economic advantages. But, the lower performance and high water sensitivity of CBEMs at early stages needs more investigation. This is the principal purpose of this research, which aims to use CBEMs as a structural binder layer after improving the mechanical properties by adding Ground-Granulated Blast-furnace Slag (GGBS). In this research the mineral filler is only materials passing (sieve No.200) replaced with (GGBS) by percentages (2, 4, 6, and 8) by total weight of aggregate with 2% of ordinary Portland cement(OPC) to enhance the mechanical properties of CBEMs and then the optimum content of GGBS was chosen based on Indirect Tensile Strength test (ITS) results. The optimum content of (GGBS) was 6% which improved significantly ITS value to more than (149%) in comparison with the control mixtures in which the mineral filler is only materials passing sieve No.200.(at age 28days). As for the other laboratory tests that represented of Marshall Stability (M.S) test and index of retained strength (IRS) test, the experimental results were very encouraging, the value of M.S of mixtures with optimum content of GGBS increased about 220% and IRS was increased about 335% in comparison with the control mixtures. Accordingly, this search has exhibited much enhanced CBEMs with GGBS in addition to the economic and ecological advantages.

Key words: cold mixture, GGBS, indirect tensile strength, durability, filler, waste materials.

1. Introduction

There are several methods to achieve the required workability of bitumen at low temperatures, i.e. decreasing bitumen viscosity; namely by using: cut back bitumen produced from mixing the bitumen with flux oil, foamed bitumen produced from foaming process of the hot bitumen with cold water, and bitumen emulsion produced from emulsification of bitumen emulsion with water [1].
This research focuses on the application of cold mixtures produced using bitumen emulsion as binder. By using cold mix asphalt, energy savings can be achieved of 95% compared to Hot Mix Asphalt (HMA) manufacturing. The weaknesses of cold mix asphalt are also apparent. Since water needs to dissipate from bitumen emulsion in order for the bitumen to develop adhesion with the aggregates, the cold mix asphalt may need several weeks to achieve its full strength [2].

This may result in lower primary strength and high porosity when compared with the conventional hot mix asphalt [3], [4] and [5]. In addition, the moisture damage is possible and the durability are of concern because of the existence of water in the mixture [6]. As a result, for heavy dusty pavements cold mix asphalt has seldom been used as a structural layer [7].

The addition of ordinary Portland cement with a proportion of (1-2) % (by mass) to cold asphalt mixture significantly improves its early mechanical properties [8] and [9], and the fully-cured material may acquire mechanical properties comparable or even better than those of an equivalent HMA [3], [10] and [11].

There are many types of by-product or waste materials which used instead of cement because they are cheaper than cement or even without cost, available, and environmentally friendly; like ground-granulated blast-furnace slag (GGBS) which is a by-product material from iron production, represented by a fine powdery material that resembles to Portland cement in its appearance.

2. Aim of the Research

To study the effect of adding Ground-granulated blast-furnace slag (GGBS) on the characteristics of CBEMs for binder course with different ages of 3, 7, 14 and 28 days.

3. Objectives of Research

The main objectives of this research are:

To encourage the use of cold bituminous emulsion mixtures CBEMs as an alternative to the conventional "HMA" in the construction of highways roads and for many causes like the simplicity of production, economic characteristic and environmental friendliness.

To investigate ways of enhancing the performance of CBEMs in particular mixtures incorporating selected by-product materials.

To encourage the use of the by-product materials such as Ground-granulated blast-furnace slag (GGBS) to improve the characteristics of mixture instead of costly materials such as cement.

4. Materials

In this research, the materials provided are commonly used in preparation of hot asphalt mixtures especially in the middle and south regions of Iraq such as aggregate, cement, bitumen emulsion.

4.1 Aggregates

The aggregates are categorized according to particle size into fine and coarse aggregates; the coarse aggregate used in this work is crushed aggregate from Al-Nibaie quarry and the fine aggregate from Najaf city. Routine tests are performed on the aggregates to evaluate their physical properties; the results are summarized in Table 1. The selected gradation with the specification limits are exhibited in Table 2 in accordance with the "general specification for roads and bridges [12], section R9.

Table 1. Physical properties of coarse and fine aggregate and mineral filler.

Material	Property	Value
Coarse aggregate	Bulk specific gravity	2.78
	Apparent specific gravity	2.82
	Water absorption, %	0.61
Fine aggregate	Bulk specific gravity	2.69
	Apparent specific gravity	2.73
	Water absorption, %	1.60
Mineral filler	Particle specific gravity	2.71

Table 2. Selected binder course gradation [12].

% passing by mass of aggregate		
Sieve opening size, mm / Standard sieves (mm)	Specification limits	Mid-limit gradation specifications
25	100	100
19	90-100	95
12.5	76-90	83
9.5	56-80	68
4.75	35-65	50
2.36	23-49	36
0.3	5-19	12
0.075	3-9	6
Filler	3-9	6

4.2 Bitumen Emulsion

Cationic emulsified asphalt is used in this research in order to prepare all specimens of CBEMs. The properties of the used bitumen emulsion are shown in Table 3.

Table 3. Asphalt Emulsion Properties.

Asphalt emulsion (K3-60)	
Property	Value
Residue by distillation, %	60
Relative Density at 15 °C, g/cm^3	1.05
Appearance	Black to dark brown liquid
Residual bitumen penetration, 1/10 mm	100

4.3 Mineral Filler

In this research, three types of fillers are used which are, material passing sieve No.200 (.075 mm), ordinary Portland cement (OPC), and Ground-granulated blast-furnace slag (GGBS).

4.3.1 Ordinary Portland Cement (OPC)

Ordinary Portland cement is supplied from AL-Kufa factory; in this research this filler (OPC) is used by various proportion i.e. 0, 2, 4 and 6% by total weight of aggregate. Chemical and Physical properties of (OPC) are shown in Table 4.

Table 4. OPC Physical and Chemical Properties.

Physical Properties	
Passing sieve No.200	95%
Density (gm./cm^3)	3.12
Specific surface area (m^2/kg)	418
Chemical testing (XRF)	
CaO	60.845%
SiO_2	24.564%
Al_2O_3	2.135%
MgO	1.625%
Na_2O	1.583%
Fe_2O_3	1.131%
K_2O	0.694%

4.3.2 Ground-granulated blast-furnace slag (GGBS)

GGBS represents a by-product material and it is used by various percentages from total weight of aggregate as partial replacement of OPC. Chemical properties of GGBS are shown in Table 5.

Table 5. Chemical Properties of GGBS.

Oxides %	GGBS
SiO_2	30.7
Al_2O_3	13.3
Fe_2O_3	0.35
CaO	42.4
MgO	6.89
Others	6.32
Specific surface area (m^2/kg)	681

4.4 CBEMs' Design Procedure

CBEMs specimens were designated depending on asphalt institute design procedure Ms-14 [13] which involves the following steps:

- Determination of Aggregate Gradation

All specimens used in this research were prepared for all tests using the total mass of aggregate =1000 gm. (for one specimen) depending on the selected aggregate gradation that was shown in Table 2 (Selected Binder Course Gradation).

- Definition of Initial Residual Bitumen Content (IRBC) and Initial Emulsion content (IEC)

The main step in the design procedure of the mixture was to calculate the estimated Initial Residual Bitumen Content (IRBC), selected as P, utilizing the Asphalt Institute empirical formula as shown below [13].

$$P = (0.05\ A + 0.1\ B + 0.5\ C) \times (0.7) \quad (1)$$

Where:

P = the percentage of Initial Residual Bitumen Content (IRBC) by mass of aggregate.
A = the percentages of coarse aggregate (retained on sieve 4.75 mm).
B = the percentages of fine aggregate (passing sieve 4.75 mm and retained on sieve 0.075 mm).
C = the percentages of filler (passing sieve 0.075 mm).

The values of (A, B) and (C) were determined based on the selected aggregate gradation that was shown in Table 2, values of (A, B) and (C) were found to be (50%, 44%, and 6%) respectively.

In an emulsion, the bitumen is suspended as tiny droplets in water. The value of IEC (Initial Emulsion Content) can be determined by dividing (P) by the percentage of the residual bitumen content in the emulsion, the content of residual bitumen in the emulsion that may be obtained by heating emulsion until whole water content evaporation, then calculation its percentage from total emulsion by using the formula as shown below:

$$IEC = \left(\frac{P}{X}\right)\ \% \quad (2)$$

Where:

IEC = Initial Emulsion Content by mass of aggregate.
X= the residual bitumen content of the emulsion

According to the selected gradation and after application of the equation (3-1), the Initial Residual Bitumen Content was 6.93% by mass of aggregate. The base bitumen content in the used emulsion was 60%, thus, the IEC is (6.93/0.6)= 11.55% by mass of aggregate.

- Determination of Pre-mixing water content

The coating degree of the aggregate and bitumen emulsion is primarily controlled by the pre-mixing water content, particularly when the aggregate gradation contains a high percentage of filler. Thanaya [16] reported that the best bitumen coating on aggregate can be achieved when the mixture is not too stiff or too sloppy. Various pre-mixing water contents were incorporated, i.e. (3–6 %) by mass of aggregate with Initial Emulsion Content (IEC), calculated from equations (1) and (2) (IEC=11.55 % by mass of aggregate), to mention the lowest content of pre-mixing water with sufficient coating. According to the visual inspection, 3% pre-mixing water (30 gram) by mass of aggregate was chosen in preparation of CBEMs.

4.5 Preparation of Cold Bituminous Emulsion Mixtures (CBEMs)
Specimens of CBEMs were mixed by using the mixing machine shown in Figure (1). The aggregate and filler material with pre-wetting water content (3 % by mass of aggregate) were added and mixed for one minute at low speed. Then bitumen emulsion (11.55 % by mass of aggregate) was gradually added during the next 1 minute of mixing at the same speed, after that the mixing process was continued for one minute at high speed (total mixing time is 3 minutes). Then the mixture was placed in the mold with hand blows, 10 blows in the center and 15 blows in the edges. Also, standard Marshall Hammer (impact compactor) shown in Figure (2) was used as the general compacting process with 75 blows to each face of the specimens.

Figure 1. Mixing Machine.

Figure 2. Marshall Hammer Compactor.

In this research, there are three specimens of cold mixture for each percentage that were designated and produced according to gradation for binder coarse that explained in Table 2 with a various percentages of filler by mass of aggregate (1000 gram), these mixture designation are shown in Table 6.

5. Testing

5.1 Marshall Stability (MS) Test
The test of stability for cold bituminous emulsion mixtures (CBEMs) utilises the same apparatus of Marshall Stability that used for testing hot mix asphalt (HMA). Generally, after preparation and compaction the sample had been modified so that the samples are tested for the curing (24 hour placed in the mold at lab temperature 30 °C) Then the samples are placed at lab temperature (30 °C) for (7days), after that tested in Marshall Apparatus shown in Figure 3.

IOP Conf. Series: Materials Science and Engineering **584** (2019) 012014 doi:10.1088/1757-899X/584/1/012014

Table 6. Mixture Designation, percentages of mineral filler in the mixture with selected test.

Mix number	Percentage of mineral filler % by total aggregate mass			Selected Test
	Materials passing sieve 0.075 mm from aggregate	OPC	GGBS	
1	6	0	0	Indirect Tensile Strength
2	4	2	0	Indirect Tensile Strength
3	2	4	0	Indirect Tensile Strength
4	0	6	0	Indirect Tensile Strength
5	2	2	2	Indirect Tensile Strength
6	0	2	4	Indirect Tensile Strength
7	0	2	6	Indirect Tensile Strength
8	0	2	8	Indirect Tensile Strength
9	6	0	0	Marshall stability
10	6	0	0	Index of Retained Strength
11	4	2	0	Marshall stability
12	4	2	0	Index of Retained Strength
13	0	2	6	Marshall stability
14	0	2	6	Index of Retained Strength

Figure 3. Marshall Stability Apparatus.

5.2 Indirect Tensile Strength (ITS) Test

Indirect Tensile Strength (ITS) test was conducted according to the method that designated by ASTM D 4123 [14]. The experimental procedure used to determine tensile, or splitting, strength of a cylindrical specimen is based on loading it diametrically in compression with a constant rate of 50.8 mm/min (2

IOP Conf. Series: Materials Science and Engineering **584** (2019) 012014 doi:10.1088/1757-899X/584/1/012014

in./min) to generate a tension zone along the specimen's loaded diameter. The expression for the maximum tensile strength generated can be stated as :

$$\sigma_t = \frac{2 \times P_{max}}{\pi H D} \quad (3)$$

Where:
σt = the ITS in kPa
P max = maximum applied load in KN
H= the height of specimen in m
D = the diameter of specimen in m
The test of indirect tensile strength (ITS) for CBEM specimen is conducted after sample compaction, and tested for curing (24 hour placed in the mold (at laboratory temperature 30 ºC)). After that placed at lab temperature (30 ºC) for (3, 7, 14 and 28) days according to the selected test age before implementing the (ITS) test by Marshall apparatus as shown in Figure (4).

Figure 4. Testing of Indirect Tensile Strength (ITS).

5.3 Index of Retained Strength (IRS) Test
Index of Retained Strength (IRS) is used to evaluate the moisture damage of the control and modified CBEMs. This test was conducted according to ASTM D 1075 which is called as "Effect of Water on Cohesion of Compacted Bituminous Mixtures". In this study, indirect tensile strength (ITS) had been used to determine the index of retained strength (IRS) for the control and modified asphalt mixtures.
In this test, two sets of samples were prepared and separated. The first set, dry samples, that after the preparation and compaction placed in the mold for 24 hour (at lab temperature 30 ºC), after extracting the samples from the mold, they were placed in the lab temperature (30 ºC) for 7 days before implementing indirect tensile strength (ITS). The second set, wet samples, with the same conditions of the first set, except after extracting the samples from the mold they were placed in the lab temperature (30 ºC) for 4 days and then placed in the water bath, at 40 ºC for three days, then the samples were extracted from water bath and placed at lab temperature (30 ºC) implementing indirect tensile strength (ITS). Figure 5 shows the water bath used for conditioning.

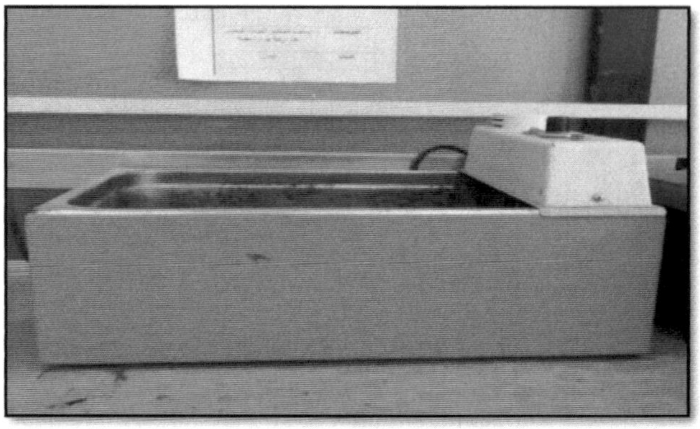

Figure 5. Water bath.

The value of index of retained strength (IRS) can be calculated by applying equation :(ASTM D 1075)

$$IRS = \frac{S_2}{S_1} \times 100\% \quad (4)$$

Where:
IRS = Index of Retained Strength, %
S_1= indirect tensile strength (ITS) of the dry specimens, kps
S_2= indirect tensile strength (ITS) of the wet specimens, kps

6. Results and Discussion

This paragraph presents the obtained results showing the extent of the impact of Ordinary Portland Cement (OPC) and GGBS on improving cold bituminous emulsion mixture (CBEMs).

6.1 Influence of OPC Addition on Indirect tensile strength (ITS) Results

Indirect Tensile Strength (ITS) is an important test that should be conducting on cold bituminous emulsion mixture (CBEMs) to indicate the tensile strength of a cylindrical specimen. Figure (6) shows the Influence of Ordinary Portland cement (OPC) addition on ITS. Different percentages where added as a replacement to the mineral filler which were 2, 4 and 6% OPC by total weight of aggregate.

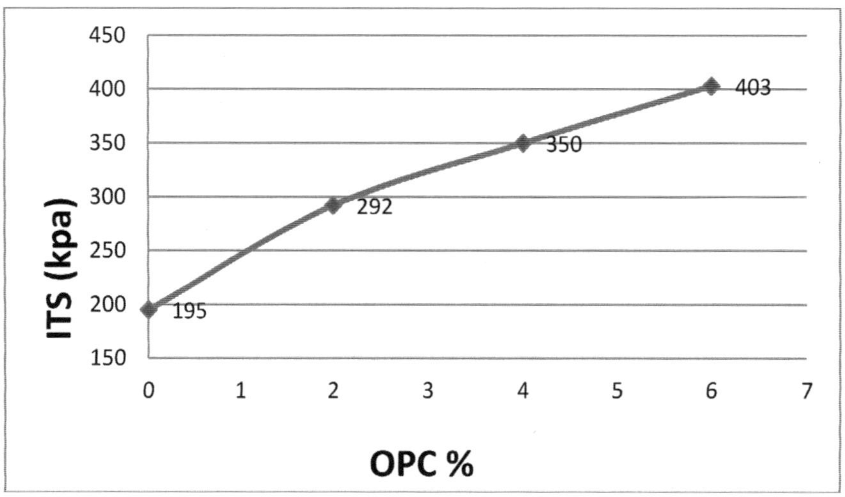

Figure 6. Influence of OPC Addition on ITS results.

6.2 Determination of Optimum GGBS content

Four percentages of GGBS were added as a filler to the cold bitumen emulsion mixtures (CBEMs) containing 2% OPC which were 2, 4, 6 and 8% by total weight of aggregate. The specimens were prepared and tested to indicate Indirect tensile strength values. ITS was adopted to indicate the optimum content of GGBS. The results are shown in Figure (7).

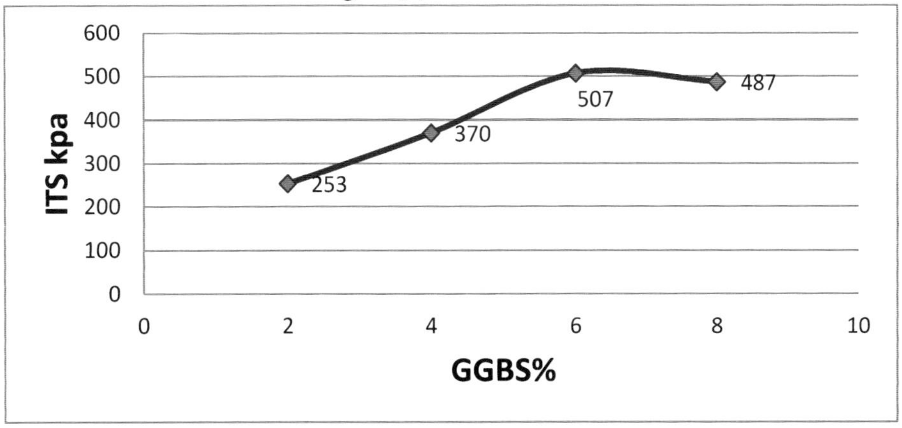

Figure 7. Indirect tensile strength for CBEMs with GGBS.

Figure 7 shows ITS values for cold mixtures with different percentage of GGBS. It is clearly shown that Indirect tensile strength of mixture increased significantly with increasing of GGBS percentage. This increment was sharply until 6% of GGBS percentage then it has been decreased slightly. The percentage of increment for the mixtures with 6% GGBS was more than 160% in comparison with the control mixtures (which used aggregate dust as filler) and almost 73% in comparison with the cold mixtures containing 2% OPC. It's worthy to say that these mixtures i.e. containing 2%OPC+6%GGBS have Indirect tensile strength more than those mixtures that have 6% OPC which is a very promising finds by means of mechanical, economical and environmental points of view. So, in accordance to these results, the optimum content of GGBS can be adopted to be 6% by total weight of aggregate.

6.3 Effect of GGBS on Indirect Tensile Strength (ITS)

Four mix used to study the effect of addition of GGBS on Indirect Tensile Strength, after different periods of time i.e. 3, 7, 14 and 28 days. The results are shown in Figure 8.

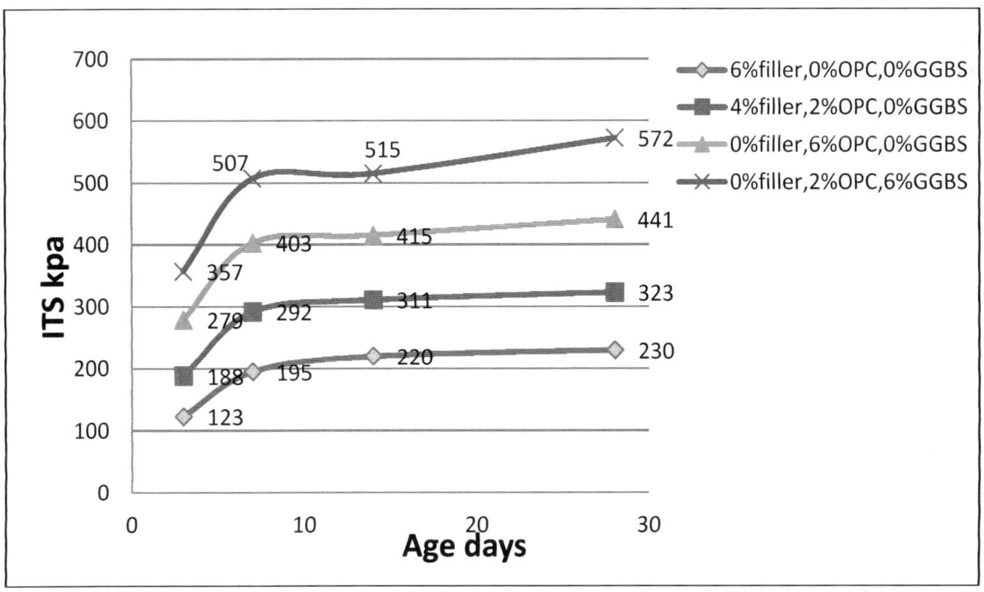

Figure 8. Effect of OPC % and GGBS % on Indirect Tensile Strength (ITS) of CBEM.

Figure 8 shows the relationship between mix type and ITS, whereas ITS equal to 195 kPa for control mixture with (6 % materials passing sieve No.200) for age (7days) increased about 50%, 107% for cold mixtures containing (2% OPC and 4% materials passing sieve No.200) and for mixtures with 6% OPC respectively. Also ITS increased about 160% for cold mixtures containing 2%OPC and 6% GGBS (Optimum GGBS content).

Figure 8 also shows that the amount of increase in ITS (kPa) is rapid in the early ages of the mixing (3 and 7) days and then continues to increase slightly, due to the evaporation of water found in the mix in early ages and gaining strength.

6.4 Influence of GGBS Addition on Marshall Stability (MS)

Marshall Stability is an important property for the performance of asphalt mixture in the binder course design. It shows the ability to resist shoving, rutting and give layer the adequate stiffness under traffic. The results of the cold mixtures with different percentages of OPC and GGBS are shown in Figure (9).

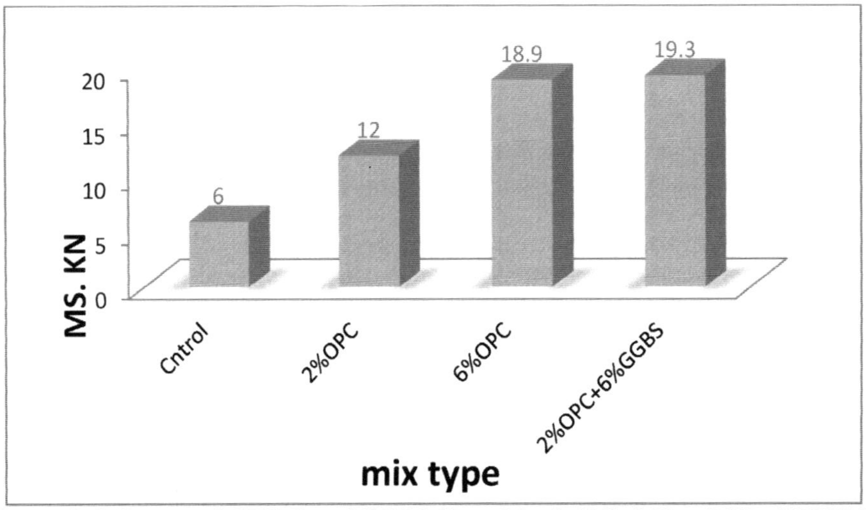

Figure 9. Influence of OPC % and GGBS % on Marshall Stability (MS) of CBEMs.

Figure 9, shows Marshall Stability (MS) values for cold mixtures with percentage of GGBS. It is clearly shown that Marshall stability of mixture increased significantly with 6% percentage of GGBS. The percentage of increment for the mixtures with 6% GGBS was more than 220% in comparison with the control mixtures (which used aggregate dust as filler) and almost 52% in comparison with the cold mixtures containing 2% OPC. It's worthy to say that these mixtures i.e. containing 2%OPC+6%GGBS have Marshall Stability more than those mixtures that have approximately 6% OPC which is a very promising finds by means of economic and environment.

6.5 Effect of GGBS on Index of Retained Strength (IRS)

Index of Retained Strength (IRS), also called water sensitivity is adopted to assess the moisture damage of the control and modified CBEMs. Figure (10) shows ITS for these mixtures before and after exposing to the water.

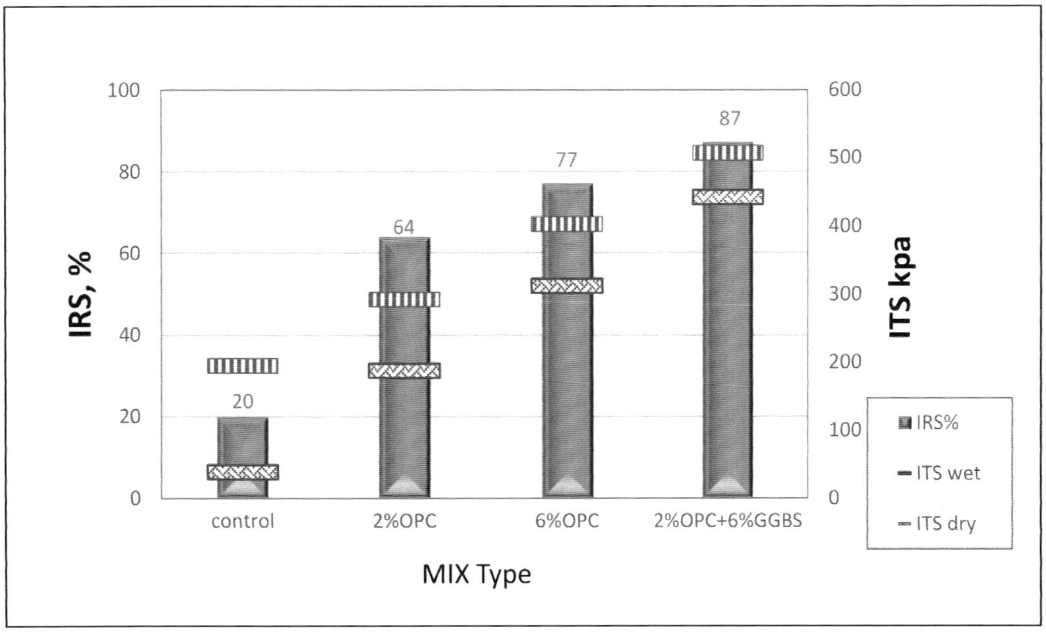

Figure 10. Effect of OPC % and GGBS % on Index of Retained Strength (IRS) of CBEMs.

Apparently in Figure (10) the IRS equals to 20% for control mix with (6% material passing sieve No.200) which is very low. IRS increased to reach (64, 77 and 87) % for mixtures with (2%, 6%) OPC and 2%OPC+6%GGBS, respectively. This finding can be attributed to that another binder has been developed which is from the hydration process between the cementitious material and the trapped water.

7. Conclusions

This study was involved in the activation of (OPC) with ground-granulated blast-furnace slag (GGBS) which can be included to produce a new binder course cold asphalt mixtures. This study led to the following conclusions:

1. Ordinary Portland cement (OPC) significantly increases Indirect tensile strength (ITS) of mixture, whereas when OPC was added by 2, 4, and 6% by total weight of aggregate, ITS (at 7 days) increased by approximately 50, 80 and 107% respectively in comparison with control mixture.
2. In accordance to the ITS test results at different ages (3, 7, 14 and 28) days, it was found that most of the strength gained for mixing was at early ages i.e. 3 and 7 days. Then, the mixtures continued to gain strength but at a lower rate. Therefore, the strength gained during the first 7 days equals 90% of the strength at the age of 28 days.
3. GGBS can be utilized as a filler instead of (OPC) to improve the mechanical properties of the mixture. When, GGBS was added by 2, 4 and 6% by total weight of aggregate, the performance of mixture improved significantly.
4. The optimum content of GGBS can be reported as 6% that corresponds high value of Indirect tensile strength (ITS) after this value there was a very slight decreased.
5. Optimum content of ground-granulated blast-furnace slag (GGBS) increases Indirect tensile strength (ITS) about 160% in comparison with control mixture and this value corresponds to those mixtures with almost 6% OPC at 7days. So, using GGBS has a very economic and environmental impact.
6. Marshall Stability (MS) for cold mixtures prepared with optimum content of ground-granulated blast-furnace slag (GGBS) increased about 220% in comparison with control mixture.
7. By means of water damage, IRS for cold mixtures containing 6% GGBS improved by 335% in comparison with the control mixtures.

References

[1] Al-Hdabi A, Al Nageim H, Ruddock F and Seton L 2014 *Laboratory Studies to Investigate the Properties of Novel Cold Rolled Asphalt Containing Cement and Waste Bottom Ash* Journal of Road Materials and Pavement Design, ISSN 1468-0629 (Print), 2164-7402 (Online), Volume 15, Issue 1 pp. 78-89.

[2] Jenkins KJ 2000 *Mix design considerations for cold and half-warm bituminous mixes with emphasis on foamed bitumen*. Ph.D. Thesis, University of Stellenbosch, South Africa.

[3] Al-Busaltan S, Al Nageim H, Atherton W and Sharples G 2012 *Green Bituminous Asphalt relevant for highway and airfield pavement* journal of construction and building materials 31.

[4] Needham D 1996 *Developments in bitumen emulsion mixtures for roads* PhD Thesis, University of Nottingham.

[5] Al-Hdabi A, Al Nageim H, Ruddock F and Seton L 2013 *Development of Sustainable Cold Rolled Surface Course Asphalt Mixtures Using Waste Fly Ash and Silica Fume* ASCE, Journal of Materials in Civil Engineering.

[6] Ibrahim H and Thom 1997 *The Effect of Emulsion Aggregate Mixture Stiffness on Both Moisture and Pavement Design* Proceeding of the second European symposium on Performance and Durability of Bituminous Materials University of Leeds.

[7] Thanaya I 2003 *Improving the performance of Cold Bituminous Emulsion Mixtures* (CAEMs) incorporating waste materials. PhD thesis. UK: University of Leeds.

[8] James A D, Needham D and Brown P S F 1996 *The benefits of using Ordinary Portland Cement in solvent free dense graded bituminous emulsion mixtures*, Proceeding International Symposium on Asphalt Technology.

[9] Thanaya I E A, Zoorob S E and Forth J P 2009 *A laboratory study on cold mix cold-lay emulsion mixtures* Proc. ICE - Transport 1: 47– 55.m https://doi.org/10.1680/tran.2009.162.1.47

[10] Brown S F, and D Needham 2000 *A Study of Cement Modified Bitumen Emulsion Mixtures* Proceeding Annual Meeting of the Association of Asphalt Paving Technologists.

[11] Al-Hdabi A, Al Nageim H, and Seton L August 2014 *Superior Cold Rolled Asphalt Mixtures Using Supplementary Cementations Materials* Journal of Construction and Building Materials, ISSN: 0950-0618, Volume 64, pp. 95-102.

[12] State Commission of Roads and Bridges (SCRB/R9) 2003, General Specification for Roads and Bridges, Republic of Iraq, Ministry of Housing and Construction, Department of Planning and Studies, Baghdad.

[13] Asphalt Institute 1989 *Asphalt cold-mix manual*, MS-14", The Asphalt Institute.

[14] ASTM Standards, Roads and paving materials 2004 in: *Annual Book of the American Society for Testing and Materials Standards*, Section 4, vols. 03–04. Washington, USA. ASTM D 4123.

Structural Behavior of Eccentrically Loaded R.C. Columns under Fire Temperature Loading

A H Kadhim*[1], M M Rasheed[2] and H K Jarallah[3]

[1]Ph.D., student, Department of Civil Engineering, College of Engineering, Mustansiriyah University, Iraq.
[2]Asst Prof., Department of Civil Engineering, College of Engineering, Mustansiriyah University, Iraq.
[3]Asst Prof., Department of Civil Engineering, College of Engineering, Mustansiriyah University, Iraq.
*Corresponding author: chelsea.alaa@icloud.com

Abstract: The main goal of this research paper is to investigate the response of the RC rectangular columns under loading simultaneously exposed to fire by using experimental study. The number of test columns were seventeenth columns. The dimension for these columns was 1600mm for length and 150mmx150mm for the cross-section. The columns were tested under axial load with two different types of eccentricity 60 mm 100mm, while the third type of loading is tested as a beam. The eccentric compression load was applied by using top and bottom cap with a column bracket. The eccentric load was applied simultaneously with fire. The test was performed under a high temperature of (400°C, 600°C, and 900°C) on the side of a compression face. At each temperature burning, cooling by two techniques of cooling, and normal cooling (by open air) and fast cooling (by direct water). The experimental results show decreasing in ultimate load capacity with increasing of temperature burning, ultimate load, load-deflection curve, strain profile, neutral axis, moment-curvature, and ductility.

Keywords: fire, an eccentric load, load capacity, eccentricity, types of cooling, strain.

1. Introduction

Concrete column is considered to be an important structural element in reinforced concrete structure because it supports the structure and carry the loads to the foundation or supports, so each damage or failure occurs in the column might cause a partial or full failure of the structure by possibly chain action [1].

The modern buildings should be provided resistance to different types of loading especially when exposing to high temperature. Always, structural engineering does not take into account the effect of high-temperature loading in the design of the structural elements. The effect of fire on structural members depends on different factors such as the amount, kind and distribution of fire, loading, the duration of exposure to fire, the cooling method practice, ventilation, and compartment size. In order to keep the human life and avoid the serviceability damage and collapse in the structure, the high-temperature fire effects need to be considered in the strength of the various main elements, such as columns, beams, slabs, shear walls [2]. The effect of fire on structural members depends on different factors such as the amount, kind and distribution of fire, loading, the duration of exposure to fire, the cooling method practice, ventilation, and compartment size. The fire could bring a chemical reaction, essentially decomposition of $Ca(OH)_2$ will cause a decrease in concrete compressive strength, which follows the development of heat. Most huge fires start from small ones, and during their development, the rate at which heat is produced exceeds the rate at which it is dissipated. It is important the structure withstand for a given time under fire effect [3]. The primary on-site investigation technique is a visual inspection, which is used to classify the degree of damage for each structural concrete member. Visually apparent damage is induced by heating includes collapse, deflection, spalling, cracking, surface crazing, colour changes, and smoke damage. A visual survey of reinforced concrete structures is performed using a classification scheme from The Concrete Society, (1990) [4]. Concrete made with siliceous or limestone aggregate shows a change in color with temperature. As this change depends on the presence of certain compounds of iron, there is some difference in the response of different concretes, so that the maximum temperature the change in color is permanent. Hence, the maximum temperature during a fire can be estimated posterior. The color sequence is approximate as follows: pink or red between 300 and 600°C, then grey up to about 900°C, and buff above 900°C. Thus, the residual strength can be approximately judged: generally, concrete whose colour has changed beyond pink is suspect, and concrete past the grey stage is probably friable and porous [5]. The column is usually vertical member and used to support an axial load and it can also resist moment, her and torsion [6].

2. Research Significant

The main objective of this research paper is to investigate the structural response of the RC rectangular columns under load simultaneously exposed to fire by using experimental study. The current research proposes a prototype RC are subjected to during fire and eccentric loading. That it will give a good idea about the rehabilitation and repair of the damaged columns due to high-temperature fire.

3. Experimental investigation

The main idea of the present research paper is to conduct an experimental test of the RC columns under high temperature. Seventeen prototype RC column were loaded under uniaxial force and fire. The dimensions of the columns are given in Figure 1. These columns were cooled normal cooling by (open air) and quick cooling by (water). The experimental programme was divided into two groups depending on the eccentricity of the applied force Table 1.

3.1. Materials Used

The normal strength concrete (NSC) are used in this experimental programme. All use mix materials (cement, gravel, and send) are confirmed with the existing used standards [7] and [8]. The mix design is given in Table 2, it was prepared in order to prepare cylinder compressive strength of about 40MPa. The mix design workability was satisfied with the requirement of [9].the main steel reinforcement (Ø12 mm) and secondary steel reinforcement (Ø8 mm) are confirmed with conforming to ASTM A615 [10]. The minimum steel reinforcement ratio of 1% and minimum ties spacing of 150mm is confirmed with the requirement of ACI-318M-14 [8]. All column specimens were cured for 28 days by using moist canvas sheets, shows Figure 2. The mechanical properties test result of the concrete at the age of 60 days before and after fire exposure are given in Table (3).

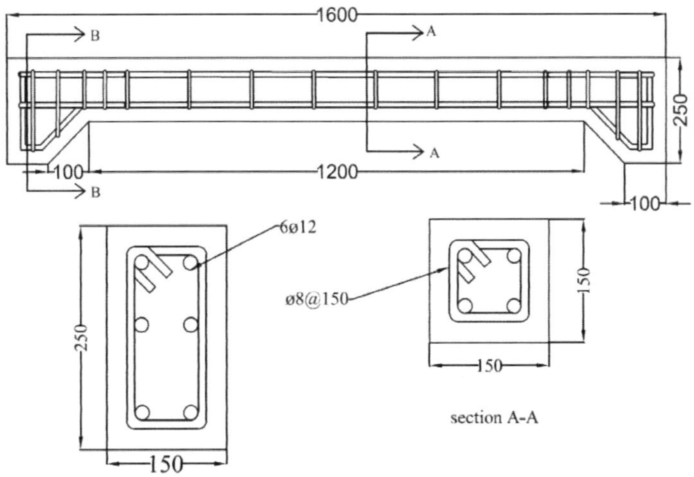

Figure 1. The tested column Dimensions.

Table 1. Summary of columns test specimens.

Group	Fire Temperature (°C)	Column ID	Main Reinforcement	Eccentricity Of Applied Load (mm)	Cooling Technique
		A	4Ø12	60	Without burring
A	400	A4	4Ø12	60	Without
		A4F	4Ø12	60	Quick
		A4N	4Ø12	60	Normal
	600	A6	4Ø12	60	Without
		A6F	4Ø12	60	Quick
		A6N	4Ø12	60	Normal
	900	A9	4Ø12	60	Without
		A9F	4Ø12	60	Quick
		A9N	4Ø12	60	Normal
B	400	B	4Ø12	100	Without
		B4F	4Ø12	100	Quick
		B4N	4Ø12	100	Normal
	600	B6F	4Ø12	100	Quick
		B6N	4Ø12	100	Normal
	900	B9F	4Ø12	100	Quick
		B9N	4Ø12	100	Normal

Table 2. Mix design for 1 m3 concrete.

Constituent Type	Mix Proportion
Cement (kg/m3)	450
Sand (kg/m3)	750
Gravel (kg/m3)	1050
Water (kg/m3)	189
W/c ratio	0.42
Slump(mm)	60

Table 3. Mechanical properties for used concrete at the age of 60 days.

Fire Temperature (°C)	25	400			600			900		
		W.O.C	N.C	F.C	W.O.C	N.C	F.C	W.O.C	N.C	F.C
Compressive Strength (MPa)	40	30	38.7	38.3	32	37	36	19.52	34	30
Splitting tensile strength (MPa)	3	2.55	2.6	2.58	1.89	2.33	2.1	1	1.84	1.34
Modulus of rupture (MPa)	3.31	2.58	2.62	2.57	2.2	2.33	2.23	1.2	1.84	1.25
Modulus of elasticity (GPa)	40.9	10.31	14.6	11.26	6.15	7.4	6.68	1.9	2.6	2.14

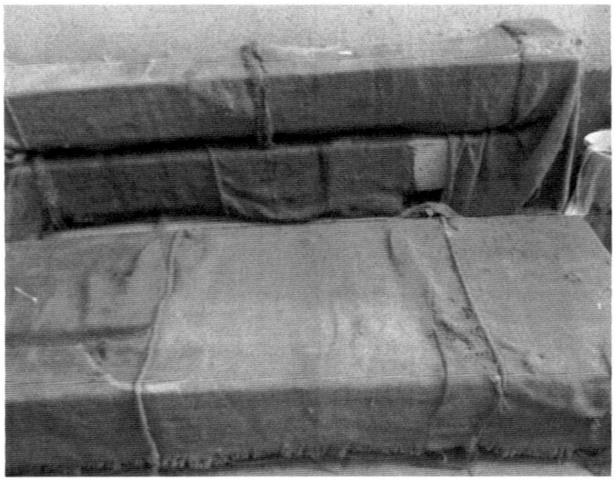

Figure 2. Casting and curing process of column specimens.

3.2. Fire test

The reinforced concrete columns were burned by testing directly subjecting a fire flame from a network of methane burners. The fire flame hits only one side of these columns. The burner network is consisting of a steel box with dimensions (800×150) mm (length × width) and contain three methane burners. While the control specimens were burned directly by fire flame from a net of methane burners in the same manner as shown in Figure 3. The fire attacked the compression side of the test RC columns as shown in Figure 4. The rate of 2C° was used to increase the temperature for fire loading in order to reach the target temperature [11]. The time duration for fire applied was 30 minutes, after that the fire flume is off and the column was left to cool down to the ambient temperature. The vertical force was applied with a rate of 1kN/s. Three dial gauge was fixed at tension face and used in order to measure the lateral deflection at two ends and mid-height of test column. The concrete strain and steel strain of the test column was measured at the tension face by using strain gauges.

3.3. Test setup and procedure

The testing machine with maximum range capacity of 3000 kN was used to apply the eccentric load on the columns (Figure 3). The age of RC tested column was 60 days. The lateral deflection was recorded at each 10kN until failure. The axial deformation of RC columns was recorded by using vertical dial gauges having a maximum needle length of 30 mm mounted at the bottom face of the columns. The cracks formation were recorded and marked at column faces. The target temperature fire flame at tension face of columns was recorded by using a digital thermometer and infrared rays thermometer. The vertical load was kept equal to 65% of the ultimate load before burning and applied fire flame. The loading was continued after burning until failure.

Figure 3. The burner network with a steel box.

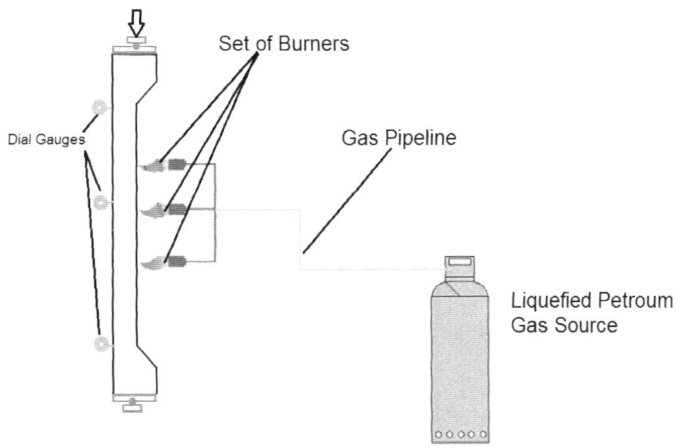

Figure 4. The furnace and other equipment.

4. Results and discussion

The following mean results are discussed for this research paper;

4.1 Effect of Burning on Load versus Deflection Results

The load-mid-height lateral deflection curve of the tested columns with the flame temperatures of 400°C, 600°C, and 900°C at the eccentricity of (60 and 100mm) are shown Figure 5, 6, 7 and 8. These columns were subjected to the vertical loads simultaneously with the fire flame. The vertically applied load was reached to 65% of the ultimate load before starting of burning. The normal cooling and fast cooling were used for burning tested columns. The load was continuously applied until failure. The significant effect was observed in the mid-height lateral deflection of columns of series A (e=60mm) and B (e=100mm) due to fire temperature. The load carrying capacity was decreased and lateral deflection was increased due to fire temperature. This can be attributed to the fact that heating causes a reduction in column stiffness, which is essentially due to the reduction in the modulus of elasticity of concrete and the reduction in the effective section due to cracking, which means that load-deflection curves for high temperatures compared with low temperatures. The load-deflection relation of the column specimens is almost linearly proportional for the two eccentricities (60 and 100mm) and for temperature exposure (400°C, 600°C and 900°) . The maximum deflection occurred at mid-height and decreased towards the support of the column specimens.

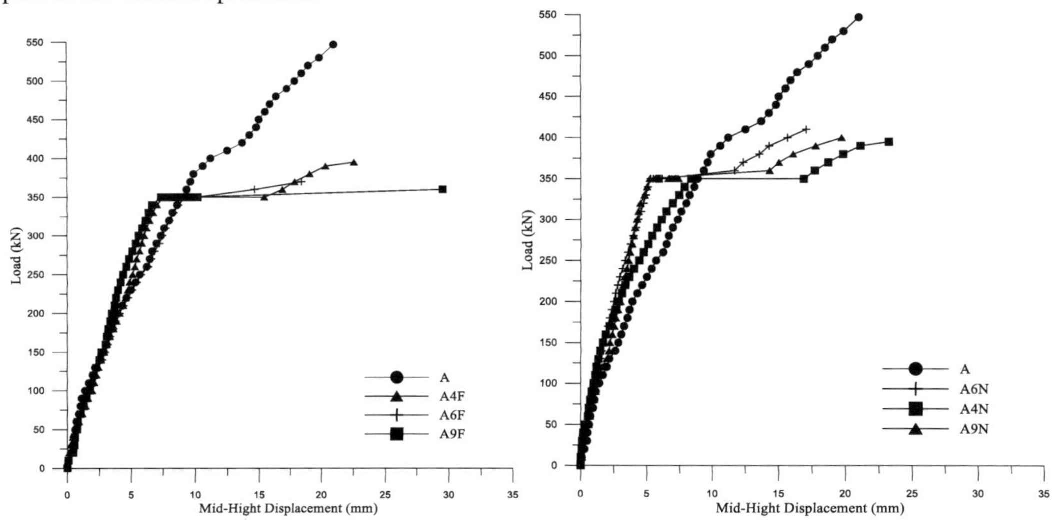

Figure 5. Load versus mid-height displacement curve of column specimen at eccentricity (e=60mm), (a-fast cooling and b-normal cooling).

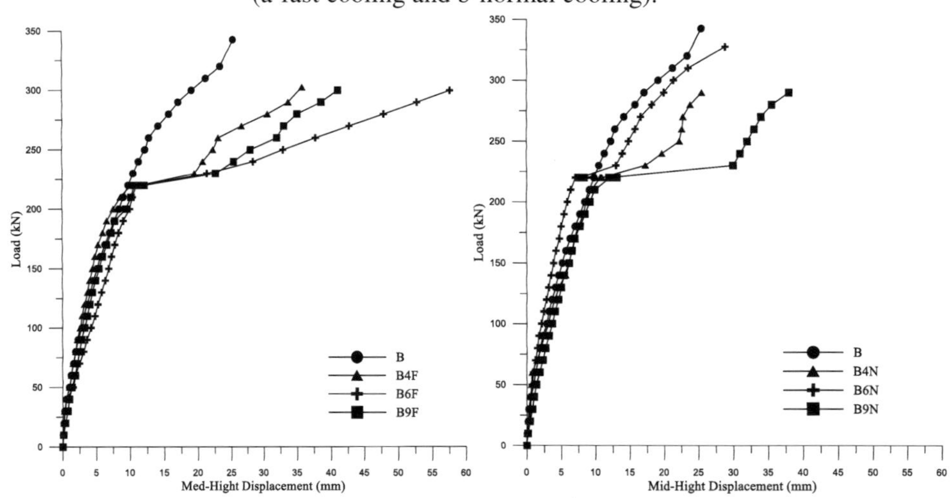

Figure 6. Load versus mid-height displacement curve of column specimen at eccentricity (e=100mm), (a-fast cooling and b-normal cooling).

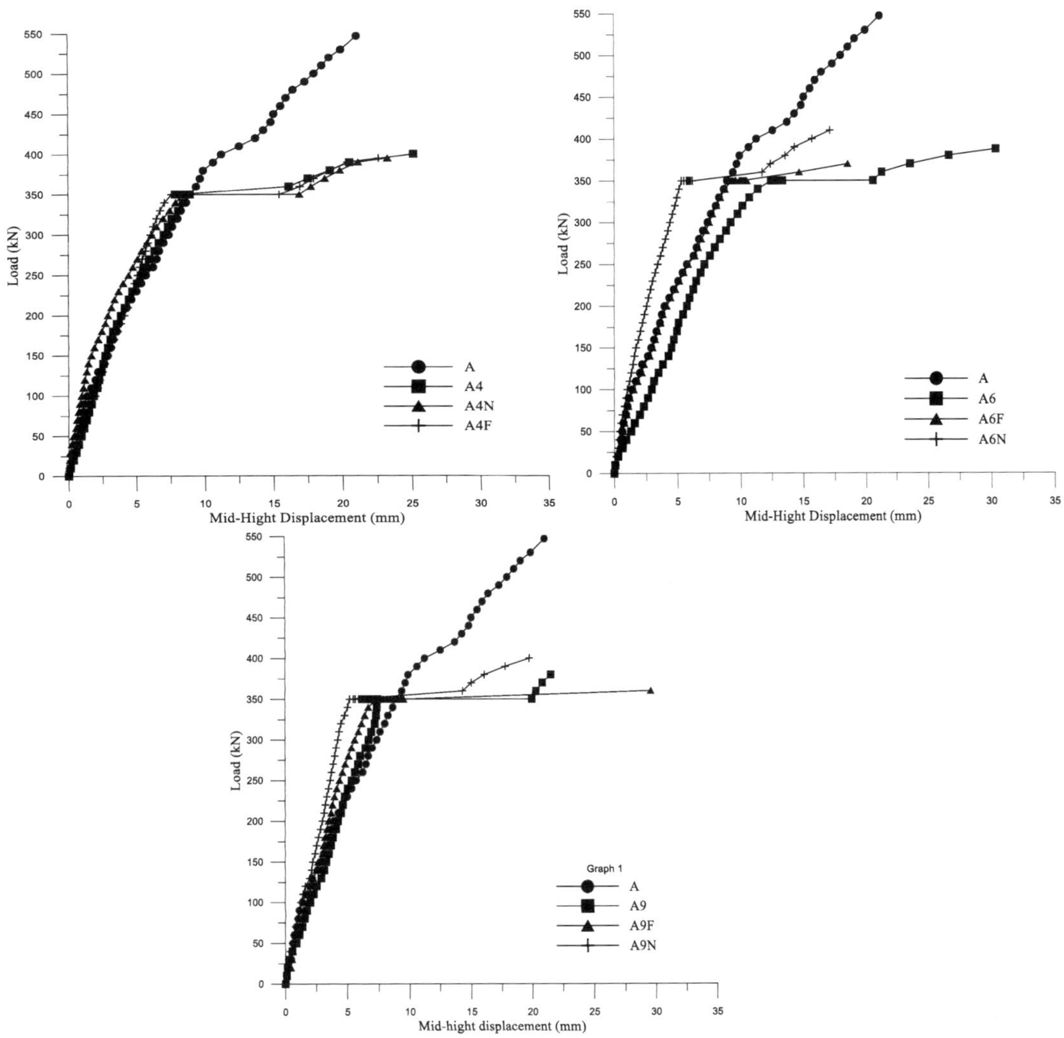

(c)

Figure 7. Load versus mid-height displacement curve of column specimen A with 60mm eccentricity (a- 400°C, b-600°C and c-900°C).

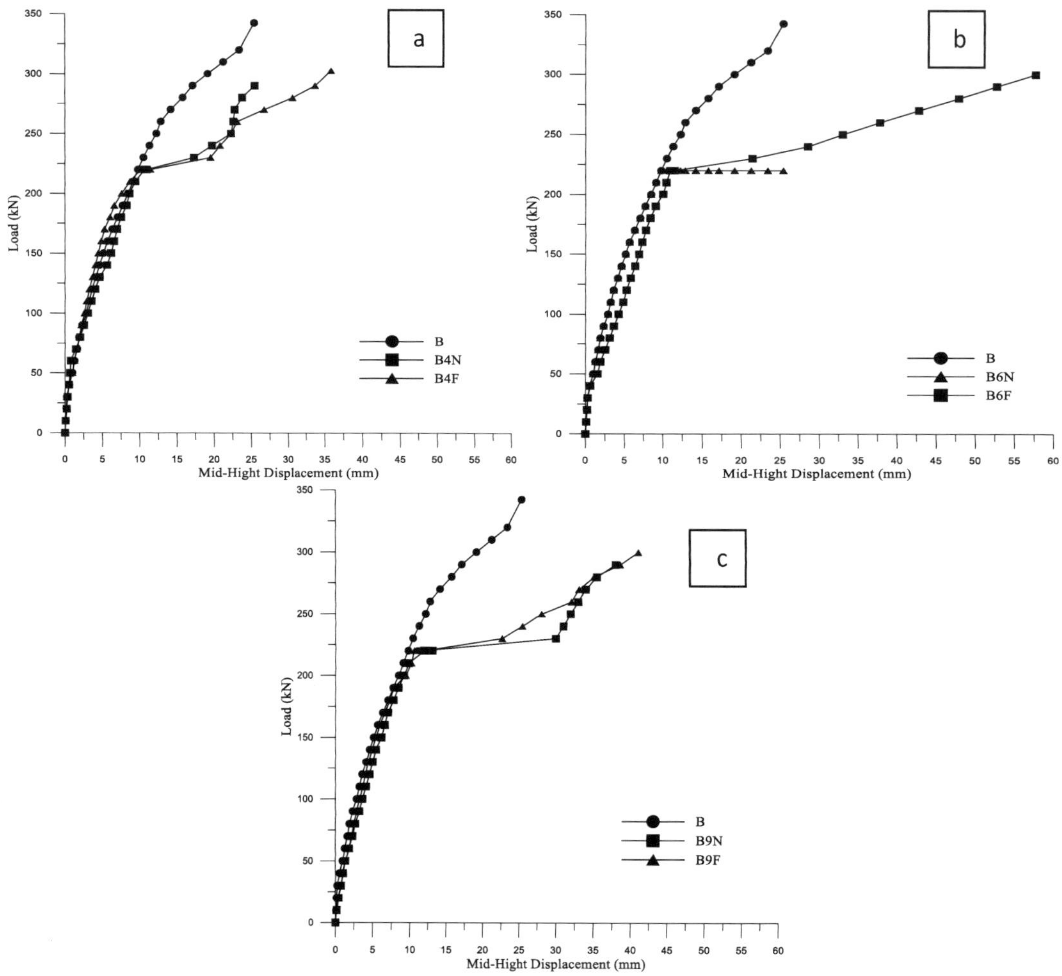

Figure 8. Load versus mid-height displacement curve of column specimen B with 100mm eccentricity (a- 400°C, b-600°C and c-900°C).

4.2. Strain Profiles

The variation of concrete strain along the depth for tested columns during different stages of loading at the constant moment region is shown in Figures (9 through 12). The strains at mid-height section of the columns in compression zone for concrete and tension zone for steel reinforcement are measured using the data logger. At the early stages of the testing process, when the column is free from cracks, the concrete resists the tensile stresses. However, by increasing the applied load, cracks appear and the concrete layer at the mid-height of the columns becomes out of concrete tensile stresses resistance and the reinforcing steel begins to resist these stresses alone. At the last stage of the test when the crack height propagates and becomes wider, the yield in the reinforcing steel becomes more eventual in the tension zone of the interaction diagram, as well as, the compression zone of the columns suffers from decreasing in its depth and high compressive stresses. That means, at the last stage of the test, the behavior of columns is inelastic and high strain values exceed the yield strain value in the reinforcing steel and concrete at the mid of the columns crushes. All the tested columns in the present study exhibit similar behavior in the property of the concrete strains, but with variations in the value of strains from one column to the other during the different stages of the applied load and at failure stage.

IOP Conf. Series: Materials Science and Engineering **584** (2019) 012015 doi:10.1088/1757-899X/584/1/012015

From figures (9 through 11) below, noted from strain profile for columns when the column is exposed to 400oC compressive and tensile strain of concrete is same than columns without exposed to fire in comparison at the same load, this behavior may be attributed to the compressive of concrete is not impact by this fire temperature.

From figures (9 through 11) below, noted from strain profile for columns when the column is exposed to 600oC compressive and tensile strain of concrete is large than columns without exposed to fire in comparison at the same load, this behavior may be attributed to the compressive strength of concrete is decrease when increased of fire temperature with fast cooling. Nevertheless, when the cooling is normal to change the failure from tension failure to compression failure.

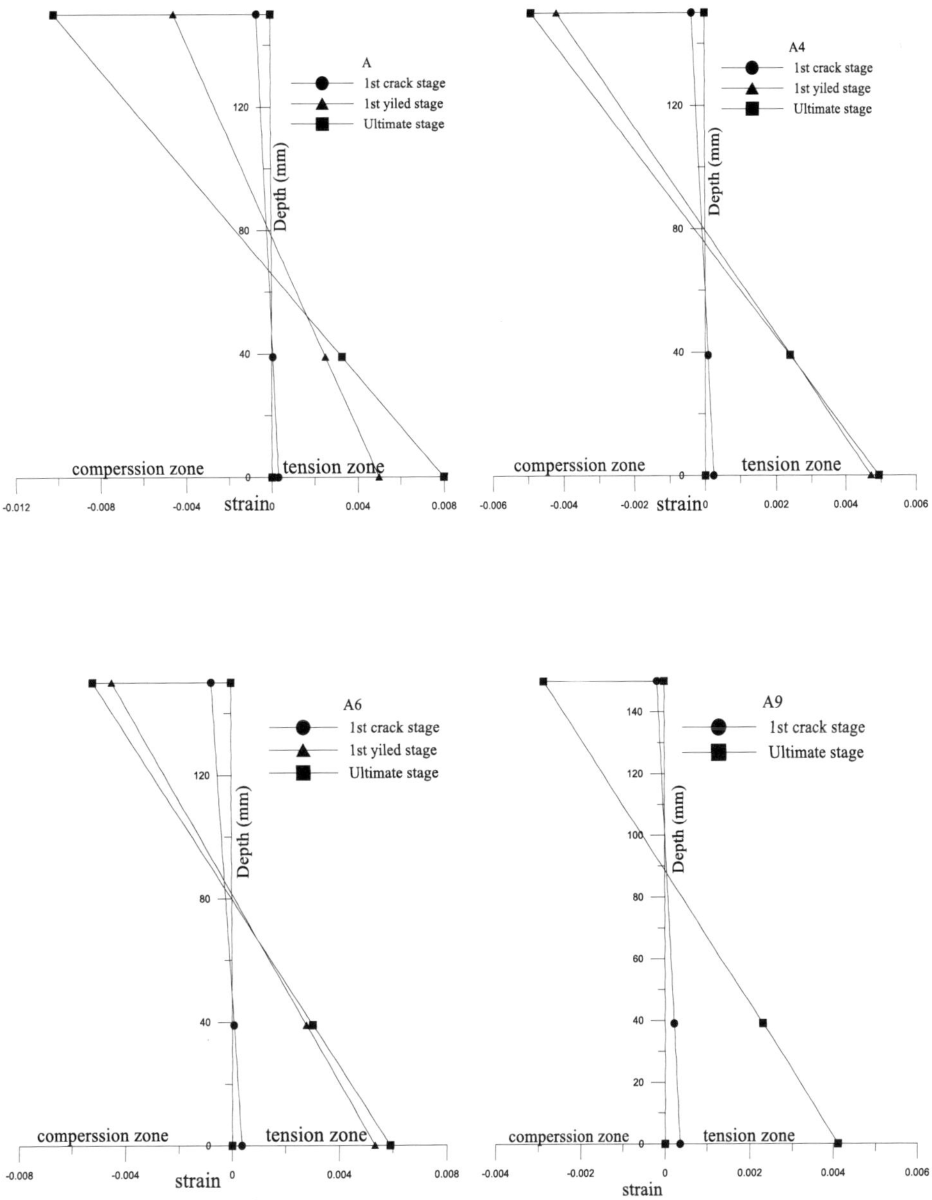

Figure 9. Strain Profile along Depth of Columns without cooling at (e=60mm).

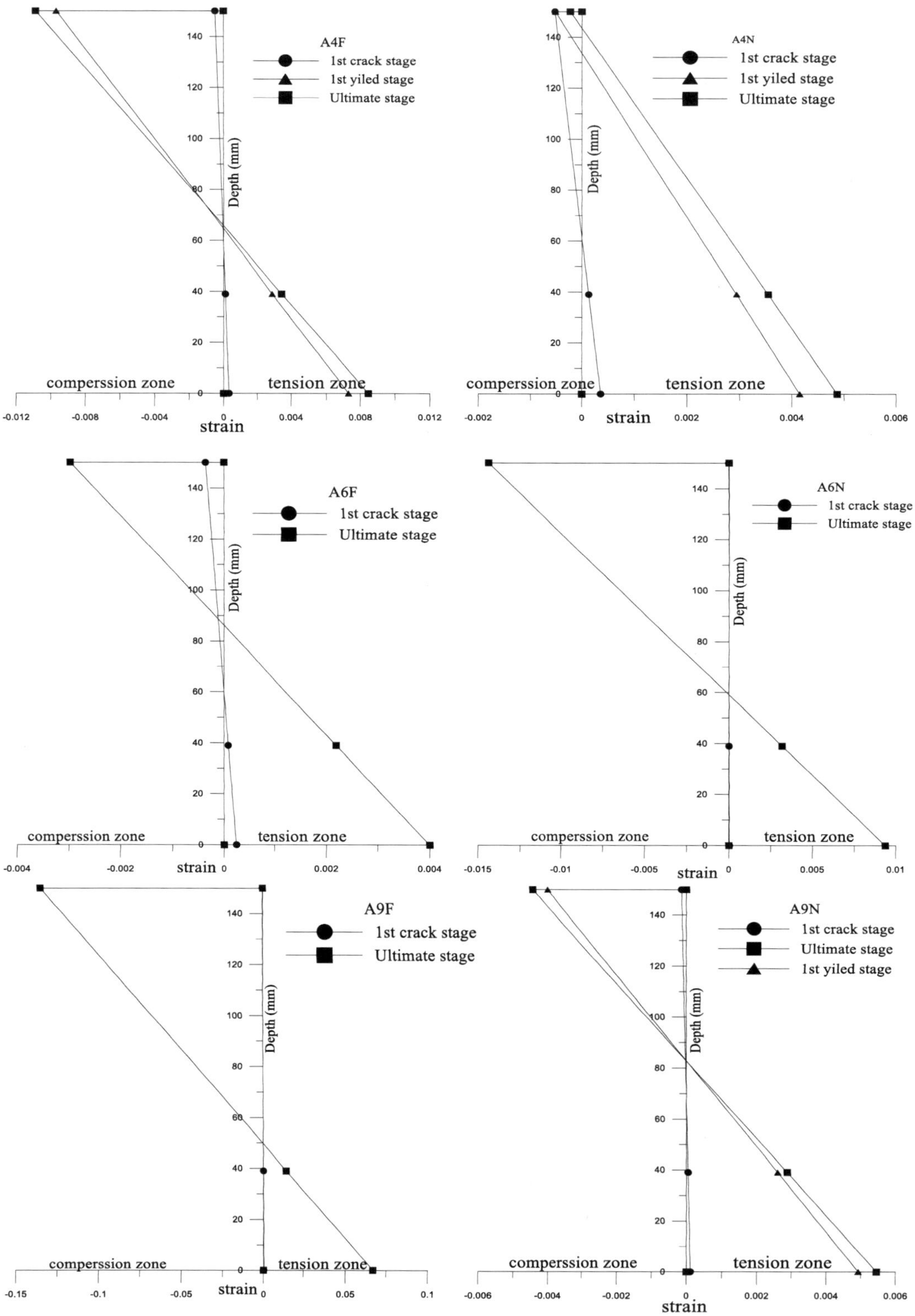

Figure 10. Strain Profile along Depth of Columns with normal and fast cooling at (e=60mm).

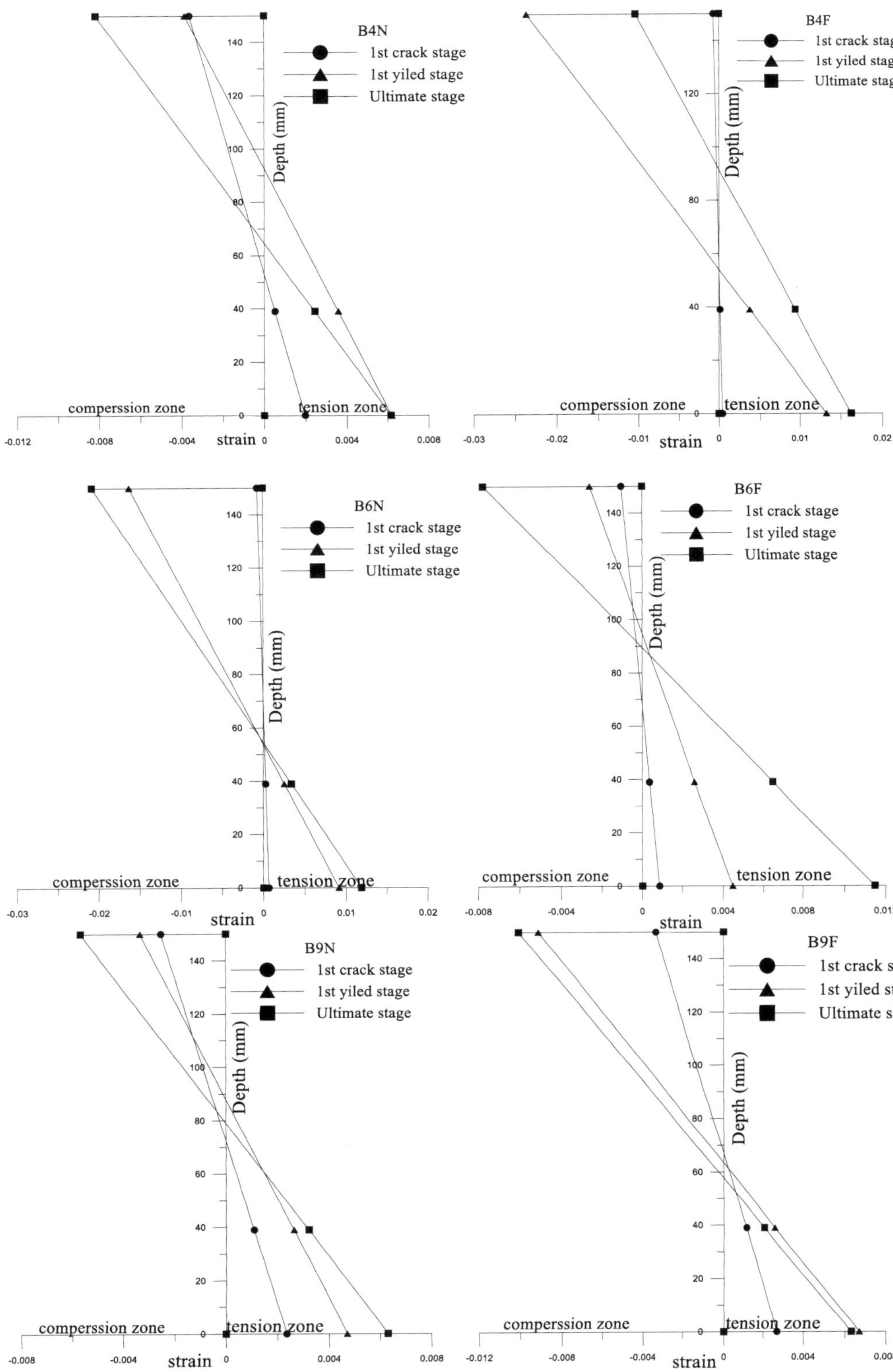

Figure 11. Strain Profile along Depth of Columns with normal and fast cooling at (e=100mm).

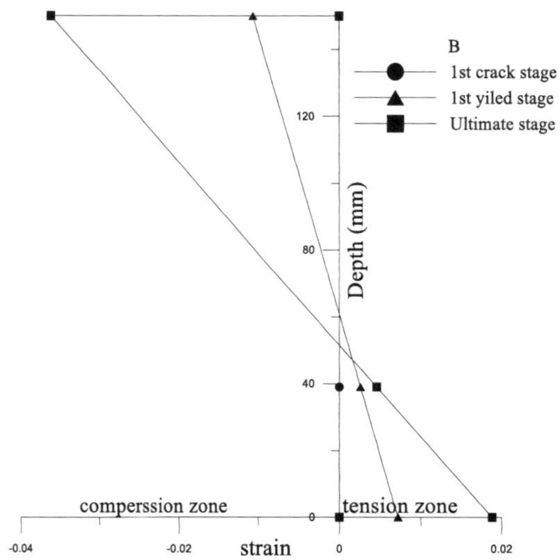

Figure 12. Strain Profile along Depth of Column without burning at (e=100mm).

4.3 Crack Patterns

The crack patterns at the failure stage of all specimen's column were shown in Figure 13. The numbers shown beside the cracks indicate the load when the crack reaches that position. The test results of load characteristics and deflection were given in Table 4.

Table 4. Test results of reference columns and columns exposed to fire flame for the two series A and B.

Specimen Identification	Temperature Level (°C)	First Crack Load (kN)	Ultimate Load (kN)	Percentage Residual Ultimate Load (%)	Maximum Deflection at Mid-height (mm)
Axial	20	50	625	100	
A	20	40	547.5	100	21
A4	400	48	400	73	25.48
A4F	400	50	392.5	71.7	22.58
A4N	400	53	395	72.15	23.22
A6	600	60	387.5	70.8	30.29
A6F	600	66	435		14.66
A6N	600	51.5	415		12.36
A9	900	100	457.5		
A9F	900	110	360	65.7	9.52
A9N	900				7.41
B	20	52.5	342.5		25.36
B4F	400	60	302.5	88.3	35.8
B4N	400	50	290	85	25.43
B6F	600	47.5	317.5		32.98
B6N	600	55	327.5		15.78
B9F	900	55	340		22.66
B9N	900	50	292.5		29.95

All columns are designed to fail in axial and uniaxial failure, which is recognized by the formation of cracks in the tensile stress zone, then yielding of steel bars and shifting the neutral axis toward the compression zone. The behavior of the tested columns under loading can be described as follows;
At early stages of loading, the tested columns are free of visible cracks and then the first crack appears at the mid-height in the tension zone. The load at which the first crack appears refers to the cracking load (Pcr). Gradually, several cracks initiate in the tension zone, with the increase in the loads, these cracks become wider. In the final stages of loading, the cracks develop and extend faster, some of which reach the compression zone until failure occurs at ultimate load capacity (Pu), crushing occurs in columns that cured with a high temperature between 600oC to 900oC, as shown in Figure 14. All cracks appeared before exposed to fire flame, but with higher temperatures and longer exposure time, the width and length of the cracks increased and extension along the columns. The uniaxial loading had effect by increasing the eccentricity so increase cracks in mid-height were observed. From this table, it can be seen that the values of ultimate load decrease when the column specimen is exposed to fire flame.

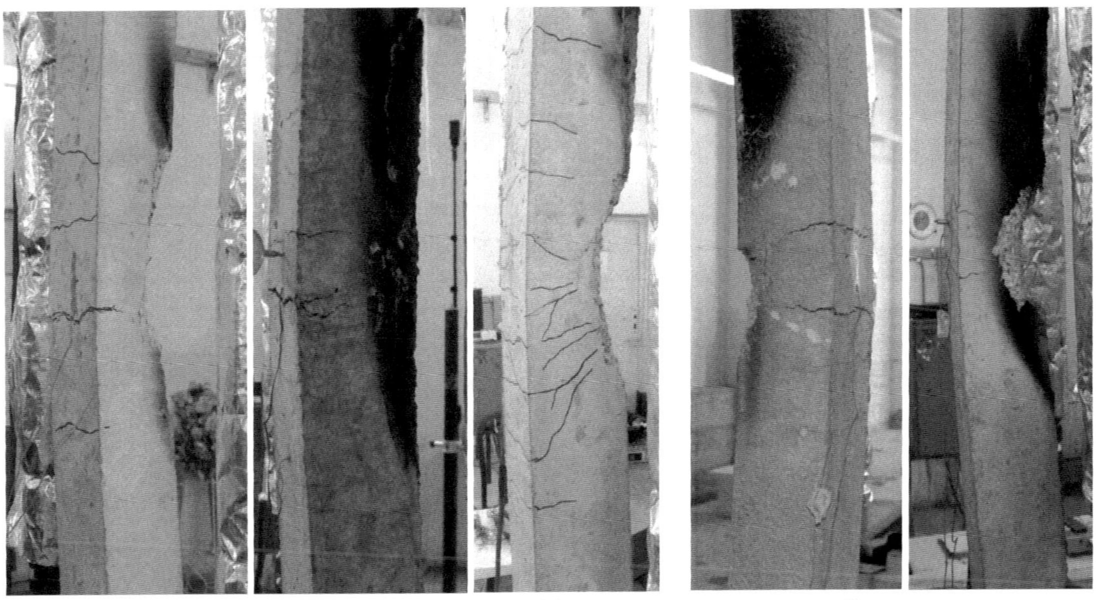

(a) (b) (c) (d)

Figure 13. Crack Patterns at Failure Stage (a-400°C, b-600°C, c-600°C and d-900°C).

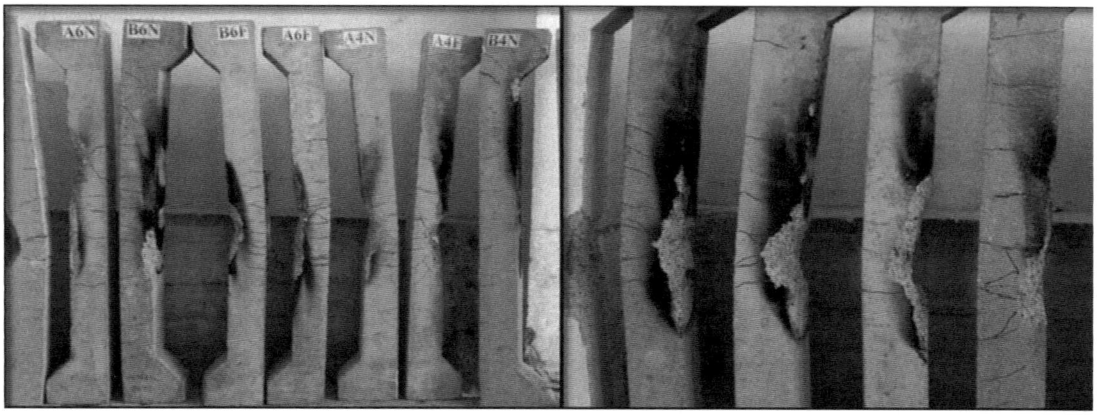

Figure 14. Columns crushing

5. Conclusions

The following conclusions can be drawn based on the aforementioned experimental investigation of the Fire Behavior of Eccentrically Loaded RC Columns:

- The compressive strength of concrete decreases with fire exposure by an amount depending on the exposure temperature when burning for a limited period (30 min.).
- The splitting tensile strength is more sensitive to fire flame exposure than the compressive strength.
- The modulus of elasticity of concrete is the most affected factor by fire flame rather than compressive strength.
- The residual ultimate load was decreased significantly due to burning by fire flame for group A and B.
- The modulus of rupture is more sensitive to fire flame exposure than the compressive strength.
- Mid- height displacement index is effective to assess the effect of fire exposure on the buckling of NSC columns in which the values of the displacement indices are highly affected after fire exposure.
- The neutral axis depth for columns after exposure to fire is lower than control columns specimens (neutral axis moves down towards the compressed face when exposed to fire).
- The compression zone and neutral axis were decreased by increase eccentricity. These results occur when columns before and after exposure to fire. The strain in the concrete increase at exposure to fire flame compared with specimens before exposure to fire flame. Where the rate of increase of strain when increasing the temperature of fire flame and when increasing the exposure time (compression, tensile).
- An increase in concrete strain with an increase in eccentricity, these results occur when columns are before and after exposure to fire.
- At a temperature equal to 400oC or less, the type of cooling does not differ in terms of its effect on concrete, the strain of concrete with normal cooling similar to strain in concrete with fast cooling.
- The experimental results show that curvature through exposure to fire flame and after cooling by two methods are more than curvature before exposure to fire flame because of the increase of strain in compression concrete zone.

References

[1] Sakai, K., and Sheikh, S.A., 1989, "*What Do We Know About Confinement in Reinforced Concrete Columns*", ACI Structural Journal, Vol.86, No.2, Mach-April, PP.192-207.

[2] Book Ng, A.H., Mirza, S.M., Lie, T.T., 1990 "Response of Direct Models of Reinforced Concrete Columns Subjected to Fire", ACI Structural Journal, Vol. 87, No.3, May-June, pp.313-325.

[3] Shettey, M. S., 1988, "*Concrete Technology Theory and Practice*", Third Edition, PP.361.

[4] The Concrete Society, 1990, "Assessment and Repair of Fire- Damaged Concrete Structures", Technical Report 33, Camberiey.

[5] Neville, A. M., 1995, "*Properties of Concrete", Longman Group, Ltd., 4th and Final Edition*", 1995, pp.388.

[6] American Concrete Institute, (2014). Building Code Requirements for Structural Concrete (ACI 318M-14) and Commentary (ACI 318R-05).

[7] ASTM C150 / C150M – 17, Standard Specification for Portland cement (2017).

[8] Iraqi Organization of Standards, IOS 45: 1984; for Aggregate.

[9] ASTM C1437 – 15, Standard Test Method for Flow of Hydraulic Cement Mortar.

[10] ASTM A615 / A615M – 09, Standard Specification for Deformed and Plain Carbon-Steel Bars for Concrete Reinforcement.

[11] Building code and requirements for structure concrete, ACI-318M-14-R10.6, p156.

[12] M.S. Abdulraheem, M.M. Kadhum, "*Experimental and numerical study on post-fire behaviour of concentrically loaded reinforced reactive powder concrete columns*", Constr. Build. Mater. 168 (2018).

Acknowledgments
The authors acknowledge the all support from the Mustansiriyah University, Iraq. The authors are also thankful to Department of Civil Engineering, College of Engineering, Mustansiriyah University for help in experimental procedure at the Structural Engineering Lab at the Mustansiriyah University.

IOP Conf. Series: Materials Science and Engineering **584** (2019) 012016 doi:10.1088/1757-899X/584/1/012016

Turbidity, Color and Chemical Oxygen Demand Removals from Synthetic Textile Wastewater Using Chitosan as a Coagulant

Hala A Abdullah[1] **and Ali J Jaeel**[2]
[1] Civil Engineering Department, M.Sc. Student, Wasit University, Iraq. Email: halaalimsc@gmail.com
[2] Civil Engineering Department, M.Sc. Student, Wasit University, Iraq. Email: alijwaid@uowasit.edu.iq

Abstract. The study investigates the performance of chitosan as a coagulant towards a treatment of synthetic wastewater by combined process of coagulation–flocculation followed by granular filtration. In this study, chitosan used due to some novel characteristics of coagulants such as non-toxic, ecosystem appropriate characteristic, biodegradability, etc. The optimum conditions for this study were at 40 mg/l of chitosan, 5 mg/l of dye, the reduction percentage is further than 98.5%, 92.5% and 94.2% for color, turbidity and chemical oxygen demand (COD) individually. In conclusion, chitosan is an effective coagulant, which can be used to remove certain pollutants with a high percentage from textile wastewater.

1. Introduction

Water is the most abundant material on earth, even though, 97% of water contains salt with percentage exceeds the threshold value and make it basically not suitable for drinking or for the various industrial purposes, 2/3 of the remaining 3% is the form of ice, then leaving only about 1% of the total water as fresh water. The environmental legislations exist to ensure the quality of the environment and these legislations involve the properties of the water drained by industrial areas [1]. Industrialized wastewater is generally handled to perform the principles for discharges to rivers or lake. Water re-use and recycling can be considered as a clear method to decrease the volume of water which must be withdrew from the surface water sources. Direct re-use and recycling require only a minimum of capital outlay and extremely low running cost [2].In the preparations and accomplishment of water reclamation and reuse, the proposed water reuse implementations affect the level of wastewater treatment desired. The quality of resulted water, and the way of allocation and implementation, water reuse may involve a completely controlled "pipe-to-pipe" system with an intermittent storage step, or it may include blending with non-reclaimed water either directly in an engineered system or indirectly through surface water supplies or groundwater recharge [3]. The principal considerations in reuse of water are (1) dependable treatment of wastewater to satisfy hard water quality needs for the designed reuse, (2) preservation of public health and (3) achievement public approval. Water reuse as well needs accurate test of framework and facilities development [4]. Textile processing production is a water intensive consumptive industry in almost every stage for processing, from sizing, desizing, scouring and bleaching, of the dyeing, finishing and printing of fabrics. Every textile plant requires large volumes of water about one hundred m3/ton and produces large quantities of effluent wastewater, that is generally described through color, salts, and organic matter [5]. The standard methods of treating textile wastewater are physical, chemical or biological methods. These methods can be adopted to reduce color and other pollutants in the wastewater with variable efficiency according to the kinds of colors and their concentration. Physical-chemical methods are always used for color removal, and other organic and inorganic impurities. Advanced filtration like (UF, NF and RO) has proved that it is more efficient and effective in removing the color , COD and other pollutants from wastewater [6] ,[7]. But these methods are relatively expensive and sensitive for impurities and colors which cause damage [8] .To achieve acceptable quality for the treated wastewaters, sand filters are widely used and the technology for design and production are more adequate and of low cost. The maintenance of the sand filter is technically easier than advanced filters. The coagulation/ flocculation process is an important part of surface water and wastewater treatment. It has a direct impact on the reliability of plant operations and final water quality together with cost control [9]. The aim of this study is to:

- The investigation using chitosan as a coagulant material to reduce the rate of discharged wastewater pollutants to the surface water sources during minimizing the pollutants concentrations of the reclaimed water.
- Determine the best dose for chitosan throughout studying their effects on the treated water characteristics (Turbidity, Color and COD removals) by physical-chemical treatment using the coagulation flocculation and filtration processes.

2. Methodology

2.1. Synthetic Wastewater

Synthetic wastewater was used to study the bench scale parameters and pilot plant model performance, the synthetic wastewater prepared to meet the original textile wastewater. For preparation synthetic wastewater, several concentrations of dye (5-10-15-20-25 mg/l) and additive (as shown in the table 1) added to a liter of deionized water.

Table 1. Composition of synthetic textile wastewater

Substance	Concentration (mg/l)	Chemical formula	Using
Starch	1000	$C_6H_{10}O_5$	Source of TSS
Acetic	200	CH_3COOH	Carbon source (COD)
Sucrose	600	$C_{12}H_{22}O_{11}$	Food of bacteria
Hydrochloric acid	300	HCL	PH adjustment
Sodium lauryl sulfate	100	$CH_3(CH_2)_{11}OSO_3Na$	Spread the dye
Sodium hydroxide	500	NaOH	PH adjustment
Sodium chloride	3000	NaCl	PH adjustment
Sodium carbonate	500	Na_2CO_3	pH Fixation
Dyes	5, 10, 15, 20, 25		Color source

2.2 Apparatus

2.2.1. Pilot Plant

The pilot Plant model includes feeding tank, mixing tank provided with mixing motor with blades turbine agitator, settling tank, sand filter, collection tank and storage tank, besides the accessories included, dosing pump, coagulant mixing tank with mixing motor and agitator, U-tube manometer, valves flow meters and drain tank, as can be seen in figure 1.

2.2.1.1. Feeding and Mixing Tanks
A steel rectangular tank of 128-liter capacity used for pumping synthetic wastewater to coagulation/ flocculation tank which consists of mixing tank with dimensions of 45cm diameter and 60cm height has been used as a coagulation/ flocculation tank. Rectangular baffle of width 4.5 cm and length 60 cm was built inside of this mixing tank. Turbine agitator with 6, blades connected with electrical motor power 80 W, use for provided a flash mixing and gentle mixing during the coagulation and flocculation processes. The stirring speed (N) was controlled by a speed regulator of the electrical mixer. The coagulation process with rapid mixing and high velocity gradient (300-350), continued for 3 min, the velocity gradient reduced to 100 then to 30 1/sec, for 20 min for the flocculation process.

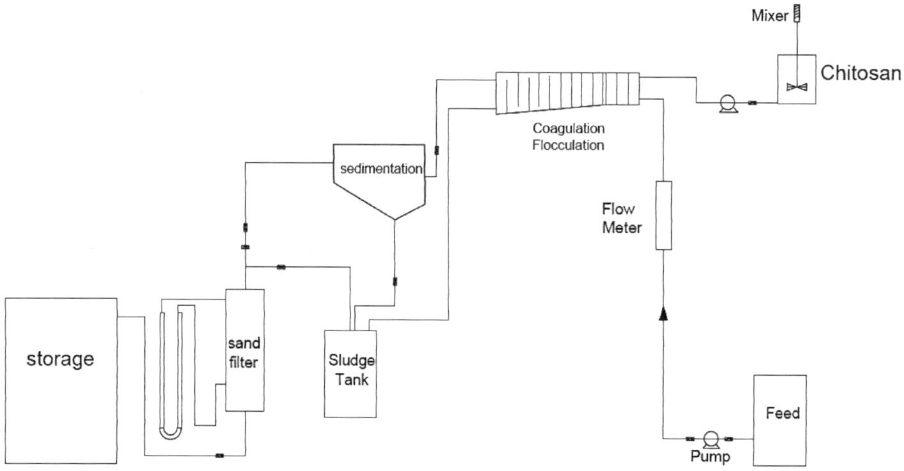

Figure 1. Schematic diagram of the treatment system (mechanical mixing tank)

2.2.1.2 Settling Tank
A rectangular steel tank of depth (D=45 cm), Length (L=90cm) and width (w=35cm) was used for settling process in a period of 30 min to remove the settling particles and flocs generated during the coagulation/ flocculation process, two steel plates of (35,40cm) height and width respectively was inserted into the sedimentation tank at the front of the inlet pipe about 10% of the projected area of the settling tank.

2.2.1.3. Sand Filter Column
Steel column of 20 cm diameter and 100 cm depth, filled with sand and gravel to use as a rapid sand filter instilled after the settling tank to remove or separate the fine particles and flocs that are not settled in settling tank from the treated wastewater. Commercial silica sand of 2.45 specific gravity and gravel was used as a filter media. A size analysis was done to the raw sand by a different mesh size sieve, the effective size d10 (Es) of the sand is (0.63) mm and d60 is (0.96) mm, the uniformity coefficient (UC) is (1.4).

2.2.1.4. Chitosan (coagulant) Tank
A steel tank of capacity 20 liter was used to prepare and supply the chitosan slurry. This tank was provided with electrical motor connected to a steel shaft with an impeller as a mixer inserted into the lime tank in order to keep continuous mixing and prevent the coagulant particles from settling at the bottom.

2.2.1.5. Drain (or sludge) and Storage Tanks
A steel tank of 300-liter capacity was used to collect the drains and the overflow from the mixer tank and hydraulic baffled channel as well as the drain exit from the settling tank and the backwash water

used to clean up the sand filter media. A steel tank of 300-liter capacity was set as a storage tank connected with a pump to supply the wastewater to be treating to the mixing tank or to the baffled mixing channel, controlled by a valve and flow meter.

2.3. Experiments and Tests

Turbidity, Color and COD testes which tested by using single beam UV spectrophotometer (SpectroDirect) further experiments were repeated 3 times for identical test results and conditions then the average values were used for calculations. The experiment procedures and tests were used according to [10] to determine the characteristics of raw and treated wastewater.

2.3.1. Chemicals and Coagulants

The coagulants and chemicals used to enhancement the removal efficiency, included chitosan. Stock solution of chitosan was prepared by adding 10 gm of chitosan to 1000 ml of tap water at 80°c to make a homogenous solution of chitosan solution strength of 1%. Thus, each 1 ml/L of stock solution contained 10 mg/L of chitosan.

3. Results and Discussion

3.1. Impact of Chitosan Concentration on Reduction of Dyes

For calculation the optimum concentration of chitosan coagulant, many chitosan concentrations (10, 20, 30, 40, and 50 mg/l) for several selections of synthetic textile wastewater were investigated and examined for calculation the sound concentration. Dye dosages of synthetic wastewater were (5, 10, 15, 20, and 25 mg/l). Figure 2 displays the impact of chitosan as a coagulant on dye reduction percentage. The concentration of 40 mg/l has great impact on color removal for a fixed dye concentration 5 mg/l of dye wastewater; the dye removal percentage was 98.5%. This optimum dosage (40 mg/l) was applied for different range of dyes concentrations as shown in figure3 the impact of optimum chitosan on dye removal adequacy for different range of dye concentrations. It is knowable that concentration of 40 mg/l has great impact on the color reduction for dye concentration 5, 10, 15, 20 and 25 mg/l the color removal efficiency were 98.5%, 96.2%, 95%, 87.4% and 82.3%, respectively.

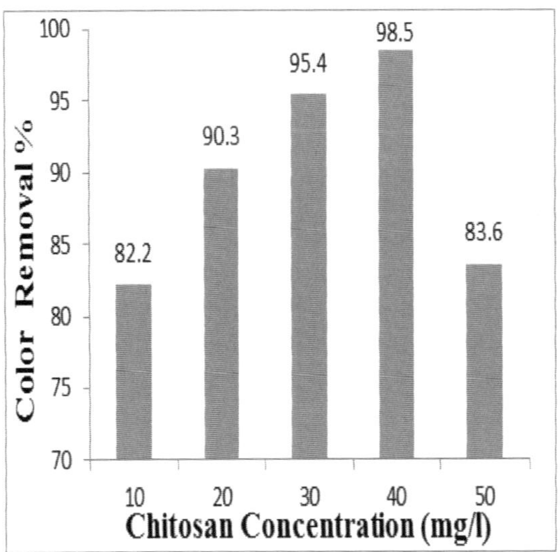

Figure 2. Optimum chitosan dosage for color removal (constant dye 5 mg/l).

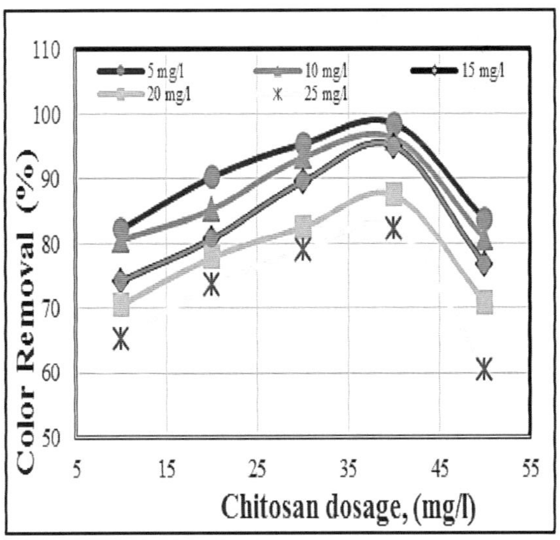

Figure 3. Color removal for different colors at optimum chitosan 40 mg/l.

3.2. Impact of Chitosan Concentration on Turbidity Reduction

Many different concentration (10, 20, 30, 40, and 50 mg/l) for several specimens of dye wastewater were experienced for selection the proper concentration of chitosan to achieve higher turbidity removal percentage. The dye dosages of synthetic textile wastewater are (5, 10, 15, 20, and 25 mg/l). Figure 4 demonstrates impact of chitosan dye removal percentage. It is knowable that concentration of 40 mg/l has valuable impact on turbidity reduction for a fixed dye concentration 5 mg/l of synthetic wastewater; the turbidity removal percentage was 92.5%. This optimum dosage (40 mg/l) was applied for different range of dyes concentrations as shown in figure 5 the impact of optimum chitosan as a coagulant on turbidity removal percentage for different range of dye concentrations. It is knowable that concentration of 40 mg/l has valuable impact on the removal of color for dye concentration 5, 10, 15, 20 and 25 mg/l the turbidity removal efficiency were 92.5%, 90%, 87%, 79% and 73.5%, respectively.

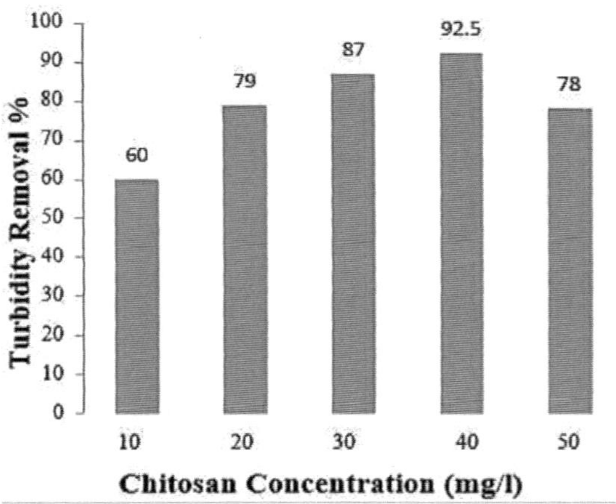

Figure 4. Optimum chitosan dosage for turbidity removal (constant dye 5 mg/l).

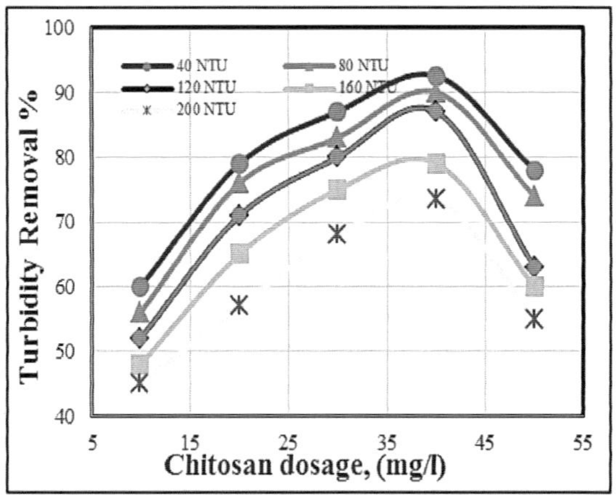

Figure 5. Turbidity removal for different colors at optimum chitosan 40 mg/l.

3.3 Impact of Chitosan Concentration on COD Reduction

To investigate the proper concentration of chitosan coagulant, various portions (10, 20, 30, 40, and 50 mg/l) for varied samples of dyes wastewater were examined to determine the best amount of coagulant. The dye dosages of synthetic dye wastewater are (5, 10, 15, 20, and 25 mg/l). Figure 6 displays impact of chitosan on COD removal percentage. It is obvious that concentration of 40 mg/l has meaningful impact on COD reduction for a fixed dye concentration 5 mg/l of dye wastewater; the COD removal efficiency was 94.2%. This optimum dosage (40 mg/l) was applied for different range of dyes concentrations as shown in figure 7 the impact of optimum chitosan on COD removal efficiency for different range of dye concentrations. It is detectable that concentration of 40 mg/l has considerable impact on COD reduction for dye concentration 5, 10, 15, 20 and 25 mg/l the COD removal efficiencies were 94.2%, 92.8%, 90.06%, 82.8% and 77.4%, respectively.

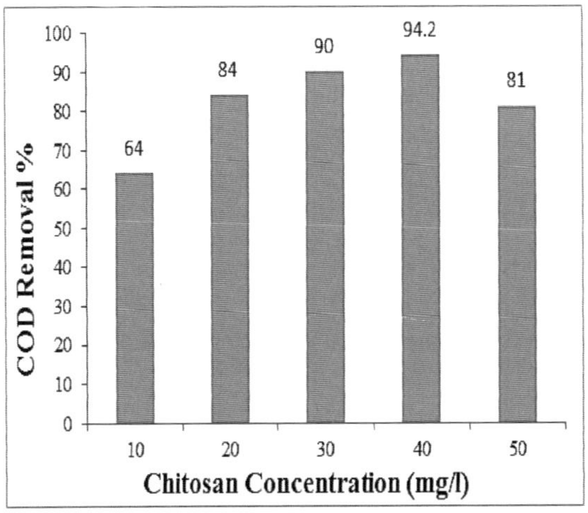

Figure 6. Optimum chitosan dosage for COD removal (constant dye 5 mg/l).

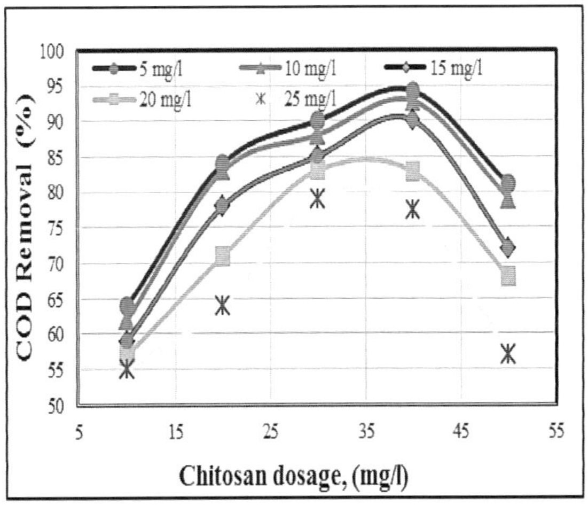

Figure 7. COD removal for different colors at optimum chitosan 40 mg/l.

3.4. Discussion

The main causes of the high percentage of color, turbidity and COD removals were due to:
1- High charge density of chitosan [11]. So, this property helped to occur rapid destabilization of the particles in less amount of chitosan.
2- The settling particles and flocs generated during the coagulation/ flocculation process were removed through the settling process in settling tank.
3- Granular filtration efficiency including removing or separate the fine particles and flocs that are not settled in settling tank.

4. Conclusions and Recommendations

The treatment of colored material in textile wastewater before discharging into rivers is important for water sources preservation using natural and chemical coagulants. Therefore, adopting chitosan powder a coagulant can be applicable at a great efficiency in dyeing and other pollutants removal from textile wastewater. The reduction percentage is further than 98.5%, 92.5% and 94.2% for color, turbidity and COD, respectively.

References

[1] Tassoula E, Diamadopoulos E and Vlachos C 2007 Tertiary physical – chemical treatment of secondary effluent from the Chaina municipal wastewater treatment plant *Global NEST J.* **9** 166-173

[2] Metcalf and Eddy 2003 *Wastewater Engineering Treatment and Reuse* 4th ed (New York: McGraw- Hill)

[3] Branes D, Bliss P J and Gould B W 1984 *Water and Wastewater Engineering System* (London: Pitman Publishing Limited London)

[4] Santoyo B A, Gómez C J L, Gomez E, MaximoM. F, Hidalgo M A M 2003 Application of reverse osmosis to reduce pollutants present in industrial wastewater *Desalination* **155** 101-108.

[5] Orhon D, Germirli B F, Kabdasli I, Insel F G, Karahan Ö, Dulkadiroglu H, Dogruel S, Sevimii F, Yediler A 2001 A Scientific approach to wastewater recovery and reuse in the textile industry *J. water Sci. and Tech.* **43** 223-231.

[6] Hatem A G and Zahraa k 2018 The Removal of terasil navy blue dye from Al-Kut textile wastewater using membrance technologies *Inter. J. of Eng. & Technol.* **7** 557-562.

[7] Mohammad F A, Zablouk M A and Abid-Alameer A M 2012 Experimental study of dye removal from industrial wastewater by membrane technologies of reverse osmosis and nanofiltration *Iranian J. of Environ. Health Sci. & Eng.* 1-9

[8] Chen X, Shen Z, Zhu X, Fan Y and Wang W 2004 Advanced Treatment of Textile Wastewater for Reuse Using Electrochemical Oxidation and Membrane Filtration. *e-mail shenzhmin@sina.com.*

[9] Rahbar M S, Alipour E and Sedighi R E 2006 Color removal from industrial wastewater with a novel coagulant flocculation *Int. J. Environ. Sci Tech.***3** 79-88

[10] APHA 1998 *Standard Methods for the Examination of Water and Wastewater* 20th edition (Washington, DC: American Public Health Association)

[11] Kumar M N 2000 A review of chitin and chitosan applications *Reactive and Functional Polymers* **46** 1-27

IOP Conf. Series: Materials Science and Engineering **584** (2019) 012017 doi:10.1088/1757-899X/584/1/012017

Effect of Embedment Depth on Raft Foundation Settlement Under Seismic Load

Abeer A A Hanash[1], Mahmoud D Ahmed[2], and Abdulmotalib I Said[3]

[1] Civil Engineering Department, University of Baghdad, Baghdad, Iraq. Email: abeer.abdulkareem@yahoo.com
[2] Civil Engineering Department, University of Baghdad, Baghdad, Iraq. Email: mahmoud_baghdad@yahoo.com
[3] Civil Engineering Department, University of Baghdad, Baghdad, Iraq. Email: Dr.Abdulmuttalib.i.Said@coenguobaghdad edu.iq

Abstract. Dynamic loads highly influence soil properties and may cause real damage to structures and buildings. This article reports the experimental results from 24 tests to study the settlement of flexible and rigid raft foundation with different embedment depth rested on dense sandy soil. A small scale building model of dimension 200*200 mm and 320 mm in height was performed with reinforced concrete raft foundation of 10 mm thickness for flexible raft and 23 mm for rigid raft, The shaking table technique was used to simulate the seismic effect, the shaker was sat to give three different excitation frequencies 1,2,and3 Hz and displacement amplitude equal to 13 mm, the foundation was placed at four different embedment depths (0,0.25B=50mm,0.5B=100mm, and B=200mm), where B is the raft side length. The results of the tests indicate that the foundation embedment have a positive effect for its ability to minimize the settlement for both flexible and rigid raft foundation and for all the excitation frequencies.

1. Introduction

Studying the effect of earthquake excitation on shallow foundations is an essential field for several decades, especially, for geotechnical engineers [1]. Generally, foundation can be classified into two major groups, shallow foundation and deep foundation. The first group contains: wall- footing, combined-footing, spread-footing and raft-foundation. The second group is: piles, piers, caissons and drilled-shaft-foundations [2]. Shallow foundations are regarded as higher safety comparing with pile foundations when subjected to heavy earthquake since design requirements involve that shallow foundations located on hard soil stratums of high bearing capacity [3]. Raft foundation is usually utilized where soil and load conditions might lead to high differential settlement for individual spread footings and for buildings with substantial overturning moments in regions of high seismicity or in high irregularity superstructure. Several factors effected the dynamic response of foundations such as the shape, dimensions, flexibility and foundation mass, depth of soil layer over bedrock, the effect of inhomogeneity, nonlinearity of soil behaviour, the embedment depth of the foundation, and anisotropy of soil properties, as stated by [4].

2. Model preparation and experimental work

Physical models are performed at much smaller scale than the actual size of proto type because the aim of physical models are to get the indicated patterns of response more rapidly and with more control over the details of the model [5]. 4 tests were performed inside a rigid steel cylindrical container of 700 mm in diameter and 600 mm in height, the inside of the container is covered by styropor sheets to prevent the wave reflection which cause additional stresses and eliminate the friction between the container faces and soil. To perform a physical model, a concrete model footings of size 200 × 200 × 10 mm for flexible raft and 200*200*23 mm for rigid raft, was used, the thickness that separates the flexible raft and rigid raft was calculated by relative stiffness factor (K) method using equation (1), it was equal to 16 mm [6], see Figure 1. The foundation model was placed centrally in the soil. A steel frame of 320 mm in height was fixed on the raft foundation to represent the building height and to hold the additional mass of 40 kg weight, this weight was calculated according to the allowable total settlement of raft foundation, building height and soil depth under raft foundation were considered to be constant 320 mm and 450 mm respectively. The used soil is dry dense sand of 70% relative density passing through sieve No.10 (2.0 mm) and retained on sieve No.200, properties of the used sand are summarized in Table 1 with the standard of the test, hygroscopic water content (≈0.5-3.0%) was mixed with the sandy soil before compaction to make sure a small cementation of soil. The soil was located inside the container in layers and then compacted to the required density which equal to (16.86) kN/m² by a steel hammer of (4.5) kg, the falling height was between (150-200) mm and the No. of blows were (4) for each layer which was founded by making a relation between the No. of drops and the developed dry density, sand-cone method was used according to [7]. to insure the required density. A vibration of (1, 2 and3) Hz in x- direction was applied to the foundation using the shaking table technique which manufactured by [8], see Figure 2.

$$k = \frac{E}{12\,Es}\left(\frac{d}{b}\right)^3 \quad (1)$$

Where:

*E, E*s = Modulus of compressibility of the foundation (26000 MPa), soil (17 MPa) respectively
b = Length of the section in the bending axis (200 mm)
d = Thickness of the raft foundation (was calculated by Eq.1 for K=0.5)
K=the relative stiffness factor (for K<5 the raft foundation is flexible)

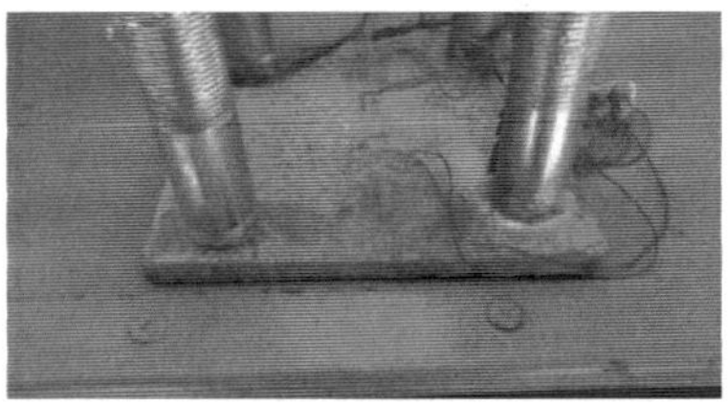

(a)

(b)

Figure 1. Modeling of raft foundation (a) flexible and (b) rigid.

3. Instrumentation and measurement of dynamic response

3.1. Linear Variable Differential Transformer (LVDT)

The maximum total settlement of raft foundation was recorded using two LVDTs placed at the edge and middle of the upper surface of raft foundation. The LVDTs transducers were calibrated by comparing the output voltage with a known deflection (using digital vernia).

3.2. Data Logger

The LVDTs were calibrated and connected to the data logger unit which provides a connection with the computer lap top. The data logger was calibrated to give direct readings for raft foundation settlement for both flexible and rigid raft. The data logger is of 2 channels for LVD readings, to attain the required accuracy, 10 reading were recorded every 1 second for each channel. The whole illustration of test components and measuring devices can be seen in Figure 3.

Table 1. Physical and mechanical properties of the used sand

Property	**Value**	**Standard of the test**
Specific Gravity, Gs	2.68	[9]
Gravel (> 4.75 mm), %	0	
Sand (4.75-0.075 mm), %	99.5	[10]
Silt and clay (< 0.075 mm),%	0.5	
Unified soil classification system (USCS), Soil Type	SP	[11]
Relative Density, DR%	70	—
Maximum dry unit weight γd min, kN / m^3	17.385	[12]
Minimum dry unit weight γd min, kN /m^3	14.365	[13]
Dry unit weight (used) γd, kN /m^3	16.85	—

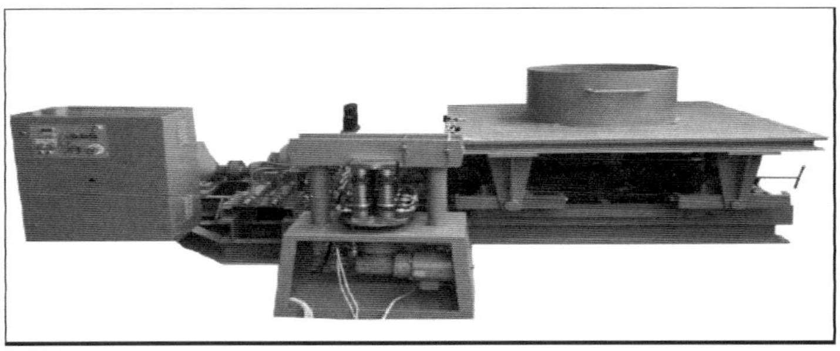

Figure 2. Shaking table photo.

(a) Testing model

(b) Testing instruments

Figure 3. General view of testing model and instruments.

4. Results and discussion

The total settlement of raft foundation with time was recorded at the edge and the middle of the raft model using LVDTs.

4.1. Effect of embedment on the settlement of raft foundation

To study the effect of increasing the embedment depth on total settlement under dynamic load, the embedment depth of flexible and rigid raft foundation was increased from zero to (0.25B, 0.5B. and B), Figures 4, 5 and 6 show the results of this variation. It can be seen that maximum total settlement decrease with increasing the embedment depth for all the excitation frequencies and for both flexible and rigid rafts, this results agrees well with [4] who stated that embedment gives noticeable decrement in dynamic response (displacement and excess pore water pressure) for all soil types but with different degrees. The summary of changing the maximum total settlement with the embedment depth for different earthquake excitation was recorded and presented in Table 2. Table 3 summaries the percentages of maximum settlement reduction comparing with surface embedded foundation. This behavior of raft foundation can be illustrated by the following points:

- More embedment depth means more contact surface area provided by basement walls gives larger side frictional forces which reverse the foundation settlement.

- Increasing embedment depth provides more confining pressure at the edges of the foundation leads to increase its stiffness and reduce the settlement.

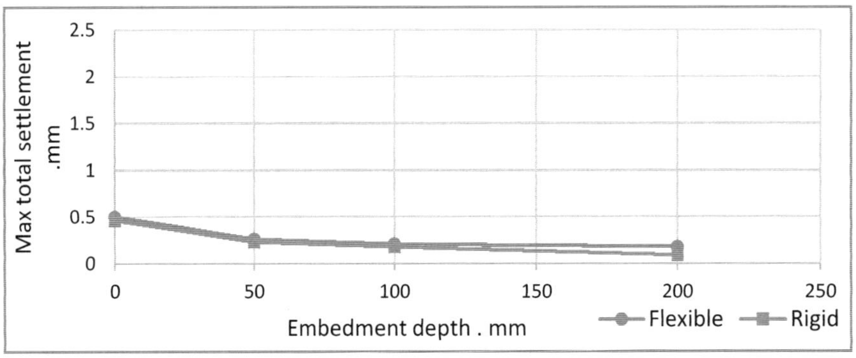

Figure 4. Variation of maximum total settlement with embedment depth under 1Hz.

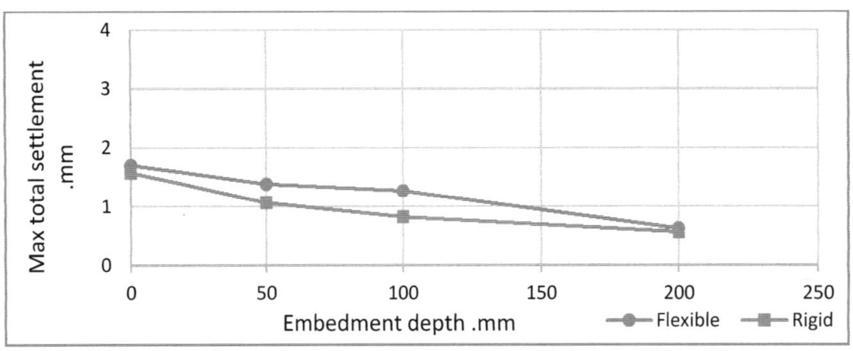

Figure 5. Variation of maximum total settlement with embedment depth under 2Hz.

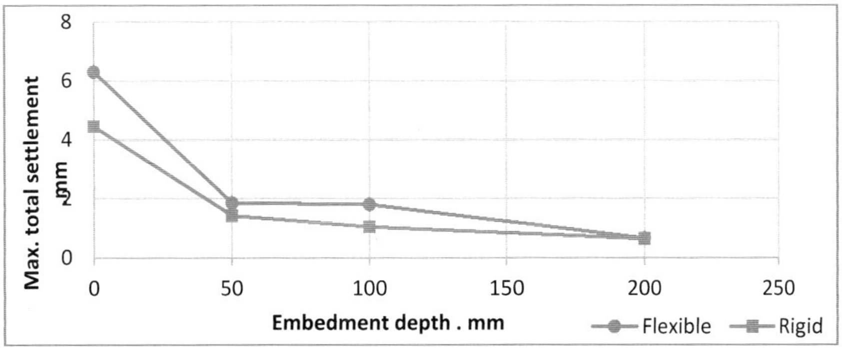

Figure 6. Variation of maximum total settlement with embedment depth under 3Hz.

Table 2. Summary of variation of the maximum total settlement with embedment depth.

Embedment Depth	Raft Type	Maximum Total Settlement(mm)		
		1Hz	2Hz	3Hz
0	Flexible	0.5	1.7	6.3
	Rigid	0.46	1.57	4.45
0.25 B	Flexible	0.26	1.38	1.85
	Rigid	0.23	1.06	1.41
0.5 B	Flexible	0.21	1.26	1.79
	Rigid	0.18	0.82	1.04
B	Flexible	0.18	0.62	0.65
	Rigid	0.09	0.56	0.63

Table 3. Summary of reduction percentages due to increasing embedment depth.

Embedment depth (mm)	Raft foundation type	Reduction percentage		
		1Hz	2Hz	3Hz
0.25B	Flexible raft	48%	19%	71%
	Rigid raft	50%	32%	68%
0.5B	Flexible raft	58%	26%	72%
	Rigid raft	61%	48%	77%
B	Flexible raft	62%	64%	90%
	Rigid raft	80%	64%	86%

- As the vertical side area of basement increased as a result of increasing embedment depth, the applied forces transmitted to the surrounding soil by shear stresses, this dissipation of applied forces leads to reduce the settlement (the settlement is a function of the force).

- Increasing the embedment depth increase in the system damping and result in reducing the footing vibration and settlement.

Figure 7 show the effect of 3Hz excitation frequency on the soil at the case of zero embedment depth, the soil seems to be failed.

(a) flexible raft foundation.

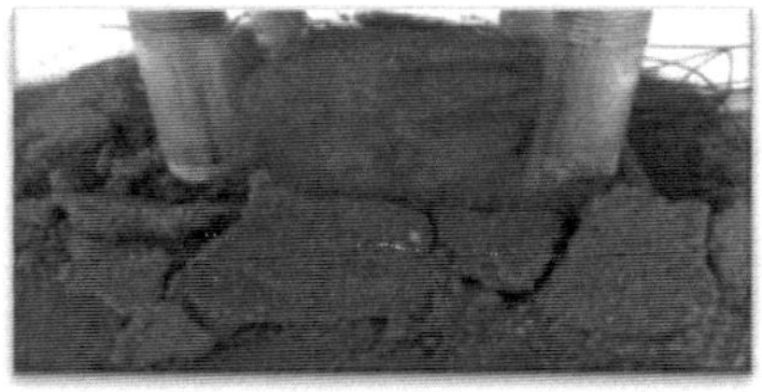

(b) rigid raft foundation.

Figure 7. The foundation failure under 3Hz excitation frequencies and zero embedment depth

4.2. Effect of raft thickness (flexible and rigid) on settlement

To investigate the dynamic behavior of flexible and rigid raft foundation, two different thicknesses are used flexible of 10 mm thickness and rigid of 23 mm thickness. The settlement – time readings had been recorded for all the tests, Figures 8, 9 and 10, represent the tests of 0.5B.

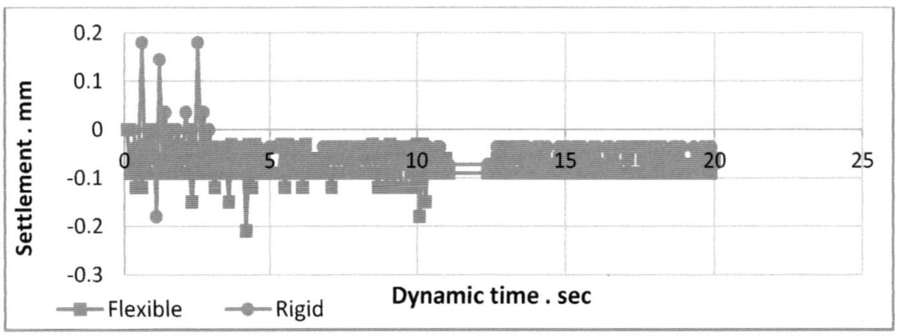

Figure 8. Settlement-time history at 0.5 B embedment depth. under 1Hz

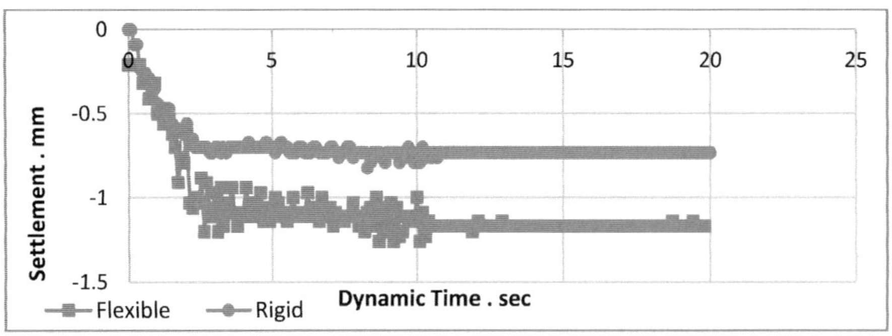

Figure 9. Settlement - time history at 0.5 B embedment depth under 2Hz.

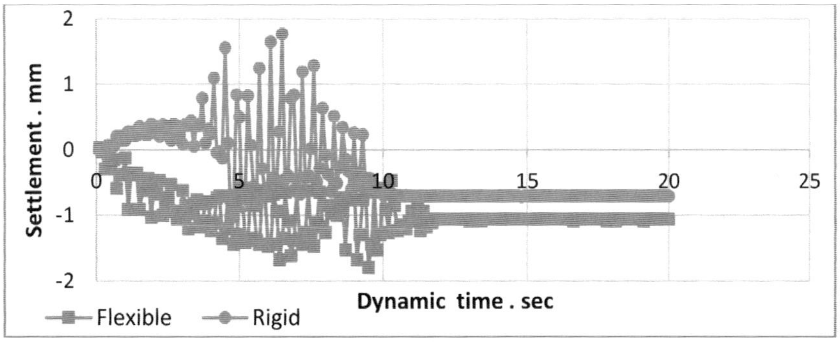

Figure 10. Settlement - time history at 0.5 B embedment depth under 3Hz.

embedment depth for (1,2and 3) Hz excitation frequency which had been chosen as a typical case to represent the effect of raft thickness on settlement with time. From figures, it can be seen that the total settlement of raft with time gave higher reading for flexible foundation than the rigid one. This response is attributed to the higher stiffness of raft foundation resulting from its higher thickness, moreover, increasing the thickness means larger side contact area and higher frictional force which resist the footing settlement. Figures 11, 12 and 13, show the effect of raft thickness for all embedment depths under 1Hz, 2Hz, and 3Hz respectively, from figures, it's clear that maximum settlement results for flexible raft were higher than those for rigid raft foundation, this conclusion agrees well with [14] who stated that the settlement of raft foundation under dynamic loading decreases with decreasing the ratio of raft diameter to raft thickness., moreover, raft thickness of 10 mm (flexible) is more economical than raft thickness 23 mm. Table 4 summaries the percentages of reduction in maximum total settlement for rigid raft comparing with flexible raft foundation.

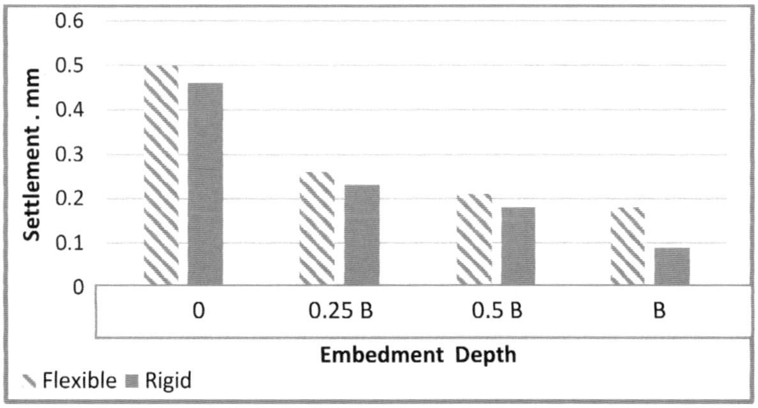

Figure 11. Variation of maximum settlement of flexible and rigged raft foundation under 1Hz.

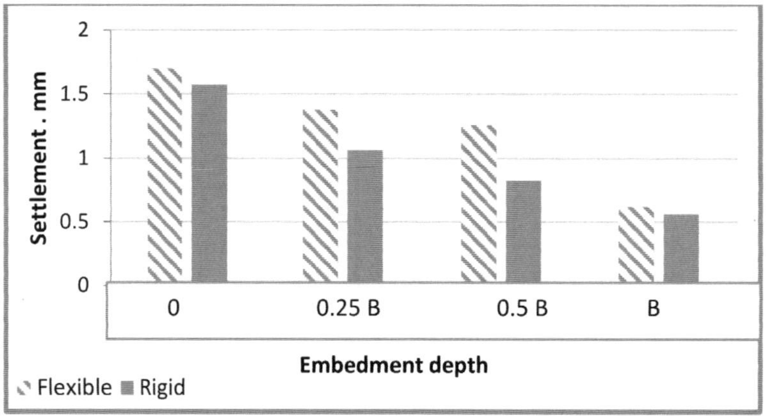

Figure 12. Variation of maximum settlement of flexible and rigged raft foundation under 2Hz.

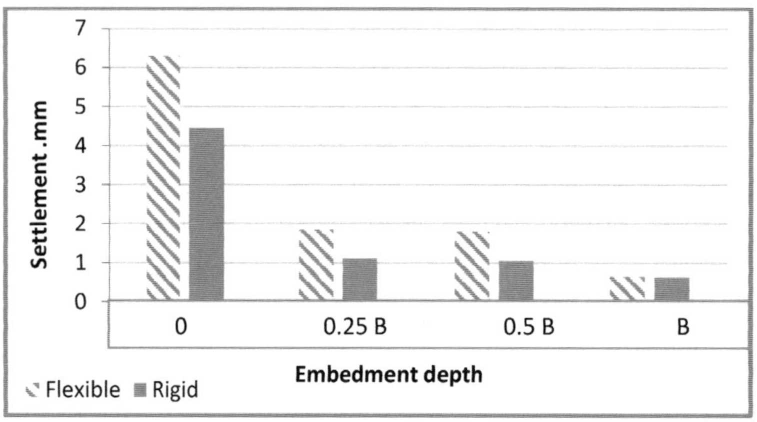

Figure 13. Variation of maximum settlement of flexible and rigged raft foundation under 3Hz.

Table 4. Summary of redaction percentages of rigid raft maximum settlement comparing with flexible raft maximum settlement.

Embedment	Reduction Percentage		
depth	1Hz	2Hz	3Hz
0	8%	8%	29%
0.25B	12%	23%	24%
0.5B	14%	35%	42%
B	50%	10%	3%

5. Conclusions

- Generally, the experimental results indicated that the foundation embedment have a positive effect on raft dynamic response because it minimizes the settlement of foundation.

- As the embedment depth increased from zero to B, the maximum settlement decreased by (62%,64%, and 90%) for flexible raft under (1,2, and 3) Hz respectively, for rigid raft the maximum settlement decreased by (80%,64%, and 86%) under (1,2, and 3) Hz, respectively.
- Generally, the larger redaction in the induced maximum settlement was when the embedment depth increased from zero to 0.25 B as the maximum settlement decreased by (48%,19%, and 71%) for flexible raft under (1,2, and 3) Hz respectively, for rigid raft the maximum settlement decreased by (50%%,32%, and 68%) under (1,2, and 3) Hz respectively. This reduction became smaller when the embedment depth increased to 0.5B comparing with raft embedded at 0.25B as the maximum settlement decreased by (19%,9%, and 3%) for flexible raft under (1,2, and 3) Hz respectively, and (22%%,23%, and 26%) for rigid raft under (1,2, and 3) Hz respectively. When the embedment depth increased to B, the percentages of reduction of maximum settlement comparing with that of raft embedded at 0.5B, were increased to be (14%,51%, and 64%) for flexible raft under (1,2, and 3) Hz respectively, and (50%%,32%, and 39%) for rigid raft under (1,2, and 3) Hz, respectively.
- The maximum settlement of rigid raft foundation was less than that for flexible raft foundation, the percentages of reduction of rigid raft maximum settlement comparing with flexible raft maximum settlement were ranged between (8% to 50%) as the embedment depth increased from zero to B under 1Hz excitation frequency. For 2Hz excitation frequency, the range was between (8% to 35%).When the excitation frequency became 3Hz, the range was between (3% to 42%).
- For 3Hz excitation frequency, in case of zero embedment depth, the soil seems to be failed for flexible and rigid raft foundations.

References

[1] Abdulrasool M A 2015 Seismic Behavior of Shallow Foundations on Some Iraqi Soils (Thesis submitted to the College of Engineering of Al-Nahrain University in a Partial Fulfillment of the Requirements for the Degree of Master of Science in Civil Engineering)

[2] Aung T Z and Tun Dr K T July 2012 Parametric Study on Foundation of Regular High-Rise R.C Building under Seismic Load (International Journal of Science- Engineering and Technology Research IJSETR Volume 1 Issue 1) pp 4-6

[3] Shirato M, Kouno T, Asai R, Nakatani S and Paolucci R 2008 Large-Scaling Experiments on Nonlinear Behavior of Shallow Foundations Subjected to Strong Earthquakes- Soil and Foundations (Japan Geotechnical Society) Volume 48- No.5,673-692, Oct.2008

[4] Al-Shammary W T S 2013 Numerical Analysis of Machine Foundation On Saturated Sandy Soil (Thesis Iraq: University of Technology) pp 99-104

[5] Wood D M 2004 Geotechnical Modelling (April) Version 2.2

[6] Gupta S C 1997 Raft Foundation Design and Analysis with a Practical Approach Advisor (Indian Buildings Congress- Former Chief Engineer- Central Public Works Department) p 25.

[7] ASTM-D7698-11a 1998 Standard Test Method for Sand-Cone Method (American Society for Testing and Materials- ASTM)

[8] Salem L A 2016 PhD. Response of Piled Raft Foundation to Earthquake Excitation (Thesis submitted to the College of Engineering of University of Baghdad- Iraq) pp 119-120

[9] ASTM-D 854 1998 Standard Test Method for Specific Gravity (American Society for Testing and Materials -ASTM)

[10] ASTM D 422-63 2001 Standard Test Method for Particle Size Analysis of Soils (American Society for Testing and Materials-ASTM)

[11] ASTM D 2487-06 2006 Standard Test Method for Classification of Soils for Engineering Purposes (Unified Soil Classification System) West Conshohocken, Pennsylvania (USAASTM)

IOP Conf. Series: Materials Science and Engineering **584** (2019) 012017 doi:10.1088/1757-899X/584/1/012017

[12] ASTM-D4253-93 2000 Standard Test Method for Maximum dry unit weight (American Society for Testing and Materials -ASTM)

[13] ASTM-D4254-93 2000 Standard Test Method for Minimum dry unit weight (American Society for Testing and Materials -ASTM)

[14] Jayalekshmi B R, Jisha S V, Shivashankar R and Narayana S S 2014 Effect of Dynamic Soil-Structure Interaction on Raft of Piled Raft Foundation of Chimney (ISRN Civil Engineering- International Scholarly Research Notices Journal Volume 2014- Article ID 790928) 11 pages.

IOP Conf. Series: Materials Science and Engineering **584** (2019) 012018 doi:10.1088/1757-899X/584/1/012018

A Conflict Index to Assess Traffic Safety at Intersections

Raghad Zeki Abdul Majeed [1, *] and Hussein A. Ewadh [2]

[1] Graduate student, Highway, and Transportation Engineering Department, Faculty of Engineering, Al-Mustansiriayah University, Baghdad, Iraq.
[2] Professor Road and Traffic Engineering, Civil Engineering Department, College of Engineering, University of Babylon. Babylon, Iraq.
*Corresponding author email: raghad.zeki@yahoo.com

Abstract. In this paper, a conflicts-based technique, using an index measure is introduced to assess traffic safety at intersections. The technique helps in identifying sites most probable to profit from safety improvement as an alternative to crashes-based analyses. As a case study, the analysis and measure of safety for 11 four-leg signalized intersections at different sites in Baghdad city, have been done independently using direct methods based on crash rate and indirect methods based on the level of conflict risk measures. A comparison is achieved between the two methods and their outputs. The study results imply that a developed method of conflicts index of signalized intersections (CISI) (based on traffic conflict frequency and severity) could serve as a viable way of measuring the safety performance of signalized intersection at urban area. CISI can be established for existing conditions at intersections as well as in evaluating different design alternatives to more accurate safety comparison and better decision-making. Application of direct and indirect method results in identifying the same sites to profit from safety improvement. In addition, the research demonstrates a promising application of CISI in; relative comparison of conflict risk at various intersections and estimating conflict risk after potential countermeasures for before-after assessment of traffic safety at intersections.

1. Introduction

In the process of assessing traffic safety at intersection sites, there are two major approaches, crash-based safety analysis (direct methods) and non–crashes based analyses (indirect methods). The decision on which method to use for the identification of sites with potential for safety improvements is a major factor in terms of the effectiveness of road safety management process. The Highway Safety Manual (HSM) provides a set of thirteen crash-based performance measures that can be applied either isolated or in conjunction with the network screening process [1].

Performance measures in HSM require different data acquisition and modeling efforts from the safety analyst which may not be available especially in developing countries. However, the crash-based approach is set back by several shortcomings such as randomness and rarity of crash occurrence.

Traffic conflict technique (TCT) is one of the non-crash based analyses represents the indirect method to assess safety condition of intersection sites. Laureshyn and Varhelyi stated, if the goals are to make a safety diagnosis of a site, TCT is the right method for that [2]. TCT is based on the detection and count of "critical incidents" occurring in real traffic situations.

TCT can be estimated through traffic simulation or observed from real traffic condition by field observation on - site or assemble conflict events from videos. Many studies applied field observation and concluded that observed conflicts on a field in some comparative studies done are more valid than conflicts found by other procedures when denoting them to crashes.

2. Analyses and assessing safety based on traffic conflicts.

The events in the traffic environment occur at different levels. The ordering of traffic events by severity is defined by Hyden as shown in figure1 in a "safety pyramid" [3].

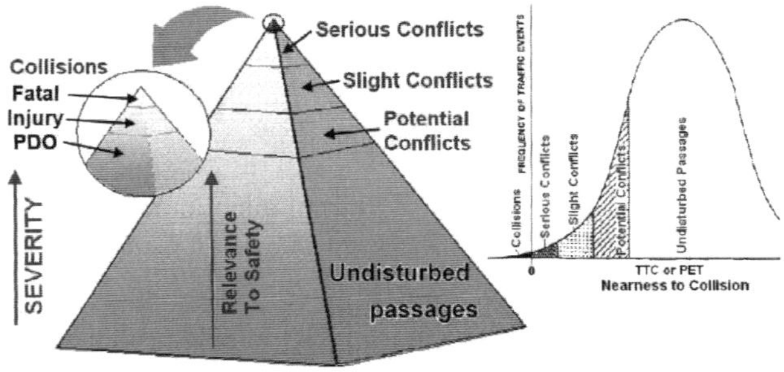

Figure 1. Safety Pyramid [3].

It is possible to group almost all of the operational definitions of the traffic conflict into two types: Evasive actions-based traffic conflict and temporal (and (or) spatial) proximity-based traffic conflict. Parker and Zegeer defined evasive actions-based traffic conflict as "… an event involving two or more road users, in which the action of one user causes the other user to make an evasive maneuver to avoid a collision" [4]. Hence, conflict according to that definition includes considerable judgment on conflict situation. Further, this definition infers that conflicts and crashes are of similar nature unless for the presence and the success of evasive action.

In a proximity-based traffic conflict, the critical events are recognized depending on temporal (and (or) spatial) proximity. It is defined as "…an observable situation in which two or more road users approach each other in space and time to such an extent that there is a risk of collision if their movements remain unchanged" [5]. The most common temporal proximity indicator family is Time to Collision (TTC) [6]. Specifically, a conflict was defined to be severe if the TTC value was less than the threshold of 1.5 sec. However, the threshold values vary in relevant studies from 1.0 to 5.0 sec.

A study conducted by Baker for investigating the relationship between crashes and traffic conflicts based on evasive action. The collected data confirm that conflicts and crashes are statistically associated [7]. Migletz carried out a conflicting study at 46 signalized and non-signalized intersections in the greater Kansas City area, the distinction of intersections by traffic volume (High, Medium and Low flow categories). The report concluded that it seems to be no observable advantage of using crash data compared to conflict data. Potentially, the combination of crash and conflict data could produce even better estimates (with lower variance) of the expected number of crashes compared to using only one data source [8]. Pietrzyk conducted comprehensive traffic conflict study by the center for urban transportation research. The study included a tabular form of observed conflicts, collected crash data and developed easy-to-use tables for Florida-based at different intersection types (4 and 3 legged signalized and unsignlized intersection). The number of lanes (instead of traffic volume) was used as an intersection classification parameter to reduce workloads in field data collection and to simplify tabulation [9].

A study conducted by Sayed and Zein described the application of traffic conflict technique to estimate traffic safety at intersections by using intersection conflict index (ICI) based on the level of conflict risk [10]. Table 1 presents the relative average conflicts risk associated with each grade of conflicts risk. Based on a study conducted Sayed and Zein, the conflict risk established ranges from LOSS A (low frequency and severity) through LOSS F (very high frequency and severity). The conflict risk, therefore, provides an indication regarding the relative risk of being involved in conflicts at intersections.

The established standards for traffic conflict frequency and severity for intersections allowing for relative comparison of conflict risk at various intersections. Further, Souleyrette and Hochstein conducted a study based on estimation conflicts (conflict frequency and severity) from simulation models and using them in assessment safety for different design alternatives [11].

Table 1. Conflict risk associated with ICI [10].

Intersection conflict index	Conflict risk
A	Negligible
B	Low
C	Moderate
D	Moderate to high
E	High
F	Extreme

3. Analysis and Assessing Safety Based on Crash Rate (CR)

CR is one of thirteen tools used in HSM as a safety performance measure. CR is used as a tool to recognize and prioritize sites as in need of modifications and for appraisal of the effectiveness of treatments. In the HSM, CR is known as the number of crashes that happen at a given site during a certain time period in relation to a specific measure of exposure (e.g., per million entering vehicles (MEV) for an intersection).

4. Crash Modification Factor (CMF)

CMFs represent the relative change in safety due to a change in one specific condition (when all other conditions and site characteristics remain constant). Therefore, a CMF may serve as an estimate of the effect of a particular geometric design (add lanes or increase the width of the lanes) or traffic control feature or the effectiveness of a particular treatment or condition.
A list of CMFs for a variety of geometric and operational treatment types for urban signalized intersection, backed by robust scientific evidence available in HSM and other references like [12,13] . The CMFs in the HSM have been developed using high-quality before/after studies that account for regression to the mean.

5. Data collection and Methodology

5.1. Study sites

This study has been used 11 four-leg signalized intersection in the urban area of Baghdad city. Figure 2 shows the locations and names of intersections in the study sites.

5.2. Crash data

Crash data are collected for three years from 2015- 2017 for 9 sites. No archived crash data are available at two sites. According to crash data there are 143 crashes occurred in 9 sites during 3 years, the average crashes for intersections was 5.29 crashes per year. Crash data in this study are represented for the crashes between vehicles only. The crash rate is estimated for each intersection by dividing the total number of crashes by MEV (total entering vehicles for major and minor streets) for the 3-year study period.

5.3. Geometric and Traffic Control data

Geometric and traffic control data needed in this study were collected from the field in each site in addition to the geometric layout of intersections and timing distribution of cycle length that obtained from relative authority (Mayoralty of Baghdad). Figure 3 presents a sample of the geometric design of Al-Masbah intersection (site No.7) in the study sites.

5.4. Traffic volume and conflict data

By the aid of positioning of surveillance cameras at each study site as well as field observation, necessary input data of traffic volume and conflicts was obtained. For each studied intersection, traffic volume and conflicts are observed for 16 hours on workdays (2 workdays for sites (1, 2, 3,4,5,6 and 7) and 3 workdays for sites (8, 9, 10, and 11)). The hours of observation are distributed to cover morning, mid-morning, noon and evening periods for each workday.

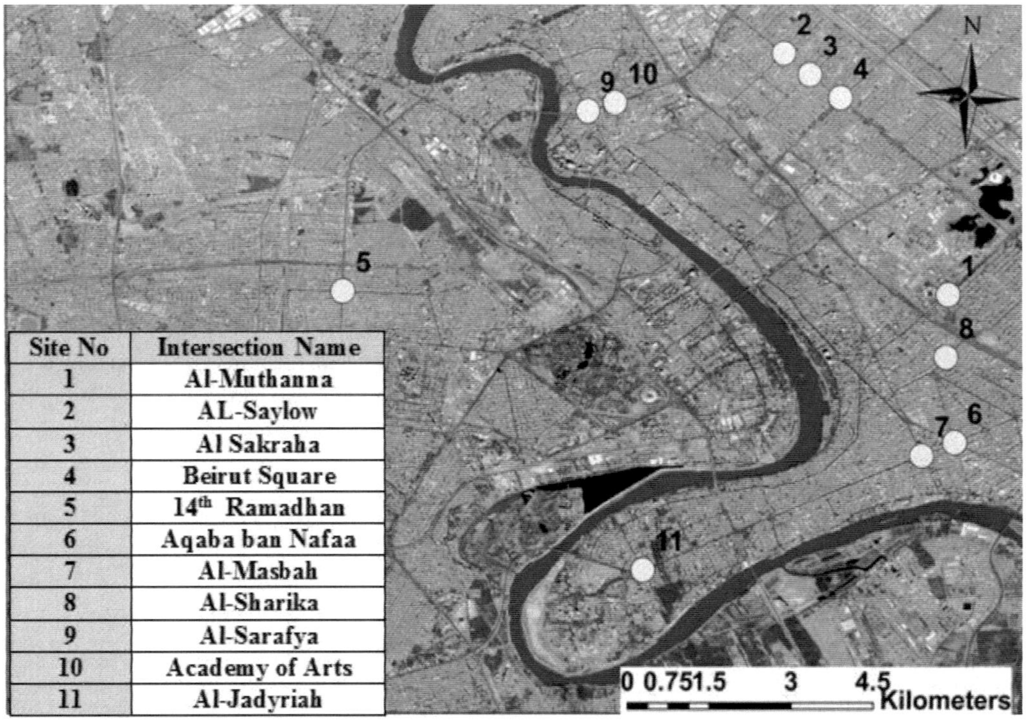

Site No	Intersection Name
1	Al-Muthanna
2	AL-Saylow
3	Al Sakraha
4	Beirut Square
5	14th Ramadhan
6	Aqaba ban Nafaa
7	Al-Masbah
8	Al-Sharika
9	Al-Sarafya
10	Academy of Arts
11	Al-Jadyriah

Figure 2. Map of Selected Sites.

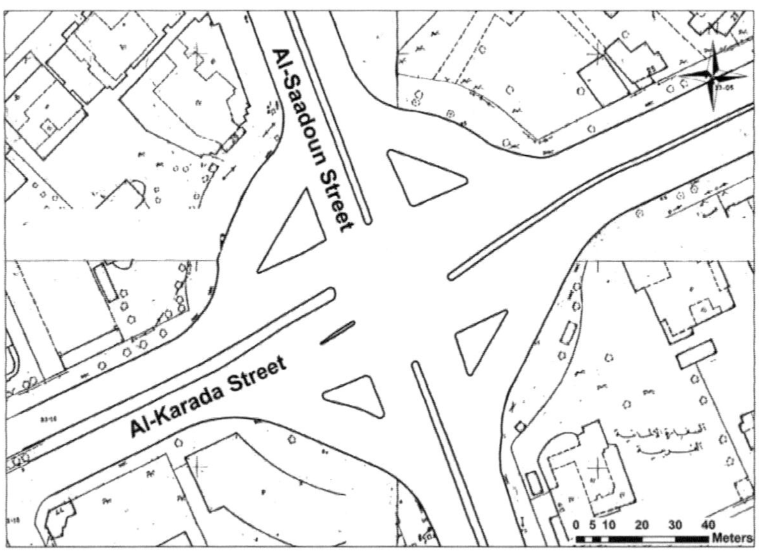

Figure 3. Geometric Layout of Al-Masbah Intersection (Site No.7).

5.4.1 Traffic volume

Two types of traffic volume measurements have been used: hourly traffic volume (HTV) and annual average daily traffic (AADT) for major and minor streets. Also, product of the hourly entrance volume (PEV) is used to represent the square root of HTV in thousand entering to intersection from major and minor streets. The AADT is estimated from account peak hour volume (PHV) for each site by using conversion factor (K) in HCM for the urbanized area to convert PHV to AADT.

The AADT is calculated based on traffic survey in 2017, AADT for 2015 and 2016 which have been estimated based on used growth factor for traffic that has been obtained from relevant authority of urban transportation study in Baghdad city.

5.4.2 Conflict data

Conflict data are collected with the help of video recording equipment. In this study, conflict data observed for 20 minutes per hour from 16 hours of video recording, therefore, the total hour of conflicts observation used in this study is 5 hours and 20 minutes for each intersection. These period fulfills requirements reported by Glauz and Migletz that the minimum number of hours needed to estimate mean hourly count for major conflict data for signalized intersection should not less than 3 hours and 25 minutes as a period of observation per each site [14]. Two types of conflicts measures are collected; total conflict frequency and serious conflict frequency. Total conflict frequency is identified based on definition introduced by Parker and Zegeer while serious conflict frequency identified based in definitions introduced by Amundsen and Hydén using 1.5 sec as threshold values for TTC [4,5]. Further, Hourly Total Conflict (HTC) is used to represent the number of total conflicts frequency at a site divided by observation hours. On the other hand, Hourly Serious Conflict (HSC) represents the number of serious conflicts frequency (conflicts with TTC less than or equal 1.5 sec) at a site divided by observation hours.

5.5. Steps of Calculation the CISI and CMF

- Calculate PEV, HTC, and HSC according to definitions in 5.4.1 and 5.4.2. for each site.
- Calculate the CISI and identifies LOS of safety for sites under study based on conflict risk and accordingly construct scatter plot.
- Calculate the crash rate for sites and compare with the results of the CISI for most probable to profit from safety improvement.
- Further reviewed for hazard sites, select potential countermeasure, apply appropriate CMF and estimate CISI for hazard locations based on proposal treatments.

6. Analysis and Results

6.1. Safety Assessment due to CISI

According to scatter plot diagram of the (HTC/PEV) on the y-axis versus the (HSC/PEV) on the x-axis, CISI is identified based on the established a CISI region boundaries using one standard deviation from the calculated mean of the overall HTC/PEV and the HSC/PEV. The methodology used in this study to identify the CISI similar to the method established by Sayed and Zein, and has been applied in various research for safety improvement [11]. Based on certain measures of the CISI, the intersections introduce a quality scale for safety called the level of service of safety (LOSS). Figure 4 shows that, there are two sites with extreme conflict risk at LOSS F (site No. 6 and 7).

These sites have the highest values of HTC/PEV and HSC/PEV across all sites under study. Seven sites have moderate conflict risk with LOSS C, but site No 8 has the highest values of HTC/PEV and HSC/PEV within this category and lay near the boundary of LOSS E. Site No.2 has the lowest value of conflict risk with LOSS A.

6.2. Safety Assessment due to Crash Rate

The proposed method in order to identify abnormal sites based on measure crash rate by using (mean plus one standard deviation) of a crash rate for sites, since in this study applied this method in the construction of conflict risk for sites. The abnormal sites for crash rate method are site No. 6 and 7 since having crash rate value more than (mean+1Std.dev.) of intersection crash rates.

Both methods used for safety analyses, crash-based methods and conflict –based method, concluded to the same result in identifying the most hazard locations.

6.3. Further review to improve extreme-risk sites

In this step, a further review of intersection characteristics (geometric layout and traffic control) is conducted for the two sites with extreme-risk according to the analyses above. Worthwhile, conflict rate (conflict per thousand vehicle passing) is estimated for two measures of conflicts for each approach. The aim of this step to identify which approach (or approaches) within an intersection may be causing in this deterioration in safety. Table 2 presents the three of CMF for four-leg signalized intersection in HSM, the CMF selected depending on deficiencies in sites and their corresponding of a higher value of conflict rate in approaches.

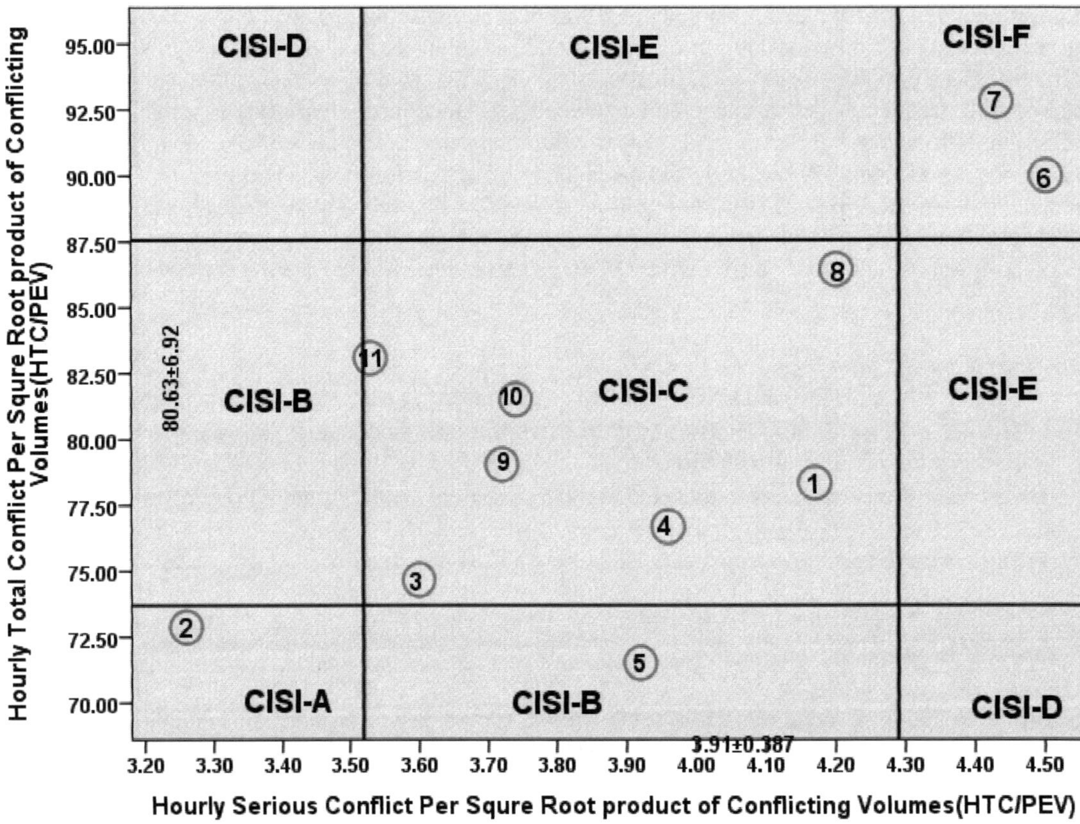

Figure 4. Conflict index for signlized intersections

Table2. CMF in HSM For four –leg signalized intersection

Treatment proposed	Value of CMFs in HSM
Add left-turn lane	0.90^N
Add right-turn lane	0.96^N
Prohibit right-turn-on-red	0.98^N

N=number of approach

Table 3 presents a summary of geometric and operation feature related to potential countermeasure in addition to the value of traffic conflicts for 8 approaches for site No 6 and 7.
This step shows that the deficiencies located in the intersection through investigating the conflict rate (conflict per thousand vehicle passing) for each approach and compared with the geometric features. Table 4 presents the countermeasures proposed for site No 6 and 7.

Table3. Summary of geometric, operation feature, and conflict rate.

Site No	Approach Direction	ELL[a]	ERL[b]	RTOR [c]	Conflict rate per 1000 vehicle (HTC/HTV)	Serious conflict rate per 1000 vehicle (HSC/HTV)
6	WB	NP[d]	NP	permitted	51.14	2.26
	NB	P[e]	NP	permitted	38.95	2.14
	EB	NP	NP	permitted	52.37	2.31
	SB	P	NP	permitted	33.48	2.23
7	WB	P	NP	permitted	40.39	2.14
	NB	NP	NP	permitted	48.28	2.15
	EB	P	NP	permitted	46.33	2.25
	SB	NP	NP	permitted	47.36	2.14

[a] Exclusive left- turn lane .
[b] Exclusive right –turn lane.
[c] right turn on red.
[d] provided.
[e]not provided.

Table 4. Conflict index measures and LOSS for potential countermeasures

Site No	Existing measures of conflict index HTC/PEV	Existing measures of conflict index HSC/PEV	Potential Countermeasures	New measures of conflict index HTC/PEV	New measures of conflict index HSC/PEV	New LOSS
			Proposal No 1			
6	90.00	4.50	Add left –turn lane at EB	86.08	4.35	E
7	92.83	4.43	Add right-turn lane at EB	91.49	4.36	F
			Proposal o 2			
6	90.00	4.50	Proposal no 1+ left-turn lane at approach	83.52	4.235	C
7	92.83	4.43	Proposal no 1+ left-turn lane at NB approach	89.5	4.28	E
			Proposal no 3			
7	92.83	4.43	Proposal no 1+ Proposal no 2 + left-turn lane at SB approach	87.46	4.18	C

7. Conclusion and Recommendation

The objective of this paper is to identify potential hazards collision sites for a sample of 11 four-leg signalized intersections in urban area of Baghdad city through developing conflict index reflect the conflict risk as an alternative measure of crashes. Also, the study conducts a comparison between the

output of the developed method and the crashes-based method in; analyses, assessing safety and proposing the potential countermeasure to improve safety at hazard areas.
The results of this study have indicated the practical value of the TCT as a quantitative method for identifying safety problems at intersections. On-site conflict observation proved also useful for gaining a valuable understanding into the practical functioning of the different intersections, and the influence of other important factors on traffic safety. With regard to the consistency of the conflicts data (conflicts frequency and severity) and the historical crash data for the period 2015-2017 it can be said that there was a reasonable level of agreement for both methods used in this study.
Although a study with more intersections is required, the study results imply that the development method (CISI based on conflict risk) could serve as a viable way of measuring the safety performance of signalized intersection at urban area. Validation of traffic conflicts against traffic crash data is a particularly difficult affair in case of limitations or unavailable of crash data. There is, however, a distinct need to develop valid and useful statistical predictive models (crashes-predication models) of the type suggested by Sayed and Zein [10].
The techniques mentioned in this research used as tools to support the effective decision. However, more studies required to explore the relation between the outputs of risk in site depending on unsafe events should be carried to explain findings and develop solutions to the problems in the future.

References

[1] AASHTO 2010 Highway Safety Manual First Edition (Washington, DC).
[2] Laureshyn A and Varhelyi A 2018 The Swedish Traffic Conflict Technique-Observer's Manua (Lund: Lund University).
[3] Hydén C 1987 The development of a method for traffic safety evaluation: the Swedish traffic conflict technique Doctoral thesis (Lund: Lund University)
[4] Parker M R and Zegeer C V 1989 Traffic conflict techniques for safety and operation Report no. FHWA-IP-88-027.
[5] Amundsen F N and C Hydén 1977 Proc., of the 1 st, int., traffic conflicts technique workshop (Institute of Transport Economics, Oslo).
[6] Laureshyn A Johnsson C De Ceunynck T Svensson Å de Goede M Saunier N and Daniels S 2016 Review of current study methods for VRU safety Report No. Deliverable 2.1 – part 4.
[7] Baker W 1972 An Evaluation of the Traffic Conflicts Technique Highway Research Record No. 384 pp 1-8
[8] Migletz D J Glauz W D and Bauer K M 1985 Relationships between traffic conflicts and accidents (Report No. FHWA/RD-84/042).
[9] Pietrzyk M 1996 Development of Expected Value Conflict Tables for Florida- Based Traffic Crashes USDOT WPI No. 0510721, Washington, D.C
[10] Sayed T and S Zein 1999 Traffic Conflict Standards for Intersections Transportation Planning and Technology vol 22 No. 4 pp. 309-323.
[11] Souleyrette R and Hochstein J 2012 Development of a Conflict Analysis Methodology Using SSAM (Iowa: Iowa Department of Transportation and FHWA)
[12] Harwood D W, Bauer K M, Richard K R ,Gilmore D K, Graham J L, Potts I B, Torbic D J, and Hauer E 2007 Methodology to predict the safety performance of urban and suburban arterials NCHRP
[13] Kolody, K., Perez-Bravo, D., Zhao, J., Neuman, T R 2014 NCHRP 17-50: Highway Safety Manual User Guise
[14] Glauz W D and Migletz D J 1980 Application of traffic conflict analysis at intersections. (Washington DC) NCHRP Report 219.

Asphalt damage reservation due to self-repair: A literature review

T Al-Mansoori[1,*], J Hussein[1], H Aziz[1]

1 Civil Engineering Dept., College of Engineering, Al-Muthanna University, Samawah, Iraq.

*corresponding author email: zzntariq@outlook.com

Abstract. In this paper, a comprehensive state-of-the-art study regarding the mechanism and enhancement of self-repair in asphalt pavement is presented. The review starts with defining the self-repair phenomenon as a natural property in asphalt materials that is associated with other properties such as the viscoelastic response. The difference between self-repair and the viscoelastic response is also presented. In addition, the factors influencing the self-repair mechanism (positively and negatively) are investigated. However, some of these factors remain vague and need to be investigated further. Furthermore, two enhancing self-repair processes, which are extrinsic technologies such as induction heating and encapsulated rejuvenators, are presented as alternative preventive maintenance applied to asphalt pavement. In conclusion, the self-repair could be achieved in an economic way in countries with hot climates, such as Iraq where temperature plays a significant positive role in enhancing the asphalt repair if enough rest period is provided, without the need to other extrinsic technologies to be applied.

1. Introduction

The worldwide use of asphalt pavement and its requirement for maintenance have urged the road agencies to increase their budget. The increased budget to keep roads safe and serving efficiently without flaws could not be reached in most countries. In the USA for instance, the American Society for Civil Engineers estimated a budget deficit of 48% in 2010 with a possibility to reach 54% by 2040 [1]. This budget deficit would lead to postponed maintenance and rehabilitation actions, causing more expensive interventions. This issue has brought the idea of the construction of long-lasting pavements [2] as a solution.

The self-repair phenomenon, which is a natural property in bituminous materials that help heal damages by themselves without external intrusion. This phenomenon is inherent in asphalt materials that can be used as an alternative to asphalt maintenance. This can hugely decrease maintenance expenses, besides extending the service life of highways and eventually promote the environment by decreasing the emissions from slowed vehicles [3]. Self-repair is an intrinsic property that happens due to the diffused molecules between the crack sides. In this way, it prevents the discontinuity of the cracked material [4,5]. Some recent research papers studied this phenomenon mainly based on the mechanical reactions of asphalt pavements [6,7,8]. Mechanically evaluated tests are able to quantify the macroscopic recovery of the damage. However, it is believed that not all damage recoveries are attributed to self-repair because of the viscoelastic responses. Ayar et al [9] and Baaj et al. [10] found that viscoelastic responses could recover damages in asphalt pavement mechanically similar to that of self-repair during the rest period. Therefore, it is crucial to present all these fundamental concepts and suggest relevant methods that can maximise the self-repair of asphalt material.

IOP Conf. Series: Materials Science and Engineering **584** (2019) 012019 doi:10.1088/1757-899X/584/1/012019

The influencing factors that cause self-repair must be studied and separated from those who lead to the mechanical recovery of asphalt during the unloading stage. In this review paper, a technical overview is presented to propose several studies concerning the asphalt self-repair ability and its possible application for clever and sustainable construction and maintenance.

2. Self-repair of asphalt

2.1. Definition

The mechanism of healing or repair in asphalt materials is still unclear because of the complex composition of asphalt with its micro-structure and potential interactions. However, for a general understanding, it can be considered as a natural reaction of asphalt material to repair cracks and eventually restore partially the original properties [11,12]. This process generally occurs in the microcracks because macrocracks cannot be wetted due to their size [13,14]. The wetting capacity, however, is affected by the type of asphalt and its properties. Figure 1 shows the schematic diagram of this mechanism.

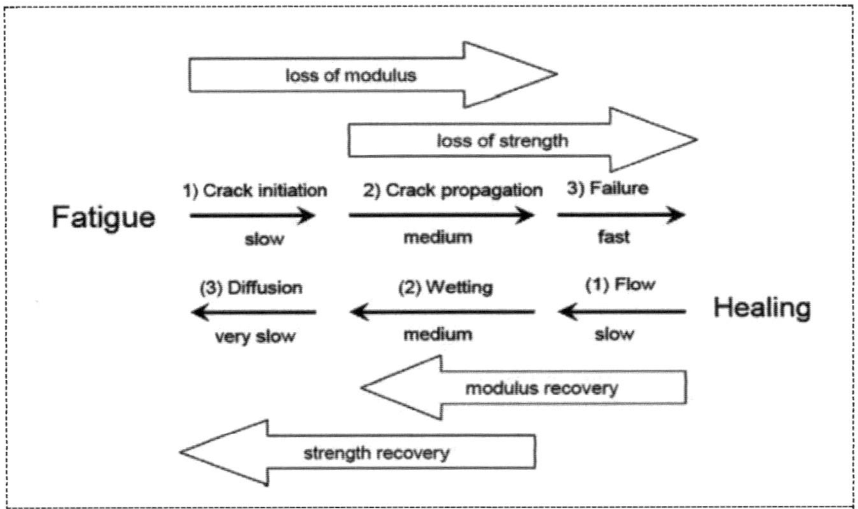

Figure 1. Schematic diagram of self-repair mechanism in asphalt [12].

2.2. The viscoelastic response

Menozzi et al. [15] stated that the asphalt self-repair cannot recover damages that result from permanent deformation, such as plastic deformation (rutting) and high viscosity failures. Healing is usually indicated as a process that occurs due to the diffusion of molecules and wetting. This process could start any time of damage or cracking, not only during rest periods [12]. The viscoelastic response and self-repair of asphalt cracks are not technically related to the damage. They can be defined as a progressive molecular rearrangement that can reverse the mechanical properties of the asphalt material, which relies on the sol-gel process. This builds stronger asphaltene connections with time [16]. The presence of rest periods leads to a progressive shift from a gel to a sol structure while restoring the gel structure [17], as shown in Figure 2.

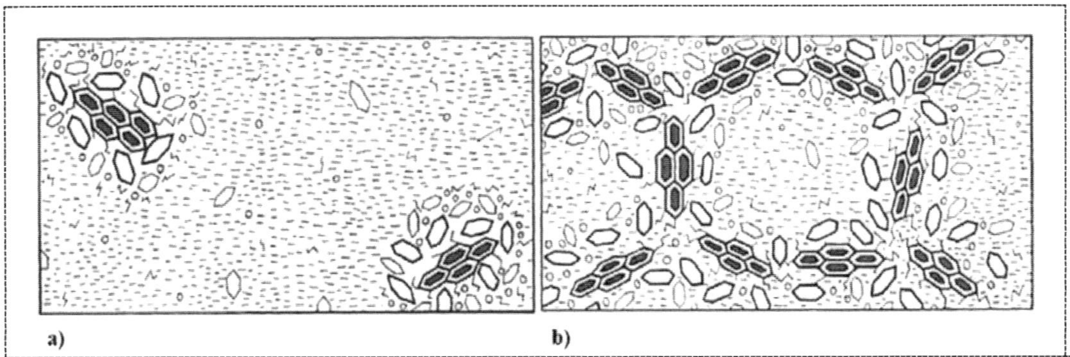

Figure 2. A schematic representation of a) SOL bitumen, and b) GEL bitumen [19].

2.3. Self-repair considerations
Due to its effect on extending the pavement life by reversing the aging of asphalt, some parameters were proposed to include the self-repair property of asphalt material in the Shell Pavement Design Manual [18]. The relationship between these parameters and the rest period were further investigated by Lytton et al. [19]. They found that those parameters could vary from 1.09 to 2.7 for a 30s of unloading. However, researchers are still not accurate on which method that could correctly quantify and predict the healing parameters for different procedures of mechanical testing. However, Gaskin [20] indicated that the compression due to traffic loading in the surface course has significantly improved the healing capability. Figure 3 below shows different types of cracks that appear at different levels of asphalt pavements.

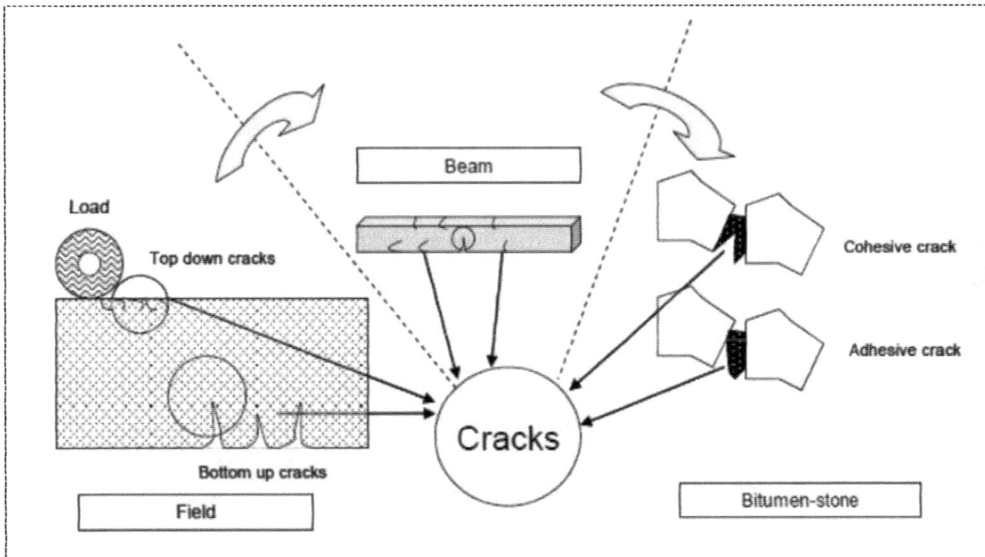

Figure 3. Type of cracks at different levels of bituminous materials [19].

3. Factors affecting self-repair

3.1. Internal forces
Such as:

- Characteristics of bitumen, many studies have found that a soft bitumen with lower penetration can flow easily into the cracks due to capillary flow, which leads to wetting the crack surfaces [15].

- Chemical composition, molecular mobility and microstructure of bitumen have considerable influence of self-repair capability of asphalt pavement [21,22].
- The surface free energy, can be considered as the surface tension of both sides of the crack, where molecules on surfaces have more energy than molecules in the core of material. This energy has a noticeable effect on the damage recovery in asphalt pavement. It may cause a negative effect on the short-term self-repair and positive effect on the long term [19].
- The self-repair effect on the filler/mastic system is higher than that of asphalt system. Al-Mansoori et al [6,8] have found that mastic asphalt system is more vulnerable to temperature effect on healing.

3.2. External forces

External factors can take part in the healing mechanism of asphalt material, such as:

- Rest period, as many studies have found that rest period can positively enhance the development of self-repair in asphalt materials [23].
- Temperature could be another essential influencer in triggering healing process. It is widely reported that healing happens rapidly with an increase of 35% at high enough temperatures [8,24]. Figure 4 below shows the effect of temperature alone in healing cracks.

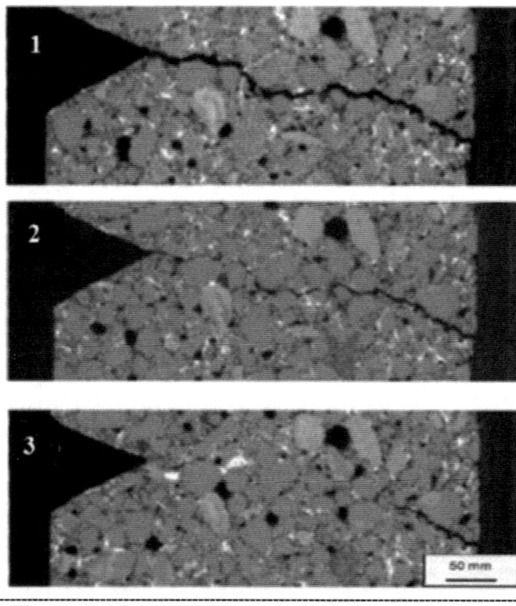

Figure 4. Natural self-repair of asphalt pavement activated by higher temperature [13].

- Traffic loading, it is observed that applying different loading magnitudes with different frequencies on asphalt samples can lead to different healing responses since they help achieve full contact of broken surfaces [23].
- Aging reports are different, some found that aging has a negative effect on the healing of asphalt [22] as the asphalt becomes harder and not easy to flow again and fill the cracks. Others reported that that laboratory aged bitumen could exhibit higher damage recovery compared with the unaged bitumen due to the difference in viscosities [25].

4. Self-repair enhancement technologies

4.1. Induction heating

Temperature and rest period are significant factors in the healing capability of asphalt materials. To increase the temperature locally on damaged asphalt, induction heating is used, which can efficiently enhance the self-repair capability [13]. However, this method is not ideal as it requires big machines to heat the asphalt and could not be applicable in the field. Besides the fact that higher temperatures may cause asphalt aging. Hot climate countries could have this natural temperature effect on asphalt repair without external intrusion as shown in Figure 5.

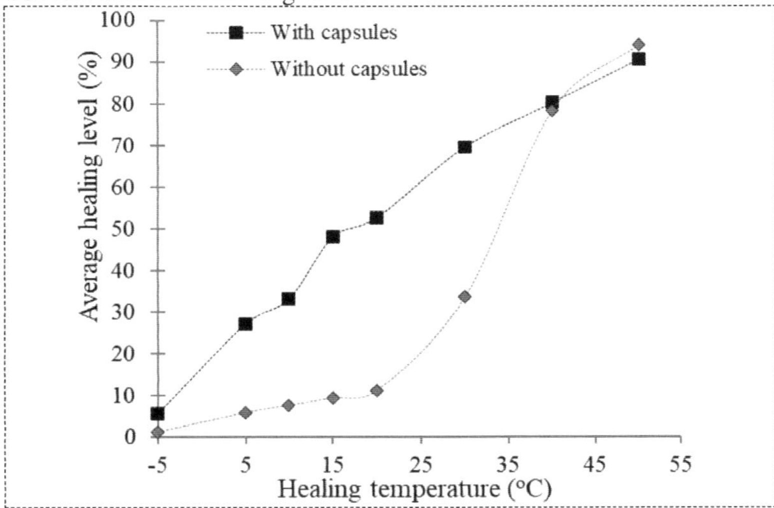

Figure 5. Different healing levels in asphalt samples with and without external intrusion at different temperatures [6].

As it appears from Figure 5, after 40°C, asphalt pavement tends to recover its strength and repair the damage without the need for external rejuvenators. This is crucial for countries such as Iraq to take this advantage.

4.2. Embedded rejuvenators

These rejuvenators are usually encapsulated and mixed with the asphalt to improve the self-repair level of asphalt materials. These rejuvenators were proved to be able to enhance the self-repair of asphalt pavement [6,7,8]. When microcracks appear, these embedded rejuvenators come out of their capsules, which will be broken by the fracture energy at the tip of the crack. Then, the rejuvenator, which should have a lower viscosity than the bitumen, is released and flows into the microcracks due to the capillary action and diffuse into the bitumen as shown in Figure 6.

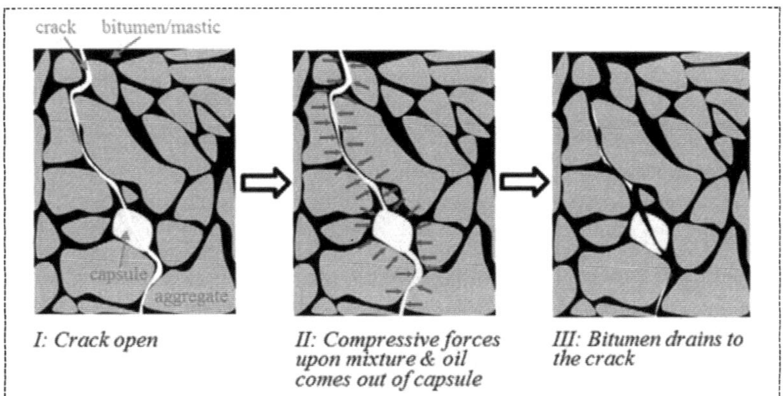

Figure 6. Schematic representation of crack healing in asphalt mixture with the presence of encapsulated rejuvenators [8].

So far, those embedded rejuvenators applied to the asphalt materials believed to be used for only one passive release and a single repair, which means they can be used once and for the first cracks only. A picture of a capsule in an asphalt sample is shown in Figure 7.

Figure 7. Image of a capsule exposed to the air in a damaged asphalt sample [8].

5. Conclusions

An enhancement process to trigger the self-repair property in asphalt is essential and has a profound help to highway agencies and the country's budget. Based on the state-of-the-art study presented in this paper. The following conclusions can be drawn:

- Self-repair of asphalt materials is a natural response to recover the original properties of the material that were damaged. This response is driven by spontaneous wetting and diffusion of molecules from both sides of the cracked asphalt.
- The self-repair process starts when the microcracks (hairline cracks) appear and continues so long as there still possible contact between the crack surfaces.
- Self-repair in asphalt is not straightforward at ambient temperature unless enhanced with enough energy, such as induction heating or added materials (encapsulated rejuvenators).
- The viscoelastic response could also cause mechanical recovery to the damaged asphalt pavement during the rest period. This is crucial if worked together with self-repair process to speed up the recovery.
- Longer rest periods, usually more than 72 hours, and high enough temperature, 40°C or more have a positive influence on the self-repair property of the asphalt materials.
- Both induction heating and embedded rejuvenators can significantly enhance the self-repair process of asphalt materials. However, based on literature, they could be unnecessary in a hot climate, in which asphalt could undergo high enough temperature to repair if enough rest period is provided.

References

[1] ASCE 2013 Report Card for America's Infrastructure American Society of Civil Engineers USA 2013
[2] Monismith C L Evolution of Long-Lasting Asphalt Pavement Design Methodology APerspective presented at the International Symposium on Design and Construction on Long- Lasting Asphalt Pavements Alabama USA 2004
[3] Chung K, Lee S, Park M, Yoo P and Hong Y Preparation and characterization of microcapsule containing self-healing asphalt J Indus Eng Chem 7 29:330, 2015
[4] Agzenai Y, Pozuelo J, Sanz J, Perez I and Baselga J. Advanced self-healing asphalt composites in the pavement performance field: mechanisms at the nano level and new repairing methodologies Recent Pat Nanotechnol 9(1):43, 2015
[5] Hager MD, Greil P, Leyens C, van der Zwaag S and Schubert U Self-healing materials. Mater Today 22(47):5424–30, 2010

[6] Al-Mansoori T, Norambuena-Contreras J and Garcia A Effect of capsule addition and healing temperature on the self-healing potential of asphalt mixtures Materials and Structures, 51:53, 2018
[7] Norambuena-Contreras J, Yalcin E, Garcia A, Al-Mansoori, T, Yilmaz M and Hudson-Griffiths, R Effect of mixing and aging on the mechanical and self-healing properties of asphalt mixtures containing polymeric capsules Construction and Building Materials 254-266, 2018
[8] Al-Mansoori T Encapsulated Healing Agents for Asphalt Self-Healing PhD thesis University of Nottingham 2018
[9] Ayar P, Moreno-Navarro F, and Gámez MCR The healing capability of asphalt pavements: a state-of-the-art review J Clen Prod 113:28 40 2015
[10] Baaj H, Mikhailenko P, Almutairi H and Di Benedetto H Recovery of asphalt mixture stiffness during fatigue loading rest periods Constr Buil Mater 158:591 600 2018
[11] Agzenai Y, Pozuelo J, Sanz J, Perez I and Baselga J. Advanced self-healing asphalt composites in the pavement performance field: mechanisms at the nano level and new repairing methodologies Recent Pat Nanotechnol 9(1):43 2015
[12] Qiu J Self-healing of asphalt mixtures: towards a better understanding of the mechanism TU Delft Delft University of Technology 2012
[13] García Á Self-healing of open cracks in asphalt mastic. Fuel 93:264 72 2012
[14] Little DN, Lytton RL, Chairl B, Williams D and Texas A An analysis of the mechanism of microdamage healing based on the application of micromechanics first principles of fracture and healing. J Assoc Asph Pav Tech 68 1999
[15] Menozzi A, Garcia A, Partl MN, Tebaldi G and Schuetz P. Induction healing of fatigue damage in asphalt test samples Construct Build Mater 74:162 8 2015
[16] Santagata E, Baglieri O, Tsantilis L and Dalmazzo D Evaluation of self-healing properties of bituminous binders taking into account steric hardening effects Construct Build Mate 41:60 7 2013
[17] Read J, Whiteoak D The shell bitumen handbook Thomas Telford 2003
[18] Lytton RL, Uzan J, Fernando EG, Roque R, Hiltunen D and Stoffels SM Development and validation of performance prediction models and specifications for asphalt binders and paving mixes Washington, DC Strategic Highway Research Program 1993
[19] Shell International Petroleum Company ltd. Shell pavement design manual: asphalt pavement and overlays for road traffic. Shell International Petroleum Company 1978
[20] Gaskin J On bitumen microstructure and the effects of crack healing University of Nottingham 2013
[21] Bhasin A, Bommavaram R, Greenfield ML and Little DN Use of molecular dynamics to investigate self-healing mechanisms in asphalt binders J Mater Civ Eng 23(4):485 92 2011
[22] Ashouri M Modelling microdamage healing in asphalt pavements using continuum damage North Carolina North Carolina State University 2014
[23] Bazin P, Saunier JB. Deformability, fatigue and healing properties of asphalt mixes Intl Conf Struct Design Asphalt Pvmts Preprint 438-451 1967
[24] Little DN, Lytton RL, Chairl B, Williams D, Texas A. An analysis of the mechanism of microdamage healing based on the application of micromechanics first principles of fracture and healing J Assoc Asph Pav Tech 68 1999
[25] Al-Balbissi A H A comparative analysis of the fracture and fatigue properties of asphalt concrete and sulphlex Texas A & M University 1983

Effect of Hybrid Polypropylene - Carbon Fiber on Self-Compacting Mortar

Muataz Ibrahim Ali[1,*], Asmaa Abdul Jabbar Jamel[2], Ahmed Mohammed Ahmad[3], Anwer Muneer Ali[4]

[1]College of Engineering, Samarra University, Iraq. Email: muitaz88@yahoo.com
[2]College of Engineering, Tikrit University, Iraq Email: ms.asmaajameel@tu.edu.iq
[3]College of Engineering, Samarra University, Iraq. Email: Ahmed_m@uosamarra.edu.iq
[4]College of Engineering, Samarra University, Iraq. Email: Eng.anwar2012@gmail.com
*Corresponding author: muitaz88@yahoo.com

Abstract. The present work, aims to improve the properties of fresh and hardened self-compact mortar (SCM) containing hybrid Polypropylene-carbon fiber. The hybrid fiber added as a percentage 0.1%, 0.5% and 1% by volume of self-compact mortar, five different mixture of 100-0%, 70-30%, 50-50%, 30-70% and 0-100% for every fibers ratios (Polypropylene to carbon). The effect of hybrid fiber on some physical and mechanical properties of (SCM) was also investigated as a compressive strength and modulus of rupture, which was examined at ages of 28 days. Results showed that the best added of hybrid fiber (50 P.P-50 C.F) % increased compressive strength with ratio 32%, so increased modulus of rupture for self-compact mortar with ratio 130% at hybrid fiber (0 P.P-100 C.F) % , also indicated that the added improve hard characteristic and decrease of fresh characteristic as a flow time, mini slump flow and specific gravity, when the volume fraction 0.5% and 1% for all hybrid fiber exceeded the limits of the standard because the largest flow time is 11 seconds.

1. Introduction

Self-compacting concrete (SCC) represents a main evolution in the constructing manufacture. Its favorable features produced from an elimination of the mechanical pressure. Compared to normal vibrated cement, SCC offers plentiful advantages in the condition of technology, economy and worker health [1]. The cracking receptivity in SCC are often happened, because Self-compacting concrete (SCC) is designed by a largest volume of mortar compared to ordinary mortar [2] also less aggregate content. The increase of volume of mortar improves flow ability but it might have unlooked-for effects on time-dependent and the mechanical properties disfigurement. Drying shrinkage is known to reduce with volume of mortar [3] and the shrinkage non- restricted where, rise nearly linearly with the increase in the volume of mortar [4].
Their problem if get cracks in concrete or mortars leading from going early saturation, damage, scaling, steel corrosion and discoloration [5]. As a result of the exposure of concrete to the loads of service and during service, also in the early stages exposed to cracks that later affect the durability of concrete [6].

Fiber reinforced concrete or mortar improve strength in the service than mortar or concrete without fiber accordingly fibers finite pervasion of the cracks inside mortar or concrete [7].
So for solve that problem some researcher suggested that added fibers (as mentioned before). That added depended on many factor as kind of fiber and its volume portion has the marked influence on properties of the fiber reinforced mortars. The fiber-reinforced composites may be classified as it is function of their fiber volume fraction like low (<1.%), mild (between 1.% and 2.%) ,so the high volume fraction (larger than 2.%) [8]. According to many report the added steel fiber (S.F) into mortars or concrete in an amount of 1 - 1.5% by volume rises flexural strength by over to 150-200%, its tensile strength by over to 100%, so the compressive strength rises by 10 - 25% [9].
In another side some of research get better the properties of (SCM) for hardened and fresh mortars such as mechanical performance and workability, by using hybrid fiber with two or more type of fiber as reduce the brittleness of cement mass, by added of carbon fiber, calcium carbonate whisker and hybrid fiber to improve mechanical prosperities of the cement mass [10].So investigates the mechanical properties such as density, compressive strength, toughness indices, flexural strength, and the impact-load test to all mixture of hybrid fiber (fibers-steel, synthetic fibers and palm fibers) such as 2 % from volume fraction. And get to increase the amount of hybrid fibers has cause the increase of resistance the impact load, increase the load cause by first crack and post-crack strength respectively [11].
This paper aims experimentally to produce self-compacting mortar SCM by using the hybrid Polypropylene-carbon fiber to improve the properties of fresh and hardened of self-compacting mortar.

2. Experimental Works

Sixteen mixes of mortar specimens are fitted and cast. At the beginning, all mixes were designed according to EFNARC [12]. Then, three ratio by volume of mortar (0.1%, 0.5% and 1.%) are used by five percentage mix of (100 P.P-0 C.F% , 70 P.P-30 C.F%, 50 P.P-50 C.F%, 30 P.P-70 C.F% and 0 P.P-100 C.F%) for every fibers ratio (Polypropylene fiber for carbon fibers).So, The fifteen mortar mixtures contain fibers and one mixture without fibers as control mixture.
Cement mortar cubes of 50 mm side length and prisms with dimensions of 40x40x160mm were used to evaluate a concrete compressive strength and modulus of elasticity, respectively. The prisms were casted and cured, according to ASTM C348 [13] and ASTM C109, to investigate the modulus of rupture.

3. Material

Cement: - Iraqi standard ordinary Portland cement was employed. The cement is tested according to Iraqi standard I.O.S No. 5/1984 [14], so the cement has achieve all a requirements. The chemical properties of this cement are showed in (Table 1).
Fine Aggregate: - Graded fine aggregate, the trading of fine aggregate as shown in (Table 2).
Fibers: - It was used two various kinds of fiber, the polypropylene fiber (P.P) and carbon fiber (C.F), the P.P made by FOSROC Company/Jordan. So the P.P fibers are 6 mm long, the physical properties as shown in (Table 3), while the second is carbon fiber manufactured by Sika Company with length of 6 mm, the physical properties of carbon fiber as shown in (Table 4).
Superplasticizer: - It was used the super-plasticizer Sikament® FFN during preparation the all mixtures in this study. It doesn't have Ion chlorides or any other element of promoting the erosion of steel reinforcement. The weight ratio (according to the cement weight) is same in all mixtures and it is 0.4%.

4. Mixing

The mix design method used in this study was according to EFNARC [12].Several trial mixes were done at the beginning in order to get the optimum admixture dosage and the required mechanical performances of SCM. The self-compacting mortar was fabricated by mixing 712 kg/m3 cement, 1212 kg/m3 fine aggregates, and 2.82 kg/m3 SP (0.4% of cement mass). The water-cement ratio was kept constant to 0.45, the same above quantities were used in casting the mortars having hybrid Polypropylene-carbon fibers.

Table 1. Chemical properties composition of cement

Oxides composition	Content %
Silica, SiO_2	13.2 %
Alumina, Al_2O_3	4.5 %
Iron oxide, Fe_2O_3	----
Magnesia, MgO	---
Sulfate, SO_3	1.11 %
Loss on Ignition, (L.O.I)	0.92 %
Insoluble material	1.03 %
Lim Saturation Factor, (L.S.F)	0.8
Silica, SiO_2	13.3 %

Table 2. Grading of fine aggregate

Sieve size	Cumulative passing %
9.5-mm (3⁄8-in.)	100
4.75-mm (No. 4)	91.32
2.36-mm (No. 8)	83.9
1.18-mm (No. 16)	67.74
600-µm (No. 30)	20.4
300-µm (No. 50)	2.46
150-µm (No. 100)	0

Table 3. Physical properties of polypropylene fibers

Form	100% Virgin Polypropylene Fiber
Specific gravity	0.91 g/cm^3
Alkali content	Nil
Sulfate content	Nil
Chloride content	Nil
Air entrainment	Air content of concrete will not be significantly increased
Young modulus	(3500-4800) MPa
Tensile strength	350 MPa
Melting point	160-170°C
Fiber length	12 mm
Ignition Point	590 °C
Cement Compatibility	Excellent

Table 4. Physical properties of carbon fibers

Details	Description
Length (mm)	6
Geometry	Fibrillated
Density (Kg/m^3)	910
Diameter (mm)	0.02
Tensile Strength (MPa)	400
Modulus of Elasticity (MPa)	3450

5. Test Methods for Fresh Mortar

5.1. Mini Slump Test

That test is one of fresh mortar tests depending on measurement of spread of paste or mortar put into the cone-shaped molding. The section of cone (diameters: 100 at bottom and 70 mm at top and height: 60mm) is put in the very smooth surface and non- sucking plate, the cone filled by mortar or paste and raised [16]. The final output results are the average the diameter of two measurements made of two perpendicular directions for the fresh paste specimens, as shown in Figure (1).

5.2. Mini V-Funnel Test

The mini V-funnel test has calculating the time required to flow the volume of paste or mortar (one liter) during the nozzle. That test is oftentimes used to degree the viscosity of paste which may be linked with properties such as pump an ability, finish ability and cohesiveness. Figure (1) shown the device used in this test.

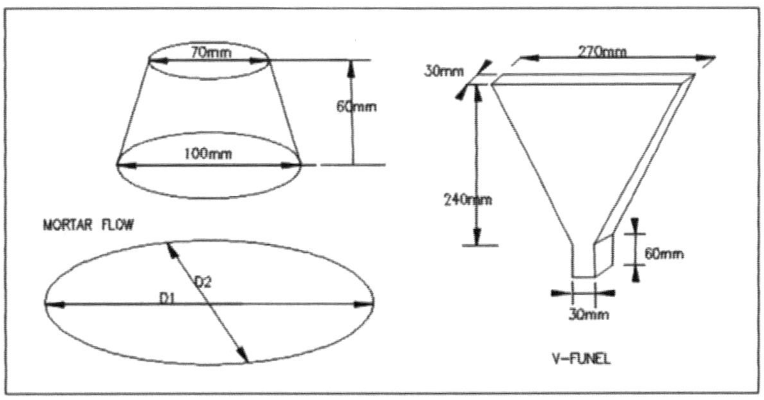

Figure 1. Dimensions of mini-slump and mini V-funnel

6. Test Methods for Hardened Mortar

6.1. Compressive Strength

According to B.S 1881: part 116: 1989 [15], compressive strength of mortar is taken by using 50 mm cubes. The electrical Compression machine are used for test with capacity of ٣000.kN by speed loading on sample 4.0.kN/ second in this test. The average value each three cubes are measured to find out the compressive strength for every mixes, so the all samples tested at age 28 days.

6.2. Flexural Strength
The flexural test worked by using prisms have dimensions 40mm*40mm*160 mm, the prisms are tested depending on the specification ASTM C78. Which states that :use Two points load on the samples. All samples are tested at age 28 days with the average of three samples to each mixes.

7. Results and Discussion

7.1. Fluidity
The data of test to mini-slump are shown in Figure (2). The control mixture achieved the highest slump flow values. The self-compacting mortars having hybrid Polypropylene-carbon fibers recorded mini-slump values smaller than that of corresponding control mixture, it can be note that the highest mini-slump value may be got for hybrid contain high C.F ratio compared to ratio for P.P.F as hybrid ratio (0 P.P-100 C.F) %, by about 255 mm, 280 mm, and 300 mm, respectively compared to control mixture, noted that the volume fraction 0.5% and 1 % is outside the limits of the European standard (EFNARC), where the limits of the standard ranging from (240 to 260) mm. Other percentage 0.1 % has within the limits of the standard ranging, so recorded mini-slump values smaller than that of corresponding the mixture composite have equal ratio for P.P.F and C.F, this may be attributed to consistency in the distribution and consensus of roughness surface for both type fibers [17].

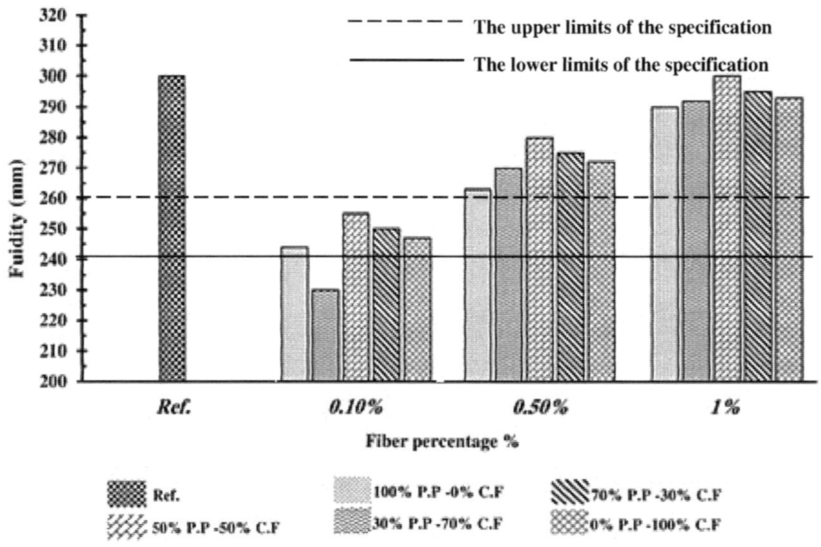

Figure 2. The relationship between mix and the fluidity

7.2. Mini V-Funnel Test
Figure (3) shows the results of time in mini V-funnel experience. The flow time of mortars containing hybrid fibers larger than that recorded to mortar without fibers and the values increased when the fibers proportions were independent as a 100% PP-0% C.F conversely from the hybrid fibers 0% PP-100% C.F, largest values when the Polypropylene fiber proportions (100% PP-0% C.F) was higher with 39%, 86% and 121%, respectively compared to control mixture because of the surface carbon fiber was more smooth from surface Polypropylene fiber, that reduction in the viscosity of self-compacting mortars [18], also the volume fraction 0.5% and 1 % is outside the limits of the European standard (EFNARC),

where the limits of the standard ranging from (7 to 11) sec, only hybrid fibers (50% PP-50% C.F) has time flow 10.5 sec that time within the standard (EFNARC) when the volume fraction equal to 0.5%. But the value of flow time for volume fraction 0.1% of all hybrid fibers has within the limits of the standard ranging.

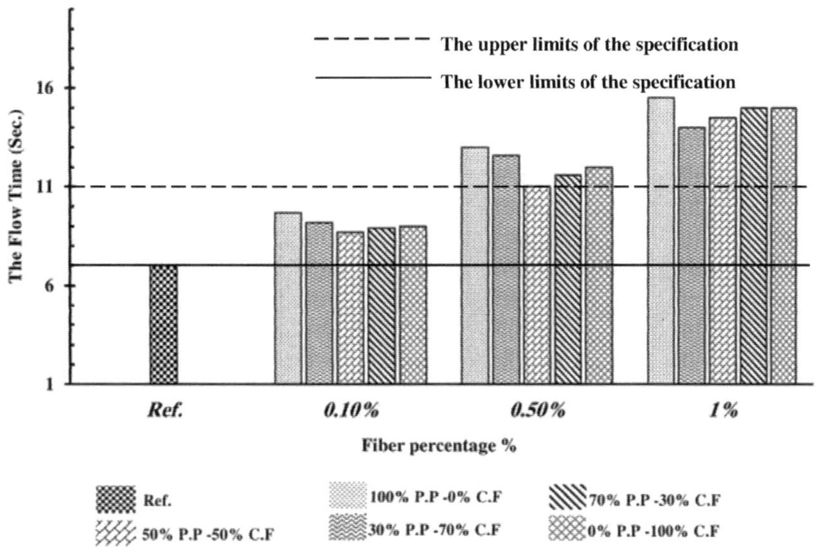

Figure 3. The relationship between mix and the flow time

7.3. Density

The hardened 28 days oven dry density of each kinds of mortar mixes are shown in Figure (4). From this figure, it can be observed that the dry density value for the reference mixture is more than the mortars containing hybrid fibers this reduced in density due to add fibers to the mixture, this added reduced unit weight for fiber percentage equal to 0.1% by 1.05%, 1.07%, 1.43%, 1.79% and 1.44% when the hybrid Polypropylene-carbon fibers are 100-0%, 70-30%, 50-50%, 30-70% and 0-100% respectively compared to control mixture so by 3.57%, 2.50%, 3.21%, 3.22% and 3.6% when the fiber percentage equal to 0.5% by volume. Finally, the fiber percentage equal to 0.5% by volume reduced the density compared to the reference mixture by 6.43%, 7.14%, 6.79%, 6.07% and 6.43% when the hybrid Polypropylene-carbon fibers are 100-0%, 70-30%, 50-50%, 30-70% and 0-100%., this result conforms to the requirement of ACI 213-R-03[19]. However, the reduction in the density in mortar containing hybrid fibers happened due to the low density for Polypropylene and carbon fibers [18].

7.4. Compressive Strength

Figure (5) Show the results of compression tests. Compressive strength is found at the age 28 days accordance to ASTM C109, as the way to quality control. The figure shows influence of fiber ratio on compressive strength, so noted that highest values compressive strength happened when the hybrid have same ratio for P.P.F and C.F on the other proportion.

The figure showed that, compressive strength amount is raised because insert the fiber to sample with many fiber volume compared to the control mixture. The addition of fibers effect on cubic hardened mortars beneath axial loads, Fissures happen in microstructure of mortars and fibers decrease crack and development. Also, the compressive strength of SCM increased. All fiber fraction mixture, the maximum raised happened at hybrid ratio (50% P.P.F- 50% C.F), it was raised with 32%, 26%, and 22% for 0.1%, 0.5% and 1% fiber fraction ratio respectively, so the peak increment is for 0.1% fiber fraction (32%).

This increased in value due to the fibrous mortars sample can develop lateral tension, then that cracks begin progress. As the progress crack approaches a fiber, the interdependence at the fiber matrix interface begins because of the tensile stresses vertical to the prospective path of the development and advancing crack. All cases mentioned before up to the interface, the head of crack meeting a process of blunting and increased in compressive strength [19].

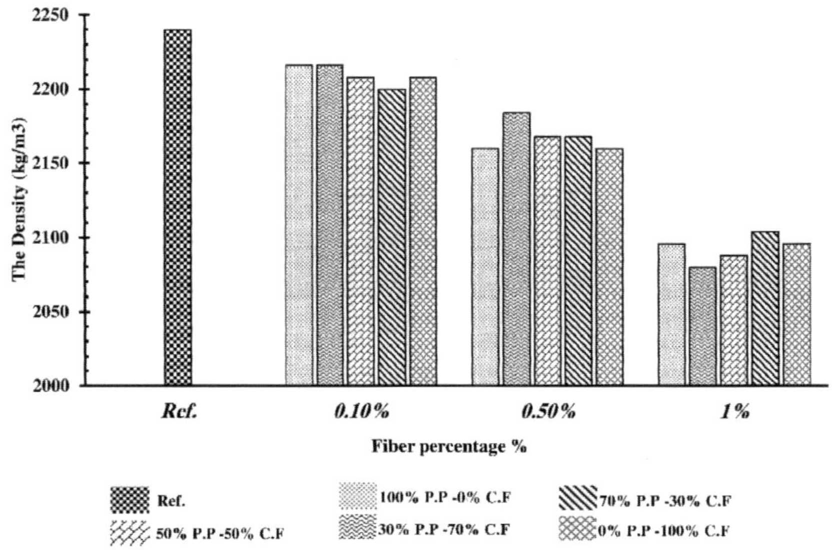

Figure 4. The relationship between mix and the density

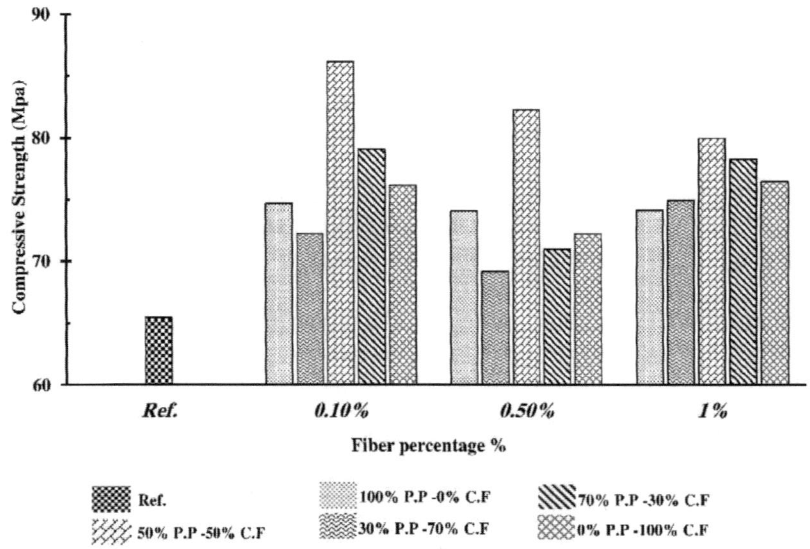

Figure 5. The relationship between mix and compressive strength

7.5. Ultrasonic Pulse Velocity

Figure (6) shows the test result of UPV test (Ultrasonic Pulse Velocity) for SCM specimens at an age of 28 days. Results showed that the mix containing fibers reduced the velocity by increased in percentage carbon fibers about 4.5, 5.5, 7.3, 6.4 and 7.3% comparing with the reference mixture when the hybrid Polypropylene-carbon fibers are 100-0%, 70-30%, 50-50%, 30-70% and 0-100%, respectively and the fiber percentage equal to 0.1% by volume. So the ratios were 15.5, 17.3, 16.4, 17.5 and 17.4%, while the hybrid fibers are 100-0%, 70-30%, 50-50%, 30-70% and 0-100%, respectively at fiber percentage equal to 0.5%. At last the increased in the velocity by ratios 21.8%, 23.6%, 22.7%, 24.5% and 23.7%, the reason to get these results is due to the few capacities for fibers to wave transfer [20].

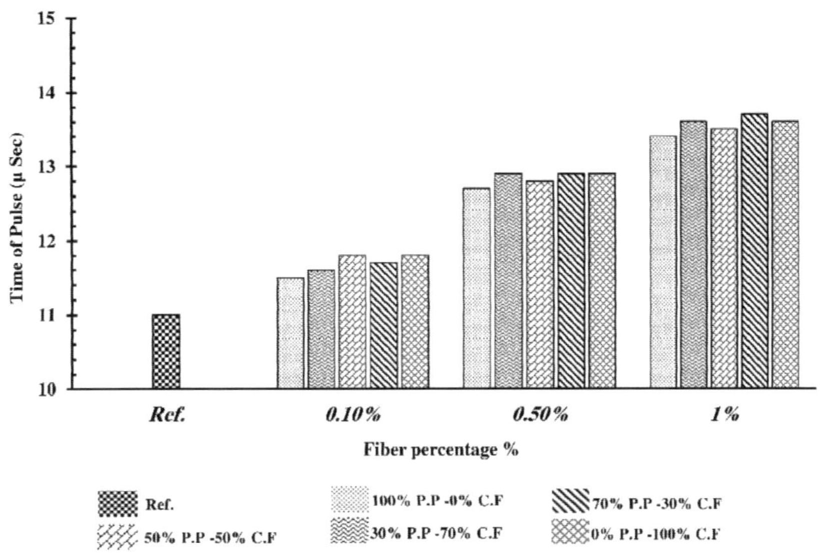

Figure 6. The relationship between mix and ultrasonic pulse velocity

7.6. Hammer Test

Figure (7) shows the hammer test results. this method gives results of the rebound number, which are linked with compressive strength, give us an initial perception of the compressive strength of the SCM without the examination of the damage, where the hammer result of control mortar highest rebound number compared with SCM containing hybrid fibers, that reduced in rebound number for hammer come back to absorption and dispersion shock resulting from the hammer by fibers [20], where the resultant reduced by increased the fiber fraction were less than when the fiber fraction equal to 1% and hybrid fibers equal to (100% P.P.F- 0% C.F).

7.7. Modulus of Rupture

The average results for the flexure experience are given in Figure (8). The flexural strength trend for carbon fiber and polypropylene fiber differs when fiber increment. In general, all fiber ratio and fiber fraction, flexure strength of SCM sample increment as carbon fiber ratio increment, it can be noted: the addition carbon or polypropylene fibers slightly raises the flexural strength compared with the reference samples, increasing the total composite fiber fraction with 0.1%, 0.5%, and 1% by volume are offers an raised in the flexural strength by about 30 % for hybrid fibers equal to (30% P.P.F- 70% C.F) when the fiber fraction 0.1%, 60 % for hybrid fibers equal to (0% P.P.F- 100% C.F) at the fiber fraction 0.5% and for hybrid fibers (0% P.P.F- 100% C.F) equal to 130% when the fiber fraction 1%. This increment results foremost from the fibers intersecting vertically with cracks in the tension half of the sample.

These fibers restrict or reduce the crack face isolate by expansion the crack, so providing the extra energy-absorbing mechanism [19].

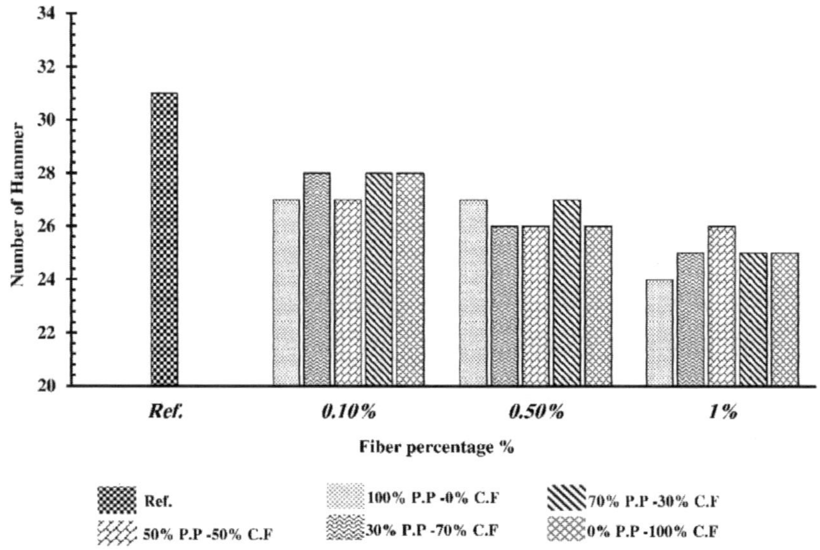

Figure 7. The relationship between mix and read of hummer

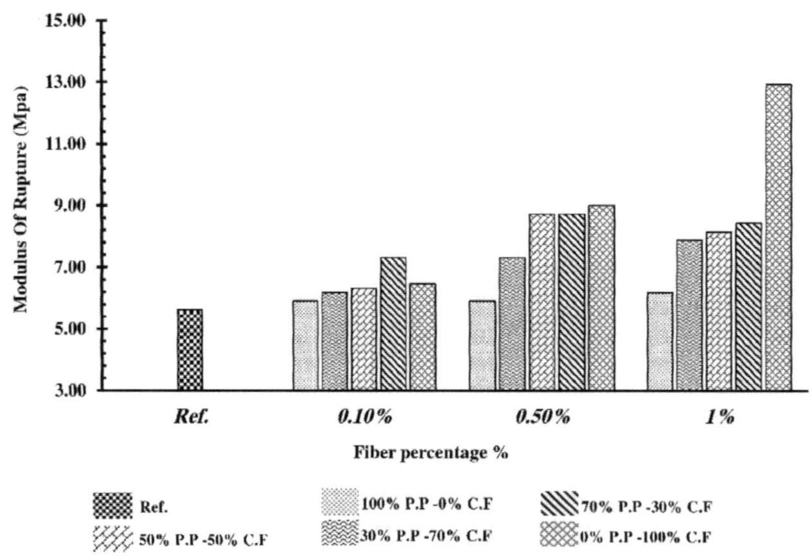

Figure 8. The relationship between mix and modulus of rupture for prism

8. Conclusions

- Added of Polypropylene and carbon fibers to the SCM mix reduced the mini-slump values, but it did not reach the minimum value for specification except when increasing the fiber percentage to 1% by volume.
- The hybrid fibers effect on flow time of SCM by increased the time, but it did not reach the maximum value for specification (12 Sec.) except when increasing the fiber percentage to 1% by volume.
- Density reduced for the hybrid fibers of SCM with increase in the proportion of fiber until it reached a ratio 7% compared to the reference mixture.
- Increased the compressive strength of SCM was contain hybrid fibers, for every fiber fraction mixtures and the maximum increment is obtained for hybridization percentage (50% P.P.F- 50% C.F), it's was equal to 32%, 26%, and 22% for 0.1%, 0.5% and 1% fiber ratio respectively, so the peak increment is for 0.1% fiber fraction (32%).
- SCM contain hybrid fibers increased the UPV (Ultrasonic Pulse Velocity) with ratio 7%, 17%, and 24% compared to the reference mixture for 0.1%, 0.5% and 1% fiber ratio respectively, and reduced the rebound number with ratio 11%, 15%, and 19% compared to the reference mixture for 0.1%, 0.5% and 1% fiber ratio respectively.
- Added the hybrid fibers on SCM increased the modulus of rupture with ratio 30%, 60%, and 130% for 0.1%, 0.5% and 1% fiber ratio respectively for the sample compared to the reference sample.

References

[1] Slater E, Moni M, Alam MS. Predicting the shear strength of steel fiber reinforced concrete beams. Constr Build Mater 2012;26:423–36.

[2] Turatsinze A, Garros M. 'On the modulus of elasticity and strain capacity of self-compacting concrete incorporating rubber aggregates'. Resour Conserv Recycl 2008;52:1209–15.

[3] Mehta PK, Monteiro PJM. Concrete; microstructure, properties, and materials. 3rd ed. New York: McGraw-Hill; 2006.

[4] Khaliq, W., and Kodur, V., "Thermal and Mechanical Properties of Fiber Reinforced High Performance Selfconsolidating Concrete at Eleavated Temperatures," Cem. Concr. Res., Vol. 41, No. 11, pp. 1112-1122, 2011.

[5] Ho AC, Turatsinze A, Hameed R, Vu DC. Effects of rubber aggregates from grinded used tyres on the concrete resistance to cracking. J Clean Prod 2012;23(1):209–15.

[6] Suhaendi SL, Horiguchi T, Saeki N. Fire resistance of fiber reinforced high strength concrete-effect of fibers on the permeability of heated concrete. In: Proc 8th int conf on concrete engineering and technology, 19–21 April 2004, Kuala Lumpur, Malaysia. p. 225–9.

[7] Iman Mehdipour, Nicolas Ali Libre, Mohammad Shekarchi. "Development of Fiber Reinforced SCM for Sustainable Construction". Journal of Structural Engineering and Geotechnics, 1 (1), 19-26, Spring 2011 19

[8] Ramadoss P, Nagamani K. Tensile strength and durability characteristics of high performance fiber-reinforced concrete. Arab J Sci Eng 2008;33(2B):307– 19.

[9] Domagała L. Modification of properties of structural lightweight concrete with steel fibres. J Civ Eng Manage 2011;17(1):36–44.

[10] Ming LI , Meng LIU , Yuanyi YANG , Zaoyuan LI , Xiaoyang GUO, "Mechanical properties of oil well cement stone reinforced with hybrid fiber of calcium carbonate whisker and carbon fiber " Petroleum Exploration and Development 42 (1) (2015) 104– 111.

[11] Eethar Thanon Dawood, , Mahyuddin Ramli, "Mechanical properties of high strength flowing concrete with hybrid fibers " Construction and Building Materials 28 (1) (2012) 193–200.

[12] Specification and guidelines for self-compacting concrete EFNARC, Association House, UK.

[13] ASTM. ASTM C348. Test method for flexural strength of hydraulic mortar. Annual Book of ASTM Standards, vol. 0401. West Conshohocken, PA, USA: American Society of Testing Materials; 1997.
[14] Iraqi Standard no. 5 for year of 1984, "Portland Cement", Central Organization for Standardization and Quality Control, 8 pages, Baghdad.
[15] BS 1881, Part 116, 1989, "Method for Determination of Compressive Strength of Concrete Cubes", British Standards Institution, 1881; PP.3.
[16] Tanyildizi H. Effect of temperature, carbon fibers, and silica fume on the mechanical properties of lightweight concretes. New Carbon Mater 2008;23(4):339–44.
[17] Bilodeau A, Kodur VKR, Hoff GC. Optimization of the type and amount of polypropylene fibres for preventing the spalling of lightweight concrete subjected to hydrocarbon fire. Cem Concr Compos 2004;26:163–74.
[18] Mazaheripour M, Ghanbarpour S, Mirmoradi SH, Hosseinpour I. The effect of polypropylene fibers on the properties of fresh and hardened lightweight self-compacting concrete. Constr. Build Mater 2011;25:351–8.
[19] Mazin Burhan, Alya'a Abbas, Sa'ad Mahmoud. 'Determination of Mechanical Properties of Hybrid Steel-Nylon Fiber Reinforced Concrete'. Modern Applied Science 2010; 4 (12).

IOP Conf. Series: Materials Science and Engineering **584** (2019) 012021 doi:10.1088/1757-899X/584/1/012021

Swelling Control of Expansive Soils Using Cement Dust

Alaa D. Almurshedi[1], Jawad K. Thajeel[2,*] and Hadeel Ch. Dekhn[3]

[1] Assistant Professor, College of Engineering, University of Baghdad – Iraq.
[2] Lecturer, College of Engineering, University of Thi-Qar – Iraq.
[3] Lecturer, College of Engineering, Al-Mustaqbal University College – Iraq.

Corresponding author email: jawad2006mm@yahoo.com

Abstract. Expansive soils exist in many regions over the world including Iraq. Expansive or swelling soil contains minerals that are capable to absorb water easily causing an increase in its volume. In this research, an expansive soil that exists in some locations in the province of Karbala in Iraq was used. Limited data is available to investigate the effect of cement dust on engineering properties of expansive soils. For this purpose, samples have been prepared by mixing the expansive soil with different percentages of cement dust. The samples have been subjected to cycles of wetting and drying to investigate the swelling behaviour. Soil samples that are used were compacted at the optimum moisture content. Five different percentages of cement dust content are used in this study include: 4, 8, 12, 16 and 20%. Fundamental tests for all soil samples. It is concluded that the addition of cement dust to the expansive soils is successful method to improve their properties such as decreasing the consistency limits, increasing the maximum dry density corresponding to lower optimum water content, decreasing the swelling pressure and swelling potential.

1. Introduction

Expansive soils are soils that experience significant volumetric changes (swell or shrink) associated with changes in moisture content. Such soil usually contains excessive amount of swelling clay minerals that have high ability to absorb water such as smectite or vermiculite. As a result, soil swells (increase in volume) as it gets wet and shrinks (decrease in volume) as it gets dry [1], [2].

Expansive soil causes considerable damages to civil engineering structures because of the pressures that are associated with heaving and/or differential settlement. Pressures are strong enough to crack foundations, floors, walls, roads, and pipelines. Many factors govern the expansion behavior of such soils like changes in water content, amount and type of clay size particles, type of soil (natural or fill), dry density, moisture content, magnitude of the surcharge pressure, and amount of non-expansive material such as gravel or cobble size particles [2].

Civil engineering projects that are constructed in areas with problematic soils are one of the most common problems over the world. This soil can be removed before construction and replaced by soil with good properties. This technique would be expensive and not economic when the soil depth is high. Therefore; researchers look for alternative methods to improve soil properties before construction. Soil stabilization is one of the most common methods that has been used to enhance the physical and mechanical properties of soils to meet the engineering requirements of specific projects [3], [4].

Kallan and Akbulut, (2004) [5], studied the effect of adding silica fume to clay soil. Permeability, swelling pressure, and compressive strength were determined for all testing samples. Silica fume-added samples were characterized with low permeability and swelling pressure and significantly high compressive strength. Silica fume was also used to investigate its effect on desiccation cracks that are developed on the surface of expansive soils. Kallan (2009) [6], concluded that silica fume is well suited for use in construction of liners in landfills.

Kolias et al., (2005) [7], investigated the effectiveness of using high calcium fly ash and cement in stabilizing high plastic (CH) and low plastic (CL) clays. Testing samples were prepared by adding various percentages of fly ash and cement to the clay soils. Uniaxial compression and indirect tension tests were carried out to determine the elasticity modulus and CBR values at 90 days. In addition, field tests were done on subgrades that have been stabilized by fly ash and cement. The results show clear improvement of mechanical properties of clay soils.

Basha, et al., (2005) [8], used rice husk ash and cement to stabilize soils and evaluate some soil properties such as plasticity, compaction, and strength and X-ray diffraction. It was found the plasticity of the treated soils was reduced. Compaction test results showed that soil samples with any percentage of rice husk and cement have lower maximum dry density associated with higher optimum moisture content compared with untreated soil samples.

2. Experimental Study

2.1. Properties of Raw Soil

Soil was collected from Karbala city Figure 1 [9] at a depth of (0.5-1m) below the ground surface. The site location and coordinates are shown on the map in Figure 1. The soil was primarily allowed to dry for 2 days and then thoroughly grinded. It was then tested to find some of the important properties that are relevant to this study as given in Table1. Sieve analysis was also conducted on the collected soil in order to determine its grain size distribution as presented in Figure 2.

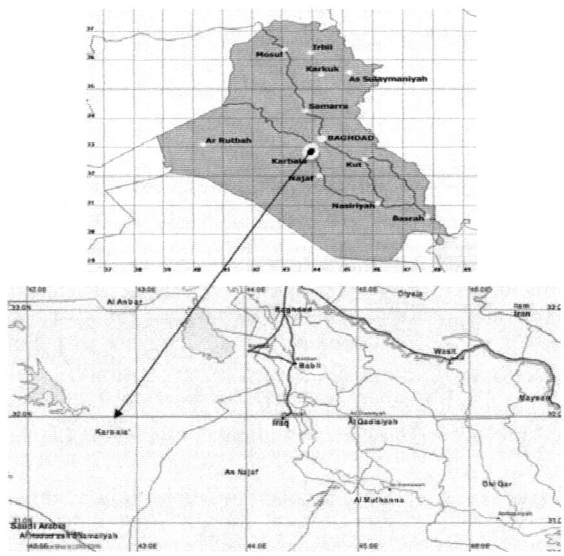

Figure 1. Location of the Karbala city [9].

Table 1. Engineering and index properties of the selected soil.

Properties	Results	Standards
Liquid Limit, LL (%)	88	ASTM 423-66[10]
Plastic Limit, PL (%)	33	ASTM D424-59[10]
Plasticity Index, PI (%)	55	
Specific gravity, Gs	2.61	ASTM D854-00[11]
Sand, S (%)	11	
Silt, M (%)	24	ASTM D 422-63[10]
Clay, C (%)	65	
Activity, A (%)	0.81	
Soil type (USCS classification)	CH	
Maximum Dry Unit Weight, γmax (kN/m^3)	14	ASTM D698-78[12]
Optimum moisture content, wopt (%)	27	

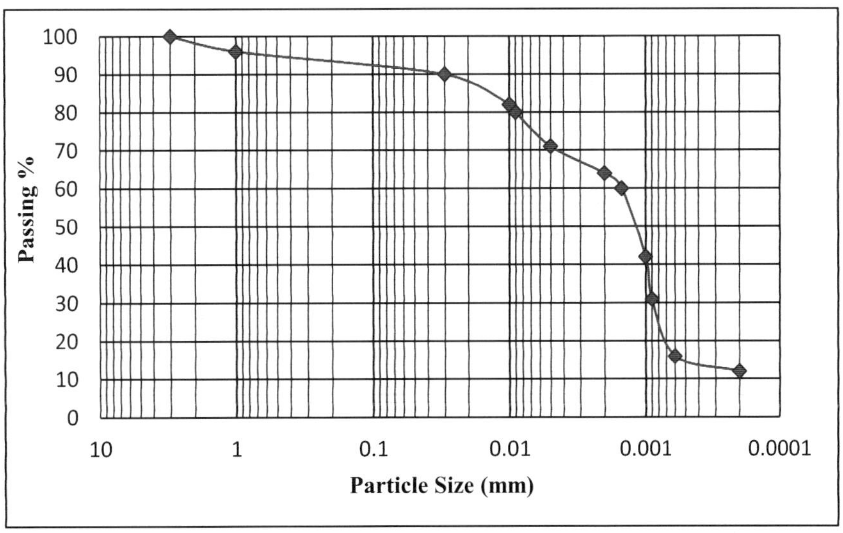

Figure 2. Grain size distribution of the selected soil.

2.2. Properties of Cement Dust

Cement dust is a kiln waste product that is accumulated during the production of Portland cement. On average approximately 1500 to 2000 ton is produced daily in Iraq [13]. It has a bad effect on the environment and needs a lot of money and effort for recycling or disposal. Cement kiln dust that was obtained from Al-Kufa Cement Factory was used in this study as a stabilizer material. Its physical properties and chemical composition are presented in Table 2.

Table 2. Physical and chemical properties of cement dust

Properties	Results
Physical Properties	
Color	Light brown
Texture	Fine
Particle shape	Semi-circle
Specific gravity, Gs	2.85
Chemical Composition by weight (%)	
Silicon dioxide (silica), SiO_2	14.1
Aluminum oxide (alumina), Al_2O_3	4.7
Iron Oxide, Fe_2O_3	1.97
Calcium Oxide (Lime), CaO	40.17
Magnesia, MgO	2.79
Sulfate, SO_3	5.85
Potassium oxide, K_2O	3.13
Sodium oxide, Na_2O	1.55
Cl	1.83
Loss of ignition (L.O.I)	24.25

2.3. Preparation of Testing Samples

In order to investigate the effect of cement dust, some of the parametric variables of soil should be considered as constants such as initial water content and dry density. The raw soil was mixed with different percentages of cement dust (4, 8, 12, 16, and 20%) by adding the required amount of water to prepare many samples with different cement dust content. The soil mixtures (7,14,21 and 28 days) were used then to prepare testing samples. Samples were compacted in molds according to Standard Proctor Compaction in order to obtain undisturbed samples. The samples were extracted from molds by pushing a Shelby tube that is 50 mm in diameter. The obtained samples were cut in to 100 mm long and used to determine the swelling, shear strength and consolidation properties.

3. Results and Discussion

3.1. Consistency

Liquid limit, plastic limit, and plasticity index were determined for samples with different cement dust content and presented in Figure 3. It can be noticed that there is a significant decrease in all indices as cement dust content increases. This can be explained by the fact that cement dust is a soft and very fine matter which can occupy the voids between soil particles instead of water. This will reduce the water content of the treated soils. Soil with low consistency limits is considered to be good soil that can be used as a base layer in construction of roads and airports because of its high shear resistance [14].

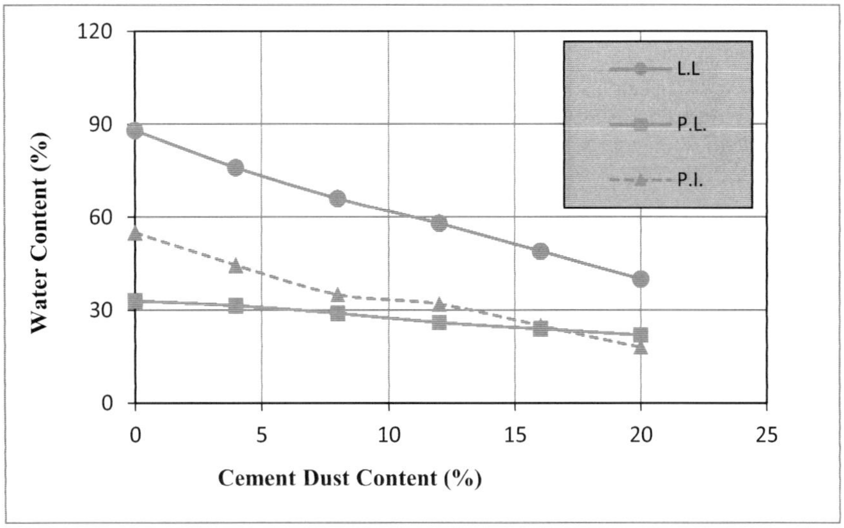

Figure 3. Effect of cement dust content on consistency limits.

The effect of curing time on plasticity index of soils with different cement dust content was studied and presented in Figure 4. It can be noticed that there is a considerable decrease in the plasticity index as samples were cured for longer time. The Figure shows that the plasticity index is decreasing at small rate when cement dust content is about 4% to 8%. Significant decrease of plasticity index can be noticed at cement dust content range of 12% to 20% for all curing periods.

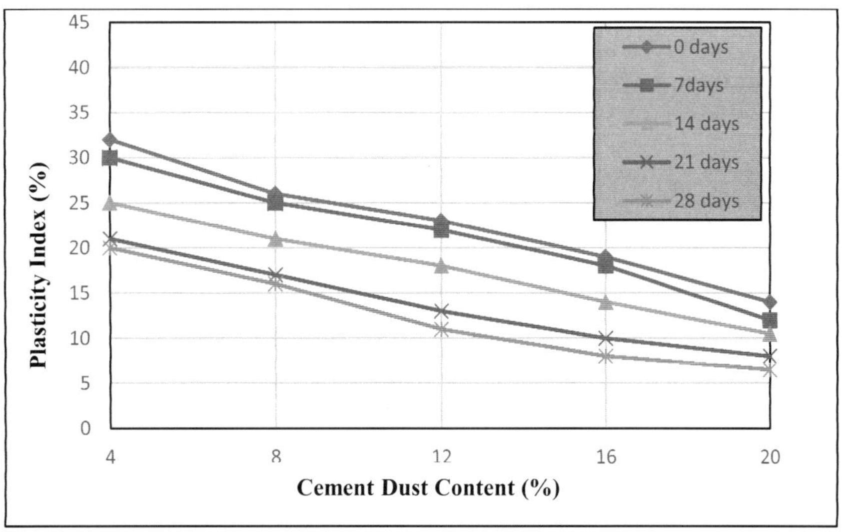

Figure 4. Variation of plasticity index with cement dust content and curing time.

3.2. Compaction Test

Compaction curves for samples with different cement dust content were determined using Standard Proctor Test as demonstrated in Figure 5. The maximum dry unit weight and the optimum moisture content were determined for each curve. It can be noticed that soil sample with higher cement dust content has the higher maximum dry unit weight with an increase of its moisture content. The increase in the optimum water content can be related to the increase in surface area of soil particles mixed with cement dust. Moreover, the increase in optimum moisture content is thought to result from the increasing desire for water, as more water is required for the dissociation of cement dust.

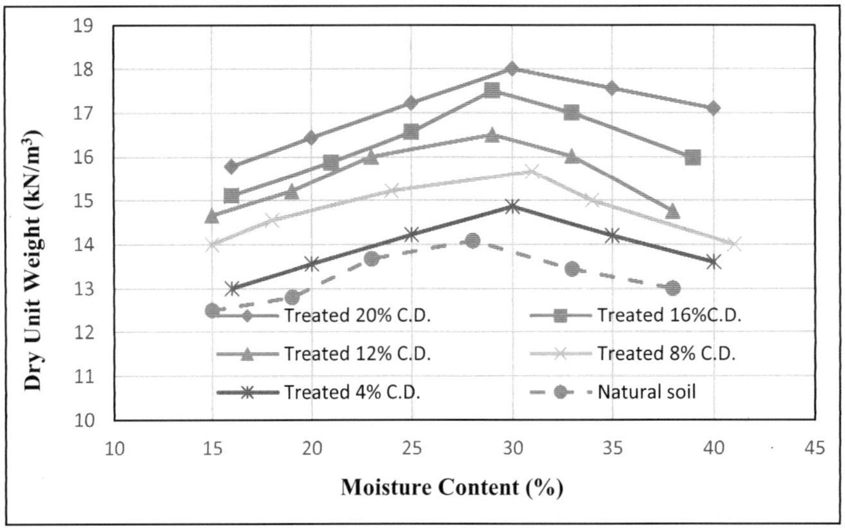

Figure 5. Compaction curve for soils with different cement dust content

3.3. Free Swelling Index

The soil-cement dust mixtures were prepared and cured for different periods. The cured samples were soaked in distilled water and kerosene for 24 hours. The height of sediment columns in water (Vw) and in kerosene (Vk) were recorded. The deferential Free Swelling Index (FSI) [15] was calculated as:

$$FCI=(V_w-V_k)/V_k \tag{1}$$

The Swelling Index was presented versus cement dust content for different curing periods as shown in Figure 6. It can be noticed that the Swelling Index decreases by 70% to 75% to for soils that are mixed with 20% of cement dust. In terms of curing period, there is a noticeable decrease in the swelling index with increasing curing days for samples that contain 4% of cement dust. There is a negligible or no effect of the curing period on the swelling index at a cement dust content of 8%.

Swelling pressure is also measured for the raw soil samples and other samples with some percentages of cement dust. The results are presented in Figure 7. A considerable reduction of 43% in the swelling pressure can be noticed for samples that contain 20% of cement dust.

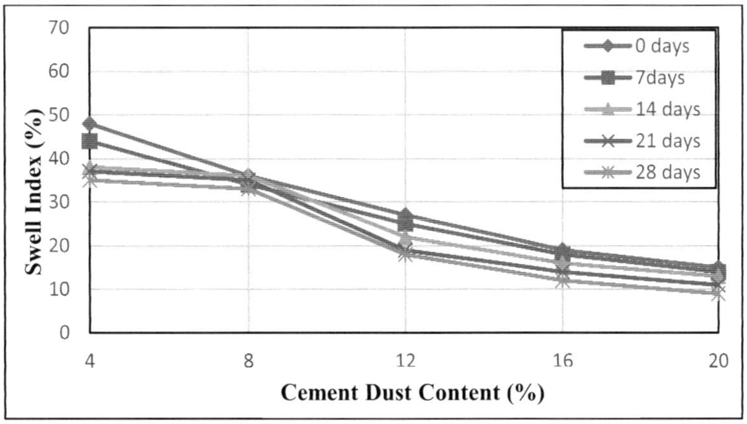

Figure 6. Variation of swelling index versus cement dust content for different curing periods

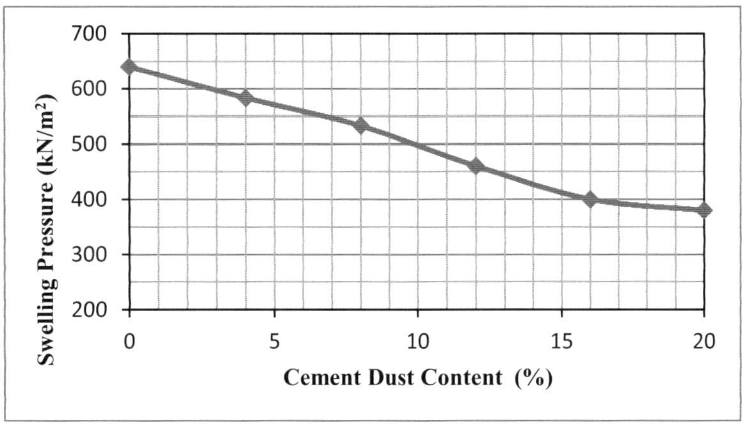

Figure 7. Effect of cement dust content on swell pressure

3.4. Consolidation Test

Results of consolidation test represented by consolidation curves (e-log σv-) for six samples with different percentages of cement dust content are shown in Figure 8. When the swelling pressure of samples assessed the dead load was applied on specimens by double increment to allow specimens to consolidate. It was noticed that as the cement dust content decreases, the initial void ratio and hence the compressibility decreases. Low compressibility values were yielded in the composite samples with 4%, 8%, 12%, and 16% cement dust content as compared with raw soil samples. When an effective stress of 25 kPa and 1000 kPa was applied, it was observed that the void ratio and compressibility gradually decrease with increasing cement dust content. While the void ratio of natural clayey soil samples decreases. This reduction in the compression and swelling can be related to the addition of low-plastic material and the interaction between clay minerals and cement dust particles.

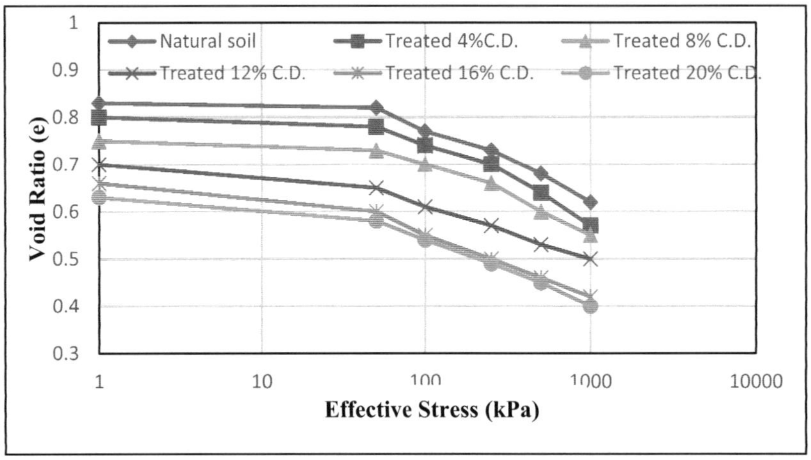

Figure 8. Void ratio versus effective stress curves using the consolidation tests

3.5. Shear Strength

Results from *Unconfined Compressive Tests (UC)* for stabilized clayey soil samples with cement dust are shown in Figure 9. The unconfined compressive strength of samples increased by 8% with increasing cement dust content from 0% to 16%. A drop down of soil strength can be noticed after increasing the cement dust to 20%. The maximum unconfined compressive strength was achieved at 16% of cement dust content, so it can be considered as the optimum cement dust content for this study. This is attributed to soil cement dust reaction, which results in the formation of cementitious materials compounds that bind soil aggregates and may be attributed to the internal friction of cement dust particles and pozzolanic reaction between cement dust and clayey soil.

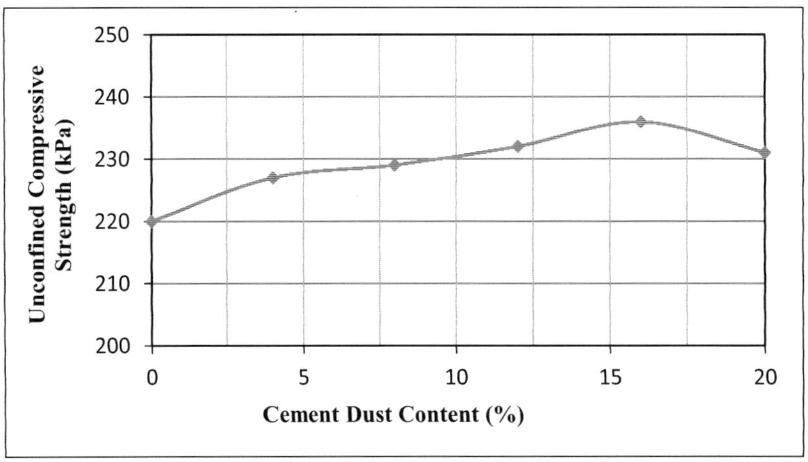

Figure 9. Effect of cement dust content on unconfined compressive strength

4. Conclusions

The present research investigated the effect of cement dust on the engineering characteristics of local expansive clayey soil. The following conclusion may be drawn:

1. The addition of cement dust up to 16% decreased the Consistency limits (liquid limits, plastic limits and the plasticity index).
2. The swelling pressure and swelling potential values steadily decreased with increasing cement dust content and the lowest values were finally reached in the stabilized samples with 16% cement dust contents. With the addition of cement dust,
3. The effect of cement dust is decreased the compressibility of clayey soil.
4. The addition of cement dust improved the unconfined compressive strength by 8% with increasing cement dust content to 16%.

References

[1] Das, B.M. (2004) "Principle of Geotechnical Engineering", 5th edition: Thomson Learning, USA.

[2] Hobart, k. (2015) " Expansive Soil and Expansive Clay: The hidden force behind basement and foundation problems". Geology.com. Accessed March 19.

[3] Al-Homoud, A.S., Basma, A.A., Malkavi, H., and Al-Basha, M.A., (1996) "Cyclic Swelling Behavior of Clays". Journal of Geotechnical Engineering Vol.121, PP.562–565

[4] Al- Murshedi, A. (2014) "Improvement of Geotechnical Properties of Expansive Soils Using Cement" Kufa Journal of Eng., Iraq Vol.5, No.2, pp.31-52.

[5] Kalkan, E., and Akbulut, S., (2004) "The Positive Effects of Silica Fume on the Permeability, Swelling Pressure and Compressive Strength of Natural Clay Liners". Engineering Geology Vol.73, PP.145–156.

[6] Kolias, S., Kasselouri-Rigopoulou V., and Karahalios, A. (2005) "Stabilization of Clayey Soils with High Calcium Fly Ash and Cement". Cement and Concrete Composites Vol.27, PP.301–313.

[7] Kalkan, E., (2009) "Influence of Silica Fume on the Desiccation Cracks of Compacted Clayey Soils". Applied Clay Science Vol.43, PP.296–302.

[8] Basha, E.A., Hashim R., Mahmud H.B., and Muntohar A.S., (2005) "Stabilization of Residual Soil with Rice Husk Ash and Cement". Construction Building Material Vol.19, PP.448–453.

[9] http://www.lib.utexas.edu/maps/iraq.html, "Maps of Iraq"

[10] Head, K.H. (1994) "Manual of Soil Laboratory Testing" .Pentech Press, London, Vol. 1, 2 &3.

[11] ASTM D 854-00 (2006) "Standard Test Method for Specific Gravity of Soils" Annual Book of ASTM Standards, ASTM, Philadelphia, USA. vol. 04.08; pp. 80-83

[12] ASTM D 698-78., (2006) "Test Method for Laboratory Compaction Characteristics of Soils" Annual Book of ASTM Standards, ASTM, Philadelphia, USA. vol. 04.08, pp. 69-76.

[13] Al- Murshedi, A. (2014) "Treatment of Marshes Soil by Cement Dust for Construction Use" Proceedings of the 1st International Engineering Conference, Spatial No. AL-Mustansiriyah University

[14] Schmitz, R.M., Schreoder, C., and Charlier, R., (2004) "Chemo-Mechanical Interactions In Clay: A Correlation Between Clay Mineralogy and Atterberg Limits". Applied Clay Science Vol.26,PP.351–358

[15] Sridharan, A. (2005) On Swelling Behaviour of Clays Proceeding, international conference on Problematic soils, North Cyprus, Vol. 2, pp. 499-516

Investigating the influence of cement replacement by high volume of GGBS and PFA on the mechanical performance of cement mortar

Ali Abdulhussein Shubbar [1,2,*], Ali Al-Shaer [1], Rasha Salah AlKizwini [3], Khalid Hashim [1], Hayder Al Hawesah [1] and Monower Sadique [1]

[1] Department of Civil Engineering, Liverpool John Moores University, UK
[2] Department of Civil Engineering, College of Engineering, University of Babylon, Iraq.
[3]Department of Environmental Engineering, College of Engineering, University of Babylon, Iraq.

* Corresponding author: alishubbar993@gmail.com , A.A.Shubbar@2014.ljmu.ac.uk

Abstract. The technique of replacing the cement with other alternative materials focus on the production of materials with similar performance and reduced environmental impacts relative to traditional cement. The main aim of this study is to investigate the effect of replacing the cement content with high volume of GGBS and PFA on the mechanical performance of cement mortar. Three mixtures were prepared with different percentages of GGBS and PFA (40%, 60% and 80%) as replacement of cement along with other mixture that made with 100% cement as a control mixture. In order to evaluate the performance of the cement mortars, compressive strength test after 7, 14 and 28 days of curing was used. The results indicated that after 7 days of curing, the increase of GGBS and PFA contents caused a reduction in the compressive strength in comparison with the control mixture. After 28 days of curing, the results indicated that the mixture incorporated 80% GGBS and PFA has higher compressive strength relative to the control mixture. Such findings will significantly contribute in reducing the cost of the produced mortar by reducing the amount of used cement and this consequently reduce the cement demands/manufacturing. Less production of cement will reduce the CO_2 emissions of the cement industry.
Keywords: GGBS; PFA; high volume; compressive strength.

1. Introduction

Cement is globally considered as the main binder in the production of concrete and mortar. However, the production of cement is associated with some negative environmental impacts such as the production of greenhouse gases and the consumption of natural resources [1-3]. The cement industry alone contribute to about 7% of the CO_2 in the atmosphere globally [4]. Therefore, there are extensive efforts around the world being undertaken with the aim of developing construction materials that are environmentally friendly and at the same time provide similar performance to traditional cement [5]. Among the known viable alternative materials to cement in different applications such as concrete and mortar production and soil stabilization are Ground Granulated Blast Furnace Slag (GGBS) and Pulverised Fuel Ash (PFA). GGBS is a by-product material of iron that is extracted from blast furnaces that are dried and ground in a rotating ball-mill into a very fine powder of GGBS [6]. PFA is a by-product produced from different coal-fired power stations [7].

IOP Conf. Series: Materials Science and Engineering **584** (2019) 012022 doi:10.1088/1757-899X/584/1/012022

Recently, many studies have been conducted to investigate the effect of using GGBS or PFA as partial replacement to cement. For example, Soriano et al. [8] investigated the effect of replacing the cement by 30% PFA on the compressive strength of cement mortars. The results indicated that the new blend provided better performance at both early and later ages relative to control mixture with 100% cement. Hawileh, et al.[9] investigated the effect of replacing the cement with up to 70% GGBS on the performance of reinforced concrete beams. The results indicated that the beams incorporated up to 70% GGBS have compressive strength similar to that of the control beams especially at later ages. The replacement of cement partially with GGBS and PFA can improve the mechanical and durability properties of cement mortar [10, 11]. Additionally, the production of mortars with lower cement content means that the negative environmental impact will be lower.
Although there are extensive research conducted by different researchers around the world to evaluate the effect of using GGBS and/or PFA on the performance of cement mortar, however, limited research investigated the replacement of cement with high volume (40-80%) of GGBS and PFA. Therefore, this paper presents the results of experimental work that has been conducted to study the effect of replacing the cement by high volume of GGBS and PFA (40, 60 and 80%) by weight on the compressive strength of cement mortar.

2. Materials and Methodology

2.1. Materials

2.1.1. Sand

The sand used in this investigation was building sand with a specific gravity of 2.62. The particle size distribution of the sand is presented in Figure 1.

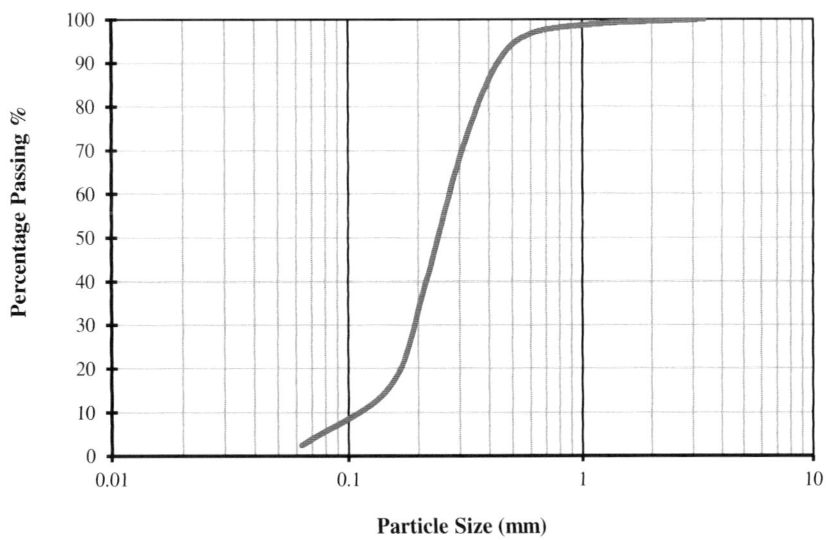

Figure 1. Particle size distribution of the sand.

2.1.2. Water
Tap water was used for all mixtures.

2.1.3. Binder Materials
The materials used in this investigation were Portland Cement (PC), Ground Granulated Blast Furnace Slag (GGBS) and Pulverised Fuel Ash (PFA). The cement was PC type CEM-II/A/LL 32.5-N. GGBS was supplied by Hanson Heidelberg Cement Group while PC and PFA supplied by CEMEX ltd Company, Warwickshire, UK.
The elemental composition of PC, GGBS and PFA was analysed by an Energy Dispersive X-ray Florescence Spectrometer (EDXRF) type Shimadzu EDX-720. Table 1 shows the chemical compositions of the PC, GGBS and PFA.

Table 1. Chemical composition of PC, GGBS and PFA.

Item	PC	GGBS	PFA
CaO %	65.21	42.51	4.81
SiO_2 %	24.56	41.06	58.83
Al_2O_3 %	1.70	5.12	18.83
Fe_2O_3 %	1.64	-	5.54
MgO %	1.30	4.25	3.86
Na_2O %	1.34	3.09	1.17
K_2O %	0.82	0.69	2.04
SO_3 %	2.62	1.27	1.06
TiO_2 %	-	0.98	1.19
LOI %	0.28	0.37	2.67
pH	12.73	11.02	10.68
Specific Gravity	2.94	2.90	2.49

2.2. Mixing proportions
During this study, the cement was replaced partially with high volume of GGBS and PFA at different percentages: 40, 60 and 80%. Table 2 provide the mixing proportions of the ternary blending mixtures. The sand to binder ratio (S/B) and water to binder ratio (W/B) were fixed during this study as 2.5 and 0.4, respectively. Additionally, all the mixtures were compared with the control one of 100% PC.

Table 2. Mixing proportions of binder materials.

Mix ID	PC (%)	GGBS (%)	PFA (%)	S/B	W/B
M1	100	0	0	2.5	0.4
M2	60	20	20	2.5	0.4
M3	40	30	30	2.5	0.4
M4	20	40	40	2.5	0.4

2.3. Samples' preparation
All required materials to produce the mortars were firstly weighted, then the binder materials and the sand were mixed using Hobart N50-110 mixer for 5 minutes. Then the water was added and the wet mixture was further mixed for 5 minutes to form the cement mortar. The produced mortar was then moulded in the prism mould with the dimensions of 40 x 40 x 160 mm. After 24 hours, the mortar prisms

were demoulded and the samples were placed in small containers for water curing until the time of testing.

2.4. Testing method

Compressive strength testing was conducted to evaluate the performance of the cement mortars according to BS EN 196-1[12]. Three samples of dimensions 40 x 40 x 160 mm were prepared for each mixing proportion and the test was carried out after 7, 14 and 28 days of curing.

3. Results and Discussion

The compressive strength of the cement mortars made from different combinations of PC, GGBS and PFA at different curing ages are presented in Figure 2.

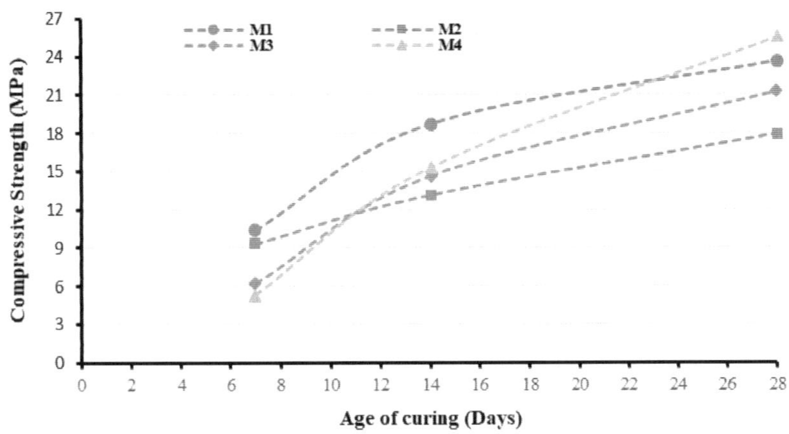

Figure 2. Compressive strength development of different ternary blends and the control cement mixture.

It can be seen from Figure 2 that increase in the period of curing resulted in improved compressive strength for all the mixtures.

At the age of 7 days, the results indicated that with increasing the GGBS and PFA content in the mixture, the compressive strength decreased. This could be due to the slow acquisition of strength for mixtures with high volume of GGBS and PFA at initial curing ages [4, 13, 14]. Additionally, all the ternary mixtures showed lower compressive strength relative to the control mixture (M1). At this age of curing, the mixtures M2, M3 and M4 showed about 90%, 60% and 51% of the compressive strength of the control mixture, respectively.

At the age of 14 days, the results showed that the compressive strength of the cement mortars incorporated GGBS and PFA has been improved with increasing the GGBS and PFA content. However, all the mixtures provided lower compressive strength relative to the control mixture.

After 28 days of curing, the results indicated that the mixture M4 with 80% GGBS and PFA has higher compressive strength relative to the control mixture with an improvement of about 8%. On the other hand, the mixtures M2 and M3 showed about 76% and 90% of the compressive strength of the control mixture, respectively.

In summary, it is essential to extend the age of curing for mixture with high volume of GGBS and PFA to ensure the development of adequate compressive strength.

4. Conclusion

The aim of this research was to investigate the effect of replacing the cement with high volume of GGBS and PFA on the compressive strength performance of cement mortar. Based on the obtained results, it can conclude that:

- Extending the period of curing resulted in enhanced compressive strength for all the mixtures. This could be due to the slow pozzolonic reaction for mortars with high percentage of GGBS and PFA that largely dependent on the formation of calcium hydroxide that requires time.
- At the age of 7 days, the results indicated that increasing the content of GGBS and PFA caused a reduction in the compressive strength of the cement mortar.
- After 14 days of curing, the results indicated that increasing the GGBS and PFA in the mixtures resulted in improved strength.
- At the age of 28 days, the results indicated that the mixture with 80% GGBS and PFA has higher compressive strength relative to the control mixture.
- Reducing the content of cement by 80% resulted in improved compressive strength of the cement mortars. Such replacement will significantly contribute to reduce cement content that consequently reduce the demand and production of cement. Therefore, the CO_2 emissions from the cement industry will be lower.

Finally, as the used by-product materials are gaining more values and the cost of these materials is increasing, therefore in order to produce sustainable binder materials with competitive financial cost, other waste materials should be investigated. For example, due to the huge production of municipal solid wastes [15-17], especially from water and wastewater facilities [18-26], the authors recommend to explore the possibility of using certain types of these wastes as cement replacement materials.

Acknowledgments
The corresponding author would like to acknowledge the financial support of this research by the Faculty of Engineering and Technology at Liverpool John Moores University (LJMU).

References

[1] Karim M, Zain M, Jamil M, and Lai F, 2013. "Fabrication of a non-cement binder using slag, palm oil fuel ash and rice husk ash with sodium hydroxide". Construction and Building Materials, vol. 49, pp. 894-902.

[2] Shubbar A, Jafer H M, Dulaimi A, Atherton W, and Al-Rifaie A, 2017. "The Development of a Low Carbon Cementitious Material Produced from Cement, Ground Granulated Blast Furnace Slag and High Calcium Fly Ash". International Journal of Civil, Environmental, Structural, Construction and Architectural Engineering, vol. 11, no. 7, pp. 905-908.

[3] Shubbar A A, Sadique M, Kot P, and Atherton W, 2019. "Future of clay-based construction materials – A review". Construction and Building Materials, vol. 210, pp. 172-187.

[4] Shubbar A A, Jafer H, Dulaimi A, Hashim K, Atherton W, and Sadique M, 2018. "The development of a low carbon binder produced from the ternary blending of cement, ground granulated blast furnace slag and high calcium fly ash: An experimental and statistical approach". Construction and Building Materials, vol. 187, pp. 1051-1060.

[5] Aprianti E, Shafigh P, Bahri S, and. Farahani J N, 2015. "Supplementary cementitious materials origin from agricultural wastes – A review". Construction and Building Materials, vol. 74, pp. 176-187.

[6] Grist E R, Paine K A, Heath A, Norman J, and Pinder H, 2015. "The environmental credentials of hydraulic lime-pozzolan concretes". Journal of Cleaner Production, vol. 93, pp. 26-37.

[7] Aprianti E, 2017. "A huge number of artificial waste material can be supplementary cementitious material (SCM) for concrete production – a review part II". Journal of Cleaner Production, vol. 142, pp. 4178-4194.

[8] Soriano L, Payá J, Monzó J, Borrachero M V, and Tashima M M, 2016. "High strength mortars using ordinary Portland cement–fly ash–fluid catalytic cracking catalyst residue ternary system (OPC/FA/FCC)". Construction and Building Materials, vol. 106, pp. 228-235.
[9] Hawileh R A, Abdalla J A, Fardmanesh F, Shahsana P, and Khalili A, 2017. "Performance of reinforced concrete beams cast with different percentages of GGBS replacement to cement". Archives of Civil and Mechanical Engineering, vol. 17, no. 3, pp. 511-519.
[10] Shubbar A, Atherton W, Jafer H M, Dulaimi A, and AL-Faluji D, 2017. "The Development of a New Cementitious Material Produced from Cement and GGBS". Presented at the The 3rd BUiD Doctoral Research Conference, the British University in Dubai, UAE, 13th May 2017.
[11] Khalil E and Anwar M, 2015. "Carbonation of ternary cementitious concrete systems containing fly ash and silica fume". Water Science, vol. 29, no. 1, pp. 36-44.
[12] BSI, " Methods of testing cement–Part 1: Determination of strength," British Standard Inistitute, London2005.
[13] Limbachiya V, Ganjian E, and Claisse P, 2016 "Strength, durability and leaching properties of concrete paving blocks incorporating GGBS and SF". Construction and Building Materials, vol. 113, pp. 273-279.
[14] Attari A, McNally C, and Richardson M G, 2016. "A probabilistic assessment of the influence of age factor on the service life of concretes with limestone cement/GGBS binders". Construction and Building Materials, vol. 111, pp. 488-494.
[15] Abdulredha M, Al Khaddar R, Jordan D, Kot P, Abdulridha A, and Hashim K, 2018 "Estimating solid waste generation by hospitality industry during major festivals: A quantification model based on multiple regression". Waste Manag, vol. 77, pp. 388-400, Jul 2018.
[16] Abdulredha M, Al Khaddar R, Jordan D, and Hashim K, 2017. "The development of a waste management system in Kerbala during major pilgrimage events: determination of solid waste composition". Procedia Engineering, vol. 196, pp. 779-784.
[17] Al-Jumeily D, Hashim K, Alkaddar R, Al-Tufaily M, and Lunn J, 2018."Sustainable and Environmental Friendly Ancient Reed Houses (Inspired by the past to motivate the future)". presented at the 11th International Conference on the Developments in eSystems Engineering (DeSE), University of Cambridge, England, UK.
[18] Alattabi A W, Harris C, Alkhaddar R, Alzeyadi A, and Hashim K, 2017. "Treatment of Residential Complexes' Wastewater using Environmentally Friendly Technology". Procedia Engineering, vol. 196, pp. 792-799.
[19] Alattabi A W, Harris C B, Alkhaddar R M, Hashim K S, Ortoneda-Pedrola M, and Phipps D, 2017. "Improving sludge settleability by introducing an innovative, two-stage settling sequencing batch reactor". Journal of Water Process Engineering, vol. 20, pp. 207-216.
[20] Hashim K S, 2017. "The innovative use of electrocoagulation-microwave techniques for the removal of pollutants from water," PhD thesis, Civil Engineering Liverpool John Moores University, UK.
[21] Hashim K S, Al-Saati N H , Hussein A H, and Al-Saati Z N, 2018. "An Investigation into The Level of Heavy Metals Leaching from Canal-Dreged Sediment: A Case Study Metals Leaching from Dreged Sediment". presented at the 1st International Conference on Materials Engineering & Science, Istanbul Aydın University (IAU), Turkey.
[22] Hashim K S et al., 2018. "Removal of phosphate from River water using a new baffle plates electrochemical reactor". MethodsX, vol. 5, pp. 1413-1418, 2018.
[23] Hashim K S et al., 2018. "Electrocoagulation as a green technology for phosphate removal from River water". Separation and Purification Technology, vol. 210, pp. 135-144.
[24] Hashim K S, Shaw S, Al Khaddar R, Ortoneda Pedrola M, and Phipps D, 2017. "Defluoridation of drinking water using a new flow column-electrocoagulation reactor (FCER) - Experimental, statistical, and economic approach". J Environ Manage, vol. 197, pp. 80-88.
[25] Hashim K S ,Shaw A, Al Khaddar R, Pedrola M O, and Phipps D, 2017. "Iron removal, energy consumption and operating cost of electrocoagulation of drinking water using a new flow column reactor". Journal of Environmental Management, vol. 189, pp. 98-108.

[26] Hashim K S, Shaw A, R. Al Khaddar R, Pedrola M O, and Phipps D, 2017. "Energy efficient electrocoagulation using a new flow column reactor to remove nitrate from drinking water - Experimental, statistical, and economic approach". J Environ Manage, vol. 196, pp. 224-233.

IOP Conf. Series: Materials Science and Engineering **584** (2019) 012023 doi:10.1088/1757-899X/584/1/012023

Solid Waste Situation in Thi-Qar Governorate

Saleem Ethaib[1,*]

[1]College of Engineering, University of Thi-Qar, 64001 Al-Nassiriya, Iraq

Email: salem_eidt@yahoo.com , dr.saleem@utq.edu.iq

Abstract. This study aims to provide an overview of the current status of solid waste in Thi-Qar governorate by a case study of five cities in this governorate. The five cities are namely: Al-Nassriya, Al-Shatra, Al-Rifai, Suk Al-Shoyokh, and Al-Chibayish. In general, the results revealed that the average generation rate of domestic solid waste was 0.85 Kg/capita/day in the study area. The organic waste was found the major components of municipal waste and it exceeded 60% of waste composition. All dumping sites in the aforementioned cities are uncontrolled dumpsite, their design not suitable, and do not work according to proper operating standards. There are no landfills or dumping sites specified for disposal the non-municipal waste. The investigations showed the presence of potential health risks and negative environmental aspects of solid waste handling in Thi-Qar governorate due to improper handling of collection, transfer, and disposal of solid waste. Thus, developing a master plan for integrated solid waste management is prerequisite to overcome the difficulties solid waste management sector in Thi-Qar governorate.

1. Introduction

Solid waste research has gained more attention in last decades; meanwhile solid waste management was one of the main concerns in developed countries particularly affected due to ineffective waste management systems [1]. The residential, commercial, industrial and institutional activities are the main sources of solid waste in urban areas [2, 3]. The generated solid waste from a community that requires collection and transport to a processing or disposal site is named municipal solid waste or refuse [2, 3]. Compared with medical waste, industrial solid waste, and construction waste, municipal solid waste has the characteristics of non-point source pollution [4]. In addition, the production of municipal solid waste widely varies depending on different parameters. For example, Abdulredha, et al. [3] found that the municipal solid waste produced by hotels, in Karbala city-Iraq, varies from 830 to 1220 g/guest-day, depending on different parameters, such as the hotels' rate. Additionally, Abdulredha, et al. [3] stated that the relationship between production of municipal solid waste and hotel features could be statistically model with a good correlation factor value (R^2 of 0.799) [5]. Poor management of solid waste has negative effects on public health and the environment due to spread vector-borne disease and increase pollution levels [6,7]. Moreover, the important physical components of municipal solid waste, such as plastics and chemical fibers, may be harmful in some contexts [8] and are suspected of being carcinogenic to humans [9]. Solid waste has become one of the global environmental issues [10].
Around the world, with rapid population growth and urbanization, 120-130 billion tons of natural resources are annually consumed and produce around 3.4-4 billion tons of municipal solid waste [11]. Iraq is one of the most populous Arab countries with population exceeding 35 million. Iraq is administratively divided into eighteen governorates. The high population growth rates, the rapid economic growth, the increase in individual income, and improper waste management practices have led to worsening solid waste management problem in the country. As well as, Iraqi waste processing and disposal infrastructure was heavily damaged after decades of conflict. A number of serious concerns

IOP Conf. Series: Materials Science and Engineering **584** (2019) 012023 doi:10.1088/1757-899X/584/1/012023

for both human health and the environment associated with the mismanagement of solid waste and lack of controls, inadequate legislation. Therefore, this country needs comprehensive plans for solid waste management.

The solid waste management is complex tasks and technical challenges [12]. Managing waste properly is essential factor to protect health of the population, promote environmental quality, for building sustainable and livable cities, but it remains a challenge for many developing countries and cities [13, 14]. Regardless of the origin, content or hazard potential, solid waste must be managed systematically to ensure environmental best practices. Thus, this study aims to identify the current solid waste situation in Thi-Qar governorate.

2. Methodology

2.1 Description of the Study Area

Thi-Qar governorate is one of the 18th Iraqi governorates and located in the south. Al-Nassriya city is the capital of Thi-Qar governorate lies on the Euphrates River about 370 km southeast of Baghdad, near the ruins of the ancient Ur city. The area of this governorate is 12900 Km^2. The population number is about 2 million people and they are live in urban and rural areas. Euphrates River passes through the governorate and supplying the water for southern cities. Meanwhile, the northern cities are supplied by steams of Tigris River. Thus, the governorate takes the advantage of these two rivers; both rivers contribute to create marsh areas. The administrative structure of Thi-Qar governorate consists of 20 units. These units were divided into 5 districts which represent main cities and 15 sub districts as shown in Figure. 1 shows the administrative map of Thi-Qar governorate. This study attempted to focus on the main five cities, Al-Nassriya, Al-Shatra, Al-Rifai, Suk Al-Shoyokh, and Al-Chibayish to identify the solid waste management process in these cities. Table 1 shows number of population and areas of these districts.

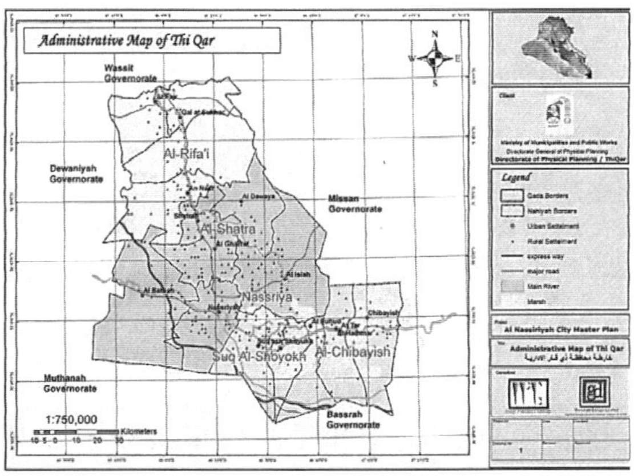

Figure 1. Administrative map of Thi-Qar governorate including selected cities (Source: Thi-Qar urban planning directorate).

Table 1: The main five cities in Thi-Qar governorate and its population

City	Number of population (2015)
Al-Nassriya	523237
Al-Shatra	170689
Al-Rifai	65905
Suk Al-Shoyokh	123381
Al-Chibayish	34922

2.2 Data collection

The information and data were collected based on published reports, questionnaire, and interviews with municipal officials and other directive authorities. This was achieved through questionnaire application, live interviews with municipal officials and other directive authorities, literature review and physical inspection. The study was accomplished during 2017.

3. Results and Discussion

3.1. Solid waste generation and composition

The Interviews with solid waste management officialdoms revealed that municipal solid waste showed that the average generation rate of domestic solid waste was 0.85 Kg/capita/ day in urban areas of Thi-Qar governorate. It is worth mention that the generation rate varies in one municipality, neighborhood based on standard living. It was around 1 kg/capita /day in high standard living neighborhoods; meanwhile, it was around 0.7 kg/capita /day in low living neighborhoods. These results are still relatively small compared to 2.1 kg/day/capita in the USA. Meanwhile is higher the production of solid waste per capita in India (less than 0.5 kg/day/capita and close to China generation rate (less than 0.9 kg/day/capita) [15]. Meanwhile, solid waste can be categorized into three types' domestic solid waste (73%), commercial and non-hazard industrial solid waste (16%), and street and green waste (11%). The average of solid waste quantities that was monthly collected by municipality cleaning departments of the five cities are Al-Nassriya, Al-Shatra, Al-Rifai, Suk Al-Shoyokh, and Al-Chibayish are depicted in Figure 2.The highest amount of solid waste was 17146 ton/month, it was found in Al-Nassriya city due to large number of population in that city.

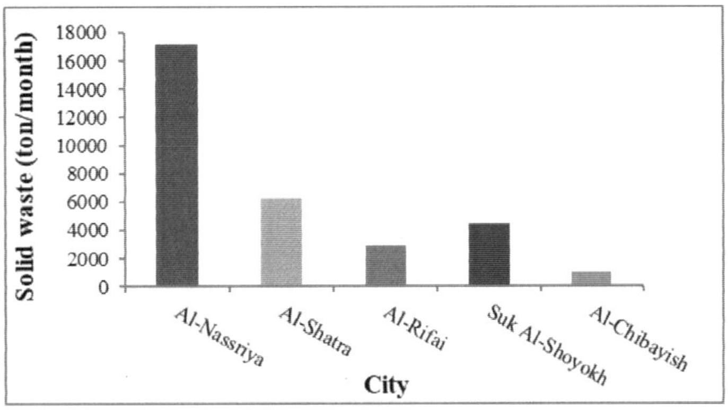

Figure 2. Monthly estimated solid waste quantities in Thi-Qar governorate.

One significant aspect of solid waste from the solid waste management that the results showed the main portion of the solid waste stream in Thi Qar governorate was an organic material. Based on Directorate of Environmental Affairs of Thi-Qar data, the composition of municipal solid waste is presented in Table 2. In general, the organic waste forms 61.3% (an average) which is biodegradable is remarkably high. Moreover, it can transform into organic fertilizer and compost materials. The rest of the solid waste stream contains considerable amounts of paper and plastic and reasonable quantities of glass and metals, and textiles that can be recovered or recycled. The other materials included inert materials (bricks, stones, dirt, etc.), wood, and electric waste. In spite of the absence of actual and real data about the amount recycled materials, it is expected that some non-organic waste are recycled or recovered by individuals and then the collected reusable materials are sold to private sector. However, there no official data about the quantity of recycled waste or the number of people who work in the sector. Waste characteristics of each community must be studied carefully before any treatment or disposal facility is designed and built.

Table 2. The composition of municipal waste in Thi Qar governorate

City	Paper (%)	Plastic (%)	Glass (%)	Metals (%)	Organic (%)	Textiles (%)	Other materials (%)
Al-Nassriya	12	13.2	1.6	2.2	65.8	2.8	2.8
Al-Shatra	10.1	12.3	2.1	2.8	60.5	2.3	8.9
Al-Rifai	9.2	13.2	2.6	3.7	56.4	1.8	7.3
Suk Al-Shoyokh	11.2	14.8	1.9	2.7	59.9	2.9	6.7
Al-Chibayish	10.2	14.3	2.4	3.9	63.8	1.5	6.4

3.2. Disposal Sites of Thi-Qar governorate

The five cities under this case study, Al-Nassriya, Al-Shatra, Al-Rifai, Suk Al-Shoyokh, and Al-Chibayish have dumping sites. Table 3 summarizes the physical and environmental conditions of the dumping sites in these cities. These sites have a negative impact on public health and the environment and do not meet the US Environmental Protection Agency (EPA) standards for landfill sites [16]. These sites located adjacent of residential area except Al-Nassriya dumping site as well as some of these sites are close to watercourses especially in Al-Shatra, Suk Al-Shoyokh, and Al-Chibayish. All dumping sites in the five cities are uncontrolled dumpsite, their design are not suitable, and doesn't work according to proper operating standards. In each location, the waste is burned and the practice is widely spread to reduce the volume of waste. Only a few equipment was employed at each site to handle waste in these sites. There is no data about the thickness of cover layer and the wastes covered without spreading and compacting. Moreover, some of these sites have shallow water table below ground surface. Therefore, the contaminated water, heavy metals and toxic elements present in the wastes contaminate the soil and ultimately seep into underground water after rainfall [17,18]. However, it is necessary to implement groundwater monitor plan, especially for the sites that have a shallow level of groundwater table.

Tables 3. The physical and environmental conditions of the dumping sites

City	Total area (m^2)	Used area (m^2)	Vacant area (m^2)	Ground water table (m)	Soil type	Equipment	Number
Al-Nassriya	960000	480000	480000	8	clay	Loader Bulldozer	1 1
Al-Shatra	20000	12000	8000	7	clay	Loader Bulldozer	1 1
Al-Rifai	25000	10000	15000	5	clay	Loader	1
Suk Al-Shoyokh	20000	7000	13000	1.5	clay	Loader	2
Al-Chibayish	150000	6500	8500	1	clay	Loader	2

3.3. Non-municipal solid waste

Non-municipal solid waste includes construction and demolition waste, medical waste, non-hazardous industrial waste, and agricultural waste.

3.3.1 Construction and demolition waste

Identification construction and demolition waste quantities are an essential requirement for the implementation of successful solid waste management [19]. Construction and demolition waste is waste that generated from construction and restoration works and building demolition. This includes building materials such as wood, bricks, concrete, insulation, electrical wiring, nails, aggregate, and etc. that are damaged or unused for various reasons during construction or generated after demolition. Construction and demolition waste may contain hazardous substances such as lead, and asbestos. On the other hand, there is the potential to recycle many components of construction and demolition waste.

Generally, these wastes in the study area are collected and disposed in open areas or near municipal dumping sites or mixed with other wastes and disposed in the landfill sites. It is worthy to mention that, sometimes, of these wastes are recycled and used in filling and dumping swamps in city area or reused in another construction project. Figure 3 presents the monthly estimated quantities of construction and demolition waste in Thi-Qar governorate.

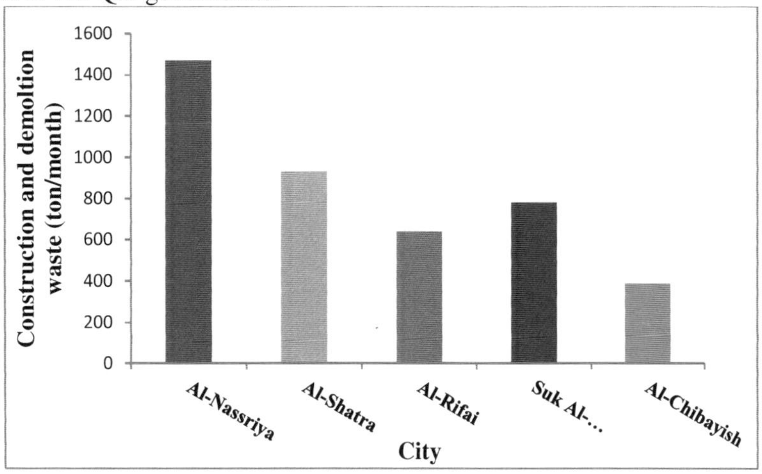

Figure 3: Estimated construction and demolition waste.

3.3.2 Medical waste

Medical waste disposal is one of the biggest day-to-day challenges faced by healthcare providers. Medical waste is a by-product of health care facilities that includes sharps, non-sharps, blood, body parts, chemicals, pharmaceutical products, medical instruments, and radioactive materials. Poor management of medical waste may expose the health care personnel, waste handlers, and the community of infectious agents, toxic materials, and increase risk of injuries. Information of concerning estimated medical waste quantities, sources, and current management practices in the main hospitals in the five cities, Al-Nassriya, Al-Shatra, Al-Rifai, Suk Al-Shoyokh, and Al-Chibayish are tabulated in Table 4. The medical wastes in these cities are processed by separation based on type, for example, the sharp materials are collected in special yellow plastic containers. Meanwhile, other wastes collected in other plastic bags and containers.

In general, the medical waste in the health care facilities in aforementioned sites are treated by incineration, the most popular method, which is still the only method used for pathological waste, such as body parts and recognizable tissues. The results revealed that the highest generation of medical waste was found in Al-Nassriya city because it has the three main hospitals. In general, the practice of treating medical wastes in the medical health facilities is meeting the environmental legislation except Alshatra hospital, whereas these wastes burned inside the incineration of hospital. The Environmental affairs office of Thi-Qar governorate closed the incinerator in Alshatra hospital due non-compliance with conditions and specifications. However, the hospital management daily sends the medical waste for burning in another hospital incinerator. After burning the medical waste, the residual material is transferred to the landfill sites by municipality cleaning staff.

Table 4. The estimated medical waste quantities and current management practices in the main hospitals of Thi-Qar governorate.

City	Hospital name	Capacity (bed)	Waste quantity (ton/month)	Action	Disposal method
Al-Nassriya	Alhussian hospital BintAlhuda hospital Alhaboby hospital Nassiriya heart center	424 285 200 120	14.2 9.33 8.51 7.911	Separation of waste based on degree of hazard in special plastic bags	Burned inside hospital incineration
Al-Shatra	Al-Shatra public hospital	168	5.66	Separation of waste based on degree of hazard in special plastic bags.	Send the waste materials to burn out side hospital.
Al-Rifai	Al-Rifai public hospital	154	5.1	Separation of waste based on degree of hazard in special plastic bags	Burned inside hospital incineration
Suk Al-Shoyokh	Suk Al-Shoyokh public hospital	180	6.21	Separation of waste based on degree of hazard in special plastic bags	Burned inside hospital incineration
Al-Chibayish	Al-Chibayish public hospital	50	3.74	Separation of waste based on degree of hazard in special plastic bags	Burned inside hospital incineration

3.3.3 Non-hazardous industrial wastes

Non-hazardous industrial wastes refer to the waste that generated from industries and cannot be potentially threatening to the health of labors and work environment. These industries include brick factories, textile, blacksmith workshop, automotive maintenance workshops and etc. The non-hazardous industrial wastes in the five cities that over mentioned are together collected with municipal wastes and disposed in the dumping sites. Table 5 shows the results of some collected data about the quantity of these wastes form some industrial activities within municipality boundaries in Al-Nassriya, Al-Shatra, Al-Rifai, Suk Al-Shoyokh, and Al-Chibayish cities.

Table 5. Estimated non-industrial waste quantities.

City	Industrial activity	Number	Estimated non-industrial waste (ton/month)
Al-Nassriya	Maintenance and blacksmith workshops	1365	661
Al-Shatra	Maintenance and blacksmith workshops	256	120
Al-Rifai	Maintenance and blacksmith workshops	101	56
Suk Al-Shoyokh	Maintenance and blacksmith workshops	150	90
Al-Chibayish	Maintenance and blacksmith workshops	20	13

3.3.4 Agriculture wastes

Agriculture waste includes garden waste, waste of poultry and livestock breeding farms, and waste from agriculture crops and trees. Figure 4 presents summary of the results of the estimated quantities of agriculture wastes in the study area. There is no special practice to handle these wastes and almost these wastes are collected and disposed with municipal solid wastes.

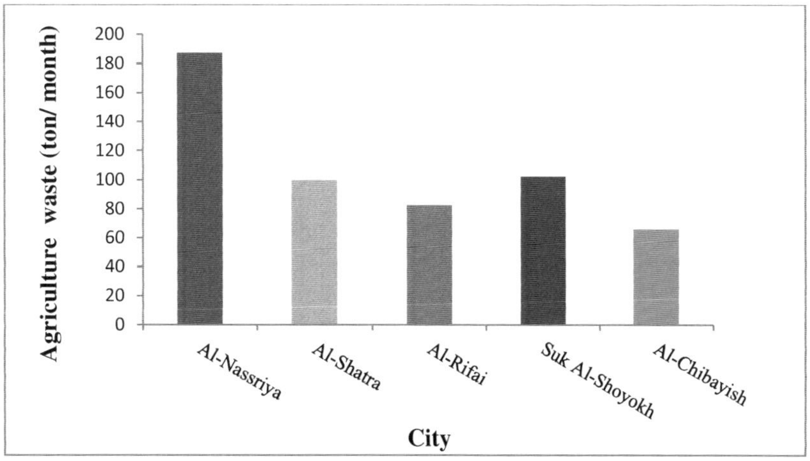

Figure 4. Estimated quantities of agriculture wastes.

3.4. Environmental aspects of solid waste handling in Thi-Qar governorate

Improper handling in the storage, collection and transfer of solid waste could lead to negative impacts on human health and the environment. During investigation and physical inspection, it was noted several points. In general, the waste collection crews are specifically facing occupational hazards such as injuries from sharp objects, airborne disease, spills of polluted materials, and traffic accidents due to a lack of health and safety facilities. Illegal burning of waste in the dumping sites in the study area is forming serious health and environmental hazards such as causing illness, changing outdoor air quality and reducing visibility.

Uncollected solid waste and spreading waste in some locations in the study area is dispiriting efforts to keep open spaces and streets clean, degrading these wastes the urban environment and emitting nuisance odor. Furthermore, presence of fraction of food waste in these locations provided an attractive shelter for flies breeding and rats which are very effective vectors that spread disease in the study area. In case, open body vehicles are being utilized for the collection and transferring of solid wastes in the study area without cover. This practice is unhygienic and spreading dust and filth, dirt from these trucks. Unsatisfied treatment of wastes in dumping sites could cause leachate and contaminating ground water resources. Iraqi Environment Law No. 27 of 2009 was issued to conserve natural resources, preserve the health of human being, support sustainable development, and increase the environmental awareness. However, the waste management problem in Iraq suffers from the lack of governmental legislation and instructions that arrange the responsibilities of recovering process between the local people and municipality administrations. As well as, the lack of legislations that regulate the investment process in Iraq caused low or absence of private sector participation in solid waste management field. Finally, due the dramatic changes in climate [20-23] which in turn could influences the production of solid wastes. Therefore, it is recommended carrying out future investigations about the relationship between the climate change and production of municipal solid wastes. Additionally, using of facilities that produce less volumes of wastes is recommended, for instance using electrocoagulation water and wastewater treatment method significantly decreases the production of sludge (solid waste) [24], which in turn enhances the management of municipal wastes.

4. Conclusions

The study showed that huge quantities of solid wastes that include more than 60 percent organic materials are generated in Thi-Qar governorate. A poor disposal infrastructure in the study area, which increases the concerns about human health and the environment due to spontaneous fires, large-scale greenhouse gas emissions and groundwater contamination probability. Based on the study results and to manage the solid waste properly in the future, the study recommended that all dumpsites that incompliant with the EPA standards should be closed. Waste separation at source should be taken place, covering at least organic and inorganic wastes. Develop suitable plan for recovery, processing, and reuse of the majority of non-hazardous industrial wastes. Establish sanitary landfill sites meets with EPA criteria's at adequate capacity to receive residual unusable wastes.

Acknowledgement
The author thanks Directorate of Thi -Qar Municipalities, Directorate of Environmental Affairs in Thi-Qar, Health Directorate in Thi-Qar governorate, Municipality offices in Al-Nassriya, Al-Shatra, Al-Rifai, Suk Al-Shoyokh, and Al-Chibayish, and Statistic Department of Thi-Qar for their corporation and support.

Reference

[1] Z. M. Bundhoo, 2018 Solid waste management in least developed countries: current status and challenges faced *Journal of Material Cycles and Waste Management* **21(1)** 1-11.
[2] J. G. Hering 2012 An end to waste? *Science* **337(6095)** 623-634.
[3] M. Abdulredha, R. Al Khaddar, D. Jordan, P. Kot, A. Abdulridha, and K. Hashim, 2018 Estimating solid waste generation by hospitality industry during major festivals: A quantification model based on multiple regression *Waste Management* **77** 388-400.
[4] D. Hou, A. Al-Tabbaa, P. Guthrie, and K. Watanabe, Sustainable waste and materials management: national policy and global perspective *Environ. Sci. Technol* **46** 2494–2495.
[5] D. Al-Jumeily, K. Hashim, R. Alkaddar, M. Al-Tufaily, and J. Lunn, Sustainable and Environmental Friendly Ancient Reed Houses (Inspired by the past to motivate the future), presented at the *The 11th International Conference on the Developments in eSystems Engineering (DeSE)*, University of Cambridge, England, UK., 2018.
[6] Erabee, I. K., and Ethaib, S. 2018 Treatment of Contaminated Landfill Leachate using Aged Refuse Biofilter Medium. *Oriental Journal of Chemistry* **34(3)**, 1441-1450.
[7] P. de Morais Lima and P. L. Paulo 2018 Solid-waste management in the rural area of Brazil: A case study in Quilombola communities *Journal of Material Cycles and Waste Management* **20 (3)** 1583-1593.
[8] M. Abdulredha, A. Rafid, D. Jordan, and K. Hashim 2017 The development of a waste management system in Kerbala during major pilgrimage events: determination of solid waste composition," Procedia Engineering **196** 779-784.
[9] O. B. Delgado, S. Ojeda-Benítez, and L. Márquez-Benavides 2007 Comparative analysis of hazardous household waste in two Mexican regions *Waste management* **27 (6)** 792-801.
[10] R. Thompson, "Classify plastic waste as hazardous," Nature, vol. 494, 2013.
[11] B. Seng, H. Kaneko, K. Hirayama, and K. Katayama-Hirayama 2011 Municipal solid waste management in Phnom Penh, capital city of Cambodia *Waste management & research* **29 (5)** 491-500..
[12] Q. Song, J. Li, and X. Zeng, 2015 Minimizing the increasing solid waste through zero waste strategy Journal *of Cleaner Production* **104** 199-210.
[13] L. Giusti 2009 A review of waste management practices and their impact on human health*Waste management* **29 (8)**2227-2239.
[14] A. M. Troschinetz and J. R. Mihelcic 2009 Sustainable recycling of municipal solid waste in developing countrie, *Waste management* **29, (2)** 915-923.
[15] R. K. Henry, Z. Yongsheng, and D. Jun 2006 Municipal solid waste management challenges in developing countries–Kenyan case study," *Waste management*, **26 (1)** 92-100.
[16] Municipal Solid Waste Landfills, Available: https://www.epa.gov/landfills/municipal-solid-waste-landfills. Accessed 25 January 2018.
[17] Erabee I. K. and Ethaib, S. 2018 Performance of Activated Carbon Adsorption in Removing of Organic Pollutants from River Water *International Journal of Engineering & Technology* **7 (4.20)** 356-358.
[18] Ethaib, S., Erabee I. K., and Abdulsahib 2018 Removal of Methylene Blue Dye from Synthetic Wastewater using Kenaf Core and Activated Carbon *International Journal of Engineering & Technology* **7 (4.19)** 909-91, 2018

[19] Z. Wu, T. Ann, L. Shen, and G. Liu 2014 Quantifying construction and demolition waste: An analytical review *Waste Management* **34 (9)** 1683-1692.

[20] S. L. Zubaidi, J. Dooley, R. M. Alkhaddar, M. Abdellatif, H. Al-Bugharbee, and S. Ortega-Martorell 2018 A Novel approach for predicting monthly water demand by combining singular spectrum analysis with neural networks *Journal of hydrology* **561** 136-145.

[21] S. L. Zubaidi, S. K. Gharghan, J. Dooley, R. M. Alkhaddar, and M. Abdellatif 2018 Short-Term Urban Water Demand Prediction Considering Weather Factors," Water Resources Management **32 (14)** 4527-4542.

[22] S. L. Zubaidi, P. Kot, R. M. Alkhaddar, M. Abdellatif, and H. Al-Bugharbee, Short-Term Water Demand Prediction in Residential Complexes: Case Study in Columbia City, USA presented at the *11th International Conference on Developments in eSystems Engineering (DeSE)*, University of Cambridge, UK, 2018.

[23] S. L. Zubaidi, P. Kot, R. M. Alkhaddar, M. Abdellatif, and H. Al-Bugharbee, Short-Term Water Demand Prediction in Residential Complexes: Case Study in Columbia City, USA presented at the 11th International Conference on Developments in eSystems Engineering (DeSE), University of Cambridge, UK, 2018.

[24] K. S. Hashim, A. Shaw, R. Al Khaddar, M. Ortoneda Pedrola, and D. Phipps, 2017 Defluoridation of drinking water using a new flow column-electrocoagulation reactor (FCER) - Experimental, statistical, and economic approach *J Environ Manage*, **197** 80-88..

Decolourization of dye solutions by electrocoagulation: an investigation of the effect of operational parameters

Khalid S. Hashim [1,2*], N H Al-Saati[3], Saif Salah Alquzweeni[4], Salah L. Zubaidi[5], Patryk Kot[1], L. Kraidi[1], Ameer H. Hussein[3], Rafid Alkhaddar[1], Andy Shaw[1], Reham Alwash[2]

[1] Department of Civil Engineering, Liverpool John Moores University, UK.
[2] Department of Environmental Engineering/ University of Babylon, Iraq.
[3] Al-Mussaib Technical Institute /Al-Furat Al-Awsat Technical University, Iraq.
[4] Department of Civil Engineering/ University of Babylon, Iraq.
[5] Department of Civil Engineering, College of Engineering, University of Wasit, Iraq.

* Corresponding author, E-mail address: k.s.hashim@ljmu.ac.uk

Abstract. This study investigates the influence of the current density on both removal of reactive black 5 dye (RB5) from water and energy consumption using aluminium-based electrocoagulation (EC) reactor. The influence of the current density was investigated by electrolysing coloured water samples containing 25 mg/l of RB5 for 60 min at three different current densities (1, 2, and 3 mA/cm2). While the initial pH, distance between electrodes, flow rate, and initial temperature of water were kept constant at 5, 4 mm, 1 L/h, and 20 ± 10C, respectively. The obtained results showed that both the removal of RB5 dye and energy consumption increased with the increase of the applied current density. For example, it has been found that the removal RB5 and the power consumption increased from about 95% to 99.5%, and the power consumption increased from 4.3 to 10.4 kWh/m3 as the current density increased from 1 to 3 mA/cm2, respectively.

1. Introduction

Water pollution is serious global problem due to the limited amount of water on planet of Earth, and due to the rapid growth in both global population and industrial activities, which in turn produce huge quantity of wastewater [2-6]. Though there is a wide range of industries that discharge huge quantities of wastewater into the surface water bodies, the textile industry is categorised as a major and serious source of water pollution as it produces large volumes of highly polluted wastewater [7, 8]. For instance, the relevant studies indicated that this industry consumes about 1-2 m3 of water per each 10 kg of textile products, and produce a similar amount of polluted wastewater [9]. Moreover, the textile effluent is categorised as serious and highly polluted wastewater due to its chemical composition, where this type of wastewater contains detergents, dyes, colour fixing agents, and organic matter [9, 10]. Furthermore, azo dyes could be decomposed in the receiving carcinogenic by-products that have long half-time life [8, 11, 12]. Discharging this highly polluted wastewater into the water bodies causes different pollution problems, for example it causes undesirable colours and odours [13]; minimises the penetration sun light that damages the aquatic life [14]; and minimises the concentration of the dissolved oxygen [11]. Thus, different treatment technologies have been used to treat the textile wastewater, such as aerobic and anaerobic degradation, oxidation, and filtration [12, 15, 16]. Unfortunately, many of the practiced methods were not efficient to remove the most dangerous pollutants such as the azo dyes [17, 18]. More badly, application of some of treatment methods to zo dyes containing wastewater results in the formation of carcinogenic compounds [19], for example, biological reduction of azo dyes under anaerobic conditions could form aromatic amines, which categorised as carcinogens [20].

Therefore, recent researches suggested different advanced treatment methods to treat the textile effluents, such as nano-filtration [1], and membrane bioreactor-ultrafiltration [21]. Amongst the recently practiced methods, electrocoagulation method (EC) received an increasing attention as an effective and affordable treatment method for textile and municipal wastewater due several merits. For instance, EC method does not required chemical additives, it requires relatively short treatment time, and it is easy to be operated, marinated, and integrated with other treatment methods. Additionally, EC method produces less sludge in comparison with traditional methods, which greatly enhances its cost-effectiveness because treatment and handling of solid wastes is a costly process [22-31].

In this context, the current study investigates the influence of applied current density on the performance of the EC method in terms of dye removal and energy consumption.

2. Aims and objectives

As it has been mentioned above, the current study investigates the influence of the applied current density on the performance of the EC method in terms of dye removal and energy consumption. It is noteworthy to mention that reactive black 5 dye (RB-5 dye) has been used, in the current study, as the model dye because it is widely used in the textile industry, and its degradation forms toxic compounds [32]. The molecular structure of RB-5 is shown in Figure (1).

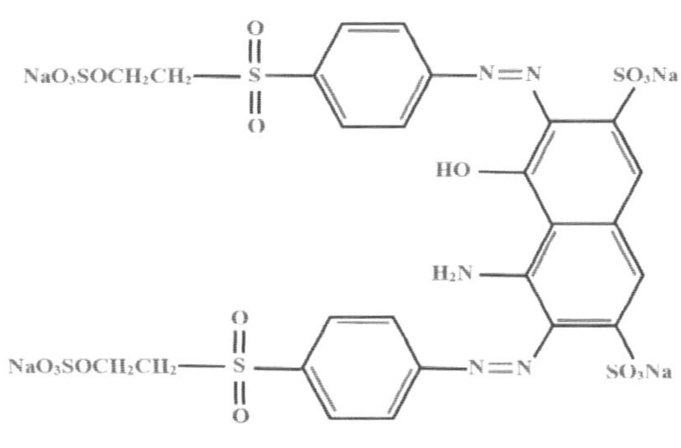

Figure 1. Molecular structure of RB-5 dye [1].

3. Methodology

3.1 EC unit

The influence of current density on the removal of RB-5 dye has been investigated by electrolysing dye containing synthetic water samples using aluminium-based EC unit, figure 2. The latter is consists of cylindrical Perspex container (25 cm in height and 10.5 cm in internal diameter), which is supplied with aluminium perforated discoid electrodes (10.4 cm in diameter and 0.1 cm in thickness), Fig. 2. These electrodes were stacked vertically within the cylindrical container with the plane of each electrode parallel and perpendicular to the direction of flow. Each electrode had the holes offset from the one above it to ensure that the water will flow in a convoluted path in order to increase mixing efficiency. A peristaltic pump (Watson Marlow, model: 504U) was used to circulate the water being treated through the EC unit, while the required current was supplied using a rectifier (HQ Power; Model: PS 3010).

Figure 2. The EC unit.

3.2 Influence of current density on dye removal and energy consumption

Initially, synthetic stock solution, 100 mg/l, was prepared by dissolving the required amount of RB-5 dye in deionised water. Samples with less concentrations of dye, 25 mg/L, were diluted from this stock solution. Influence of current density on the removal of RB-5 dye has been investigated by subjecting diluted water samples to different current densities (1, 2, and 3 mA/cm2) for 60 min at flow rate of 1 L/h, inter-electrode distance of 4 mm, initial pH of 5, and ambient temperature of (20 ±10C).

5 mL water samples were periodically collected, each 5 min, from the EC unit to monitor the residual concentration of RB-5 dye. The collected samples were filtered at 0.45 μm filters (Whatman filters); the residual dye concentration in the filtrate was measured using a spectrophotometer (Hach Lange DR 2800). The removal efficiency (RE %) was determined using the following equation [33]:

$$RE\% = \frac{C_0 - C_t}{C_0} \times 100\% \quad (1)$$

Where C_0 and C_t represent the initial and the measured concentrations of RB5, in mg/L, respectively.
While the energy consumption (E), during the course of each set of experiments, was measured using the following formula [34]:

$$E = \frac{I*V*T}{Vol.} \tag{2}$$

Where I is the current (A), V is the potential (V), t is the treatment time (H), and Vol. is the volume of treated water (m3).

4. Results and discussion

4.1. Influence of current density on dye removal

As it has been mentioned above, the influence of current density on the removal of RB-5 dye has been investigated by electrolysing coloured water samples (25 mg/L of RB-5 dye) for 60 min at three current densities (1, 2, and 3 mA/cm2).

The obtained results showed a direct proportion between the removal of the RB-5 dye and the magnitude of the applied current density. It can be seen from figure 3 that 30 min was enough to achieve 100% removal of RB-5 dye at current density of 3 mA/cm2, but 60 min was required at current density of 2 mA/cm2 to reach a removal percentage of 98.6%. %. This increase in removal efficiency with the increase of current density could be explained by the fact that the dissolved aluminium ions from the anode increases as the current density increase, which enhances RB-5 dye removal as a consequence [35].

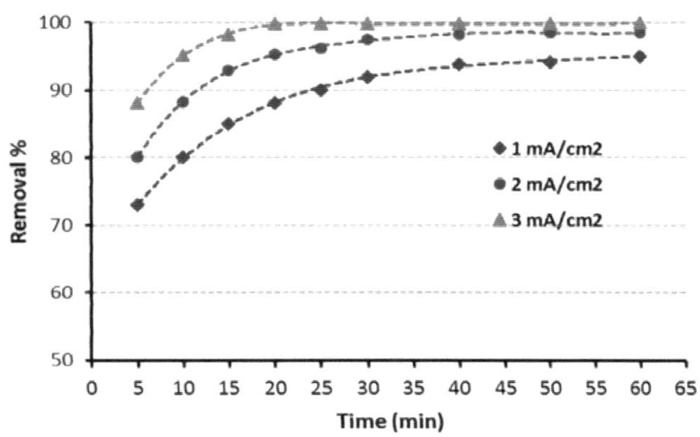

Figure 3. RB-5 dye removal at different current densities.

4.2. Influence of current density on energy consumption

Though it has been found that increasing the current density positively influences the removal of BR-5 dye, the obtained results showed that increasing the current density negatively influenced the energy consumption, where Figure 4 shows that as current density increased from 1 to 3 mA/cm2, the power consumption increased from 4.3 to 10.4 kWh/m3, respectively.

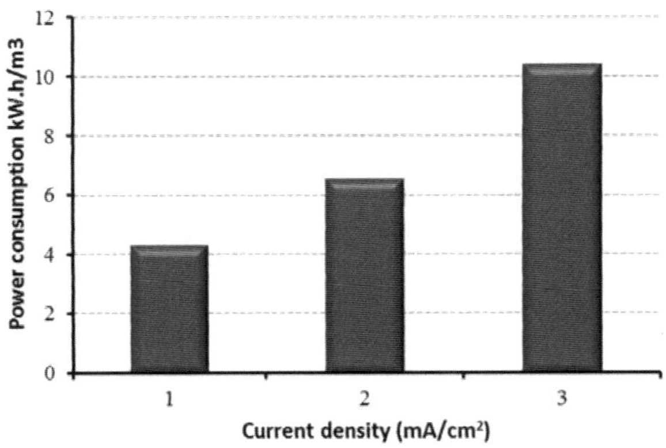

Figure 4. Power consumption at different current densities.

5. Conclusions

According to the outcomes of the current study, the removal of RB-5 dye can be effectively enhanced by increasing the generation rate of fresh coagulants from the metallic anode, which in turn could be enhanced by increasing the magnitude of the applied current density. However, the increase in the current density is beneficial to a certain limit as it increases the power consumption, which negatively influences the applicability of this method. Therefore, the magnitude of the applied current must be optimised to achieve the highest removal efficiency at a reasonable operational cost.

It is recommended to carry out further investigations about the influence of other operating parameters, such as water conductivity, ability of the EC method to remove dyes from water and wastewater. Additionally, the ability of the EC method to remove other pollutants from water and wastewater should be studied.

Finally, due the dramatic changes in climate [36-39] that greatly influencing the global water, the authors recommend to carry out future investigations about the development of new water treatment methods, such as integrating EC method with other methods.

References

[1] Y. K. Ong, F. Y. Li, S.-P. Sun, B.-W. Zhao, C.-Z. Liang, and T.-S. Chung, "Nanofiltration hollow fiber membranes for textile wastewater treatment: Lab-scale and pilot-scale studies," *Chemical Engineering Science,* vol. 114, pp. 51-57, 2014.

[2] K. S. Hashim, A. Shaw, R. Al Khaddar, M. Ortoneda Pedrola, and D. Phipps, "Defluoridation of drinking water using a new flow column-electrocoagulation reactor (FCER) - Experimental, statistical, and economic approach," *J Environ Manage,* vol. 197, pp. 80-88, Jul 15 2017.

[3] K. S. Hashim, A. Shaw, R. Al Khaddar, M. O. Pedrola, and D. Phipps, "Iron removal, energy consumption and operating cost of electrocoagulation of drinking water using a new flow column reactor," *Journal of Environmental Management,* vol. 189, pp. 98-108, 2017.

[4] K. S. Hashim, A. Shaw, R. Al Khaddar, M. O. Pedrola, and D. Phipps, "Energy efficient electrocoagulation using a new flow column reactor to remove nitrate from drinking water - Experimental, statistical, and economic approach," *J Environ Manage,* vol. 196, pp. 224-233, Jul 01 2017.

[5] K. S. Hashim, N. H. Al-Saati, A. H. Hussein, and Z. N. Al-Saati, "An Investigation into The Level of Heavy Metals Leaching from Canal-Dreged Sediment: A Case Study Metals Leaching from Dreged Sediment," presented at the 1st International Conference on Materials Engineering & Science, Istanbul Aydın University (IAU), Turkey, 2018.

[6] N. H. Al-Saati *et al.*, "Statistical modelling of turbidity removal applied to non-toxic natural coagulants in water treatment: a case study," *Desalination and Water Treatment,* vol. 150, pp. 406-412, 2019.

[7] A. M. Lotito, M. De Sanctis, C. Di Iaconi, and G. Bergna, "Textile wastewater treatment: aerobic granular sludge vs activated sludge systems," *Water Res,* vol. 54, pp. 337-46, May 1 2014.

[8] P. A. Joshi and K. J. Mhatre, "Microbial efficiency to degrade Carbol fuchsin and Malachite green dyes," *Advances in Applied Science Research,* vol. 6, no. 12, pp. 85-88, 2015.

[9] H. Yin *et al.*, "Textile Wastewater Treatment for Water Reuse: A Case Study," *Processes,* vol. 7, no. 1, p. 34, 2019.

[10] V. S. Munagapati and D.-S. Kim, "Adsorption of anionic azo dye Congo Red from aqueous solution by Cationic Modified Orange Peel Powder," *Journal of Molecular Liquids,* vol. 220, pp. 540-548, 2016.

[11] P. I. Firmino, M. E. da Silva, F. J. Cervantes, and A. B. dos Santos, "Colour removal of dyes from synthetic and real textile wastewaters in one- and two-stage anaerobic systems," *Bioresour Technol,* vol. 101, no. 20, pp. 7773-9, Oct 2010.

[12] V. L. Gole and P. R. Gogate, "Degradation of brilliant green dye using combined treatment strategies based on different irradiations," *Separation and Purification Technology,* vol. 133, pp. 212-220, 2014.

[13] J. Fu *et al.*, "Degradation of Active Brilliant Red X-3B by a microwave discharge electrodeless lamp in the presence of activated carbon," *Environ Technol,* vol. 31, no. 7, pp. 771-9, Jun 2010.

[14] S. C. Santos and R. A. Boaventura, "Treatment of a simulated textile wastewater in a sequencing batch reactor (SBR) with addition of a low-cost adsorbent," *J Hazard Mater,* vol. 291, pp. 74-82, Jun 30 2015.

[15] D. K. Mahmoud, M. A. M. Salleh, W. A. W. A. Karim, A. Idris, and Z. Z. Abidin, "Batch adsorption of basic dye using acid treated kenaf fibre char: Equilibrium, kinetic and thermodynamic studies," *Chemical Engineering Journal,* vol. 181-182, pp. 449-457, 2012.

[16] D. Al-Jumeily, K. Hashim, R. Alkaddar, M. Al-Tufaily, and J. Lunn, "Sustainable and Environmental Friendly Ancient Reed Houses (Inspired by the Past to Motivate the Future)," in *2018 11th International Conference on Developments in eSystems Engineering (DeSE)*, 2018, pp. 214-219: IEEE.

[17] Y. Aşçı, E. A. Demirtas, C. F. Iscen, and A. S. Anagun, "A statistical experimental design to determine the azo dye decolorization and degradation by the heterogeneous fenton process," *Fresenius Environmental Bulletin,* vol. 24, no. 11a, pp. 3717-3726, 2015.

[18] H. Hayat, Q. Mahmood, A. Pervez, Z. A. Bhatti, and S. A. Baig, "Comparative decolorization of dyes in textile wastewater using biological and chemical treatment," *Separation and Purification Technology,* vol. 154, pp. 149-153, 2015.

[19] P. Aravind, V. Subramanyan, S. Ferro, and R. Gopalakrishnan, "Eco-friendly and facile integrated biological-cum-photo assisted electrooxidation process for degradation of textile wastewater," *Water Res,* vol. 93, pp. 230-41, Apr 15 2016.

[20] N. D. Lourenço, R. D. G. Franca, M. A. Moreira, F. N. Gil, C. A. Viegas, and H. M. Pinheiro, "Comparing aerobic granular sludge and flocculent sequencing batch reactor technologies for textile wastewater treatment," *Biochemical Engineering Journal,* vol. 104, pp. 57-63, 2015.

[21] D. De Jager, M. S. Sheldon, and W. Edwards, "Colour removal from textile wastewater using a pilot-scale dual-stage MBR and subsequent RO system," *Separation and Purification Technology,* vol. 135, pp. 135-144, 2014.

[22] S. Safari, M. Azadi Aghdam, and H. R. Kariminia, "Electrocoagulation for COD and diesel removal from oily wastewater," *International Journal of Environmental Science and Technology,* vol. 13, no. 1, pp. 231-242, 2015.

[23] S. Elabbas *et al.*, "Treatment of highly concentrated tannery wastewater using electrocoagulation: Influence of the quality of aluminium used for the electrode," *J Hazard Mater,* Jan 4 2016.

[24] K. S. Hashim, A. Shaw, R. Alkhaddar, and M. O. Pedrola, "Controlling of Water Temperature during the Electrocoagulation Process Using an Innovative Flow Columns - Electrocoagulation Reactor," *International Journal of Environmental, Chemical, Ecological, Geological and Geophysical Engineering,* vol. 9, no. 8, pp. 869-872, 2015.

[25] K. S. Hashim, A. Shaw, R. Alkhaddar, M. O. Pedrola, and D. Phipps, "Effect of the supporting electrolyte concentration on energy consumption and defluoridation of drinking water in the electrocoagulation (EC) method," in *The 2nd BUiD Doctoral Research Conference*, The British University in Dubai, 2016.

[26] A. Shaw, K. S. Hashim, R. Alkhaddar, M. O. Pedrola, and D. Phipps, "Influence of electrodes spacing on internal temperature of electrocoagulation (EC) cells during the removal (Fe II) from drinking water," in *The 3rd BUiD Annual Doctoral Research Conference*, The British University, Dubai, 2017.

[27] M. Abdulredha, R. Al Khaddar, D. Jordan, P. Kot, A. Abdulridha, and K. Hashim, "Estimating solid waste generation by hospitality industry during major festivals: A quantification model based on multiple regression," *Waste Manag,* vol. 77, pp. 388-400, Jul 2018.

[28] M. Abdulredha, A. Rafid, D. Jordan, and K. Hashim, "The development of a waste management system in Kerbala during major pilgrimage events: determination of solid waste composition," *Procedia Engineering,* vol. 196, pp. 779-784, 2017.

[29] A. W. Alattabi, C. Harris, R. Alkhaddar, A. Alzeyadi, and K. Hashim, "Treatment of Residential Complexes' Wastewater using Environmentally Friendly Technology," *Procedia Engineering,* vol. 196, pp. 792-799, 2017.

[30] A. W. Alattabi, C. B. Harris, R. M. Alkhaddar, K. S. Hashim, M. Ortoneda-Pedrola, and D. Phipps, "Improving sludge settleability by introducing an innovative, two-stage settling sequencing batch reactor," *Journal of Water Process Engineering,* vol. 20, pp. 207-216, 2017.

[31] I. A. Idowu *et al.*, "An analyses of the status of landfill classification systems in developing countries: Sub Saharan Africa landfill experiences," *Waste Management,* vol. 87, pp. 761-771, 2019.

[32] S. H. Chang *et al.*, "Treatment of Reactive Black 5 by combined electrocoagulation-granular activated carbon adsorption-microwave regeneration process," *J Hazard Mater,* vol. 175, no. 1-3, pp. 850-7, Mar 15 2010.

[33] K. S. Hashim *et al.*, "Removal of phosphate from River water using a new baffle plates electrochemical reactor," *MethodsX,* vol. 5, pp. 1413-1418, 2018.

[34] U. T. Un, A. S. Koparal, and U. B. Ogutveren, "Fluoride removal from water and wastewater with a bach cylindrical electrode using electrocoagulation," *Chemical Engineering Journal,* vol. 223, pp. 110-115, 2013.

[35] S. L. Zubaidi, J. Dooley, R. M. Alkhaddar, M. Abdellatif, H. Al-Bugharbee, and S. Ortega-Martorell, "A Novel approach for predicting monthly water demand by combining singular spectrum analysis with neural networks," *Journal of hydrology,* vol. 561, pp. 136-145, 2018.

IOP Conf. Series: Materials Science and Engineering **584** (2019) 012024 doi:10.1088/1757-899X/584/1/012024

[36] S. L. Zubaidi, S. K. Gharghan, J. Dooley, R. M. Alkhaddar, and M. Abdellatif, "Short-Term Urban Water Demand Prediction Considering Weather Factors," *Water Resources Management,* vol. 32, no. 14, pp. 4527-4542, 2018.

[37] S. L. Zubaidi, P. Kot, R. M. Alkhaddar, M. Abdellatif, and H. Al-Bugharbee, "Short-Term Water Demand Prediction in Residential Complexes: Case Study in Columbia City, USA," presented at the 11th International Conference on Developments in eSystems Engineering (DeSE), University of Cambridge, UK, 2018.

[38] S. L. Zubaidi, P. Kot, R. M. Alkhaddar, M. Abdellatif, and H. Al-Bugharbee, "Short-Term Water Demand Prediction in Residential Complexes: Case Study in Columbia City, USA," in *2018 11th International Conference on Developments in eSystems Engineering (DeSE)*, 2018, pp. 31-35: IEEE.

Development of Risk Optimisation Model for Oil and Gas Transportation Pipeline Routes

L Kraidi [1, 2,*], R Shah [1], W Matipa [1], F Borthwick [1], Khalid Hashim [3]

[1] Department of the Built Environment, Faculty of Engineering and Technology, Liverpool John Moores University, UK.
[2] PhD Student, Henry Cotton Building, Faculty of Engineering and Technology, Liverpool John Moores University, UK.
[3] Department of Civil Engineering, Faculty of Engineering and Technology, Liverpool John Moores University, UK.

* Corresponding Author: L.A.Kraidi@2016.ljmu.ac.uk

Abstract. Identifying safe routes for the pipelines that transport Oil and Gas (O&G) products is a challenging topic in the current environment; particularly in the insure countries. Because the relevant data about the probability and severity levels of the Risk Factors (RFs) that affect the safety of these pipelines are rare. Which makes the existing risk assessment tools ineffective to analyse these RFs and identify safe route for these pipelines. Hence, this paper aims to develop a risk assessment tool that can identify safe routes for the new O&G pipelines in Iraq in a systematic way using the following steps. Firstly, an industry-wide questionnaire survey was conducted to gather the data about the probability and severity levels of the RFs in such projects in Iraq. Secondly, the Fuzzy Inference System (FIS) in MATLAB was used to analyse and rank the RFs. Because the FIS can reduce the uncertainty in risk analysis, which results from the lack of data and the biasness of stakeholder's judgments about the RFs. Thirdly, the existing information from the new pipelines projects were analysed to identify the potential RFs in the proposed routes for these projects. As the O&G pipeline network in Iraq is above-the-ground, this paper focused on the RFs that affect this type of pipelines. Fourthly, the safest route for the new pipeline was identified by optimising the risk index value for each route. While, the route that has less value of risk index is the safest route. This paper analysed the five routes that were suggested to build a new gas export pipeline in Waist in Iraq. The pipeline will transport the extracted gas from Badra filed to the shipping points in Iraq. It was found that route number 4 is the safest route for this pipeline.

1. Introduction

Oil and Gas Pipelines (OGPs) must be planned, designed, installed, operated and maintained regarding the safety requirements to transport the petroleum products safely. However, several Risks Factors (RFs) are threatening the safety of these projects, such as terrorism, sabotage, thefts, corrosion, design and construction defects, natural hazards, operational errors and many more. Meanwhile, the current risk assessment tools are inaccurate to analyse the RFs in OGP projects in the developing countries due to the data scarcity and lack of research about them in these countries. As stated by Kraidi et al., [1], the risk management system in OGP projects in Iraq suffers from the scarcity of data about the probability and severity levels of the RFs in these projects. The alternative way of identifying and analysing the RFs in such a situation is via conducting a literature review about the RFs in OGP projects and collecting the stakeholders' perceptions about them [2]. Nevertheless, analysing the RFs based on the stakeholders' perceptions results in uncertain results. Because the stakeholders have different perceptions about the probability and severity levels of the RFs [3]. Therefore, the RFs in this paper will be analysed using the Fuzzy Inference System (FIS) in MATLAB. Because the FIS uses linguistics terms (e.g. very low, low, moderate, high and very high) to analyse the RFs, which is useful to calculate the Risk Index (RI) of the factors when there are neither sharp boundaries nor precise values of their probability and severity levels [4].

2. Aims and objectives

The purpose of the paper, therefore, is to develop a risk assessment tool that helps in analysing the RFs in the pipelines' projects and choosing safe routes for the new projects in a systematic way. This tool will analyse the OGP projects in two cities southern of Iraq, which are Waist (Al Kut) and Basra. Because the current risk management system in these projects is inadequate, which obstructs gas export activities. Figure 1 shows the flowchart of the risk assessment tool.

3. Research approach

Figure 1 explains the flowchart of the risk assessment tool.

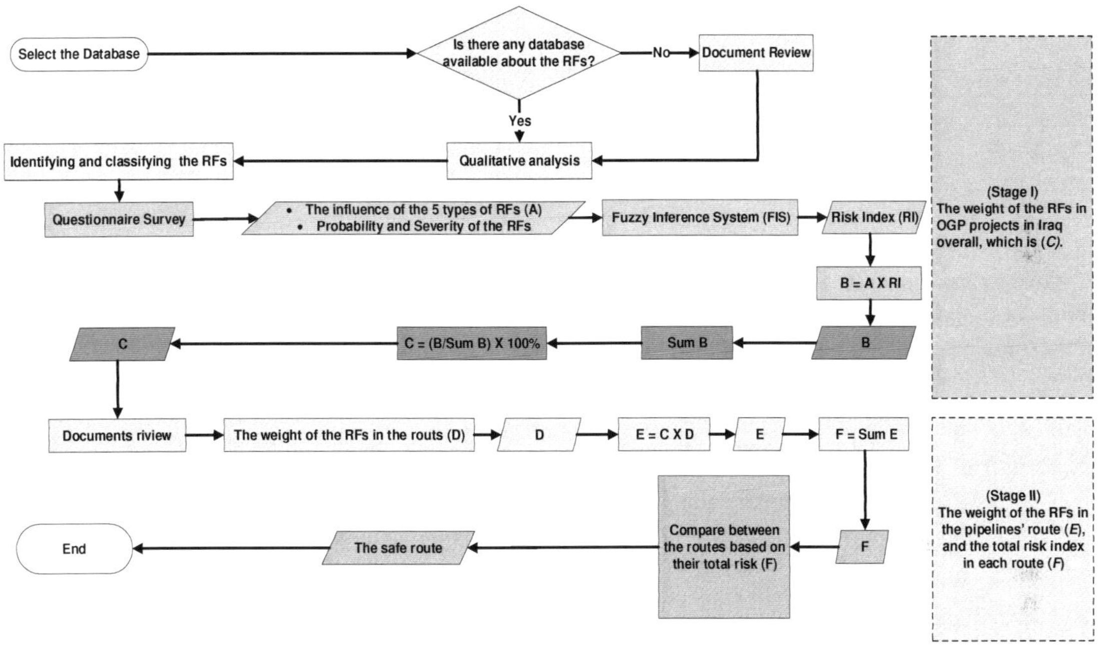

Figure 1. The flowchart of the risk assessment tool.

Following figure 1, the risk assessment tool works in two stages. In stage I, extensive investigations were carried out to identify the RFs in OGP projects in different countries and circumstances to overcome the problem of data scarcity about them in Iraq. The identified RFs were classified into five groups based on their type. Then, the RFs were evaluated via a questionnaire survey that was distributed amongst the stakeholders in OGP projects in Iraq using an online survey tool. The outputs of the survey were the weight of each group of the RFs (A). As well as the probability and severity levels of the RFs, which were used as inputs for the FIS in MATLAB to calculate the RI of the RFs, see Figure 2. Appendix A shows examples of the questionnaire survey. Appendix A shows the size of the sample and the response rat. B is the weight of the RF that considers the weight of its groups (A) and its value of RI. The results of (Stage I) of risk analysis are shown in **Error! Reference source not found.**.

$$B = A \times RI \quad (1)$$

C is the weight of the RFs from 100%,

$$C = (B / (Sum\ B)) \times 100\% \quad (2)$$

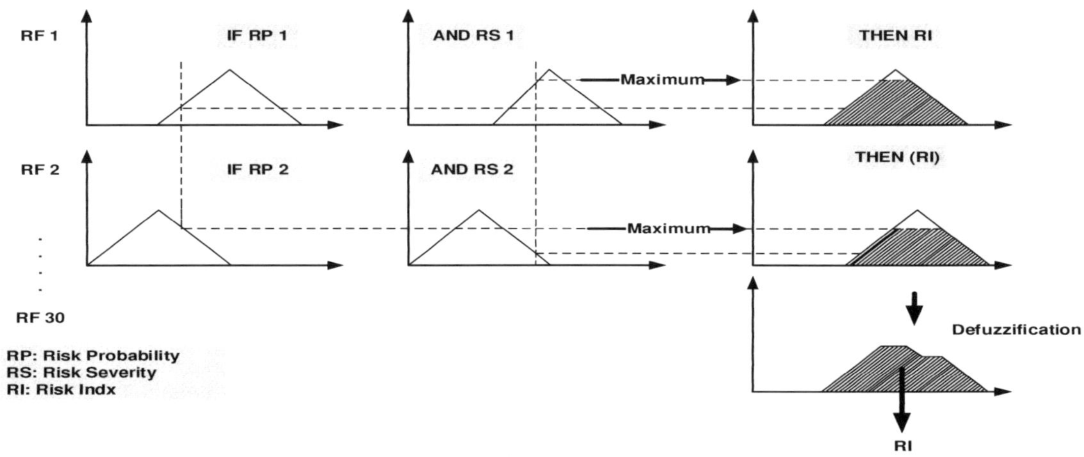

Figure 2. The diagram of the Fuzzy Inference System (FIS).

Stage II of the risk assessment tool is about analysing the RFs in specific routes of OGPs. This paper has analysed the five proposed routes for a new gas pipeline, which will be built from Badra gas field in Waist to Basra in order to export the extracted gas from that field via the sea. The available documents about these five routes were subjectively analysed to identify the RFs that might threaten the pipelines in these routes. D is the weight of the RFs in the route. Based on the document analysis, in the case that the RF is threatening the pipeline, then $\boldsymbol{D} = \mathbf{1}$; otherwise, $= \mathbf{2}$. E is the final weight of the RFs in the route.

$$E = C \times D \quad (3)$$

Where ***C*** is the weight of the RFs in OGP projects in Iraq overall and ***D*** is the weight of the RFs within the specific route. ***F*** is the total risk index in the route.

$$F = SUM(C) \quad (4)$$

The route that has less value of ***F*** is the safest route. Because it has the less total impact of the RFs. The results of analysing the OGPs' routes are shown in **Table 2**.

Table 1. The identified RFs from the literature review and their values of probability, severity and index.

RFs [1-3 and 5,6]	Type*	A	Probability	Severity	RI	B	C
Terrorism, sabotage and the security risk	S&S	2.84	3.995	4.490	3.99	11.33	5.43
Stealing the products	S&S		3.692	4.081	3.75	10.65	5.10
Public awareness	S&S		3.712	4.106	3.80	10.79	5.17
Staff threats	S&S		3.323	3.571	3.35	9.51	4.55
Socio-political effects	S&S		3.449	3.611	3.49	9.91	4.75
Leakage of sensitive information	S&S		2.980	3.399	3.38	9.60	4.60
Corruption	R&L	1.45	3.980	4.323	3.87	5.61	2.69
The absence of the law on TPD	R&L		3.606	3.682	3.54	5.13	2.46
Lack of risk management practice	R&L		3.530	3.652	3.51	5.09	2.44
Lack of proper training	R&L		3.646	3.859	3.71	5.38	2.58
Lack of risk registration	R&L		3.566	3.662	3.60	5.22	2.50
Little research on this topic	R&L		3.621	3.697	3.55	5.15	2.46
The geographical location	PL	2.37	3.717	4.192	3.76	8.91	4.27
The pipeline is easy to access	PL		3.631	3.773	3.57	8.46	4.05
Land ownership conflicts	PL		3.495	3.646	3.68	8.72	4.18
Geological risks	PL		2.747	3.182	3.17	7.51	3.60
Vehicles accidents	PL		2.465	2.970	2.80	6.64	3.18
Animals accidents	PL		1.894	2.020	1.95	4.62	2.21
Improper safety regulations	HSE	1.89	3.687	3.960	3.70	6.99	3.35
Improper inspection and maintenance	HSE		3.657	3.899	3.69	6.97	3.34
The risk related to the aboveground pipeline	HSE		3.667	3.949	3.70	6.99	3.35
Limited warning signs	HSE		3.626	3.732	3.56	6.73	3.22
Inadequate risk management	HSE		3.227	3.505	3.48	6.58	3.15
Natural disasters	HSE		2.652	3.066	3.10	5.86	2.81
Corrosion	OC	1.45	3.687	3.990	3.72	5.39	2.58
The weak ability to manage the risk	OC		3.631	3.848	3.67	5.32	2.55
Shortage of modern equipment	OC		3.667	3.924	3.68	5.34	2.55
Design, construction and material defects	OC		3.333	3.611	3.64	5.28	2.53
Operational errors	OC		3.101	3.409	3.30	4.79	2.29
Hacker attacks on the system	OC		3.066	3.066	3.03	4.39	2.10

*Security and Safety (S&S), Rules and Regulations (R&R), Pipeline Location (PL), Health Safety and Environment (HSE) and Operations Consent (OC)

IOP Conf. Series: Materials Science and Engineering **584** (2019) 012025 doi:10.1088/1757-899X/584/1/012025

Table 2. The results of analysing the RFs and testing the pipelines routs.

RFs	C	Route 1		Route 2		Route 3		Route 4		Route 5	
		D1	E1	D2	E2	D3	E3	D4	E4	D5	E5
Terrorism, sabotage and the security risk	5.43	0	0.00	1	5.43	1	5.43	0	0.00	0	0.00
Stealing the products	5.10	1	5.10	1	3.20	1	5.10	0	0.00	0	0.00
Public awareness	5.17	1	5.17	1	3.24	1	5.17	1	5.17	1	5.17
Staff threats	4.55	0	0.00	1	2.86	0	0.00	1	4.55	0	0.00
Socio-political effects	4.75	0	0.00	1	2.98	1	4.75	1	4.75	0	0.00
Leakage of sensitive information	4.60	1	4.60	0	0.00	1	4.60	0	0.00	1	4.60
Corruption	2.69	1	2.69	1	3.30	1	2.69	1	2.69	1	2.69
The absence of the law on TPD	2.46	1	2.46	1	3.02	1	2.46	1	2.46	0	0.00
Lack of risk management practice	2.44	1	2.44	1	2.99	1	2.44	1	2.44	1	2.44
Lack of proper training	2.58	1	2.58	1	3.16	1	2.58	1	2.58	1	2.58
Lack of risk registration	2.50	1	2.50	1	3.07	1	2.50	1	2.50	1	2.50
Little research on this topic	2.46	1	2.46	1	3.03	1	2.46	1	2.46	1	2.46
The geographical location	4.27	0	0.00	1	3.21	1	4.27	0	0.00	0	0.00
The pipeline is easy to access	4.05	1	4.05	1	3.05	1	4.05	0	0.00	1	4.05
Land ownership conflicts	4.18	1	4.18	1	3.14	0	0.00	1	4.18	1	4.18
Geological risks	3.60	0	0.00	1	2.70	1	3.60	0	0.00	1	3.60
Vehicles accidents	3.18	0	0.00	1	2.39	1	3.18	0	0.00	1	3.18
Animals accidents	2.21	0	0.00	0	0.00	0	0.00	0	0.00	0	0.00
Improper safety regulations	3.35	1	3.35	1	3.16	1	3.35	1	3.35	1	3.35
Improper inspection and maintenance	3.34	1	3.34	1	3.15	0	0.00	0	0.00	1	3.34
The risk related to the aboveground pipeline	3.35	1	3.35	1	3.16	0	0.00	1	3.35	1	3.35
Limited warning signs	3.22	0	0.00	1	3.04	1	3.22	0	0.00	1	3.22
Inadequate risk management	3.15	1	3.15	1	2.97	1	3.15	1	3.15	1	3.15
Natural disasters	2.81	1	2.81	0	0.00	1	2.81	0	0.00	1	2.81
Corrosion	2.58	0	0.00	1	3.17	1	2.58	1	2.58	0	0.00
The weak ability to manage the risk	2.55	1	2.55	1	3.13	1	2.55	1	2.55	1	2.55
Shortage of modern equipment	2.55	1	2.55	1	3.14	1	2.55	1	2.55	1	2.55
Design, construction and material defects	2.53	1	2.53	0	0.00	1	2.53	0	0.00	1	2.53
Operational errors	2.29	1	2.29	1	2.81	1	2.29	0	0.00	1	2.29
Hacker attacks on the system	2.10	0	0.00	0	0.00	0	0.00	0	0.00	0	0.00
Sum =	100.00	F1=	64.12	F2=	78.50	F3=	80.27	F4=	51.29	F4=	66.55

4. Discussion

The initial list of RFs that affect the OGPs was identified based on the literature review and it was evacuated via a questionnaire survey. The participants were asked to add RFs that have not mentioned in the survey. After analysing their comments, the list of the affective RFs was revised. For instance, some of the RFs have been deleted from the list like the hacker attacks on the operating or control system and animals' accidents as these RFs have a very low impact on the OGPs in Iraq based on the results of the survey. Some of the RFs like construction and material defects were split up into three RFs, which are design defects, construction defects and material defects. Some of the RFs were merged as one RFs like geographical location like "insecure areas" and the pipeline is easy to access. Some of the RFs added to the list like the pipes are older than the design age. Table 3 shows the final lists of RFs by their type.

Table 3. The final lists of RFs by their type.

Groups	The RFs before the survey	The RFs after the survey	RF number
Security and Safety (S&S)	Terrorism and sabotage	Terrorism, sabotage and the security	1
	Thieves	Thieves	2
	Public's Low legal and moral	Public's Low legal and moral awareness	3
	Staff threats	*Deleted*	
	Socio-political such as poverty and	Socio-political effects such as poverty and	4
	Leakage of sensitive information	Leakage of sensitive information	5
Rules and Regulations (R&R)	Corruption	Corruption	6
	The law does not apply on the	The law does not apply on the saboteurs	7
	Stakeholders are not paying proper attention	Not paying proper attention to risk management (e.g. not following scheduled	8
	Lack of proper training	Lack of proper training	9
	Lack of the accidents database and	Lack of the accidents database and	10
	Limited researchers about this	Limited researchers about this subject	11
Pipeline Location (PL)	Geographical location like	Geographical location e.g. insecure zones	12
	The pipeline is easy to access	*Deleted*	
	Conflicts over land ownership	Conflicts over land ownership	13
	Geological risks such as	Geological risks such as groundwater and	14
	Vehicles accidents	*Deleted*	
	Animals attacks on the pipeline	*Deleted*	
Health Safety and Environment (HSE)	Improper safety regulations	Improper safety regulations	15
	Improper inspection and	Improper inspection and maintenance	16
	The above-the-ground pipeline increases sabotage and thefts	The above-the-ground pipeline increases sabotage and thefts opportunities	17
	Limited warning signs	*Deleted*	
	Inadequate risk management	*Deleted*	
	Natural disasters and weather	Natural disasters and weather conditions	18
Operations Consent (OC)	Corrosion and lack of corrosive	Corrosion and lack of corrosive protection	19
	The weak ability to identify and	The weak ability to identify and monitor the	20
	Shortage of the IT services and	Shortage of the IT services and modern	21
	Design, construction and material defects	Construction defects (e.g. welding defects and damage the pipes during the	22
		Design defects	23
	Operational errors	Operational errors	24
	Hacker attacks on the operating or	*Deleted*	
The added RFs after the survey		The unqualified staff, lack of experience and not well educated about risk	25
		Pumping more than one type of petroleum product and crude oil from different fields	26
		Salts and metals contents in the transported	27
		External oil spots that negatively affect the	28
		Not taking the future urban planning into	29
		Poor quality pipes and material defects	30
		The pipes are older than the design age	31

5. Conclusion

- The developed risk assessment tool in this paper provides a systematic approach of choosing safe routes for OGP projects, specifically for the organisations that just began analysing the RFs in OGPs more effectively, which is the case in OGP projects in Iraq.
- Using the FIS in risk assessment remedies the problems of the traditional approaches to risk analysis and ranking.
- It was found in Table 2 that Route 3 is the riskiest route (F3 =80). Meanwhile, Route 4 is the safest route (F = 51). Therefore, the export gas pipeline should be built on this route.
- The initial list of the RFs has been identified from the literature review. This list has been revised after the survey based on analysing the participants' comments about adding the RFs that affect the safety of OGPs in Iraq and did not mention in the survey.
- The future work of this paper is to estimate the consequences of OGPs failures. Moreover, evaluate the cost and time impact of the RFs; as well as, the cost and time impact of the risk mitigation methods that should be applied to mitigate them.
- The future work also includes analysing the probability, severity and RI of the added RFs. As well as, investigating their impact in the pipelines' routes.

Acknowledgement
The financial support from the Ministry of Higher Education and Scientific Research, Iraq and AL-Muthanna University is highly appreciated.

References

[1] Kraidi L, Shah R, Matipa W, and Borthwick, F. 2019, Analyzing the critical risk factors associated with oil and gas pipeline projects in Iraq. International Journal of Critical Infrastructure Protection, vol. 24, pp. 14-22. https://doi.org/10.1016/j.ijcip.2018.10.010

[2] Kraidi L, Shah R, Matipa W, and Borthwick, F. 2017. Analysing the Critical Risk Factors of Oil and Gas Pipeline Projects in Iraq. presented at the The 3rd BUiD Doctoral Research Conference, Faculty of Engineering & IT, the British University in Dubai. URL:http://content.buid.ac.ae/bdrc/BDRC2017-Full-Proceedings-Faculty-of-Engineering-and-IT.pdf

[3] Lavasani S, Yang Z, Finlay J, and Wang J. 2011. Fuzzy risk assessment of oil and gas offshore wells. Process Safety and Environmental Protection, vol. 89, no. 5, pp. 277-294. https://doi.org/10.1016/j.psep.2011.06.006

[4] Biezma V, Agudo D, Barron G. 2018. A Fuzzy Logic method: Predicting pipeline external corrosion rate. International Journal Press of Pressure Vessels and Piping. Vol. 163(June) pp. 55–62. https://doi.org/10.1016/j.ijpvp.2018.05.001

[5] Kraidi L, Shah R, Matipa W, and Borthwick, F. 2018. An Analysis of the Critical Risk Factors in Oil and Gas Pipeline Projects Using a Comprehensive Risk Management Framework. Presented at the ARCOM 34th Conference, Belfast, UK, 2018. URL: http://www.arcom.ac.uk/-docs/archive/2018-Working-Papers.pdf

[6] Kraidi L, Shah R, Matipa W, and Borthwick, F. 2018. Analyzing the critical risk factors in oil and gas pipelines projects regarding the perceptions of the stakeholders. In Creative Construction Conference 2018 - Proceedings. Ljubljana, Slovenia, Budapest University of Technology and Economics. (pp. 304–311). https://doi.org/10.3311/CCC2018-041

Appendix A

Table 4: Questionnaire design (example)

RFs	What is the probability scale that the RFs threating the OGPs?					What is the severity scale of the flowing RFs?				
	Rare	**Unlikely**	**Possible**	**Likely**	**Almost Certain**	**Negligible**	**Minor**	**Moderate**	**Major**	**Catastrophi**
Terrorism and sabotage										
Thieves										
Public's Low legal and moral										
Staff threats										
Socio-political such as poverty and education										

Appendix B

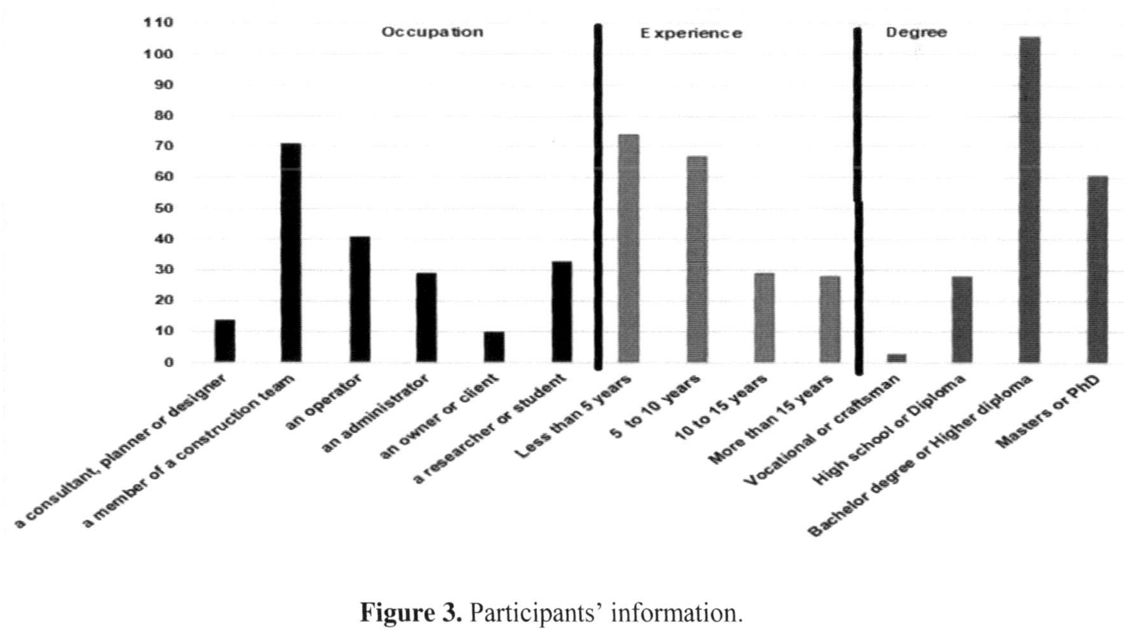

Figure 3. Participants' information.

The response rate was 199 out of 400 (49.75%).

Using LARS –WG model for prediction of temperature in Columbia City, USA

Salah L. Zubaidi[1*], Patryk Kot[2], Khalid Hashim[2,3], Rafid Alkhaddar[2], Mawada Abdellatif[2], Yousif Raad Muhsin[1]

[1]Department of Civil Engineering, College of Engineering, University of Wasit, Iraq.
[2]Department of Civil Engineering, Liverpool John Moores University, UK.
[3]Department of Environmental Engineering/ University of Babylon, Iraq.

* Corresponding author, E-mail address: salahlafta@uowasit.edu.iq

Abstract: Climate change has placed considerable pressure on the residential environment in different areas of the world. These issues have increased the motivation of researchers to analyse and forecast the changes in critical climatic factors, such as temperature, in order to offer valuable reference outcomes for management and planning in the future. This study set out to determine to what extent global warming would affect Columbia City, Missouri, USA. The Long Ashton Research Station Weather Generator (LARS-WG) model is used for downscaling daily maximum temperatures based on the SRA1B scenario. Seven General Circulation Models (GCMs) outputs are employed for three selected periods, 2011–2030, 2046–2065 and 2080–2099. The findings show that (1) statistical analysis confirmed the skill and reliability of the LARS-WG model to downscale maximum temperature time series; (2) the ensemble mean of seven GCMs exhibited an increasing based on yearly and monthly data for all periods compared with baseline period 1980-1999. The findings can contribute to a better understanding of the impacts of climate change on the urban environment and encourage planners and stakeholders to find the best solution for mitigation of these impacts.

Keywords: Climate change; Columbia City; LARS-WG; statistical downscaling.

1. Introduction

Humanity has faced uncommonly from global challenges, and one of the most critical issues is global warming. Global warming, which is often named climate change, is the phenomenon of rising in temperature of atmospheric that leads to modifications in the ecosystem. So, these changes are impacting, directly and indirectly, both people and their surrounding environment for short and long term [1, 2]. UNFCCC [3] stated that climate change has several harmful influences such as 1) shortages in quantity and quality of agricultural production; 2) spread many diseases, e.g. cholera and malaria; 3) depilation the freshwater resources which in turn exerts extra pressure on water and wastewater treatment facilities [4-14]. In addition, several researchers proved that temperature driven water consumption such as [15, 16, 17].

Many studies have applied downscale approaches to simulate future climate factors, such as Abdellatif, et al. [18], who investigated the spilling volume of urban drainage catchment in northwest England under

IOP Conf. Series: Materials Science and Engineering **584** (2019) 012026 doi:10.1088/1757-899X/584/1/012026

different climate change scenarios and climate models using a generalised linear model and an ANN (hybrid GLM-ANN) model. In addition, Räisänen [19] used 12 regional model simulations (RCM) and baseline (1980–2010) monthly mean temperature to investigate the changes in snowfall in northern Europe. Moreover, Masanganise, et al. [20] utilised 10 global climate models (RCM) and monthly temperature and rainfall to describe climate changes by mid-century in Zimbabwe, 2040-2070 period, relative to the 1980-2010 baseline.

This study endeavours to achieve three aims include firstly, to evolve a downscaling model that has the capability to simulate current and future maximum temperature in Columbia City. Next, to employ the developed downscale model together with information about climate change to simulate maximum temperature for three future periods. Finally, to analyse the obtained outcomes to assess the future maximum temperature in the studied area.

2. Area of study and data set

Columbia City locates in northern mid-Missouri state that lay middle of the United States of America (Figure 1). Columbia City has an area of around 164.5 km², and its coordinates are 38.9517° N, 92.3341° W. The city has a growth in population from 62,061 to 108,500 capita between 1980-2010. The groundwater represents the primary source of the freshwater resource. The water company serve about 44,360 different customers inside the city, and 92% of them is residential customers [21].

The weather in the city is hot, muggy and wet in summer and the warmest period between 31st May to the 17th September, with an average daily high temperature above 26 °C. While winter is very cold with an average daily high temperature that reaches no higher than 10 °C and the coldest period is between 28th November to 26th February [22]. Historical daily maximum temperature data (°C) were gathered from Columbia City's council for the period from 1980-2010.

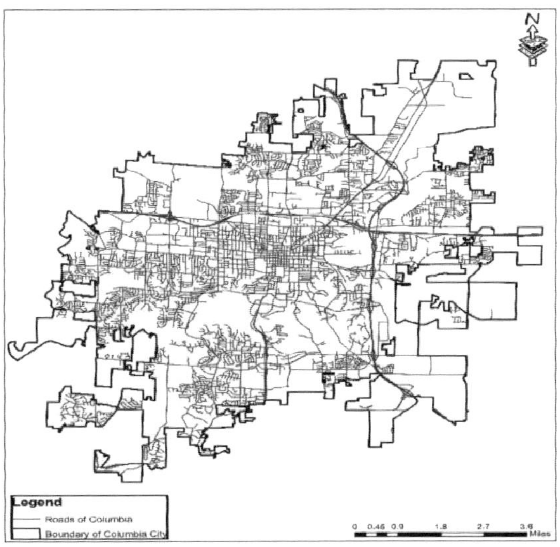

Figure 1. Location map of Columbia

3. Downscaling Technique

Several approaches have been utilised to assess global climate simulations and to downscale different global climate scenarios, resulting from different emissions scenarios, for the evaluation of climate influences on hydrologic systems [18, 19, 23]. Daniels, et al. [24] mentioned that downscaling techniques could be divided into two main types: dynamic and statistical. In addition, downscaling methods are based on several General Circulation Models (GCMs) and the Intergovernmental Panel on Climate Change (IPCC) scenarios.

A GCM is the main tool applied to project the influences of emissions on climate in the future. It provides information at a scale of 100-500 kilometres for one grid size and temporal scales of monthly means and longer. Accordingly, it is considered too coarse for planning and assessment of impact for almost all decision makers. However, the downscaling technique has been used to obtain information on climate change at a scale more relevant to stakeholders. It is utilised spatially, with a resolution of 20 kilometres, or even a specific location, and temporally (e.g., daily temperature sequences from monthly or seasonal temperature amounts) [25]. Roy and Majumder [26] point out that the IPCC offers various climate change scenarios and attempts to forecast the result of that variation on different related natural phenomena. These scenarios are based on pollution and land use dynamics, population and the influence of climate change on these activities. These scenarios – such as A2 that depicts a very heterogeneous world, continuously growing global population and economic development – are oriented regionally. B1 describes a convergent world, where global population will peak mid-century and clean technologies will be introduced. A1B is balanced between using fossil and non-fossil energy across all sources. More details about IPCC scenarios can be found in [27].

Daniels, et al. [24] stated that the statistical downscaling method has advantages such as being efficient, computationally inexpensive and offering several various emissions scenarios and GCM pairings, while the dynamic downscaling method needs high computational resources and expertise and a high volume of data inputs. Accordingly, this study will apply the statistical downscaling method and, more precisely, The Long Ashton Research Station Weather Generator (LARS-WG) model.

3.1. The Long Ashton Research Station Weather Generator (LARS-WG)

LARS-WG is a stochastic weather generator utilised to generate a long series of synthetic data in three periods (2011-2030, 2046-2065 and 2080-2099). In addition, it has been designed to study the impact of climate change. Moreover, it has been tested for various sites across the world, and the results showed that the model is able to simulate climatic factors with reasonable skill [28]. Semenov [28] inspected and stated that the period from 1980-2010 is valid as a baseline period for generating future climatic factors. Additionally, several researchers have adopted this period in studies [19, 20].

For this research, version (5.5) of LARS-WG and 7 GCMs will be used to reduce the uncertainty for three different periods. Additionally, the A1B emissions scenario will be applied in the LARS-WG model. This scenario will be investigated to show its influence on residential complexes. Usage of more than one GCM and the A1B scenario is to eliminate the uncertainty that results from the assumption in each approach. More precisely, it is utilised for long-term period projection. More details about the seven GCMs that were selected from IPCC AR4 can be found in [29].

4. Results and discussion

4.1. LARS-WG Calibration and Validation

The daily maximum temperature data for Columbia City for the period 1980-2010 (31 years) was employed for calibration and validation of the LARS-WG model. Graphic comparison of mean and standard deviation between observed and simulated data was used to assess the ability and reliability of the model to downscale future temperature data. Figure 2 shows the model had a very good performance in fitting mean and standard deviation in terms of maximum temperature.

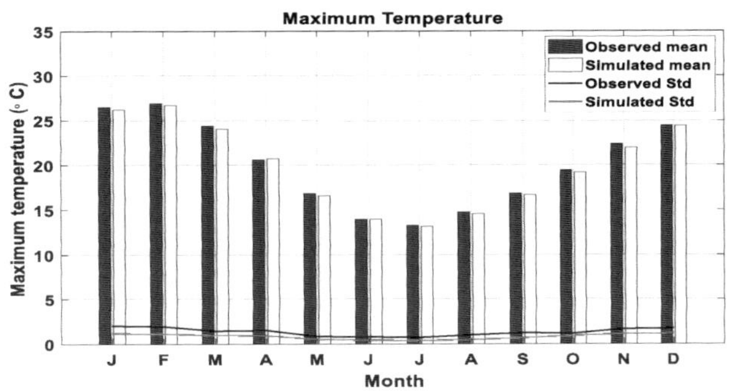

Figure 2. Measured and simulated mean and standard deviation of Tmax at study area

The LARS-WG model reveals an adequate performance in replicating the observed maximum temperature in this study, which obviously emphasises that the model is appropriate for this region. Similar reliable performances in generating maximum temperature have been found for the LARS-WG model in various locations around the world, as mentioned in [29]. Accordingly, confidence was increased to employ the downscaling model in this study.

4.2. Projection of Future Temperature

The developed and calibrated LARS-WG model for Columbia City was then employed to forecast maximum temperature for three periods, 2011-2030, 2046-2065 and 2080-2099, dependent on the SRA1B scenarios generated from seven GCMs. The results of the yearly maximum temperature forecasting data for all GCMs and their ensemble mean for the three periods are plotted in Figure 3. This figure shows that all the GCMs' projected data over the period 2011-2030, in general, have little variation. The pattern of HADCM3 and NCCCSM models are closer to each other, while GFCM21 and MPEH5 models have the same pattern but are not close to each other. INCM3 and IPCM4 models have an irregular pattern in some years. The limitation of ensemble projected between around 18.4-19.6 (i.e. it means the maximum and minimum values over the 20 years). The results for the second period, 2046-2065, present that all GCMs show approximately the same pattern, except the IPCM4 and MPEH5 models, which have considerable variation. In addition, the GCMs exhibit anomalous behaviour in the end quarter of the period. The limitation of ensemble projected approximately 20.3-21.3. The third period, 2080-2099, shows that the variation between the GCMs has increased, compared with the previous two periods. HADCM3 and IPCM4 models have almost the same pattern and offer highest temperature predictions degree compared with the other models, which demonstrate approximately the same trend, except for MPEH5, which shows the lowest one. The limitation of ensemble projected around 21.4-22.4.

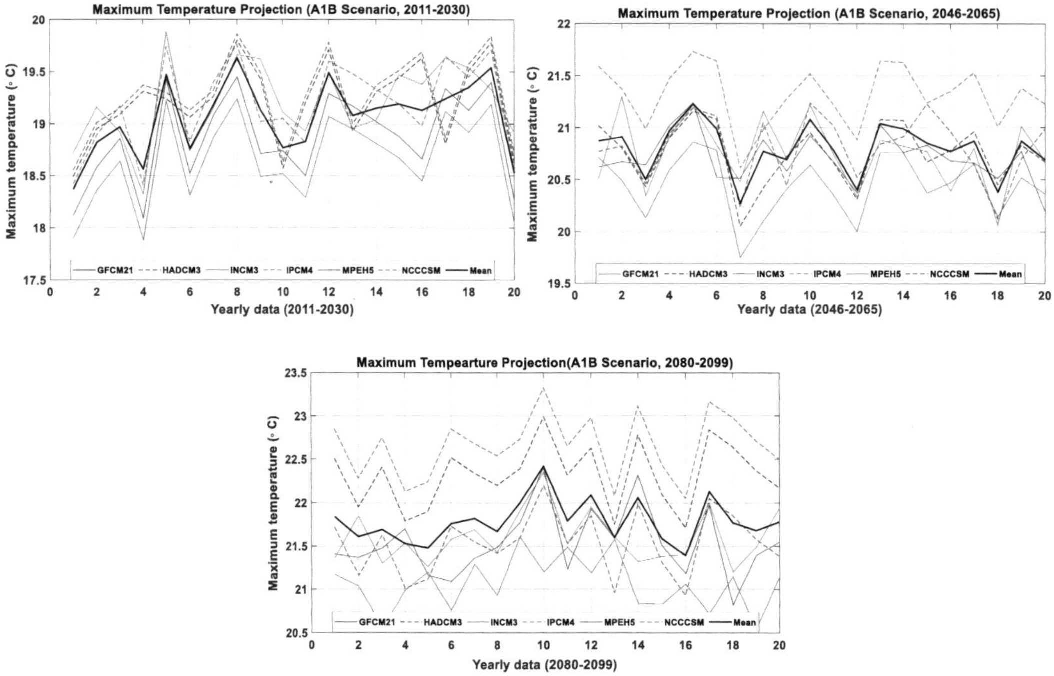

Figure 3. Projected yearly maximum temperature data for all periods

Additional results based on a monthly basis can increase our knowledge about climate change. Therefore, the ensemble means of maximum temperature forecasting from the seven GCMs were considered and the variations between baseline (1980-1999) and ensemble means for all periods were plotted, as shown in Figure 4. It can be seen that temperature will increase based on a monthly basis for all periods except June, July, August and September in the first period. The temperature will increase more as the prediction moves into the future (i.e., 2080-2099 is the hottest period).

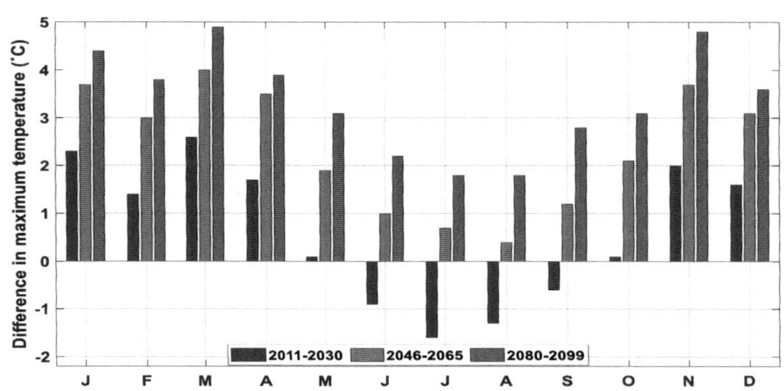

Figure 4. The differences in maximum temperature between the periods (2011-2030, 2046-265 and 2080-2099) and the current

5. Conclusions

In this study, the applicability of the LARS-WG model in downscaling daily maximum temperature was tested in Columbia City. Historical daily maximum temperature data, seven GCMs and SRA1B scenario were employed in the LARS-WG model to downscale maximum temperature in three periods of 2011-2030, 2046-2065 and 2080-2099. Historical data for the period of 1980-2010 were utilised to calibrate the downscaling model and to compare with future scenarios. From the study, it is concluded that:

- The LARS-WG model has the ability to perform excellently in downscaling daily maximum temperature data in the study area.
- The downscaled maximum temperature from a simulation of seven GCMs has various changing oscillating trends in all future periods. This also clarifies that more GCMs should be utilised in climate change research to minimise the models' uncertainty.
- The monthly differences between the baseline value and the ensemble means for all future periods show that maximum temperature was increased in all months except June, July, August and September. In addition, maximum temperature value increased as the prediction moved into the future.

These findings contribute in several ways to our understanding of climate change and provide a basis for further research. Additionally, they enhance the knowledge of policymakers to select the right solutions – those which mitigate the impacts of global warming in residential complexes and support sustainability.

Acknowledgements

The authors thank the city council of Columbia City, USA for the data they provided.

References

[1] J. Borbora and A. K. Das, "Summertime Urban Heat Island study for Guwahati City, India," Sustainable Cities and Society, vol. 11, pp. 61-66, 2014.

[2] D. Jewitt, B. F. Erasmus, P. S. Goodman, T. G. O'Connor, W. W. Hargrove, D. M. Maddalena, and E. T. Witkowski, "Climate-induced change of environmentally defined floristic domains: A conservation based vulnerability framework," Applied Geography, vol. 63, pp. 33-42, 2015.

[3] UNFCCC, "United Nations Framework Convention on Climate Change, Climate Change: Impacts, Vulnerabilities and Adaptation in Developing Countries," Bonn: UNFCCC 2007.

[4] K. S. Hashim, N. H. Al-Saati, A. H. Hussein, and Z. N. Al-Saati, "An Investigation into The Level of Heavy Metals Leaching from Canal-Dreged Sediment: A Case Study Metals Leaching from Dreged Sediment," presented at the 1st International Conference on Materials Engineering & Science, Istanbul Aydın University (IAU), Turkey, 2018.

[5] K. S. Hashim, I. A. Idowu, N. Jasim, R. Al Khaddar, A. Shaw, D. Phipps, P. Kot, M. O. Pedrola, A. W. Alattabi, and M. Abdulredha, "Removal of phosphate from River water using a new baffle plates electrochemical reactor," MethodsX, vol. 5, pp. 1413-1418, 2018.

[6] Ethaib, S., Erabee I. K., and Abdulsahib, " Removal of Methylene Blue Dye from Synthetic Wastewater using Kenaf Core and Activated Carbon" International Journal of Engineering & Technology, vol. 7, no. 4.19 pp. 909-91, 2018.

[7] K. S. Hashim, R. A. Khaddar, N. Jasim, A. Shaw, D. Phipps, P. Kota, M. O. Pedrola, A. W. Alattabi, M. Abdulredha, and R. Alawsh, "Electrocoagulation as a green technology for phosphate removal from River water," Separation and Purification Technology, vol. 210, pp. 135-144, 2018.

[8] S. Ethaib, R. Omar, M. K. S. Mazlina, A. B. D., Radiah and S. Syafiie, "Development of a hybrid PSO–ANN model for estimating glucose and xylose yields for microwave-assisted pretreatment and the enzymatic hydrolysis of lignocellulosic biomass," Neural Computing and Applications, vol.30, pp., 1111-1121, 2016.

[9] K. S. Hashim, A. Shaw, R. Al Khaddar, M. Ortoneda Pedrola, and D. Phipps, "Defluoridation of drinking water using a new flow column-electrocoagulation reactor (FCER) - Experimental, statistical, and economic approach," J Environ Manage, vol. 197, pp. 80-88, Jul 15 2017.

[10] Erabee I. K., and Ethaib, S. M., " Performane of Activated Carbon Adsorption in Removing of Organic Pollutants from River Water" International Journal of Engineering & Technology, vol. 7, no. 4.20 pp. 356-358, 2018.

[11] K. S. Hashim, A. Shaw, R. Al Khaddar, M. O. Pedrola, and D. Phipps, "Iron removal, energy consumption and operating cost of electrocoagulation of drinking water using a new flow column reactor," Journal of Environmental Management, vol. 189, pp. 98-108, 2017.

[12] A. W. Alattabi, C. Harris, R. Alkhaddar, A. Alzeyadi, and K. Hashim, "Treatment of Residential Complexes' Wastewater using Environmentally Friendly Technology," Procedia Engineering, vol. 196, pp. 792-799, 2017.

[13] K. S. Hashim, A. Shaw, R. Al Khaddar, M. O. Pedrola, and D. Phipps, "Energy efficient electrocoagulation using a new flow column reactor to remove nitrate from drinking water - Experimental, statistical, and economic approach," J Environ Manage, vol. 196, pp. 224-233, Jul 01 2017.

[14] D. Al-Jumeily, K. Hashim, R. Alkaddar, M. Al-Tufaily, and J. Lunn, Sustainable and Environmental Friendly Ancient Reed Houses (Inspired by the past to motivate the future), presented at the The 11th

International Conference on the Developments in eSystems Engineering (DeSE), University of Cambridge, England, UK., 2018.

[15] S. L. Zubaidi, J. Dooley, R. M. Alkhaddar, M. Abdellatif, H. Al-Bugharbee, and S. Ortega-Martorell, "A Novel approach for predicting monthly water demand by combining singular spectrum analysis with neural networks," Journal of hydrology, vol. 561, pp. 136-145, 2018.

[16] S. L. Zubaidi, P. Kot, R. M. Alkhaddar, M. Abdellatif, and H. Al-Bugharbee, "Short-Term Water Demand Prediction in Residential Complexes: Case Study in Columbia City, USA," presented at the 11th International Conference on Developments in eSystems Engineering (DeSE), University of Cambridge, UK, 2018.

[17] S. L. Zubaidi, S. K. Gharghan, J. Dooley, R. M. Alkhaddar, and M. Abdellatif, "Short-Term Urban Water Demand Prediction Considering Weather Factors," Water Resources Management, vol. 32, no. 14, pp. 4527-4542, 2018.

[18] M. Abdellatif, W. Atherton, R. Alkhaddar, and Y. Osman, "Quantitative assessment of sewer overflow performance with climate change in northwest England," Hydrological Sciences Journal, vol. 60, no. 4, pp. 636-650, 2015.

[19] J. Räisänen, "Twenty-first century changes in snowfall climate in Northern Europe in ENSEMBLES regional climate models," Climate dynamics, vol. 46, no. 1-2, pp. 339-353, 2016.

[20] J. Masanganise, T. Mapuwei, M. Magodora, and K. Basira, "Multi-model projections of temperature and rainfall under representative concentration pathways in Zimbabwe," 2014.

[21] Jacobs and M. St. Louis, "Long Range Water System Study," Water & Light Department, USA2015.

[22] W. Spark, "Average Weather in Columbia Missouri," 2013, Available: https://weatherspark.com/y/10940/Average-Weather-in-Columbia-Missouri-United-States-Year-Round, Accessed on: 20/02/2019.

[23] E. P. Salathe, P. W. Mote, and M. W. Wiley, "Review of scenario selection and downscaling methods for the assessment of climate change impacts on hydrology in the United States Pacific Northwest," International Journal of Climatology, vol. 27, no. 12, pp. 1611-1621, 2007.

[24] A. E. Daniels, J. F. Morrison, L. A. Joyce, N. L. Crookston, S.-C. Chen, and S. G. McNulty, "Climate Projections FAQ," U.S. Department of Agriculture, Forest Service, Rocky Mountain Research Station.2012.

[25] S. Trzaska and E. Schnarr, "A review of downscaling methods for climate change projections," Center for International Earth Science Information Network (CIESIN).2014.

[26] U. Roy and M. Majumder, Vulnerability of Watersheds to Climate Change Assessed by Neural Network and Analytical Hierarchy Process. Springer, 2015.

[27] M. Parry, O. Canziani, J. Palutikof, P. v. der, and L. C. Hanson, "AR4 Climate Change 2007: Impacts, Adaptation, and Vulnerability," Cambridge University Press, USA.2007.

[28] M. A. Semenov, "Validation of ELPIS 1980−2010 baseline scenarios using the observed European Climate assessment data set," Clim Res, vol. 57, pp. 1-9, 2013.

[29] Y. Osman, M. Abdellatif, N. Al-Ansari, S. Knutsson, and S. Jawad, "Climate Change and Future Precipitation in an Arid Environment of the Middle East: Case Study of Iraq," Journal of Environmental Hydrology, vol. 20, pp. 1-8, 2017.

Effect of void Ratio of Inner Steel Tube on Compression Behavior of Double Skin Tubular Column

Hanaa A. Ali [1] and Wissam D. Salman [2]
1 Graduate Student Department of Civil Engineering, College of Engineering, University of Diyala, Iraq.
2 Assistant Professor, Department of Civil Engineering, College of Engineering, University of Diyala, Iraq.
Corresponding author: dr_wissam80@yahoo.com

Abstract. Hybrid double skin tubular columns (DSTCs) are a composite members that consist of outer FRP tube and inner steel tube and concrete filled in between. These three materials combination in one element have several advantages do not found in common columns. The total 12 circular DSTCs were poured and tested under axial load. Every specimen is average of a pair of samples of the variables, diameter and thickness of steel tube, to investigate effect of inner steel tube void ratio on compression behavior of hybrid DSTC. The void ratio was defined as a ratio between the two inner steel and outer FRP tube diameter. The test results show that the stress-strain behavior of concrete material in the hybrid DSTCs has a better confined by two tubes. While the DSTC with small diameter has similar behavior to concrete filled FRP tube (CFFT). Thickness of inner steel tube is slightly effect on concrete stress inside DSTC.

Keywords: Double skin tubular columns, FRP tube, Steel tube and void ratio.

1. Introduction

Over the past two decades, the fiber reinforced polymer (FRP) composites confining concrete. At recent years excessive research on the hybrid composites fiber-reinforced polymer (FRP) confined concrete because of unique property of excellent corrosion resistance, hybrid FRP concrete tubular columns have great potential for using to be exposed to a harsh environment in outward. Concrete filled FRP tube (CFFT) by **Yu et. al. 2014**[1], **Xie et. al. 2015**[2]. Concrete filled steel tube (CFST), **Shi et. al. 2015**[3]. Hybrid double skin tubular column (DSTC) composite section was proposed by **Teng et. al. 2004[4]** consists of outer FRP tube, inner steel tube and concrete filled in between. This composite section is perfectly used for compression members when steel tube centricity. Also it is used for tension members when steel tube is eccentricity (but in this case shear connecters is needed). **Yu et. al. 2007** studied concept and behavior of DSTC [4], **Yu et. al. 2012** studied compressive behavior of circular and square DSTC [5], **Ozbakkaloglu et. al. 2013** investigated the effect of high and normal strength concrete on compressive behavior of DSTC **[6]**, **Fanggi et. al. 2014** studied effect of inner steel tube cross section on DSTC [7], **Ozbakkaloglu el. at. 2014** investigated seismic behavior of DSTC **[8]**. These studies were illustrate a better ductile behavior of the DSTCs under axial load, combined axial and flexure load and seismic load. FRP tube thickness was significant effect on stress and strain curve and inner steel tube prevent concrete spalling from inward. Concrete material inside the hybrid DSTCs restrained by two tubes leads to increasing its compressive strength. The experimental procedure of this paper includes investigation of the effect of inner steel tube void ratio on compressive behavior of carbon FRP –concrete –steel double skin tubular columns. The study was aimed to investigating the circular DSTCs with cross section of dimension (100 mm*310 mm) to provide properties of column according to ACI CODE [9].

2. Experimental Testing Description

All 12 DSTCs have 100 mm in diameter (D) and 310 mm in height, the outer FRP tubes were made by embedded carbon fiber sheet on polymer resin Table 1 illustrates details of specimens. The first letter of specimens name refers to double skin tubular column (D) and the letter (C) refer to concrete compressive strength of cubic fcu (70mm* 70mm) C60. The letters (ds) refer to diameter of inner steel tube diameter 48mm ds48, diameter 38mm ds38, diameter 32mm ds32 and diameter 25mm ds25. The letters (ts) refer to thickness of inner steel tube thickness 1.5mm ts1.5, thickness 2mm ts2 and thickness 3mm ts3. The roman numeral (I) refer to number of carbon fiber reinforced polymer layers I: one layers.

Table 1. Details of specimens

Specimens Designation	Cubic Compressive Strength of Concrete strength fcu (MPa)	Cylinder Compressive Strength of Concrete strength fć (MPa)	Diameter of Inner Steel Tube (mm)	Thicknes s of Inner Steel Tube (mm)	Number of layers of FRP Tube
DC60ds48ts1.5I	60	50	48	1.5	1
DC60ds48ts2I	60	50	48	2	1
DC60ds48ts3I	60	50	48	3	1
DC60ds38ts2I	60	50	38	1.5	1
DC60ds32ts2I	60	50	32	1.5	1
DC60ds25ts2I	60	50	25	1.5	1

3. Column Materials

3.1. Concrete material

The specimens were poured with high strength self-compact concrete (mortar) with cement: sand 1:1.5, water/cementituos ratio 0.35. The silica fume was added by 15% to increase concrete compressive strength and to attain workable of mix super plasticizer with 2.17% was added.

3.2Fiber Reinforced Polymer (FRP)

Fiber reinforced polymer (FRP) includes the imbedded fibers in polymeric resin and has several advantages compared with steel. These advantages are good corrosion resistance and high strength to its weight ratio. Carbon fiber reinforced polymer tube (CFRP) was used in this work consists of two components, unidirectional carbon fiber sheet and epoxy resin. Table 2 illustrate the properties of carbon fiber sheet from data sheet. The carbon sheet was cut into three pieces, two small with 50 mm width and third pieces with 310mm width, with overlapping zone 100mm, that pieces embedded with epoxy and rounded on PVC tube with diameter 100 mm after 48 hours carbon fiber reinforced polymer (CFRP) tube was removed from PVC tube and ready to use.

Table 2. Properties of Carbon Fiber

Nominal thickness t f (mm-ply)	Tensile strength ff (MPa)(nominal)	Ultimate tensile strain,6 ε f (%)	Elastic modulus Ef (GPa)(nominal)
0.12	4100	1.7	231000

3.3 Steel Tubes

Six types of circular hollow steel tubes (HST) was set up inside of DSTCs specimens in center, three specimens for each type of steel tubes were prepared to compression test, test result include average of these three specimens, ds refer to diameter and ts refer to thickness of steel tube. Specimens prepared to compression test by sit up, four leaner variable displacement transducer (LVDT) were placed on two opposite sides between loading steel plate and support to measure axial displacement, axial shortening was gain of average of four reading, four strain gauges with 5mm was installed in mid height of steel tube two vertically to measured axial strain and two horizontal to measured lateral strain, capacity of testing machine 2000 kN, with load control rate 1kN/sec. testing was occur in civil engineering lap at university of diyala. Data obtained from test include that was axial load-strain curve, axial stress–strain curve, peak load, yield load, yield stress and max strain were shown in Table 3.

Table 3. Properties of Steel Tube

Specimens Designation	Peak load (kN)	Yield load (kN)	yieling stress (MPa)	Axial strain (mm/mm)	lateral strain (mm/mm)	Modes Failure
HSTds25ts2	55	40	315	-1.500	2.380	G.B.[1] and E.F.[2]
HSTds32ts2	67	55	324	-1.548	1.823	G.B.[1] and E.F.[2]
HSTds38ts2	70	65	325	-1.624	2.224	G.B.[1] and E.F.[2]
HSTds48ts2	97	80	327	-1.557	2.308	G.B.[1] and E.F.[2]
HSTds48ts1.5	75	60	327	-1.550	2.034	G.B.[1], E.F.[2] and R.[3]
HSTds48ts3	175	130	350	-1.666	1.167	G.B.[1] & E.F.[2]

[1] G.B. global buckling, [2] E.F. elephant footing, [3] R. rippling

4. Test of Specimens

Concrete was poured between two tubes with no compaction in order to not made any movement of tubes. Every specimen was poured in a pair of columns, after 48 hours columns was remove from mold, double skin tubular columns were covered them with nylon, at day of 30 columns were subjected to axial compressive load until failure. The specimens were prepared for testing by setting specimen at the positions between loading and supporting in the beginning of test loading plate touch steel plate (steel disc with diameter 95 mm willed on rectangular steel plate with dimension 140*200 mm and thickness 3mm), Figure 1 illustrate test machine, data logger read axial compression load number (1), and four L.V.D.T. were fixed on two opposite sides of column its end touch the steel plate and its wired to the logger number 2 to read axial shortening, strain gauges were reach to the data logger number 3 to read strains of two tubes.

Figure 1. Testing machine.

4.1 Axial Load Capacity of Specimens

Table 4 shown the axial load capacity of double skin tubular columns (DSTC), the values of $\frac{Pu}{Pc+Ps}$ ratio define as percentage between ultimate load of specimens (Pu) to ultimate load capacity of concrete (Pc) and ultimate load of steel test (Ps), Pc was get from multiplying strength of unconfined concrete (fć of cylinder) by cross section of concrete in side DSTC. Ps was ultimate load of steel get from compression test of steel tubes. The $\frac{Pu}{Pc+Ps}$ ratio shown the axial load was decrease by (4% and 8%) in DSTCs with ts (2 and 3) mm respectively, compared with DSTC of ts (1.5) mm, that's refer to inner steel tube thickness was not clear effect on axial load because small cross section of DSTC. While the axial load was increase by (5%, 7% and 16%) in DSTCs with ds (38, 32 and 25) mm respectively, compared with DSTC of ds (48) mm, because of better restrained of concrete by two tubes, steel tube inside and FRP outside.

4.2 Load- Shortening Behavior of Specimens

The load-shortening of DSTCs with ts (1.5, 2 and 3) mm was shown in Figure 2, the behavior of curve at first part was bilinear and in DSTCs, while drop in load was occur after transition point in specimens with ts (2 and 3) mm, that's happen because redistribution stress of concrete after yielding of inner steel tube had been happen. Table 4 shown the test results, it can be seen that the service shortening at (0.7Pu) [10,11,12,13] decreased by (27% and 55%), and ultimate shortening slightly increase by (1.4% and 5.3%), in DSCTs with ts (2 and 3) mm, respectively, compared with those of ts (1.5) mm. Figure 3 illustrate load -shortening relationship of DSTCs with ds (48, 38, 32 and 25) mm, when decrease inner steel tube diameter ultimate load was increase, behavior of the DSTC approximately became similar to behavior of concrete filled fiber tube column, the service shortening decreasing by (12%, 31% and 46%) and the ultimate shortening was increase by (1.6%, 2.5% and 17%) in DSTC with ds (38, 32 and 25) mm, respectively compared with reference DSTCs of ds (48) mm.

4.3 Axial Stress-Axial Strain Behavior

Axial stress of unconfined concrete (fć) was obtained from cylinder of dimension of 100 mm × 200 mm, strain was measured by strain gauges were installed axially on mid height of cylinder, the axial stress of concrete inside DSTC (fc) evaluated by ultimate load of DSTC subtraction from ultimate load of steel tube test to same strain, axial strain of columns was get from average of two strain gauges at mid of height of specimen, enhancement ratio (ER), (fc/fć) was percentage between ultimate stress of concrete confined by FRP tube to ultimate stress of unconfinement concrete.

Table 4. Summery of Test Result

Specimens Designation	Pu (kN)	Δs (mm)	Δu (mm)	Pc (kN)	Ps (kN)	$\frac{Pu}{Pc+Ps}$	fc (MPa)	CR	ER	εa (mm/mm)	K1	εh (mm/mm)
DC60ds48ts1.5I	518	2.8	5.90	303	75	1.37	74.0	0.188	1.48	-3.5	2.14	3.20
DC60ds48ts2I	527	2.2	6.20	303	97	1.32	69.0	0.188	1.38	-3.70	2.26	3.35
DC60ds48ts3I	603	1.8	6.77	303	175	1.26	70.5	0.188	1.41	-3.82	2.34	3.50
DC60ds38ts2I	561	2.3	6.30	335	70.0	1.39	75.05	0.188	1.50	-3.75	2.30	2.65
DC60ds32ts2I	590	1.8	6.36	352	67.0	1.41	76.50	0.188	1.53	-4.35	2.67	3.30
DC60ds25ts2I	646	1.4	6.70	368	55.6	1.53	80.50	0.188	1.61	-4.75	2.91	2.80

The nominal confinement ratio (CR) of designing outer FRP tube was percentage between confining pressure to concrete compressive strength, calculated from Eq. (1) with different compressive strengths of concrete assemblage as uniform distribution of confinement DSTCs. (T. Ozbakkaloglu 2013)

$$\frac{f_1}{f_{\acute{c}}} = \frac{2t_f E_f \varepsilon_f}{f_{\acute{c}} D} \qquad (1)$$

Where: fl:confining pressure, fć: concrete compressive strength, Ef: modulus of elasticity of FRP, tf : total nominal thickness of FRP, εf : ultimate tensile strain of the fibers, Df: internal diameter of the FRP tube.

4.4 Effect of Inner Steel Tube Thickness

Figure 4 to 6 illustrate the stress–strain relationship of DSTCs with ts (1.5, 2 and 3) mm, from Table 4, it can be seen that the stress of concrete inside DSTC was decrease and enhancement strength ratio, while axial strain increased with increase thickness of inner steel tube, with ts (1.5, 2 and 3) mm respectively, that's because small cross section of specimens and effect of inner steel tube thickness did not appear clearly.

4.5 Effect of Inner Steel Tube Diameter

Ultimate stress of concrete in DSTC effected by change diameter of inner steel tube, void area ratio was increase by increase diameter of inner steel tube, stress-strain curve of DSTC with ds (38, 32, 25) mm as illustrated in Figure 7 to 9, ultimate stress of concrete were increase because of core of concrete, by confined concrete by two tubes of DSTC the ER was increase in DSTCs with ds (48, 38, 32 and 25) mm, axial strain was increase in DSTCs with decreased inner steel tube diameter, because reduce inner void and increase amount of concrete.

IOP Conf. Series: Materials Science and Engineering **584** (2019) 012027 doi:10.1088/1757-899X/584/1/012027

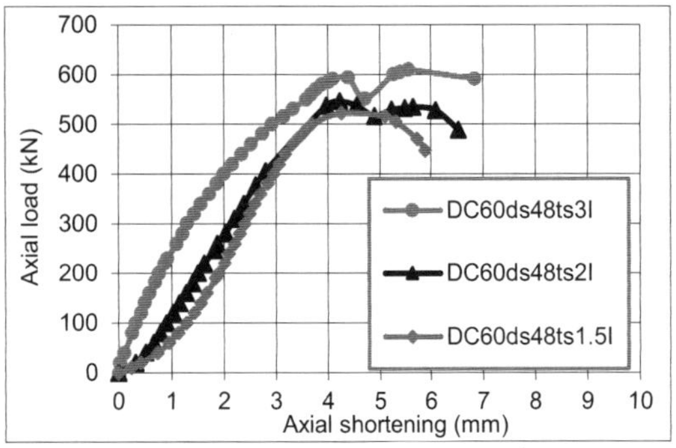

Figure 2. Load -Shortening relationship of DSTCs with ts (2 and 3) mm.

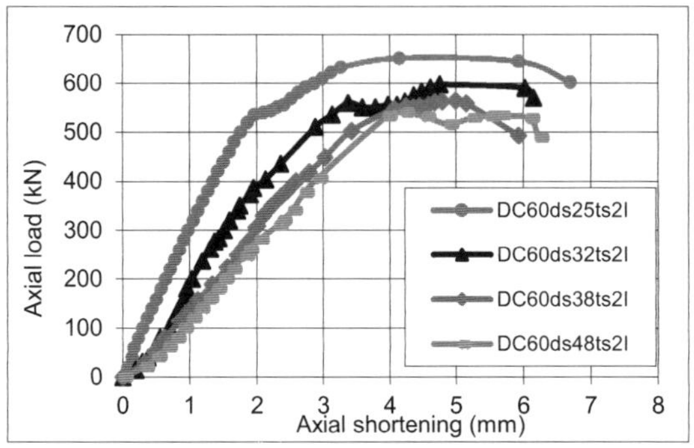

Figure 3. Load –shortening relationship of DSTCs with ds (48, 38, 32 and 25) mm.

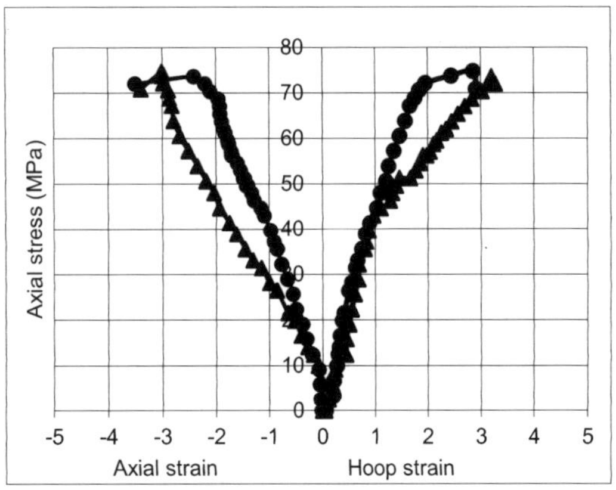

Figure 4. Stress-strain relationship of specimen DC60ds48ts1.5I.

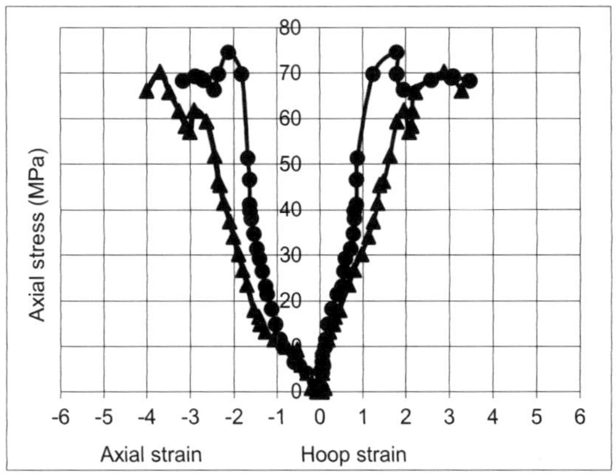

Figure 5. Stress-strain relationship of specimens DC60ds48ts2I.

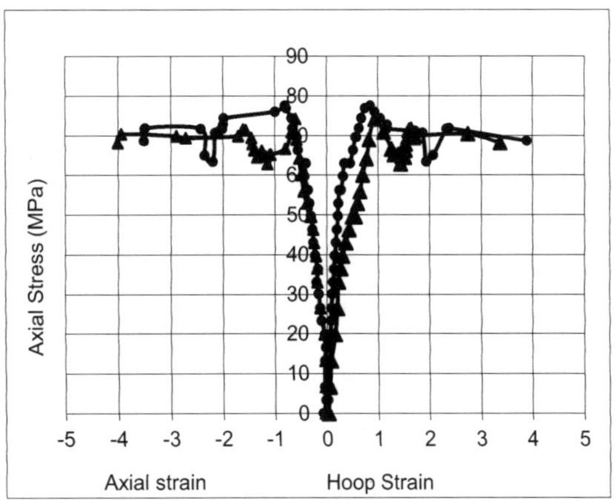

Figure 6. Stress-strain relationship of specimens DC60ds48ts3I.

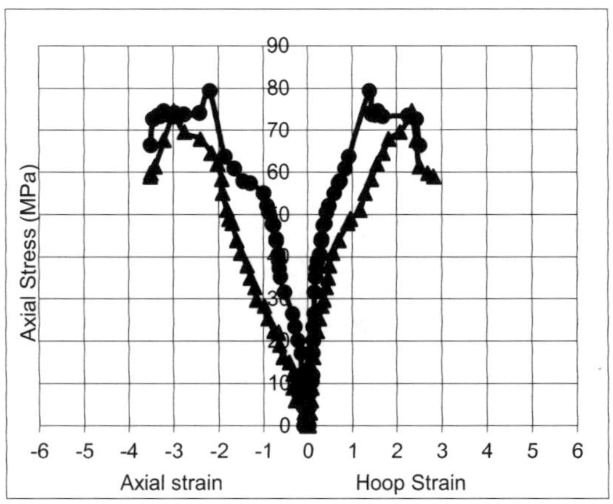

Figure 7. Stress-strain relationship of specimens DC60ds38ts2I.

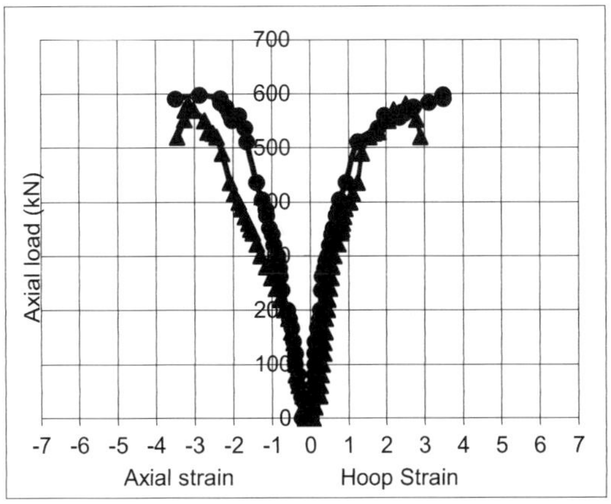

Figure 8. Stress-strain relationship of specimens in DC60ds32ts2I.

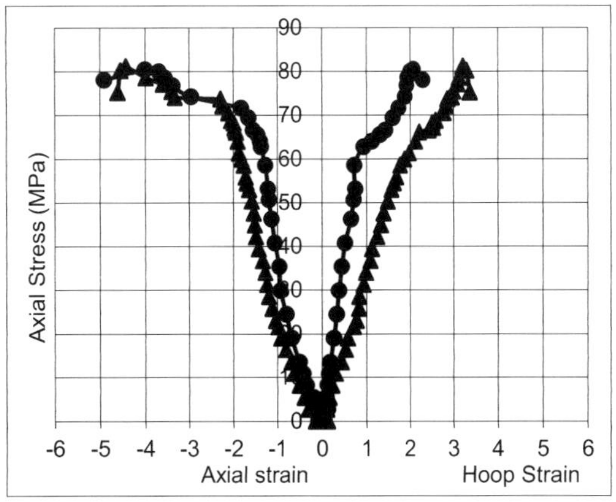

Figure 9. Stress-strain relationship of specimens in DC60ds25ts2I.

4.6 Hoop Strain

Hoop strain of FRP tube of DSTCs (εh) was get from average of three strain gauges distribution along overlap region. The hoop strain of specimens was illustrated in Table 4, it can be seen that the εh slightly increased by (5% and 9%) in DSTCs with ts (2 and 3) mm, respectively compared with DSTCs of ts (1.5) mm and the εh was decreased by (20%, 1.4% and 16%) in DSTCs with ds (38, 32 and 25) mm compared with DSTC of ds (48) mm.

4.7 Modes of Failure

Figure 10 shown modes failure of DSTCs, all specimen in this study were failed by hoop rupture at top height of outer FRP tube because under high ultimate load movement of concrete was occur, while specimens DC60ds48ts1.5I and DC60ds48ts3I failed by hoop rupture at the mid height of specimens, the position of failure of FRP was corresponded to the location of steel tube deformation.

Figure 10. Failure modes of specimens.

5. Conclusion

This paper has presented the results of an experimental study on the behavior of FRP-concrete-steel double-skin tubular columns under axial load. The experimental producer involved 12 high strength concrete specimens with DSTCs designed and manufactured. Those experimental testing are implemented in this paper to investigate the influence of the void ratio on the concrete stress-strain relationship.
The following conclusions can be drawn:

1- High strength self-compact concrete was effective confined by two tubes, steel tube inward and FRP outward lead to high ductile behavior and strength of concrete inside DSTCs.
2- When increased thickness of inner steel tube from (1.5 to 2 and 3) mm, the nominal confinement ratio didn't effected by properties of inner steel tube, the ultimate load increased by (2% and 16%), the service shortening decreased by (27% and 55%), the ultimate shortening slightly increase by (1.4% and 5.3%), the stress of concrete inside DSTCs and strength enhancement ratio decreases by (7% and 5%), the enhancement axial strain increased by (6% and 9%) and hoop strain slightly increased by (5% and 9%) in DSTCs.
3- When increase diameter of inner steel tube from (25 to 32, 38 and 48) mm, the ultimate load decreased by (9% to 18%), the service shortening increasing by (39% to 88%) and the ultimate shortening was decrease by (5% to 7%), the stress of concrete inside DSTCs decreased by (5% to 14%), the strength enhancement ratio decreased by (2% to14%), enhancement of axial strain decreased by (8% to 22%) and hoop strain decreased by (5% to 20%).

References

[1] Yu T, Fangi XL, Teng JG. FRP-confined self-compacting concrete under axial compression. Journal of materials in civil engineering. 2013 Nov 29;26(11):04014082.
[2] Xie T, Ozbakkaloglu T. Behavior of steel fiber-reinforced high-strength concrete-filled FRP tube columns under axial compression. Engineering Structures. 2015 May 1;90:158-71.

[3] Shi XS, Wang QY, Zhao XL, Collins FG. Structural behaviour of geopolymeric recycled concrete filled steel tubular columns under axial loading. Construction and Building Materials. 2015 Apr 15;81:187-97.
[4] Teng JG, Yu T, Wong YL, Dong SL. Hybrid FRP–concrete–steel tubular columns: concept and behavior. Construction and building materials. 2007 Apr 1;21(4):846-54.
[5] Yu T, Teng JG. Behavior of hybrid FRP-concrete-steel double-skin tubular columns with a square outer tube and a circular inner tube subjected to axial compression. Journal of composites for construction. 2012 Sep 10;17(2):271-9.
[6] Ozbakkaloglu T, Fanggi BL. Axial compressive behavior of FRP-concrete-steel double-skin tubular columns made of normal-and high-strength concrete. Journal of Composites for Construction. 2013 Jun 7;18(1):04013027.
[7] Fanggi L, Alfonsius B, Ozbakkloglu T. Effect of Inner Steel Tube Cross-Sectional Shape on Compressive Behavior of Square FRP-Concrete-Steel Double-Skin Tubular Columns. In Applied Mechanics and Materials 2015 (Vol. 752, pp. 578-583). Trans Tech Publications.
[8] Ozbakkaloglu T, Idris Y. Seismic behavior of FRP-high-strength concrete–steel double-skin tubular columns. Journal of Structural Engineering. 2014 Mar 5;140(6):04014019.
[9] ACI Committee. Building code requirements for structural concrete (ACI 318-05) and commentary (ACI 318R-05). American Concrete Institute.
[10] Ibrahim AM, Oukaili NK, Salman WD. Flexural behavior and sustainable analysis of polymer bubbuled reinforced concrete slabs. *Fourth Asia Pacific Conference on FRP in Structures (APFIS 2013)* 11-13 December, Melbourne, Australia © 2013 International Institute for FRP in Construction.
[11] Salman WD, Mansor AA, Mahmood M. Behavior of Reinforced Concrete One-Way Slabs Strengthened by CFRP sheets in Flexural Zone. International Journal of Civil Engineering and Technology (IJCIET) Volume. 2018;9:1872-81.
[12] Salman WD. STRENGTH AND BEHAVIOR OF COMPOSITE STEEL TUBE CONCRETE BEAM. DIYALA JOURNAL OF ENGINEERING SCIENCES. 2015;8 (2):163-81.
[13] Salman WD. Strengthening of Reinforced Concrete One-Way Slabs with Opening using CFRP Strip in Flexural. International Journal of Science and Research (IJSR). 2015 May;4 (5):980-91.

Acknowledgement

The authors would like to gratefully acknowledge to the staff in structural laboratory and civil Department of engineering college, Diyala University.

IOP Conf. Series: Materials Science and Engineering **584** (2019) 012028 doi:10.1088/1757-899X/584/1/012028

Effect of Wasit Thermal Power Plant on Water Quality of Tigris River Downstream to Al Zubaidiayh City

Hiba Alaa[1] and Ali Jwied Jaeel[2]

[1] Civil Engineering Department, M.Sc. Student, University of Wasit, Iraq. Email: hibaalaa340@gmail.com
[2] Civil Engineering Department, PhD, University of Wasit, Iraq. Email: alijwaid@uowasit.edu.iq

Abstract. This study was directed at Wasit Thermal Power Plant (WTPP) to determine the water quality status. Nine sampling stations established along the reach of Tigris River, the first station was 500m upstream (US) to WTPP and it was a reference station, the second station at WTPP and seven stations downstream (DS) to WTPP for collection of water samples. Water samples were prepared for their physico-chemical characteristics at all sampling stations were compared with the reference station. Sampling of all stations (downstream to WTPP) showed higher water pH, total dissolved solids (TDS), electrical conductivity, and chemical oxygen demand (COD) than that of sampling station upstream to WTPP and lower water dissolved oxygen (DO). The results showed increasing of pH, TDS, EC, and COD by 68%, 166%, 60% and 478% respectively, and decreasing the value of DO by 66% indicated that discharge from WTPP increased pollution load in river Tigris.

1. Introduction

Water is a basic and primary need for all vital life processes, and it has now been established that life first arose in the aquatic environment. Since prehistoric times, man has been closely linked to water, and the evidence of the earlier civilization that developed all historical human settlements on inland water resources has proved conclusively, according to the World Water Commission in the twenty-first century, more than half of the world's major rivers are so depleted and contaminated that they endanger human health and toxins surrounding ecosystems. When exhaust water finds its way into any water body that pollutes water, water pollution is primarily related to household and industrial waste. Both types of wastewater pose threats to water quality that can be classified as health risks and health hazards. Many freshwater resources are polluted through human activities. Some 25,000 people (in developing countries) are said to die every day because of their daily use of sewage. Many more millions suffer from recurrent and devastating water-borne diseases, **[1]**. About half of people living in developing countries do not have access to safe drinking water and 73% do not have sanitation facilities. Some of their waste eventually pollutes drinking water supply, resulting in high levels of suffering. Any human use of water, whether for drinking, irrigation, industrial processes or recreation, has some quality requirements to make it acceptable. This quality standard can be described in terms of the physical, chemical and biological properties of such water **[2]**. In many places, surface and underground water is exposed to industrial, agricultural and municipal waste. Water is contaminated with organic and inorganic chemicals as it passes through urban settlements, agricultural land and industries. It has been a common practice around the world for waste disposal in streams / rivers. Organic matter disposed of in the river dissolved oxygen is used in water through its aerobic decomposition. DO low levels may interfere with water uses for a variety of purposes. Water pollution is the contamination of water by natural and foreign matter such as microorganisms, chemicals, industrial wastes or other waste or sewage in quantities that are likely to cause damage to living organism. The US Environmental Protection Agency (EPA) identifies pollution due to the presence of undesirable substances that cause its chemical composition, quantity, inhibiting or obstructing natural processes, and causing undesirable environmental and health effects. Water pollution not only effects on water quality but also it threatens human health, economic development and social prosperity so that it considered as environmental problem that requires an effective and quick solution [**3**]. Furthermore, assessment of water quality by using accurate techniques such as Water Quality Index (WQI) to estimate the status of water and maps by GIS to represent the spatial distribution of parameters have become an important experimental and practical. Water quality studies have focused on situations where serious pollution problems arise, particularly in densely populated urban areas. For example, the city of Baghdad is populated and produces a large amount of wastewater from different sources disposed of directly in the Tigris River or after treatment. In the past few years, the increase in wastewater discharged directly into the river by pumping stations has caused the storm sewage network to increase pollution levels in the river water [**4**]. Most of the potential negative environmental impacts of recycled water use in the environment come from the origin of recycled water as liquid water. These impacts include other water resources and potential pollution of surface and groundwater sources. Public health risks and other environmental impacts that may directly or indirectly affect the public. Fortunately, no significant adverse effects have ever occurred. It is important that all public water systems serve the water with the best possible quality for their customers [**5**]. Tigris is an important river of the central part of Iraq supplies water to a large command area of Wasit province. Ali 1978 [**6**] concluded that Tigris River in Baghdad area is highly polluted comparing with Euphrates and Abu- Graib stream. Al-Masri and Ali 1985 [**7**] confirmed that discharging treated and untreated wastewater into river limited the use of water for different uses as it flows downstream. For evaluating water quality parameters Al-Masri and Ali 1985 [**7**] were studied some water parameters such as: sulfate, Alkalinity, chloride, calcium, hardness, and magnesium in Tigris River through Baghdad city. Ghadban 1993 [**8**], concluded that after the war on Iraq in 1991, most water intakes of treatment plants, were heavily polluted by sewage discharges causing ecological damage and public health hazards. Abd-Ali 1993 [**1**], also concluded that war on Iraq in 1991, affected the bacteriological quality of the Tigris river in Baghdad especially down river at Dora site, as a result of discharging the material sewage to the river. In Wasit Governorate (central Iraq), the Tigris River is facing the impact of conservative pollution due to the continuous discharge of domestic wastewater during the current construction phase (2010-2016) for Wasit Thermal Power Plant (WTPP). The present study aims to study the impact of contaminated industrial wastewater from this plant directly in the river on water quality during the operation period from September 2018 to March 2019.

2. Experimental Works

2.1. Wasit Thermal Power Plant (WTPP)

Wasit Thermal Power Station (WTPP) is located in Wasit Governorate. The station is 120 km southeast of Baghdad as shown in figure 1. The source of water upon which the project is the Tigris River, which is located on the right bank of the river about 100 meters from the project power station is one of the projects that have significant environmental impacts on the region began to work the first unit of this station in 2012 and a production capacity of 3301 MW and the station AL Zubaydiyah one of the largest electrical stations in Iraq, where the site was selected on the right side of the Tigris River area (800) and in 2015 the total units 6 units card output of 2550 MW The station works with natural gas, crude oil and heavy fuel [**9**].

IOP Conf. Series: Materials Science and Engineering **584** (2019) 012028 doi:10.1088/1757-899X/584/1/012028

Figure 1. Location of WTPP in Wasit Province (Middle of Iraq)

2.2. Wastewater sampling

Prior to sampling, first, the sampling stations were established on the water resource map in Wasit. Sampling stations were selected in the study area at a distance slightly equal to each other. Second, the GPS receiver (Garmin 62S) was used to identify each station and take the station coordinates in the study area to collect water samples as shown in table 1. Water samples were collected in the morning between 9 am and 10 am. The samples were collected using the Grab method according to the [**10**]; samples were collected in 2 liters' plastic bottles and stored in an ice box of 4°C and transferred to the laboratory within twenty-four hours for analysis. Water samples were collected by lowering the plastic bottles before cleaning them at the bottom of the water body, and a depth of 30 cm, allowed for excessive flow before pulling out. Ten sampling and sampling points were used 500 meters from each other. The first station was up stream of 500 meters; the second was at the drainage point (WTPP point source). The third location was to the eighth sites at 500m, 1000m, 1500m, 2000m, 2500m, 3000m downstream. The study was conducted within six months.

Table 1. Samples location.

Distance (Km)	Station (#)	Longitude	Latitude
-500	1	45.08803	32.78939
0 (WTPP)	2	45.0876	32.77943
500	3	45.09476	32.77386
1000	4	45.10042	32.77261
1500	5	45.10546	32.77105
2000	6	45.10942	32.7695
2500	7	45.11427	32.76878
3000	8	45.11899	32.76915
3500	9	45.1291	32.77269

2.3. Experimental Works

For each collected sample, the following parameters as shown in (table 2) were investigated.

Table 2. Physical-chemical parameters analyzed to assess water quality of Tigris River

Parameter	Symbol	Unit	Method
Acidity function	pH	-	pH-meter
Dissolved Oxygen	DO	mg/l	Winkler's Method
Total Dissolved Solid	TDS	mg/l	Filtration papers
Electrical Conductivity	EC	µs/cm	EC-meter
Chemical Oxygen Demand	COD	mg/l	SpectroDirect

2.3.1. Acidity Function (pH)
pH is a crucial issue that affects the survival of aquatic organisms, solubility and toxicity of the many metals within the waters. The acidity of waters will increase the solubility of the many metals that cause adverse effects on aquatic organisms. pH was measured with the help of a pH meter. Decreasing pH scale price throughout heat months is also due to elevation of temperature and water temperature due to increase the biological activity of microorganisms and rise the amount of CO_2 in the water, whereas augmented level of pH scale throughout cold months is also due to increase the density of plant and augmented level of dissolved oxygen within the water and consumption of CO_2 that cause elevating the pH scale.

2.3.2. Dissolved Oxygen (DO)
Dissolved oxygen level of the samples was measured with the D.O meter. The electrode was dipped into each sample, after it had been rinsed, then readings were taken and the description was given in [**11**].

2.3.3. Total Dissolved Solids (TDS)
The amount of solids that pass through the filter paper when filtering a certain volume of water is the total dissolved solids, so TDS can be measured as a concentration (mg/l). Total soluble solids consist of inorganic salts (calcium, magnesium, sodium, chloride, and potassium) and small amounts of organic substances dissolved in water. Total soluble solids in drinking water come from natural sources, sewage and runoff to rural and urban areas. The high dissolved solids are considered to be unpalatable to the consumer so 500 mg / l can be acceptable. But more than its, it becomes less palatability, especially for drinking purposes and less dependent on some industries. Concentration of dissolved solids greater than 1200 mg / L causes taste and salinity. TDS test by filtration followed by oven drying 200 ml (V) of sample was filtered by filter paper to separate the suspended solid, then the beaker was weighted (A) and filled by the filtered sample. The sample was dried in the oven at 105°C for 5 hours, the beaker with the dissolved solid was weighted again (B), then TDS was calculated by:
$TDS=(B-A)/V$
A=weight of beaker filled by filtered sample, (mg).
B= weight of beaker with the dissolved solid, (mg).
V= volume of sample (L)

2.3.4. Electrical Conductivity (EC)
The solution's capacity to carry an electrical charge is called electrical conductivity. This capacity influenced by the presence of ions, their total concentration, mobility, valence, and liquid's temperature. Conductivity measurement is very important for identifying certain water properties such as alkalinity, total dissolved solids, hardness and salinity. It is important to note that there is a reciprocal relationship between conductivity and temperature, where the conductivity increases with increasing temperature. The conductivity is the opposite of the electrical resistance. Most inorganic acids, inorganic bases and

IOP Conf. Series: Materials Science and Engineering **584** (2019) 012028 doi:10.1088/1757-899X/584/1/012028

inorganic salts have a very good conductivity in their solutions. When the water is ongoing to be purer, its conductivity will be decreased.

2.3.5. Chemical Oxygen Demand (COD)

In environmental chemistry, the chemical oxygen demand (COD) is an indicative measure of the amount of oxygen that can be consumed by reactions in a measured solution. It is commonly expressed in mass of oxygen consumed over volume of solution which in SI units is milligrams per liter (mg/l). A COD test can be used to easily quantify the amount of organics in water. The most common application of COD is in quantifying the amount of oxidizable pollutants found in wastewater. COD is useful in terms of water quality by providing a metric to determine the effect an effluent will have on the receiving body. The test was done using SpectroDirect which is a modern single-beam spectrophotometer with an excellent performance that is specifically designed for water and wastewater testing. The instrument is equipped with a wide range of pre-programmed methods based on the proven range of Lovibond® tube tests, liquid reagents.

3. Results and Discussion

3.1. Impact of WTPP on pH value

As shown in figure 2 the pH value of water was within the standard and acceptable value of river water at station #1 (500m before WTPP), then after that at the mixing point (station #2 at WTPP) which is considered the point source of pollution, the value of pH jumped to 10.6 yielding increasing about 68% from original normal value (7.2) before reaching WTPP. The obtained result indicated the notable impact of this power plant on the water quality of the river concerning this important parameter.

3.2. Impact of WTPP on DO concentrations

As shown in figure 3 the DO concentration (7.5 mg/l) of water was within the standard and acceptable value of river water at station #1 (500m before WTPP), then after that at the mixing point source of pollution, the value of DO lowered to its minimum value 2.55 mg/l yielding decreasing about 66% from original normal value (7.5 mg/l) before reaching WTPP. The obtained result indicated the notable impact of this power plant on the water quality of the river concerning dissolved oxygen parameter.

3.3. Impact of WTPP on TDS concentrations

As shown in figure 4 the TDS concentration (750 mg/l) of water was at station #1 (500m before WTPP), then after that at the mixing point source of pollution, the value of TDS increased to a maximum value of 2000 mg/l yielding decreasing about 166% from original normal value (750 mg/l) before reaching WTPP. The obtained result indicated the notable impact of this power plant on the water quality of the river concerning total dissolved solids.

3.4. Impact of WTPP on EC values

As shown in figure 5 the EC values (1500 μs/cm) of water was at station #1 (500m before WTPP), then after that at the mixing point source of pollution, the value of EC increased to 2410 μs/cm yielding increasing about 60% from original normal value (1500 μs/cm) before reaching WTPP. The obtained result indicated the notable impact of this power plant on the water quality of the river concerning electrical conductivity.

3.5. Impact of WTPP on COD concentrations

As shown in figure 6 the COD concentration (55 mg/l) of water was within the standard and acceptable value of river water at station #1 (500m before WTPP), then after that at the mixing point source of pollution, the value of COD increased to its maximum value 318 mg/l yielding increasing about 4.78 times its original normal value (55 mg/l) before reaching WTPP. The obtained result indicated the notable impact of this power plant on the water quality of the river concerning dissolved oxygen parameter.

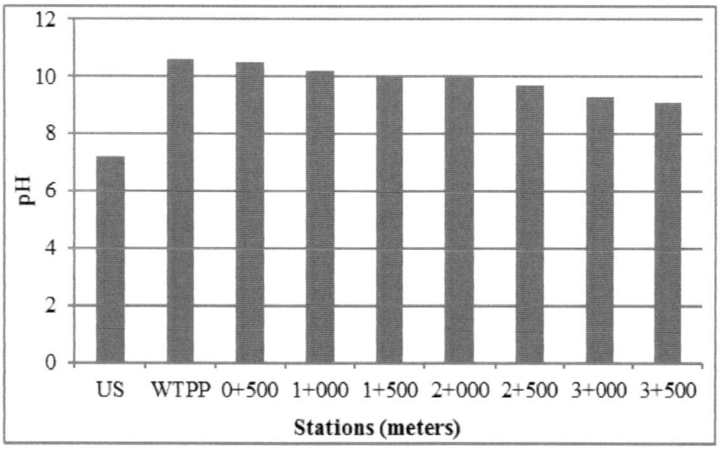

Figure 2. Impact of WTPP on pH values

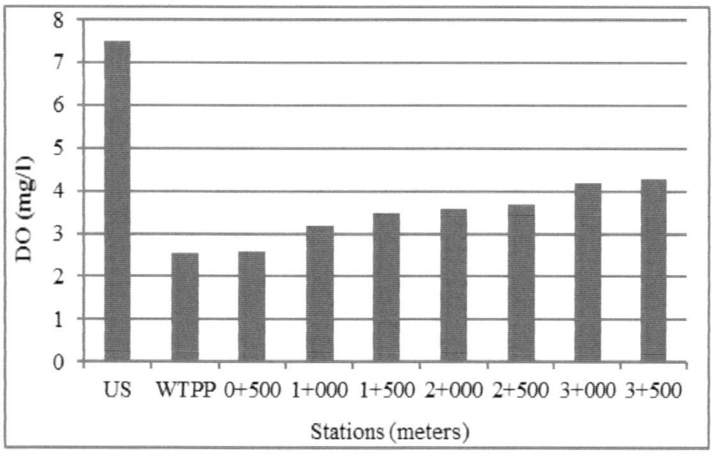

Figure 3. Impact of WTPP on DO concentrations

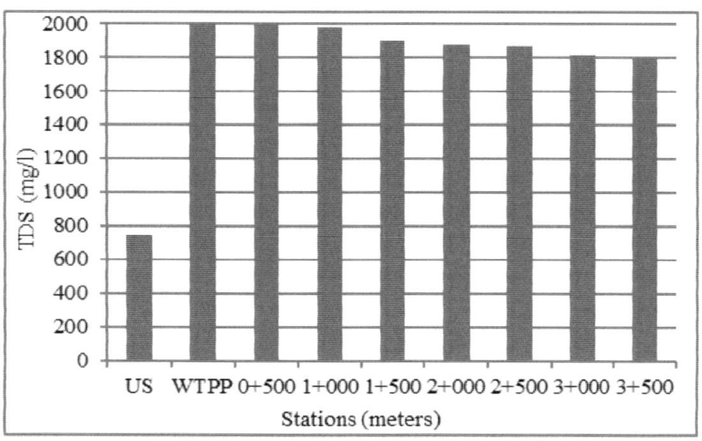

Figure 4. Impact of WTPP on TDS values

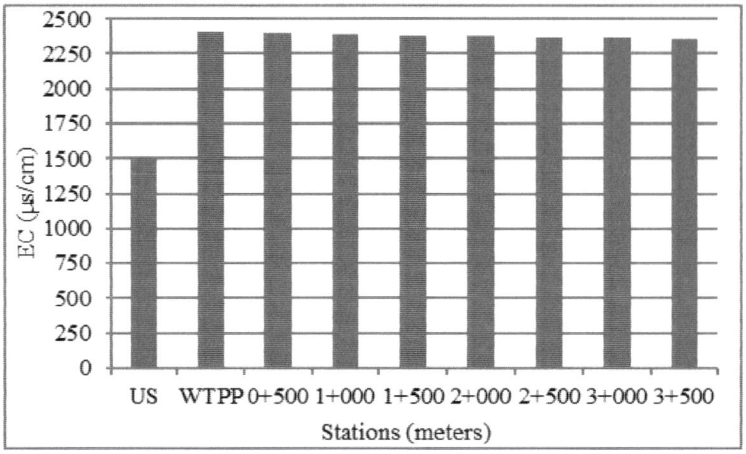

Figure 5. Impact of WTPP on EC values

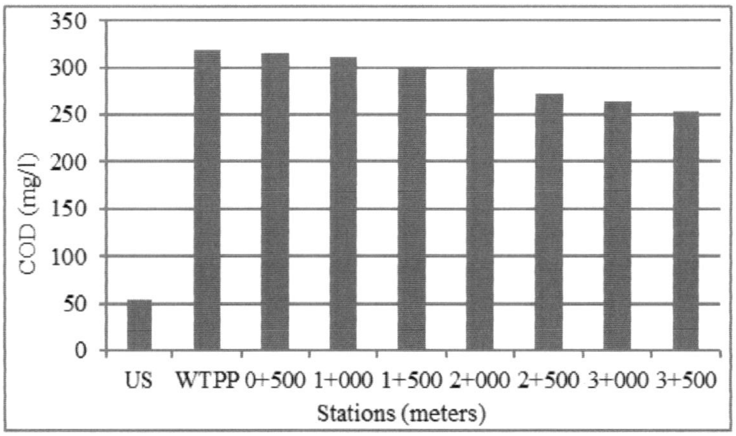

Figure 6. Impact of WTPP on COD concentrations

4. Conclusions

Tigris is an important river in Iraq it provides water to a huge land of Wasit province. The present study was intended to assess of water quality and so that an evaluation of pollution masses on the river just close to its origin. The study imparts that the water quality is deficient and contaminants of both organic and inorganic origin are coming into the river, thereby spoiling its water quality. The high concentrations of pH, TDS, EC, and COD are due to high concentration of organic and inorganic matter in the industrial wastewater flowing from the WTPP. The high concentration of TDS generates high concentration of dissolved inorganic solids. Also the discharged water from WTPP has a very low dissolved oxygen and that makes an adverse effect on the dissolved oxygen of water in the river at the mixing point and downstream of WTPP. The severely pollution of water of Tigris river has rendered its water unfit for human consumption especially there are many villages downstream about 1km away from the project. Advanced treatment of industrial wastewater before drain in the river is the lone aims by which the problematic of water pollution could be diminished. During the present investigation, the water quality of Tigris river in the upstream of WTPP was observed quite well in comparison to downstream of the river, which was found to be highly polluted due to discharge of industrial wastewater through surface drains.

IOP Conf. Series: Materials Science and Engineering **584** (2019) 012028 doi:10.1088/1757-899X/584/1/012028

References

[1] Abd-Ali. A. H., 1993, A Statistical Study on Some Effects of the Aggressive War upon the Quality of Water in the City Baghdad, M.Sc. Thesis, Build. And Const. Dept. University of Technology.

[2] WHO (2003) The World Health Report: Shaping the Future: World Health Organization 1211 Geneva 27, Switzerland.

[3] Al-Aadhamy, S.A. (1966). Evaluation of the chlorine ion concentrations in Baghdad (AlRusafa) city and their environment effect. M.Sc. Thesis. Baghdad Univ. Iraq.

[4] Dahr, J., (2004), "Pollution Chokes The Tigris; A Main Source of Baghdad's Drinking Water.

[5] William, O. Maddous, (2001), "Water Resources Planning ", First Edition, p.98

[6] Ali, A. I., 1978, Study the Bacterial Quality of Water in Iraq, M.Sc. Thesis, Agr. College, University of Baghdad.

[7] Al-Masri N.A. and Ali A.A. ,1985, Effect of Domestic and Industrial Wastewater Discharge on Pollutant Variation in River Tigris, Proceeding of Iraq Conference on Engineering, IDE, pp.467-474.

[8] Ghadban, F.D. 1993, Evaluation of Some Environmental Effects in Iraq (Water and Wastewater Treatment Plants), M.Sc. Thesis, College of Eng., University of Baghdad.

[9] Salman, Khalid Jawed**.** The negative effects of the absence of studies on environmental impact assessment and sustainable development in the province of Wasit (electrical energy in a model. *Journal of the Faculty of Basic Education for Educational and Human Sciences / Babylon University*. 2018, Vol. 38, p. 15.

[10] World Bank (1988). Pollution study on clean up river Kaduna. Part C report. KEPA/ABUCONS.

[11] APHA, AWWA, WPCF (1995). "Standard Methods for the Examination of Water and Wastewater. "American Public Health Association, Washington, DC

IOP Conf. Series: Materials Science and Engineering **584** (2019) 012029 doi:10.1088/1757-899X/584/1/012029

Water quality index of Tigris River on Waist Governorate for aquatic life

Huda Yousif Enad[1] and Ali Jwied Jaeel[2]

[1]Civil Engineering Department ,M.Sc.Student, Wasit University , Iraq.
hudayous.94@gmail.com

[2]Civil Engineering Department, PhD, Wasit University, Iraq.
alijwaid@uowasit.edu.iq

Abstract. In this study, a mathematical model Canadian Council of Ministers of the Environment Water Quality Index (CCME WQI) has been used. Several physical and chemical parameters are chosen to calculate the water quality index as follows: temperature, pH, total dissolved solids (TDS), turbidity, dissolved oxygen (DO), total nitrogen (TN) and total phosphorus (TP). The results showed that all stations received a poor assessment of the water quality index, indicating that the water of the Tigris River is highly polluted for aquatic life due to discharging of poorly treated wastewater from Al Azziziyah Wastewater Treatment Plant (AWWTP).
Keywords—River pollution; domestic wastewater; Turbidity; Dissolved Oxygen; Total Nitrogen; Total Phosphorus.

1. Introduction

The main sources affecting surface water quality are human activities and natural influences. Water quality can be determined without these sources through atmospheric processes, rock mineralization, nutrient spills, soil organic matter, hydrological factors and biological processes within the aquatic body that alter the physical and chemical components of water [1]. In recent years, water quality assessment has become very important in rivers affected by pollutant disposal. Industrial, municipal and household waste disposal are the main source of water pollution [2]. Stream pollution is any weakness to the local water properties through the disposal of industrial contaminants to the extent that it is no more useful for drinking purposes or backup the biotic societies living on it [3]. Disposal of polluted water from agricultural, industrial and municipal sources without treatment or partial treatment before discharge is a common practice in developing countries [4]. Physical and chemical properties of water determine the level of safety of the water system [5]. The physical and chemical properties are compared to determine water quality according to specific criteria [6]. The simplest and easiest way to evaluate and determine the state of water quality is the water quality index [7]. The water quality indicator can be expressed using a color, symbol, number or description for silver (water). This indicator consists of simple and less complex criteria that are used as water quality calculations [8]. The water quality index consists of four common stages: selection of parameter, conversion of the coefficient to a single scale, selection of the weight of the coefficients and the appropriate assembly function [9]. The water quality index may be used as an environmental indicator and has been developed to be used as an indicator of chemical and physical water quality [10]. The Canadian Water Quality Index compares the value of observations to an objective value as this objective value reflects water quality [11, 12, and 13]. The aim of this study

IOP Conf. Series: Materials Science and Engineering **584** (2019) 012029 doi:10.1088/1757-899X/584/1/012029

is to apply the water quality index to assess water quality in the Tigris River within the boundaries of Al-Azzizyah by measuring the physical and chemical parameters that determine the level of water quality. Sites that do not exceed standard water quality have high scores of water quality index (close to 100), and sites zero). The advantages of CCME WQI can be summarized as represent measurements of a variety of variables in a single number; flexibility in the selection of input parameters and objectives; adaptability to different legal requirements and different water uses; statistical simplification of complex multivariate data and combine various measurements in a variety of different measurement units in a single metric.

2. Experimental Works

2.1. Study area

Wasit Governorate locates in the southern part of the central region of Iraq (figure 1). The city of Aziziyah locates in the north of Wasit province and southern the capital Baghdad, about 85 km and, an area of 2252 km^2 and represents 3% of the area of the province area, and Tigris River passes through its land.

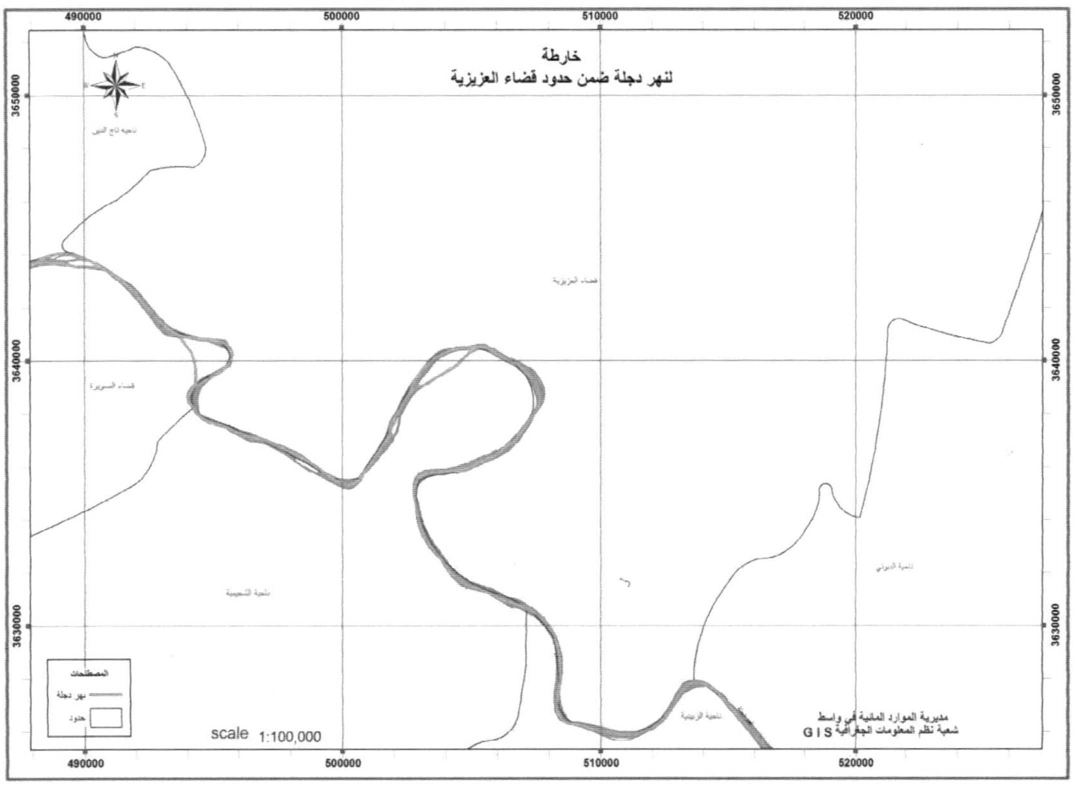

Figure 1. The image showing study area

2.2. Sample collection

Samples were collected from five sites distributed within the study area. The locations of these stations were identified using the German GPS 60 (table 1). These sites cover areas where there are contaminants due to the discharge of the waste water treatment plant at the first site, and the agricultural and industrial activities and human waste in the river in other locations. Samples were taken during January 2019 (rainy month). Four samples were collected at each point in five different locations for one month on a weekly basis using a 1000 ml plastic bottle, which was first washed using hydrochloric acid to avoid contamination and then distilled water. Chemical and physical tests were conducted at the health laboratory of the Department of Civil Engineering at Wasit University. Water quality index was calculated through a set of physical and chemical parameters based on data availability and importance [14]. The parameters are pH, temperature, total dissolved solids (TDS), turbidity, dissolved oxygen (DO), total nitrates (TN) and total phosphates (TP). The water quality index for all sites was calculated on the basis of standards for the protection of aquatic life (table 2) [11].

Table 1. Samples location.

Station (#)	Distance (Km)	Longitude	Latitude
1	0	N36°31.334'	E05°05.764'
2	1.5	N36°30.820'	E05°07.145'
3	3	N36°20.820'	E05°08.276'
4	5.5	N36°20.820'	E05°09.023'
5	9.5	N36°20.820'	E05°12.143'

Table 2. Water quality standard according to the Canadian Guideline.

Parameter	Unit	CCME guideline
Temperature	°C	15
pH	-	6.5-9
TDS	mg/l	500
Turbidity	NTU	5
DO	mg/l	5.5-9
NO_2	mg/l	0.06
PO_4	mg/l	0.3

2.3. Calculation and classification of water quality index (WQI)

The water quality index was calculated and classified in two phases, firstly a mathematical formula expressing the index to determine the amount of deviation and compliance with the water quality standards is used. The formula consists of three important components; the first component called the scope and represents the percentage of the number of parameters that are out of compliance with the total number of measured parameters [11].

$$F_1\ (Scope) = \left[\frac{Number\ of\ Failed\ Variables}{Total\ Number\ of\ Variables}\right] \times 100 \quad \ldots\ldots\ldots\ldots\ldots\ldots (1)$$

The second component is called frequency and refers to as the percentage of failed tests (do not meet the guidelines)

$$F_2\ (\text{Frequency}) = \left[\frac{Number\ of\ Failed\ Tests}{Tolal\ Number\ of\ Tests}\right] \times 100 \quad \ldots\ldots\ldots\ldots\ldots\ldots. (2)$$

The third component is called amplitude and represents the number of failed tests that do not comply with the guidelines and is calculated in two steps:

1. $excursion = [\frac{Objective}{Failed\ Test\ Value}] - 1$ .... (3)

(When the test value should not fall below the target)

$excursion = [\frac{Failed\ Test\ Value}{Objective}] - 1$ .... (4)

(When the test value should not exceed the target)

2. $nse = \frac{\sum excursion}{Number\ of\ Tests}$ .... (5)

F_3 (Amplitude) $= \left[\frac{nse}{0.01\ nse + 0.01}\right]$ .... (6)

The following equation gives a numeric value between 0 and 100 for water quality status

$\boldsymbol{CWQI} = 100 - \left[\frac{\sqrt{F\ 1^2 + F\ 2^2 + F\ 3^2}}{1.732}\right]$ .... (7)

Water quality is classified into five categories according to CCME WQI (Table 3).

Table 3. Water quality classification.

Parameter	CCME WQI value
Excellent	95-100
Good	80-94
Fair	60-79
Marginal	45-59
Poor	0-44

3. Results and Discussion

The water quality index (WQI) gives one value that reflects water quality. Rating standards allow for easy and simple quality assessment [15]. The chemical and physical properties of water greatly help to save aquatic life. The results of the study showed that the values of the water quality index are 23, 26, 31, 29 and 33 in stations 1, 2, 3, 4 and 5, respectively (Figure 2).

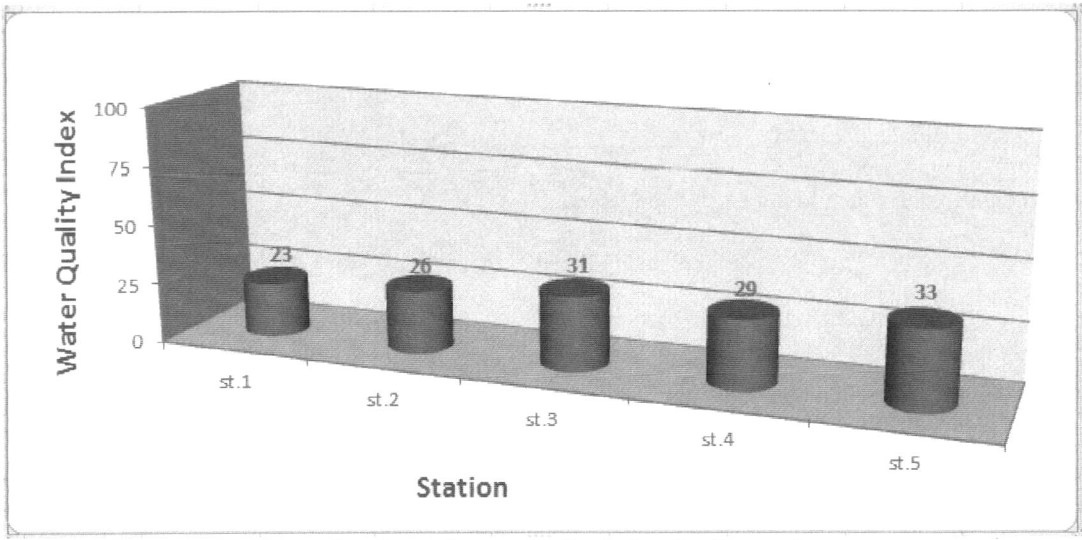

Figure 2 Values of water quality index for five stations

It can be said that the water of the Tigris within the study area is much polluted for aquatic life, because all stations indicated a poor evaluation of the water quality index. This is due to many Industrial, municipal and agricultural pollutants into the river without treatment [16]. The results showed that the first station exhibited the lowest assessment because of the discharges of Al- Azziziyah wastewater treatment plant that adversely affects the quality of the river.

The decreasing and increasing value of the water quality index is due to measured physical and chemical processes. Water temperature values ranged from (15-18.3) °C, all within the limits. The temperature is an important factor because it affects the characteristics of the river such as photosynthesis of plants, the amount of oxygen dissolved in water and the amount of metabolism of organisms [17]. pH is an important factor for water quality The values of pH changed from 7.8 to 8.3, all within permissible limits. Water is almost alkaline [18].

The values of total dissolved solids varied between (503 – 1120) mg/L exceeding in all stations the permissible limits and recommended by the CCME. The main source of high concentration of TDS is water pollution from sewage and industrial sources [19]. The turbidity values ranged between 88.5 to 305 NTU and exceeded the limit in all points. High values of turbidity are resulted from the amount of mud, silt and organic and inorganic materials during winter [14]. Dissolved oxygen is used as a measure of water quality, a very important factor in aquatic life [5]. The values of dissolved oxygen at all locations within the recommended limits for CCME. The normal values of dissolved oxygen concentrations are due to several factors, including low temperature, oxygen mixing with water and low salinity [18]. The total nitrogen level in all stations exceeds the permissible limit. The decrease in total nitrogen concentration in the Tigris reflects a decrease in biological processes in winter [20]. Sources of nitrogen compounds in the ecosystem are animal waste, wastewater, fertilizers, decomposition and disintegration of organic matter and nitrogen stabilization processes [21, 22]. The concentration of phosphorus exceeded the maximum allowed (0.3 mg/ L) at all sites and ranged between (0.5-1.23) mg/l. High phosphorus concentrations adversely affect the aquatic environment and act on lack oxygen and excessive growth of plants and algae [23].

4. Conclusion

It can be concluded from this study that the water quality index (WQI) gives a clear and simple assessment and can reduce water quality data to only one value. As the water quality index (WQI) is able to calculate water quality clearly and realistically, it can also be used to help water quality management and decision making.

5. References

[1] UNEP/GEMS, United Nations Environment Programme Global Environment Monitoring System/Water Programme. Water Quality for Ecosystem and Human Health. UN GEMS/Water Programme Office c/o National Water Research Institute 867 Lakeshore Road Burlington, Ontario, L7R 4A6 CANADA available on-Line, 2006.

[2] Campbell, L. M. Mercury in Lake Victoria (East Africa): Another emerging issue for a Beleaguered Lake. PhD, Thesis. Waterloo, Ontario, Canada, 2001.

[3] Agrawal, A., Pandey, S. R. and Sharma, B. 2010. Water pollu¬tion with special reference to pesticide contamination in In¬dia. Journal of Water Resources and Protection, 2, 432-448. https://doi.org/10.4236/jwarp.2010.25050

[4] E. J. Abdullah, "Evaluation of surface water quality indices for heavy metals of Diyala River-Iraq," Evaluation, Baghdad: University of Baghdad, 2013.

[5] Lomniczi, I.; Boemo, A. and Musso, H. Location and Characterization of Pollution Sites by Principal Component Analysis of Trace Contaminants in A Slightly Polluted Seasonal River: A Case Study of the Arenales River (Salta, Argentina) J. Water SA., Vol. 33, No. 4, pp. 479-485, 2007.

[6] Robertson, D.M.; Saad, D.A. and Heisey, D.M. A regional Classification Scheme for Estimating Reference Water Quality in Streams Using Land-Use-Adjusted Spatial Regression-Tree Analysis J. Envir. Manag. Vol. 37, No. 2, pp. 209-229, 2006.

[7] W.K. Dodds, J.R. Jones and E.B. Welch. Suggested Classification of Stream Trophic State: Distributions of Temperate Stream Types by Chlorophyll, Total Nitrogen, and Phosphorus. J. of Water Research, Vol. 32, pp.1455-1462, 1998.

[8] Fernández, N.; Ramìrez A. and Solanon, F. Phsico-Chemical Water Quality Indices- A Comparative Review. Revista Bista, Unversidad de Pamplona, Bucaramanga,Colombia, ISSN(Versiόn impresa) 0120-4211, pp.19-30, 2004.

[9] House, M.A. Water Quality Indices. Unpublished Ph.D. Thesis. Middlesex Polytechnic, Queensway, Enfield, England, U.K. 1986,

[10] Salim, B.J.; Gholamreza, N.B.; Amir, S.; Masoud, T. and Mojtaba, A. Water Quality Assessment of Gheshlagh River Using Water Quality Indices. J. Environmental Science, Vol. 6, No.4, pp.19-28, 2009.

[11] CCME, Canadian Council of Ministers of the Environment. Canadian Water Quality Guidelines for the Protection of Aquatic Life: Canadian Water Quality Index 1.0 Technical Report. In Canadian Environmental Quality Guidelines, Winnipeg, Manitoba, 2001. (http://www.ccme.ca/assets/pdf/wqi_usermanualfctsht_e.pdf).

[12] Khan, F.; Husain, T. and Lumb, A. Water Quality Evaluation and Trend Analysis in Selected Watersheds of the Atlantic Region of Canada" J. Environmental Monitoring and Assessment, Vol. 88, pp. 221–242, 2003.

[13] Lumb, A.; Doug, H. and Tribeni, S. Application of CCME Water Quality Index to Monitor Water Quality: A Case of the Mackenzie River Basin, Canada J. Environmental Monitoring and Assessment., Vol. 113, pp. 411-429, 2006.

[14] APHA, American Public Health Association. Standard Methods for the Examination of Water and Wastewater, 21st Edition Washington, DC. 22621, pp., 2005.

[15] Crabtree, R. W.; Cluckie, and Forster, C. F. A comparison of two quality models. Water Research, Vol. 20, pp. 53-61, 1986.

[16] WHO, World Health Organisation and United Nations Children's Fund Joint Monitoring Programme for Water Supply and Sanitation. Progress on Drinking-Water and Sanitation: Special Focus on Sanitation.UNICER, NewYork, NY, USA, and WHO, Geneva, Switzerland, 2008.

[17] Suski, C. D.; Killen, S.S.; Keiffer, J.D. and Tufts, B.L. The influence of environmental temperature and oxygen concentration on the recovery of largemouth bass from exercise. Implications for live release angling tournaments. J. of Fish Biol. Vol. 68, pp.120-136, 2006.

[18] Ahipathy, M.V. and Puttaiah, E.T. Ecological Characteristics of Vrishabhavathy River in Bangalore (India). j. Environmental geology, Vol. 49, No. 8, pp.1217-1222, 2006.

[19] Boyd, C.E... Water Quality an Introduction. Kluwer Acadamic Publi-shers, Boston, USA, 330pp. 2000.
[20] WHO, World Health Organization. Nitrate and nitrite in drinking-water: Background document for development of WHO Guidelines for Drinking-water Quality. WHO/SDE/WSH/07.01/16/Rev/1, World Health Organization, 20 Avenue Appia, 1211 Geneva 27, Switzerland, 2011.
[21] Environment Canada. Canadian water quality guidelines for the protection of aquatic life: Ammonia. In: Canadian environmental quality guidelines, 1999. Canadian Council of Ministers of the Environment, Winnipeg, 2010.
[22] McKenzie, J.M.; Siege, D.I.; Patterson, W. and McKenzie, D.J. A Geochemical Survey of Spring Water from the Main Ethiopian Rift Valley, Southern Ethiopia: Implications for Well-Head Protection. J. Hydrogeology Vol. 9, pp. 265-272, 2001.
[23] Fadiran, A.O.; Dlamini, S.C. and A. Mavuso. A Comparative Study of the Phosphate Levels in Some Surface and Ground Water Bodies of Swaziland. Bull. Chemical Society of Ethiopia, vol. 22, NO. 2, pp.197-206, 2008.

Effect of increasing floors number on the natural frequency compared with further masses added to steel building models

Ahmed A. Alalikhan [1,*], Mohsen A. Habelalmateen [1] and Haider A. Al-Tameemi [1]

[1] Civil Engineering Department, Faculty of Engineering, University of Kufa, Najf, Iraq.

* Corresponding author: ahmeda.jasim@uokufa.edu.iq

Abstract. This paper investigates the change in the natural frequency of whole building due to the increase or decrease of the number of floors of a steel building model. Five steel building models were adopted in the present study to investigate the natural frequency required to be identified in each model. The first model represents a one floor building model, the second building model contains two floors and so on. The first four models were subjected to further masses distributed upon four points within the last floor. Each model consists of the same structural details that are proposed in the other models such as details of the beams, columns and slabs. The herein parameters are the number of floors and the further added masses upon the models up to four floors. All models were analyzed using ANSYS software to identify the natural frequencies for the extracted modes of vibration about X, Y, and Z building axes. The study is aimed to produce a relationship between the change of floors number and the further mass addition in the field of structural frequency attenuation based on the indication of the first identified mode. Results show that higher masses required to be added to attenuate value of natural frequency for the one floor building model than the further masses required for building with two floors to achieve the same purpose. This relationship continues for the other models with higher number of floors but with less further required masses.

1. Introduction

The attention towards the role of floors vibration in the structural design was noticed since the early of 19th century [1]. At that time, the researchers declared that the stiffening of the structural elements is required to minimize the floor vibration produced by the human activities or other sources as much as possible [1]. Other researchers [2] investigated the full-height non-structural partitions effect on the dynamic properties of the composite floor. Both of the previous techniques are related with the vibration of the floor itself as a local behaviour for a certain part vibration in the building while present study focuses on the global vibration behaviour of the building due to the increase in floors numbers as aforementioned by some researchers in literature [3-4].

The natural frequency is an important measure of vibration rate of any structural system in which either be in cycles per second (Hertz) or in radians per second. It is important to the designers to identify the natural frequency for each structure in order to insure that it is not coincide with the frequency obtained by the excitation forces to avoid the resonance phenomena [5]. In this way, attenuating frequency of a certain structural system is an important aim in which the system will be with an insufficient time that had been already changed due to the frequency attenuation to react with the cyclic load [5]. Here, cyclic load is the causative of excitation which could be due to different reasons such as human activities, traffic vibration or any other reason. As an example, if the cyclic load has a sufficient time to complete one cycle, then this will allow the building to react and response to the excitation load sufficiently, and

vice versa [5]. Hence, the attenuation of frequency should be taken into account by the designers to produce acceptable dynamical designed systems.
Present study investigates the relationship between the increasing floors number of steel structures and the natural frequencies produced in these structures. This relationship is compared with the frequency attenuated based on the addition of further masses as a specified attenuation technique [6-8] according to the indication of first identified mode.

2. Adopted models

Due to the widely use of steel structures in the societies [9], five steel structural models were adopted in the present study. Starting with a one floor building model, additional floor was added to each one of the other four models to produce models with two, three, four and five floors. Later, models up to four floors were subjected to further masses added on the last floor distributed on four points so that the total adopted models become nine models. All the details of the beams, columns and slabs are the same for each floor. Thus, only the one floor building model is shown in Figure 1 and other models with additional floors are similar.

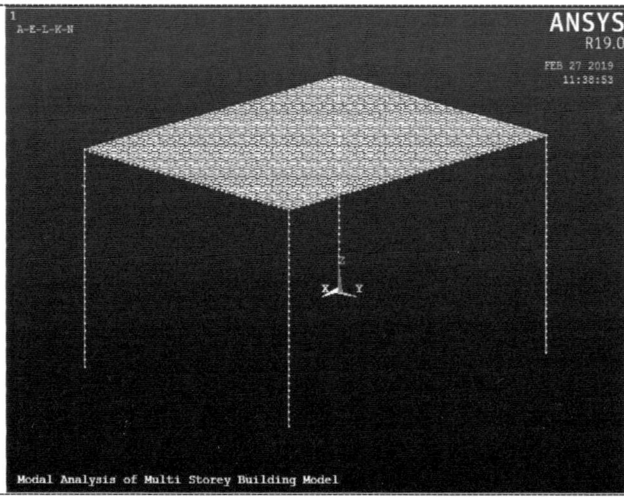

Figure 1. One floor steel model.

As shown in Figure 1, four beams are existed in each floor with I-shape and the dimensions details shown in Figure 2a. Also, there are four steel columns with a hollow square cross section shown in Figure 2b carry a steel slab with50 mm thickness.

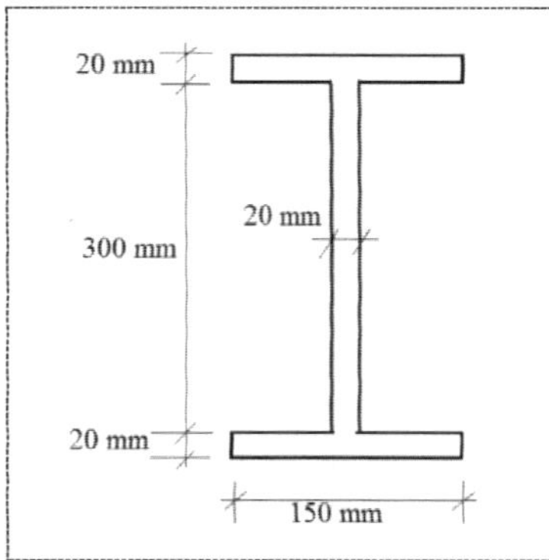

Figure 2a. Beam dimensions.

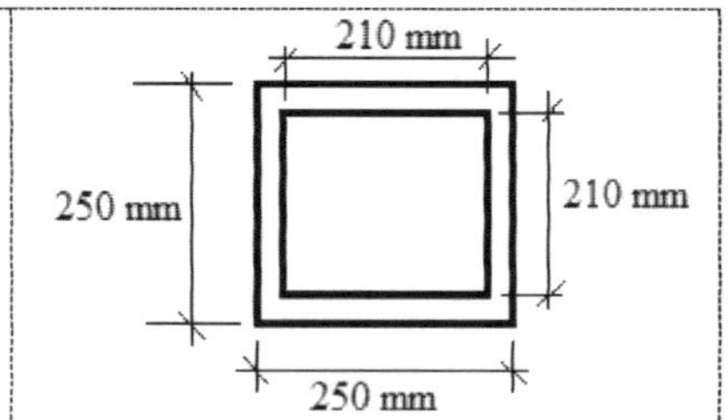

Figure 2b. Column dimensions.

Each model was simulated using ANSYS software based on the following details, the height of each floor (in Z-direction) is 3000 mm, distance between columns centre to centre is 4000 mm in Y-direction while it is 5000 mm in X-direction. The Element Type (ET) was used as Beam4 for the beams and columns, element type of Shell63 was used to simulate the slab and element type of Mass21 to represent the further masses. Only one Material Property (MP) was considered for all elements in the structure represented by the steel material that has modulus of elasticity of 200 GPa, Poisson's ratio of 0.3 and density of 7860 kg/m^3. The columns are completely fixed in all directions in their ends within the ground level during the analyzing process. Each one meter length, of the beams or columns, has been divided into 10 elements in the finite element model so that the total elements for the slab will be 40×50 total existed elements.

3. Application of modal analysis

Each model was analyzed based on modal analysis via ANSYS software to identify natural frequencies and the shape for each of the three considered modes of: first bending about X-axis, first bending about Y-axis and first torsion about Z-axis. Figure 3 shows the mode shapes and values of natural frequencies for the models without further added masses based on the number of floors.

With respect to models with further masses that were added to the models up to four floors, the shape of adopted modes were the same as in the case of models without further added masses. So that only values of natural frequencies were changed for the models up to four floors. The comparison between natural frequencies values identified form models without further masses and the corresponding values obtained by models with further masses was executed based on indication of the first mode only which is the bending about X-axis mode and therefore they are listed in Table 1. Also, Table 1 shows values of masses that were added to the models based on their floors and the attenuated frequency value due to the mass addition for each model.

Table 1. Values of natural frequencies (Hz) for models with further added masses.

Number of floors	1	2	3	4
Value of additional mass (Ton)	80	30	20	10
Attenuated frequency of the first bending about X- axis mode(Hz)	2.67	1.53	1.03	0.85

IOP Conf. Series: Materials Science and Engineering **584** (2019) 012030 doi:10.1088/1757-899X/584/1/012030

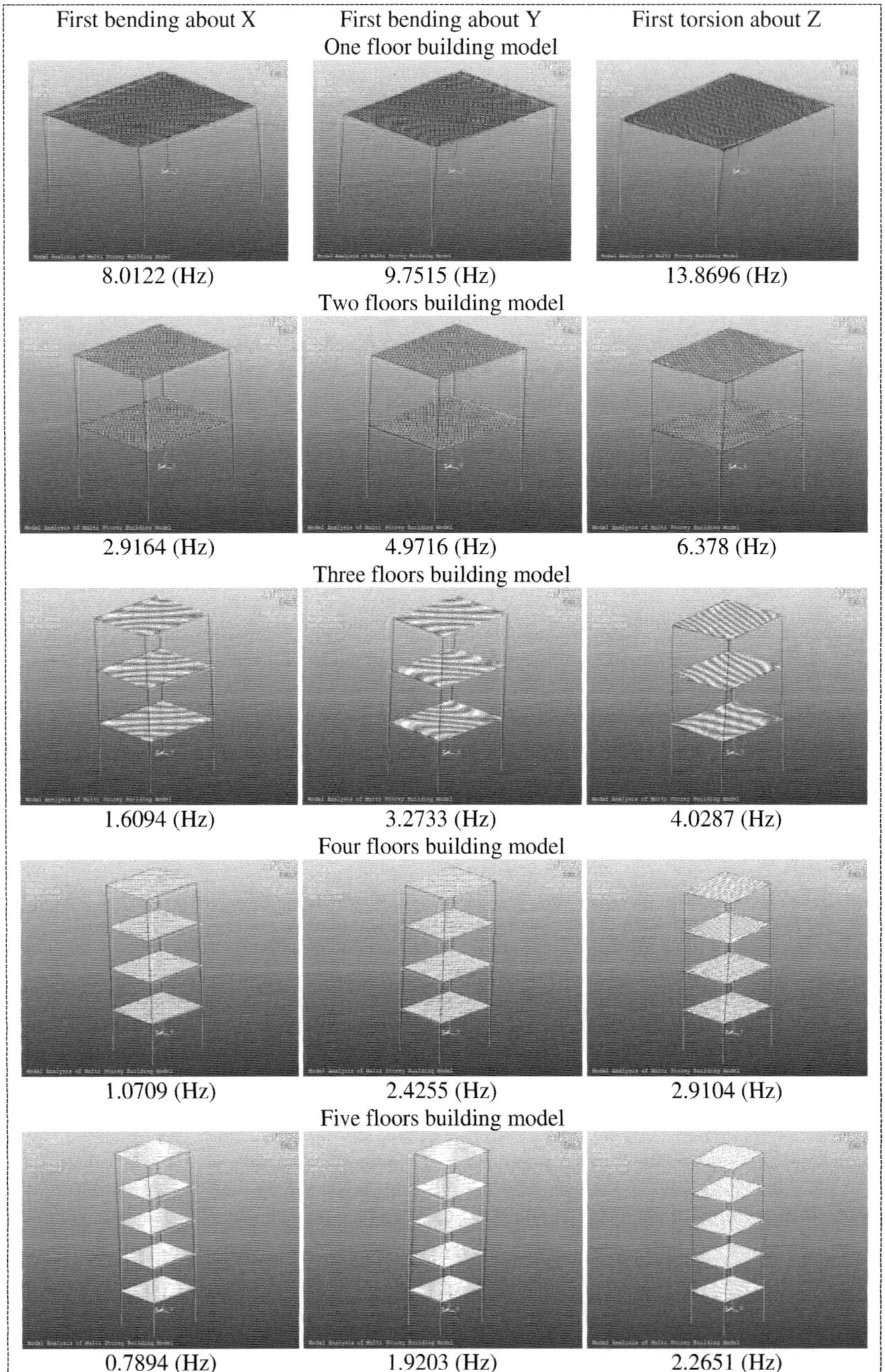

Figure 3. Shape of modes and values of natural frequencies for the adopted modes.

IOP Conf. Series: Materials Science and Engineering **584** (2019) 012030 doi:10.1088/1757-899X/584/1/012030

4. Results and Discussion

Values of natural frequencies were obtained based on two response:

4.1. Frequency response to additional floors

For the adopted mode and based on the modal analysis, the relationship between value of the natural frequency and the number of floors for the models without further added masses are plotted in Figure 4. It is clearly shown that the values of natural frequencies were decreased as the number of floors was increased as an adversely relationship. The curve of first mode (bending moment about X-axis) shown in Figure 4 decreases sharply for the models with one and two floors while it decreases with lesser sharpness when the number of floors are increased.

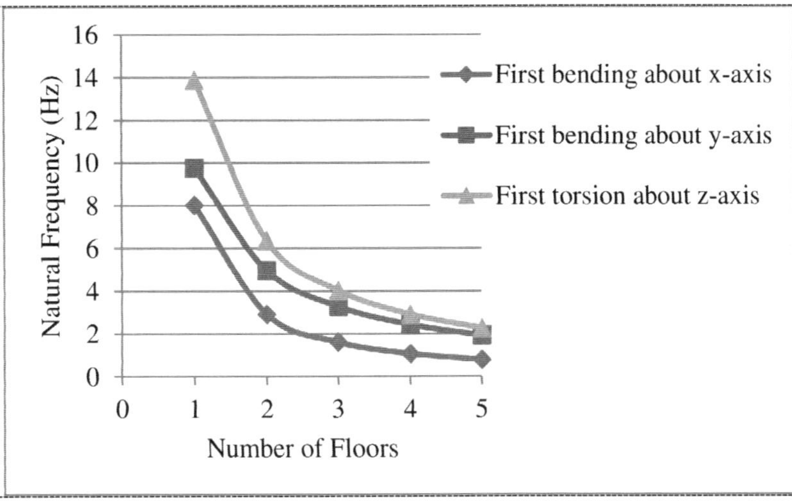

Figure 4. Values of natural frequencies for the first three modes with respect to floors number.

The relationship is similar for the comparison between the adopted three modes which are decreased sharply for one and two floors models and become lesser sharpness for other models. Nevertheless, when the building with more than two floors the first mode have the smallest change value compared with other modes.

4.2. Frequency response to further masses

For each model, further masses were added to attenuate its frequency to a value nearly equal to that of the model with one additional floor. Here, the aim is to reach a low frequency value for each model without additional floor. In this case, only the first mode (dominant mode) was investigated to be as the indicator for the frequency attenuation. Starting with the model of one floor, further masses were added gradually to reach a value of frequency nearly equal to that of two floor model. The same procedure was repeated then for the model with two floors to reach a value of frequency close to that of the three floor model and so on. Results of the procedure are plotted in the bar chart shown in Figure 5.

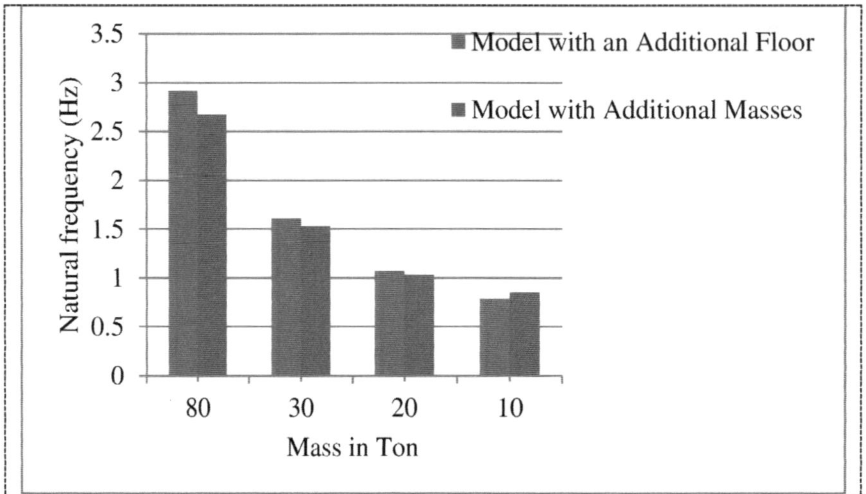

Figure 5. Attenuating frequency by additional masses compared with increasing floors number.

Based on Table 1 and Figure 5, further masses of 80 ton was required to attenuate the first mode in the one floor building model to reach a frequency value approaching to that of the corresponding first mode in the two floors building model. Next, further masses of 30 ton was required to attenuate the frequency value of the two floor model to produce a frequency value close to that of the corresponding three floors model without further masses. The repeated procedures applied upon the three and four floors building model produce 20 and 10 ton, respectively, to attenuate natural frequencies such that they respectively approach to frequencies values of the four and five floors. Both groups' results of natural frequencies obtained by models without further masses and the corresponding group values obtained by models with further masses are plotted in Figure 6 to recognize the relationship based on the results of the first bending about X-axis mode. The relationship exhibits high difference (5.33 Hz) between values of natural frequency for models with further masses and the corresponding natural frequency values of models without further masses. The difference is gradually decreased to 0.219 Hz as the number of floors is increased, as shown in Table 2.

Table 2. Differences between values of the first mode natural frequencies for models without further masses and the corresponding values of models with further masses.

Number of floors	1	2	3	4
Further masses	80	30	20	10
Frequency (Hz) for models without further masses	8.0122	2.9164	1.6094	1.0709
Frequency (Hz) for models with further masses	2.6729	1.5338	1.0349	0.8517
Difference (Hz) (frequency value without mass - frequency value with mass)	5.3393	1.3826	0.5745	0.2192

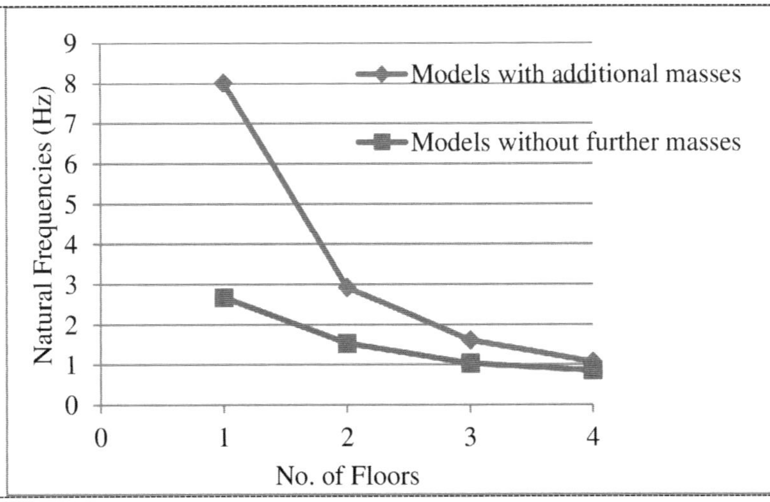

Figure 6. Differences of natural frequency values for model without and with further masses.

5. Conclusions

In the field of frequency attenuation, present study exhibits the following conclusions:

1. Increasing the number of floors in the building model decreases natural frequency values as the same action of further added masses.
2. Attenuating values of natural frequencies for the one floor building model requires high value of further mass (80 ton) to approximately become as the frequency value of two floor building model without further masses. The required value of further masses becomes smaller (30 ton) to make the frequency values of two floors model with further masses close to the frequency value of the three floors model without further masses. Required masses decrease to values of 20 ton and 10 ton added to the three and four floors models, respectively, to approach the frequency values of the corresponding four and five floors models without additional masses.
3. For both cases of increasing floors number or further mass addition, the relationship of reduction in values of the natural frequencies starts sharply for the models with one and two floors and becomes lesser sharpness for the models of higher floors number.
4. The differences between the values of natural frequencies of models without further masses and that of models with further masses start highly, with difference of 5.33 Hz, and gradually decreased to 0.219 Hz as the number of floors is increased.

References

[1] Sinin H, Mohammad N and Norsuzailina M 2011 Dynamic property analysis and development of composite concrete floor (CCF) and vibration serviceability: a review Int. J. of the Physical Sciences 6 1 p 26

[2] Petrovic S and Pavic A 2011 Effects of non-structural partitions on vibration performance of floor structures: a literature review Int. Conf. of Structural Dynamics (Leuven) (Belgium) pp 1014-20

[3] Goel R and Chopra A 1997 Dynamic period formulas for moment-resisting frame buildings Int. J. of Structural Engineering 123 11 pp 1454-61

[4] Naeim F 1998 Performance of 20 extensively-instrumented buildings during the 1994 Northridge earthquake The Structural Design of Tall Buildings 7 3 pp 179-194

[5] Andrew L, Stephan J and Paul J 2009 Design of floors for vibration: a new approach, revised edition The Steel Construction Institute (Silwood Park) p 128

[6] Mišković Z, Al-Wazni S and Alalikhan A 2018 Damage detection for civil structural

health monitoring application - a case study of the steel grid bridge structural model Technical Gazette 25 2 pp 266-275

[7] Alalikhan A, Al-Wazni S and Mišković Z 2016 Tests of heuristic optimization techniques in vibration based damage detection Gradevinar 68 7 pp 543-557

[8] Mišković Z, Al-Wazni S, Alalikhan A and Mišković L 2016 FE model updating based on dynamic properties by heuristic optimization 6th Int. Conf. of Civil Eng. - Science and Practic (Montenegro) p 8

[9] Muiris C and Julian M 2014 Utilization of structural steel in buildings Proceedings of the Royal Society a Mathematical, Physical and Engineering Sciences p 14

Recycling of disposal Polypropylene Blister Tablets and Strapping ties as fiber reinforcement for Pervious concrete

Ahmed M Ahmed[1, *], Ahmed H Hussein[1], and Mohammed T Hammood[2]

[1] University of Samarra, College of Engineering, Iraq.
[2] Tikrit University, College of Engineering, Iraq

*Corresponding Author: ahmed_m@uosamarra.edu.iq

Abstract. Pervious concrete (PC) was introduced as a solution for the runoff from rainfall. To insure high porosity, pervious concrete should contain limited quantities of fine aggregate, which adversely effects concrete mechanical properties. As an attempt to improve PC mechanical properties, fibers were added to concrete matrix. As PC is allows water to pass through it, metal fibers are uncourageous to be added to it. Plastic fibers were selected as an alternative to metal fibers to be added to PC. Concrete physical and mechanical properties were studied for mixtures containing different quantities of fibers. Tests results showed a slight improvement in concrete compressive strength (0.48% to 1.44%) as fibers were added. The addition of fibers improved splitting tensile strength by (6.36% to 27.67%). As adding fibers, concrete flexural strength increased by (11.37% to 24.17%) compared to the reference mixtures. of PC. Concrete physical properties (Density, Porosity and permeability) were also investigated. The addition of fibers to concrete has a slight effect on (density, porosity and permeability).

1. Introduction

The permeability of most materials used in construction is low [1]. Due to the use of these materials, the hydrological cycle in the cities is radically changed, as rainwater no more infiltrating to the soil and lead to run-off. Responding to that, researchers are searching for new materials with high permeability characteristics without altering their mechanical properties. In that sense, Pervious Concrete (PC) is among these materials [2].

Pervious concrete pavements (PCP) are structures permeable enough to serve storm water management and bear pedestrian/traffic loads, depending on the application. In this pavement system, a pervious concrete layer is placed on a highly voided stone bed as the base layer, to insure a rapid percolation of runoff through the pavement system instead of allowing it to run on the surface [3, 4].

Pervious concrete (PC) has acquired renewed interest in the past decade due to its positive environmental impacts. Widespread researches employing different strategies has been carried out to improve the overall performance of PC. Lots of literatures have been published. As high performance pervious concrete (HPPC) advanced, extensive application of this material has been made possible. [5]

However, this type of pavement cannot be used for heavy traffic due to a high amount of voids and consequently low strength of pervious concrete. [6]

Several research efforts focused on improving the mechanical and durability properties of PC by using various supplementary materials, fibre reinforcement can be used. Fiber reinforcement should be selected carefully depending on its interaction with a given pervious concrete mix. To avoid corrosion,

IOP Conf. Series: Materials Science and Engineering **584** (2019) 012031 doi:10.1088/1757-899X/584/1/012031

Metallic reinforcement should not be used for pervious concrete. Instead of metallic fibers, fibers made of carbon, glass, synthetic or natural materials may be added to pervious concrete [7-13].
Saeid Hesami et al (2014) [6], studied the effect of adding different weight percentages of rice husk ash as a cement replacement in concrete mixtures on the mechanical properties. They also added 0.2% volume fraction of glass and 0.5% volume fraction of steel and 0.3% volume fraction of polyphenylene sulfide (PPS) fibers to improve the mechanical properties of the pervious concrete. Different water to cement (w/c) ratios were used in their study. They investigated concrete physical and mechanical properties including permeability, porosity, compressive strength, flexural strength, and tensile strength. Due to the results, concrete mechanical properties were improved as fibers and rice husk ash were added. [6]
Also, the effect of adding volume fractions of cured carbon fibre composite to pervious concrete on its abrasion resistance was studied by Rangelov et al. [14]. They noticed that mixtures containing cured carbon fibre composite have better performance than the control mixture.
Fiber reinforcement enhances the strength of the matrix and delays the generation of the cracks. However, the bonding mechanism between fiber and concrete matrix is controversial in literature and needs better explanation. Due to surface smoothness and inert chemical nature of commercially available fibers, researchers studied several treatment techniques to improve the fiber-matrix bonding properties. The use of fibers in pervious concrete is even more challenging due to high porosity and insufficient fiber-matrix bonding interface [15].

2. Materials and Methods

2.1. Cement
Ordinary Portland cement (OPC. Type I) was used in this investigation confirming to (ASTM C 150) [16].

2.2. Gravel
Two types of gravel were used in this experimental work, both of them was rounded single sized gravel. The first type of gravel passed through (9.5 mm) sieve and retained on (4.75 mm) sieve. The second type of gravel passed through 12 and retained on 9.5 mm sieve.

2.3. Fibers
PVC strapping ties were cut to pieces (3 cm in length, 2 mm in width and 1 mm in thickness) as shown in fig.1. Two volumetric ratios of this fibers were added to concrete (0.5 %) and (1 %) by total volume of concrete.
PVC Blister tablets were cut to pieces (3 cm in length, 2 mm in width and 1 mm in thickness) as shown in fig.2. Two volumetric ratios of this fibers were added to concrete (0.5 %) and (1 %) by total volume of concrete.

A. Before cut B. After cut

Figure 1. Strapping ties.

A. Before cut B. After cut

Figure 2. Blister tablets.

2.4. Water

Portable water was added to the dry mixture.

3. Mix proportion

To reach the best consistency and achieve an appropriate porosity for the matrix, trial batches were done. Figure (3) and table (1) shows the details of trial mixes and its results.

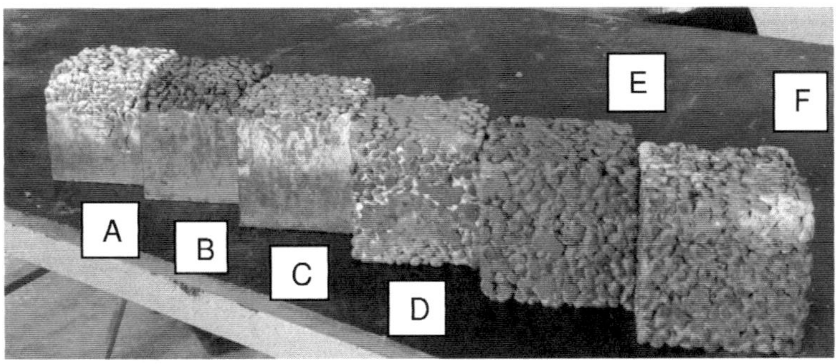

Figure 3. Trial Cubes.

Table 1. Trial mixes.

Cube	Cement	Gravel kg / m3	Sand	W/C ratio	Superplasticizer
A	kg / m3	1400	kg / m3	0.25	%
B	425	1400	400	0.2	1
C	425	1400	400	0.35	1
D	425	1400	400	0.4	0
E	425	1600	400	0.3	0
F	425	1700	200	0.3	0

The best proportioning discovered through trial mixes (F) was followed in the experiential work. Table 2. describes the selected symbols for each of the mixtures and table 3. Contains the proportions of each of the investigated mixtures.

Table 2. Mixtures symbolling

symbol	Description
R1	is the refers to the reference mixture, casted using (9 - 4.75 mm) aggregate.
R2	is the refers to the reference mixture, casted using (12.5 – 9 mm) aggregate
M1F-0.5%	is the matrix casted using (9 - 4.75 mm) aggregate and reinforced with (0.5% Strapping ties fibers).
M1F-1%	is the matrix casted using (9 - 4.75 mm) and reinforced with (1% Strapping ties fibers).
M2F-0.5%	is the matrix casted using (12.5 – 9 mm) and reinforced with (0.5% Strapping ties fibers).
M2F-1%	is the matrix casted using (12.5 – 9 mm) and reinforced with (1% Strapping ties fibers).
B1F-0.5%	is the matrix casted using (9 - 4.75 mm) aggregate and reinforced with (0.5% Blister tablets fibers).
B1F-1%	is the matrix casted using (9 - 4.75 mm) aggregate and reinforced with (1% Blister tablets fibers).
B2F-0.5%	is the matrix casted using (12.5 – 9 mm) aggregate and reinforced with (0.5% Blister tablets fibers).
B2F-1%	is the matrix casted using (12.5 – 9 mm) aggregate and reinforced with (1% Blister tablets fibers).

Table 3. Mixtures symbolling

symbol	Cement kg / m^3	Gravel kg / m^3	Fibers kg / m^3	Water kg / m^3
R1	425	1700	0	127.5
R2	425	1700	0	127.5
M1F-0.5%	425	1700	7.5	127.5
M1F-1%	425	1700	15	127.5
M2F-0.5%	425	1700	7.5	127.5
M2F-1%	425	1700	15	127.5
B1F-0.5%	425	1700	7.5	127.5
B1F-1%	425	1700	15	127.5
B2F-0.5%	425	1700	7.5	127.5
B2F-1%	425	1700	15	127.5

4. Mix procedure

The proportioned ingredients were mixed and batched using a mechanical drum mixer of 125 liter in volume. The specimens were molded and compacted according to (ASTM C 192) [17]. The specimens were immersed in water 24 hours after casting and until tested at 28 days age.

Figure 4. Pervious concrete specimens.

5. Tests.

The investigated properties were (compressive strength, Splitting tensile strength, flexural strength, Porosity, Density and permeability).

5.1. Concrete Compressive Strength Test

Concrete compressive strength was tested according to (BS 1881, Part 116) [18]. For compressive strength test, three (15 cm) cubes of concrete were casted for each mix. Concrete specimens were immersed in water (24 HRs) after they were casted, and stayed in water until testing it at 28 days age.

5.2. Splitting Tensile Strength Test

According to (ASTM C 496) [19], three cylinders (15 cm in dimeter and 30 cm in height) were casted for each mix for the Splitting Tensile strength test. Concrete specimens were cured for 28 days before tested.

5.3. Flexural Strength Test

Concrete flexural strength was tested through casting three (10*10*50 cm) prisms for each of the five mixtures. Curing started (24 Hrs) after casting and continued until testing at 28 days age. The test was conducted according to (ASTM C 78) [20].

5.4. Density

Three cylinders were casted for density measurement of each mixture. Concrete density was measured according to (ASTM C 1754) [21], as listed in the procedure below:

- Determine the dimensions of the test specimen (Diameter and Length).
- Dry the specimen.
- Determine the constant dry mass of the specimen A
- Determine the submerged mass, B.

$$\text{Density} = (K*A)/(D^2*L) \tag{1}$$

Where:
A = dry mass of the specimen, g [lb],
D = average diameter of the specimen, mm [in.],
L = average length of the specimen, mm [in.], and
K = 1 273 240 in SI units or 2 200 in [inch-pound] units

5.5. Porosity

Concrete porosity was also tested through measuring the volume of water could be held by voids of a specimen with respect to the gross volume of the specimen its self. Porosity test was conducted according to (ASTM C1754) [21].

Figure 5. Porosity test procedure.

5.6. Permeability

To measure concrete permeability, tow test procedures could be followed (Constant head permeability test) and (Falling head permeability test) according to previous studies, Constant head permeability test is preferred to investigate the permeability of pervious concrete [22]. Three cylinders were casted to measure concrete permeability. Each cylinder was installed in the test instrument described in figure (6). The coefficient of permeability was calculated for each specimen.

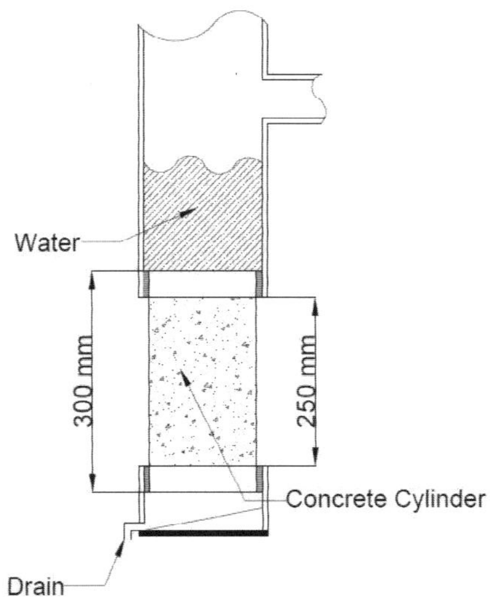

Figure 6. Permeability test.

6. Test results

6.1. Total Void in aggregate based on Angularity number

The single sized gravel allows the creation of continuous voids which is the base concept of pervious concrete. Total void percent and aggregate angularity number are shown in table 4.

Table 4 shows the angularity number for two sizes of aggregate used in this study. The percentage of voids present in the aggregate increases as the angularity number increases. It is observed that the size of aggregate has pronounced effect on the voids.

Table 4. Total void percent and aggregate angularity number

Gravel	Total void percent	aggregate angularity number
Grading 1	37.5%	4
Grading 2	39.5%	6

6.2. Compressive Strength

Table 5 shows the compressive with various mixture proportions. The specimens with all the two selected size of aggregate *followed* the same trend. It is observed that compressive strength is slightly effected as the mixture proportions changed.

Table 5. Compressive strength test results

Mix	R1	R2	M1F-0.5%	M1F-1%	M2F-0.5%	M2F-1%	B1F-0.5%	B1F-1%	B2F-0.5%	B2F-1%
Concrete compressive strength	11.1	10.4	11.2	11.15	10.45	10.48	11.25	11.26	10.45	10.47
Difference compared with R1 and R2	0	0	+0.9	+0.45	+0.48	+0.77	+1.35	+1.44	+0.48	+0.67

6.3. Splitting tensile strength

The splitting tensile test results are shown in table 4. The tests results indicate that the splitting tensile strength was increased by (6.36 % to 27.67 %) as fibers were added.

Table 6. Splitting tensile strength test results.

Mix	R1	R2	M1F-0.5%	M1F-1%	M2F-0.5%	M2F-1%	B1F-0.5%	B1F-1%	B2F-0.5%	B2F-1%
Concrete compressive strength	2.53	2.05	2.83	3.21	2.32	2.52	2.71	2.23	2.25	2.48
Difference compared with R1 and R2	0	0	+11.9	+26.9	+13.2	+22.9	+6.36	+27.67	+9.76	+20.97

6.4. Flexural strength

The flexural strength of pervious concrete was increased by (11.37 % to 24.17 %) when adding fibers as shown in table 7.

Table 7. Flexural strength test results.

Mix	R1	R2	M1F-0.5%	M1F-1%	M2F-0.5%	M2F-1%	B1F-0.5%	B1F-1%	B2F-0.5%	B2F-1%
Concrete compressive strength	2.11	1.82	2.4	2.62	2.15	2.25	2.35	2.61	2.08	2.19
Difference compared with R1 and R2	0	0	+13.75	+24.17	+18.13	+23.63	+11.37	+23.7	+14.29	+20.33

6.5. Permeability, porosity and density

The permeability of PC is increased as the size of gravel increased. This is due to the increase in volume of voids presented in *aggregate*. The increase of PC density is a logical result of the increase of porosity as the aggregate size increased. Density, Porosity and permeability is slightly effected by fiber type an volume fraction. The results of density, porosity and permeability are presented in table 8.

Table 8. Permeability, porosity and density

Mix	R1	R2	M1F-0.5%	M1F-1%	M2F-0.5%	M2F-1%	B1F-0.5%	B1F-1%	B2F-0.5%	B2F-1%
Density kg/m3	2125	2100	2128	2130	2102	2104	2127	2129	2102	2103
Porosity %	19.51	20.84	19.33	19.18	20.53	20.09	19.58	19.81	21.12	21.18
Permeability (cm/s)	0.907	1.094	0.905	0.904	1.093	1.09	0.906	0.904	1.092	1.091

7. Conclusion

- According to the permeability of Pervious concrete, it presents an efficient solution for most of runoff cases.
- PC permeability increases as the size of aggregate increases.
- The use of (superplasticizers) in pervious concrete results in mix segregation.
- The mixture proportion for pervious should be carefully selected to insure uniform distribution of matrix components.
- Plastic fiber has slight effect on PC compressive strength.
- The addition of recycled plastic fibers to PC increase its tensile strength by (6.36% to 27.67%).
- The addition of recycled plastic fibers to PC increase its flexural strength by (11.37% to 24.17%).

References

[1] Rui Z, Man X, Roberto V N and Kay W 2016 Influence of pore tortuosity on hydraulic conductivity of pervious concrete: Characterization and modeling Cons. and Build. Materials, Vol 125, p 1158-1168.

[2] Anush K C and Krishna P B 2016 Pervious concrete as a sustainable pavement material Research findings and future prospects: A state-of-the-art review Cons. and Build. Materials, Vol 111, p 262-274.

[3] ACI. 522R–10. Report on Pervious Concrete 2010 ACI Committee.

[4] Bethany Eisenberg; Kelly C L and David R. S Permeable Pavements, 2015 ASCE.

[5] Rui Z, Zhen L and Chi-sun P 2018 Research and application of pervious concrete as a sustainable pavement material: A state-of-the-art and state-of-the-practice review Cons. and Build. Materials Vol 183, p 544-553.

[6] Saeid H, Saeed A and Mahdi N 2014 Effects of rice husk ash and fiber on mechanical properties of pervious concrete pavement Cons. and Build. Materials Vol 53 p 680-691.

[7] Elnaz K, Mohd R S, Jahangir M, Mohd W H and Mahdi R 2016 Properties of sustainable lightweight pervious concrete containing oil palm kernel shell as coarse aggregate Cons. and Build. Materials Vol 126, p 1054-1065.

[8] Dang H N, Mohamed B, Nassim S, Lydia L and Fabienne B 2013 Valorization of seashell by-products in pervious concrete pavers Cons. and Build. Materials Vol 49 p 151-160.

[9] Benjamin R, Kingsten B and Narayanan N 2014 Fracture behavior of pervious concretes: The effects of pore structure and fibers Eng. Frac. Mechanics Vol 118 p 1-16.

[10] Kevern J T, Biddle D and Cao Q 2015 Effects of Macrosynthetic Fibers on Pervious Concrete Properties Journal of Materials in Civil Engineering, Vol 27.

[11] Milena R, Somayeh N, Liv H and Karl E 2016 Using carbon fiber composites for reinforcing pervious concrete Cons. and Build. Materials Vol 126 p 875-885.

[12] Ćosić K, Korat L, Ducman V and Netingera I 2015 Influence of aggregate type and size on properties of pervious concrete Cons. and Build. Materials Vol 78, p 69-76.

[13] Hussein A I and Hashim AR 2016 Effect of palm oil clinker incorporation on properties of pervious concrete Cons. and Build. Materials, Vol 115 p 70-77.

[14] Milena R, Somayeh N, Liv H and Karl E 2016 Using carbon fiber composites for reinforcing pervious concrete Cons. and Build. Materials, Vol 126 p 875-885.

[15] Lutfur A, Mijia Y and Xinnan W Effectiveness of chemical treatment on polypropylene fibers as reinforcement in pervious concrete Cons. and Build. Materials Vol 163 p 32-39.

[16] ASTM C150 (2015), Portland Cement, Annual Book of ASTM Vol 04.02.

[17] ASTM C192 (2015), Standard Practice for Making and Curing Concrete Test Specimens in the Laboratory Annual Book of ASTM Vol 04.02.

[18] BS 1881, Part 116 (2004) Method for Determination of Compressive Strength Of Concrete Cubes, British Standards Institution, BS 1881, Part 116.

[19] ASTM C496 (2015), Standard Test Method for Splitting Tensile Strength of Cylindrical Concrete Specimens, Annual Book of ASTM Vol 04.02.

[20] ASTM C 78 (2015), Standard Test Method for Flexural Strength of Concrete (Using Simple Beam with Third-Point Loading, Annual Book of ASTM Vol 04.02.

[21] ASTM C 1754 (2015), Standard Test Method for Density and Void Content of Hardened Pervious Concrete, Annual Book of ASTM Vol 04.02.

[22] Gersson F B S, Isaac G, Raquel S. T and Berenice M. Toralles 2017Comparison between the falling head and the constant head permeability tests to assess the permeability coefficient of sustainable Pervious Concretes Case Studies in Construction Materials, Vol 7 p 317-328.

Influence of using alum sludge as an aggregate on hot mix Asphalt

Sahar A. Ali Dawood[1, *], Hayder Abbas Obaid[2], Tameem Mohammed Hashim[3] and Hadeel Kareem Jasim[4]

[1,3,4] Department of Civil Engineering, Al-Mustaqbal University Collage, Babil, Iraq.
[2] Department of Civil Engineering, Faculty of Engineering, University of Babylon, Iraq.

*Corresponding author: *E-mail:*sahar.ali31@yahoo.com

Abstract: Alum sludge (AS) is one of the final products from water treatment processes. Handling and disposal of alum sludge has become one of the major economic and environmental concerns. An experimental investigation was examined to study the possible utilization of alum sludge as aggregate substitute in asphalt mixtures for pavement construction applications. Five asphalt mixtures containing various contents of the AS, 0% (reference mix), 30%, 50%, 70% and 100%, as a replacement to the fine aggregate were prepared in this study. The results conducted that alum sludge is a suitable material for producing asphalt mixes that contains of AS ranging from 30% to 50% as partial replacement for fine aggregates to achieve almost all the standard requirements for aggregates in asphalt mixtures..

Keywords: Alum sludge, Mechanical Properties, Marshall, Stability, Flow.

1. Introduction

The hot-mix asphalt (HMA) mixtures are complex material consisted of asphalt binder and mineral aggregates. The asphalt binder is used as birdlime in asphalt pavement [1], whilst the aggregate acts as the structural frame work. HMA contains 90–95% by weight and 75% up to 85% by volume of mineral aggregates [2]. The asphalt mixtures contain high percentages of aggregates; thus, use types of aggregates are needed for the construction and maintain roads. Use of waste material as replacement of aggregates in the pavement making lead to prevent the accumulation of these wastes, environmental pollution and as well as reduce primary production costs [3]. The sludge is one of the important water treatment plants in the water industry. In fact, sludge-treated water sludge (alum sludge) is an unavoidable by-product obtained as a result of water treatment, such as coagulation, flocculation, clarification and filtration, where the use of aluminum sulfate as the primary coagulant [4-10]. It became the disposal of sludge alum (AS) resulting from drinking water treatment plants gradually a threat to the environment and requires complex and expensive handling and disposal procedures [11-18]. Generally, alum sludge contains useful chemical compounds such as silicon dioxide (SiO_2) and aluminum oxide (Al_2O_3), which are the main components of cement [19].

In European countries, the sludge management method comprises the several stages such as sludge gathering and storage, pumping to thickening area, thickening, storage of thickened sludge, pumping to dehydration area, dehydration, atomization, and final storage. In contrast, the discharge of sludge directly into the nearby stream. This traditional practice of the performance of sludge directly to a nearby stream became less acceptable because this discharge could violate the permitted current standards [20].

This has been identified globally as the most feasible alternatives to solve the sludge discharge problem including several reuse options such as soil barriers, reduction of nutrients in the soil laden spin-off and run in brick making, in the ceramics industry, the pavement and structural improvement in the soil, cement and adhesives industry, lightweight aggregate.
Sludge used as supplementary cementitious material and fine aggregate substitute in preparation of cement mortar and concrete. Bearing in mind that the calcination conditions i.e., temperature and duration affect the microstructure and the pozzolanic properties of industrial solid wastes and as a result control the mechanical properties of pozzolanic cements [21]. Previous studies were investigated to study the potential of using sludge and sludge ash in the production of building and construction materials, artificial aggregate, cement, and ceramics production. Typical examples of the direct use of sludge ash from wastewater treatment plants are as filler in concrete. Owaid et al. 2014; Owaid et al. 2016 [22, 23] showed that the incorporation of alum sludge as a partial replacement to ordinary Portland cement in concrete is an effective means for improving the fresh state and hardened properties of high performance concrete (HPC).

2. Materials and experimental works

2.1 Materials

2.1.1 Asphalt binder

An asphalt binder of (40-50) penetration grade brought from Al Hilla refinery, Iraq, and it was used in this study. The physical properties and related standards are summarized in Table 1, which meet the requirements of the specifications [24].

Table 1. Physical Properties of Asphalt Binder

Test	No.Standards	Requirements	Results
Penetration	ASTM D5	40-50	49
Flash Point	ASTM D92	≥260	237
Ductility	ASTM 113	≥100	102
Specific Gravity	ASTM D70	-	-
Softening Point	ASTM D36	≥46	47.5

2.1.2 Aggregates

The source of the fine and coarse aggregates used in this study were obtained from Al-Najaf city. The graduation of aggregates is very important in hot mix design to check its suitability for hot mix asphalt. Table 2 shows the physical properties of the aggregates. Gradation of aggregates, is provided in Table 3. The gradation curves of the mixture design are shown in Figure 1.

Table 2. Physical properties of Aggregates.

Test	No. specification	Results
Coarse aggregate		
Bulk specific gravity (g/cm³)	ASTM C 127	2.632
SSDspecific gravity(g/cm³)		2.64
Apparent specific gravity (g/cm³)		2.663
Fine aggregates		
Bulk specific gravity(g/cm³)	ASTM C 128	2.637
SSD specific gravity "(g/cm³)		2.645
Apparent specific gravity(g/cm³)		2.662
Abrasion loss (%)(LOS Angeloes)	ASTM C 131	24.1
Flat and elongated particles (%)	ASTM D 4791	17.1

Table 3.Graduation of Aggregates and Specifications.

Sieve SIZE	% Passing	Specification
19	100	100
12.5	90	90-100
9.5	87	76-90
4.75	60	44-74
2.36	40	25-58
0.3	18	5-21
0.075	5	4-10

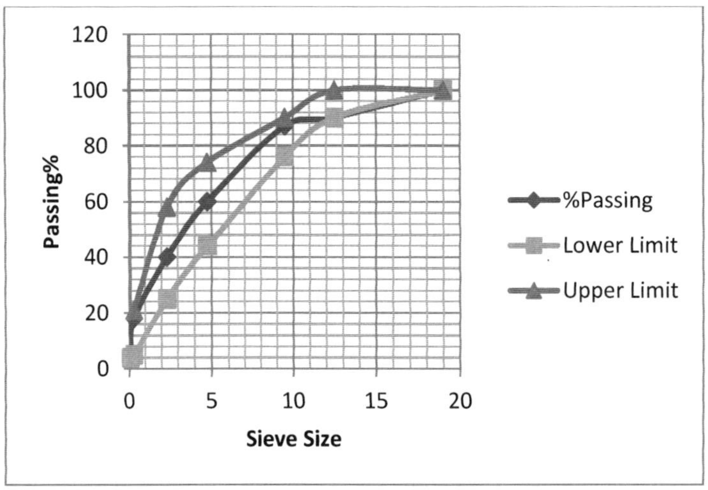

Figure 1. Grading curves of aggregates and specification limits.

2.1.3 Mineral filler

The physical properties and chemical compositions of mineral filler are presented in Tables (4) and (5), respectively.

Table 4. Physical properties of Mineral Filler

Test items		**Results**
Specific gravity		2.706
Sieve Size (mm)	No.30(0.6)	100
	No.50(0.3)	100
	No.200(0.075)	97.5

Table 5. Chemical Compositions of Mineral Filler

Chemical Compositions	**CaO**	**SiO_2**	**Al_2O3**	**Fe_2O_3**	**SO_3**	**MgO**
Results	62.74	22.68	5.06	3.24	2.2	2.36

2.1.4 Waste material (alum sludge)

The waste materials used in this study was alum sludge (AS) that was collected at Al Hilla city water treatment plant, then oven-dried at 105 ℃ for 24 h as shown in Figure 2. The dried AS was crushed and sieved the graduation same of requirements for fine aggregate which that used in this study. The physical characteristics of AS was light brown. X-ray fluorescence (XRF) was used to analyse the chemical composition of alum sludge as seen in Table (6).

Figure 2. Waste materials (Alum sludge) oven-dried at 110 °C

Table 6. Chemical composition of waste material (alum sludge).

Chemical Analysis, (%)	CaO	SiO_2	Al_2O_3	Fe_2O_3	MgO	SO_3	K_2O	Na_2O
Waste material (alum sludge)	0.42	41.38	33.03	3.84	0.31	0.27	1.53	0.14

2.2 Mix Design

The gradation was adopted according to the aggregate conditions, and the details were shown in Figure 1. According to the hot mix asphalt (HMA) mixture gradation design in relevant specifications [24]. Marshall Gradation Design Method was utilized to determine the asphalt-aggregate ratio. For the results, reveal that the optimal asphalt-aggregate ratio was 4.8 %.

2.2.1 Marshall test

Marshall test was conducted to study the mechanical character of prepared asphalt mixtures with a loading rate of 50 mm/min at 60 °C, and the Marshall stability and Flow value were determined according to ASTM D1559 [25].

2.2.2 Indirect Tensile Strength Test

The indirect tensile strength test is used to measure the creep compliance and strength of asphalt mixture .the specimen is loaded until failure-peak load is measured throughout the test, (AASHTO, 2005).

3. Result and Discussion

3.1. Marshall Properties

The results of all Marshall Stability tests are summarized in Table 7. All results shown for each specimen are the average value of three tests. It was found that it can be concluded that there is possible replacing (30-50) % of fine aggregate by alum sludge at hot asphalt concrete mixtures.

Table 7. Marshall Stability test results for Mixtures with different alum sludge content.

Fine aggregate		stability	flow	%Air	Density
sand%	alum sludge%	(kN)	mm	voids	(gm/cm^3)
100%	**0%**	8	4.1	4	2.16
70%	**30%**	8.1	2.6	3.8	2.10
50%	**50%**	8	2.7	4.2	2.13
30%	**70%**	7.5	2.3	4.5	2.08
0%	**100%**	7	2.3	4.5	2.0

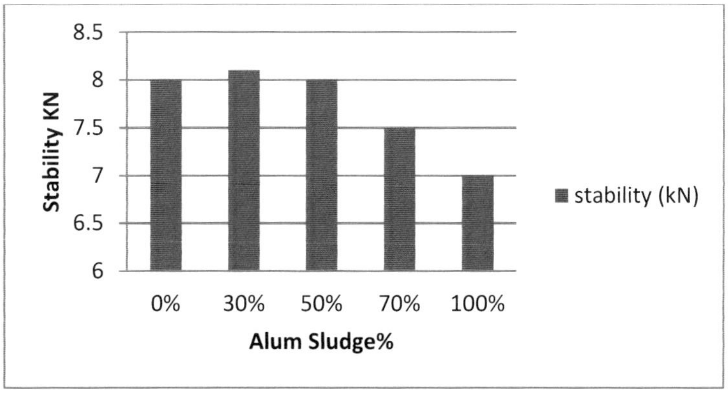

Figure 3. Results of Marshall Stability of different asphalt mixtures.

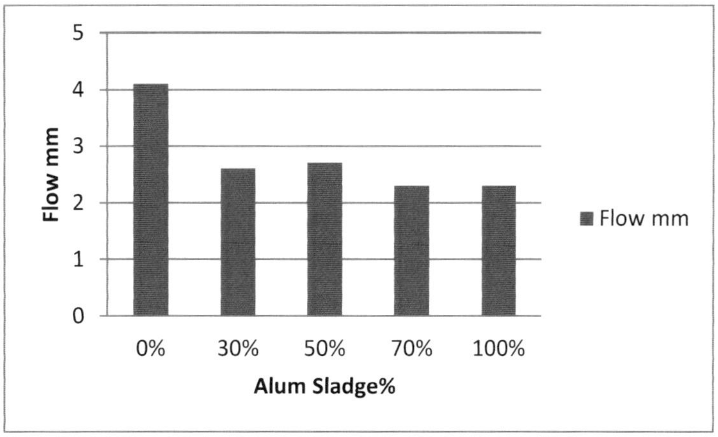

Figure 4. Flow of different asphalt mixtures.

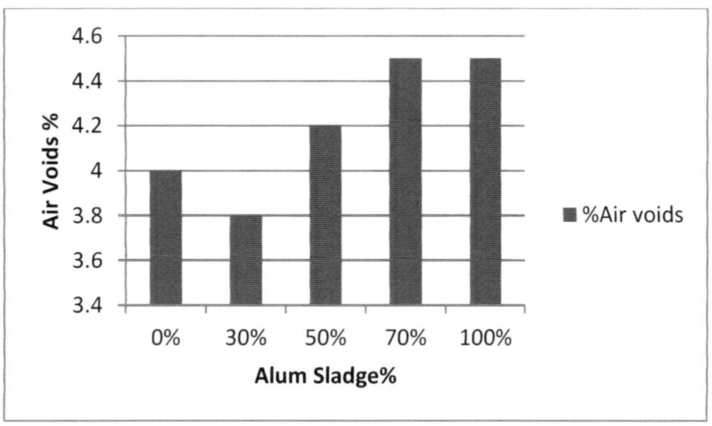

Figure 5. Results of Air Voids of different asphalt mixtures.

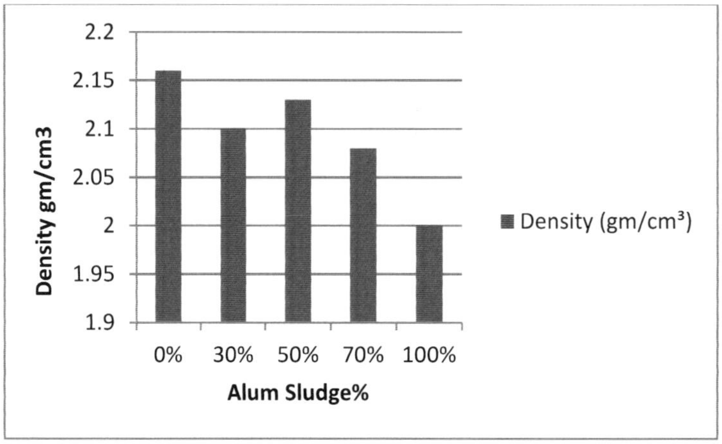

Figure 6. Results of Density of different asphalt mixtures.

3.2. Indirect Tensile Strength Test results

The effects of replacing fine aggregate with AS in asphalt mixtures are presented in Table 8. The results revealed that the indirect tensile strength increased with an increase in AS content for all mixes up to 50%. Beyond this replacement level, the tensile strength shows a decline.

Table 8. The ITST Results of conditioning and un conditioning samples

Mixture	P Max load (KN)	ITST
Alum Sludge (0%)	24.9	2.480
Alum Sludge (30%)	26.9	2.707
Alum Sludge (50%)	28.92	2.865
Alum Sludge (70%)	23.88	2.377
Alum Sludge (100%)	23.6	2.349

4. Conclusions

The mixtures of the alum sludge and conventional aggregates from natural sources has satisfied all the properties required for asphalt mixtures and produced expected mix design properties in accordance with the standard specification requirements at (30-50)% replacing aggregate. For ITST test results, it indicated that replacement of alum sludge by weight of fine aggregate lead to improve the asphalt mixtures using up to 50%, but they decreased with a further increase in alum sludge contents because of the waste material effect on the mixture density.

References

[1] White TD, Haddock JE, Rismantojo E. 2006. Aggregate tests for hot-mix asphalt mixtures used in pavements. NCHRP report 557.

[2] Topal A, Sengoz B. 2005. Determination of fine aggregate angularity in relation with the resistance to rutting of hot-mix asphalt. Constriction& Building Material 19:155–63.

[3] C. Sullivan,M.Tyrer,C.R.Cheeseman and N.J.D.Graham.2010.Disposal of water treatment wastes containing arsenic–areview,Sci.TotalEnviron.4081770–1778.

[4] K. S. Hashim, A. Shaw, R. Al Khaddar, M. Ortoneda Pedrola, and D. Phipps, "Defluoridation of drinking water using a new flow column-electrocoagulation reactor (FCER) - Experimental, statistical, and economic approach," *J Environ Manage,* vol. 197, pp. 80-88, Jul 15 2017.

[5] K. S. Hashim, A. Shaw, R. Al Khaddar, M. O. Pedrola, and D. Phipps, "Iron removal, energy consumption and operating cost of electrocoagulation of drinking water using a new flow column reactor," *Journal of Environmental Management,* vol. 189, pp. 98-108, 2017.

[6] K. S. Hashim, A. Shaw, R. Al Khaddar, M. O. Pedrola, and D. Phipps, "Energy efficient electrocoagulation using a new flow column reactor to remove nitrate from drinking water - Experimental, statistical, and economic approach," *J Environ Manage,* vol. 196, pp. 224-233, Jul 01 2017.

[7] K. S. Hashim, N. H. Al-Saati, A. H. Hussein, and Z. N. Al-Saati, "An Investigation into The Level of Heavy Metals Leaching from Canal-Dreged Sediment: A Case Study Metals Leaching from Dreged Sediment," presented at the 1st International Conference on Materials Engineering & Science, Istanbul Aydın University (IAU), Turkey, 2018.

[8] N. H. Al-Saati et al., "Statistical modelling of turbidity removal applied to non-toxic natural coagulants in water treatment: a case study," Desalination and Water Treatment, vol. 150, pp. 406-412, 2019.

[9] D. Al-Jumeily, K. Hashim, R. Alkaddar, M. Al-Tufaily, and J. Lunn, "Sustainable and Environmental Friendly Ancient Reed Houses (Inspired by the Past to Motivate the Future)," in *2018 11th International Conference on Developments in eSystems Engineering (DeSE)*, 2018, pp. 214-219: IEEE.

[10] A. Shaw, K. S. Hashim, R. Alkhaddar, M. O. Pedrola, and D. Phipps, "Influence of electrodes spacing on internal temperature of electrocoagulation (EC) cells during the removal (Fe II) from drinking water," in *The 3rd BUiD Annual Doctoral Research Conference*, The British University, Dubai, 2017.

[11] K. S. Hashim, A. Shaw, R. Alkhaddar, and M. O. Pedrola, "Controlling of Water Temperature during the Electrocoagulation Process Using an Innovative Flow Columns - Electrocoagulation Reactor," *International Journal of Environmental, Chemical, Ecological, Geological and Geophysical Engineering,* vol. 9, no. 8, pp. 869-872, 2015.

[12] K. S. Hashim, A. Shaw, R. Alkhaddar, M. O. Pedrola, and D. Phipps, "Effect of the supporting electrolyte concentration on energy consumption and defluoridation of drinking water in the electrocoagulation (EC) method," in *The 2nd BUiD Doctoral Research Conference*, The British University in Dubai, 2016.

[13] M. Abdulredha, R. Al Khaddar, D. Jordan, P. Kot, A. Abdulridha, and K. Hashim, "Estimating solid waste generation by hospitality industry during major festivals: A quantification model based on multiple regression," *Waste Manag,* vol. 77, pp. 388-400, Jul 2018.

[14] M. Abdulredha, A. Rafid, D. Jordan, and K. Hashim, "The development of a waste management system in Kerbala during major pilgrimage events: determination of solid waste composition," *Procedia Engineering,* vol. 196, pp. 779-784, 2017.

[15] A. W. Alattabi, C. Harris, R. Alkhaddar, A. Alzeyadi, and K. Hashim, "Treatment of Residential Complexes' Wastewater using Environmentally Friendly Technology," *Procedia Engineering,* vol. 196, pp. 792-799, 2017.

[16] A. W. Alattabi, C. B. Harris, R. M. Alkhaddar, K. S. Hashim, M. Ortoneda-Pedrola, and D. Phipps, "Improving sludge settleability by introducing an innovative, two-stage settling sequencing batch reactor," *Journal of Water Process Engineering,* vol. 20, pp. 207-216, 2017.

[17] I. A. Idowu *et al.*, "An analyses of the status of landfill classification systems in developing countries: Sub Saharan Africa landfill experiences," *Waste Management,* vol. 87, pp. 761-771, 2019.

[18] K. S. Hashim *et al.*, "Removal of phosphate from River water using a new baffle plates electrochemical reactor," *MethodsX,* vol. 5, pp. 1413-1418, 2018.

[19] A. B. M. A. Kaish, a, Khalid Mohammed Breesem and Manal Mohsen Abood,. 2018.Influence of Pre-treated Alum Sludge on Properties of High-Strength Self-compacting Concrete, journal of Cleaner Production.

[20] M. Arabani and A.R. Azarhoosh .2012. The effect of recycled concrete aggregate and steel slag on the dynamic properties of asphalt mixtures, *Construction and Building Materials*, 35 (1–7).

[21] A.Tantawy.2015.Characterization and pozzolanic properties of calcined alum sludge, *Materials Research Bulletin*, Volume 61, January, Pages 415-421.

[22] H.M Owaid, R Hamid, MR Taha. 2014. Strength-ultrasonic pulse velocity relationship of thermally activated alum sludge multiple blended high performance concretes. Key engineering materials 594, 521-526

[23] H.M Owaid, R Hamid, MR Taha. 2016. Variation of ultrasonic pulse velocity of multiple-blended binders concretes incorporating thermally activated alum sludge ash. KSCE Journal of Civil Engineering 21 (4), 1235-1246.

[24] P.R. China, Ministry of Communications.2005. Technical Specification for Construction of Highway Asphalt Pavements, China Communications Press. ASTM. ASTM D1559–89, Standard Test Method for Resistance to Plastic Flow of Bituminous Mixtures using Marshall Apparatus, Annual Book of ASTM Standards USA, 1995.

[25] ASTM. ASTM D1559–89, Standard Test Method for Resistance to Plastic Flow of Bituminous Mixtures using Marshall Apparatus, Annual Book of ASTM Standards USA, 1995.

Enhancement of Rigid Pavement Capacity Using Synthetic Discrete Fibers

Ahmed Alsabbagh[1], Salam Wtaife[2], Alaa Shaban[3], Nakin Suksawang[4] and Emad Alshammari[5]

[1] Lecturer, Civil Engineering Department/College of Engineering, University of Babylon, Iraq.
[2] Lecturer, University of Misan, Iraq.
[3] Lecturer, Civil Engineering Department/ College of Engineering, University of Kerbala, Iraq.
[4] Associate Professor, Mechanical and Civil Engineering Department, Florida Institute of Technology, United States.
[5] Lecturer, University of Hail, Kingdom of Saudi Arabia.
*corresponding author: aalsabbagh2015@my.fit.edu

Abstract. Adding synthetic discrete fibers to unreinforced rigid pavements has been significantly increased during the last decades. Synthetic fibers improve flexural capacity, toughness, fatigue resistance, and durability, and reduce crack width. Most design methods of rigid pavements adopted modulus of rupture (i.e., flexural strength) as the primary parameter in the input design process. The elastic modulus of rupture, which has calculated depending on ASTM C78 formula does not reflect the benefits of the synthetic fiber. The effective modulus of rupture (MOR') has been proposed to quantify the added capacity of fibers over unreinforced concrete, which is calculated depending on the equivalent flexural strength. This study aimed to determine the flexural capacity of rigid pavement, which included filtrated polypropylene discrete fibers. The concept of the effective modulus of rupture was adopted to calculate the capacity of rigid pavements. Adding Polypropylene in concrete reduced the compressive strength, modulus of elasticity, elastic modulus of rupture, while improves effective modulus of rupture and flexural toughness.

1. Introduction

Although the rigid pavements are more expensive than flexible pavements, it has longer service duration and lowers required maintenance process [1, 2]. The design purpose of rigid pavements is to provide enough reinforcement to control cracks, which are generated from applied traffic load and environment variation [3]. Therefore, over several decades welded steel meshes have been utilized as a traditional option for the reinforcement of slabs on the ground or rigid pavements. Last three decades, the utilize of discrete fibers in concrete pavements has increased. Adding any discrete fibers to concrete does not prevent cracks appearance but delay the amount and intensity of cracks [4]. Fiber-reinforced concrete (FRC) has been widely utilized to improve the performance of unreinforced concrete and uses in new infrastructure, rehabilitation, repair, and retrofits. FRC is a composite mixture composed of discrete fibers and concrete matrix, which composed of cement, sand, aggregates, and additives. Discrete fibers can significantly improve flexural performance, impact resistance, fatigue properties, shrinkage resistance, and ductility, and decrease width and intensity of cracks [5, 6, 7, 8, 9, 10, 11].

The primary role of discrete fibers is bridging forces across initiated cracks due to distributing within a concrete matrix, which leads to make these cracks closer together [12, 13, 14, 15]. According to several experimental works of rigid pavements or slabs on ground, discrete fibers improve slab capacity, toughness, and reflective cracking resistance, and reduce the cracks width and the required slab thickness [12, 16, 17]. Two main factors control of FRC performance, type and amount content of discrete fiber. Each fiber kind with specific content has a specific behavior within a concrete matrix that is different from other concrete matrix or fiber kind. Fiber amount contributes to determining the fiber role within the concrete matrix that is ranged from few amounts (0.01%) to high amount (3%) [18, 19].
Steel fibers have a successful history in paving applications, but over the last two decades synthetic fibers have become predominant due to their lightweight, low cost, and resistance to rust damage [12, 20]. Synthetic fibers are human-made fibers resulting from petrochemical processes [21]. There are many synthetic fibers types have been used in concrete applications such as acrylic, aramid, carbon, nylon, polyester, polyethylene, polyvinyl alcohol (PVA) and polypropylene. Synthetic fibers have high elongation properties, low density, and low modulus of elasticity [22]. Polypropylene (PP) is the most common type used for concrete pavements applications [23, 24]. Polypropylene is available in two kinds, monofilament polypropylene (FPP) and fibrillated polypropylene (MPP) [25]. FPP is microfiber (length <25 mm) and is effective within the concrete matrix mechanically (not chemically) by a cohesive mechanism [26]. FPP fibers improve the properties of the plastic concrete stage by increasing the homogeneity of the mixture. Also, they decrease the bleeding rate and plastic settlement [25]. FPP fibers have adverse or little effect on mechanical properties of concrete such as compressive strength, modulus of elasticity, and flexural strength (modulus of rupture). On the other hand, flexural toughness has been significantly increased by adding FPP fibers.

2. Rigid Pavement Capacity

Many testing approaches have been proposed to determine several properties of FRC including compression, tension, shear, and bending tests. All design methods of rigid pavement have been adopted used the flexural strength (i.e., modulus of rupture) to determine the flexural capacity of pavement, which calculated depending on ASTM C78 formula. Moreover, adding synthetic fibers to concrete does not improve the mechanical properties like a modulus of rupture (MOR) especially for synthetic microfibers [27, 28, 29, 30]. Hence, using the elastic formula of ASTM C78, 2010 [31] does not reflect the benefits of the fiber [32, 33, 34]. Therefore, Moens and Nemegeer [50] first recommended the post-cracking concept (toughness) to take into account the benefits of the discrete fiber in the concrete slab on the ground. Altoubat et al. [33] proposed using the equivalent flexural strength ratio (Re,3) to quantify the added flexural capacity of discrete fibers over unreinforced concrete. Re,3 is defined as the ratio between the equivalent flexural strength (fe,3) value and the elastic modulus of rupture (MOR). They also used an effective modulus of rupture (MOR') to determine the slab capacity, which presents a percentage increase to the (MOR). The fe,3, Re,3, MOR, and MOR' are calculated as follows:

$$MOR = \frac{PL}{bd^2} \tag{1}$$

$$fe,3 = \frac{TL}{3bd^2} \tag{2}$$

$$Re,3 = \frac{fe,3}{MOR} \tag{3}$$

$$MOR' = \frac{fe,3}{MOR} \tag{4}$$

*Where **MOR** is the modulus of rupture (MPa), **fe,3** is the equivalent flexural strength (MPa), **Re,3** is the equivalent flexural strength ratio (%), **MOR'** is an effective modulus of rupture (MPa), **P** is the maximum flexural load recorded (N), **T** is the are under load-deflection curve up to an ultimate point (Joule), **L** is the beam span (mm), **b** is the beam width (mm), and **d** is the beam depth (mm).*

3. Aim of the Study

This research aimed to determine the flexural capacity of rigid pavement, which included filtrated polypropylene discrete fibers. The concept of the effective modulus of rupture was adopted to calculate the capacity of rigid pavements.

4. Experimental Work

In this study, Type I/II ordinary Portland cement of ASTM C150 [36] with a specific gravity of 3.15 was used. Natural fine sand with a specific gravity of 2.63 was used. A crushed limestone with a maximum size of 19 mm was used as coarse aggregate for all mixtures. Both coarse and fine aggregates conformed to the ASTM C33 [37] specification. All FRC mixtures had the same water-cement ratio (w/c) of 0.45. Fibrillated polypropylene (FPP) discrete fibers were used in this study and added with different contents. The length and diameter of FPP were 38 and 0.9 mm, respectively. FPP had 1138 MPa in tensile strength and 7.8 in specific gravity. The mixing procedure involved mixing the coarse and fine aggregates in the mixer for 1 minute. Portland cement was added then to the mixer, and all the materials were dry-mixed for more 1 minute. FPP was spread into the mixture, and the materials were mixed again for 2 minutes. Finally, superplasticizer was poured into the total amount of water outside the mixer, and the solution was added gradually and mixed for 2 minutes. For compressive strength specimens, the concrete mixture was placed in the plastic cylinder molds in two layers and vibrated for 30 seconds. For flexural testing, the mix was cast using in steel prism mold in two later and vibrated for 1 minute. All cylinders and prisms samples were stored in a water tank saturated for 27 days. All compressive and flexural tests were tested at 28 days. Table 1 presents the mix proportions.

Table 1. Summary of Mix Proportions (% by volume)

Case	Cement	Aggregates	Sand	Water	Polypropylene
Plain1	12.8	40.1	28.0	18.1	0
0.5PFRC	12.7	39.9	27.9	18.0	0.5
0.8PFRC	12.6	39.7	27.9	17.9	0.8
Plain2	29.2	0	33.0	36.8	0
1.5PFRCC	28.7	0	32.5	36.3	1.5
2PFRCC	28.6	0	32.3	36.1	2.0

5. Results and Discussion

5.1. Compressive Strength

Figure 1 shows the compressive strength results for plain, PFRC, and PFRCC. The plain mixture showed the highest compressive strength compared with all FRC mixtures. PP fibers had a little adverse effect on the compressive strength especially for concrete mixtures without aggregates (PFRCC). Figure 2 shows that PP reduced modulus of elasticity with increase dosage amount. The results of compressive parameters agree with previous studies [17, 19, 38]. On the other hand, PP fibers had a significant effect on the failure mode of the compressive failure from brittle to ductile mode.

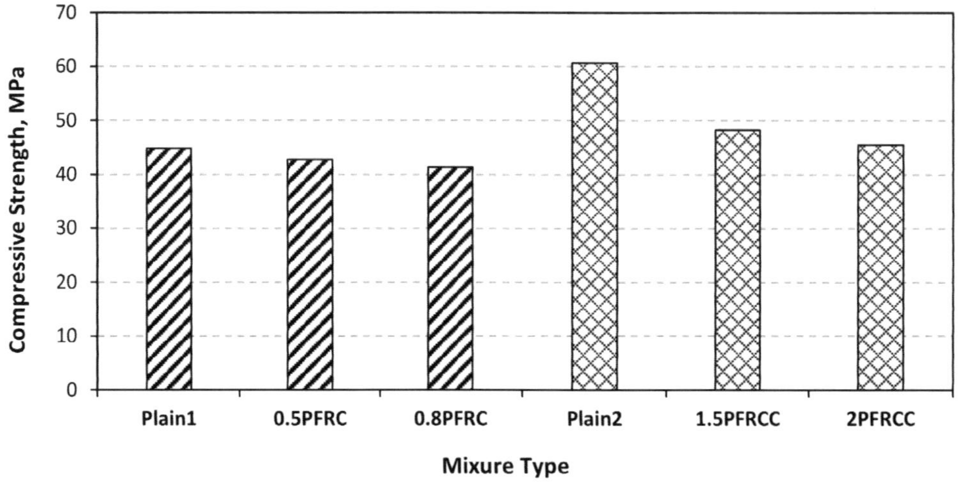

Figure 1: Results of Compressive Strength.

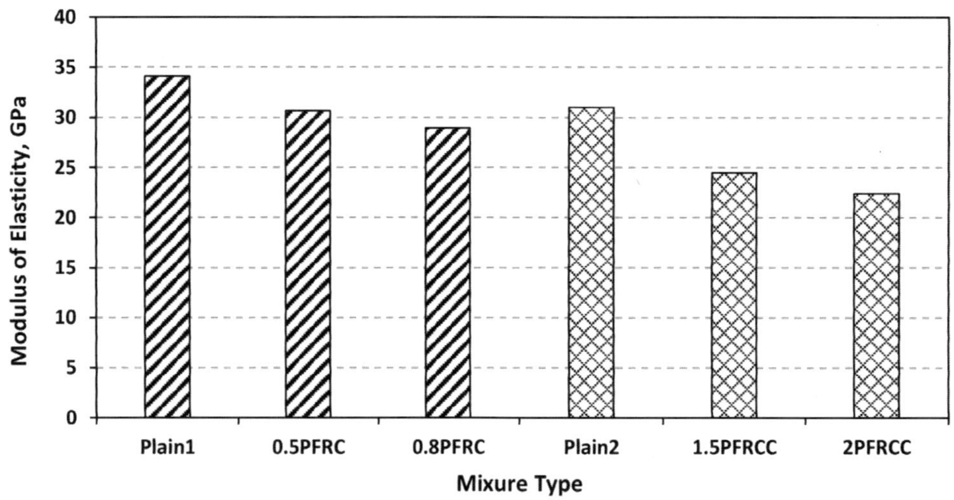

Figure 2. Results of Modulus of Elasticity.

5.2. Flexural Strength

Load-deflection curve, which calculated based on the average value from two LVDTs for all the specimens are shown in Figure 3. These load-deflection curves were utilized to calculate all flexural parameters such as modulus of rupture (MOR), residual strength, (R150), effective modulus of rupture (MOR'), and flexural toughness. From Figure 3, polypropylene fiber improves the flexural deflection at failure significantly by more than 25 times of plain mix. All FRC exhibits drop loading which immediately occurs when a plastic hinge was formed. The PFRC mix showed a highly drop in load after peak point by more than 90%, while the PFRCC mix dropped by 50%. The peak load for PFRC mixtures was higher than PFRCC mixtures due to the effect of aggregates in mixtures.

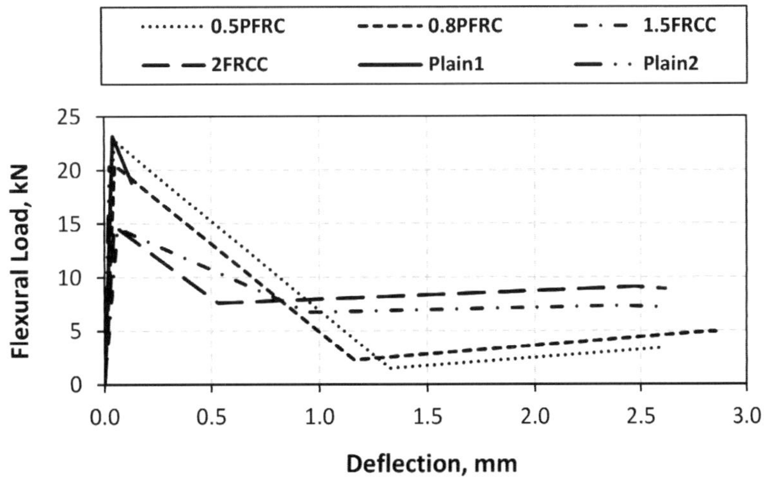

Figure 3. Load-Deflection Curves of Plain, PFRC and PFRCC.

The results of the tests for flexural capacity (elastic modulus of rupture and effective modulus of rupture) are given in Figure 4. It was observed that elastic modulus of rupture (MOR) decreases with increasing PP content, while effective modulus of rupture increases with the same PP content. Therefore, using the residual strength concept to determine the flexural capacity of FRC reflect the fibers benefits. Figure 5 shows the flexural toughness of plain, PFRC, and PFRCC mixture. The incorporation of PP fibers improved the flexural toughness by 20 times of plain concrete.

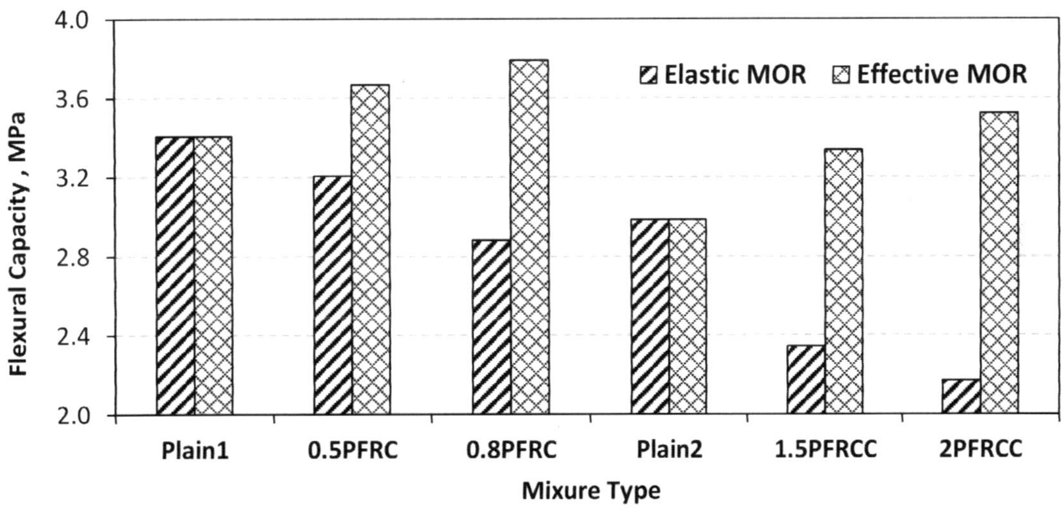

Figure 4. Effect of Fibrillated PP Content on Flexural Capacity of Concrete.

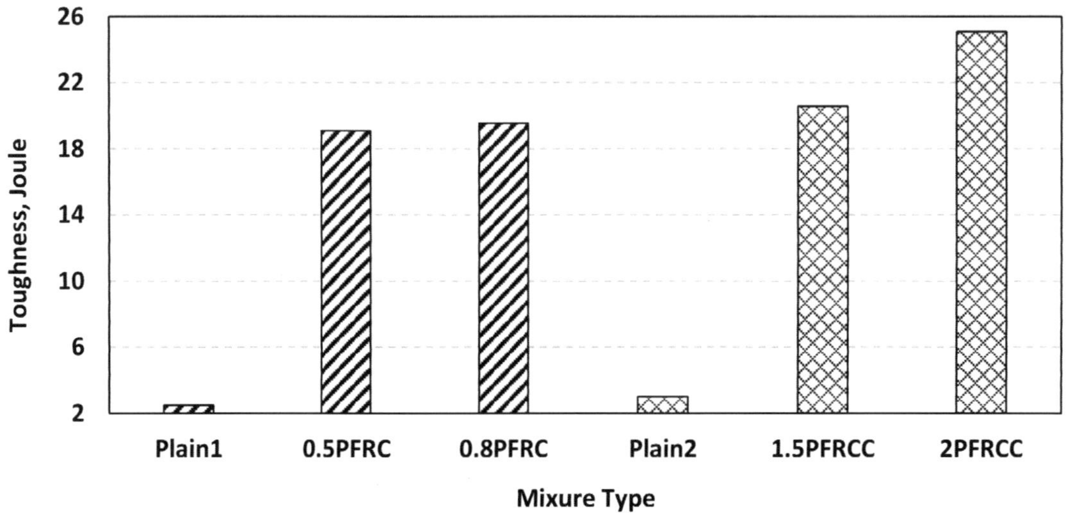

Figure 5: Effect of Fibrillated PP Content on Flexural Toughness of Concrete

6. Conclusion

Based on the results of the experimental work, the following conclusions can be drawn:

1. The value of the compressive strength and modulus of elasticity decreased by adding polypropylene fibers in concrete, especially for the concrete mix without aggregates.
2. The mode of compressive failure was changed from a brittle to a ductile failure by adding polypropylene in concrete.
3. The addition of polypropylene fibers had decreased the modulus of rupture, which is calculated based on ASTM formula. The obtained results indicated that the use of 0.5, 0.8, 1.5, and 2.0% PFRC led in 6, 18, 27, and 37% reduce in MOR, respectively, especially for a concrete mix without aggregates.
4. Results of flexural toughness showed that PFRC samples acquire a high flexural toughness compared to plain concrete samples.
5. Effective modulus of rupture had significantly improved by adding polypropylene fibers in concrete. Adding 0.5, 0.8, 1.5, and 2.0% polypropylene fibers in concrete increased the MOReff. by more than 8, 11, 12, 18%, respectively.

References

[1] Mohr, P., Hansen, W., Jensen, E., & Pane, I. (2000). Transport properties of concrete pavements with excellent long-term in-service performance. Cement and Concrete Research, 30(12), 1903-1910.

[2] Bognacki, C. J., Marsano, J., & Baumann, W. (2000). Spending concrete dollars effectively. Concrete International, 22(9), 50-56.

[3] Deacon, C. (1991). Welded steel fabric in industrial ground floor construction. Concrete, 25(7), 41-44.

[4] Wtaife, S., Shaban, A., Alsabbagh, A., Alshammari, E., & Suksawang, N. (2018). Effect of Hybrid Discrete Fibers on Mechanical Properties of FRC. IOP Conference Series: Materials Science and Engineering, 454 (2018), 012170.

[5] Ramakrishnan, V., Wu, G. Y., & Hosalli, G. (1989). Flexural fatigue strength, endurance limit and impact strength of fiber reinforced concretes. Transportation Research Record, 1226, 17-24. [5]

[6] Bentur, A., & Mindess, S. (1990). Fiber Reinforced Cementitious Composites, II Elsevier Applied Science. London and New York. [6]
[7] Wang, Y., Li, V. C., & Backer, S. (1990). Experimental Determination of Tensile Behavior of Fiber. ACI Materials Journal, 87(5).
[8] Balaguru, P. N., & Shah, S. P. (1992). Fiber-reinforced cement composites.
[9] Shah, S. P., Daniel, J. I., Ahmad, S. H., Arockiasamy, M., Balaguru, P. N., Ball, C. G Ball Jr, H.P., Banthia, N., Batson, G.B., Bentur, A. and Casamatta, D. (1993). Guide for specifying, proportioning, mixing, placing, and finishing steel fiber reinforced concrete. ACI Materials Journal, 90(1), 94-101.
[10] Jun, Z., & Stang, H. (1998). Fatigue performance in flexure of fiber reinforced concrete. ACI Materials Journal-American Concrete Institute, 95(1), 58-67. [10]
[11] Mindess, S., Young, J. F., & Darwin, D. (2003). Concrete, 2003. Prentice Hall, Upper Saddle River, NJ.
[12] Harrington, D. S., & Riley, R. C. (2012). Guide to concrete overlays of asphalt parking lots.
[13] Kim, M. O., & Bordelon, A. C. (2017). Age-dependent properties of fiber-reinforced concrete for thin concrete overlays. Construction and Building Materials, 137, 288-299.
[14] Kelly, J. (1990). Fibres for floors. Concrete, 24(4), 19-20.
[15] Banthia, N., Yan, C., & Mindess, S. (1996). Restrained shrinkage cracking in fiber reinforced concrete: a novel test technique. Cement and Concrete Research, 26(1), 9-14. 1[5]
[16] Beckett, D., Van De Woestyne, T., & Callens, S. (1999). Corner & edge loading: on concrete industrial ground floors reinforced with steel fibres. Concrete, 33(3), 22-24. [16]
[17] Cervantes, V., & Roesler, J. (2009). Performance of concrete pavements with optimized slab geometry. Urbana, 51, 61801.
[18] Roesler, J. R., Bordelon, A. C., Ioannides, A., Beyer, M., & Wang, D. (2008). Design and concrete material requirements for ultra-thin whitetopping. Illinois Center for Transportation.
[19] Suksawang, N., Wtaife, S., & Alsabbagh, A. (2018). Evaluation of elastic modulus of fiber-reinforced concrete. ACI Materials Journal 115(2).
[20] Labib, W., & Eden, N. (2006). An investigation into the use of fibres in concrete industrial ground-floor slabs. Liverpool John Moores University, Liverpool.
[21] ACI Committee 544. (2001). Report on Fiber Reinforced Concrete. American Concrete Institute.
[22] Manolis, G. D., Gareis, P. J., Tsonos, A. D., & Neal, J. A. (1997). Dynamic properties of polypropylene fiber-reinforced concrete slabs. Cement and Concrete Composites, 19(4), 341-349.
[23] Rajan, S., & Olek, J. (2002). Concrete Overlay as a Rehabilitation Option for Distressed Asphalt Pavements. Joint Transportation Research Program, 37.
[24] Winkelman, T. J. (2005b). The Illinois White-topping Experience: A Practical Approach. In International Conference on Best Practices for Ultrathin and Thin White-toppings Federal Highway Administration American Concrete Pavement Association Colorado Department of Transportation Research Board International Society for Concrete Pavements.
[25] Knapton, J., (2003) Ground bearing concrete slabs, Thomas Telford, London.
[26] Banthia, N., & Gupta, R. (2006). Influence of polypropylene fiber geometry on plastic

shrinkage cracking in concrete. Cement and concrete Research, 36(7), 1263-1267.
[27] Abbas, A. A. (2002). Analysis and design of industrial ground floor slabs using the finite element method (Doctoral dissertation, Imperial College London (University of London).
[28] Hulett, T., & Clarke, J. (2003). Technical report 34 concrete industrial ground floors: A guide to design and construction: Overview of the new report. Concrete, 37(2), 22-29.
[29] Brandt, A. M. (2008). Fibre reinforced cement-based (FRC) composites after over 40 years of development in building and civil engineering. Composite structures, 86(1-3), 3-9.
[30] Bothma, J. (2013). The structural use of synthetic fibres: thickness design of concrete slabs on grade (Doctoral dissertation, Stellenbosch: Stellenbosch University).
[31] ASTM C78. (2010). Standard test method for flexural strength of concrete (using simple beam with third-point loading). American Society for Testing and Materials.
[32] Bache, H. H., & Vinding, I. (1990, October). Fracture mechanics in design of concrete pavements. In Proceedings of the second international workshop on the design and evaluation of concrete pavements, CROW/PIARC, Siguenza, Spain (pp. 139-164).
[33] Altoubat, S. A., Roesler, J. R., Lange, D. A., & Rieder, K. A. (2008). Simplified method for concrete pavement design with discrete structural fibers. Construction and Building Materials, 22(3), 384-393.
[34] Gaedicke, C., & Roesler, J. (2009). Fracture-based method to determine the flexural load capacity of concrete slabs. FAA COE Rep (31).
[35] Moens, J., & Nemegeer, D. (1991). Designing fiber reinforced concrete based on toughness characteristics. Concrete International, 13(11), 38-43.
[36] ASTM C150. (2001): Standard specification for Portland cement. American Society for Testing and Materials.
[37] ASTM C33. (2001): Standard specification for coarse aggregates. American Society for Testing and Materials.
[38] Pliya, P., Beaucour, A. L., & Noumowé, A. (2011). Contribution of cocktail of polypropylene and steel fibres in improving the behaviour of high strength concrete subjected to high temperature. Construction and Building Materials, 25(4), 1926-1934.

Improvement Properties of Self -Healing Concrete by Using Bacteria

Z M Hussein [1*], A H Abedali [1] and A S Ahmead [2]

[1] Highway and Transportation Engineering Department, Collage of Engineering, Mustansiriya University, Iraq.
[2] Food science Department, College of Agricultural Engineering Science, Baghdad University, Iraq.
*Corresponding author E-Mail: zainabali@uomustansiriyah.edu.iq

Abstract. The aim of this investigation, self-healing of cracks in concrete by using named Bacillus Subtilis bacteria isolated from agricultural soils and dry soil. Four concrete mixes include, one plain mix without bacteria, other mixes containing a different concentrations of bacteria 10^3, 10^6 and 10^9 cell/ml respectively were prepared. Hardened properties of concrete water absorption, ultrasonic plus velocity, and compressive strength at 7, 28, 60 and 90 days. Also concrete specimens applied to load 10% of compressive strength at 28 days were studied. The result indicate there is decrease in water absorption of (15-38)%, as well as the decrease in the concrete specimens of loading 10% of compressive strength of (23- 41%) relative to control concrete. Increase in compressive resistance of concretes specimens of (6- 20)%, (6-17)%, (8-22)% and (7-20)% for ages 7, 28, 60 and 90 days respectively, concrete specimens loaded by 10% of compressive strength, increased by (25-34%) and (34-39%) for 60 and 90 days relative to control concrete . It was observed that the UPV test of concrete specimens was increased. Scanning Electron Micrographic gave an increased in density, deposition of calcite carbonates in the voids and closure of the crack by precipitation of calcium carbonate layers was observed.

Key words: Bacteria concrete, self-healing concrete, bio-concrete, Micro-crack.

1. Introduction

Concrete is one of the most commonly used mixture materials, the concrete mixture suffers from many problems that lead to the occurrence of a micro-cracks or in motion cracks propagated quickly to create different types of failure may not be processed. The main defects of the concrete are the appearance of micro-crack resulting from relatively low tensile strength due to high external loads, high tensile stresses due to irregular deflection, restrain of shrinkage, plastic deformation and temperature variations. Thus affecting the durability of concrete. The appearance of micro- cracks has a wider impact than other cracks that make concrete more porosity and permeable to enter liquids and chemicals such as sulphates and chlorides for long periods that lead to the fracture and damage of concrete as well as the oxidation of reinforce steel and so, affect the durability of concrete [1].

However, most approach treatment concrete is conventional methods because have relied on main materials which are epoxy materials, acrylic resins, and polymers. These materials are not compatible to repair concrete for coefficient of thermal expansion and their produce production leads to pollution to the environment [2]. In order to overcome the defects of conventional methods of repair concrete, it can be used self-healing ability as effective material since it involves urea of producing bacteria. The bacteria play an important role to grow inside the concrete micro-structure which assists in mineralization of calcium carbonate, by through hydrolyzing urea existing in the surrounding environment. Consequently, carbon dioxide of urea emits with collecting calcium ions and deposit calcium carbonate in the formation of calcite.

Bio- Concrete or Self-Healing Concrete have a new type of concrete developed by researchers called Bio-Concrete or Self-Healing Concrete, can be used to treat the inanition treatment cracks by adding a special type of bacteria to a concrete mixture that has the ability to produce and precipitate of calcium limestone (calcium carbonate) will have the ability to repair and treatment hair cracks [1].

The first study was conducted on calcium deposition properties in 1995 by researcher Gollaspuid and others in 1995. Researchers used a technique to reduce porosity and soil permeability by injecting bacteria [3]. Also, several studies were also conducted to develop different methods of using bacteria to improve properties concrete and repair cracks Ravindranatha, N. Kannan, Likhit [4], R A. B. Depaa and T. Felix Kala [5], Chithra P, BaiShibi Varghese [6] and Abhishek Thakur, Akshay Phogat and Khushpreet Singh [7]. Locally, a study was conducted to treatment concrete cracks by using bacteria (Bacillus Subtilis), silica powder and sodium chloride. Silica, chloric acid and bacteria were also used for the treatment of cracks in by making cracks in the concrete with a width of 1, 2 mm. Ultrasonic plus velocity and the scanning electron microscopy were examined. The researchers concluded that the crack with the width of the 1mm was better and easier than the width of the 2 mm and using the silica and bacterial treatment material gave the best results [8].

The research idea involves adding different concentrations of the prepared solution bacteria type Bacillus Subtilis (B.S), adding to the concrete mixture, conducting the water absorption tests, compressive strength, ultrasonic plus velocity test comparing of the control specimens. Also, concrete specimens at the age of 28 days was applied load 10% of the compressive strength of the specimens examined for the purpose of healing micro- cracks in the concrete specimens to and this return to curing and tested with ages 60 and 90 days to demonstration the effect of the effectiveness of bacteria in concrete cracks self-healing.

2. Research objectives

The main objectives of present research work are;
1- Studying the effect of various concentration of Bacillus Subtilis bacteria on mechanical properties of concrete with different age.
2- Predicted self-healing of micro-crack of micro structure of concrete.

3. Experimental work

In this research, ordinary Portland cement type (I) was used. The tests result demonstrate that the physical and chemical properties of cement used conform with Iraq standards No.5/ 1984 [9]. Natural sand passing sieve No. 4.75 mm was used. The physical properties for sieve analysis shows that the sand used is combatable with the Iraqi Standard No. 45/1984 [10]. Nominal maximum size of 14 mm of natural crushed coarse aggregate was used. The sulphate content and grading of the coarse aggregate satisfy the requests for Iraqi Standard (No. 45/ 1984) [10]. Condensed Silica fume (SF) is a highly active pozzolanic materials. It is a by-product from the produce from silicon or ferro- silicon metal. Tables 1 and 2 demonstrate the physical and chemical properties of condensed silica fume utilize in this study as partial replacement of cement. The results prove that SF utilized meets the requirements of ASTM C1240 – 06 [11]. The tap water used in mixing and curing of concrete was potable water from the water supply network. Also, High Range Water-

Reducing Admixture (HRWRA) with a commercial Sika-Viscocrete- 5930 was used. It is the third generation of superplasticizer meets the satisfies of ASTM C494M/ 04 types F [12]. The mechanism of HRWRA contributes to flow able concrete with to decrease the of w/c ratio in the mixture. Table 3 specifies the technical explanation of the hydrous solution of superplasticizer utilized during this study.

Table 1. Physical properties of condensed silica fume

Physical properties	Results, SF	ASTM C, 1240-05 Limits
Specific surface area, min, (m^2/kg)	20000	15000
Strength active Index with Portland cement at 7 days, min. percentage of control	122	≥ 105
Percent retained on 45μm (No.325), max, %	9	≥10

Table 2. Chemical analysis of condensed silica fume*

Oxides Composition	Oxide Content (%)	ASTM C1240-5 Limitations
SiO_2	90.51	≥ 85
Al_2O_3	0.60	-
Fe_2O_3	2.32	-
Na_2O	0.15	-
CaO	0.58	-
MgO	0.30	-
TiO_2	0.01	-
K_2O	1.26	-
P_2O_5	0.10	-
SO_3	0.35	≤ 4
L.O.I	3.82	≤ 6

* Test were accomplished at by the National Center for Geological Survey and Mines.

Table 3. Technical explanation Sika -Viscocrete-(5930)*

Main Action	Results
Form	Turbid liquid
Colour	Light brown
Relative density	1.080 kg / litre
PH value	8.0 ± 1.0
Viscosity	128±1.30 cps @ 20 °C
Transport	Not classified as dangerous
Labeling	No hazard label required

* According to the manufacturer (The Chemical Company BASF)

3.1 Prepared of bacteria used

The concrete is characterized by high alkalinity between (10-13), so the bacteria used in this research should be suitable for alkaline concrete. The isolated bacteria used from agricultural soils are Bacillus Subtilis (BS). The bacteria were prepared first by preparing the Nutrient Broth Urea (NBU) by dissolving the following feed in each liter of Nutrient Both 5g / l Urea 20g / l and calcium chloride g / l^3. The bacteria were grown at a temperature of 37 ° C in the incubator for 24 hours in incubator 150 rpm. After that bacterial growth, phenotypic tests were performed to confirm the bacterial form and the identification keys in Holt et al. 200 Berges manual. The bacteria returned colony / ml unit by casting in the dishes and the different concentrations of solution bacteria are 10^3, 10^6 and 10^9 cell / ml as shown in Figure 1 were prepared.

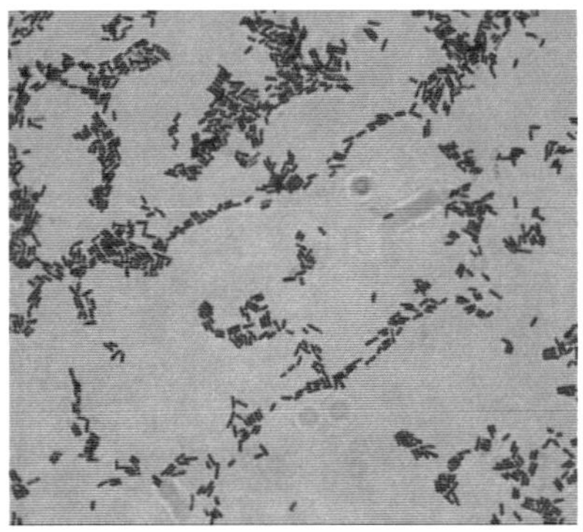

Figure 1. Bacteria Bacillus Subtilis under the micrographic.

3.2 Concrete mixes

Four concrete mixes were intended in this research included, control concrete (reference mix) and three concrete mixes including, the different concentration of solution bacteria (10^3, 10^6 and 10^9) cell / mm.

3.3 Mix proportion

Control Concrete mixture was designed to conformity to British method for concrete mixes design [13] to achieve compressive strength 41 MPa at 28 days, without any admixtures. The mix proportioning were (1: 1.5: 2), (Cement: Sand: Coarse aggregate) by weight, with a cement content of 450 kg/m^3, w/c 0.42 and slump test (100 ± 5 mm). Several trials of concrete mixes were accomplished to choose the optimum dosage of superplasticizer (HRWRA). The w/c ratio was adapted to have the similar workability of the control mix with the slump of (100 ±5 mm). The major function of using superplasticizer HRWRA is to decrease the quantity of mixture water while maintaining a similar workability of the control mixture. The details of the designed control concrete mixture including different dosages of (HRWRA) are presented in Table 4. The empirical results in this examination show that the best dosage of superplasticizer HRWRA is 2 liter per 100 kg of cement, which leads to a water decrease of about 44% and the maximum value of compressive strength of 56.4MPa at 28 days.

Table 4. Details of the trial mixes with various dosages of HRWRA

Mix Proportions By Weight	w/c Ratio	HRWRA (L/100kg of Cement)	Slump (mm)	Water Reduction (%)	Compressive Strength MPa	
					7 Days	28 Days
1: 1.5: 2 Cement: Sand: Gravel With cement content of 450 kg/ m^3	0.42	0	95	0	34.7	45
	0.38	1	100	9.5	36.3	47.8
	0.32	1.5	98	23.8	42.7	53
	0.28	2	100	33.3	45.5	56.3
	0.27	2.5	105	35.7	43.1	54.3

3.4 Concrete mixing producer

The material mixing procedure was carried out by an electrical rotating container mixer of 0.1m^3 volume. All materials mixed by mixture for 5 minute and 70 % of blending water was added to the dry materials then mixed for 1 minute. The superplasticizer was mixed with the 30% remaining mixing water, next it is addition to the mixture and mixed for 2 minutes. Finally, the Bacterial solution added to the mix and mixed 1 minute.

3.5 Preparation, casting and, curing of concrete

Steel molds 100 mm were prepared to clean and oiled before placed of fresh concrete to avoids to adhesion of fresh concrete to the inner surfaces of the molds. Fresh concrete is adding in molds of three layers satisfy the requirements of standard specifications and positioned on the vibration table according to ASTM C-192/C192M [14]. The upper layer of concrete was softened by a steel shovel, and the specimens were covered with a plastic cover for 24 hours to avoid water drying. Subsequent that the specimens were de-molded and entirely immersed curing water until the time of the test.

4. Experimental Tests

4.1 Water Absorption

. Water absorption test was carried out according to ASTM C642 [15] at the age of 28 days.

4.2 Ultrasonic Pulse Velocity Test

Ultrasonic pulse velocity apparatus test (UPV) was accomplished according to ASTM C597 [16]. The rate result for three 100 mm cube samples is recorded.

43 Compressive Strength

The compressive strength test was done on concrete cubes samples of 100 mm according to BS 1881: part 116 [17].

44 Scanning Electronic Micrographic (SEM)

The concrete specimen was served for this examination by taking apart from the fracture surface of the concrete. Apart of concrete specimens were dried in a furnace at 60°C for 7days, to remove any moisture in the specimens, after that, these specimens were covered with a light layer of conductive material to prevent the formalization of electric charge when the electron ray scans the specimen.

5. Results and discussion

51 Water absorption

The effect of adding different concentrations of bacteria to concrete specimens on water absorption is illustrated in Table 5 and Figure 2. The results showed that, water absorption of concrete decreased with increased concentration of bacterial for all ages of concrete specimens of (15-38) %. The highest reduction in water absorption for concrete specimens was the concentration of 10^6 cells / mm with a decrease of water absorption (28%) compared to the control specimen. The reducing water absorption was because of the effectiveness of bacteria in depositing layers of calcium carbonate into the cavities, pores and micro- cracks of concrete specimens that seal pores and micro cracks lead improved permeability. The examination of the concrete samples after loading 10% of the compressive strength, the results showed a clear reduction of water absorption relative to control specimen of (23- 41%) due to the presence of bacteria that contributing to $CaCO_3$ production, which has reduced the percentage of pours [5].

Table 5. Influence the concentration of bacteria on water absorption

Symbol Specimens	water Absorption % @ 28 days	Water absorption % after loading 30% of compressive strength
MR	13	22
MB3	11	17
MB6	8	13
MB9	10	15

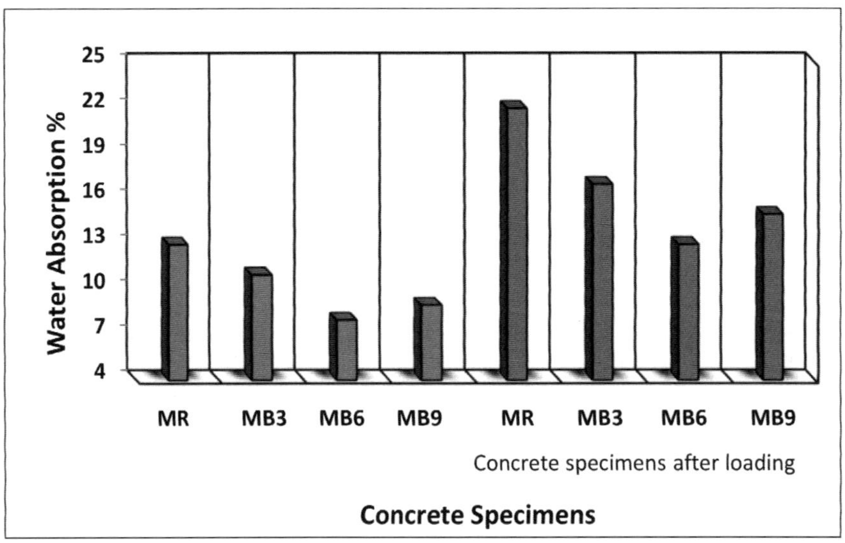

Figure 2. Effect of the different of concentration of bacteria on water absorption

5.2 Compressive strength

Table 6 and Figure 3 show the influence of the concentration of bacteria on the compressive strength of concrete for all ages. Concrete samples containing different concentrations of bacteria increases of compressive strength compared of to the control specimen are (6- 20), (6-17), (8-22) % and (7-20%) for ages 7, 28, 60 and 90 days respectively. The results showed that the highest improvement was for the concentration of bacteria 10^6 cells / mm relative of the control specimen. Also, the results showed that the concentration of bacteria increased to 10^9 cells / mm, which reduced the compressive strength relative to 10^6 cells / mm. The results showed a remarkable improvement of the concrete specimens after loading 10% of compressive strength of ages 60 and 90 are (25-34%) and (34-39%) to relative the control specimens loaded with 10% of compressive strength. The improvement in compressive strength by biochemistry is due to the effectiveness of microorganisms in the deposition of calcium carbonate in the micro-cracks and voids that act to healing of cracks subjected to loading on the concrete specimens.

Table 6. The effect of concentration of bacteria on compressive strength

Symbol specimens	Compressive Strength MPa				Compressive strength after loading 10%of compressive strength at 28 MPa	
	7 days	28 days	60 days	90 days	60 days	90 days
MR	34.7	45.5	48	50.5	33	34.5
MB3	38.2	47.7	51.8	53.9	41.3	46.2
MB6	43.3	52.5	58.3	60.7	44.3	48.0
MB9	41.5	50.0	54.2	57.4	43.5	47.6

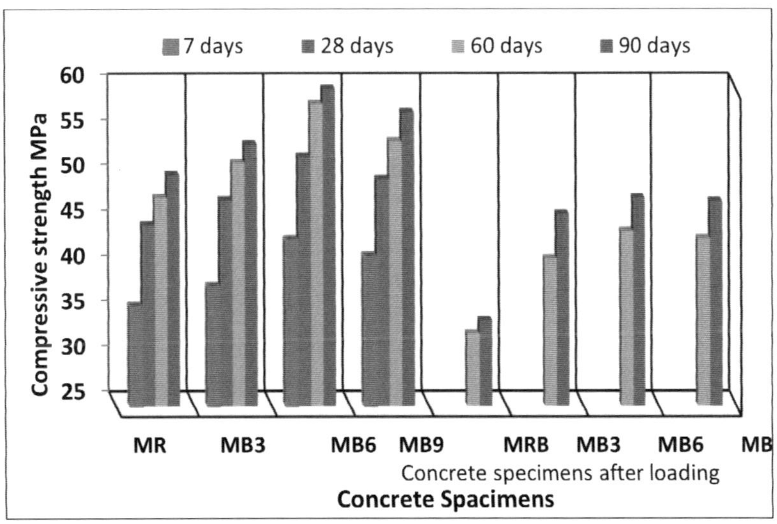

Figure 3. Influence the concentration of bacteria on compressive strength

5.3 Ultrasonic Pulse Velocity (UPV)

The (UPV) test is used to estimation the uniformity and quality of bacterial concrete, showed the existence of voids and cracks, and develop the depth of the cracks. The completion of UPV of concrete samples containing concentration of bacteria are shown in Table 7 and Figure 4. Concrete specimens containing the concentration of bacteria 10^6 has less time of transmission than other concentrations of bacteria. And also, for concrete specimens loaded with 10% of compressive resistance, it a clear improvement in the examination. This is due to the deposition of layers of calcium carbonate, filling the voids and closing the fine cracks inside the concrete.

Table 7. The results of UVP for concrete specimens

Symbol specimen	UVP (km/sec)				UVP (km/sec) after loading 10% of compressive strength at 28 MPa	
	7 days	28 days	60 days	90 days	60 days	90 days
MR	4.12	4.22	4.27	4.33	3. 5	3.12
MB3	4.14	4.32	4.36	4.39	3.60	3.71
MB6	4.23	4.38	4.40	4.51	3.75	3.85
MB9	4.19	4.35	4.38	4.47	3.60	3.78

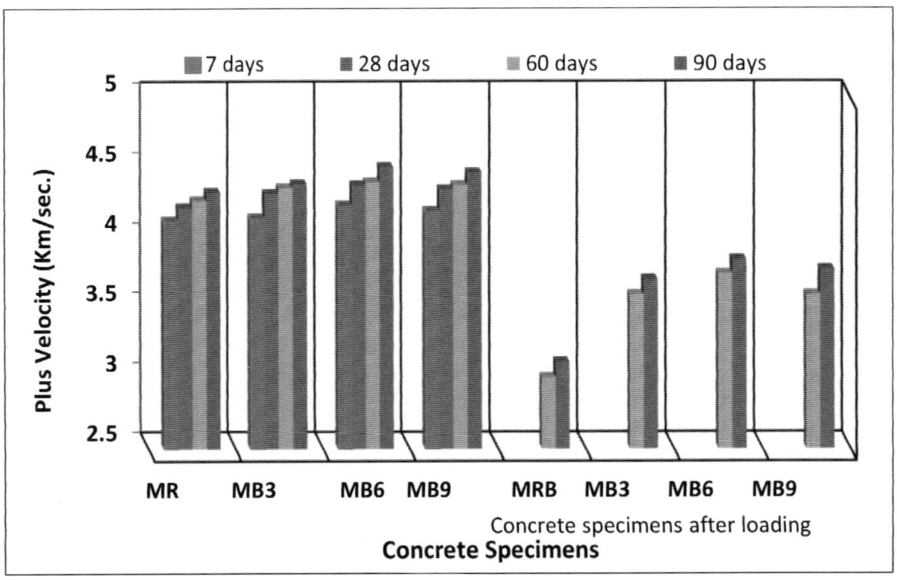

Figure 4. Ultrasonic pulse velocity of concrete specimens

5. 4 Scanning Electron Micrographic (SEM)

Scanning electronic micrograph of control specimen and concrete specimen containing concentration bacteria solution 10^6 cell/mm after loading 10% of compressive strength at age 28 days of curing prepared in this research was achieved to study the microstructure of that concrete specimens.

Figure 5 (a-b-c) displays the image of control concrete without bacteria, where, pores and micro-cracks can be easily seen in the image. Figure 6 (a- b-c)' demonstrates the SEM analysis of specimen concrete with 10^6 cell / mm of concentration of bacteria has appeared different calcite crystals formed in concrete. High calcium amounts in it prove that calcite was present in the form of calcium carbonate due to bacteria. Figure 6.-b shows the presence of crystalline calcium carbonate correlating with bacteria. The deposition of calcite

IOP Conf. Series: Materials Science and Engineering **584** (2019) 012034 doi:10.1088/1757-899X/584/1/012034

used achieving as barrier to deleterious materials and subsequently improves properties concrete and permeability.

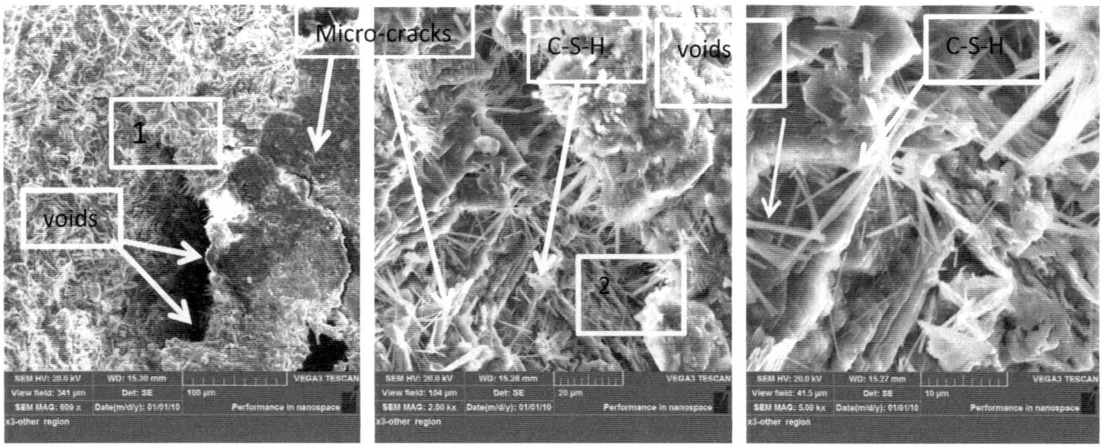

a. SEM Image of control concrete b. SEM Image at zone 1 c. SEM Image cracks of control concrete

Figure 5. Scanning electron microscopy (SEM) images of fracture surface control concrete without bacteria

a. SEM image of concrete b. SEM image at zone1 c. SEM image at zone 2

Figure 6. SEM images of fracture surface of concrete specimen containing concentration 10^6cell/mm of bacteria.

6. Conclusions

The main conclusions that can be extracted from this research are:

1-The results showed a decrease in water absorption of (15-38)%, as well as decrease in the concrete specimens of loading 10% of compressive strength of (23- 41%) relative to control concrete. The highest reduction was for concrete specimens containing a concentration of 10^6 cell/mm.

2 - Increase in compressive resistance of concretes specimens of (6- 20) %, (6-17) %, (8-22)% and (7-20)% for ages 7, 28, 60 and 90 days respectively. concrete specimen loading 10% of compressive strength

increases are (25-34%) and (34-39%) for 60 and 90 days, relative to compressive strength of control specimens that do not containing bacteria.
3 - It was observed that UPV test of concrete specimens increased because of the time of the transmission of the least wave with the presence of calcium carbonate, which fill the voids and cracks- micro.
4- Scanning Electron Micrographic gave increased in density, deposition of calcite carbonates in the voids and closure of the crack by precipitation of calcium carbonate layers was observed.

Acknowledgments

The authors would like to thank Mustansiriyah University to support this work.

Reference

[1] Jonkers H M 2011 Bacteria – based self- healing concrete HERON vol 56(1) p 1-12.
[2] Wasim Kh and Basit E M 2016 Crack healing in concrete using various bio influenced self-healing techniques Construction Building Material vol 102 (1) p349–357.
[3] Gollapudi U U Knutson C I Bang S and Islam M R 1995 A new method for controlling leaching through permeable channels Chemosphere 30 p 695–705.
[4] Ravindranatha Kannan N and Likhit M L 2014 Self-healing material bacterial concrete International Journal of Research in Engineering and Technology vol 3(3) p656-659.
[5] RA B Depaa T and Felix K 2015 Experimental Investigation of Self-Healing Behavior of Concrete using Silica Fume and GGBFS as Mineral Admixtures, Indian Journal of Science and Technology vol 8(36).
[6] Chithra P and BaiShibi V 2016 An experimental investigation on the strength properties of fly ash based bacterial concrete International Journal of Innovative Research in Advanced Engineering vol3(8).
[7] Thakur A Phogat A and Singh K 2016 Bacterial Concrete and Effect of Different Bacteria on the Strength and Water Absorption Characteristics of Concrete: A Review. International Journal of Civil Engineering and Technology vol 7(5) p43-56.
[8] Ibrahim A Redha A and Sana'a M 2014 Use of bacteria to self- healing concrete cracks, Diyala journal for engineering sciences vol 4(4) p27-36.
[9] Iraqi Standard No. 5 1984 Portland Cement the Central Organization for Standardization and Quality Control (in Arabic).
[10] Iraqi Specification, No.45 1984 Aggregate from Natural Sources for Concrete and Construction (in Arabic).
[11] ASTM C1240 2007 Standard Specifications for Silica Fume Used in Cementitious Mixtures Annual Book of ASTM Standards vol 04 p 200-208.
[12] ASTM C494/C494M 2004 Standard and Specification for Chemical admixtures for Concrete Annual Book of ASTM Standards vol 4 (2) p 271–279.
[13] Building Research Establishment 1997 Design of Normal Concrete Mixes Second Edition London.
[14] ASTM C 192/C192M 2015 Standard Practice for Making and Curing Concrete Test Specimens in the Laboratory Annual Book of ASTM Standards American Society for Testing and Materials vol 4(2).
[15] ASTM C642 2013 Standard Test Method for Density Absorption and Voids in Hardened Concrete Annual Book of ASTM Standards vol 4(2).
[16] ASTM C597 2004 Standard Test Methods for Pulse Velocity through Concrete, Annual Book of ASTM Standards vol 4 (2) p 310-313.
[17] BS 1881 Part 116 1989 Method for Determination of Compressive Strength of Concrete Cubes British Standards Institution p3.

IOP Conf. Series: Materials Science and Engineering **584** (2019) 012035 doi:10.1088/1757-899X/584/1/012035

Influence of current density and electrodes spacing on reactive red 120 dye removal from dyed water using electrocoagulation/electroflotation (EC/EF) process

B A Abdulhadi[1,2,*], P Kot1, K S Hashim[1], A Shaw1 and R Al Khaddar[1]

[1] Department of Civil Engineering, Liverpool John Moores University, UK.
[2] Department of Environmental engineering, Babylon University, Iraq
*Corresponding author: B.A.Abdulhadi@2017.ljmu.ac.uk

Abstract. The current study investigates the influence of the current density (CD) and electrodes gap (EG) on the removal of reactive red 120 dye (RR-120) from drinking water using electrocoagulation (EC) process. The influence of CD was studied by treating the dyed water at three different CDs (2, 4 and 6 mA/cm2). While the influence of EG was investigated at three EGs (5, 10, and 15 mm). The Results obtained showed that increasing the CD enhanced the removal of the dye. However, increasing the EG significantly decreased the dye removal efficiency. It was found that dye removal increased from 87% to 98%, as the CD increased from 2 to 6 mA/cm2, respectively. While increasing the EG from 5 to 15 mm decreased the dye removal from 96% to 80%, respectively.

1. Introduction

Reactive Red 120 dye (RR-120) is a component of azo dyes and it is widely used in textile industries [1, 2]. RR-120 is categorised as azo dye as it contains azo bond ($-N = N -$) in its molecular structure [3], figure 1. The textile industry released up to 60% of the entire dyestuff used in the tinting process into the environment, which causes serious problems due to their likely toxicity to both human and aquatic life [4, 5]. For example, discharging dyestuff into the water bodies creates unwanted water odours and colours ; limits sun light penetration; and decreases the concentration of the dissolved oxygen [6]. Furthermore, the final results of degradation of azo dies, under certain conditions, could be carcinogenic oxygen [6]. A wide range of techniques therefore has been applied to remove dyes from water and wastewater, such as chemical oxidation, reverse osmosis and aerobic and anaerobic degradation [6, 7]. Unfortunately, some dyes have complex and stable molecular structures that make their removal by these tradition treatment methods very difficult [8]. Therefore, different traditional and advanced treatment methods have been used recently for removing these dyes, and other pollutants, from wastewater and drinking waters such as filtration, Nano-filtration, bioreactor-ultrafiltration and electrocoagulation [9-11]. The electrocoagulation (EC) has been suggested as an effective alternative process to the coagulation process in water and wastewater treatment over the last century because it tolerates many environmental and economic attractive advantages [12-15]. The EC process, in comparison with other water treatment processes, can be considered a cost effective process as it does not require any chemical additions and easy to operate [15, 16]. Additionally, its efficiency in treating different type of pollutants from drinking waters is well documented [17]. Furthermore, the EC method produces small volumes of sludge [18-21], the latter categorised as solid waste that required complex and expensive handling and treatment procedures, which enhances the cost-effectiveness of the EC method. Finally, the performance of the EC method has been modelled to predict the removal of different pollutants under different operational conditions. For example, Hashim, Shaw [16] found that the relationship between the operating parameters and nitrate removal, using aluminium-based EC reactor, could be modelled with R^2 of 0.848, the latter is a statistical tool to evaluate the statistical relationship between two variables or events [22]. Additionally, Hashim, Khaddar [10] stated that the performance of aluminium-based EC units, in terms of phosphate removal, could be modelled with an R^2 of 0.882. It is noteworthy to highlight that these values of R^2 indicate good agreement between the predicted and measured removals [11, 22, 23].

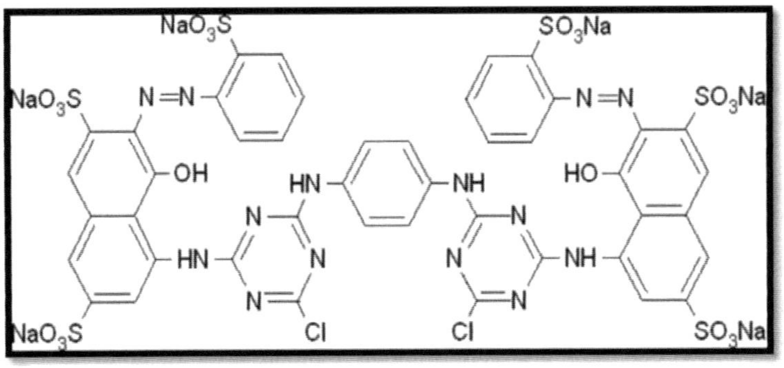

Figure 1. The molecular structure of RR-120 dye.

2. Experimental

2.1. EC reactor

Mollah, Morkovsky [12] demonstrated that the design of the EC cells have not significantly changed during the last few decades. Generally, rectangular cells with plane rectangular electrodes are still the predominant design. Thus, the first stage of the current study was devoted to develop a new EC reactor. This new reactor consists of 1.0 L rectangular Perspex container, figure 2-A. The Perspex has been used as it is an inert and affordable material. The container is supplied with four rectangular perforated aluminium electrodes (total effective area of 280 cm^2). Each electrode has 35 hole each 5mm diameter drilled in lines of 3 and 4 holes to ensure each hole is offset from the one before. This configuration is

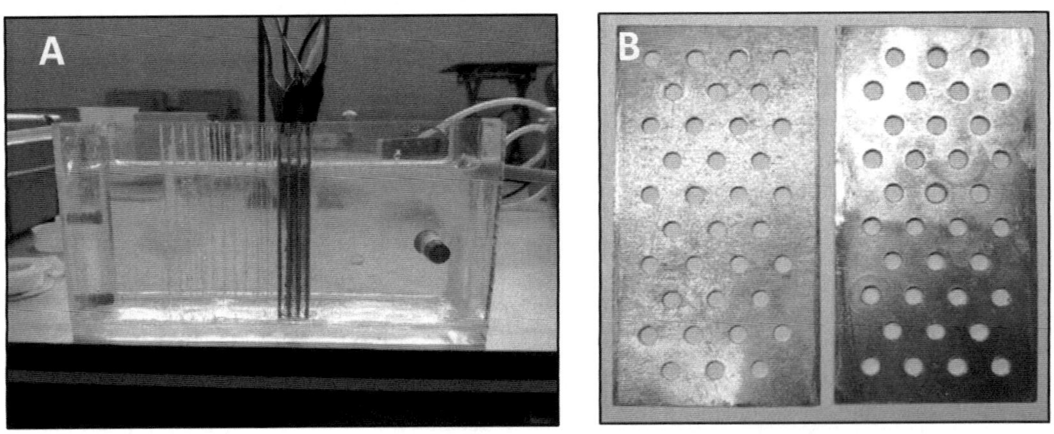

Figure 2. A) The EC reactor, B) Aluminium electrode.

to ensure that the water will flow in a convoluted path in order to increase solution mixing efficiency, figure 2-B. Aluminium has been used in the current study as it is an effective electrode materials [24]. This reactor was connected to both a bench top DC power supply (HQ Power; 0–30 V) and a peristaltic pump (Watson Marlow type, model: 504U) to supply the required electrical current and to circulate flow water through the reactor. The whole unit is shown in Figure 2.

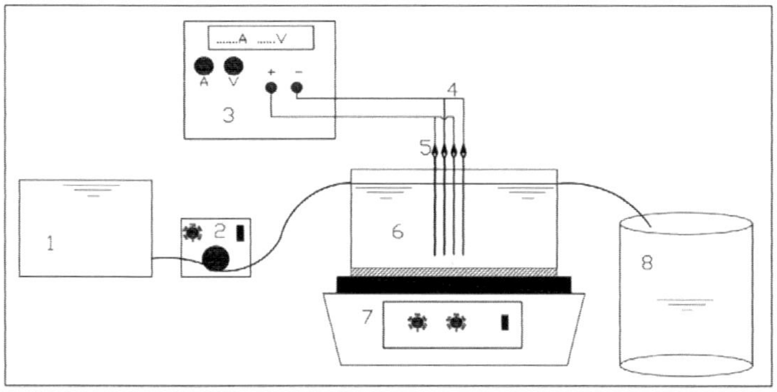

Figure 3. EC setup: 1.influent tank (polluted water), 2. Peristaltic pump, 3. Power supply, 4. Electrical wires, 5. Electrodes, 6. EC reactor, 7. Magnetic stirrer, 8. Effluent tank (treated water)

2.2. Material and methods

According to the literature, the concentration of reactive dyes receiving rivers is normally ranging between 1.2 to 3 mg/L (average value, 25 days each year) or 5 and 10 mg/L (average value, 50 days each year), and could increase up to 300 mg/L in some cases at 1 or 2 days each year (the worst case) [25]. Thus, in the current study, samples with 20 mg/L (as the worst case of the 50 days average value) of RR-120 have been used to investigate the influence of the current density (CD) and electrodes gap (EG) on the decolourization process.

Initially, a stock solution, having 1000 mg/L of RR-120 dye, was prepared by dissolving 1000 mg of RR-120 in one litter of deionised water. Then, samples with less concentrations of RR-120 (20 mg/L) were diluted from this stock solution. The experimental work was initiated by treating the diluted samples at three different CDs (2, 4 and 6 mA/cm^2) for 12 minutes, based on preliminary design experiment, at flow rate 100 ml/min, keeping the EG, initial pH, water conductivity, and initial water temperature constant at 5 mm, 8, 0.4 mS, and 20±1^{0}C, respectively. The initial pH and the initial conductivity were controlled using HCl or NaOh, and NaCl, respectively. While, the influence of the EG was investigated by changing distance between electrodes from 5, to 10, and 15 mm keeping the CD at the optimum value that obtained from the previous step. The initial pH, water conductivity, and initial water temperature constant at 8, 0.4 ms, and 20±1^{0}C, respectively.

The progress of decolourization process was monitored by collecting 5 mL samples at each 2 minutes for 12 minutes. The collected samples were filtered at 0.45 µm filter papers (Whatman filters)[26], the filtrate was analysed using a spectrophotometer (Hach Lange DR 2800) at 512 nm. All chemicals were supplied by Sigma Aldrich and used as supplied.

Dye removal efficiency (RE %) was determined as follows:

$$RE\% = \frac{C_{\circ}-C}{C_{\circ}} \times 100\% \qquad \textbf{(1)}$$

Where C_0 and C are the influent and effluent concentrations of RR-120 dye, in mg/L, respectively.

3. Results and discussion

3.1. Influence of CD

This part of the study was conducted to study the influence of CD (2, 4 and 6 mA/cm^2) on the removal of RR-120 dye from drinking water using EC process. The results obtained in showed that the removal of RR-120 dye from the synthetic water samples increases with the increase of the applied current density, figure 4. For example, it has been found that he removal efficiency, after 8 minutes of treatment, has increased from 80% to 88% and 96% as the CD increased from 2 to 4, and 6 mA/cm^2, respectively. This could be explained by the fact that increasing the CD increases the dissolving rate of aluminium ions from the anode, which enhances the dye removal [27, 28]. However, increasing the applied CD results in an increase in power consumption, thus CD of 4 mA/cm^2 will be used to carry out the rest of experiments.

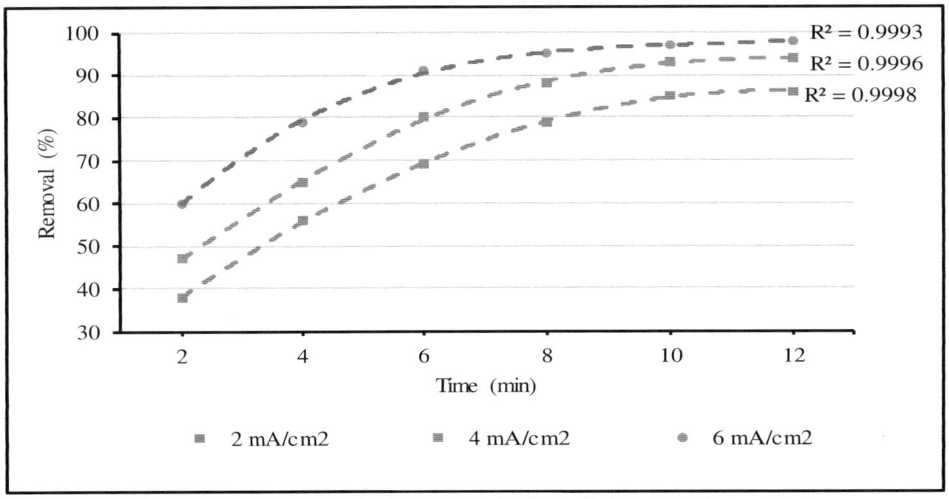

Figure 4. The removal efficiency of RR-120 dye at different CDs.

3.2. Influence of EG

In order to explore the influence of the gap between electrodes on the removal of RR-120 dye, coloured water samples were treated at three different EGs (5, 10, and 15 mm), keeping the CD, initial pH, water conductivity, and initial water temperature constant at 4 mA/cm^2, 5 mm, 8, 0.4 ms, and 20±1^{0}C, respectively.

The results obtained indicated that the wider the EG was, the lower the removal efficiency was, figure 5. For instance, after 12 min of electrolysing, the dye removal decreased from 96%, to 89%, and 80% as the electrode distance increased from 5 to 10, and 15mm, respectively. This behaviour could be attributed to the fact that increasing the gap between electrodes results in decreasing the required attraction force for flocs formation, and thus reduces the dye removal. In addition, as the electrodes gap increase, the produced gas bubbles will raise fast toward the surface of the polluted solution then reduces the mixing effect required to attract the different ions and form the adsorbent as a consequence [26].

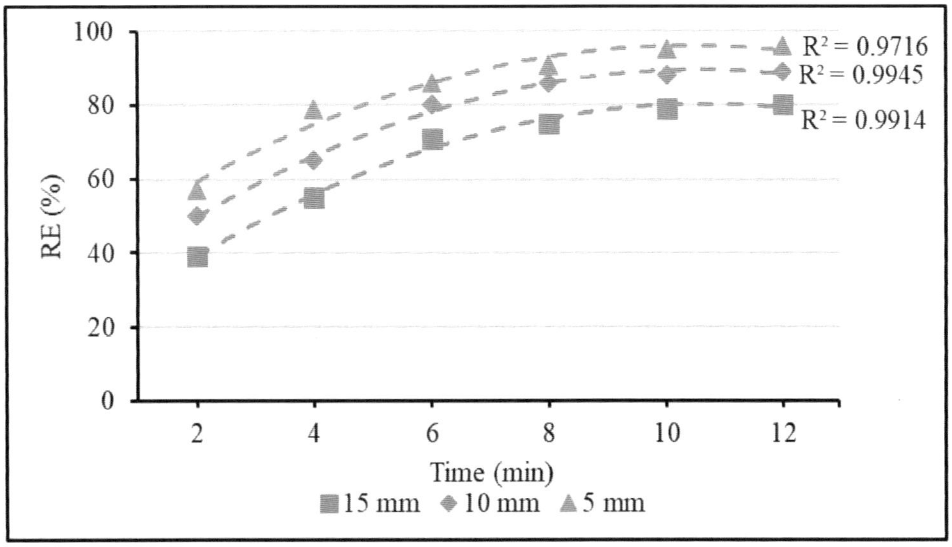

Figure 5. The removal efficiency of RR-120 dye at different EGs.

4. Conclusion

The outcomes of the current study indicated that the removal of RR-120 dye, using aluminium-based electrocoagulation method, is positively influenced by the increase of current density and negatively by the gap between electrodes. Therefore, to enhance the performance of EC units, in terms of dye removal from water or wastewater, it is recommended to keep the gap between electrodes as short as possible, and increase the applied current density to the level that does not influence the cost-effectiveness of the treatment method. Moreover, the obtained coefficient of correlation (R^2) for the relationship between the operational variables and the responses is more than 0.97 for all operation conditions. These statistical values indicate a very good relationship between the predicted and measured removals and can be used for modelling the performance of the EC method[29].

Finally, due to the recent development in sensing technology [30], the authors recommend the application of this technology in the development of smart EC units. For example, sensors could be used to control the applied current density according to the concentration of the targeted pollutants.

5. References

[1] Neamtu, M., et al., Oxidation of commercial reactive azo dye aqueous solutions by the photo-Fenton and Fenton-like processes. Journal of Photochemistry and Photobiology A: Chemistry, 2003. 161(1): p. 87-93.

[2] Hashim, K.S., The innovative use of electrocoagulation-microwave techniques for the removal of pollutants from water, in Civil Engineering 2017, Liverpool John Moores University, UK.

[3] Cho, I.-H. and K.-D. Zoh, Photocatalytic degradation of azo dye (Reactive Red 120) in TiO2/UV system: Optimization and modeling using a response surface methodology (RSM) based on the central composite design. Dyes and Pigments, 2007. 75(3): p. 533-543.

[4] Saharan, V.K., M.P. Badve, and A.B. Pandit, Degradation of Reactive Red 120 dye using hydrodynamic cavitation. Chemical Engineering Journal, 2011. 178: p. 100-107.

[5] Robinson, T., et al., Remediation of dyes in textile effluent: a critical review on current treatment technologies with a proposed alternative. Bioresource Technology, 2001. 77(3): p. 247-255.

[6] Firmino, P.I., et al., Colour removal of dyes from synthetic and real textile wastewaters in one- and two-stage anaerobic systems. Bioresour Technol, 2010. 101(20): p. 7773-9.

[7] Gole, V.L. and P.R. Gogate, Degradation of brilliant green dye using combined treatment strategies based on different irradiations. Separation and Purification Technology, 2014. 133: p. 212-220.

[8] Cardoso, N.F., et al., Comparison of Spirulina platensis microalgae and commercial activated carbon as adsorbents for the removal of Reactive Red 120 dye from aqueous effluents. Journal of Hazardous Materials, 2012. 241-242: p. 146-153.

[9] Yagub, M.T., et al., Dye and its removal from aqueous solution by adsorption: A review. Advances in Colloid and Interface Science, 2014. 209: p. 172-184.

[10] Hashim, K.S., et al., Electrocoagulation as a green technology for phosphate removal from River water. Separation and Purification Technology, 2018. 210: p. 135-144.

[11] Hashim, K.S., et al., Defluoridation of drinking water using a new flow column-electrocoagulation reactor (FCER) - Experimental, statistical, and economic approach. J Environ Manage, 2017. 197: p. 80-88.

[12] Mollah, M.Y., et al., Fundamentals, present and future perspectives of electrocoagulation. J Hazard Mater, 2004. 114(1-3): p. 199-210.

[13] Yasir Aljaberi, F. and W. Mohammed, Novel method for electrocoagulation removal of lead from simulated wastewater by using concentric tubes electrodes reactor. Vol. 101. 2018.

[14] Hashim, K.S., et al., Removal of phosphate from River water using a new baffle plates electrochemical reactor. MethodsX, 2018. 5: p. 1413-1418.

[15] Hashim, K.S., et al., Iron removal, energy consumption and operating cost of electrocoagulation of drinking water using a new flow column reactor. Journal of

a. Environmental Management, 2017. 189: p. 98-108.

[16] Hashim, K.S., et al., Energy efficient electrocoagulation using a new flow column reactor to remove nitrate from drinking water - Experimental, statistical, and economic approach. J Environ Manage, 2017. 196: p. 224-233.

[17] Zhu, B., D.A. Clifford, and S. Chellam, Comparison of electrocoagulation and chemical coagulation pretreatment for enhanced virus removal using microfiltration membranes. Water Research, 2005. 39(13): p. 3098-3108.

[18] Hashim, K.S., A. Shaw, and R. Alkhaddar, Enhancement of Dissolved Oxygen Concentration during the Electrocoagulation Process Using an Innovative Flow Column: Electrocoagulation Reactor. International Journal of Civil and Environmental Engineering, 2016. 3(7).

[19] Hashim, K.S., et al., Controlling of Water Temperature during the Electrocoagulation Process Using an Innovative Flow Columns - Electrocoagulation Reactor. International Journal of Environmental, Chemical, Ecological, Geological and Geophysical Engineering, 2015. 9(8): p. 869-872.

[20] Hashim, K.S., et al. Effect of the supporting electrolyte concentration on energy consumption and defluoridation of drinking water in the electrocoagulation (EC) method. in The 2nd BUiD Doctoral Research Conference. 2016. The British University in Dubai.

[21] Shaw, A., et al. Influence of electrodes spacing on internal temperature of electrocoagulation (EC) cells during the removal (Fe II) from drinking water. in The 3rd BUiD Annual Doctoral Research Conference. 2017. The British University, Dubai.

[22] Pallant, J., SPSS survival manual. 2005, Sydney: McGraw-Hill Inc.

[23] Hashim, K.S., et al., An Investigation into The Level of Heavy Metals Leaching from Canal-Dreged Sediment: A Case Study Metals Leaching from Dreged Sediment, in 1st International Conference on Materials Engineering & Science. 2018: Istanbul Aydın University (IAU), Turkey.

[24] Chaturvedi, S. and P.N. Dave, Removal of iron for safe drinking water. Desalination, 2012. 303: p. 1-11.

[25] Carmen, Z. and S. Daniela. Textile organic dyes–characteristics, polluting effects and separation/elimination procedures from industrial effluents–a critical overview. in Organic

IOP Conf. Series: Materials Science and Engineering **584** (2019) 012035 doi:10.1088/1757-899X/584/1/012035

pollutants ten years after the Stockholm convention-environmental and analytical update. 2012. InTech.

[26] Khandegar, V. and A.K. Saroha, Electrocoagulation for the treatment of textile industry effluent--a review. J Environ Manage, 2013. 128: p. 949-63.

[27] Un, U.T., A.S. Koparal, and U. Bakir Ogutveren, Fluoride removal from water and wastewater with a bach cylindrical electrode using electrocoagulation. Chemical Engineering Journal, 2013. 223: p. 110-115.

[28] Yasir Aljaberi, F. and W. Mohammed, Adsorption of lead from simulated wastewater via electrocoagulation process: kinetics and Isotherm Studies. Vol. 4. 2018.

[29] Yasir Aljaberi, F., Studies of Autocatalytic Electrocoagulation Reactor for Lead Removal from Simulated Wastewater. Vol. 6. 2018.

[30] Teng, K.H., et al., Embedded Smart Antenna for Non-Destructive Testing and Evaluation (NDT&E) of Moisture Content and Deterioration in Concrete. Sensor2019. 19(3): p. 547.

IOP Conf. Series: Materials Science and Engineering **584** (2019) 012036 doi:10.1088/1757-899X/584/1/012036

Chitosan as a Widely Used Coagulant to Reduce Turbidity and Color of Model Textile Wastewater Containing an Anionic Dye (Acid Blue)

Hala A Abdullah[1] and Ali J Jaeel[2]

[1]Civil Engineering Department ,M.Sc.Student, Wasit University , Iraq.
halaalimsc@gmail.com

[2]Civil Engineering Department, PhD, Wasit University, Iraq.
alijwaid@uowasit.edu.iq

Abstract. The study focuses on using chitosan as a widely used coagulant (a biopolymer) to treat model textile wastewater containing an anionic dye (acid blue). Jar test experiments had been used in the coagulation-flocculation process. Acid Blue was selected as a model dye for testing chitosan's ability to treat model textile wastewater. Chitosan is found from chitin by deacetylation. The biopolymers, because of biodegradable properties, availability and low cost have been used as coagulants. The dosage impact, pH, settling time, decolorization and turbidity reduction of simulated textile wastewater were studied. Chitosan showed great performance to remove turbidity and color. The usage of chitosan as a primary coagulant would decrease the cost of wastewater treatment. Besides, it is eco-friendly coagulant without harmful impacts.
Keywords: Coagulant; chitosan; model textile wastewater; coagulation- flocculation; biopolymer; turbidity; color

1. Introduction

Textile industries have a highly positive effect on the economic development countries. However, the unacceptable heterogeneous aspects in composition effluent are one of the basic problems associated with textile factories [1]. Textile finishing's wastewater, especially dyeing process effluent, contains a wide range of chemicals, organic dyes, and auxiliaries. Color, inorganic salts, chemical oxygen demand (COD), total dissolved solids (TDS) and heavy metals are sorted as the main contents in the effluent of the textile industry. So, dye effluent must be treated before being discharged [2]. Some techniques such as coagulation, adsorption, or advanced oxidation can be used for the decolorization and reduction the level of pollutants from textile wastewater. Electrocoagulation is another process to remove pollutants from water and wastewater [3] . K.S. Hisham et al, investigated the impact of the electrocoagulation method on the analysis of the surface of aluminum electrodes for phosphate reduction process, findings of the study demonstrated that 99% of phosphate was eliminated within 1 hour of electrolysis at an initial pH of 6, primary dose of phosphate of 100 mg/L [4].While Omwenea et al, investigated the impact of initial pH, current density initial phosphorus dosage and electrocoagulation period on phosphorus reduction from domestic wastewater. Phosphorus removal from domestic wastewater was 99.99% [5] .Among the available treatment techniques, Coagulation- flocculation can be considered as one of the most important solid-liquid separation processes due to simplicity, cost-effective, easy technique for treating the textile effluent [6] and it can be used as a pre-treatment, a post-treatment or even a main treatment [7] [8].Chemical coagulants are commercially available with wide usage .Negative impacts on human health and the possible contribution to Alzheimer's disease and percentile dementia have been recorded [9] [10] [11] [12] [13]. Therefore, special importance has been focused on a new generation of coagulants classified as environmental friendly coagulant such as chitosan.

Chitosan is an amino polysaccharide obtained in the process of deacetylation of chitin, which, besides cellulose, are the most common natural polymers in the world [14].The main sources of chitosan are organisms living in seawater – shrimps, crabs, and lobster shells. Chitosan has a structure similar to those of chitin and cellulose and has unique properties as compared to other biopolymers. However, chitosan has greater solubility in an aqueous acid medium than its precursor polymer (chitin) and cellulose, result from the presence of amine groups and primary and secondary hydroxyl groups, which are responsible for its application in multiple functions [15] [16]. The main advantages of chitosan are non-toxicity, no harmful effect on human health, linear cationic polymer of high molecular weight and biodegradability [17] [18].

This study examines chitosan capability as a widely used coagulant to remove color and the turbidity of simulated textile wastewater in coagulation-flocculation process. Initial pH, chitosan dosage and settling time were considered to know the optimum conditions to achieve the best performance of chitosan.

2. Materials and methods

2.1. Wastewater

For the preparation of model textile wastewater, a varying concentration of acid blue dye in range of (5 -10 and 15) mg which was supplied from Al-Kut Textile Factory /Iraq as a source of color and other additive shown in table 1, were added into a liter deionized water. The dye solution was suspended at a fast speed (120 rpm) for 2 minutes followed by slow speed (20 rpm) for 20 minutes. Then, the mixture was left for 30 minutes to deposit. When the settling period has been ended, samples were taken to measure the initial parameters.

Table 1. Model textile wastewater component.

Substance	Concentration (mg/l)	Chemical formula	Using	Supplied from/ Purchased from
Starch	1000	$C_6H_{10}O_5$	Source of TSS	Macclesfield, Cheshire, UK
Hydrochloric acid	300	HCL	pH adjustment	BDH Chemicals Ltd Pool, England
Sodium lauryl sulfate	100	$CH_3(CH2)_{11}OSO_3Na$	Spread the dye	CDH Ltd. India
Sodium chloride	3000	NaCl	pH adjustment	Macclesfield, Cheshire, UK
Sodium carbonate	500	Na_2CO_3	pH Fixation	HiMedia Lab. Pvt. Ltd. India

2.2. Chitosan

Chitosan was supplied by CDH Ltd. (India) as shown in figure 1(a). The chitosan (1 g) was added after being ground and sieved to (1L) of demineralized water, after which 1 g of acetic acid was added under agitation for 1 hour to dissolve the chitosan powder. The acetic acid was being used in chitosan preparation due to chitosan solubility in the aqueous acid medium [15] [16]. The solution (figure 1 (b)) with pH 4 was used as a coagulant after leaving it for 48 hours.

(a) (b)

Figure 1. (a) Chitosan before ground **(b)** Chitosan solution.

2.3. Coagulation–flocculation experiments
Experiments on model textile wastewater coagulation and flocculation were performed using a jar test apparatus (VELP Scientific, Italy) which consists of four beakers, four stainless steel paddles as shown in figure 2. This test considered the best way to assess the coagulation process by finding the best dosage of coagulant. Each 1000ml beaker was filled with synthetic textile wastewater with notable masses of dyes. In every beaker, a divergent coagulant concentration is adapted with dissimilar concentrations fluctuating from (10, 20, 30, 40, 50, 60 and 70) mg/l. In the jar test device, the beakers faced to fast mixing speed 120 rpm for 120 seconds to disperse chitosan solution in all solution, and after that to quit mixing speed 20 rpm for 1200 seconds to form the flocs. The mixers are switched for 1800 seconds for settling the flocs. At the end of the settling period, the samples were picked from a distance 3-4 cm under the surface of each beaker by syringe. These samples tested by using single beam UV spectrophotometer (SpectroDirect) turbidity and color. All the experiments were conducted at room temperature of 20 ±2. The experiments were performed in duplicates and the average values were considered.

Figure 2. Jar test apparatus used in the experimentation.

3. Result and discussion

3.1 Chitosan dosage effect

The impact of chitosan concentration was studied at pH 6, 2 min with 120 rpm then 20 min with 20 rpm and 30 min of settling time for dosage (10 -20 -30 -40 -50 -60 and 70 mg/l) of chitosan. Figure 3 shows the impact of chitosan dosage on color and turbidity removals when dye concentration 5 mg/l. It is clear from the results that when chitosan dosage equal to 40 mg/l, 99% and 76.5 % for turbidity and color respectively were recorded. From the dosage 10mg/l to 40 mg/l the removal of turbidity and color was increased. Charge density phenomenon could be explained in this case. Chitosan has a high charge density [19].Therefore; this helped to occur rapid destabilization of the particles in less amount of chitosan. When increased chitosan dosage from (40-70) mg/l, the removals of turbidity and color decreased, explained by colloidal surfaces have been filled with adsorbent and no more places founded on it to format bridges between them. Figure 4 displays the reduction of turbidity and color when using optimum dosage of chitosan (40 mg/l) to treat dye with concentration (10 and 15 mg/l) with (90.6% turbidity, 70.1 % color) for dye 10 mg/l and (85.4% turbidity, 66.3 % color) for dye 15mg/l.

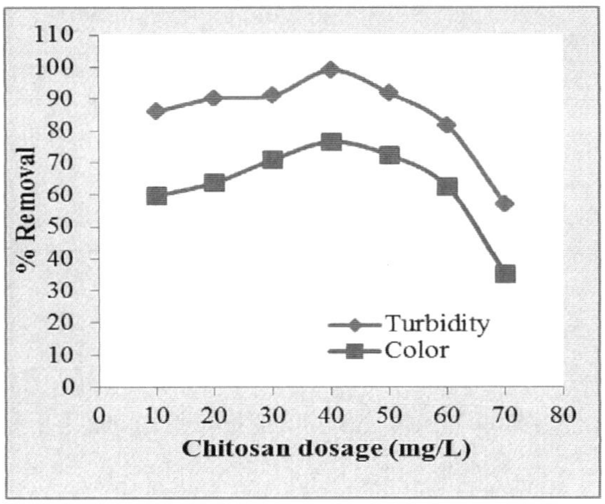

Figure 3. Chitosan dosage effect on turbidity and color removals (constant dye 5 mg/l).

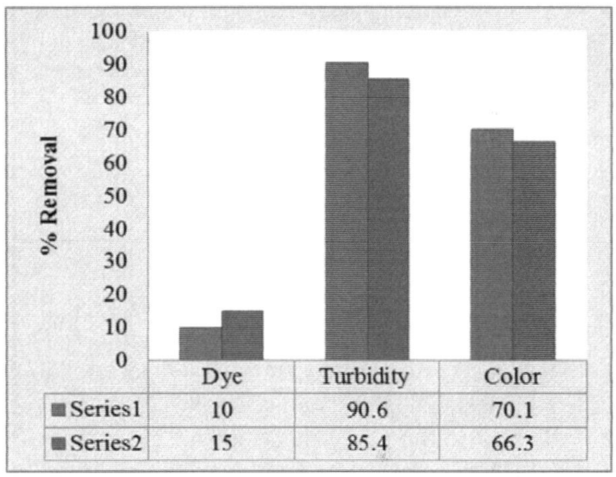

Figure 4. Effect of optimum chitosan dosage 40 mg/l on turbidity and color removals for 10 and 15 dye concentration.

3.2 pH effect

pH value is an important parameter influence on the solubility of chitosan in aqueous solution. pH impact was studied for pH which started from pH 2 till pH 12 using dilute HCl or NaOH to adjust pH value, optimum dosage of chitosan 40 mg/l, 5 mg/l of dye, 2 min with 120 rpm, 20 min with 20 rpm and 30 min of settling time. Figure 5 demonstrates the pH impact on turbidity and color removal, pH 5-6 recorded the highest reduction with 99% and 76% of turbidity and color reduction respectively. This can be supported due to chitosan activity in the acid range of pH and indicating that the pH can be considered as a critical parameter affecting on chitosan ability to reduce the level of pollutants. Figure 6 represents the pH effect on reducing the level of certain pollutants when dye with concentration (10 and 15) mg/l and pH value 5.5 (average value). The highest removals were (91.2 % turbidity, 72 % color) for dye 10 mg/l and (89.4% turbidity, 69.1 % color) for dye 15mg/l.

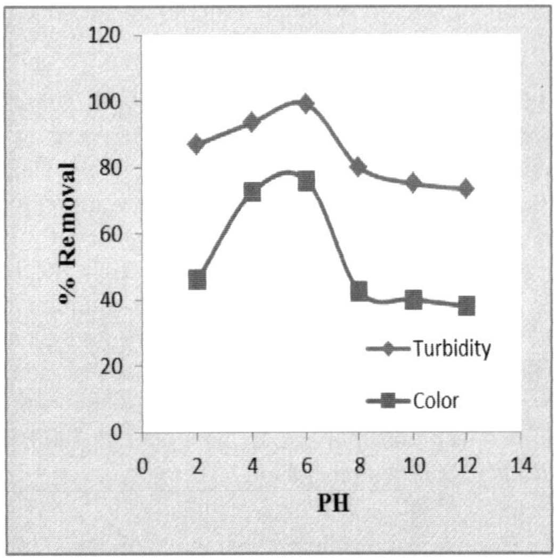

Figure 5. PH effect on turbidity and color removals. Dye 5 mg/l.

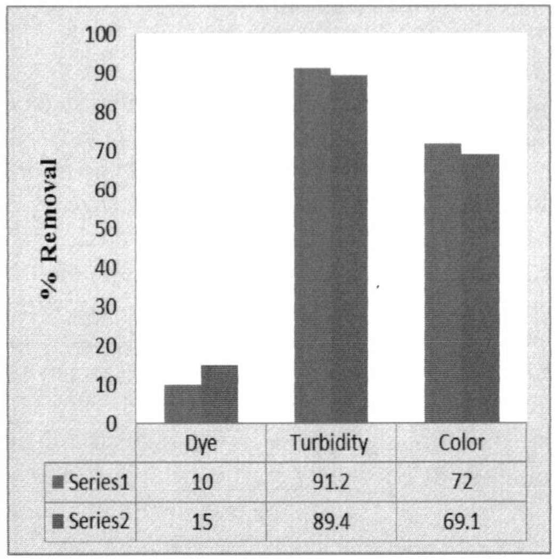

Figure 6. Turbidity and color removals when PH = 5.5 for dye (10 and 15) mg/l.

3.3 Settling time effect

The impact of settling time on turbidity and color reduction was investigated by testing several experiments using optimum chitosan dosage (40 mg/l), dye (5 mg/l) ,pH value (5.5) , 2 min with 120 rpm , 20 min with 20 rpm and settling periods of (10-20-30-40-50 and 60 min). Figure 7 displays the effect of settling time on pollutants reduction. It can be stated; a direct correlation between settling period and percentage removals where the turbidity and color removal efficiency reaches to 99.7% and 78% respectively in time 50 min and after time 50 there no change observed in the removal percentage of certain pollutants. When dye concentration (10 and 15) mg/l at the same condition and settling time 50 min, the percentage reduction were (96 % turbidity, 74.1% color) for dye 10 mg/l and (92.8% turbidity, 70.3 % color) for dye 15mg/l as shown in figure 8.

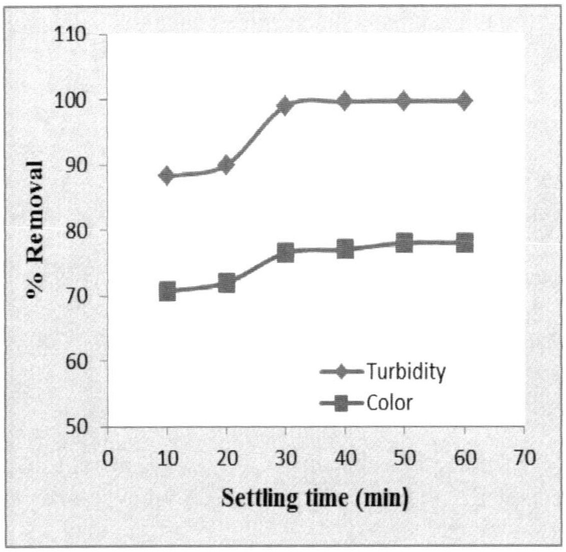

Figure 7. Effects of settling time on turbidity and color removals. Dye concentration 5mg/l, PH=5.5.

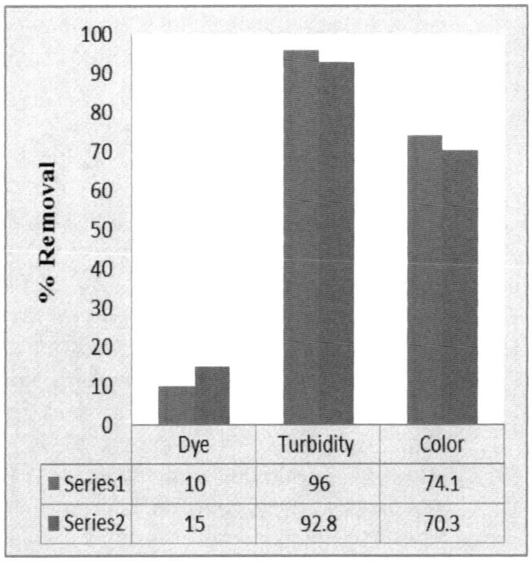

Figure 8. Turbidity and color removal settling time 50 min, PH=5.5. Dye concentration 10 and 15 mg/l.

4. Conclusions

Among different technologies available to treat textile wastewater such as physical, chemical, biological, and advanced methods, chemical coagulation-flocculation is still a suitable technology due to its cost-effectiveness and simplicity. In this study, chitosan used as a widely used coagulant to treat model textile wastewater containing an anionic dye (acid blue). Chitosan recorded 99.7% and 78% of turbidity and color removal for dye concentration 5 mg/l under an optimum condition.

References

[1] Yaseen D A and Scholz M 2019 Textile dye wastewater characteristics and constituents of synthetic effluents: a critical review *Int. J. of Environ. Sci. and Technol.* 2 1193-1226

[2] Vera G, Aleksandra V and Marjana S 2005 Efficiency of the coagulation/flocculation method for the treatment of dye bath effluents *Dyes and Pigments* 67 93-97

[3] Khalid S H, Ibijoke A I, Nisreen J, Rafid A , Andy S, David P, Kot P, Montserrat O P, Ali W A, Muhammad A, Reham A, Teng K H, Keyur H J and Mohammed H A 2018 Removal of phosphate from river water using a new baffle plates electrochemical reactor *J. of MethodsX* 5 1413-1418

[4] Khalid S H, Rafid A, Nisreen J, Andy S, David P, KotaP,Montserrat O P, Ali W A, Muhammad A and Reham A 2019 Electrocoagulation as a green technology for phosphate removal from river water *J. of Separation and Purification Technol.* 210 135-144

[5] Omwenea P I, Kobyaa M, and Canb O T 2018 Phosphorus removal from domestic wastewater in electrocoagulation reactor using aluminium and iron plate hybrid anodes *J. of Ecological Eng.* 123 65-73

[6] Rao L N 2015 Coagulation and flocculation of industrial wastewater by chitosan *Int. J. of Eng. and Applied Sci.* 2 50-52

[7] Dos S A B, Cervantes F J and Van L J B 2007 Review paper on current technologies for decolourisation of textile wastewaters: perspectives for anaerobic biotechnology *Bioresour. Technol.* 98 2369–2385

[8] Ga¨hr F, Hermanutz F and Oppermann W 1994 Ozonation: an important technique to comply with new German laws for textile wastewater treatment *Water Sci.Technol.* 30 255-263
[9] Divakaran R and Pillai V N S 2001 Flocculation of kaolinite suspensions in water by chitosan *Water Res.* 35 3904-3908
[10] Ali S A, Pal S, and Singh R P 2010 Flocculation performance of modified chitosan in an aqueous suspension *J. Appl. Polym. Sci.* 118 2592-2600
[11] Crapper D R, Krishnan S S and Dalton A J 1973 Brain aluminium in Alzheimer's disease experiment at neurofibrillary degeneration *Sci.* 180 511-573
[12] Jadhav M V and Mahajan Y S 2011 Consequences of residual aluminium from alum coagulant on humans *Int. J. Res. Chem. Environ.* 1 22–27
[13] Al-Mutairi and Nayef Z 2006 Coagulant toxicity and effectiveness in a slaughterhouse wastewater treatment plant *Ecotoxicology and Environ. Safety* 65 74–83
[14] Kurita K 2006 Chitin and chitosan: functional biopolymers from marine crustaceans *Mar. Biotechnol.* 8 203-206
[15] Ahmad A L, Sumathi S and Hameed B H 2005 Residual oil and suspended solid removal using natural adsorbents chitosan, bentonite and activated carbon: a comparative study *Chem. Eng. J.* 108 179–185
[16] Pillai C K S, Paul W and Sharma C P 2009 Chitin and chitosan polymers:chemistry, solubility and fiber formation *Prog. Polym. Sci.* 34 641-678
[17] Nayab A K M , Salma K S, Soumen S, Ayshi D, Farhana S M and Mohammad N H 2018 Treatment of textile waste water using natural catalyst (chitosan and microorganism) *J. of Textile Eng. and Fashion Technol.* 4 320–325
[18] Kos L 2018 Use of Chitosan for textile wastewater decolourization *Fibres and textiles in Eastern Europe* 24 130-135
[19] Kumar M N 2000 A review of chitin and chitosan applications *Reactive and Functional Polymers* 46 1-27

IOP Conf. Series: Materials Science and Engineering **584** (2019) 012037 doi:10.1088/1757-899X/584/1/012037

Comparison of Compressibility Behaviour of Lime-Cement Stabilized Lateritic Clay Soil Contaminated by Heavy Metals

K A Saeed[1], K A Kassim[2], H Nur[3], N Z M Yunus[4]

[1] Lecturer, Water Resources Engineering Department, Al-Mustansiriyah University, Baghdad, Iraq: khitamhussein@ uomustansiriyah.edu.iq
[2] Professor, Faculty of Civil Engineering, Universiti Teknologi Malaysia, 81310 Skudai, Malaysia, e-mail: kanuar@utm.my

[3] Professor, Ibnu Sina Institute for Fundamental Science Studies,Universiti Teknologi Malaysia, 81310 Skudai, Malaysia e-mail: hadinur@utm.my

[4] Senior Lecturer, Faculty of Civil Engineering, Universiti Teknologi Malaysia, 81310 Skudai, Malaysia e-mail: nzurairahetty@utm.my

Abstract. The type of stabilizer, heavy metal and soil can typically affect the strength and compressibility development in the treatment of contaminated soils. This is due to the difference in chemical reactions and interferences occurring among the stabilizers, metals, and clay minerals. A standard consolidation test using an oedometer was carried out to investigate the cement–lime induced changes in the compressibility of clay soils contaminated with either copper or zinc. In this study, laterite clay soil obtained from the hilly area at Universiti Teknologi Malaysia was used. Control samples of cement-and lime-stabilized soils without heavy metals were compared with the treated contaminated stabilized soil at various contaminants contents. Such percentage of contaminants means that the soil falls under industrially contaminated soil. To highlight the effect of the two heavy metals on the consolidation behavior of clay samples treated with cement and lime, a comparison has been made at a curing period of 7, 28, 100, and 200 days respectively. It was found that the interference of heavy metals during the process of hydration (cement and lime) contributed directly to the inhibition of the compression development of the tested soil samples. Therefore, it is crucial to identify the type of stabilizer, heavy metal, and soil involved to ensure successfully and remediate treatment.

1. Introduction

Nowadays, due to heavy industrialization, it is important to protect the environment from hazardous pollutants associated with waste generation and disposal. Several methods have been developed to reduce the potential release of toxic species into the environment or to transform the hazardous wastes into nontoxic substances. By using a chemical treatment, the stabilization and solidification processes can bind the hazardous waste compound to an established unsolvable System and catch the waste inside a compact cementitious medium, respectively [1]. The chemical reaction that arises from the interaction of waste components with a solidifying binder plays an important role that controls the quality of binder-based solidified waste products [2]. This method is commercially used in situ or ex. situ for treatment of contaminated soil because it is relatively cheap and results in well qualified long-term physical and chemical stability, which can increase the mechanical and structural characteristics [3]. Therefore, the end products of such stabilized wastes can be reused for engineering fills, highway construction, pavement subgrade, backfill, base material, and so on.

The effectiveness of the stabilization and solidification process is dependent on the types of wastes and heavy metals. Portland cement using fly ash, lime, slag, and soluble silicates is among the solutions used in the chemical treatment of soil contaminated by heavy metal waste [4]. Ordinary Portland Cement (OPC) is the most commercially used solution as it is readily available, cost- effective, and also compatible with a variety of wastes [5-8]. However, several studies reported that the addition of heavy metals to the stabilized waste could retard the hydration process and, in some cases, inhibition of the cement hydration process occurred [9]. Cement. hydration can also be delayed in. the existence of zinc (Zn.), lead (Pb.), and copper (Cu.) when double salts are formed between these heavy metals and calcium ions that dissolve early from the cement [10 and 11]. Meanwhile, Tashiro and Oba, (1980) [12] reported that the heavy metal oxides of chromium (Cr), Cu, Zn, arsenic (As), and Pb influence the cement's physical properties due to reactions with cement paste, which influence the cement paste desensitization and strengthening development during early stages of cement hydration.

Generally, the existence of heavy metals in the polluted cement-stabilized soil can retard the cement reactions. However, their effect on the reaction with lime has not yet been comprehensively studied. It should be noted that the lime in the form of (hydrated. Lime) is rather high. in Malaysia and has commonly been used in the treatment of tropical soils [13]. Therefore, in this study, lime was used as an alternative stabilizer in order to compare the efficiency and suitability of lime and cement as binders for stabilization/solidification (S/S) treatment of tropical laterite soils contaminated with heavy metals. The unconfined compressive strength of stabilized soil at different curing period intervals was determined to investigate the strength development for both cement and lime in contaminated stabilized soils. In addition, the one-dimensional consolidation (odometer) test was conducted to examine the compression index parameter for contaminated samples.

2. Materials and specimen preparation

The laterite residual soil used in this study was a reddish clay soil with a large quantity of iron oxides usually establish in humid areas. Table 1 shows (the chemical and physical properties) of the untreated soil. Hydrated lime was utilized in this research is It works well with clayey soils, particularly those with a reasonable to high plasticity index (PI > 15). In addition, OPC has been used as the second stabilizer in this work. Both substances were in powder form, and they were obtained from a factory in Pasir Gudang, Johor. The chemical compositions of the hydrated lime and Portland cement are shown in Table 2.

Table 1. The physical properties and chemical composition of laterite.

Engineering and physical properties (laterite)	**Values**	**Chemical composition oxides**	**Values (%)**
CEC (meq/100 g)	22.7	SiO_2	21.55
pH (L/S = 2.5)	4.86	Al_2O_3	24.31
Specific gravity	2.75	Fe_2O_3	29.4
Liquid limit, LL (%)	72.8	Na_2O	0.07
Plastic limit, PL (%)	35.6	K_2O	0.11
Plasticity index, PI (%)	37.2	P_2O_5	16.71
BS classification	MH	SO_3	3.98
ICL (%) 2.00 soluble	5	CO_2	3.65
Maximum dry density (Mg/m^3)	1.33		
Optimum moisture content (%)	34		
Unconfined compressive strength (kPa)	288		
Cc	0.24		

Note: ICL: Initial Consumption of Lime, Cc: Compression Index

Table 2. Chemical compositions of the hydrated lime and Portland cement.

Hydrated lime		**Portland cement**	
Chemical composition	**Values (%)**	**Chemical composition**	**Values (%)**
Calcium oxide, CaO	74.23	Calcium oxide, CaO	64.73
Phosphorus oxide, P_2O_5	0.08	Potassium oxide (K_2O)	0.57
Magnesium oxide, MgO	0.74	Magnesium oxide, MgO	1.1
Calcium sulfate, $CaSO^4$	0.12	Sulfur trioxide (SO_3)	2.01
Ferric oxide, Fe_2O_3	0.17	Ferric oxide, Fe_2O_3	3.1
Aluminium oxide, Al_2O_3	0.11	Aluminium oxide, Al_2O_3	5.2
Silica, SiO_2	0.14	Silica, SiO_2	21.1
Loss on ignition, LOI	24.35	Loss on ignition, LOI	0.8
		Insoluble residue	0.1

The components that were added to reproduce inorganic metal contamination were copper nitrate trihydrate (Cu (NO3)2 * 3 H2O) to provide the copper (II) ions (Cu2+) and zinc nitrate tetrahydrate (Zn (NO3)2 * 4 H2O) to provide zinc (II) ions (Zn+2). Both of them were from Merck (KGaA), Germany. The two substances were selected due to their occurrence at many polluted situates; their chemical and physical properties are as listed below in Table 3.

Table 3. The chemical and physical properties of the selected reagents

Metal	Chemical and physical properties	Metal	Chemical and physical properties
Copper dinitrate trihydrate	$Cu(NO_3)_2 * 3\ H_2O$	Zinc nitrate tetrahydrate	$Zn(NO_3)_2 * 4\ H_2O$
Solubility	2670 g/l (20 °C)	Solubility	(20 °C) soluble
Melting point	114 °C	Melting point	45 °C
Molar mass	241.60 g/mol	Molar mass	261.44 g/mol
Density	2.05 g/cm^3 (20 °C)	Density	1.0 g/m^3
Bulk density	1050 kg/m^3	pH value	5 (50 g/l, H_2O, 20 °C)
pH value	3–4 (50 g/l, H_2O, 20 °C)		

Preceding studies accompanied on lateritic soil have shown that oven drying can significantly modify the plasticity and compaction properties of the soil [14]. Consequently, in this study, all the soil samples were prepared from air-dried soil cracked into lesser sizes and sieved through a 2 mm sieve. The sieving was complete to confirm that the soil was of uniform grade.

The concentration of stabilizing ions has a great influence on the short and long term reactions respectively during the initial modification and the cementation of the treated samples. Based on [5 and 16], concentration of lime between 2 and 3% of the soil's dry weight is adequate for amendment purposes, and for stabilization purposes, the concentration typically ranges from 5 to 10%. The latter range is more prominent for pozzolanic reactions. Kassim and Chern, (2004) [13] discovered that, for typical Malaysian cohesive soils, the optimum lime content for stabilization should be around 3 to 6%. Due to the studies on improving the contaminated soils are relatively less, thus, the suitable percentages of contamination-stabilizers were based on the percentage of pozzolanic reactions. For comparison purposes, two equivalent contents of stabilizers were used (5 and 10%) for both lime and cement in this study.

The essential quantity of water or optimum water .content. (OWC) depends on nature of the soil and stabilizer content In BS.1377: Part 4: 1990 (Clause 3.3.4.1) [17 and 18], five sequence of standard Proctor compaction tests are mentioned to ascertain OWC for several lime and cement mixtures. In this study, the mix proportions (percentage lime, percentage cement) were (0, 0), (5, 0), (10, 0) and (0, 5), (0, 10) as percentages by weight of dry soil. Depending on these percentages the labels of samples for lime and cement treated laterite clay soil are (LCLT5, LCLT10, LCCT5, and LCCT10) respectively. Should be noted that (LC) mentions to lateritic clay, and (LT and CT) mention to lime and cement treated one after another. The optimum moisture contents were adopted in accordance with these mix proportions as the calculated water content of contaminated soil samples. Soil samples artificially polluted with heavy metals were equipped by mixing zinc and copper nitrate as the sources of contaminants into the natural laterite clay. Three percentages of heavy metals were nominated, namely 0.1, 5, and 10% (dry soil weight basis). These different percentages were based respectively on the level of pollutants commonly found in naturally contaminated soil, industrially contaminated soil, and highly contaminated soil. Control samples (cement- and lime-treated soils without heavy metals) were as well set for evaluation purposes.

To taste the contaminated the soil, all the laboratory instruments and the sampling device were previously immersed in a 5% nitric acid solution, followed by deionized water for a day before sampling in order to remove traces of metals. Through the preparation of the sample, lime and cement were first individually mixed with the dry medium to certify a homogeneous distribution of the two stabilizers in the soil. Then, de-ionized water mixed with a pre-set quantity of copper nitrate or zinc nitrate was added to the mixture. Should be noted, only pure de-ionized water was used for the control samples. Based on laboratory trials, an additional quantity of water (2% of OWC) was added to all mix designs to take evaporation losses into account. After the mixing process, the soils were compacted in 38 mm × 76 mm cylindrical moulds. The preparation process was concluded by making sure that the designed (O.W.C) and maximum dry Density according to Clause (4.1.5 of BS 1924: Part 2: 1990b) were achieved.

After that, the specimen was cured for two hours, removed from the mold, inserted into a thin-walled P.V.C tube, wrapped in a thin plastic lid and sealed with a tight rubber lid. Then, samples were kept to cure in a temperature controlled room (27 ± 2 °C) till the experiments were performed after 7, 14, 28, 100, and 200 days.

3. RESULTS AND DISCUSSION

Oedometer tests were carried out to evaluate the compressibility characteristics of cement- and lime-based lateritic clay soils contaminated with either 5% copper or 5% zinc. Such a contaminant percentage (5%) is commonly indicative of industrially contaminated soil. The curing time was considered to estimate the effect of heavy metals on the soil compression index. Here, a higher content of both stabilizers (10%) was selected to assess the effect of curing time on contaminated soil compressibility. To underscore the effect of both heavy metals on the consolidation behaviour of clay samples treated with cement, comparisons were made as shown in Figure 1(a–d) at curing intervals of 7, 28, 100, and 200 days, respectively. From the void ratio–axial pressure relationship of uncontaminated samples treated with cement (controls) and contaminated treated samples at 7 days, it is clear that the former group's compression began at a lower void ratio than that of the corresponding contaminated samples. However, the difference in void ratio between contaminated and uncontaminated samples decreased with curing time, as shown in Figure 1(b–d). This was most likely due to cation exchange reactions, where an increase in both flocculation and aggregation caused a chemically induced 'pre-consolidation' effect, which increased the vertical effective yield stress and reduced the compressibility characteristics [19 and 20].

Figure 2 reveals variations in the compression index of cement-treated clay soil with and without heavy metals over time. It is obvious from this figure that the compression index of clay samples decreased markedly over time. This is in agreement with [21], who reported that the compressibility of treated homogeneous specimens was much less than that of untreated clay. This implies that cement-treated clay undergoes structuration due to the effect of cementation up to its apparent yield stress level, after which restructuration occurs. As for heavy metals' effects on the compressibility of treated contaminated clay samples, it was observed that their presence increased the compressibility of contaminated clay soils and that this increase was dependent on the heavy metal type. Zinc had a greater effect on the increase of compressibility than did copper, most likely due to its chemical reactions with calcium hydroxide and clay minerals. A gradual reduction in Cc of contaminated clay over time was most likely the result of soil hardening, mainly due to an increase in the amount of pozzolanic reaction compounds over time, which, in turn, reduced heavy metal inhibition.

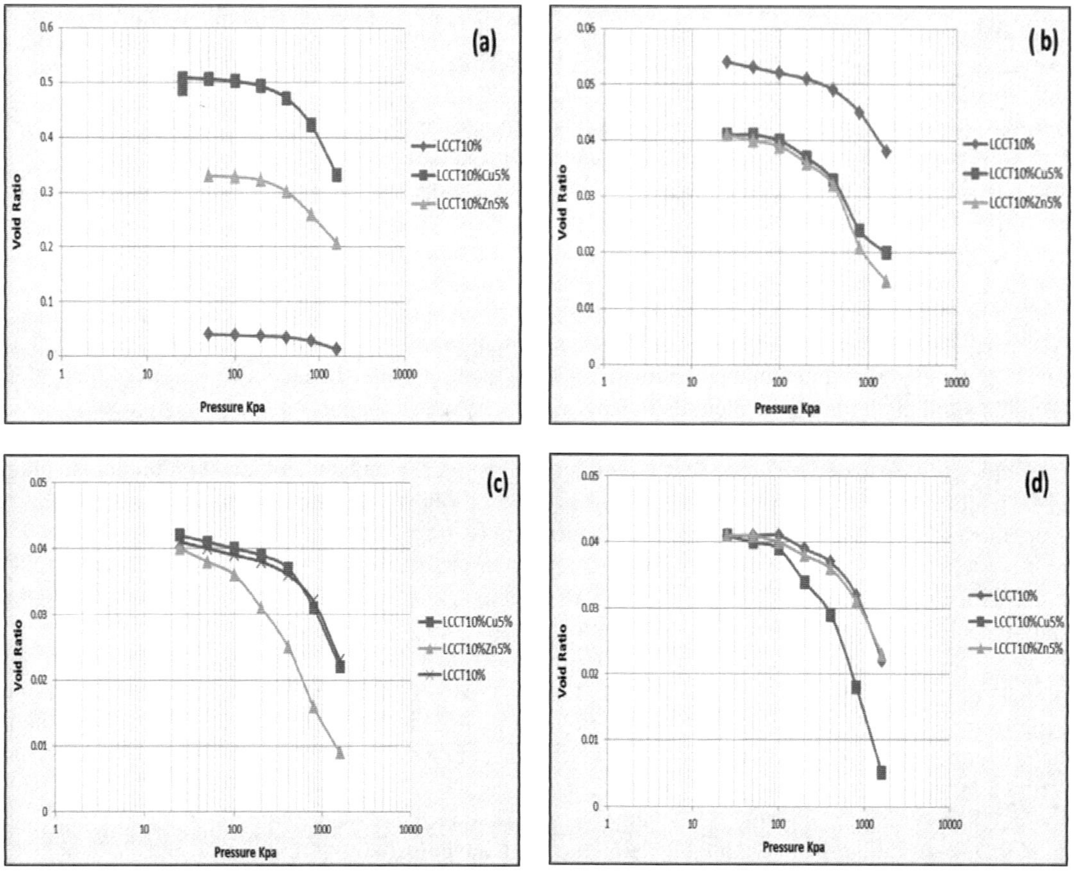

Figure 1. Relationship of void ratio with axial pressure of uncontaminated and copper/zinc-contaminated cement-stabilized laterite clay at different curing times: (a) 7 days, (b) 28 days, (c) 100 days, and (d) 200 days.

Figure 3 illustrates the consolidation behaviour for lime-stabilized lateritic clay soil in comparison with heavy-metals-contaminated lateritic clay treated by the same stabilizer. The comparison has been applied with respect to changes in the curing time, as shown in Figure 3(a–d), at 7, 28, 100, and 200 days respectively. Figure 3(a) shows that the consolidation curve of lime-treated soil started at a higher void ratio than that of cement-treated clay samples, as observed in Figure 1(a), which means that cement is more suitable for improving the compressibility characteristics of lateritic clay soil.

IOP Conf. Series: Materials Science and Engineering **584** (2019) 012037 doi:10.1088/1757-899X/584/1/012037

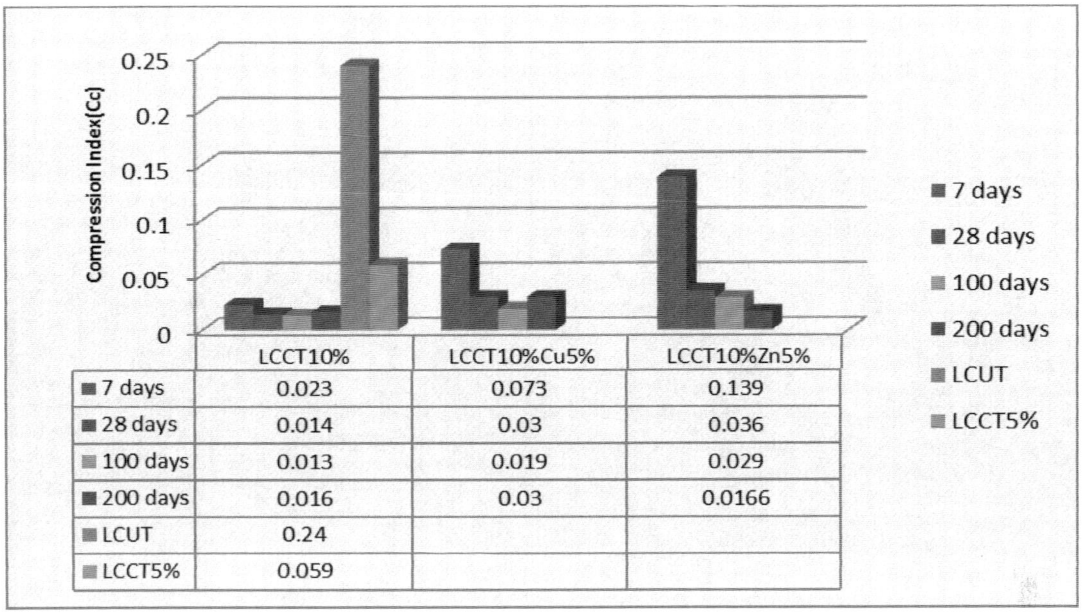

	LCCT10%	LCCT10%Cu5%	LCCT10%Zn5%
7 days	0.023	0.073	0.139
28 days	0.014	0.03	0.036
100 days	0.013	0.019	0.029
200 days	0.016	0.03	0.0166
LCUT	0.24		
LCCT5%	0.059		

Figure 2. The variation of the compression index of uncontaminated and copper/zinc-contaminated cement-stabilized laterite clay at different curing times.

Figure 4 depicts the evaluation using the compression index of lime-treated clay to estimate the effect of heavy metals on lime stabilization. As shown in this figure, it is obvious that the effect of heavy metals on the compressibility of contaminated lime-treated clay soil is different from the effect on cement-treated clay soil. Irrespective of the type of heavy metal, the compressibility of lime-treated clay decreased after both 7- and 28-day curing periods. The compressibility of uncontaminated treated clay and contaminated treated clay increased after 100 days due to an increase in the percentage void ratio resulting from the agglomeration and rearrangement of soil particles over time. However, a reduction in the compression index occurred at 200 days of curing time. This was attributed to the production of pozzolanic reactions, which formed a hardened skeleton matrix that enclosed the unaltered soil particles and led to strengthening of the soil and reduction of its compressibility. As such, it can be concluded that although heavy metals have less retarding effect on lime with regard to the compressibility of soil, cement is preferable for improving the compressibility of the contaminated soil.

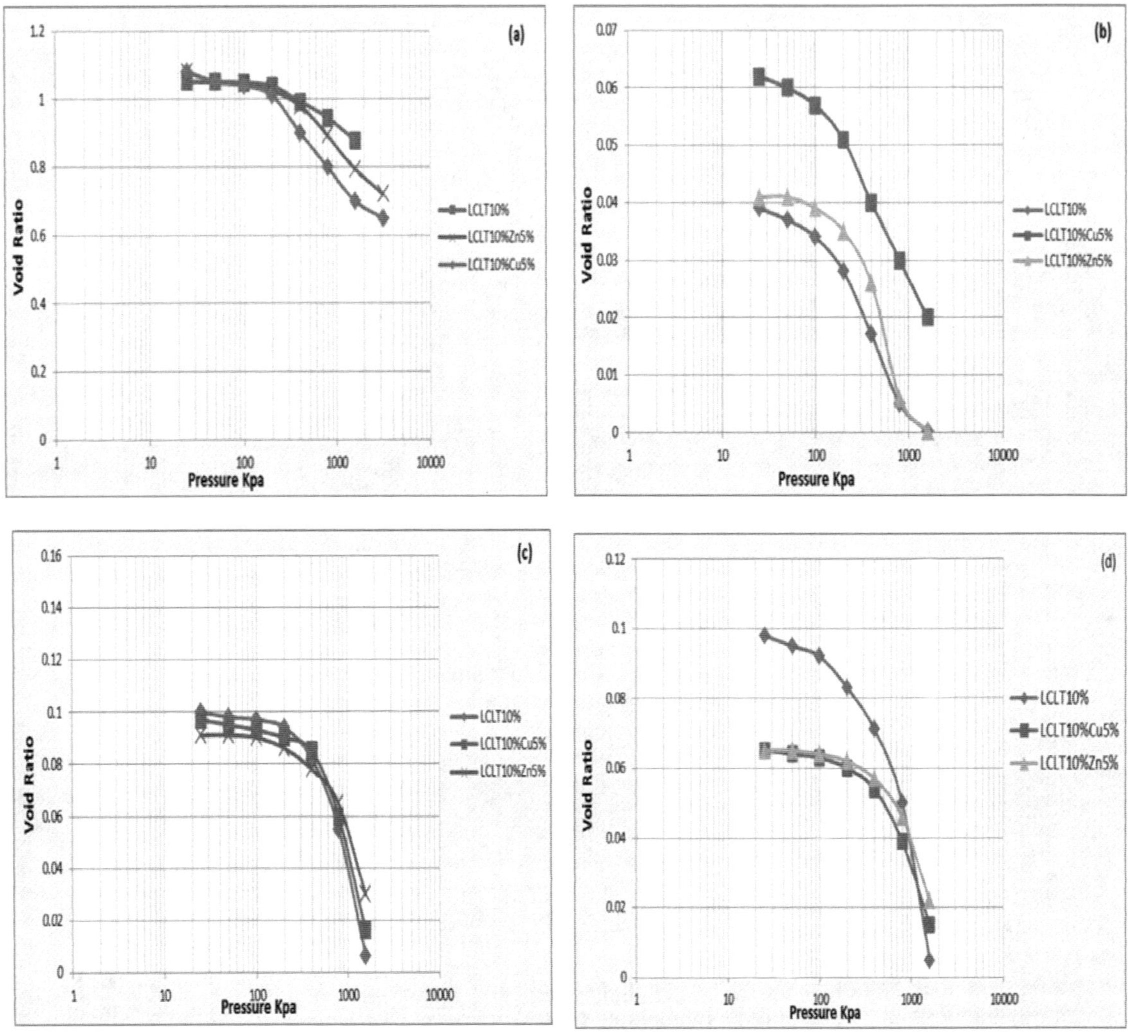

Figure 3. Relationship of the void ratio with axial pressure of uncontaminated and copper/zinc-contaminated lime-stabilized laterite clay at different curing times: (a) 7 days (b) 28 days, (c) 100 days, and (d) 200 days.

IOP Conf. Series: Materials Science and Engineering **584** (2019) 012037 doi:10.1088/1757-899X/584/1/012037

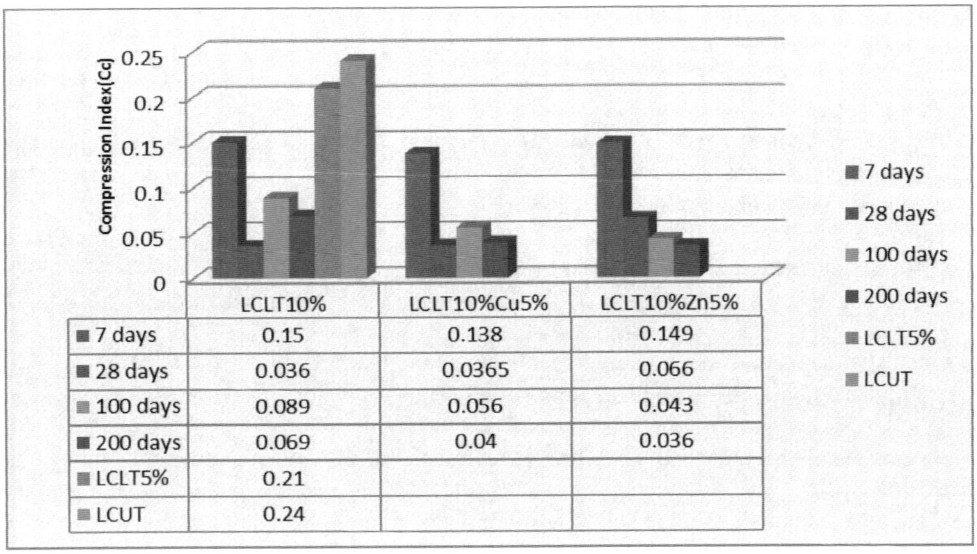

	LCLT10%	LCLT10%Cu5%	LCLT10%Zn5%
7 days	0.15	0.138	0.149
28 days	0.036	0.0365	0.066
100 days	0.089	0.056	0.043
200 days	0.069	0.04	0.036
LCLT5%	0.21		
LCUT	0.24		

Figure 4. Variation of the compression index of uncontaminated and copper –zinc-contaminated lime-stabilized laterite clay at different curing times.

4. CONCLUSIONS

On the basis of laboratory examinations, the following conclusions can be drawn:

- The results from the one-dimensional oedometer test showed that cement is more active than lime in reducing the compressibility of contaminated soil.
- Zinc has more effect on the increase of compressibility than copper due to its chemical reaction with calcium hydroxide and clay minerals. Moreover, zinc metal has an amphoteric nature which is exhibit maximum solubility and then may affect the compressibility properties.
- Irrespective of the type of heavy metal, the compressibility of lime treated laterite clay has decreased after both 7 and 28 curing periods. The compressibility of uncontaminated treated clay and contaminated treated clay increased up to 100 days. The increasing in the compressibility may attribute the mechanism interparticulate forces in the soil particles which are occurred in the presence of non-expanding mineral such us kaolinite mineral

References

[1] Wiles CC (1987) A review of solidification/stabilization technology. Journal of Hazardous Materials, Vol. 14: 5-21.

[2] Asavapisit S, Naksrichum S, and Harnwajanawong N (2005) Strength, leachability and microstructure characteristics of cement-based solidified plating sludge. *Cement and Concrete Research*, 35:1042-1049.

[3] Conner JR (1990) Chemical fixation and solidification of hazardous wastes. Van Nostrand Reinhold, New York.

[4] Choon, KP (1999). Hydration and solidification of hazardous wastes containing heavy metals using modified cementitious materials. Cement and Concrete Research. 30, 429-435.

[5] Spence RD and Shi C (2004) Stabilization and Solidification of Hazardous, Radioactive and Mixed Wastes. CRC Press, Boca Raton, FL, US.

[6] USEPA (1999) Solidification/Stabilization resource guide. EPA542-B-99-002, Adapted from http://www.epa.gov.

[7] USEPA (2001) Treatment technologies for site cleanup. Annual status report (10th E.)", EPA542-R-01-004, Adapted from http://www.epa.gov.
[8] USEPA (2004) Treatment technologies for site cleanup. Annual from http://www.epa.gov.
[9] Hills CD and Sollar CJ (1993) Ordinary Portland cement based solidification: the role of OPC reviewed. Cement. Concrete. Res.23, 196-212.
[10] Trussell S and Spence RD (1994) A review of solidification/stabilization interferences. Waste Mangmant.14:507-519.
[11] Tashiro C, Takahashi H, Kanaya M, Hirakida I and Yoshida R (1977(Hardening properties of cement mortar adding heavy metal compound and solubility of heavy metal from hardened mortar. Cement Concrete Res7 (3), 283-290.
[12] Tashiro C and Oba J (1980) The effect of Cu(OH)2 on the hydration of C3A. in: Proceedings of the seventh international congress on chemistry of cement,2,59-63.
[13] Kassim KA, and Chern KK (2004) Lime stabilized Malaysian cohesive soils. Journal Kejuruteraan Awam 16(1), 13-23.
[14] Goswami, R. K. and Mahanta, C. (2007). Leaching Characteristics of Residual Lateritic Soils Stabilised with Fly Ash and Lime for Geotechnical Applications. *Waste Management*. 27, 466-481.
[15] Das, B. M. (1990). Principle of foundation engineering. PWS-KENT publishing company, Boston.
[16] Maher M, Marshall C, Harrison F and Baumgaertner K (2005) Context sensitive roadway surfacing selection guide. FHWA-CFL/TD-05-004
[17] British Standards Institution (1990) British Standard methods of test for soils for civil engineering purposes. Part 2, Classification tests. London, BS1377.
[18] British Standards Institution (1990) British Standard methods of test for soils for civil engineering purposes. Part 4, Compaction-related tests. London, BS1377.
[19] Nalbantoglu Z and Tuncer ER (2001) Compressibility and hydraulic conductivity of a chemically treated expansive clay. Canadian Geotech.Vol.38:154-160
[20] Ho MH, Chan CM and Bakar I (2010) One dimensional compressibility characteristics of clay stabilized with cement-rubber chips. International Journal of Sustainable Construction Engineering & Technology, Vol 1.No.2
[21] Ho MH and Chan CM (2011). Some Mechanical Properties of Cement Stabilized Malaysian Soft Clay. World Academy of Science, Engineering and Technology.

Dissipation of Energy of Flow by Conventional Type of Gabion Weir

A M Al- Fawzy[1] and F M Al- Mohammed[2]

[1] Directorate of Water Resources in Kerbala, 56001, Kerbala, Iraq. E-mail: ali_meckiwre85@yahoo.com
[2] Kerbala Technical Institute, Al-Furat Al-Awsat Technical University, 56001, Karbala, Iraq. E-mail: fmhmjme@yahoo.com

Abstract: The present study aimed to investigate the effect of using the conventional (rectangular) shape of gabion weir in dissipating the energy of flow. For that a series of laboratory tests (25 experiments and 194 test runs) were conducted on the gabion weir by using of a laboratory flume of 12 m long by a cross section of 0.3 m width, and 0.5 m height to test five different lengths of the weir, (40 cm, 60 cm, 80 cm, 100 cm, and 120 cm), with five respective monosized samples of natural quarry gravel with diameters (9.5-14) mm, (14-19) mm, (19-25) mm, (25-37.5) mm, and (37.5-50) mm. The tests were conducted for a range of discharge (0.7-15.0) l/s. Dimensionless analysis was used to analyze the data by using Buckingham Pi-Theorem to find the relation between the difference in energy of gabion ends and the unit discharge. Beside, a comparison discharge was chosen to find the relation between the difference in energy of gabion ends and the other variables. Results of experiments showed that the difference in energy increase by increasing both the unit discharge and the length of the gabion weir and decrease by increasing the equivalent diameter of the gravel sample.

1. Introduction

Weirs consider as one of the hydraulic structures that can widely used in water works such as energy dissipation, regulation and control of water, flood mitigation, etc. Weirs differ according to the material which the body of the weir made from. Some of them have a solid body constructed from an impervious material such as concrete, while other types have a pervious body constructed from non-cohesive agglomerations of natural or broken particles such as rockfill. The weir type that has porous body constructed from pervious material contained in a metal mesh call as "gabion weir". Gabion weirs are structures having a stable texture, flexible nature, and easy way to build. This way illustrates by using a porous medium, such as rockfill, as filling material to form the body of the weir and putting it in a mesh grid. Many studies were carried out on the weir with different types and materials to discuss multi-subjects, and multi-situations in each subject. Some of these studies were deal with the effect of this type of hydraulic structures on the hydraulic characteristics of flow in different states such as, the behaviour of stepped chutes in transition-flow situation [1], the behaviour of stepped spillways in through-flow state [2], the seepage of flow through rockfill structures, and its theoretical solution during overtopping-flow condition [3], the attitude of permeable rubble mound river structures in cases of determination of discharge, and evaluation of flow behaviour through and around the body of the structure [4], [5], and the effect of broad-crested weir on design of hydraulic control structures and their measurements by varying the weir side slopes [6]. While other studies took in consider the behaviour of the pervious weir on energy dissipation, and the energy losses through it or on both of it sides such as, the energy loss through stepped gabion spillways [7], the gabion weir as energy dissipater with through and over-flow conditions [8], [9], [10], and [11], where it was concluded generally that the gabion weirs in stepped shape are good dissipaters of energy of flow with range of (25% - 85%) distributed unequally between the through-flow, and the over-flow conditions, where the through-flow have (80% - 85%) of the later percentage, and the rest is for over-flow. This goodness is up to 3 steps, but the total energy dissipation decreases with more steps. Besides that, the energy dissipation is inversely proportional with the depth of critical flow over the gabion weir, and the slope of step has no change on energy dissipation amount. The present study aims to investigate the dissipated amount of energy by using of the conventional, rectangular, type of the gabion weir.

2. Laboratory Work

In this study, twenty five laboratory experiments were carried out on five physical models by using five different gravel samples with each physical model. The total number of test runs was one hundred and ninety four test runs, and every physical model graphically named as configuration. The physical models used were fabricated from steel rods rapped with a grid sheet with an opening less than the size of the filling material to form the gabion basket then filled with the gravel material. Figure 1 shows definition sketch of the physical models and Table 1 presents the dimension of these models. The gravel samples used were five monosized gravel samples taken from the typical stone quarry of region of Bahr AnNajaf in AnNajaf Governorate, graded according to the American Society of Testing and Materials, ASTM, and numbered as gravel sample number1, gravel sample number2, gravel sample number3, gravel sample number4, and gravel sample number5 respectively. Table 2 presents the details of the used gravel samples in this study, and the value of porosity for each one of them. The laboratory flume used in this study was a glass sides tilting flume with dimensions of 12 m long by cross section of 0.3 m wide with 0.5 m height. The flume had a slope of (+0.006) during the test runs. The discharge was measured in the flume by two ways, the first way was by using a flow-meter device, and the other was the volumetric way, the second way used to verify the readings of the flow-meter. The discharge values were in range of (0.7-15.0) l/s. Carrying out the test runs illustrated firstly by stabilizing the physical model of gabion weir inside the flume, filling the model with the gravel, covering the model, and operating the flume pump.

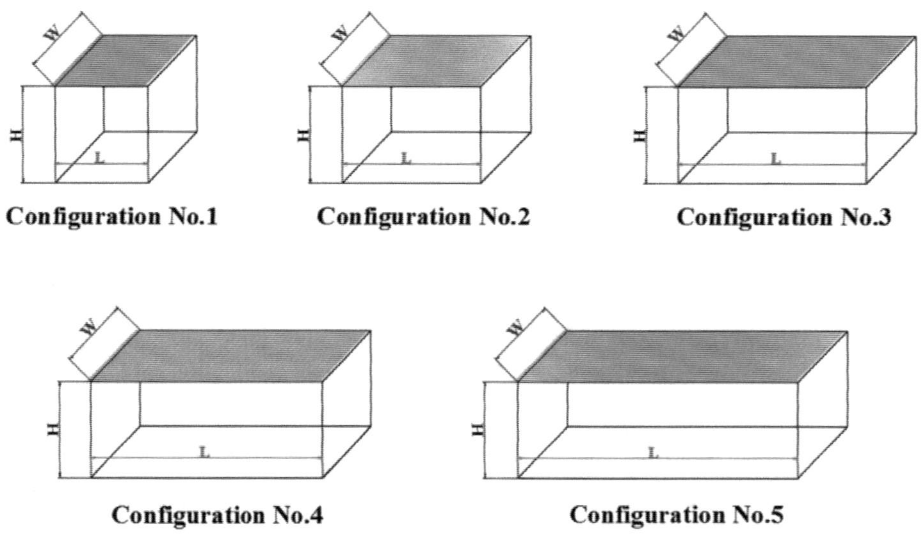

Figure 1. The physical models used in this study.

Table 1. Dimensions of the physical models.

Physical model number, (Configuration number)	Physical model dimensions		
	Height, H (cm)	Length, L (cm)	Width, W (cm)
1	40	40	30
2	40	60	30
3	40	80	30
4	40	100	30
5	40	120	30

Table 2. Details of the used gravel samples, and the value of porosity for each gravel sample.

Gravel Sample No.	Diameter of used gravel sample, (mm)	Porosity, n
1	9.5 – 14.0	48.8
2	14.0 – 19.0	51.6
3	19.0 – 25.0	52.3
4	25.0 – 37.5	52.8
5	37.5 – 50.0	53.5

Secondly by measuring the discharge for the flow, depths of water upstream and downstream the model, the Phreatic line of water through the model, and finally by repeating the steps above for seven times after reaching the study state of the flow.

3. Dimensional Analysis

Generally, energy dissipation depends on hydraulic and geometric variables. For the rectangular shape of gabion weir, these variables can be expressed functionally as:

$$f\{q, y_u, y_1, y_2, LT, dm, \rho, g, n, i_s\} = 0 \tag{1}$$

Where q is discharge per unit width measured by square length unit per time unit, (L^2/T), y_u is the depth of water at the upstream side of gabion weir (L), y_1 and y_2 are the depths of water at downstream side of gabion weir before and after the hydraulic jump respectively (L), LT is the gabion weir length (L), dm is the equivalent diameter of the used gravel sample (L), g is the gravitational acceleration (L/T^2), ρ is Mass density (M/L^3), n is porosity of the used gravel samples, and i_s is the slope of the laboratory flume. The energy at the upstream side of gabion weir was calculated by equation (2):

$$E_u = Z_u + y_u + (v_u^2/2g) = Z_u + y_u + (q^2/2gy_u^2) \tag{2}$$

Where E_u is the energy at the upstream side of gabion weir (L), Z_u is the elevation head (L), and v_u is the velocity of flow at the upstream side of gabion weir (L/T). While, the energy at the downstream side of gabion weir was calculated by equation (3):

$$E_1 = y_1 + (v_1^2/2g) = y_1 + (q^2/2gy_1^2) \tag{3}$$

Where E_1 is the energy at the downstream side of gabion weir (L), and v_1 is the velocity of flow at the downstream side of gabion weir (L/T). The difference in energy between the upstream and downstream sides of gabion weir, ΔE, was calculated by equation (4):

$$\Delta E = E_u - E_l \quad (4)$$

From equations (2), (3), and (4), it can be noticed that E_u depends on Z_u and y_u, Z_u depends on the slope of the laboratory flume, i_s, and E_l depends on y_1. So, equation (1) can be re-written as:

$$f\{q, \Delta E, LT, dm, \rho, g, n\} = 0 \quad (5)$$

By using of dimensional analysis (Buckingham Pi – Theorem), variables of equation (5) can be expressed by non-dimensional parameters as follows:

$$\Delta E/LT = f\{(dm/LT), (q/(g^{0.5} \,.\, LT^{1.5})), n\} \quad (6)$$

4. Results and Discussion

The In order to represent the relationship between the discharge and the energy dissipation, a non-dimensional data set was used. The dimensionless parameter $(q*(g^{0.5} \,.\, LT^{1.5})^{-1})$ was used for discharge representation, and the parameter (ΔE/LT) was used for representation of energy dissipation. Accordingly, the relationship above will be written as:

$$(\Delta E/LT) = f_1\{q*(g^{0.5} \,.\, LT^{1.5})^{-1}\} \quad (7)$$

By using of "Microsoft – Excel" computer program, it was found that the best form of equation of trend line which represents the relationship between energy dissipation dimensionless parameter, (ΔE/LT), and unit discharge dimensionless parameter, $(q*(g^{0.5} \,.\, LT^{1.5})^{-1})$ was the power form. Accordingly, equation (7) becomes :

$$\Delta E/LT = c_1\{q*(g^{0.5} \,.\, LT^{1.5})^{-1}\}^{k1} \quad (8)$$

Where c_1 and k1 are constants. Figures 2, 3, 4, 5, and 6 show the variation of values of energy dissipation dimensionless parameter with values of unit discharge dimensionless parameter for all test runs carried out in this study. In these figures the values of energy dissipation dimensionless parameter is directly proportional with the values of unit discharge dimensionless parameter, and values of former decreases by increasing the dm.

Tables 3, 4, 5, 6, and 7 respectively present the values of constants c_1 and k1 of energy dissipation dimensionless parameter-unit discharge dimensionless parameter relationship for configurations of this study.

Tables 8, 9, 10, 11, and 12 respectively summarize the values of difference in energy ΔE at minimum 0.7 l/s, and 3.6 l/s applied discharges where the discharge of 3.6 l/s was selected for comparison purposes. In these values ΔE at the minimum applied discharge was varied between 0.024 m and 0.225 m, and ΔE at 3.6 l/s applied discharge was varied between 0.112 m and 0.331 m. The latter group of tables show that generally the values of ΔE at the minimum and 3.6 l/s applied discharges decrease by increasing the value of dm. While the values of ΔE at the minimum and 3.6 l/s applied discharges increase by increasing the length of the gabion weir.

By using of multi-linear regression for the data set to correlate the dependent parameters with other independent ones, [12], [13], [14], [15], [16], [17], [18], [19], and [20], equation (6) can be re-written as:

$$\Delta E = 1.0538*10^{12}\{(q^{0.6387} \,.\, LT^{0.1629})*(dm^{0.1210} \,.\, g^{0.3194} \,.\, n^{6.6063})^{-1}\} \qquad R^2 = 0.92 \quad (9)$$

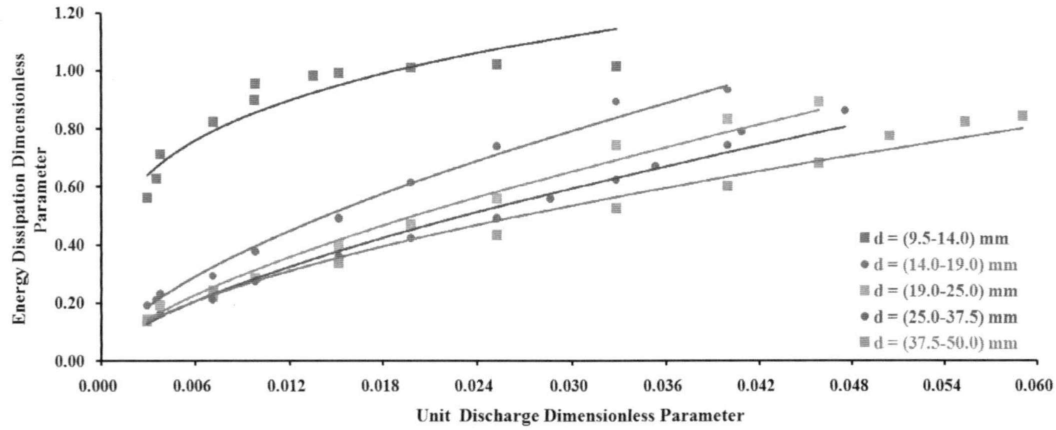

Figure 2. Variation of energy dissipation dimensionless parameter values with unit discharge dimensionless parameter values for tests runs of configuration number 1.

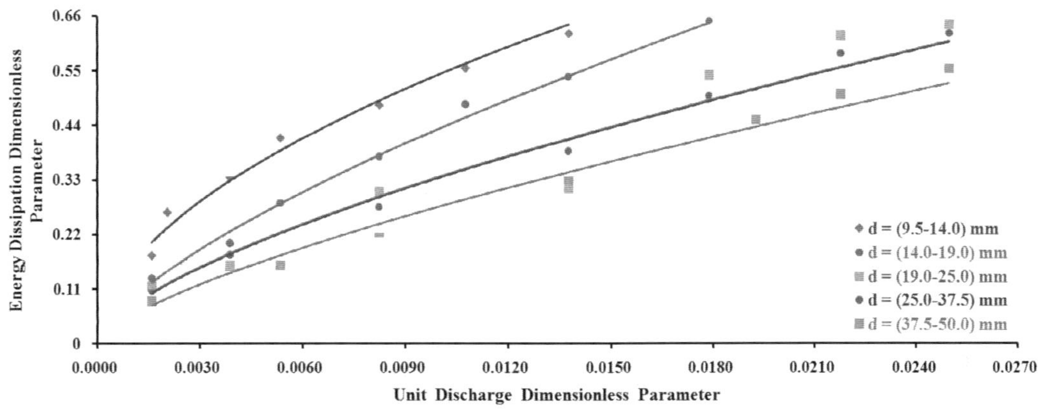

Figure 3. Variation of energy dissipation dimensionless parameter values with unit discharge dimensionless parameter values for tests runs of configuration number 2.

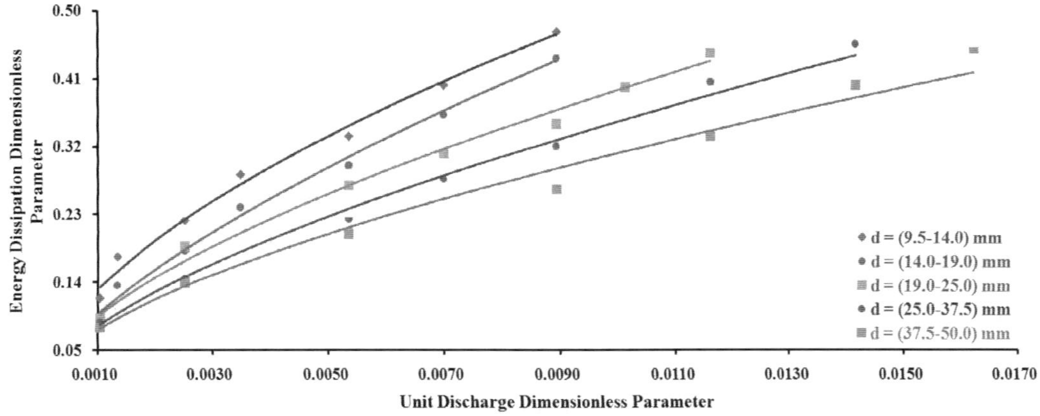

Figure 4. Variation of energy dissipation dimensionless parameter values with unit discharge dimensionless parameter values for tests runs of configuration number 3.

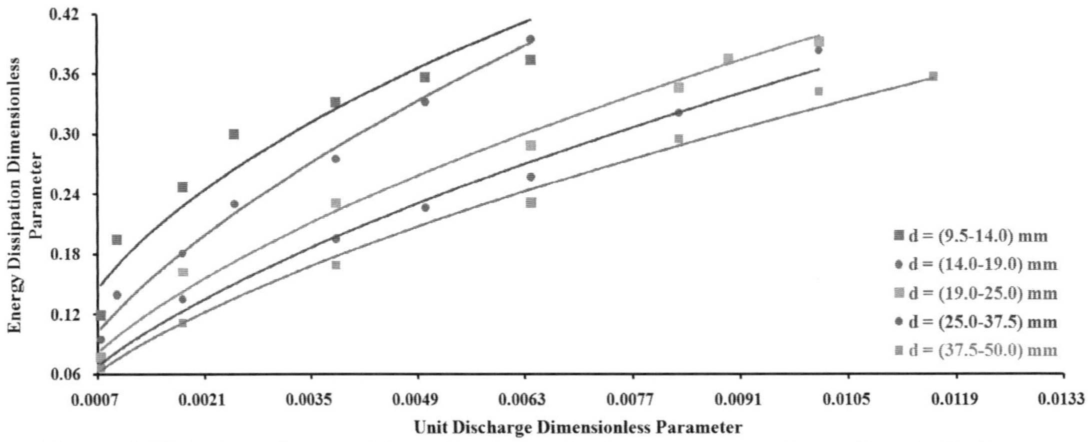

Figure 5. Variation of energy dissipation dimensionless parameter values with unit discharge dimensionless parameter values for tests runs of configuration number 4.

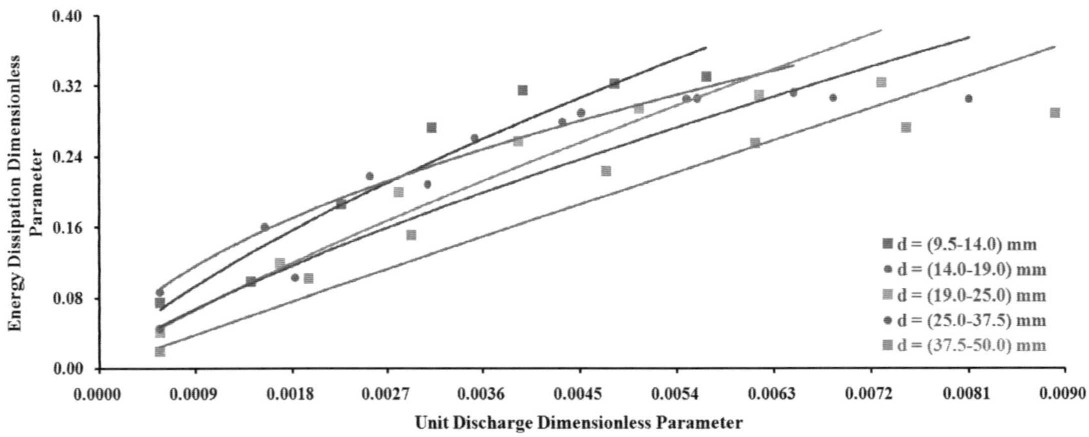

Figure 6. Variation of energy dissipation dimensionless parameter values with unit discharge dimensionless parameter values for tests runs of configuration number 5.

Table 3. Values of c_1, k1, and R^2 of trend line equation of energy dissipation dimensionless parameter – unit discharge dimensionless parameter relationship for configuration number 1.

dm (mm)	c_1	k1	R^2
9.5-14	2.6152	0.2418	0.87
14-19	7.0435	0.6230	0.99
19-25	6.5543	0.6581	0.98
25-37.5	6.0787	0.6632	0.99
37.5-50	4.3063	0.5949	0.99

Table 4. Values of c_1, k1, and R^2 of trend line equation of energy dissipation dimensionless parameter – unit discharge dimensionless parameter relationship for configuration number 2.

dm (mm)	c_1	k1	R^2
9.5-14	6.2872	0.5319	0.97
14-19	10.249	0.6867	0.99
19-25	6.7195	0.6507	0.95
25-37.5	6.7451	0.6521	1.00
37.5-50	6.9118	0.6991	0.99

Table 5. Values of c_1, k1, and R^2 of trend line equation of energy dissipation dimensionless parameter – unit discharge dimensionless parameter relationship for configuration number 3.

dm (mm)	c_1	k1	R^2
9.5-14	7.4574	0.5862	0.98
14-19	10.882	0.6825	0.98
19-25	6.8343	0.6190	0.99
25-37.5	6.6507	0.6374	1.00
37.5-50	5.1675	0.6101	0.99

Table 6. Values of c_1, k1, and R^2 of trend line equation of energy dissipation dimensionless parameter – unit discharge dimensionless parameter relationship for configuration number 4.

dm (mm)	c_1	k1	R^2
9.5-14	4.5409	0.4740	0.90
14-19	8.4444	0.6080	0.98
19-25	6.1932	0.5973	0.99
25-37.5	6.6308	0.6316	0.99
37.5-50	5.7588	0.6253	1.00

Table 7. Values of c_1, k1, and R^2 of trend line equation of energy dissipation dimensionless parameter – unit discharge dimensionless parameter relationship for configuration number 5.

dm (mm)	c_1	k1	R^2
9.5-14	16.259	0.7345	0.94
14-19	5.3307	0.5449	0.99
19-25	23.0600	0.8325	0.98
25-37.5	15.885	0.7783	0.98
37.5-50	37.630	0.9830	0.96

Table 8. Values of ΔE at minimum and 3.6 l/s applied discharges measured for configuration number 1.

dm (mm)	ΔE at minimum applied discharge (m)	ΔE at 3.6 l/s applied discharge (m)
9.5-14	0.225	-
14-19	0.077	0.179
19-25	0.058	0.160
25-37.5	0.055	0.145
37.5-50	0.055	0.135

Table 9. Values of ΔE at minimum and 3.6 l/s applied discharges measured for configuration number 2.

dm (mm)	ΔE at minimum applied discharge (m)	ΔE at 3.6 l/s applied discharge (m)
9.5-14	0.107	0.284
14-19	0.088	0.235
19-25	0.070	0.184
25-37.5	0.055	0.165
37.5-50	0.051	0.134

Table 10. Values of ΔE at minimum and 3.6 l/s applied discharges measured for configuration number 3.

dm (mm)	ΔE at minimum applied discharge (m)	ΔE at 3.6 l/s applied discharge (m)
9.5-14	0.095	0.266
14-19	0.071	0.236
19-25	0.074	0.214
25-37.5	0.069	0.179
37.5-50	0.064	0.163

Table 11. Values of ΔE at minimum and 3.6 l/s applied discharges measured for configuration number 4.

dm (mm)	ΔE at minimum applied discharge (m)	ΔE at 3.6 l/s applied discharge (m)
9.5-14	0.119	0.331
14-19	0.094	0.275
19-25	0.076	0.230
25-37.5	0.067	0.195
37.5-50	0.066	0.168

Table 12. Values of ΔE at minimum and 3.6 l/s applied discharges measured for configuration number 5.

dm (mm)	ΔE at minimum applied discharge (m)	ΔE at 3.6 l/s applied discharge (m)
9.5-14	0.094	0.308
14-19	0.104	0.297
19-25	0.049	0.262
25-37.5	0.053	0.224
37.5-50	0.024	0.112

5. Conclusion

The flow of water through and over rockfill materials is a wide important subject, especially in management of water resources and as an ecological friendly solution in various domains. Due to the porous nature of the gabion weirs, it allow the migration of some aquatic lives in the water path that leads to preservation of the water environment. Within the limitations of this study, it have been concluded that:

1- the dimensionless equation is possible to use for design purposes.
2- the energy dissipation is directly proportional with unit discharge.
3- the energy dissipation decreases with the increase of the equivalent diameter of the used gravel sample.
4- at the same gravel sample and discharge, the value of energy dissipation increases by increasing the length of the gabion weir.

References

[1] Chanson H and Toombes L 2004 Hydraulics of stepped chutes: the transition flow *J. Hydr. Res.* **24** 43-54.

[2] Kazemi-Nasaban Gh 1996 Flow characteristics through gabion stepped spillways M. Eng. Thesis (Shahid - Chamran University – Ahwaz - Iran).

[3] Li B and Garga V K 1998 Theoretical solution for seepage flow in overtopped rockfill J. Hydraul. *Eng.***124** 213-217.

[4] Michioku K, Maeno S, Furusawa T and Haneda M 2005 Discharge through a permeable rubble mound weir *J. Hydraul. Eng.* **131** 1-10.
[5] Michioku K, Takehara K and Etoh T 2007 An experimental study on flow field in and around rubble mound river structures *Journal of Hydro-Science and Hydraulics* **25** 37-45.
[6] Sargison J E and Percy A 2009 Hydraulics of broad-crested weirs with varying side slopes *J. Irrig. Drain Eng.* **135** 115-118.
[7] Chinnarasri Ch 2003 Energy loss through stepped gabion spillways *Proc. Energy Technology Toward a Clean Environment 2nd Regional Conf.* pp 436-441.
[8] Chanson H 2001 Hydraulic design of stepped spillway and downstream energy dissipaters *Dam Engineering* **11** 205-242.
[9] Salmasi F, Taghi-Sattari M and Mahesh-Pal 2012 Application of data mining on evaluation of energy dissipation over low gabion-stepped weir *Turk. J. Agric.* **36** 95-106.
[10] Kells J A 1994 Energy dissipation at a gabion weir with through and over flow *Can. Soc. Civ. Eng. Annul. Conf.* pp 26-35.
[11] Peyras L, Royet P and Degoutte G 1992 Flow and energy dissipation over stepped gabion weirs *J. Hydraul. Eng.* **118** 707-717.
[12] Azizi A, Meftah-Halghie M, Ahmedi M Z and Golmayie S A 2008 Evaluation the affection of used material porosity on energy dissipation in gabion stepped weirs *J. Agric. Sci. Natur. Resour.***15** 1.
[13] Vashisth A 2017 Energy dissipation over stepped gabion weir *International Journal of Dynamics of Fluids* **13** 153-159.
[14] Stephenson D 1979 Gabion weir energy dissipaters in *Proc. Congress 13th ICOLD* q50 r3 pp 33-43.
[15] Kang J, Kim S, Yeo H and Lee N 2014 Experimental study on flow characteristic in sloping weir *Engineering* **6** 329-337.
[16] Ashour M A, Sayed T and El-Attar S 2014 A new water energy dissipater for efficient energy dissipation and enriching the flow with dissolved oxygen content *Limnol, Rev.* **14** 3-11.
[17] Ashour M A, Sayed T and El-Attar S 2015 Impact of curved shaped energy dissipaters downstream diversion head structures to the dissolved oxygen content in irrigation canals and enhancement of irrigation water quality *International Journal of Scientific Research and Innovative Technology* **2** 14-26.
[18] Shafai-Bajestan M and Kazemi-Nasaban Gh 2011 Experimental study on gabion stepped spillway *Experimental Methods in Hydraulic Research, Geoplanet : Earth and Planetary Science*, pp 267-274.
[19] Gupta S K, Mehta R C and Dwivedi V K 2012 Modelling of relative length and relative energy loss of free hydraulic jump in horizontal prismatic channel *Proc. Nirma University 3rd Int. Conf.* vol 1 pp 529-537.
[20] Moussa Y A, Ali A M and Saleh Y K 2016 Performance of Sills Over Aprons under The Effect of Submerged Hydraulic Jump, (Case Study : Naga Hammadi Barrage) *Ain Shams Eng J* **9** (*Preprint 10.1016*).

Acknowledgments

The authors would like to express their gratitude to Babylon university/college of engineering and Al-Furat Al-Awsat technical university for their valuable assistance and support.

IOP Conf. Series: Materials Science and Engineering **584** (2019) 012039 doi:10.1088/1757-899X/584/1/012039

Experimental analysis of static capacity of fibrous concrete cambered slabs after impact loading effect

Al Kulabi Ahmed Kamil[1,*] and Al Zahid Ali Adnan[2]

[1, 2] Department of Civil Engineering, University of Kufa, Najaf, 54001, Iraq
*Corresponding author: ahmedk.kadhim@uokufa.edu.iq

Abstract. Fibrous concrete is an enhanced material by adding fibers to the ingredients of normal concrete. This addition improved some properties of the regular concrete, which are growing concrete ductility, decreasing concrete shrinkage, and raising concrete resistance to the dynamic attacks. The aim of this paper is more investigating the properties of fibrous concrete by finding out the effect of impact loading induced by a free fall ball on flexural capacity of fibrous concrete from one side, and finding out the importance of camber construction in reducing the influence of impact loading on the flexural capacity of concrete slabs from another side. For that purpose, three concrete slabs of (80X80X4) cm dimensions were dynamically and then statically loaded. The three specimens had the same steel fiber dosage of 0.5%, but they had different slab camber at the mid span where one specimen had no camber, the second specimen had 10 mm camber, and the third specimen had 20 mm camber. In addition to the difference in the slabs′ camber, the three slabs were subjected to a different number of impacts and from various heights to get a better explanation. Tests results showed that presence of fibers and increasing slab camber caused an improvement in the specimen resistance to both of impact and flexural loads, decreased the specimen crack width and number, and provided a safer residence for tenants.

1. Introduction

What is called researchers all over the world for improving the weak properties of the normal concrete is to achieve the building' structural integrity. Normal concrete is renowned in providing enough strength for compression loading, but it is not good in resisting impact loading, increasing the rate of causing structures collapse and disasters.

In the last years, the world suffered some actions that threated both of structures and people. Most affecting actions on buildings comprise terrorist threats and global worming actions which generate sudden dynamic loads in addition to the static ones; the thing that inspired specialists in concrete to produce a new type of concrete with enough capability in absorbing the created energy on structures by dynamic loads [1, 2].

Published studies categorized dynamic loads into two sets depending on the source that are generated from: naturally and humanly dynamic loads [3, 4]. Naturally dynamic loads include loads produced by natural world sources, such as winds, tornados, earthquakes, etc. while humanly loads comprise loads created by man-made sources, such as rockets, and blasts.

Fibrous concrete is the most preferred concrete for situations like these where it can provide 50 Mpa strength when it is compared with normal concrete that can achieve less than 50 Mpa strength [5]. And, it is more durable and can provide more strength for tensile loading comparing with normal concrete [6, 7].

Hence, to have a clear idea about how dynamic loading will influence the behaviour of fibrous concrete, how presence of camber will stand the dynamic loading, how height of impacting weight and number of impacts will reduce the flexural strength of the slabs, and how the failure type will be under these fluctuant parameters, authors of this paper intended to test three concrete slabs with different characteristics.
The three concrete slabs had the same dimensions of (80X80X4) cm and had the same content of steel fiber of 0.5%. Next, one of the slabs had no camber, which is named T4Ftest, where the bottom surface was plane, while the other two slabs had cambers of 10 mm, which is named T4C1, and 20 mm, T4C2. Moreover, the flat slab, T4Ftest was subjected to three impacts from a 90 cm height, T4C1 was subjected to one impact from a height of 120 cm, and T4C2 was subjected to four impacts; one from a 120 cm height and the others from 90 cm height as performed by the author Al Zahid [8]. After testing the three slabs dynamically, the same slabs were loaded and tested statically to get answers for all the questions listed in the previous paragraphs and to have a better explanation regarding structures loading capacity and behaviour when they experience unexpected dynamic loads plus static loads.

2. Materials and method

In this section, a brief description for all the used ingredients for casting the three fibrous concrete slabs is included, and the way of performing dynamic and static loading is summarized; comprising a description for the testing machines and the way of installing the slabs.

2.1. Materials

Fibrous concrete studied in this paper is produced by mixing all of: cement, sand, silica fume, steel fiber, and glenium.

- Cement: Ordinary Portland cement is the type of cement that is used for producing the three slabs, and it is manufactured locally by Al Najaf Al Shraf factory in Najaf, Iraq. This type of cement meets the Iraqi standard specification No. 5- 1984.
- Fine Aggregate: The particle sizes used for these slabs are within the range (600 μm) to (150 μm), and the physical properties are 2.65 specific gravity and 1.3% absorption.
- Silica Fume: Sika-Fume S92D is used for this work as illustrated in figure 1. It is useful as a binder and filler in concrete [9].
- Steel Fiber: WSF 0213 is used to enhance concrete ductility [10]. This type of fiber is made in China and meets (ISO 9001/2008) specifications. Fibers are with brass coated surface, with 2300 MPa tensile strength, straight, of 13 mm average length, of 0.2 mm diameter, and with 65 aspect ratio. figure 2. shows sample of steel fiber that was used.
- Glenium: The type of superplasticizer that was used is Glenium 54 as shown in figure 3 to improve the performance of concrete where it works on reducing water.

Figure 1. Silica fume.

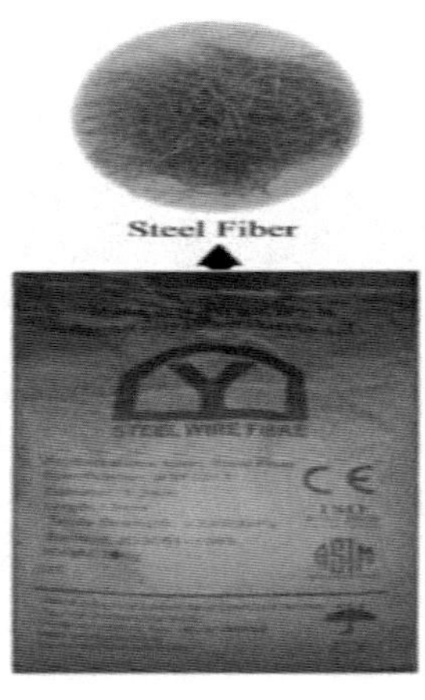

Figure 2. Steel fiber.

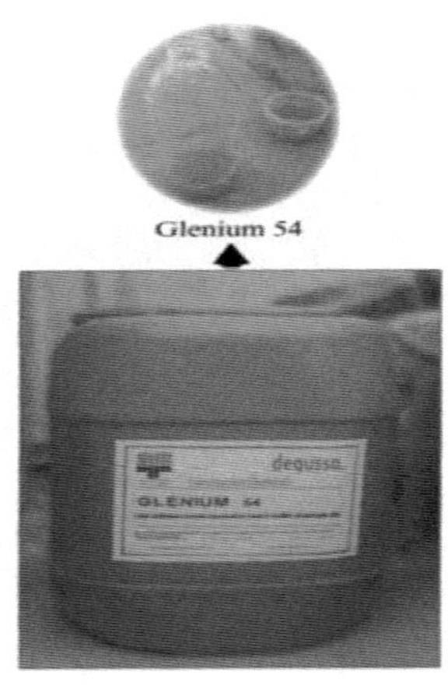

Figure 3. Glenium 54.

IOP Conf. Series: Materials Science and Engineering **584** (2019) 012039 doi:10.1088/1757-899X/584/1/012039

2.2. Impact testing machine

The drop-weight testing equipment was fabricated locally by the author Al Zahid [8] to find out the strength of the fibrous concrete slabs for impact loading as illustrated in figure 4.

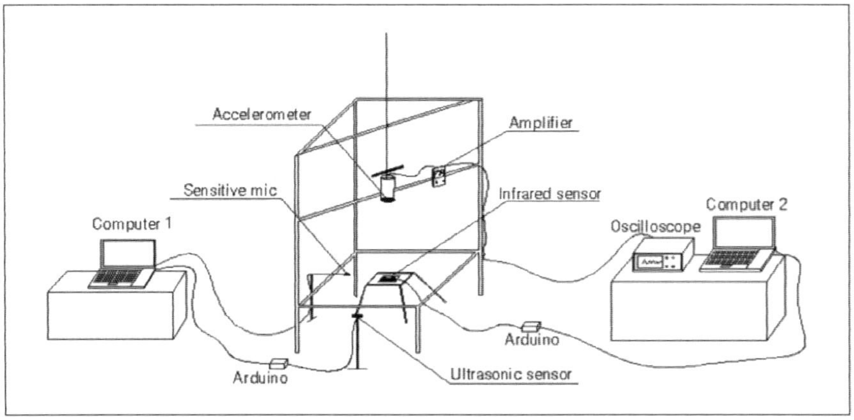

Figure 4. Drop-weight testing equipment.

The testing equipment was made up by using I and rectangular steel sections, and other supplements were needed as well to perform the test and get the required data, such as a dropping weight, an electronic ultrasonic sensor, an electronic accelerometer sensor, and an electronic infrared sensor.
A 5 Kilograms drop-weight was required to apply impacts on the set up slabs on the equipment where a steel ball was used for that purpose. Next, the reason for setting up the electronic accelerometer sensor is to know the acceleration of the drop weight by recording the impacting force made by the drop weight with time. And, the electronic infrared sensor was necessary to record the specimens´ displacement with time. While, the electronic ultrasonic sensor was set up to measure the specimens´ lateral displacement with time.

2.3. Methodology of performing impact loading test

After installing the fibrous concrete slabs on the drop-weight testing machine by the researcher Al Zahid [8], the impacting weight, steel ball, was fallen freely by the effect of its own weight on the top faces of the cambered slabs and at the center from different heights (90 and 120 cm) as illustrated previously. While the electronic sensors were put at the following positions: the electronic accelerometer sensor was put on the drop-weight to record the impacting force and time data, electronic infrared sensor was installed at the center of the bottom faces of the cambered faces to provide the author with the slabs deflection with time data, and the electronic ultrasonic sensor was fixed in a position where the lateral displacements of the cambered slabs can be obtained.

2.4. Static load testing methodology

Some researchers all around the world conducted a number of studies to define the characteristics and behaviors of both fibrous concrete and normal weight concrete from one side, and to well specify the differences between them. For instance, Aravind et al. [11] performed experimental research about the static capacity of reinforced fibrous concrete beams with dimensions of (1.5X1.5X0.18) m. Moreover, Jia et al. [12] conducted a laboratory investigation regarding the behavior of fibrous concrete under impact loads influence by subjecting two-ways reinforced concrete slabs to blast loading.
However, these experimentally conducted searches and other studies did not investigate deeply the properties and behaving ways of concrete after applying concrete to dynamic loads. The reason that inspired the authors of this paper to perform a laboratory study and then provide a complete

interpretation concerning the flexural capacity and behavior of fibrous concrete after applying impacts forces to those who are interested in designing structures.
Therefore, the same slabs that were applied to impact loading where installed one more time on the hydraulic flexural strength testing machine, which is shown in figure 5.

Figure 5. The hydraulic static testing machine.

A solid square cube with dimensions of (15X15) cm was set up at the center of the top face of slabs to distribute the applied load by the hydraulic machine in two directions of slabs and realize the condition of two-way slabs.

3. Results and discussions

In this section, the results of the three tests: compression test, impact loading test, and flexural loading test are included. In addition, the outcomes of the tests are discussed in details.

3.1. Compression test results

Three cubes for each slab were cast on the same day of casting the slabs and tested for knowing their compression strength by the author Al Zahid [8]. The average of each three cubes and for each slab is calculated and included in table 1. The cubes set of the slabs were tested at different ages as it can be noticed from table 1.

Table 1. Average compressive strength test of three cubes.

Slab	Compression Strength (MPa)	Age (Day)
T4Ftest	84.265	47
T4C1	89.313	62
T4C2	85.280	63

3.2. Impact loading test results

Table 2 comprises the following information: the maximum impact loading capacity in Kilo-newton for each slab, the number of impacts, the height of dropping weight, and the corresponding deflection for the maximum load in mm.

Table 2. Impact test results.

No. of Impacts	T4Ftest		T4C1		T4C2			
	90 cm	Deflection	120 cm	Deflection	90 cm	Deflection	120 cm	Deflection
1st Impact	4.45	3.40	8.91	2.71	N/A	N/A	11.87	1.79
2nd Impact	3.71	4.3	N/A	N/A	7.92	2.07	N/A	N/A
3rd Impact	3.36	3.21	N/A	N/A	7.17	2.25	N/A	N/A
4th Impact	N/A	N/A	N/A	N/A	6.68	4.28	N/A	N/A

As it is clear from table 2, the slab with the biggest camber, T4C2, stood the highest impacting load in the first loading for approximately 11.87 kN with the least corresponding deflection of 1.79 mm when it is compared with the other two slabs. While, the slab with a less camber, T4C1, provided less strength for the impacting force which is 8.91 kN with a higher corresponding deflection of 2.71 than T4C1. And, the slab with zero camber, T4Ftest, endured the least impact loading for nearly 4.45 kN with the highest corresponding deflection of 3.4 mm comparing with the other two slabs, although it was subjected to the dropping weight from a lower height of 90 cm when it is compared with the height of the drop-weight of the T4C1 and T4C2 that is 120 cm in the first loading.
Next, the cracks progression after each impact loading is photographed for all the slabs. Figures. 6 (a through c) show the cracks growth of the slab T4Ftest where the crack width was not specified by the author Al Zahid [8] because of an electronic problem. The figures illustrate the slab's stiffness is decreasing after each dynamic loading.

(a) (b) (c)

Figure 6. The crack pattern of the slab T4Ftest after a) the first impact loading, b) the second impact loading, and (c): the third impact loading.

Moreover, figure 7 shows the created cracks after freely applying a drop-weight on the slab T4C1. The average cracks width was 0.0362 mm.

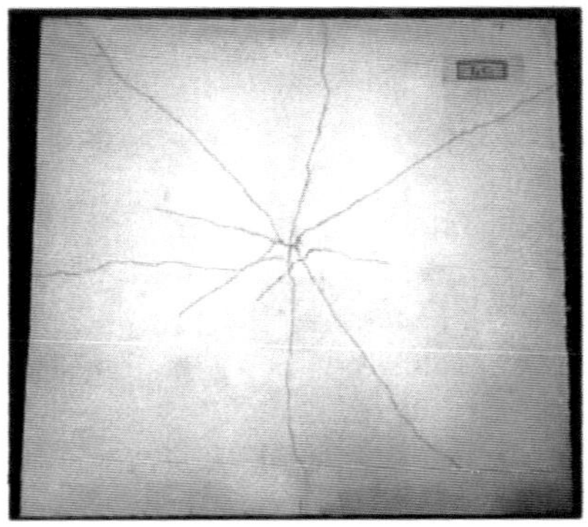

Figure 7. The crack pattern of the slab T4C1 after applying an impacting load.

Finally, figures 8 (a through d) exhibit the cracks development of the slab T4C2 after freely dropping a weight from a 120 cm height in the first time and from a 90 cm height in the other three subsequent stages of dynamic loading. The average cracks width was 0.0175, 0.02, 0.02, and 0.025 mm in the 1st, 2nd, 3rd, and 4th loading stage, respectively.

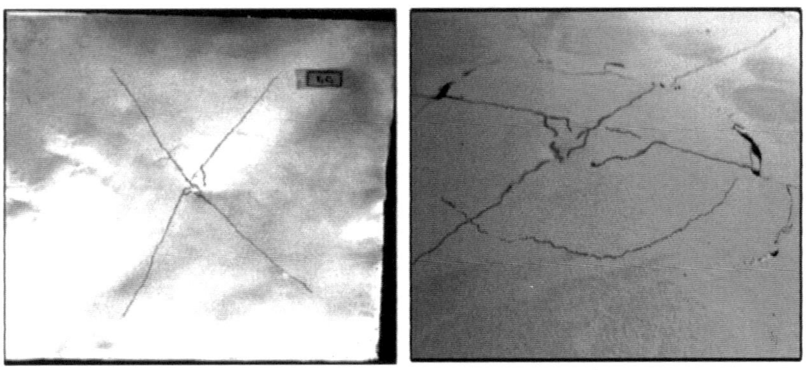

(a) (b)

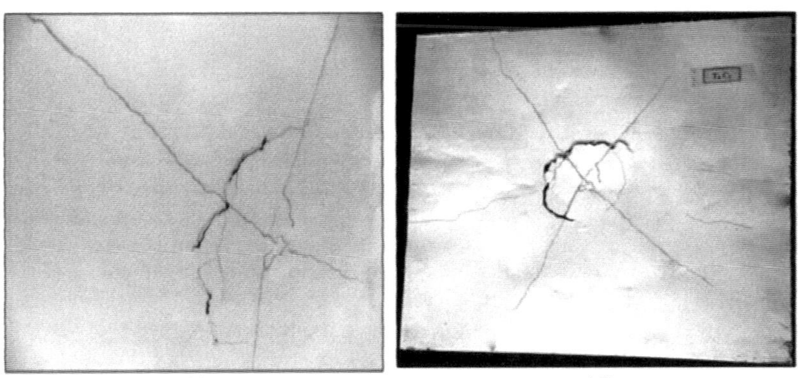

(c) (d)

Figure 8. Crack pattern of the slab T4C2 after (a): the 1st stage of loading, (b): the 2nd stage of loading, (c): the 3rd stage of loading, and (d): the 4th stage of loading.

3.3. static load testing results

As specified in the previous sections, the three impacts loaded fibrous concrete slabs were loaded one more time statically. The slabs were progressively statically loaded where the static load was increased gradually by 0.5 kN till slabs failure, and the corresponding deflection for each load increment was recorded. Table 3 includes the maximum flexural capacity of the slab T4Ftest after being impacted for three times from a height of 90 cm, the maximum flexural capacity of the slab T4C1 after being impacted one time from a height of 120 cm, and the maximum flexural capacity of the slab T4C2 after impacting it from a height of 120 cm for one time and from a height of 90 cm for three times. Besides, table 3 comprises the deflection corresponding to the maximum flexural loading of each slab.

Table 3. Static test results.

Slab	Static Load (kN)	Deflection (mm)
T4Ftest	19.5	13.6
T4C1	16.5	4.91
T4C2	17.0	4.21

From the results of static loading in table 3, the relationship among the maximum flexural strength, deflection, and camber height of the slabs was a non-uniform relation since the slabs were subjected to a different number of impacts and from different heights. And so, the slab with no camber, T4Ftest, provided the highest flexural strength of 19.5 kN, and it was the most deflected slab for about 13.6 mm, the slab with 1 cm camber, T4C1, stood a less static loading than T4Ftest which was 16.5 kN and deflected less for approximately 4.91 mm, and the slab with 2 cm camber provided a higher flexural strength than T4C1 but less than T4Ftest, and it was the least deflected slab than the other two slabs for nearly 4.21 mm. Figure 9 illustrates the develop of load-deflection curves of the three fibrous concrete slabs.

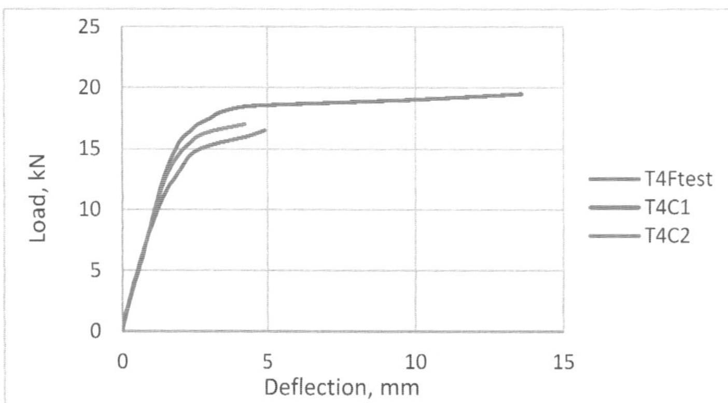

Figure 9. Load-deflection curves of the fibrous concrete slabs.

The average crack width and number generated by flexural loading were specified and compared for all slabs. For that purpose, the cracks created during impact loading were drawn using a blue marker, and the cracks caused by flexural loading were marked using a red marker to recognize the two generated cracks.

Loading the slab T4Ftest statically led to more expanding the impact loading cracks and to creating new cracks. The average cracks width generated by the flexural loading is 1.718 mm. Moreover, new five main cracks showed up; the cracks started from the loading point toward the slabs edges. As it is shown from figure 10 a, the new cracks were spread out on the bottom face of the slab through the four quarters of the slab.

Figure 10 b shows the cracks distribution and number of the slab T4C1 after the flexural loading. For this slab, some of the new cracks tracked the impact cracks and then continued their way to the slab edges starting from the loading point. While, the other new cracks started from the center of the slab deriving new paths to the slab edges. The average cracks width is 1.023 mm.

Finally, the slab T4C2 had a different cracks pattern and distribution as shown in figure 10 c. This slab had a circular crack extending around the center of the slab with a number of diagonal cracks where some of them started from the center of the slab, and the others developed from the boundaries of the circular crack. Almost all the diagonal cracks extended through different paths of the impact loading cracks paths on the bottom face of the slab. The flexural loading cracks have an average width of 1.69 mm.

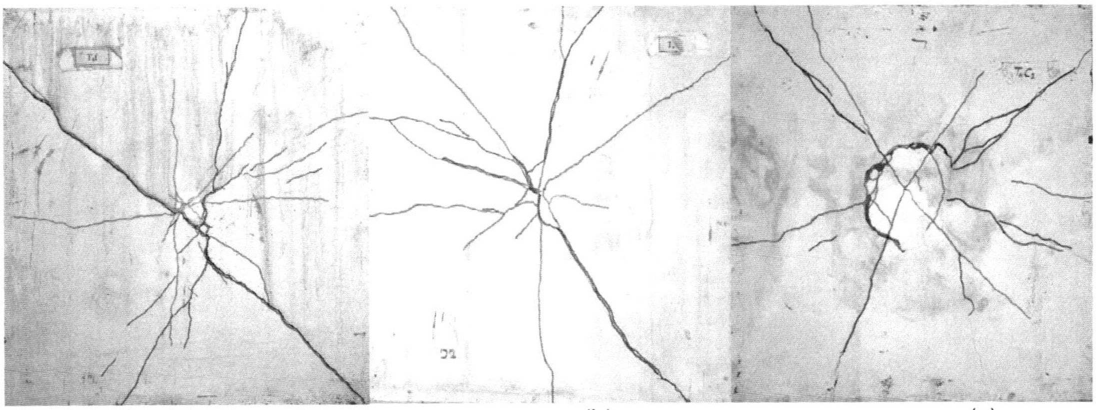

(a) (b) (c)

Figure 10. Impact-Flexural cracks pattern of the slab a): T4Ftest, b): T4C1, and c): T4C2.

3.4. Results discussion

Based on the experimental observations and outputs, discussion the results of dynamic and static tests of the fibrous concrete slabs is performed.

As it can be seen from the previous sections, the height of camber represents the most affecting parameter on specifying the maximum impact loading, maximum static loading, deflection, width of cracks and numbers, and failure type.

Concerning maximum loading capacity, increasing slabs camber height led to increasing the slabs resistance to the applied loads. For example, increasing slab camber by 10 mm (T4Ftest and T4C1) led to increasing impact resistance by 100.25% although the free weight was dropped from a higher elevation on the slab T4C1 which is 120 cm, and increasing the slab thickness by 20mm (T4Ftest and T4C2) led to increasing impact resistance by 166.74% from a higher elevation (120) cm in the first dynamic loading comparing with impacting distance of the slab T4Ftest which is 90 cm. Figures 11 through 14 show the influence of increasing slabs camber on the dynamic loading capacity of fibrous concrete slabs in the 1st, 2nd, and 3rd loading stages. Since, the slabs T4C1 was impacted from a distance of 120 cm for one time and the slab T4C2 from the same distance for one time and 90 cm for three times, these slabs withstood less static loading than the slab T4Ftest as illustrated in figure 14, but they deflected less as it would be explained in the deflection section.

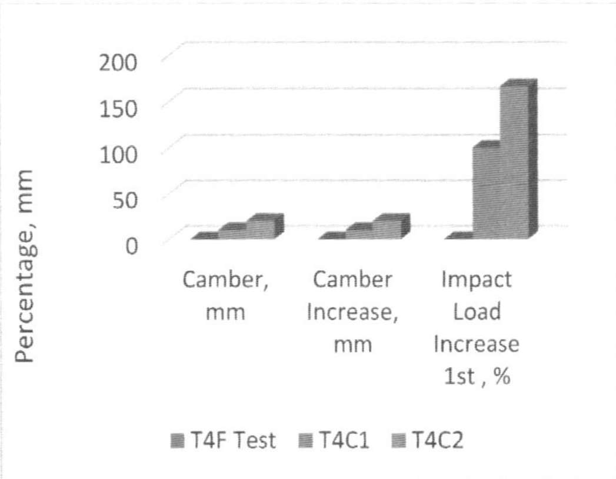

Figure 11. The relation between camber height and dynamic loading capacity in the first loading stage.

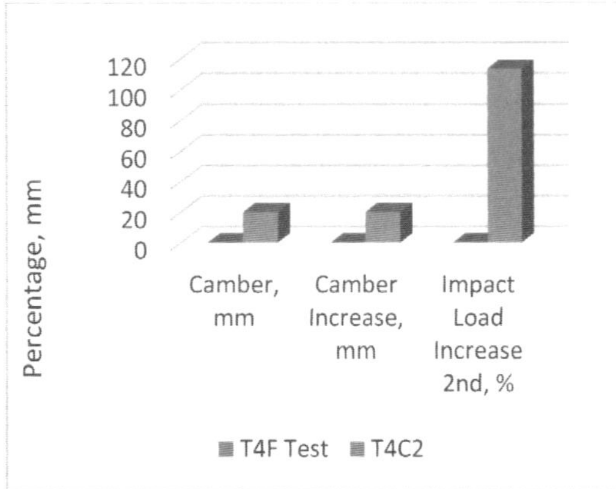

Figure 12. The relation between camber height and dynamic loading capacity in the second loading stage.

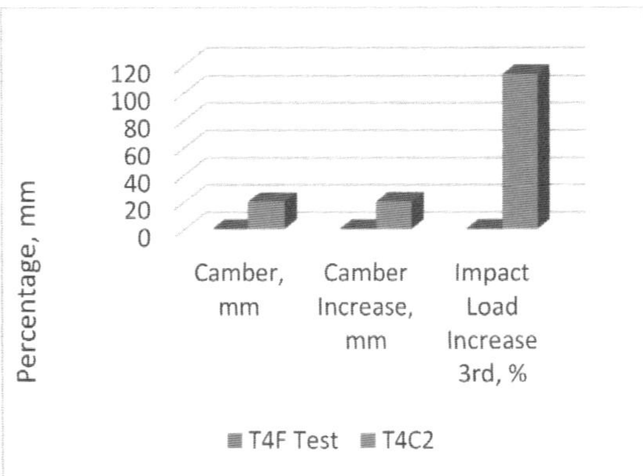

Figure 13. The relation between camber height and dynamic loading capacity in the third loading stage.

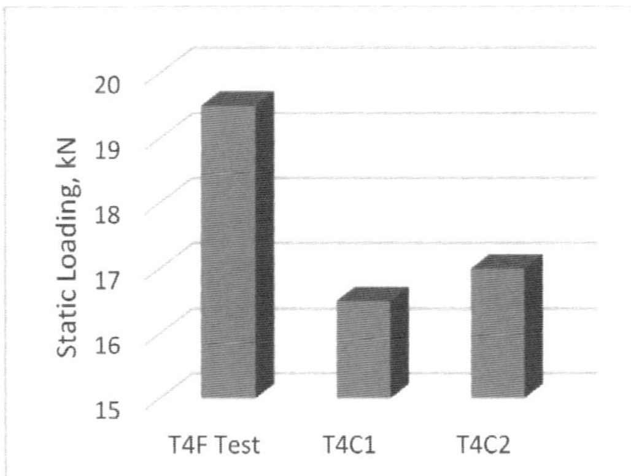

Figure 14. Static loading capacity of the three after dynamic loading stages.

From figure 14, T4Ftest resisted the highest flexural load while the flexural load of the slabs T4C1 and T4C2 was 15.4% and 12.8% less than the load of the slab T4Ftest, respectively; the impacting load was dropped from a higher distance than the impacting distance of the slab T4Ftest.

Regarding deflection, what was found from the results that the relation between slabs deflection and camber is an inverse relation where increasing the slabs camber leads to decreasing slabs deflection. For instance, in the 1st dynamic loading stage increasing the slab camber by 10 mm (T4Ftest and T4C1) and 20mm (T4Ftest and T4C2) led to decreasing deflection by 20.3% and 47.4%, respectively, although these slabs, T4C1 and T4C2, were impacted from a higher distance and withstood more loading as explained previously. Figures 15 (a through c) illustrate the importance of raising up slabs camber in decreasing slabs deflection. Moreover, deflection of slabs caused by static loading was decreasing with increasing slab camber, see figure 15 d.

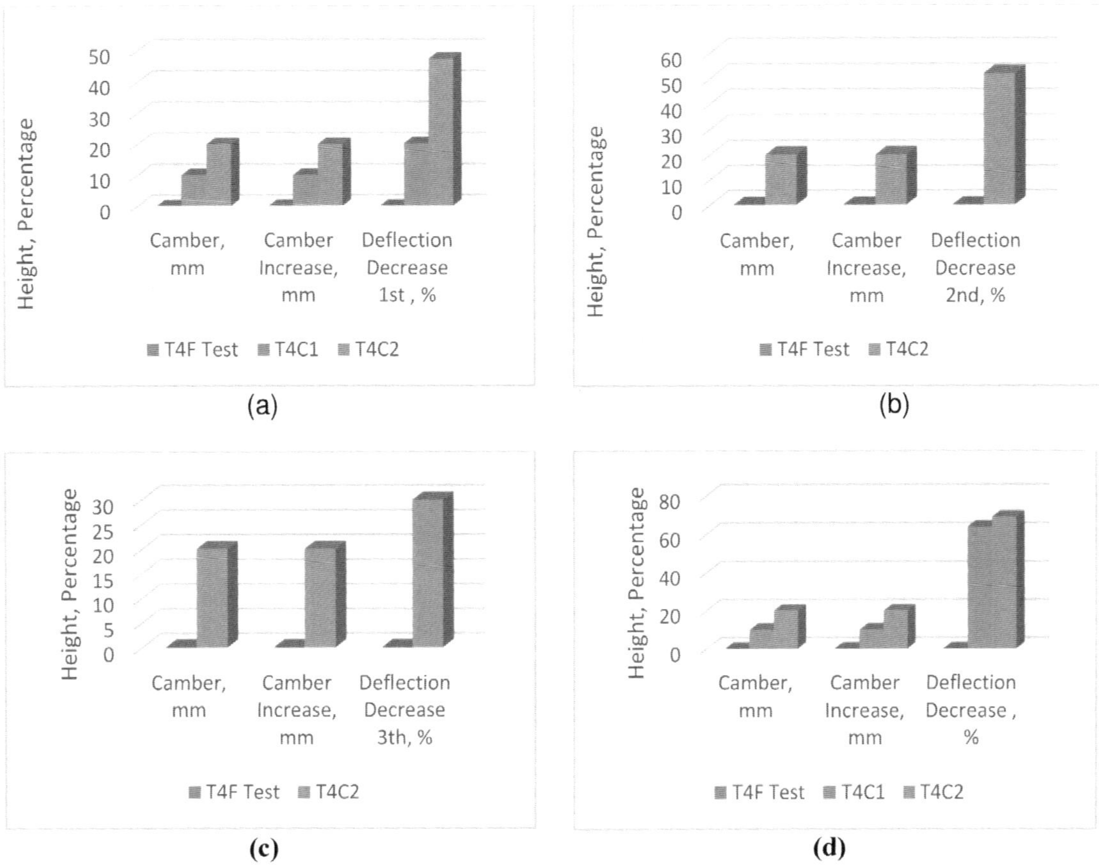

Figure 15. The relation between slab camber and deflection in the (a): 1st stage of dynamic loading, (b): 2nd stage of dynamic loading, (c): 3rd stage of dynamic loading, and (d): static loading.

Respecting crack width, the developed cracks during dynamic loading is a dependent function of three variables, which are height of impacting weight, number of impacts, and slab's camber. Figure 16 shows how camber height affects the width of the developed crack where increasing slab's camber by 100% led to decreasing the average crack width by 51.7% when they impacted from the same height and by the same weight. Moreover, the average crack width of the slab T4C2 was increasing gradually by repeating the dynamic loading in the following loading stages where crack width was 0.02, 0.02, and 0.025 mm in the 2nd, 3rd, and 4th loading stage, respectively. The same concept is applicable for the static loading case where slab camber is the key parameter in controlling the width of the developed cracks; increasing slab camber by 10 mm (T4Ftest and T4C1) led to decreasing crack width by 40.5%, and increasing crack width by 20 mm (T4Ftest and T4C2) led to decreasing crack width by 1.6%. For the slab T4C2, the decrease percentage was less because as explained previously this slab was impacted for three more times than the slab T4C1.

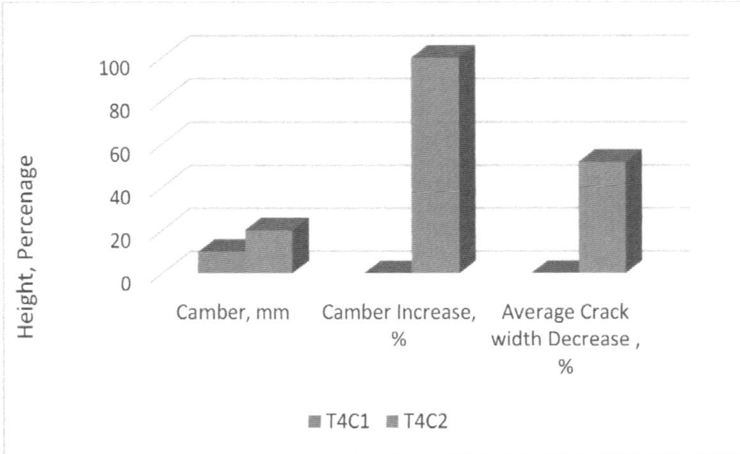

Figure 16. Effect of raising slab camber on reducing crack width developed from dynamic loading.

Failure type, a number of factors specifies failure type of concrete slabs, such as concrete type: normal weight or fibrous concrete, loading type: dynamic or static, position of loading, loading amount, etc. For dynamic loading, Jia et al. [12] found that failure type depends mainly on weights of explosives and hitting points or positions of the slabs; whether explosives will impact slabs at their borders or center. Moreover, they proved that degree of damage of slabs will increase by increasing explosive weight and/or by moving the impacting position toward slab borders away from center, and the type of failure will change gradually from flexural to flexural shear failure. Consequently, these factors are the key factors in specifying slabs failure when they will be statically loaded after being dynamically loaded. For this study, different slabs failure was observed; that is because of differences in height of impacting loading, number of impacts, and camber height. Hence, failure type of the slabs T4Ftest and T4C1 is flexural, while the slab T4C2 failed in punching shear.

Finally, buildings collapse is the main concern of designers. Buildings that may be exposed to explosions threaten their residents' life to the risk of sudden collapse. Therefore, structures which are constructed by slabs with cambers as it was noticed from the previous results are the safest structures where more loads can be withstood and less crack number and width would be caused; consequently, the cost of repairing of this kind of structures is less.

4. Conclusions

Researchers all around the world are inspired to more investigate and improve concrete properties attempting by that providing safe structures after increasing the risky rates on residents because of collapse cases as a result to the terrorism attacks. Hence, three concrete slabs with dimensions of (80X80X4) cm, steel fiber of 0.5%, and with different camber height: 0, 10, and 20 mm were dynamically and statically loaded. The results showed that concrete slabs with the highest camber are the most resisting slabs to impact and static loads and the safest slabs for tenants where deflection and cracks number and width are less. Moreover, slabs with the highest camber can be maintained and with a less cost after being exposed to attacks.

References

[1] Chen J.J., Ng P.L., Li L.G., and Kwan A.K.H. 2017 Modern building materials, structures and technologies, MBMST 2016, ELSEVIER, Vol. 172, pp. 165-171.

[2] Vinayagam P. 2012 Experimental investigation on high performance concrete using silica fume and superplasticizer, International Journal of Computer and Communication Engineering, Vol. 1, pp. 168-171.

[3] Patel P. B., Thakur L. S., Maroliya M. K., and Patel B. A. 2017 Shear and impact behavior of reactive powder concrete with varying powder content, *Kalpa*

IOP Conf. Series: Materials Science and Engineering **584** (2019) 012039 doi:10.1088/1757-899X/584/1/012039

publications in CE, Vol. 1, pp. 307-314.

[4] Koccaz Z., Sutcu F., and Torunbalci N. 2008 Architectural and structural design for blast resistance buildings, in *The 14th WCEE*, 12-17.

[5] Al-Jubory N. H. 2013 Mechanical properties of reactive powder concrete (RPC) with mineral admixture, *Al-Rafadain Engineering Journal*, Vol. 21, pp. 92-101.

[6] Wu Z., Shi C., and Wu L. 2016 Effects of steel fiber content and shape on mechanical properties of ultra-high performance concrete, *ELSEVIER*, Vol. 103, pp. 8-14.

[7] Singh S. and Ritu 2016 Natural reactive powder concrete (Rpc) – in form of a superplasticized portland cement mixture with silica fume, steel fibers and ground fine quartz – a review *International Journal of Technical Research (IJTR)*, Vol. 5, pp. 217-222.

[8] AL Zahid M. A. A. 2016 Impact resistance of cambered reactive powder concrete slabs with steel stiffeners, PHD Eng. thesis, Babylon University, Babylon, Iraq.

[9] Rasol M. A. 2015 Effect of silica fume on concrete properties and advantages for Kurdistan region, Iraq, International Journal of Scientific and Engineering Research, Vol. 6, pp. 170-173.

[10] Baarimah A. O. and Mohsin S. M. S. 2017 Behaviour of reinforced concrete slabs with steel fibers, in *IOP*, 271-012099.

[11] Aravind M., Senthil K. V., and Manikandan G. 2017 Flexural behavior of high strength reactive powder concrete, *SSRG International Journal of Civil Engineering*, pp. 350-354.

[12] Jia H., Yu L., and Wu G. 2014 Damage assessment of two-way bending RC slabs subjected to blast loadings, *The Scientific World Journal,* Vol. 2014, pp. 1-12.

The behavior of Lightweight Aggregate Concrete Made with Different Types of Crushed Bricks

H Adem [1], E Athab [2], S Thamer [3], AT Jasim[4]

[1] BSc graduate, Department of civil engineering, University of Kufa, Iraq. Email: hadiadem95@gmail.com

[2] BSc graduate, Department of civil engineering, University of Kufa, Iraq. Email: runrun9626@gmail.com

[3] BSc graduate, Department of civil engineering, University of Kufa, Iraq. Email: SaraAlhmedawi@gmail.com

[4] Asst. Prof., Department of civil engineering, University of Kufa, Iraq. Email: alit.albozwaida@uokufa.edu.iq

Abstract. Lightweight concrete can significantly reduce a dead load of structural concrete elements compared to normal weight concrete. The reuse of construction and demolition wastes, especially crushed clay bricks, represents a major contribution to the environment. Due to the nature of clay bricks, it can be considered as source of fine and coarse aggregate to produce lightweight structural concrete .In this study, used three types of crushed clay brick as coarse aggregate for concrete aiming to produce structural lightweight concrete. Density, compressive and flexural strengths were tested. Results show that the 28-day air-dry density, compressive and flexural strengths values of the concrete made with Iso crushed brick were higher than that for concrete made with crushed Iranian and Nahrawan bricks. Results of 28 air-dry density and compressive strength show that the lightweight concretes used were complied with the requirements for lightweight structural concrete.

Keywords: Lightweight concrete, Crushed bricks aggregate, Compressive strength, Flexural strength, Density.

1. Introduction

The use of waste aggregate in concrete is not new, it used widely since World War II in Germany [1]. 400-600 Million cubic meters was used in concrete construction , the use of waste clay bricks is a benefit as it reduces environmental pollution and decreases the cost of concrete since the volume of aggregate in the concrete of about 60 -75 % . Using waste bricks lead to reduce the density of concrete and also to reduce compressive strength. Hiremath and Sanjay investigation [2], use waste clay brick aggregates as replacement of coarse aggregate with the different ratio from 0 to 100%, compressive strength decreased from about 31 MPa to 11 MPa, but tensile strength increased by 25 % replacement with 20% increment in splitting tensile strength. Subramani [3] showed that the using 50% crushed brick as coarse aggregate replacement increases the value of flexure strength by about 5% compared to reference mix. Waste thermostone also used as coarse and fine aggregate to provide lightweight concrete [4], and lead to reduce compressive strength by 75% by using 100% replacement. Padmini, et al. [5] showed that the density of brick aggregate concrete is 15% lower as compared conventional concrete, and it increases with cement content and reduces with increase in water cement ratio. This investigation aims to study the possibility of produce lightweight concrete using crushed clay bricks as coarse aggregates. In this research, the density, compressive strength and flexural strength are studied. The main variables in this investigation are; types of crushed bricks used *(Iranian crushed brick, Nahrawan crushed brick and Iso crushed brick as coarse aggregates)*, and W/C (0.45 and 0.30)

2. Experimental program

2.1 Materials

2.1.1 Cement

Ordinary Portland cement manufactured by KAR Group was used in all mixes throughout this study. Tables 1 and 2 show the chemical analysis and physical properties of the used cement. The results show that the cement conformed to Iraqi specifications (I.S.O) No. 5/1984.

Table 1. Chemical properties of used cement*.

Oxides composition	Content %	I.Q.S.NO.5,1984[50]
CaO	56	
SiO_2	19	
AL_2O_3	3.98	
Fe_2O_3	3.76	
Lime saturation Factor(L.S.F)	0.962	0.66-1.02
MgO	2.79	< 5%
C_3A	4.187	
SO_3	2.32	< 2.8%
Loss on ignition(L.O.I)	3.04	< 4%
Insoluble Residue	1.01	< 1.5%

*Chemical analysis tests were carried out by the Engineering Consulting Bureau. Laboratories-University of Kufa

Table 2. Physical Properties of Cement*.

Physical properties	Test result	limits of IOS No.5:1984
Fineness: specific surface, Blaine cm^2/gm	2760	2300 (minimum)
Setting time, Vicat's method:-		
Initial (hrs: min.)	1:45	00:45 (minimum)
Final (hrs :min)	4:10	10: 00 (maximum)
Compressive strength of cement mortar using cubes (70.7mm) MPa		
3 days	20.3	15 (minimum)
7 days	28.6	23 (minimum)

* Physical tests were carried out by the Engineering Consulting Bureau. Laboratories- University of Kufa

2.1.2 water

Ordinary tap water was used for mixing and curing for all concrete mixes of this study.

2.1.3 High Range Water Reducing Admixture (HRWRA)

A high performance concrete superplasticizer based on modified polycarboxylic ether, which is known commercially (SP 90), was used throughout this investigation as a (HRWRA). It is the third generation of superplasticizers and it complies with ASTM C 494-2003 Type F

2.1.4 Fine aggregate

Normal weight natural sand, brought from Al-Najaf quarry was used as fine aggregate in this work. The grading is given in Table 3. The physical and chemical tests on sand used throughout this work are shown in Table 4. Results indicated that the fine aggregate grading and sulfate content are within the requirements of the Iraqi specification No.45/1984.

Table 3. Grading of the used sand and the requirements of I.O.S 45/1984.

Sieve size	% Passing	% Passing limits of I.O.S 45/1984 (Zone 2)
10 mm	100	100
4.75 mm	100	90-100
2.36 mm	97	75-100
1.18 mm	80	55-90
600 μm	53	35-59
300 μm	12	8-30
150 μm	6	0-10

Table 4. Chemical and physical properties of the used sand.

Property	Specification	Result	Limit of I.O.S No. 45/1984
Bulk Specific gravity	ASTM C128-2003	2.53	
Absorption, %	ASTM C128-2003	1.9	
Dry loose unit weight, kg/m^3	ASTM C29-2003	1580	
Sulfate content (as SO_3), %	I.O.S No.45-84	0.08	0.5 (Max.)
Materials finer than 0.075 mm sieve, %	I.O.S No.45-84	1.1	5.0 (Max.)

2.1.5 Coarse aggregate

Three types of crushed bricks with maximum size (19 mm) were used in this study as shown in Figure 1. Table 5 shows the compressive strength of bricks and the density and absorption of crushed bricks. Table 6 shows the grading of crushed brick aggregate.

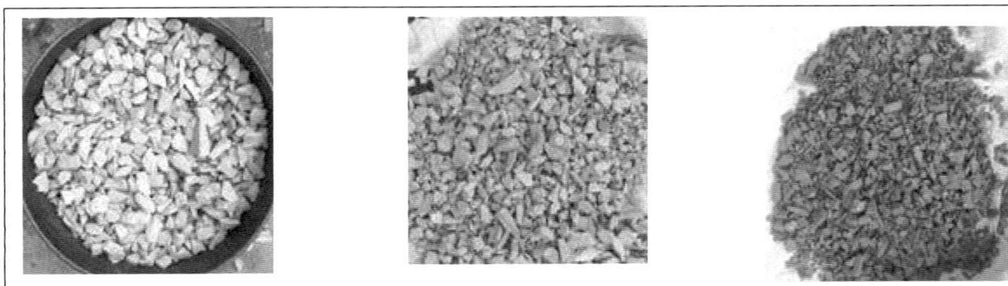

(a) (b) (c)

Figure 1. Crushed bricks (a) Iranian brick (b) Nahrawan brick (c) Iso brick

Table 5. Properties of three types of crushed bricks used as coarse aggregate.

Brick type	Density, kg/m^3	Absorption %	Compressive strength of brick (MPa)
Iranian Bricks	816	32.30	10.10
Nahrawan Brick	835	23.14	17.08
Iso-Bricks	889	15.87	30.21

Table 6. Selected grading of coarse lightweight aggregate.

Sieve size (mm)	Selected % passing	% Retained	% Passing ASTM C330
19	100	0	90-100
12.5			
9.5	50	50	10-50
4.75	10	40	0.15
2.36	0	10	-

2.2 Mix concrete design

Six different LWAC (Lightweight Aggregate Concerte) mixes (identified as MIR, MNH, MIS, MIR-H, MNH-H, and MIS-H), were used in this study. [MIR-mix made with Iranian crushed bricks, MNH- mix made with Nahrawan crushed bricks , MIS- mix made with Iso crushed bricks] and the other three mixes containing High Range Water Reducing Admixtures. The six concrete mixes differed by water/cement ratio and the type of crushed brick used. These mixes were designed in accordance with ACI committee 211-2[6] to produce a structural LWAC. The details of the mixes used throughout this investigation are given in Table 7.

Table 7. Details of the mixes used throughout this study.

Type of mix	Cement Kg/m^3	Fine agg,	Crushed brick as Coarse Agg.	Type of crushed brick	w/c	HRWRA% by wt. of cement	Slump mm
MIR	350 kg	960 kg	480 kg	Iranian	0.45	-	100
MNH	350 kg	960 kg	480 kg	Nahrawan	0.45	-	105
M IS	350 kg	960 kg	480 kg	Iso	0.45	-	110
MIR -H	350 kg	960 kg	480 kg	Iranian	0.30	2.00	110
MNH-H	350 kg	960 kg	480 kg	Nahrawan	0.30	2.00	105
M IS-H	350 kg	960 kg	480 kg	Iso	0.30	2.00	110

2.3 Preparation and Curing of Test Specimens

All concrete specimens were cast in steel moulds in layers. Each layer was compacted by a steel rod. The specimens were kept in their molds at room temperature with a plastic sheet cover to minimize water losses for 24 hours. After that they were demolded and immersed in water tank up to the age of 28 days at laboratory temperature of about (23±2 °C).

2.4 Tests

2.4.1 28-Day Air Dry Density

The test was performed in accordance with ASTM C567 (2003) [7] using 150×300 mm cylindrical specimens. After demolding the specimens were left for 6 days in air - tight polyethylene sheets. After that, they were removed and immersed in water for one day. The next day, saturated surface dry and suspended- immersed weight were taken. The specimens were left in the laboratory for 21 days, then the dried specimens were weighed and the 28-day air-dry density was calculated

2.4.2 Compressive strength test

The compressive strength test was determined according to B.S. 1881 Part 116 [8]. This test was conducted on 150 mm cubes using an electrical testing machine with a capacity of 2000 kN at a loading rate of 15 MPa per minute.

2.4.3 The flexural strength

The flexural strength test (two point load) was determined according to B.S. 1881 Part 118 **[9].** This test was conducted on 100x100x400 mm prisms using an electrical testing machine with a capacity of 2000 kN.

3- Results and discussion

3-1 28 air-dry density

The 28 day air dry density of all types of LWAC mixtures are presented in Figure 2. Results show that the 28 day air dry density of lightweight concretes produced from deferent types of crushed bricks aggregate ranged between (1935-1990) kg/m^3. However, all LWAC mixes conform to the requirements of structural LWAC, according to RILEM (1975) [10] classification, which limits the maximum density to 2000 kg/m^3. As expected, the 28-day air-dry unit weight values increase with the decrease of w/c ratio. The values of air-dry unit weight ranged between (1988-1990) kg/m^3. This is attributed to the low water content and consequently, low air voids.

3.2 Compressive strength

The results of the compressive strength for each mix tested at 28 days are shown in Figure 3. It is clear from these results that the compressive strength value of the concrete made with Iso crushed brick was higher than that for concrete with crushed Iranian and Nahrawan bricks by about 42.5% and 18.6 % respectively. This may be because of the high porosity and low strength of Iranian and Nahrawan bricks comparing to Iso, which affects negatively on the strength.

Test results also showed that using an HRWRA increases the compressive strength with the decrease in w/c ratio. This can be first attributed to the better dispersion of the cement particles throughout the concrete mix, and second to the lower initial volume of voids that are needed to be filled by the products of hydration due to the reduction in the volume of mixing water. The compressive strength were 17.70, 22.36 and 25.49 MPa for concrete mixes made with Iranian, Nahrawan and Iso crushed bricks respectively. These results show that the lightweight concretes used were complied with the requirements of ACI Committee 213R for structural lightweight concrete [11].

3.3 Flexural strength

The results of flexure strength tests are shown in Figure 4. The test results showed that the flexure strength values were 2.29, 3.63 and 3.98 MPa for concrete mixes with Iranian, Nahrawan and Iso crushed brick respectively. It is clear that mix with Iranian crushed brick give smaller values than Nahrawan and Iso crushed Bricks due to its low weight and high water absorption.

Results also indicate that the incorporation of HRWRA in LWAC leads to higher flexural strength compared to their corresponding reference concrete. This behavior is mainly attributed to the significant reduction in the capillary porosity of the cement matrix as well as to the good dispersion of the cement grains throughout the mix. The flexural strength values were 3.89, 4.08 and 4.44 MPa for mixes with Iranian, Nahrawan and Iso crushed brick respectively

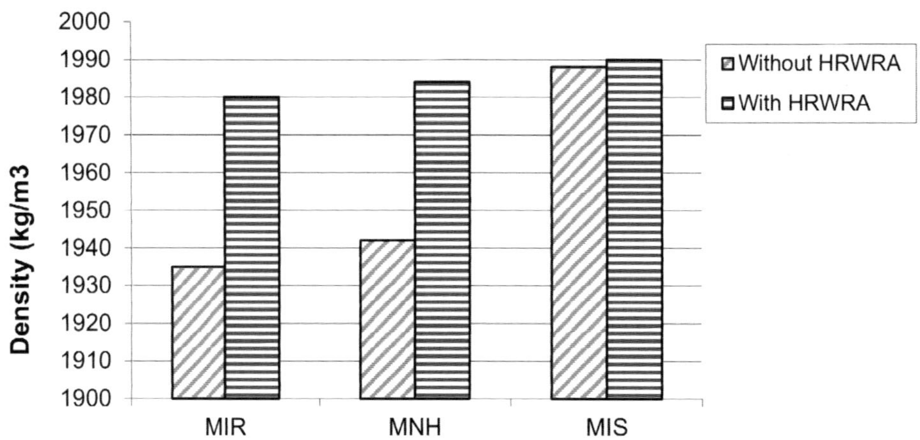

Figure 2. Air dry desity of concrete with defferent crushed bricks.

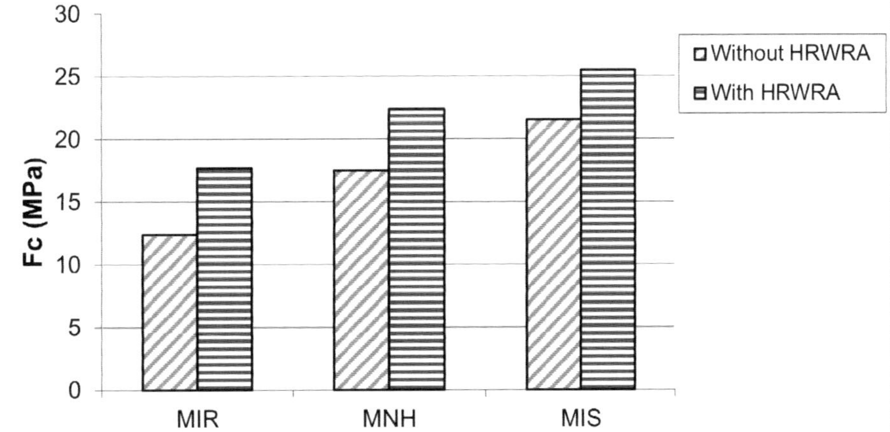

Figure 3. Compressive strength of concrete with defferent crushed bricks.

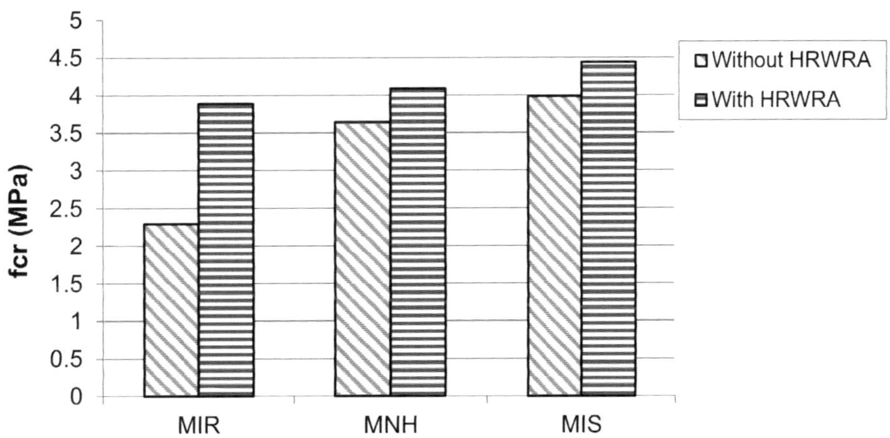

Figure 4. Flexural strength of concrete with defferent crushed bricks.

4. Conclusions

1- The 28-day air-dry density of lightweight concretes produced from deferent types of crushed bricks aggregate ranged between (1935-1990) kg/m^3.
2- The compressive strength value of the concrete made with Iso crushed brick was higher than that for concrete with crushed Iranian and Nahrawan bricks by about 42.5% and 18.6 % respectively. Test results also showed that using an HRWRA increases the compressive strength values.
3- The flexure strength of the lightweight concrete cast using Iso crushed brick is higher than lightweight concretes cast using Iranian and Nahrawan crushed brick, as aggregates by about 42%,and 9 %, respectively. Results also indicate that the incorporation of HRWRA in LWAC leads to higher flexural strength compared to their corresponding reference concrete
4- The results of 28 air-dry density and compressive strength show that the lightweight concretes used were complied with the requirements for lightweight structural concrete.

References

[1] Otoko G 2004 Used of crushed clay bricks as aggregate in concrete ,*International journal of engineering and technology research* , vol 2 NO.4, ISSN 2327 – 0349 pp. 1-9 .
[2] Hiremath M and Sanjay S 2017 Replacement of coarse aggregate by demolished Brick waste in concrete. *International Journal of science technology and engineering*, vol4 issue 2 pp. 31-36.
[3] Subramani, T.2015 Experimental investigation of using concrete waste and brick waste as coarse aggregate. *Intern jour of application &innovation in engineering & management* vol4 issue 5 pp. 294 – 303.
[4] Borhan T 2015 Effect of using recycled lightweight aggregate on the properties of concrete. *Journal of Babylon University, engineering sciences* vol 23 No.2 pp-1-12.
[5] Padmini AK, Ramamurthy K and Mathews MS 2001 Behavior of concrete with low-strength bricks as lightweight coarse aggregate. *Magazine of Concrete Research* vol.53 No.6 pp. 367-375.
[6] ACI Committee 211.2 2003 Standard Practice for Selecting Proportion for Structural Lightweight Concrete ACI Manual of Concrete Practice, Part 1 pp.1-16.
[7] ASTM C567 2003 Standard Test Methods for Determination Density of Structural Lightweight Concrete Annual Book of ASTM Standard, Vol. 04-02 pp302-304.
[8] B.S. 1881 Part 116 1983 Method for Determination of Compressive Strength of Concrete Cubes British Standard Institute 1983 pp. 1-3
[9] BS 1881 Part 118: 1983 Method for Determination of flexural Strength of Concrete ", British Standard Institute, pp 1-6
[10] RILEM 1975 Terminology and Definitions of Lightweight Concrete. Recommendation LCI. 1st edition.
[11] ACI 213 R 2003 Guide for Structural Lightweight-Aggregate Concrete' ACI Manual of Concrete Practice "2003

Acknowledgments
The authors wish to pay their considerate thanks to Concrete Laboratory, Department of Civil Engineering, University of Kufa, for their assistance in this research work.

Hot Asphalt Mixture Characteristics with Nano-Metakaolin Materials

Mustafa A Kareem[1], Abbas Al-Hdabi[2], Nibras A Al-Sahaf [3] and Laith F Hasan[4]

[1] Post graduate student, Department of Civil Engineering, Faculty of Engineering, University of Kufa, Iraq. Email: kareemmustafa41@gmail.com
[2] Asst. Prof., Department of Civil Engineering, Faculty of Engineering, University of Kufa, Iraq. Email: abbas.alhadabi@uokufa.edu.iq
[3] Lecturer, Department of Civil Engineering, Faculty of Engineering, University of Kufa, Iraq. Email: nibrasa.alsahhaf@uokufa.edu.iq
[4] B.Sc. Civil Engineering, University of Kufa, Iraq. Email: laithf.hasan@uokufa.edu.iq.

Abstract. In this research, Nano-metakaolin was used to investigate the achievable prospects to enhance asphalt mixture properties. The aim of this study was to evaluate the effect of adding Nano-Metakaolin on the properties of asphalt mix and comparing it with the local and international requirements, besides identifying the optimum percent of NMK to be incorporated in the hot mix asphalt. As a modifier to the bitumen several percentages of Nano-Metakaolin was used i.e. 1, 2, 4, 8, 16 and 20% by weight of bitumen. Modified asphalt binders were subjected to physical and conventional tests which were penetration, kinematic viscosity, softening point, and ductility. From these tests, optimum Nano-Metakaolin was 16%. After obtaining the optimum Nano-Metakaolin, two sets of asphalt concrete mixtures were prepared, the first was with the pure asphalt (control mixture) and the second was with the optimum Nano-Metakaolin i.e. 16% (modified mixes). The control and modified mixtures have been evaluated based on experimental tests, such as Marshall Stability (MS), Marshall flow, Marshall stiffness, Indirect Tensile Strength (ITS) and Index of Retained Strength (IRS). The results for these tests indicated an improvement in the mechanical proprieties i.e. increased Marshall Stability, Indirect Tensile Strength (ITS), and index of retained strength. As a final result, the use of (16%) Nano-metakaolin by weight of asphalt enhanced asphalt concrete properties and produce durable mixtures for highway construction.

Keywords: Nanotechnology; Asphalt mixtures; Pure asphalt; Nano-metakaolin; mechanical properties.

1. Introduction

Asphalt concrete is one of the most common types of pavement surface materials used in the world. Asphalt pavements must undergo heavy loads and unfavorable environmental conditions and their resulting distresses such as high-temperature rutting and low-temperature cracking for an acceptable period of time. Therefore, modification of asphalt layers is necessary.

It is a composite material consisting of mineral aggregates, asphalt binder, and air voids. The load-carrying behavior and resulting failure of such material depend on many mechanisms that are strongly related to the local load transfer between aggregate particles [1].

Due to the rapid urbanization and industrialization of the world over the last century, the construction and maintenance of transportation roadways is a constant demand in both urban and rural areas. Furthermore, due to excessive traffic loads and environmental factors, many existing pavements have already reached the end of their service life and other pavements will soon require maintenance [2].

Fatigue cracking and rutting (permanent deformation) distresses are the most significant challenges that can affect the functional performance of asphalt highways. The reducing of the problems would be more economical for recovery while reconstructing defects require more costs. One approach to enhance asphalt characteristics is by using a variety of modifiers including rubbers and polymers. Lately, several effective solutions have been proposed such as the adoption of nanomaterials to modify asphalt and/or asphalt mixtures characteristics performance [3].

Nanomaterials in present days are being utilized widely in a spectrum of fields. Nanotechnologies work on modifying base materials on the molecular level to produce new materials with better properties. For example, modifying the properties of Portland cement using Nanomaterials have been examined by several studies. Lately, Nanomaterials have been used in road paving. Many studies have been carried out to investigate the potential advantage of adding different Nanomaterials to asphalt binders such as "Nano clay, Nano silica, Nano-hydrated lime, Nano-sized plastic powders, or polymerized powders, Nanofibers, and Nanotubes" [4].

The most common distresses that could face unmodified paving asphalt is the rutting deformation (due to high temperature) and cracking (due to low temperature). Modifying and strengthening asphalt hence should be considered in the asphalt pavement industry [5].

As the world continues to urbanize and construct more transportation roadways, the need for high-quality and sustainable pavement is a constant need. Due to these demands, transportation experts and engineers focus on improving the performance and service-life of pavements. Such modifications aim to improve the mechanical properties and durability of asphalt mixtures by adding nano-metacaolin to the asphalt cement.

2. Methodology

2.1 Selected Materials

In this research, the materials provided are commonly used in preparation of hot asphalt mixtures especially in the middle and south regions of Iraq such as aggregate, cement, asphalt binder.

2.1.1 Aggregates

The coarse aggregate has been brought from a hot-mix asphalt plant in Al-Najaf city (Iraq) and it was originally brought from Al-Nibaee quarry (North of Baghdad). Tables 1 and 2 list the main chemical and physical characteristics of the used aggregate which were tested in accordance to ASTM, (2004) [6]. While Table 3 shows the selected gradation for type IIIA surface course mixture in accordance to Iraqi specification for roads and bridges (SCRB, 2003) [7].

Table 1. Chemical composition of Nibaee aggregate.

Chemical compound	Results, %
L.O.I.	6.55
MgO	0.78
SiO_2	82.52
CaO	5.37
SO_3	2.7
Fe_2O_3	0.69
Al_2O_3	0.48
Mineral composition	
Calcite	10.92
Quartz	80.03

Table 2. Physical properties of aggregates

Property	ASTM designation	Test Results	SCRB specifications
Coarse aggregate			
Bulk specific gravity	C127	2.610	…….
Apparent specific gravity	C127	2.641	…….
Percent wear by Los Angeles abrasion, %	C131	22.7	30 Max
Soundness loss by sodium sulfate Solution, %	C88	3.4	12 Max
Flat and elongated particles, %	C4791	5	10 Max
The degree of crushing, %	D5821	96	90 Max
Fine aggregate			
Apparent specific gravity	C128	2.68	…….
Sand equivalent, %	D2419	57	45 Min
Bulk specific gravity	C128	2.631	…….
Clay lumps and friable particles, %	C142	1.85	3 Max
Angularity, %	C1252	54	…….

Table 3. Aggregate size separation by using sieve analysis.

Sieve size, mm	Specification limit	Selected gradation
	% passing by total weight of aggregate	
19	100	100
12.5	90-100	95
9.5	76-90	83
4.75	44-74	59
2.36	28-58	43
0.300	5-21	13
0.075	4-10	7

IOP Conf. Series: Materials Science and Engineering **584** (2019) 012041 doi:10.1088/1757-899X/584/1/012041

2.1.2 Asphalt Cement

One type of asphalt cement is used with (40-50) penetration grade brought from AL Nasiriya refinery. Table (4) shows the physical properties of asphalt cement which were evaluated according to ASTM standards ASTM, (2004) [6] and compared with Iraqi Specification known as a general specification for roads and bridges [7].

Table 4. Properties of (40-50) base asphalt binder

Property	ASTM Designation	Test Results	Requirements for penetration graded asphalt cement
Penetration at 25 C°, 100 gm,5 sec (1/10 mm)	D5	40	40-50
Softening point, Ring and ball °C	D36	50	....
Ductility at 25 °C, cm	D113	105	>100
Flash point, °C	D92	273	>232
Specific gravity at 25 °C	D70	1.03	....
Solubility in trichloroethylene,% wt	D2042	99.31	>99
Residue from thin-film oven test	D1754		
-Ductility at 25 °C, cm	D113	33	>25
Retained penetration,% of original	D5	59	>55

2.1.3 Filler

Filler materials represent mineral particles that pass Sieve No.200 (0.075mm). The mineral filler used in this research was ordinary Portland cement (OPC) obtained from Al Kufa factory, at 7 percent content, which represents the average value of SCRB specifications [7]. Table (5) shows the Physical properties of this cement.

2.1.4 Nano-metakaolin (NMK)

The nano-metakaolin used in this study is manufactured by Iranian Nanomaterials Pioneers Company in Iran [8]. The properties of nano metakaolin were with surface area of 480000 cm^2/g and of average dimensions of 200*100*20 nm. Table (6) gives the chemical composition of NMK. Practical steps of the preparation process of nano-metakaolin (NMK) is shown in Figure (1).

Table 5. OPC Physical properties of fillers.

Property	Portland cement
Passing sieve No. 200, (%)	94.76
Specific surface area (m2/kg)	418
Specific gravity (ASTMC188-95)	3.1

Table 6. Chemical properties of NMK

Chemical content	%
SiO_2	45.5
Al_2O_3	37
Fe_2O_3	0.2
TiO_2	1.5
CaO	0.01
MgO	0.02
Na_2O	0.03
K_2O	0.07
L.O.I	12.5

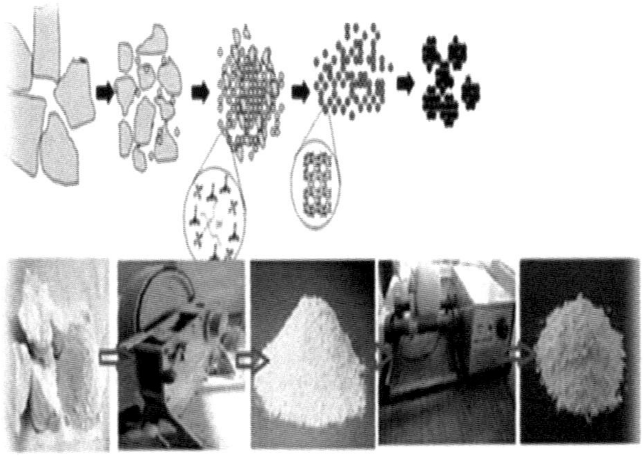

Figure 1. Practical steps of the preparation process of Nano-metakaolin

3. Experimental works

3.1 Testing on Asphalt Binder

Asphalt binder (40-50) was used in this research. In order to evaluate asphalt cement properties, several tests were conducted on both unmodified asphalt binder and modified asphalt specimens that have been prepared. The modified specimens are prepared by adding various percent of Nano-metakaolin 1, 2, 4, 8, 16, and 20%. Different laboratory tests have been performed such as the penetration, softening point, kinematic viscosity, and ductility.

3.2 Testing Program

All materials and mixture were prepared under ASTM specification D1559-89. The mixture was compacted according to the Marshall mix design method standards. Three important tests were carried out on the prepared asphalt mixture; these are Marshall stability, indirect tensile and index of retained strength tests.

3.2.1 Marshall Test

Marshall Method for designing hot asphalt mixtures was used to determine the optimum asphalt content. According to standard 75-blows on each face. Marshall design method designated as ASTM D1559-89, (AASHTO T 245-13) [9] a number of specimens each of 1200 gm in weight was prepared using five different asphalt contents (from 4 - 6% with 0.5 % incremental). Three specimens were used to prepare asphalt mixture with one- asphalt content to have an average value of Marshall Flow, bulk density, and stability.

3.2.2 Indirect Tensile Strength (ITS) Test

A mechanical displacement control testing frame was used to conduct the indirect tensile tests in accordance with ASTM D4123 (ASTM, 2004) in order to evaluate the tensile strength of asphalt concrete mixtures.

3.2.3 Index of Retained Strength (IRS) Test

Index of retained strength (IRS) test is used to evaluate moisture damage of asphalt pavement in accordance with method ASTM D1075 (ASTM, 2004). It is one of the tests required by (SCRB, R/9, 2003) [8] specifications to be performed on asphalt mixes used in a surface course in addition to Marshall Tests [10].

4. Results and discussion

The results of laboratory work have been obtained and analyzed in order to identify the optimum percent of Nano-metakaolin to be added to the asphalt cement; in addition to study the effect of adding the optimum percent of Nano-metakaolin on mechanical properties of asphalt concrete mixtures. Laboratory work results are presented in this chapter in two stages. First, several sorts of testing have been conducted to investigate the impact of adding Nano-metakaolin material to the base asphalt (40-50 penetration grade). The tests include kinematic viscosity, penetration, softening point and ductility to find the optimum Nano-metakaolin content. The second stage, Marshall Test was carried out with different percentages of asphalt which are (4.0, 4.5, 5.0, 5.5, and 6.0%) and the results are analyzed in order to obtain the optimum asphalt content. After obtaining OAC and optimum NMK, two sets of asphalt concrete specimens with and without NMK were prepared. The results are analyzed on several experimental tests including Marshall Test, Marshall Stiffness, indirect tensile strength, Index of retained strength and Volumetric Properties. Finally, the results were compared with the Iraqi SCRB specifications [8].

4.1 Effect of Nano-metakaolin on Penetration Test Results

Four tests were carried out to study the effect, Penetration tests were carried out for the base and modified asphalt with different percentages of Nano-metakaolin i.e. 1, 2, 4, 8, 16, and 20% by weight of asphalt at different temperatures which were (15, 25 and 35°C). Figure (2) shows the penetration test results for different percentages of Nano-metakaolin. According to the results, with the addition of different percentages of Nano-metakaolin, the penetration value at high temperature decreased dramatically while there is a little decrease at low temperature in comparison with unmodified asphalt. The best results for the asphalt binder used in this study obtained in the asphalt modified with 16% Nano-metakaolin.

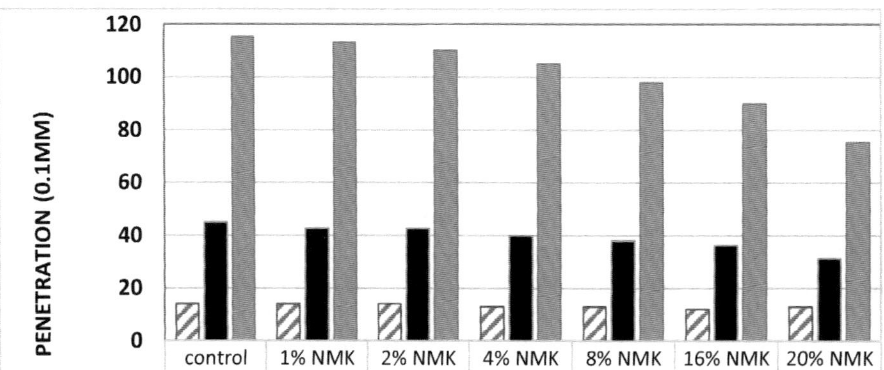

	control	1% NMK	2% NMK	4% NMK	8% NMK	16% NMK	20% NMK
penet. 15 c°	14	14	14	13	13	12	13
penet. 25 c°	45	43	43	40	38	36	31
penet. 35 c°	115	113	110	105	98	90	75

Figure 2. Variation of penetration values versus Nano NMK content

4.2 Effect of Nano-metakaolin on Viscosity Test Results

The Kinematic viscosity test has been carried out at percentage of 4, 8, and 16% by weight of asphalt at different temperatures which were 90, 110, and 130°C. Figure (3) shows the effect of Nano-metakaolin content on the Kinematic viscosity of asphalt with change temperature. By adding Nano-metakaolin to the base asphalt binder, the viscosity increased. The maximum viscosity was related to the case of adding 16% (NMK) which occurred at 130 °C.

4.3 Effect of Nano-metakaolin on Softening Point Test Results

The results of softening point test for base asphalt binder and asphalt binder modified with different percentages of Nano-metakaolin are shown in Figure (4). The figure demonstrates a change in softening point from adding 1% NMK to 20% NMK. There is a significant increase in softening point after adding 4, 8, and 16% Nano-metakaolin by weight of asphalt.

4.4 Effect of Nano-metakaolin on Ductility Test Results

Figure (5) demonstrates the results of ductility test for both unmodified asphalt and modified asphalt with different percentages of Nano-metakaolin.

In accordance with the above testing, the results showed that the optimal percent for Nano-metakaolin is 16%.

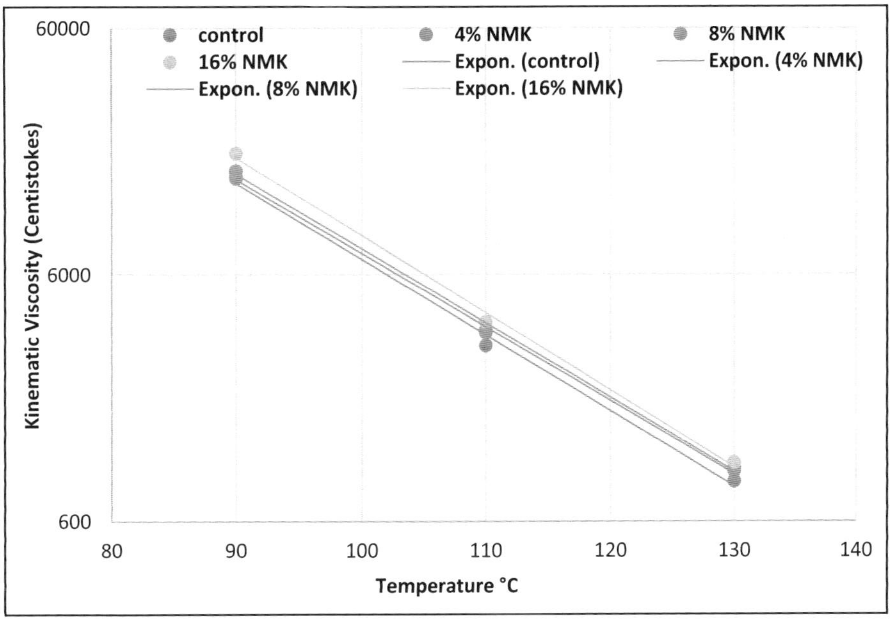

Figure 3. Variation of kinematic viscosity versus Nano NMK content

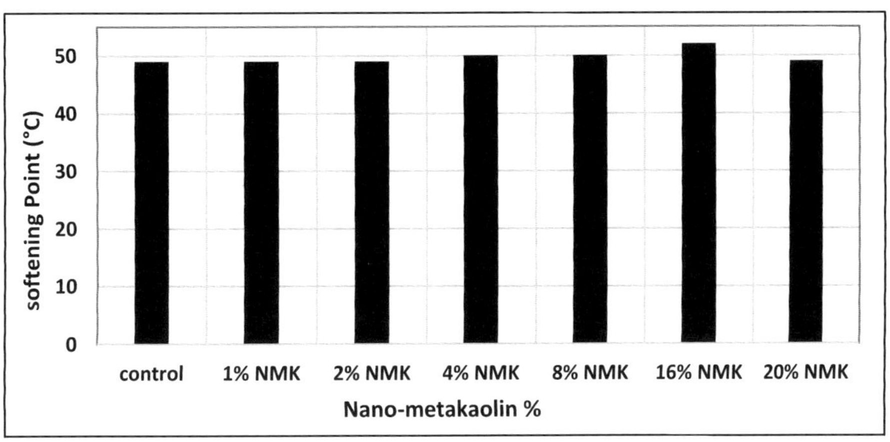

Figure 4. Variation of softening point versus Nano NMK content

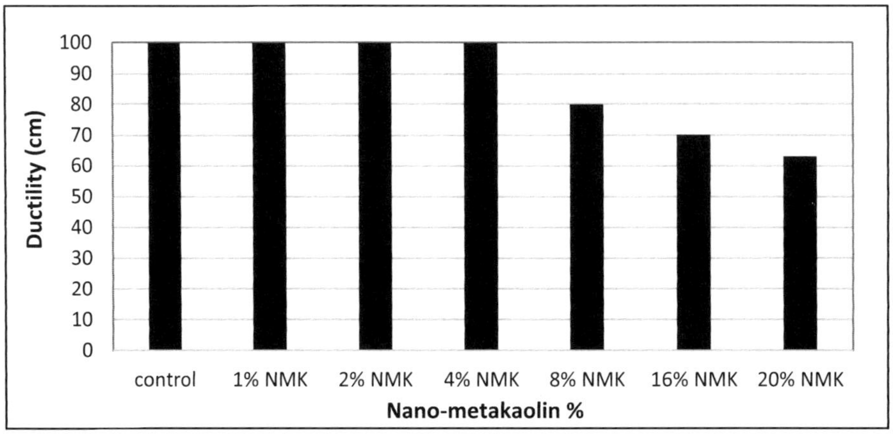

Figure 5. Variation of ductility versus Nano NMK content

4.5 Optimum Asphalt Content

1200 gm in weight were prepared using five different asphalt contents (4 - 6% with 0.5 % incremental) in order to find optimum asphalt content (OAC). Figure (6) shows a summary of Marshall test outcomes.

4.5.1 Determination of optimum asphalt content (OAC)

Figure 6 is utilized to obtain three AC values as follows:
The asphalt content at the highest MS = 4.6%
Asphalt content at the highest (% GMB) bulk density = 5.5%
Asphalt content at the median of allowed percentages of air voids (% AV) i.e. 4% for surface course mixtures = (5.35) %

Optimum asphalt content (OAC) $= \frac{4.6\%+5.5\%+5.35\%}{3} = 5.15\%$

The result of OAC satisfies with Iraqi specifications (SCRB, R/9, 2003) [7] and (ASTM, 2004) [6].

4.6 Effect of NMK on Mechanical Properties of HMA

Two sets of the specimens were prepared; asphalt (40-50) penetration grade and the second mixture is prepared as a modified binder with Nano-metakaolin of 16% metakaolin by weight. They have been evaluated based on experimental tests which are Marshall Stability MS, Marshall Flow, Marshall Stiffness, ITS, and IRS. Figure (7) illustrates the influence of the addition of Nano-metakaolin on Marshall Stability test. The figure implies that Marshall Stability increased significantly when 16% of Nano-metakaolin has been added to the base asphalt binder. Figure (8) represents the influence of the addition of Nano-metakaolin on Marshall Flow test. There is a slight decrease in Marshall Flow when 16 percent of NMK has been added to the base asphalt binder.

The indirect tensile strength test results are shown in Figure (9). The value has been increased by adding Nano-metakaolin to the base asphalt binder. High tensile strength at failure is a desirable property for stiff mixtures. ITS increased almost by 8% for modified mixtures in comparison with the control mixtures. While, Figure (10) shows the influence of the addition of NMK on the index of retained strength test for modified and unmodified mixtures.

Due to the results above, the Using of Nano Metakaolin enhanced the Characteristics Hot Asphalt Mixture with the optimum ration of 16% by weight of asphalt material.

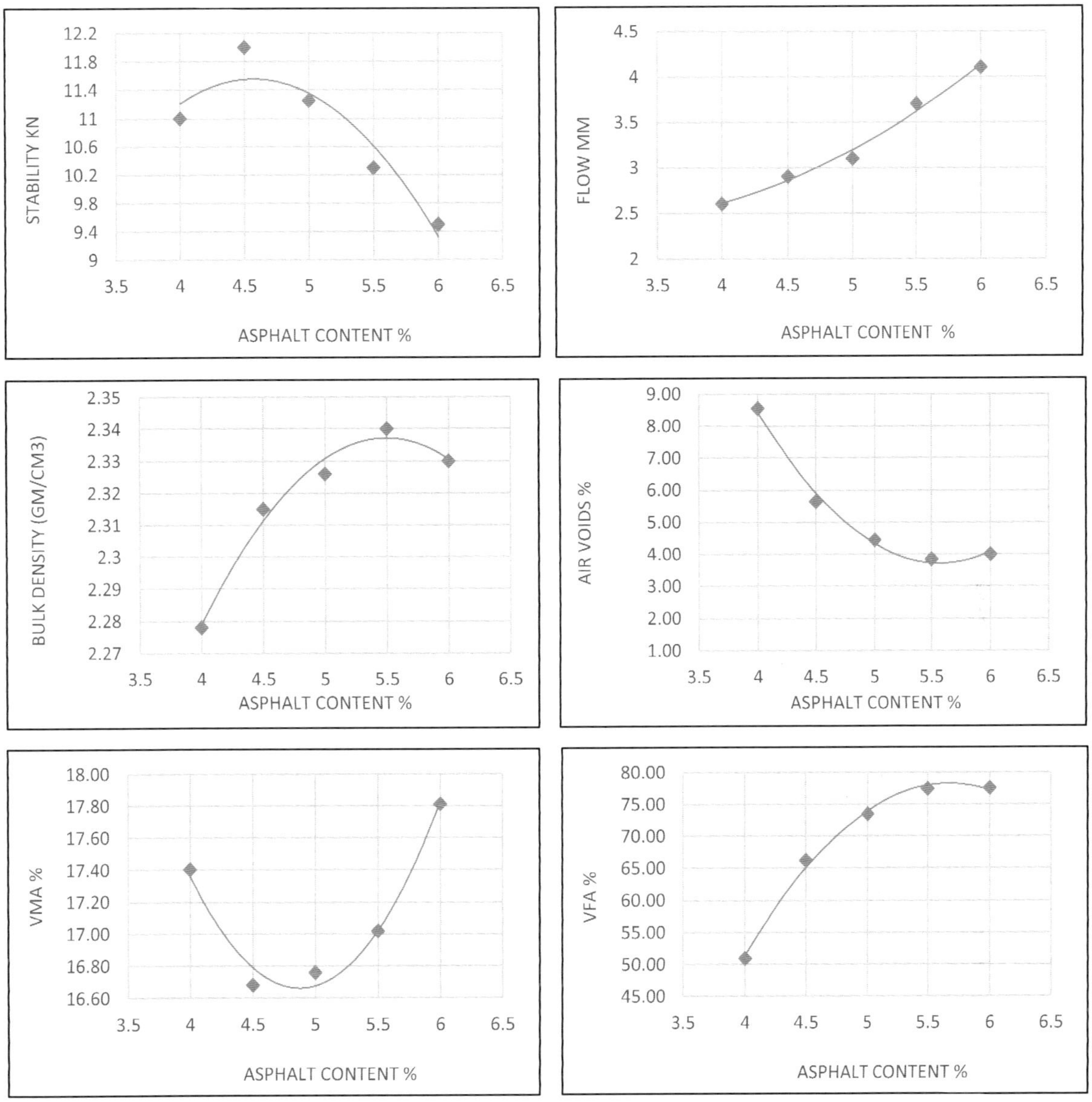

Figure 6. The relationships between asphalt cement and the properties of control mixtures

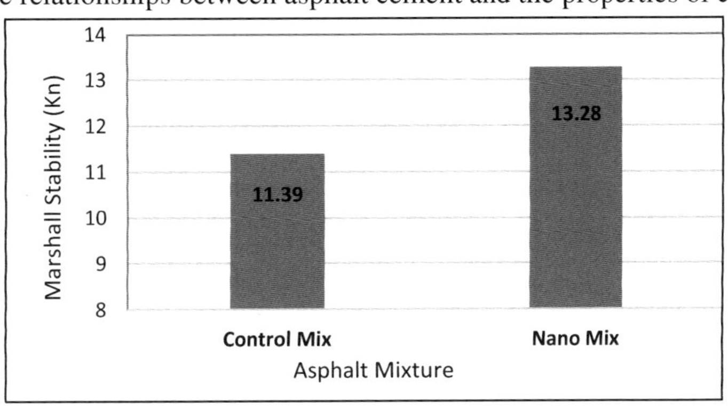

Figure 7. Effect of Nano metakaolin on Marshall Stability

IOP Conf. Series: Materials Science and Engineering **584** (2019) 012041 doi:10.1088/1757-899X/584/1/012041

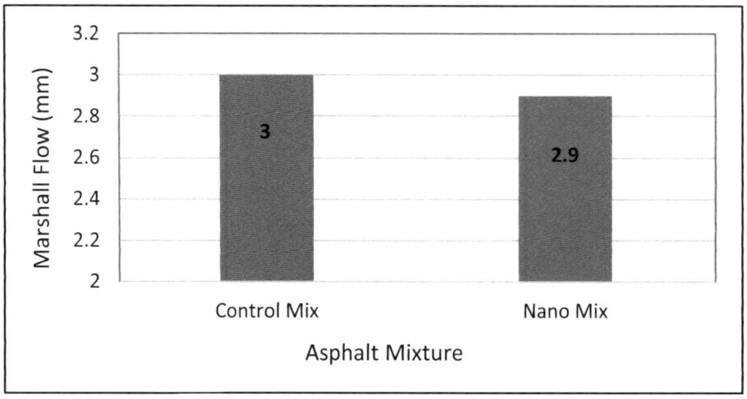

Figure 8. Effect of Nano metakaolin on Marshall Flow

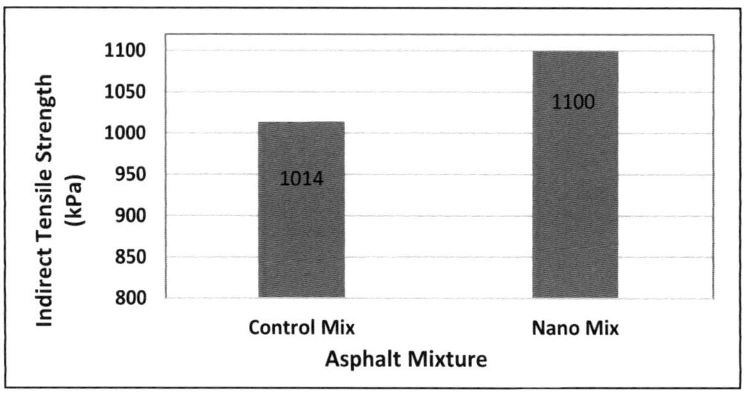

Figure 9. Effect of Nano metakaolin on indirect tensile strengt

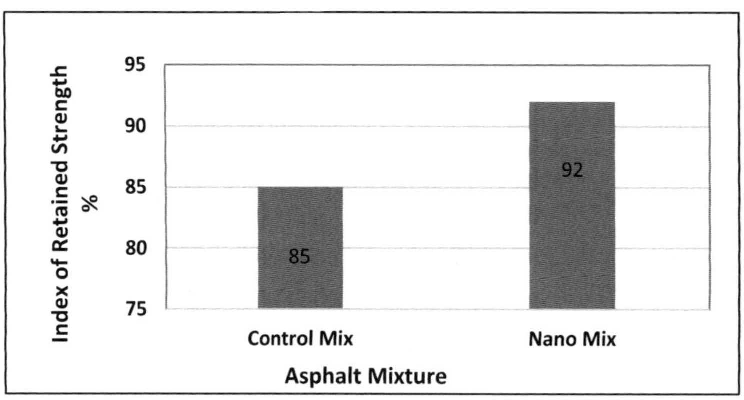

Figure 10. Effect of Nano metakaolin on the index of retained strength

IOP Conf. Series: Materials Science and Engineering **584** (2019) 012041 doi:10.1088/1757-899X/584/1/012041

5. Conclusions

The aim of this research was to evaluate the effects of Nano-metakaolin as an asphalt binder modifier on the asphalt mixtures. Several laboratory tests were conducted to evaluate the characteristics of HMA by different contents of Nano-metakaolin added to the base asphalt binder. Based on the limited test results obtained in this research, the following conclusions can be drawn.

1. Regarding kinematic viscosity, penetration, softening point, and ductility test. The optimum amount of Nano-metakaolin to be added as a modifier of asphalt mix was obtained to be 16% by the total weight of asphalt mix.
2. The results of the penetration test and ductility test represent decreased penetration degree and the ductility also declined after adding Nano-metakaolin to the base asphalt binder. This indicates that modified asphalt with NMK is more stiffer in comparison with base asphalt binder.
3. Marshall stability is increased by adding 16% of NMK to the base mixture. Also, Marshall stability is increased by almost 17% for the modified mixes in comparison with the control mixtures.
4. By means of Indirect Tensile Strength. ITS increased by almost 7% for modified mixtures in comparison with the unmodified asphalt concrete mixtures.
5. In terms of the Index of Retained Strength test. IRS is increased from 85 to 92%. IRS reduced performance and increased maintenance for asphalt pavements. Thus, enhance IRS rise performance and reduce maintenance.

References

[1] Sadd M H, Dai Q, Parameswaran Vand Shukla A 2004 Microstructural Simulation of Asphalt Materials: Modeling and Experimental Studies Journal of Materials in Civil Engineering, 16(2), 107-115.

[2] Siriwardane H, Gondle R and Kutuk B 2010 Analysis of Flexible Pavements Reinforced with Geogrids Geotechnical and Geological Engineering, 28(3), 287-297.

[3] Shafabakhsh G H, Mirabdolazimi S M, and Sadeghnejad M 2014 Evaluation the effect of nano-TiO2 on the rutting and fatigue behavior of asphalt mixtures Construction and Building Materials, 54, 566-571.

[4] Sadeghnejad M and Shafabakhsh G 2017 Experimental Study on the Physical and Rheological Properties of Bitumen Modified with Different Nano Materials (Nano SiO2 & Nano TiO_2) International Journal of Nanoscience and Nanotechnology, 13(3), 253-263.

[5] Lewandowski L H 1994 Polymer modification of paving asphalt binders Rubber Chemistry and Technology, 67(3), 447-480.

[6] ASTM Standards 2004 Roads and Paving Materials Annual Book of the American Society for Testing and Materials Standards, Section 4, Vol. 04-03.

[7] State Commission of Roads and Bridges (SCRB/R9) 2003 General Specification for Roads and Bridges Republic of Iraq, Ministry of Housing and Construction, Department of Planning and Studies, Baghdad, Revised Edition, Addendum No.3.

[8] Iranian Nanomaterials Pioneers Company http://www.irananotech.com

[9] AASHTO 2007 Standard Specifications for Transportation Materials and Methods of Sampling and Testing. 5th edition, American Association of State Highway and Transportation Officials, Washington, D.C., USA.

[10] Al-Jumaily M A H 2008 Evaluation of Hydrated Lime Filler in Asphalt.

The Effect of the Shear Connectors Layout on the Sustainable Lightweight Concrete Sandwich

Laith Sh. Rasheed[1], and Suad Abbas Ali[2]

[1] Dept. of Civil Eng., University of Karbala, Iraq. Email: laith.alqarawee@uokerbala.edu.iq
[2] Graduate student, Dept. Of Civil Eng., University of Karbala, Iraq. Email: suadabbass3@gmail.com

Abstract. Sandwich Slab is a modern shape of structural construction. It contains a low density layer known as the core layer confined by two strong layers. In the present study, the two layers of light weight concrete [LWC] were tied together by a cage of steel reinforcement named as shear connectors, sandwiched between them a layer of polystyrene. The experimental investigation was conducted on six slabs. Two parameters were investigated (the steel area of the shear connectors which is depended on the number of W-Shaped shear connectors and the layout of W-shaped shear connectors). The main reason for using crushed bricks as a lightweight aggregate was to the benefit from the recycling broken bricks as a sustainability material. The slabs were tested as simply supported under two line loads. Continuous truss-shaped shear connecter showed better behaviour than discrete W-shapes shear connectors. Experimental test results indicated that the presence of discrete shear connectors at the ends of the specimens in distance of one-fourth of span had significant better performance than at the center of the specimence.

Keyword: The Sandwich slab, Sustainable lightweight concrete, Crushed bricks, Shear connectors.

1. Introduction

Concrete sandwich slabs are a layered system composed of two thin, and high strength layers held together by shear connectors [1-4]. Those layers usually separated by a lightweight insulating layer, such as extruded polystyrene (XPS) or expanded polystyrene (EPS) [5]. For more than 50 years concrete sandwich slabs were used for many different structures as exterior and interior walls [6]. The presence of the low-density core layer provides high thermal resistance and structural performance [6]. The manufacturers are looking for new, applicable product lines and because of energy performance and general aesthetics of the panels so the urgent need to develop concrete sandwich panels [3]. Concrete sandwich slab may be regarded as fully composite, a semi-composite or non-composite panel, Full composite panel is accomplished by providing full shear transfer between the two layers on another meaning Full composite panel are designed and manufactured so that the two concrete layers act together as a single unit to resist applied loads till failure [2]. When sandwich slab was used as a structural member, the slab must be designed as a full composite slab in order to take full advantage of the strength of the two layers and to prevent individual layer buckling [7]. The main aim of this work is producing structural lightweight concrete sandwich slabs with high strength over weight ratio. The over-exploitation of natural materials raises controversy about protection natural resources so that the researchers studied this problem and tried to develop a solution to balance the lack of natural future resources [8]. The reuse of broken bricks to produce new and good concrete is a possible solution.

Previous studies showed that the crushed bricks are possible option to use as a structural material in terms of mechanical properties and structural behavior [9]. One of the important properties of the crushed bricks is the low density. it's density is about 40% less than the normal coarse aggregate [9]. Panels with crushed bricks as coarse lightweight aggregate exhibit better structural performance than panels with normal aggregate [10]. Lower density means more advantages to reduce the labor, reduce the construction's time, and contribute to green building ideology. From the previous researches, it is clear that studies on sandwich slabs are still limited and there is a weakness. Most of concrete wythes in the sandwich panels were made of conventional concrete. So these sandwich panels are strong but they have lower strength over weight ratio. Therefore, further investigation on using both lightweight concrete in the wythes and polystyrene core layer was required.

2. Experimental Work

The experimental program involves testing one solid slab, one sandwich specimen with normal concrete, and four sandwich slabs with lightweight concrete in the wythe layers. Three of the lightweight sandwich specimens were built with different shear connectors layout, but with same shear reinforcement ratio. All specimens have been tested under two-line load. In current experiment, three parameters were investigated. The first parameter was the steel area of the shear connectors; the second one was the effect of different shear connectors layouts which arrangement gives better composite ratio. The first slab specimen denoted by (RN). RN is a solid slab with normal weight concrete reinforced with longitudinal steel bars (Ø6 @ 6mm) and it considered as a control slab for this experiment.
The second slab specimen (SN-C45) was sandwich slab with normal weight concrete in the skin layers. SN-C45 was reinforced with continuous steel truss as shear connectors. The inclined angle for each bent in this specimen was 45°.
The third slab specimen (SB-45) was sandwich slab with crushed bricks as coarse lightweight aggregate in skin layers. SB-45 was reinforced with continuous truss as a shear connector. The inclined angle for each bent in this specimen was 45°.
The fourth specimen (SB-L1), the fifth specimen (SB-L2), and the sixth specimen (SB-L3) were sandwich slabs with crushed bricks as coarse aggregate in skin layers, and they were with different layouts of discrete W-shape shear connectors as illustrated in the Figure 1.

2.1 Specimens Details

All specimens were designed to act in one-way action and had the same total dimensions which they were (1100mm total length × 400 mm width × 90mm thickness). The sandwich slabs consist of two 30 mm thick skin layers of lightweight concrete, and 30 mm thickness cork layer between them. Figure 1 shows details of the reinforcement for each slab.

IOP Conf. Series: Materials Science and Engineering **584** (2019) 012042 doi:10.1088/1757-899X/584/1/012042

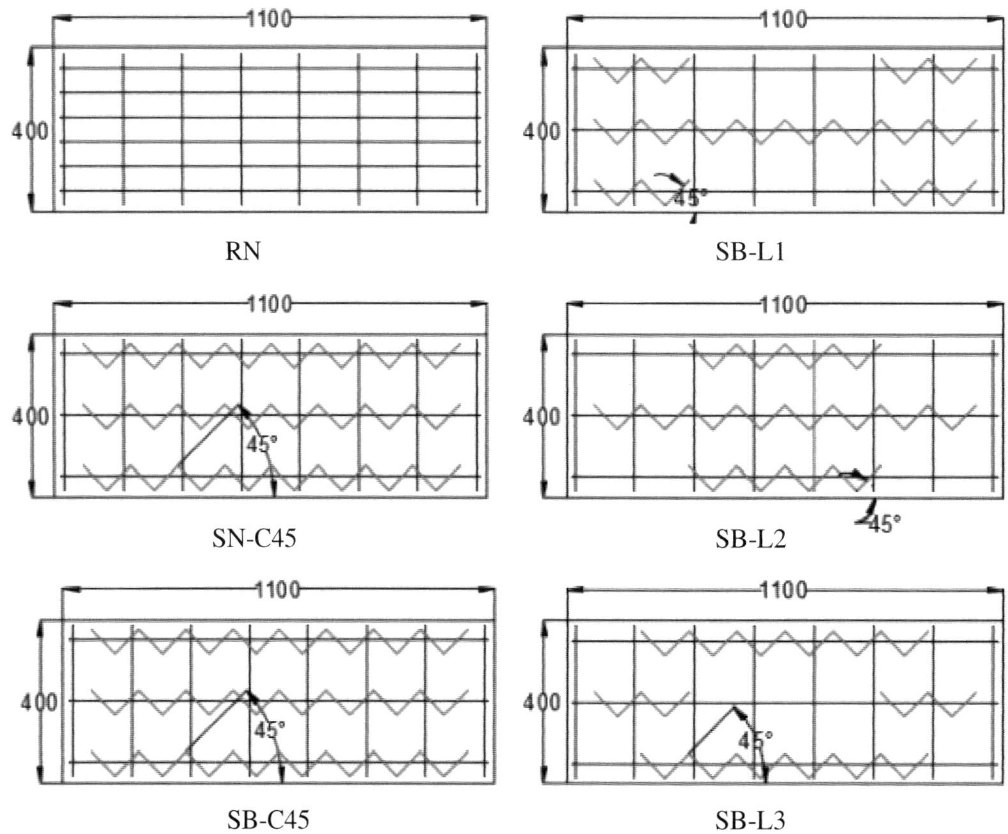

Figure 1. Details Reinforcement of the Tested Slabs.

2.2 Materials

Sulfate resistance cement (Type V) was used in the mix to prepare all specimens. The labrotory tests of the physical and chemical properties were performed according to the specification No.5 1984 of Iraqi Specification. Crushed coarse aggregate within the maximum size of 9.5 mm was used in the mix of the normal weight concrete, and crushed bricks with maximum size of 9.5 mm and bulk density of 805 kg/m^3 was used in the mix as a lightweight coarse aggregate. The grading of the crushed bricks and normal coarse aggregate was same and conforming the ASTM C330-05. The mixing proportions of the conventional normal concrete and lightweight concrete were (1:2.1:1.2) with w/c = 0.4 and 0.5%super plasticizer from the weight of the cement. Hyperplast PC 200 was used as high Performance Super-Plasticizer Concrete Admixture (known commercially as Flocrete PC200). All sandwich specimens were reinforced with a mesh of 6 mm diameter bars of spacing 150 mm deformed steel reinforcement. The steel mesh had yielding strength of 533 MPa. The meshes centered at both of the concrete layers of the slabs as shown in Figure 2. The outer concrete layers tied together with steel truss shear connector of 4mm diameter deform bars. After one day of casting, the forms were removed and the slabs were submerged in water at room temperature for 28 days of curing.

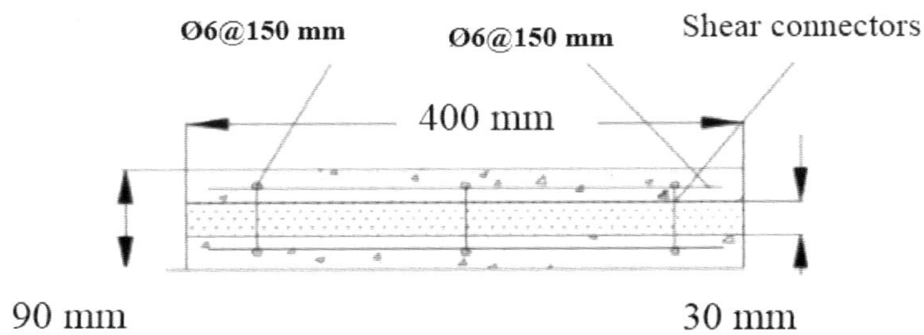

Figure 2. Cross Section Details of the Sandwich Slabs.

2.3 Tests Procedure and Equipment

The test setup was simply supported under two lines load applied equally at (1/3 of the span). Effective span was (990mm). The Test was conducted by using universal machine with load cell capacity equal to (1000kN) as illustrated in Figure 3. For top and bottom concrete layers, Strains were measured at mid span by using demec points. At one end of all specimens, the end slip was measured at the center of width for the upper concrete layer by using LVDT with 100mm range, and the end slip for the lower concrete layer was measured at the center of width by using LVDT with 10mm range. At mid-span LVDT with 100-mm range was used to measure the deflection. Figure 3 displays the supports and the location of the applied load arrangement. Note (1,2&3 represent the LVDTs). When the load was zero, first readings of LVDT were recorded, then the load gradually increased in stages while deflection and end slip measurements simultaneously recorded up to failure.

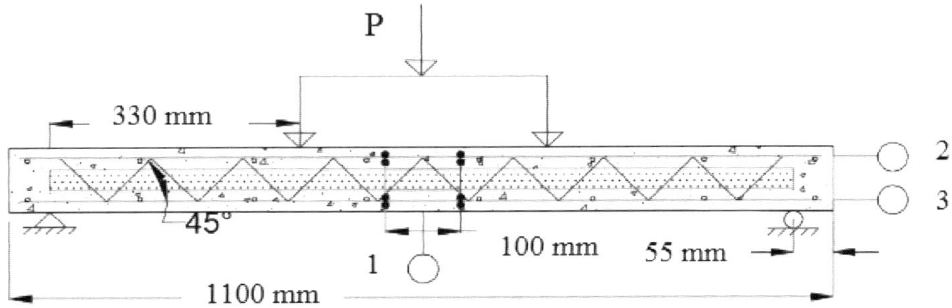

Figure 3. The Load Arrangement of the Tested Slabs.

3. Results and discussion

The experimental results were analyzed in the context of Load-Deflection curves, Load- Strain relationship and cracks pattern at the failure. The results indicate that hardened density for lightweight aggregate concrete when using crushed bricks are below 1945 kg/m^3 with cylinder compressive strength 25.2 MPa. And tensile strength 2.83 MPa.

3.1 The Load Deflection Relationship

To understand the effect of the layout of the shear connectors on the flexural strength of the sandwich slabs, load- deflection curves for the solid and sandwich slabs at mid span were showed in the Figure4. All tests of the slabs were conducted under load increment 3 kN. Table (1) displays total weight, failure loads and deflection of tested slabs, and a comparison with a reference slab as appropriate.

Table 1. Failure Loads and Deflection of Tested Slab.

Specimen symbol	Total Weight of specimen (Kg)	First Cracks Load (kN)	Failure load (kN)	Mid-span Deflection mm	Reference slab	The Difference in the ultimate load compared with the reference slab %	The Difference in the deflection compared with the reference slab %
RN	94.5	11.86	33	11.00	_	_	_
SN-C45	65	10	30	13.50	RN	-9.09	22.72
SB-C45	57.4	12	27.84	22	SN-C45	-7.20	62.96
SB-L1	56.4	6	21.65	15.42	SB-C45	- 22.23	-29.91
SB-L2	56.3	3	17.60	18.00	SB-C45	-36.78	-18.18
SB-L3	55.8	6.6	19.89	16.23	SB-C45	-28.55	-26.23

From the table 1 can be noticed that:

- For the same type of coarse aggregate (normal weight aggregate) used in the concrete mix, the flexural load capacity for the SN-C45 slab was decreased by about 9.09% from the flexural load capacity for the solid slab. From another point of view using a sandwich slab will increase the deflection value by about 22.72% and ductility index will increase by 2.27 % as compared with the RN slab. Also, the total weight will decrease by about 31.21%.

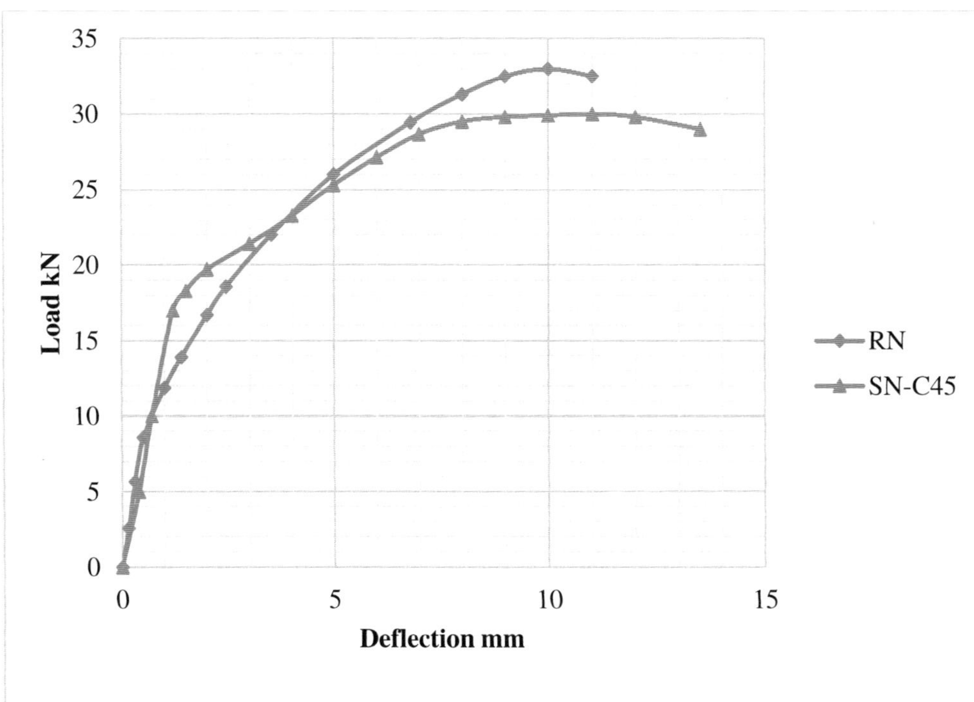

Figure 4. Load Deflection curves for the RN & SN-C45 slabs.

- Flexural strength was decreased about 7.20% when using crushed bricks as a coarse aggregate in the outer concrete layers of the SB-C45 slab instead of normal coarse aggregate SN-C45 slab. This decease accompanies by an increase in deflection value and ductility index by 69.23 %, and 39.64%; respectively. Also, the total weight will decrease by about 39.25% when compared with RN slab.

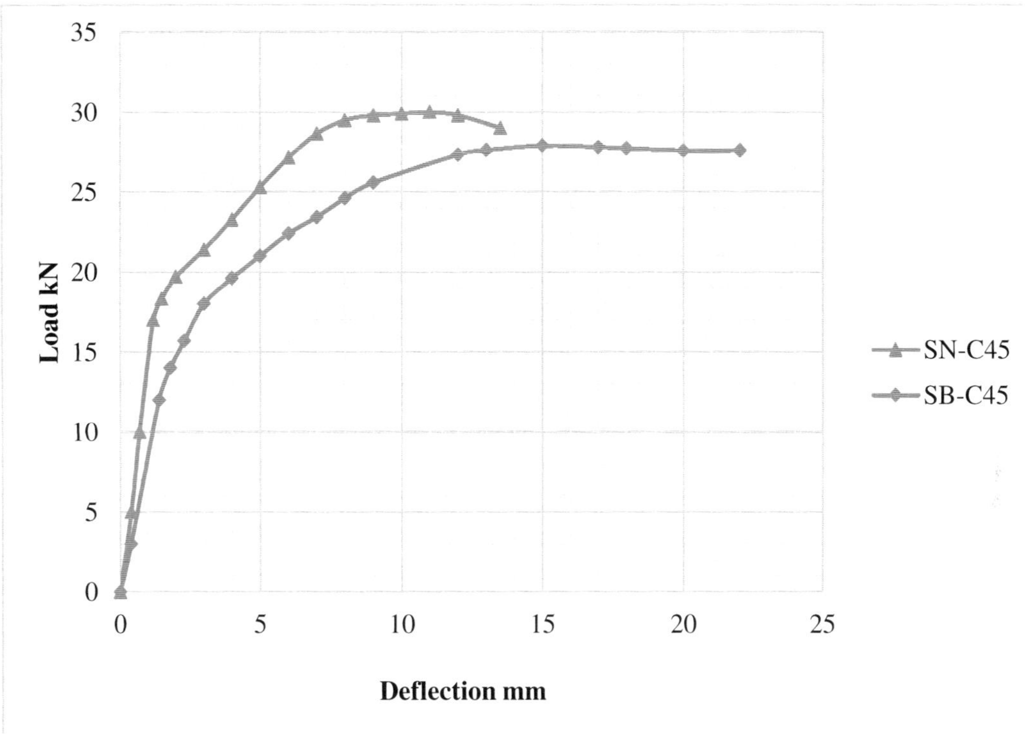

Figure 5. Load Deflection curves for the SN-C45 & SB-C45 slabs.

- When discrete shear connector with a steel area (2/3) of the continuous truss steel area was used: -
  I. Flexural strength was decreased about 22.23% when using SB-L1 slab, compared with ultimate strength when SB-C45 slab. So that, max deflection value and ductility index were decreased by about 29.91 %, 1.97 %: respectively.
  II. When using SB-L2 slab, flexural strength was decreased about 36.78% compared with ultimate strength when SB-C45 slab, and max deflection will decrease by 18.18 %. While ductility index will increase by 14.57%.
  III. Ultimate strength and max deflection value will decrease by about 28.55%, 26.22%; respectively, when using SB-L3 slab as compared with SB-C45 slab with an increase in ductility index by 3.31%.

Also, when using sandwich slabs with discrete shear connectors, the total weight of reached more than 40% when compared of RN slab.

IOP Conf. Series: Materials Science and Engineering **584** (2019) 012042 doi:10.1088/1757-899X/584/1/012042

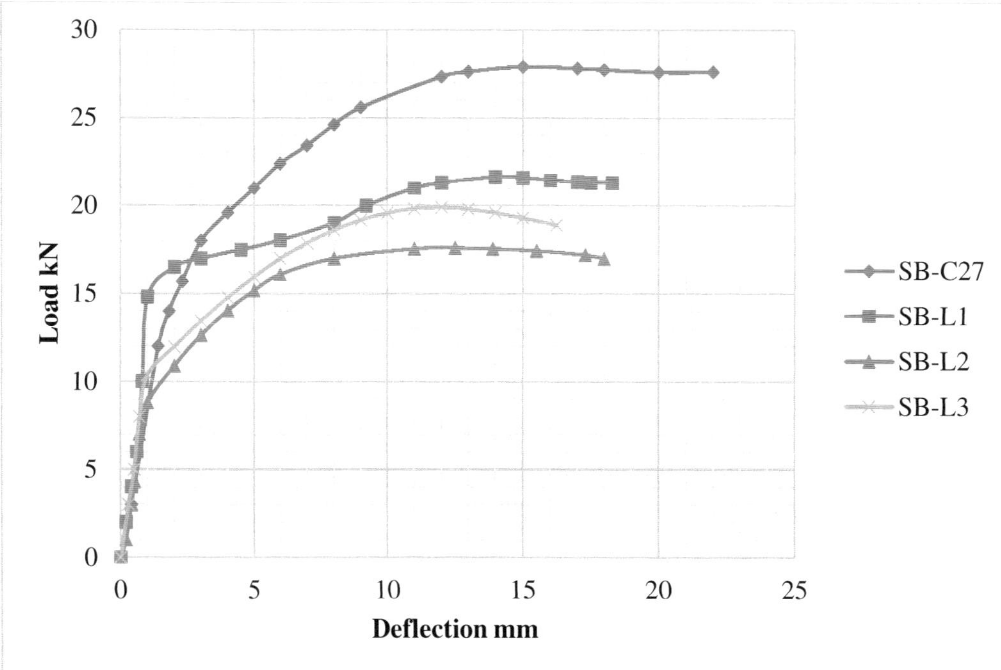

Figure 6. Load Deflection curves for the SB-C45, SB-L2, SB-L2, SB-L3 slabs.

3.2 Load-Strain Relationship

The composite behaviour of the concrete sandwich slabs was predicted by using load strain profiles When each layer acts independently accompany with a relative slip occurring at the interface, this means the panel behaved as non-composite. While when the two concrete layers are connected with full or efficient shear transformation, the panel behaved as full-composite [11] . From Figures 7-11, It is clear that SN-C45 slab exhibits full composite action till the load was 78.26% of the ultimate load, while SB-C45 & SB-L2, SB-L3 slabs show full action till the load reach 72.46%, 73.5%, 71.79% of the ultimat load. SB-L1 slab exhibits minimum composite action when the load was 40% of the ultimate load.

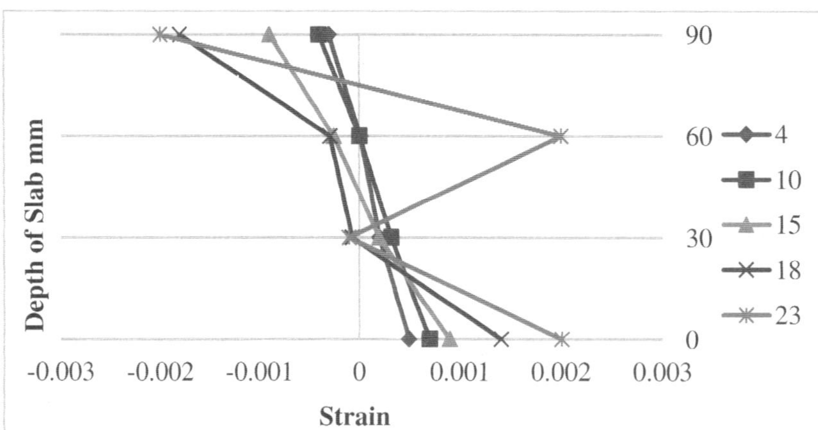

Figure 7. Strain Distribution along Depth of **SN-C45** slab with a Different load stages.

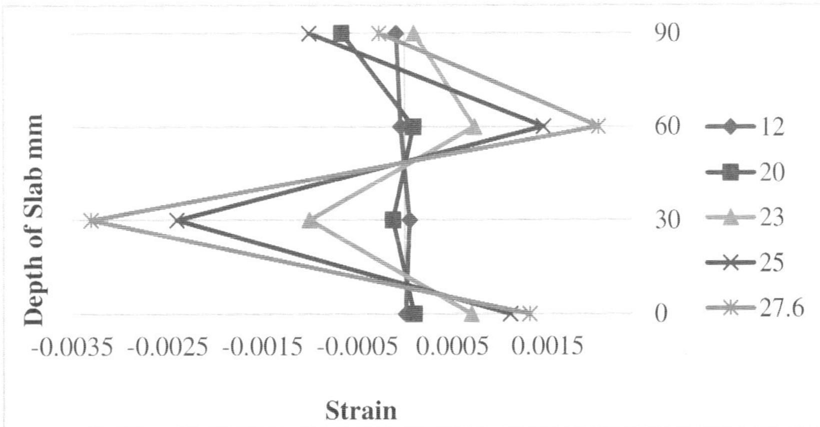

Figure 8. Strain Distribution along Depth of **SB-C45** slab with a Different load stages.

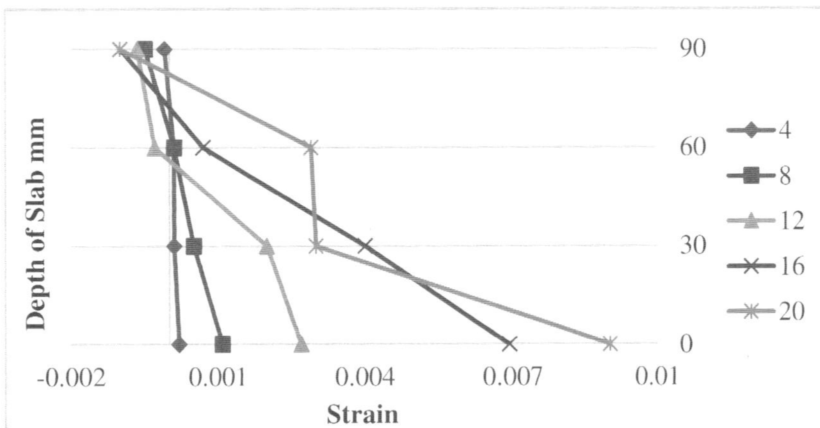

Figure 9. Strain Distribution along Depth of **SB-L1** slab with a Different load stages.

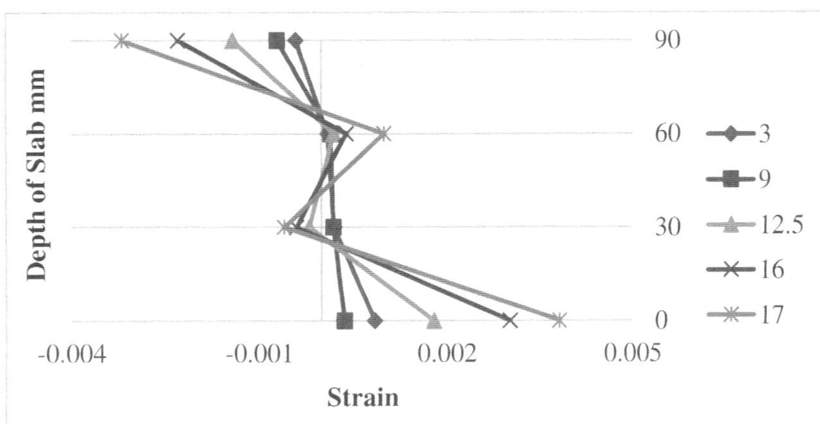

Figure 10. Strain Distribution along Depth of SB-L2 slab with a Different load stages.

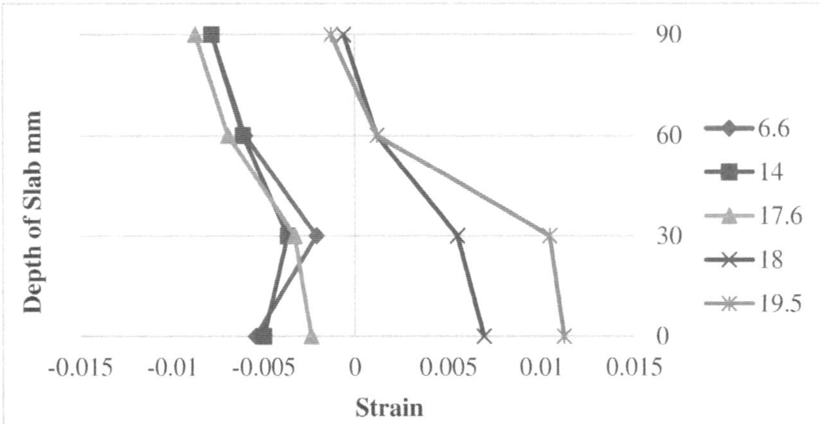

Figure 11. Strain Distribution along Depth of **SB-L3** slab with a Different load stages.

3.3 Crack Pattern

Figure (1) shows cracks patterns that appeared in the lower face of tested slabs. All slabs exhibited flexural cracks which developed along the width of the panel. Small cracks were noticed in the top concrete layer. This behavior is agreed with previous researchers [2, 8]. All sandwich slabs show ductile behaviour, exhibiting large deformation before the failure, more than the solid conventional panel. All sandwich slabs failed due to the tension failure at the bottom layer except the solid conventional panel which failed in the tension zone accompanied by concrete crushing.

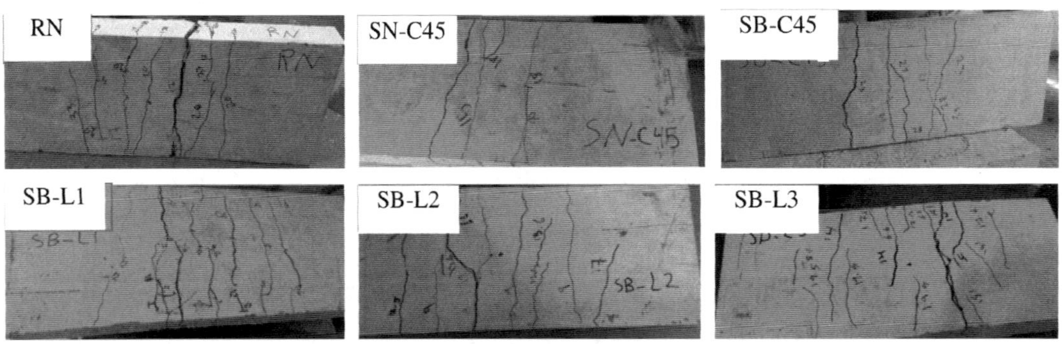

Figure 12. Cracks Distribution of the Different Tested slabs.

4. Conclusions

Structural lightweight aggregate concrete, with a cylinder compressive strength about 25.2 MPa and air dry density of 1954 kg/m could be produced from waste crushed bricks as a coarse lightweight aggregate, natural sand, and high-performance superplasticizer (PC-200).

I. Solid slab with normal aggregate showed an increase in flexural ultimate strength capacity as compared with the sandwich slab with normal aggregate while SN-C45 slab had less hardened density than solid slab. Also, sandwich slab will increase the deflection value, and ductility index.

II. The sandwich slab that contains crushed bricks as a coarse aggregate with 45 inclined angle continuous shear connectors bent had less total weight compared with normal aggregate SN-C45 sandwich slab, this reduction was accompanied by reduce in load capacity while the ductility and toughness will increase.

III. SB-C45 slab was represented the best layout of shear connectors than the other tested sandwich slabs.

IV. The presence of shear connectors at the ends of the panel section had significant effect on the flexural strength and the composite ratio of the tested slabs. when using discrete truss shear connectors. (SB-L1), (SB-L2), (SB-L3) slabs exhibited decreasing in load capacity as compared with SB-C45 slab. This reduction accompanies by a reduction in total weight. The area of steel shear connectors (which is considered an expensive material) has reduced.

V. It was clear that SN-C45 slab show better composite action and less slip value when compared with SB-C45 due to the density of concrete which increases when the compressive strength increase.

References

[1] Benayoune, A. Samad, A. A. Ali, A. A. & Trikha, D. (2007) Response of pre-cast reinforced composite sandwich panels to axial loading, *Construction and Building materials*. 21, 677-685.

[2] Benayoune, A. Samad, A. A. Trikha, D. Ali, A. A. & Ellinna, S. (2008) Flexural behaviour of pre-cast concrete sandwich composite panel– experimental and theoretical investigations, *Construction and Building Materials*. 22, 580-592.

[3] Samsudin, S. & Mohamad, N. (2013) Structural behaviour of precast lightweight foamed concrete sandwich panel under axial load: an overview, the *International Journal of Integrated Engineering*. 4.

[4] PCI Committee on Precast Sandwich Wall Panels (2011) State of the art of precast/prestressed concrete sandwich wall panels, *PCI Journal*. 56, 131-176.

[5] Jiang, H. Guo, Z. Liu, J. & Liu, H. (2018) The Shear Behavior of Precast Concrete Sandwich Panels with W-shaped SGFRP Shear Connectors, the *KSCE Journal of Civil Engineering*,

[6] Kang, W.H. & Kim, J. (2016) Development of statistical design models for concrete sandwich panels with continuous glass-fiber-reinforced polymer shear connectors, *Advances in Structural Engineering*. 19, 239-254.

[7] Benayoune, A. Samad, A. A. A. Trikha, D. Ali, A. A. A. & Ashrabov, A. (2006) Structural behaviour of eccentrically loaded precast sandwich panels, *Construction and building Materials*. 20, 713-724.

[8] Mohamad, N. Khalifa, H. Samad, A. A. A. Mendis, P. & Goh, W. (2016) Structural performance of recycled aggregate in CSP slab subjected to flexure load, *Construction and Building Materials*. 115, 669-680.

[9] Al-Mamoori, F. H. N. (2015) Structural Behavior of Light Weight Concrete Slab Panels Reinforced with CFRP Bars, University of Babylon, 264.

[10] Adheem, A. H, Rasheed, L. S. & Ali, I. M. (2018) Flexural behavior aggregate concrete, the *International Journal of Civil Engineering*. 9.

[11] Zhi, Q. & Guo, Z. (2016) Experimental evaluation of precast concrete sandwich wall panels with steel–glass fiber–reinforced polymer shear connectors, *Advances in Structural Engineering*.

IOP Conf. Series: Materials Science and Engineering **584** (2019) 012043 doi:10.1088/1757-899X/584/1/012043

Tire Rubber Waste for Improving Gypseous Soil

L N Snodi[1,*] and I S Hussein[2]

[1,2] Department of Civil Engineering, College of Engineering, University of Tikrit, Iraq.

*Corresponding author, Email: dr.lamyaanajah@tu.edu.iq

Abstract. Each year large quantities of tire wastes are released to the environment in an undesirable way. Those waste materials cause many environmental and economic problems. This study introduces an attempt to improve Gypseous soil properties using rubber tires waste.

This paper presents an experimental study that explores the possibility of using rubber tires waste as an additive to improve the strength of gypseous soils. Different percentages of the waste material (2, 4, 6 and 8%) were mixed with the designated soil. The obtained results show that the angle of internal friction increases significantly with the addition of a different percentage of tire residues, unlike cohesion which was barely affected. In addition, the waste tires rubber minimizes the maximum dry density and optimum moisture content for this type of soil. The difference in optimum moisture content and maximum dry weight with the addition of waste fraction is linear.

1. Introduction

Scrap tires are increasing every year which is causing a major environmental problem [1]. Rubber tires particles are using in landfill engineering as subgrade reinforcement for construction roads over soft soil or for improvement some physical and engineering properties of collapsible soil. Cetin et al. (2006) suggested using waste tires rubber application in civil engineering due to the [2]:

1. Low density.

2. High durability.

3. High thermal insulation.

4. Low cost compared with the other fill materials.

Gypseous soil is a rich soil with salt deposited which is one of the many types of collapsible soils. The problems are associated with water seepage through the soil; the gypsum dissolves thus causing subsidence of the ground level. In order to understand this problem, it should be prevented or reduce the dissolution of gypsum in the soil [3]. Many researchers was studied during the last 3 decades the treatment of that problem, some of them were physical treatment like cement addition, earth reinforcements and stone piles while the other were chemical treatment like treatment by Lime, Clinker, Polycoat, Emulsified Asphalt or fuel oil [4,5,6,7]. Other researchers studied the effect of adding waste tires as an improvement material, [8] study the effect of replacing sand aggregate grain size with wastes rubber and plastics at different mixing ratio. Also [9] made a numerical analysis for a reinforced wall using rubber tire chips-sand mixtures as backfill material. [10] Presents a laboratory study showed the effect of granular rubber waste tire on the physical properties of a clayey soil.

This paper presents an experimental study for improvement properties of Gypseous soil by mixing it with rubber tire particles which is one of the most little cost material.

2. Materials

2.1 Waste tires rubber

The material used in this study was the waste of rubber tires for bicycles Figure (1) and was manually cut by scissors with squares (1cm^2) and thickness (3 mm).

Figure 1. Tires Rubber Waste.

2.2 Soil

This study performed on natural Gypseous soil which it had highly gypsum content (41.24%). The soil was taken from Tikrit city in Salah-Aldeen governorate (Tikrit University). Samples were taken from depth range (1.0-3.0) meters below the ground surface after excavating the upper soils strata.

3. Results and Discussions

3.1 Properties Tests

The physical tests are computed for samples taken from depth 2 meters below the ground surface. The obtained samples (disturbed and undisturbed) were taken to the soil mechanic laboratory, Civil Engineering Department, College of Engineering at the University of Tikrit. The grain size distribution result as shown in Figure (2), the soil is poorly graded sand (SP) according to the United soil classification (USC). The properties of soil are shown in Table (1). All the above tests conducted according to the ASTM specification [11] except the specific gravity which was conducted according to the British Standard [12].

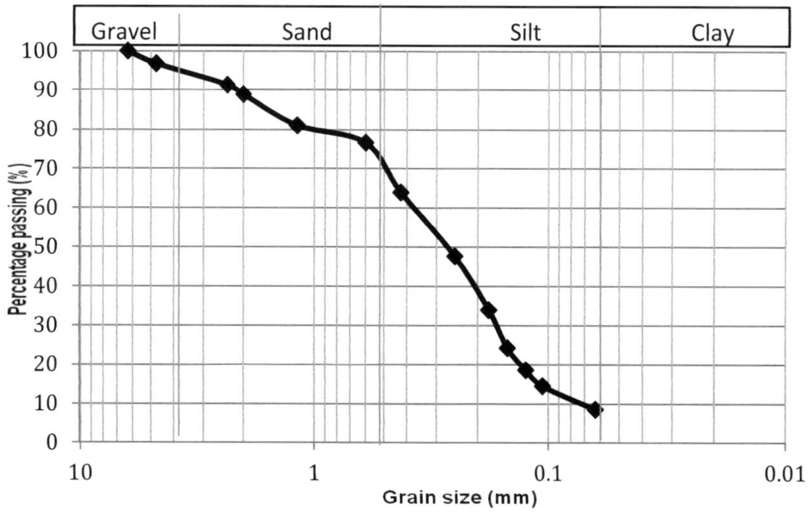

Figure 2. Grain size distribution curves for the soils used in the study.

Table 1. Properties of the Gypseous soil.

Soil Properties	Value
Water Content (OMC) (%)	4.9
Specific Gravity	2.75
Liquid Limit (%)	26
Plastic Limit (%)	N.P
Plasticity Index (%)	-------
Maximum Dry Density (MDD) (gm/cm^3)	1.44
Gypsum Content (%)	41.24

3.2 Direct Shear Test

The direct shear test was conducted according to the ASTM (D3080-72), which was used to determine the shear strength parameters, the cohesion (c) and the angle of internal friction (φ). Direct shear test conducted first on the undisturbed sample, and the value of the cohesion (c) and angle of internal friction (φ) was (3.1N/mm^2) (28.9°) respectively. Direct shear test conducted on a disturbed sample which re-molded with adding a different percentage of waste tire rubber content (0, 2, 4, 6, 8%) as shown in Table (2). It could be observed that the angle of internal friction (φ) of gypseous soil increased significantly by adding a different percentage of waste tire rubber. The reason of this behavior may be related to increasing in the surface of friction between soil particle and the rubber particle, which is the surface area of the rubber particle greater than that for the soil particle.

Table .2 Results of Direct Shear Test for Gypseous Soil.

Rubber waste content (%)	**Cohesion (N/mm²)**	**Friction angle (φ °)**
0	3.1	28.9
2	3.4	34.6
4	3.6	37.5
6	4.1	41.2
8	4.8	47.8

3.3 Compaction Test

Standard Procter Compaction test was used to obtain maximum dry unit weight (MDD) and optimum moisture content (OMC) of untreated and treated soil samples according to the procedure of ASTM designated as D1557-79 [13].

As shown in Table (3) which related moisture content and dry unit weight of gypseous soil with adding a different percentage of waste tire rubber content (0, 2, 4, 6 and 8%). It could be observed that rubber particle decreased the maximum dry density for the soil because that the specific gravity of the rubber is less than that for the soil. It could benefit from this point by using the rubber particle in the construction of embankments or as reinforcement material in geotechnical works [10]. Also, it could be observed that adding rubber particle decreased the optimum moisture content because rubber particle does not suck water.

The difference in optimum moisture content and maximum dry density with the percentage of waste tire rubber ratios is linear as shown in Figure (3), and Figure(4) and the shape of the curving curves is similar to that of the rubber-free samples added.

Table 3. Results of Compaction Test.

Rubber Content (%)	OMC (%)	MDD (gm/cm³)
0	18.65	1.44
2	18.33	1.40
4	18.11	1.37
6	17.92	1.34
8	17.73	1.32

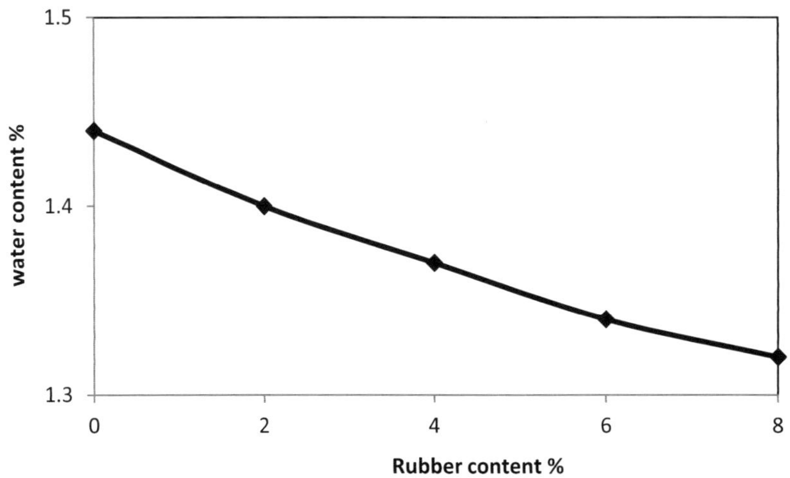

Figure 3. Effect of waste tire rubber content on the optimum moisture content (OMC %).

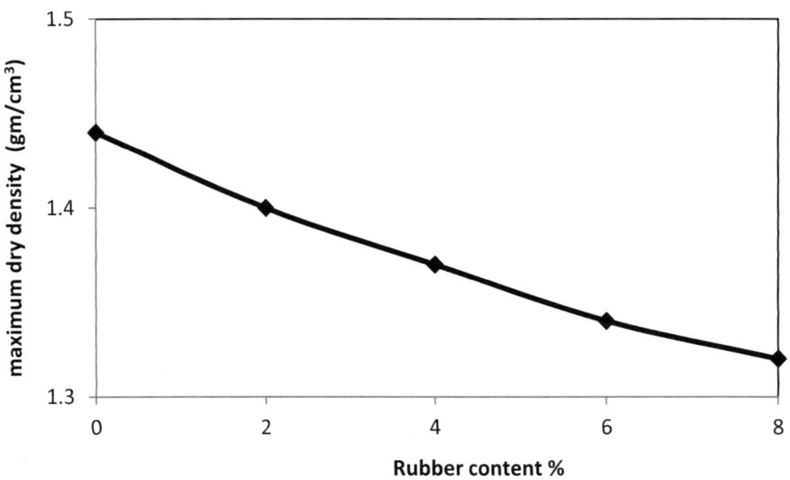

Figure 4. Effect of waste tire rubber content on the maximum dry density (MDD).

4. Conclusion

Based on the testing results of effect tired rubber for improvement Gypseous soil the following conclusions can be drawn

1. Addition waste tires rubber to Gypseous soil increases the value of cohesion but not significantly.

2. The angle of internal friction for the Gypseous soil increases significantly with the addition of different proportions of tire residues.

3. The waste tires rubber cut minimizes the maximum dry density of the soil and optimum - moisture content. The difference in optimum moisture content and maximum dry weight with the addition of waste fraction is linear

4. The shape of the curves of the soil containing the rubber pieces is similar to the models without additives

References

[1] Sivakumar Babu, G.L. Bali,R.S. (2017). Geoenvironmental Practices and Sustainability (Part of the Developments in Geotechnical Engineering book series (DGE)). DOI 10.1007/978-981 -10-4077-1.

[2] Cetin, H., Fener, M. and Gunaydin, O., (2006). Geotechnical properties of tire-cohesive clayey soil mixtures as a fill material. *Engineering Geology* No.88, p. 110-120.

[3] Al-Abdullah, S.F., Dalaly, N.K. and Saleam, S.N. (2000). A Proposed Approach on Improvement of Gypseous Sandy Soil. Journal of Engineering, University of Mustansiriya.

[4] Al-Janabi, M.D. (1997). Compressibility of Gypseous Soils Stabilized with Lime. M.Sc. thesis, Department of Civil Engineering, University of Mustansiriya.

[5] Al-Neami, M.M. (2010). Improvement of Gypseous Soil by Clinker Additive. *Journal of Engineering*, University of Technology. Vol.28, No.19.

[6] Abbas, H.O. (2012). Improvement of Sabkha Soil by Using Geomesh and Addition of Polycoat. *Journal of Engineering*, University of Technology. Vol.30, No.4.

[7] Aziz, H.Y. (2011). Gypseous Soil Improvement using Fuel Oil", *World Academy of Science, Engineering and Technology.*

[8] Abbas, F.M., Abbas, M.O., Abbas, R.O. and Shymaa K.G., (2011). Improvement of Soil by Waste Tires Addition. *Journal of Engineering*, University of Technology. Vol.29, No.16.

[9] Youwai, S. and Bergado, D.T. (2004). Numerical analysis of reinforced wall using rubber tire chips-sand mixtures as backfill material. *Computers and Geotechnics* 31, 103-114.

[10] Ramireza, G.G.D, Casagrandea, M.D.T., Folleb, D., Pereirab, A. and Paulonb, V.A., (2015). Behavior of granular rubber waste tire reinforced soil for application in geosynethetic reinforced soil wall. *Revista IBRACON de Estuturas e Material*, Vol. 8, No.4.

[11] ASTM, (1989). Soil and Rock, Building, Stones, Geotextile. Sections 4, Vol. 04-08.

[12] British Standard Institution (1975). Method of Testing Soils for Civil Engineering Purposes. B.S.1377.

[13] ASTM D1557 - 09 Standard Test Methods for Laboratory Compaction Characteristics of Soil Using Modified Effort.

The Behaviour of Reinforced Concrete Bridge Piers Using Different Types of Concrete Mix

Ali H Al-Mamoori[1], Wajde S Alyhya[1], and Hajir A Al-Hussainy[1, 2]
[1] Department of Civil Engineering, University of Kerbala, Iraq

[2] Corresponding author, email: hajerahmad204@gmail.com

Abstract. This paper investigates the behaviour of reinforced concrete piers produced from four different concrete mixes. All samples have identical dimensions of 200 mm in width, 200 mm in depth, and 600 mm in length with a column cross-section of 200 × 300 mm and 200 mm in depth. The four various concrete mixes were: Normal concrete (NC), Green concrete (GC) which used a recycled aggregate with a replacement ratio of 50% instead of normal aggregate, GC in addition to 2% volumetric ratio of recycled steel fibre (TRSF) and high strength concrete mix at pier cap region (THSC). Experimental tests have been conducted to determine the behaviour of the piers in terms of first cracking loading, load failure, crack width and deflection. Results showed that the ultimate load value of GC was less than the NC pier by 2.41%. Furthermore, there was a significant increase in the ultimate load of TRSF pier by 4.76% as compared with GC pier because of the recycled steel fibre in the pier cap that affects positively on crack width. The results also showed that the use of HSC in the pier cap region improved the failure load by 29.7 % when compared with GC pier.

Keywords: Green Concrete, High Strength, Reinforced Pier, Steel fibre

1. Introduction

Piers are the parts of the bridge that provides vertical supports at intermediate points. The two main functions of piers are: transferring superstructure vertical loads to the foundations and resisting horizontal forces acting on the bridge[1]. With increasing the rate of demolition, it is imperative to reuse waste materials effectively to conserve non-renewable natural resources [2]. Several research studies were conducted on the use of recycled aggregate in order to be used in concrete to investigate the properties of the produced concrete. Most of the previous studies investigated the mechanical properties of the resulting concrete. However, limited works were concentrated to understand the durability aspects of emerging building materials most likely with concrete made from natural aggregates [3]. Recycled aggregates are largely used as fillers in low-level applications and road construction, whereas it is little used in structural construction due to its irregular shape, which can affect the feasibility of concrete as well as the large water absorption capacity [4]. The high paste content and high water absorption of the recycled aggregate resulted in poor operating capacity and significant loss of concrete strength [5]. Previous research indicated that the addition of an extra 5% of cement to a concrete mix containing recycled aggregates can compensate for the decrease in strength [6]. Furthermore, the variation in concrete properties, which used recycled aggregates were greater than those with normal aggregates because of its inconsistent surface [7].

Many researchers have studied the mechanical properties of recycled concrete. However, no researchers have studied the effect of using recycled concrete on structural units, especially piers. Therefore, the aim of this research is to investigate the behaviour of reinforced concrete piers using various concrete mixes.

2. Experimental Work

2.1. Details of the Tested Piers

All specimens have an identical geometry and reinforcement pattern as shown in Fig. 1 and Table 1.

The cross-section of each reinforcement concrete piers is shown in Fig.2

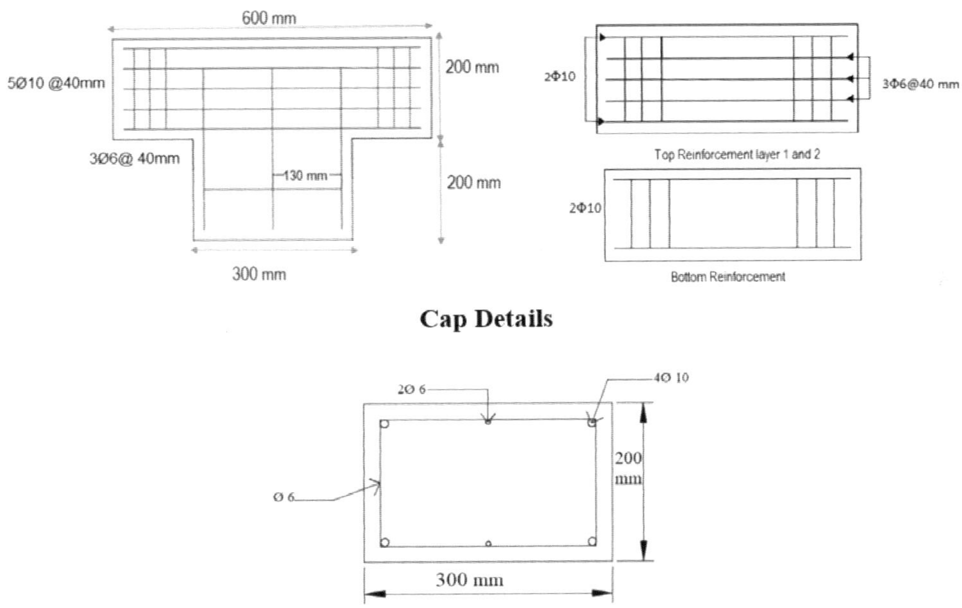

Figure 1. The cross section details of the pier.

Table 1. Description of Tested Pier Specimens.

Pier Case	Concrete type	Cap Details		Column Details	
		Normal Aggregate %	Recycled Aggregate %	Normal Aggregate %	Recycled Aggregate %
NC	Normal concrete	100	0	100	0
GC	Green concrete	50	50	50	50
TRSF	Green concrete +2% by vol. ratio of the recycled steel fibres	50	50	50	50
THSC	High strength concrete	50	50	50	50

2.1.1. Case No. 1 (NC)

It represents the reference which is cast by using normal concrete (NC) and reinforced with steel bar as shown in Fig. 2.

2.1.2. Case No. 2 (GC)

This case investigates the effect of using green concrete (GC), which is produced by replacing half natural aggregate (NA) by recycled coarse aggregate (RCA) and reinforcing with steel bar as shown in Fig. 2.

2.1.3. Case No. 3 (TRSF)

This case investigates the behaviour of green concrete that has recycled steel fibres of 2% by volume in the pier cap part with GC in the column part as shown in Fig.2.

2.1.4. Case No. 4 (THSC)

This case investigates the effect of using high strength concrete (HSC) in the pier cap while the green concrete mix is used for the column as shown in Fig.2.

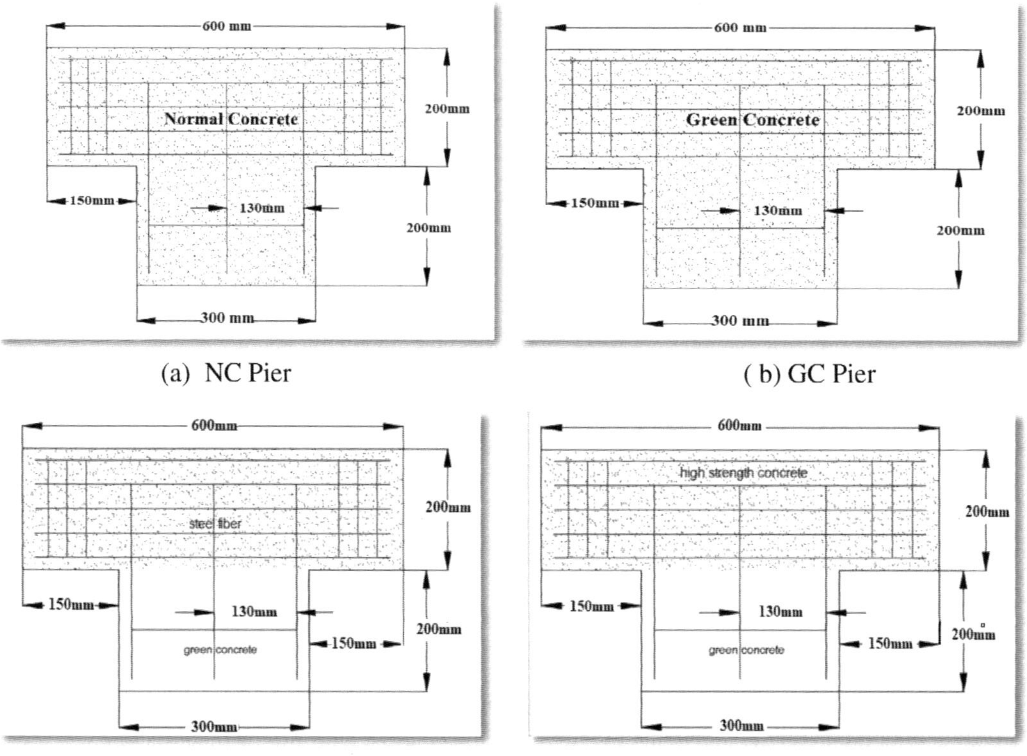

(a) NC Pier (b) GC Pier

(c) TRSF Pier (d) THSC Pier

Figure 2. Details of the tested piers.

2.2. Materials

2.2.1. Cement

CIM V is used, which is produced by Karbala factory according to the Iraqi specification No. 5/1984 [8].

2.2.2. Fine Aggregate

Natural sand from Al-Ukhaidher is used for producing various concrete mixes with a maximum particle size of (4.75 mm). The physical and chemical properties were according to the Iraqi specification No. 45/1984 [9].

2.2.3. Coarse aggregate
Natural aggregate is used with a (19 mm) as maximum particle size, which is washed by water and then left to dry in the air before using [9].

2.2.4 Recycled coarse aggregate
Old concrete samples, which are available in the laboratory are collected to be used as a recycled coarse aggregate (RCA is broken down into small size particles by hand hummer and sieved according to the Iraqi specifications' limits No. 45 /1984 [9].

2.2.5. Recycled steel fibres
The steel fibres used in this investigation was a recycled one produced and extracted from old cars tires. The tires have been burned and then the steel fibres were extracted and cut into small pieces of 2 cm length. The diameter and length of the recycled steel fibre were 1 mm and 20 mm respectively with a dry density of 6740 kg/m^3.

2.2.6. Steel reinforcement
Ukrainian deformed bars with two sizes of 6- and 10-mm diameter was used in reinforcing all concrete piers. From each size, three samples have been tested. The tensile strength was tested according to ASTM A- 615-15[10].

2.2.7 Superplasticizer
High-range water-reducing admixture was used to produce high strength concrete (HSC) mix only, which is produced by Sika Company and known commercially as SikaViscoCrete®-5930.

2.3. Concrete mix proportions
Normal and high strength concrete mixes were used for casting pier specimens using recycled coarse aggregates at 50% replacement ratio by normal aggregates as green concrete. All concrete mixes consisted of cement, fine aggregate, coarse aggregate, recycled aggregate concrete, and water. Several empirical mixes have been carried out in the laboratory to select the final mix of (1: 1.4: 2.2) with w/c ratio of 0.43 for the normal and green concrete mix and (1: 0.97: 1.23) with w/c ratio of 0.38 for high strength mix. The green concrete mix was produced by replacement of 50% of natural coarse aggregate with recycled coarse aggregate. The mixing process was performed by using a rotary mixer of 0.1 m^3 and the mix proportions are shown in Table 2.

Table 2. Mix proportions of different types of concrete.

Materials	Normal	Green concrete	High-strength
Cement	465	465	500
Fine Aggregate	651	651	485
Coarse Aggregate	1023	511.5	615
Recycled Coarse Aggregate	0	511.5	615
w/c	0.43	0.43	0.38
Superplasticizer	0	0	1

2.4. Casting of Concrete
All moulds were prepared and all their internal surfaces were oiled. After mixing, the concrete was poured into the moulds and the compaction was applied using an electric vibrator. After one day, the moulds were opened and the pier specimens were placed in special treatment basins.

2.5. Testing
After curing, the pier specimens transferred from the water tank. In order to clarify cracks, the surfaces have been dried and painted by white colour. To measure the deflection, the load was applied by using two points load on the specimens in which the distance from centre to centre was 450 mm.

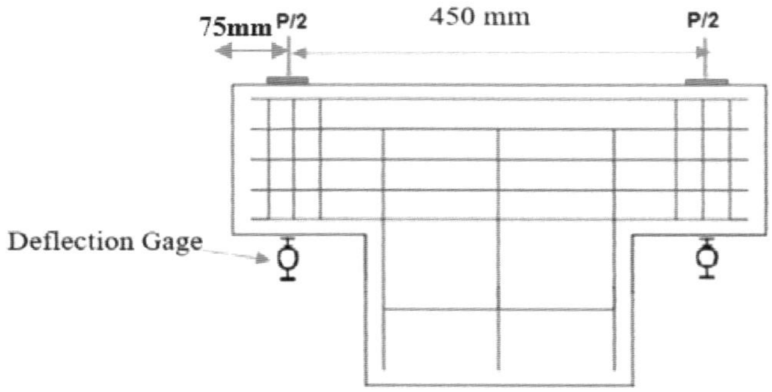

Figure 3. The deflection gages and point loads for pier specimens.

Figure 4. The hydraulic testing machine of concrete pier specimens.

2.6. Concrete tests

2.6.1. Compressive strength test
The compressive strength test was carried out according to ASTM C39M-05 standard test method using cylindrical specimens [11]. The cylinder was with a diameter of 100 mm and 200 mm in length to determinate the compressive strength at age 28 days.

2.6.2. Splitting tensile strength test
The splitting tensile strength was carried out according to the ASTM C496 [12]. For each mix, three cylinders with a 100 mm in diameter and 200 mm in length were used to determinate the splitting tensile strength at age 28 days.

IOP Conf. Series: Materials Science and Engineering **584** (2019) 012044 doi:10.1088/1757-899X/584/1/012044

3. Experimental results

3.1. Mechanical properties of concrete

The mechanical properties of concrete samples tested in this study included compressive strength, splitting tensile strength, absorption, and density. The average values of three samples were recorded to represent every single property as shown in Table 3.

Table 3. Mechanical properties of concrete.

Mix Notation	*fc'* (MPa)	*ft* (MPa)	Absorption (%)	Density (g/cm^3)
NC	29.5	3.٥	3.7	2.485
GC	29.1	3.٣	4.4	2.343
RSF	35.0	٤,٠	٥,٠	2.343
HSC	47.2	4.5	1.٣	2.412

3.2. Crack pattern and first cracking load

When the tensile stress of concrete reaches the ultimate strength, the cracks occurred in the reinforced concrete piers. Due to the applied load, different types of cracks appeared in pier specimens. These cracks include flexural, flexural shear and shear cracks. The results of all piers are presented in Table 4, which includes the first cracking load, crack width, ultimate load and the deflection at failure stage.

Table 4. The First Crack Load, Ultimate Load, Crack Width, and the Deflection at Failure Stage.

Piers Notation	First crack Pcr (kN)	Ultimate load Pu (kN)	Crack width (mm)	Deflection (mm)
NC	200	538	2.3	6.6
GC	180	525	3.5	7.8
TRSF	238	550	2.2	5.3
THSC	320	681	3.7	4.5

From the results, it can be noticed that the cracks started at the top of the pier then propagated downward and become wider with increasing the applied load. Fig. 5 shows the cracks patterns of the pier specimen.

IOP Conf. Series: Materials Science and Engineering **584** (2019) 012044 doi:10.1088/1757-899X/584/1/012044

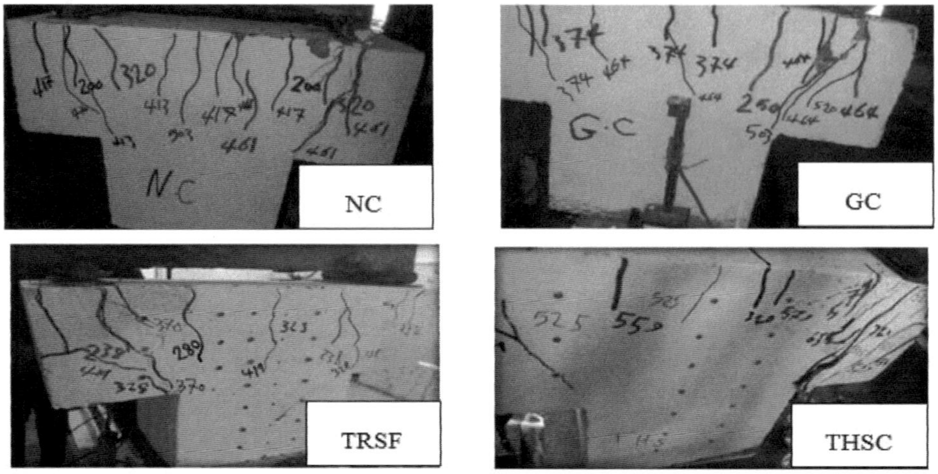

Figure 5. The Crack pattern at the failure stage of all tested piers.

3.3. Crack width

The propagation of cracks has been observed and the crack width was recorded for each 40 KN of specimens loading. This monitoring was continued until reaching the failure loads. Using a microscope type AEM40X with an accuracy of 0.05 mm, the crack width was determined. For NC pier, the first crack load was 200 kN with 0.1 mm wide flexural in type, which was close to the middle pier cap and it propagated from the top surface. At service load, 70% of the ultimate load NC specimen recorded a crack width of 1.22 mm at a load of 417 kN which increased to 1.33 mm at 461 kN load. Further loading to nearly 530 KN resulted in further cracks with a maximum crack width of 2.3 mm.

On the other hand, the GC specimen recorded the first crack at 180 kN with a crack width of 0.1 mm which was increased to 3.5 mm at 520kN. For TRSF specimen, the first crack width was 0.08 mm which increased to 1.4 mm at 361 kN. The maximum crack was 2.3 mm at 550 kN. It has been noticed that the use of recycled steel fibre has minimized the crack width. HSC pier found to have a first crack width of 0.1 mm. Moreover, at 70% of the ultimate load, the width of the cracks developed to reach a value of 1.5 mm. The maximum crack width of 3.65 mm at 680.7 kN as shown in Fig. 6.

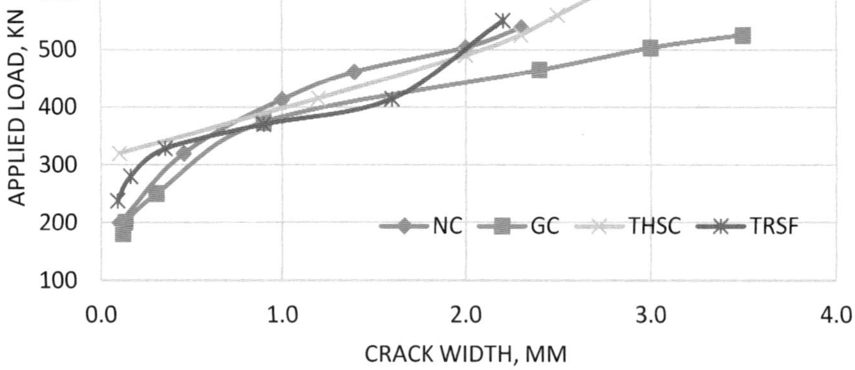

Figure 6. Load-crack width curve of various tested piers.

3.4. Ultimate load and deflection

The load-deflection curves for all tested specimens are shown in Fig. 7. Deflections have been measured at the cantilever part of the pier cap for each load increment. Generally, piers behaved elastically at early loading stages, with no visible cracks. With increasing the load, the behaviour of the samples modified from elastic to non-linear and the cracks became clear. At the third stage shear and the flexural shear, cracks continued to propagate downward and they behaved plastically, after yielding of steel reinforcement.

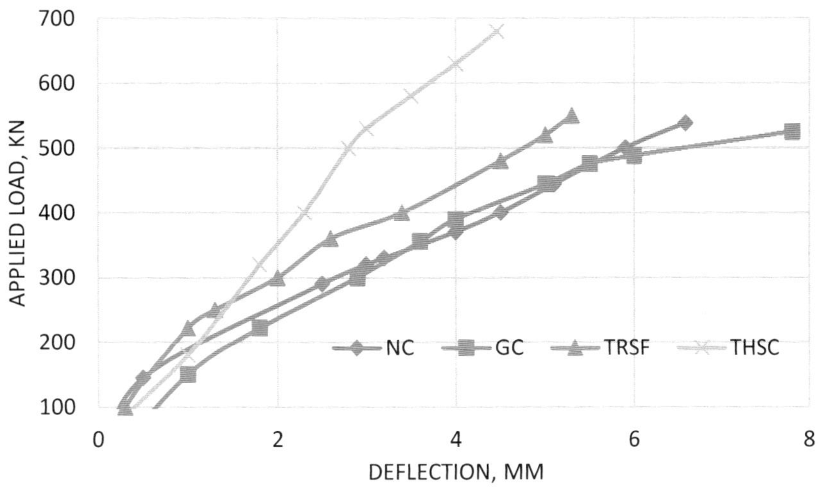

Figure 7. Load-deflection behaviour of all tested piers.

Results showed that the failure load of the NC pier was greater than that of GC pier by 2.4% and the deflection of NC pier was lower than GC specimen by 18.5% this reduction due to the higher air content of recycled coarse aggregate in the green concrete mix. By adding recycled steel fibre the load-deflection curve and the ultimate load increased significantly due to increased stiffness. The results of such improvement showed that the ultimate load of TRSF pier was greater than those without recycled steel fibre by 4.7 %, and the deflection was less by 29.6% in comparison with GC pier. Using high strength concrete mix (HSC) has improved the mechanical and physical properties of the specimen. For THSC, (pier with high strength concrete at pier cap only), the ultimate strength was greater than GC pier by 29.7%. Also, the deflection decreased by about 43% in comparison with the GC pier. These results may be due to the effect of using high strength concrete.

4. Conclusions

This study revealed several important points in the philosophy of green concrete piers as follows:

1. The ultimate strength of the pier cast using green concrete (GC) was less than those for NC by 2.4%, which means that GC is an effective alternative to be used for structural purpose.
2. The deflection under the ultimate load of the GC pier was higher by 18.5% than that of NC pier.
3. Adding recycled steel fibre to the pier cap increased the ultimate load by 4.8%. However, it caused a decrease in the deflection by 29.6 compared to GC. Use of high strength concrete (HSC) increased the ultimate load by 29.7% while the deflection decreased by 42.9% when compared with GC pier.
4. Generally, the increase in the ultimate current capacity of all pier model was conjugated by an increase in the first crack load and a decrease in the crack width corresponding to the applied load.

References

[1] G. Fu, *Bridge Design LRFD and LRFR*. John Wiley & Sons, Inc., 2013.

[2] A. K. Padmini, K. Ramamurthy, and M. S. Mathews, "Influence of parent concrete on the properties of recycled aggregate concrete," *Constr. Build. Mater.*, vol. 23, no. 2, pp. 829–836, 2009.

[3] F. T. Olorunsogo and N. Padayachee, "Performance of recycled aggregate concrete monitored by durability indexes," *Cem. Concr. Res.*, vol. 32, pp. 179–185, 2002.

[4] S. Hasaba, M. Kawamura, K. Toriik, and K. Takemoto, "Drying shrinkage and durability of the concrete made of recycled concrete aggregate," *Trans. Japan Concr. Inst.*, vol. 3, pp. 55–60, 1981.

[5] N. Su and B. L. Wang, "Study on the engineering properties of recycled aggregate concrete and recovered aggregate from demolished concrete," *J. Chinese Inst. Civ. Hydraul. Eng.*, vol. 12, no. 3, pp. 435–444, 2000.

[6] S. Frondistou-Yannas, "Waste concrete as aggregate for new concrete," in *Journal Proceedings*, 1977, vol. 74, no. 8, pp. 373–376.

[7] C. De Pauw, "Fragmentation and recycling of reinforced concrete some research results," in *Adhesion problems in the recycling of concrete*, Springer, pp. 311–319,1981.

[8] Iraqi Specification, No. 5/1984, "*Portland Cement*".

[9] Iraqi Specification, No. 45/1984, "*Aggregate from Natural Sources for Concrete and Construction*".

[10] ASTM A615/A615M-15a, "Standard Specification for Deformed and Plain Billet-Steel Bars for Concrete Reinforcement," *ASTM Int.*, 2015.

[11] ASTM C39/C39M-05, "Standard Test Method for Compressive Strength of Cylindrical Concrete Specimens," *ASTM Int.*, p. 8, 2004.

[12] ASTM Standard C496, "Standard Test Method for Splitting Tensile Strength of Cylindrical Concrete Specimens," *ASTM Int.*, pp. 1–5, 2004.

Performance of piled raft foundation supported by either connected or unconnected piles

Athraa M J Alhassani[1,3,*], Ala Nasir Aljorany [2]

[1]PhD Student, Department of Civil Engineering, University of Baghdad, Iraq
[2]Professor, Department of Civil Engineering, University of Baghdad, Iraq
[3] Lecturer, Department of Civil Engineering, University of Kufa, Iraq
*Corresponding author, email: athraa_alhassani@yahoo.com

Abstract. Piled raft is commonly used as foundation for high rise buildings. The design concept of piled raft foundation is to minimize the number of piles, and to utilize the entire bearing capacity. High axial stresses are therefore, concentrated at the region of connection between the piles and raft. Recently, an alternative technique is proposed to disconnect the piles from the raft in a so called unconnected piled raft foundation UCPR, in which a compacted soil layer (cushion) beneath the raft, is usually introduced. The piles of the new system are considered as reinforcement members for the subsoil. In the current study, the behavior of both connected and unconnected piled rafts systems has been studied experimentally. The effects of different factors, such as; the piles number and configuration, on the load settlement behaviour and pile load sharing are also investigated. The results indicate that when unconnected piles are used, the total settlement of the piled raft system is significantly reduced, and the pile load sharing considerably decrease, e.g., for piled raft with nine piles the total settlement has decreased by 35.6% and pile load sharing has decreased by 20.9%.

1. Introduction

The piled raft is a composite foundation that comprises three parts; raft, piles and soil. The design philosophy of piled raft differs from that in conventional foundation where it is assumed as either a raft or pile group that carry the superstructure loads and insuring a certain value of safety factor. Basically, piled raft concept is proposed in order to obtain an economic design compared to conventional pile foundation by using the required number of piles that are important to minimize settlement to an acceptable limit and the transmitted loads are shared between the raft and piles. However, very large bending moments may be developed when these small number of piles are structurally connected to the raft as well as cracks in the raft and relatively high axial stress concentration at the pile head may be occur. Thus, the probability of structural failure of the foundation is greater than that related to the bearing capacity failure of the supporting soil. The dimensions and number of piles may consequently be increased in order to prevent the possibility of structural failure. To overcome the problem of connection reaction an alternative approach has been proposed where unconnected piles are used and provided an interposed cushion layer between the raft and the piles. The piles in such condition will behave as soil stiffeners instead of as structural members [1-3].

Recently, many numerical studies have been presented to investigate the behavior of unconnected piled raft [1], [4-7]. In addition, many experimental studies [8-10] are dealing with the UCPR.

Liang et al. reported that a granular cushion layer plays an important role in improving the bearing capacity of the subsoil and in changing the load transfer mechanism of piles. An elastoplastic numerical analysis was performed by Eslami and Malekshah, showed that the thickness and the stiffness of cushion affects the location of the maximum axial stress along piles.

Cao et al. [8] reported that the settlement and the bending moments at the piles head are considerably reduced by using unconnected piled raft. Fioravante and Giretti [10] studied the stiffness of the foundation when the granular cushion layer was interposed between the raft and piles. They found that the relative stiffness of pile and soil underneath raft affect the load distribution mechanism in the foundation. Also, it is found that the initial stiffness of foundation was essentially a function of the piles stiffness.

In the current study, the behavior of piled rafts with connected and unconnected piles is investigated experimentally. The influence of the cushion layer and number of piles used with connected and unconnected piled rafts is also investigated.

2. Testing equipment

The set up used to perform the experimental work of the current study comprises a loading frame and steel container measured 800 mm deep and (750mmx750mm) in plan as shown in Figure 1. A square steel plate, with side dimensions 210 mm and thickness 3mm is used as a model raft. The modulus of elasticity and Poisson's ratio of the steel plates were 210×10^6 kPa and 0.30, respectively. According to raft-subsoil stiffness ratio, k_{rs}, which is defined in Eq. (1) below, the model raft can be considered as flexible since its value is as low as ($k_{rs} = 0.2$) [11]:

$$k_{rs} = 5.57 \frac{E_r}{E_s} \frac{(1 - \nu_s^2)}{(1 - \nu_r^2)} \left(\frac{B_r}{L_r}\right)^{0.5} \left(\frac{t_r}{L_r}\right)^3 \quad (1)$$

An aluminum alloy model piles of hollow square section with outside dimension of 14 mm and inside circular hole of diameter 11 mm and a length, *L*, of 420 mm were used. The piles were closed-ended with modulus of elasticity of about 50×10^6 kPa and Poisson's ratio 0.3, the piles were treated as flexible in the test interpretation. Three model piles were instrumented by strain gauges to measure axial stresses at pile head and pile end. The side dimension of the container for all tests was greater than three times of the raft side and the end tip of piles located about 20D above the bottom of the container.

The settlement of the foundation system was recorded by two LVDTs (Linear Variable Differential Transformer) of 0.01 mm accuracy positioned at the middle edges of the foundation system. The application of the Load was done by an electrical hydraulic jack tied to the top of the main loading frame. Load cell of 20 kN capacity was connected to the lower end of the jack to measure the applied load. All measurements devices were connected to the data acquisition to record and save all the readings automatically.

3. Materials

A dry sand was used as a foundation soil. The specific gravity of the solid particles is about 2.64. From the grain size distribution curve of the soil shown in Figure 2, the coefficient of curvature (*Cc*) is 0.91, the coefficient of uniformity (*Cu*) is 2.05, and the mean grain size (*D50*) is 0.34 mm. According to the Unified Soil Classification System (USCS) the foundation soil can be classified as poorly graded sand (SP). The foundation soil is characterized by maximum dry unit weight 17.49 kN/m^3 and a minimum dry unit weight 15.04 kN/m^3. The soil was prepared for each model test to a relative density of about 50% which correspond to 16.81 kN/m^3 dry unit weight.

In the unconnected piled raft tests, the cushion layer, prepared by using a well graded soil. The grain size distribution curve of the cushion soil is shown in figure 2. All unconnected piled raft tests were performed with cushion layer compacted to a maximum standard Procter dry unit weight and optimum moisture content. The maximum Procter dry unit weight and optimum moisture content were 18.65 kN/m^3, and 12.4, respectively. The thickness of cushion layer was taken 5 cm for all model tests.

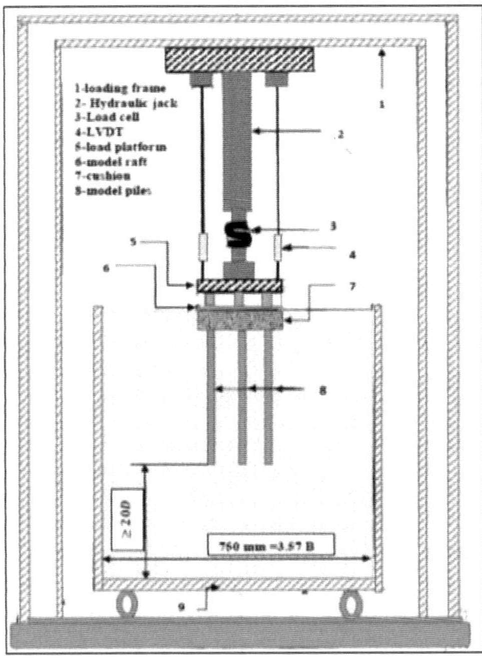

Figure 1. Loading model test for piled raft with cushion layer.

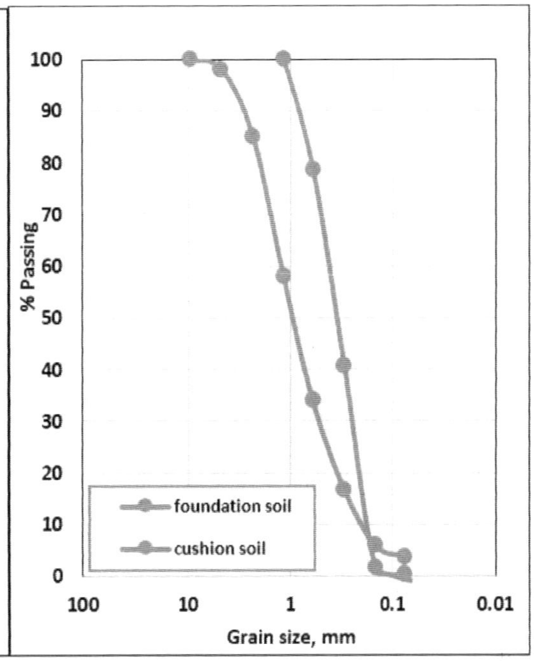

Figure 2. Grain size distribution curves of foundation and cushion soil.

4. Results and Discussions

Several model tests on connected and unconnected piled raft were performed in this study. The interposed cushion layer that was used with the unconnected piled raft models, was 5 cm in thickness and about 60 MPa modulus of elasticity. To evaluate the relative improvement of the raft performance with either connected or unconnected piles, a nondimensional factor, called the bearing pressure improvement BPI has been introduced. BPI can be defined as:

$$BPI = \frac{q_{Piled\ Raft}}{q_{Raft}} \quad (2)$$

Where

$q_{piled\ Raft}$= bearing stress of a connected or unconnected piled raft.

q_{Raft}= bearing stress of raft only, at the same settlement level.

The settlement of the raft, *S*, is expressed in terms of a nondimensional form as relative settlement ratio, *(S/B)* %, which it is the ratio of maximum settlement to raft width. The comparisons of the piled raft response considering the different studied parameters were taken at two levels of settlement ratios *S/B*, at 1% and 5%. Figure 3 and figure 4 show the relationship between applied pressure and settlement ratio for different number of piles of connected and unconnected piled raft. The stress–settlement relationships of the CPR1, CPR4, CPR5, CPR9, UNPR1, UNPR4, UNPR5 and UNPR9 tests are almost linear up to a certain stress level. This trend of the initial part of the curve was also reported by Poulos [12], as shown in Fig. 11.

The variation of BPI with pile number of connected piled raft at two settlement ratios, 1%, and 5% is shown in Figure 5. It can be concluded from the figure that, initially the rate of BPI increases with increasing pile number. However, the increasing rate of BPI becomes quite small when the number of piles reaches a certain value and it seems that the effect on reducing the settlement of raft becomes insignificant. This pattern of behavior is consistent with the conclusion by Poulos and Davis [13] "that the number of piles required to reduce settlements under a raft to a tolerable limit is usually small and any further addition of piles may result in only marginal further reductions in settlements". Figure 6 shows that the BPI for unconnected piled raft with different number of piles. From this figure it is clear that as the number of piles increases, the BPI increase approximately at the same rate. This may be attributed to the fact that the purpose of the unconnected piles is to reinforce the subsoil. In addition, the bearing pressure improvement ratio for connected and unconnected piled raft increase as the relative settlement ratio, S/B increase from 1% to 5%. Increasing BPI with increasing S/B ratio could be because of the influence of the densification of the sand produced due to load progressing and probable interactions between the piles. Figure 7 shows the comparison of BPI for connected and unconnected piled raft at relative settlement 1%.

V. Fioravante [10] stated that the performance of the settlement of an unconnected piled raft can be assessed by means of the settlement efficiency ratio, as it defined in following:

$$\eta = \frac{w_{unpiled} - w_{piled}}{w_{unpiled}} \tag{3}$$

Where $\boldsymbol{w_{unpiled}}$ and $\boldsymbol{w_{piled}}$ are the settlements amount of the unpiled raft and the UCPR, respectively. Since (and $0 \leq \eta \leq 1$), the UCPR be more efficient as the value of η be larger, at a given load level.
The settlement efficiency of the UCPRs with the applied vertical load is presented in Figure 8. It is shown that, at experienced range of displacement the settlement efficiency is approximately constant. This may be attributed to the reason that the piled raft did not reach the ultimate load. It can be concluded that the number of piles affect the settlement efficiency e.g.: η = 0.3, 0.4, and 0.68 for UNPR4, UNPR5, and UNPR9, respectively.
Figure 9 shows the Influence of number of piles on percentage of load shared by the piles. Generally, the percentage of load carried by piles increases with increasing the number of piles either for connected or unconnected piles. The comparison of pile load sharing between connected and unconnected piled raft for similar number of piles shown in Figure 10. From this figure it can be noticed that the load shared by unconnected piles is lower than that shared by connected piles except for piled raft with one pile, the unconnected pile carried load more than connected pile.

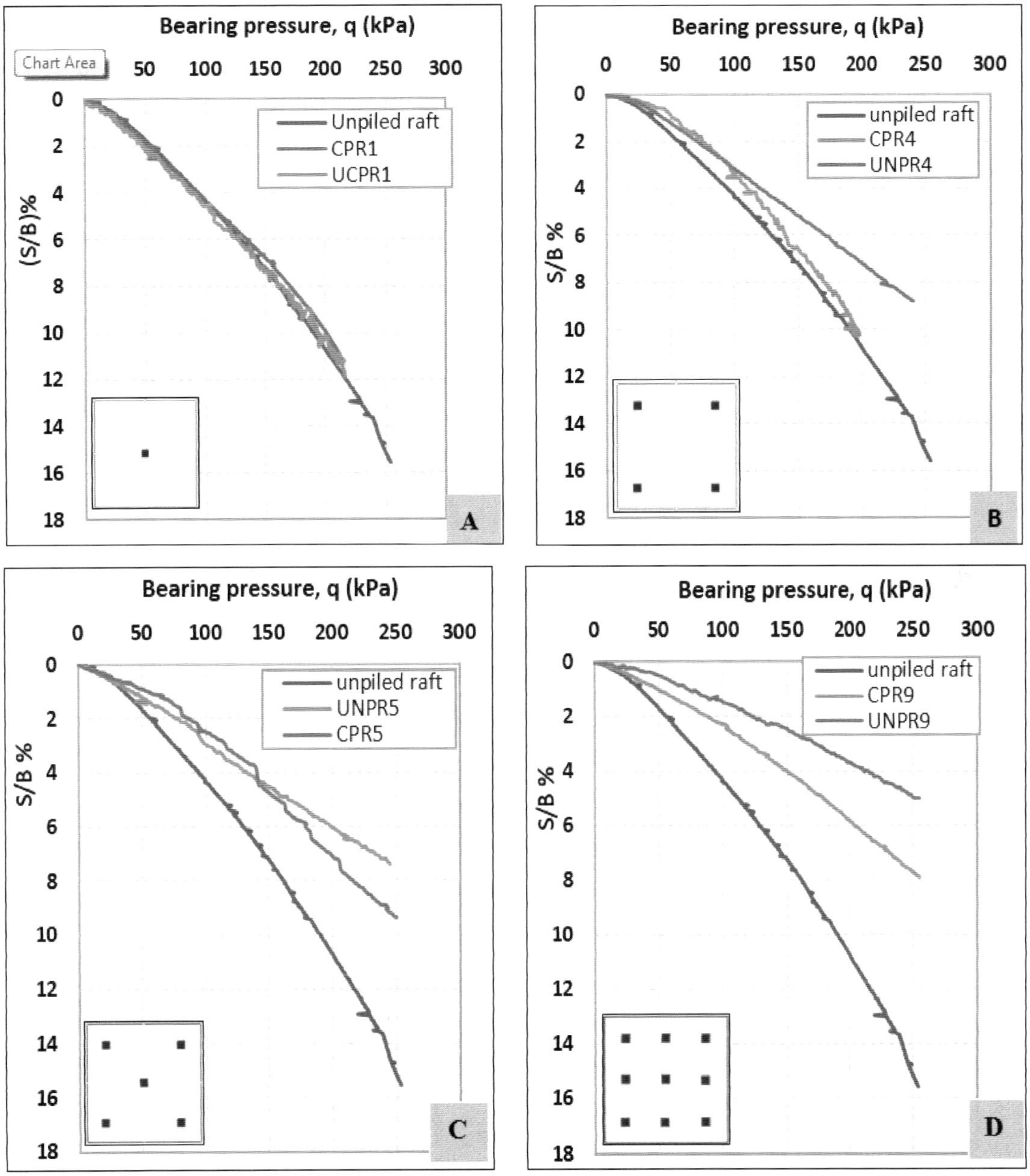

Figure 3. Relative settlement vs. bearing pressure for different number of connected and unconnected piled raft, A- one pile, B- four piles, C- five piles, D- nine piles.

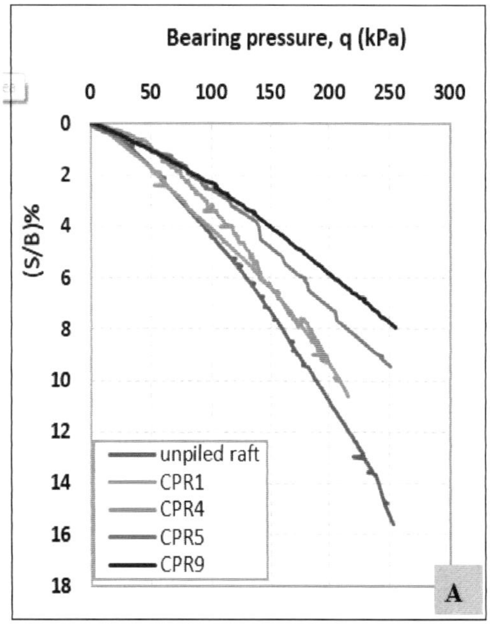

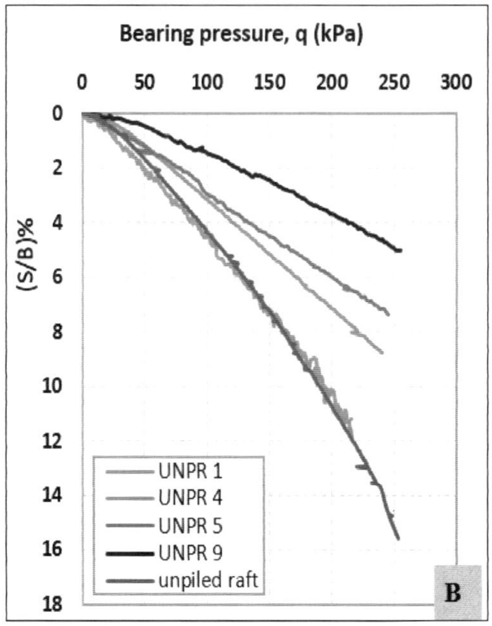

Figure 4. Variations of pressures with relative settlements different number of piles, A- Connected piles, B- Unconnected piles.

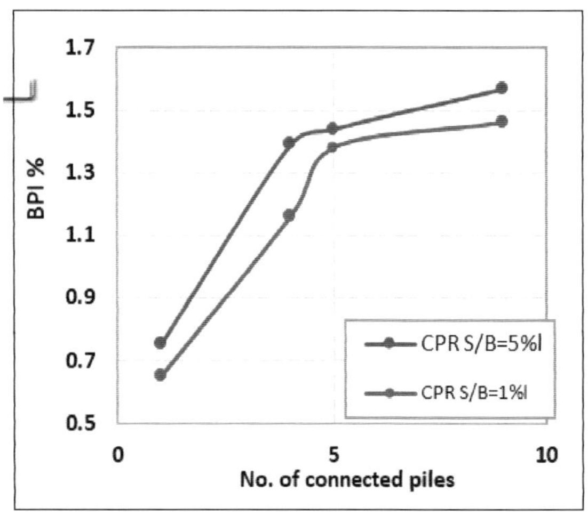

Figure 5. Variation of bearing pressure improvement with different number of connected piled raft at two S/B ratios

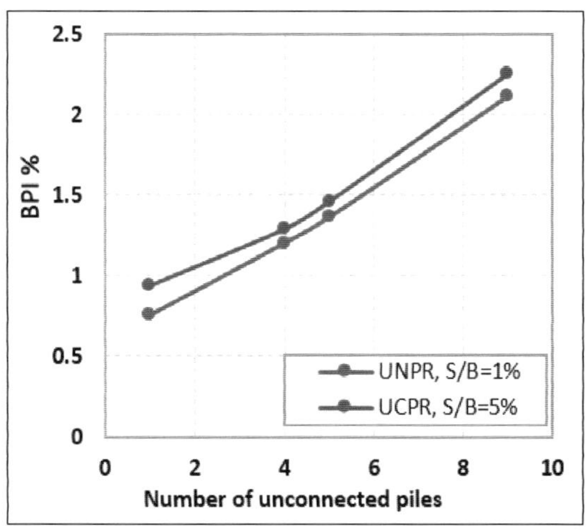

Figure 6. Variation of bearing pressure improvement with different number of unconnected piled raft at two S/B ratios

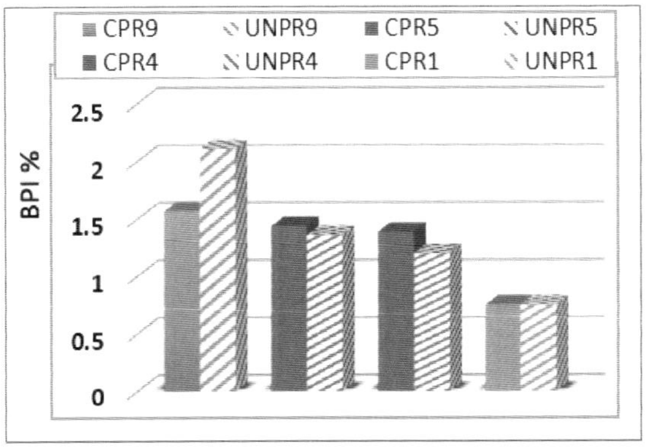

Figure 7. Comparison of BPI for connected and unconnected piled raft

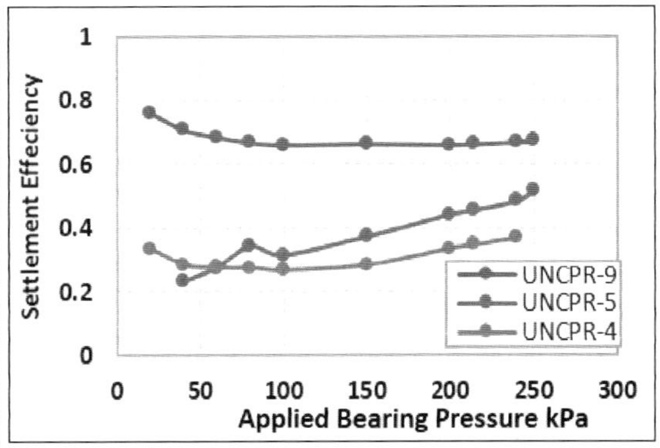

Figure 8. Settlement efficiency of unconnected piled raft.

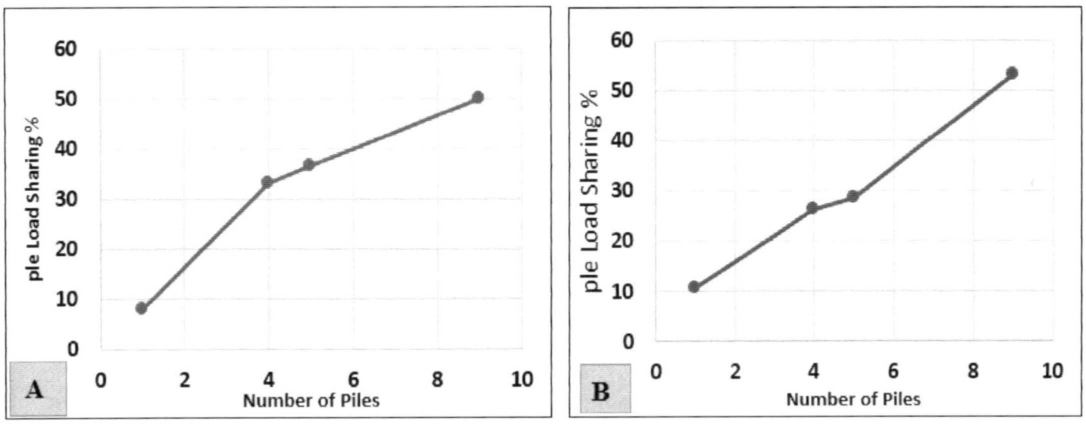

Figure 9. Pile load sharing of piled raft, A- connected piles, B- unconnected piles

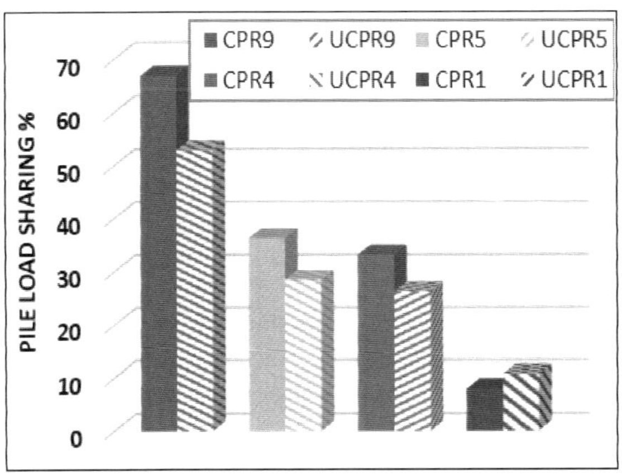

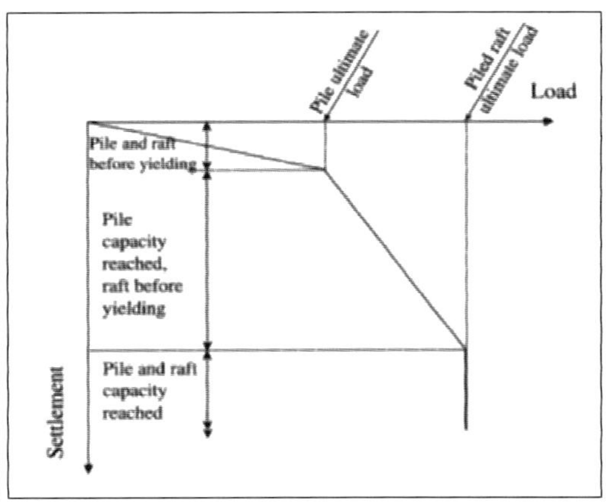

Figure 10. Comparison of pile load sharing for

Figure 11. Load-Settlement curve (adapted from Poulos 2001) [11].

5. Conclusions

Within the scope of the experimental model tests performed in this investigation, the following conclusions can be stated:

1. The amount of settlement relevant to unconnected piled raft system is less than that of connected piled raft.
2. The bearing pressure improvement BPI with connected piled raft initially increases as the pile number increase then becomes quite small when the number of piles reaches a certain value. For unconnected piled raft, as the number of piles increase the BPI increase almost at the same rate.
3. In unconnected piled raft the settlement efficiency is approximately constant in the experienced range of displacement.
4. The percentage of load carried by piles increase with increasing the number of piles either for connected or unconnected piled raft .

6. References

1. Wong I, Chang M, Cao X. 2000. Raft foundations with disconnected settlement-reducing piles, In: Hemsley JA, editor. Design applications of raft foundations, London: Thomas Telford Ltd, pp. 469.
2. Burghignoli A, Jamiolkowski M.B and Viggiani C. (2007). Geotechnics for the Preservation of Historic Cities and Monuments: Components of a Multidisciplinary Approach. In Proceedings of the 14th European Conference on Soil Mechanics and Geotechnical Engineering, Madrid, Spain. pp. 3-38
3. Jamiolkowski M B, Ricceri G, Simonini, P (2009). Safeguarding Venice from high tides: site characterization & geotechnical problems. In Proceedings of the 17th International Conference on Soil Mechanics and Geotechnical Engineering, Alexandria, Vol. 4, pp. 3209-3230.
4. Liang FY, Chen LZ, Shi XG. 2003. Numerical analysis of composite piled raft with cushion subjected to vertical load, Computers and Geotechnics, No. 6, Vol. 30, pp. 443-453
5. Eslami A, Veiskarami M, Eslami M. 2012. Study on optimized piled-raft foundations (PRF) performance with connected and non-connected piles-three case histories, International Journal of Civil Engineering, No. 2, Vol. 10, pp. 100-111 .
6. Sharma V, Vasanvala S, Solanki C. 2011. Effect of cushion on composite piled–raft foundation, Journal of Engineering Research and Studies, Vol. 2.
7. Hor B, Song M J, Jung M H, Song Y H and Park Y H. 2015. A 3D FEM analysis on the performance of disconnected piled raft foundation, The 15th Asian Regional Conference on Soil Mechanics and Geotechnical Engineering, Japanese Geotechnical Society Special Publication.
8. Cao X, Wong I, Chang M. 2004. Behavior of model rafts resting on pile-reinforced sand, Journal of Geotechnical and Geoenvironmental Engineering, 2004, No. 2, Vol. 130, pp. 129-138.
9. El Sawwaf M. 2010. Experimental study of eccentrically loaded raft with connected and unconnected short piles, Journal of Geotechnical and Geoenvironmental Engineering, No. 10, Vol. 136, pp. 1394-1402.
10. Fioravante V, Giretti D. 2010. Contact versus noncontact piled raft foundations, Canadian Geotechnical Journal, No. 11, Vol. 47, pp. 1271-1287.

IOP Conf. Series: Materials Science and Engineering **584** (2019) 012046 doi:10.1088/1757-899X/584/1/012046

Intelligent non-destructive technique for crack existence inspection in a structure using Particle Swarm Optimization (PSO)

Saad Al-Wazni[1] and Ahmed Y AZainul-abideen[2]

[1,2] Lecturer at Civil Eng. Dept., Faculty of Eng. at University of Kufa, Najaf, Iraq

* Corresponding author: Email address: saad.alwazni@uokufa.edu.iq

Abstract. Increasing cost of erection and maintenance of modern structures becomes need new method of crack investigations before catastrophic failure is happened. A proposed technique for non-destructive inspection of crack in a structure is presented using standard Particle Swarm Optimization (PSO) method. The vibration properties, natural frequencies and mode shapes, of the structure are included in the non-destructive proposed technique. The new technique of crack inspection is verified in a steel clamped beam structural model. The crack position and size are considered and two cases of crack scenarios are adopted in this study. The crack is presented in the FE model as a reduction in the stiffness of specified elements. The dynamic analysis of the FE model is implemented using ANSYS-APDL software. The proposed technique includes a subroutine created by high technical level language in MATLAB software. This new technique exhibits robust and satisfactory performance of inspection the position and size of crack in the adopted structural model.

Keywords: Vibrations of structures, intelligent nondestructive techniques, crack investigation, Particle Swarm Optimization.

1. Introduction

Intelligent nondestructive techniques are modern methods to evaluate the integrity and reliability of the structural system in many engineering disciplines. The information of crack coordinates and size in the structure is main skill of the intelligent nondestructive techniques. The crack in the structure changes the stiffness of the system. These changes of the structural properties alter the vibration characteristics of the structure. By using the vibration properties (modal frequencies and mode shapes) of the structure, the nondestructive technique is effective. In this technique, the operational system data of structure could be extracted without the need for dissociation of the structure [1].

This study developed new technique using Particle Swarm Optimization (PSO) to extract the information of crack in the structure. The PSO optimization method has solved complex non-convex problems in hyperspace [2]. Nowdays, The PSO is implemented to include large scale problems with significant efficiency in solve difficult optimal. In the specified complex problems, the PSO exhibits better than Genetic Algorithms [1]. Kennedy and Eberhart explored the idea of the PSO method during were modeling the social behavior of artificial life, like as birds flocks, fish schooling and bee collections..etc [2].

In the PSO method, the target function is very important to get the global optimal in the convex and nonlinear problems. Nanda et al. (2012) adopted technique to investigate the crack in a cantilever steel structural model using PSO with incremental swarm size [3]. The change in the natural frequencies of the structure was only included in their target function without and with noise. They improved that the PSO effective method in damage investigation. Vaez and Fallah (2017) presented a hybrid technique for crack detection using PSO with Genetic Algorithm (GA) and carried out it on three types of steel thin plate structural numerical model only in their study [4]. They used fitness function included the error in the modal frequencies and mode shapes and verified their approach with several crack scenarios without and with noise. They concluded that PSO method improved their hybrid technique with GA to investigate the crack in the structure. As well as, Wei et al (2018) proposed technique for crack investigating using modified PSO [5]. They implemented their technique on three types of structures; beam, truss and plate and reported that PSO is robust to detect crack in the structure.

In this study, the standard PSO with all parameters designed in different values is adopted in the new proposed technique. Also,the modal mode shapes were included in the adopted target function as well as the modal frequencies of the structure. Also, the weights of the parameters in the target function are studied. Those important modifications of the PSO technique and fitness function are main contributions in this research.

2. Particle Swarm Optimization (PSO)

A real simulation of the Particle Swarm Optimization method is a flock of birds in a searching for food. The target of the swarms is to find the position of the more food. In fact, each bird is going to do random search and it will keep the bushy location of food it experiences. Furthermore, all birds can memorize the acquaintance about their best exploration of positions, then each bird is leaded by its detection and by best position notified by the others. So, by changing the direction of their course to travel somewhere between two positions, the birds will detect the region by overflying positions of most focus and finally is going to dragged to the bushy position [2].

The PSO algorithm is implemented for all type of optimization problems with low effort of calculations. The Genetic Algorithm (GA) method is close to the PSO method, but, the PSO is more robust [1]. As well as, The GA optimization has binary system counter to the PSO method has real parameters to simplify the computations.

The PSO mimics the behavior of birds during searching for food. Here each bird indicates to a particle (solution from searching space) and the group of particles indicates to swarm. Each particle has location and velocity which are modified continuously. That's mean, the particle moves by two vectors, the first accelerates it to the best locally and the second to the best globally of the entire swarm. These two vectors of the particles are shown in the figure 1. Eventually, design the number of particle is more significant for exploring the precise optimal solution. In this study, the number of particle is designed to ten particles.

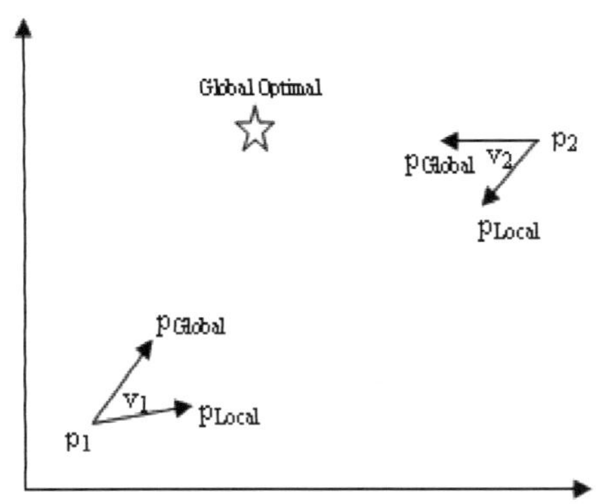

Figure 1. Concept of modification of a particle in PSO searching space.

The equation of modified velocity of the particles during the search is written in equation (1) [2]:

$$v_{i+1} = \omega \times v_i + c_1 \times r_1 \times (p_{Local} - p_i) + c_2 \times r_2 \times (p_{Global} - p_i) \tag{1}$$

where ω is a variable changed dynamically which generates acceleration for the particle depends on the last iteration. The p_i is the particle in the current time (solution). The p_{Local}, p_{Global} are the local and global optimal solution during the iterations. The c_1, c_2 are constants represent the directions of the particle to the best weighted by random values locally and globally, respectively. Those constants are design within range between (0 4]. r_1, r_2 are random number ranged between (0 1). The v_i is the current velocity of the particle. The next better step of the particle is given by kinematic equation as equation (2):

$$p_{i+1} = p_i + \Delta t . v_{i+1} \tag{2}$$

where Δt is the time incremental step of the particle movement in the searching space.
The proposed technique begins the iteration with random swarm of the particle. The convergence criterion controls the new generations of particles to get the global target. In this work, the convergence criterion is a specified number of iteration and the standard PSO was implemented.

3. Finite element analysis of the structural model

This study adopted Nanda et al (2012) structural model, the model was a steel clamped beam [3]. The structural model has length of 600 mm and material properties 200 GPa, 7800 kg/m^3, respectively. The finite element model was divided into 20 elements, as shown in figure 2.

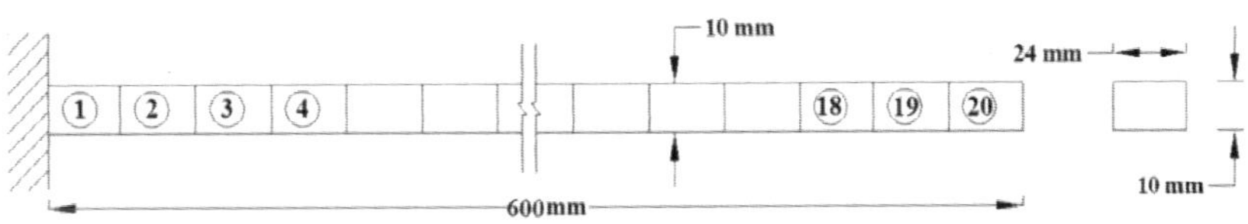

Figure 2. The adopted structural model [3].

Their study was included six natural frequencies extracted from the finite element analysis. This study carried out the numerical analysis using ANSYS-APDL software to extract the vibration properties [6], as shown in figure 3.
From figure 3, the first six natural frequencies and mode shapes are included and represented the translations of model nodes in z-axis without noise. The results of the analysis are very close and compared with the reference, as show in figure 4. The differences percentages in natural frequencies values are ranged (-0.000044) in the 1st mode and (-0.00383) in the 6th mode.
The single cracked state is included and the comparison of crack investigation in convergence of target function, between the study and the reference, is conducted to verify and exhibit the efficiency and powerful of the proposed technique.

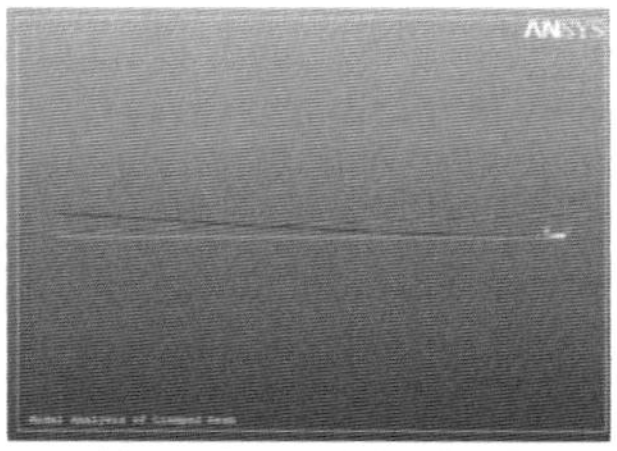

1st Bending =22.721 Hz

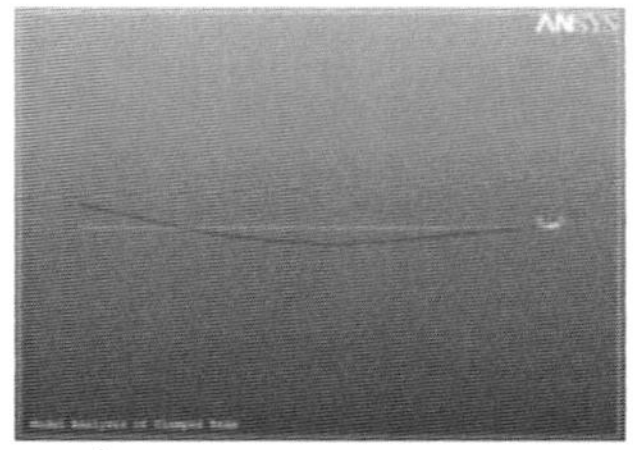

2nd Bending =142.343 Hz

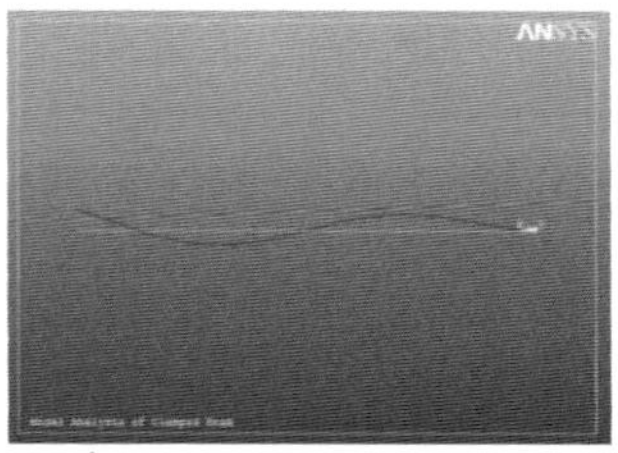

3rd Bending =398.363 Hz

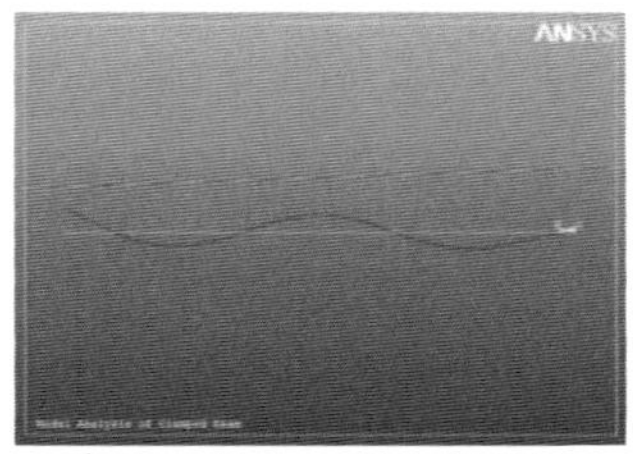

4th Bending =780.087 Hz

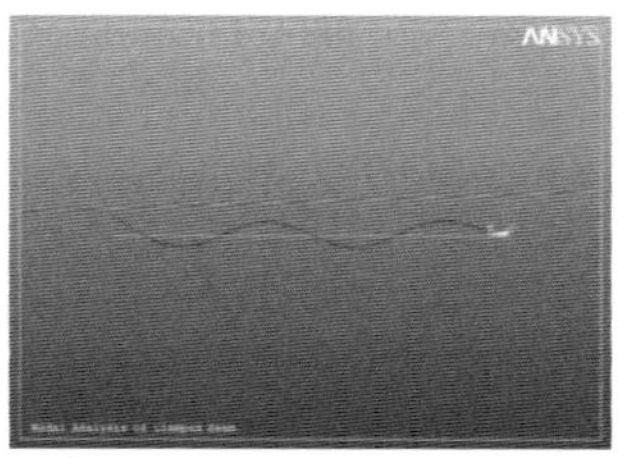

5th Bending =1288.396 Hz

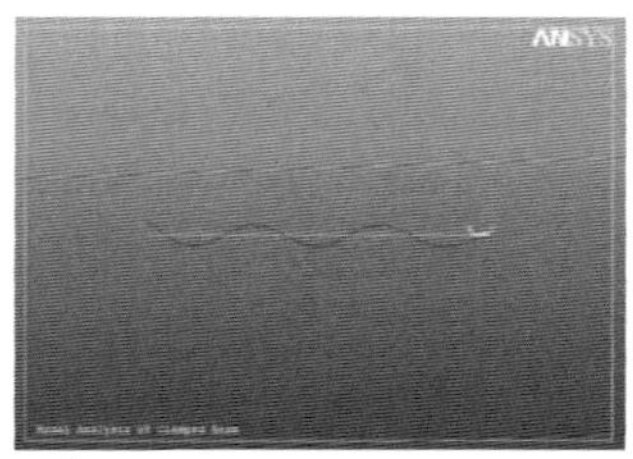

6th Bending =1922.711 Hz

Figure 3. Modal frequencies and mode shapes of the clamped beam.

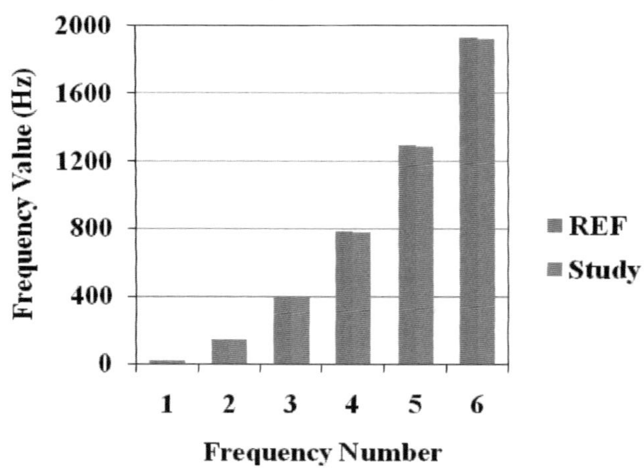

Figure 4. The comparison in natural frequencies between the study and the reference [3].

IOP Conf. Series: Materials Science and Engineering **584** (2019) 012046 doi:10.1088/1757-899X/584/1/012046

4. Performance of new proposed technique

As mentioned previously, the existence of cracks in the structure changes its properties. This changes effects on the vibration properties of the structure. The comparison between the intact and cracked structure state could supply appropriate diagnosis.

The general dynamic response equation for the intact structure could be given in equation (3):

$$M.\ddot{u}(t) + K.u(t) = 0 \tag{3}$$

where M, K are the mass and stiffness matrices, respectively and $\ddot{u}(t)$, $u(t)$ are the acceleration and displacement, respectively.

When crack occurred, the reduction in the stiffness of the local element and global structure is obtained. The changes in the eigenvalue problems are acquired, as shown in equations (4) and (5):

$$(K - \omega_i^2 M)u_i = 0 \tag{4}$$

$$(K_c - \omega_{ci}^2 M)u_{ci} = 0 \tag{5}$$

where ω_i, u_i represent the ith modal natural frequency and mode shape, respectively. The symbol $'c'$ represents the cracked structure and $i = 1,2,3,..., n$ is equal the few first frequencies.

These small changes in the properties need sensitive target function to gives accuracy and can gain the best global or very close to the optimum. Current study adopts fitness function includes two parts. The first one is the square of percentage differences between cracked and examined natural frequencies of the structure. The second term is the sum of the norm of the differences between cracked and examined relative nodes translation of the structure. The second term was not included in the target function which was used in the study of Nanda et al (2012) [3]. Those altering and adding in the fitness function are one of the important differences between this study and the reference. To verify the proposed new technique, two crack scenarios were adopted to simulate the cracks in the structural model. These two scenarios were carried out in the study of Nanda et al (2012) [3], and adopted here for comparative and verification.

4.1. First crack scenario

In this case, a single crack created in the element no.3 at 75 mm close to the fixed support, as shown in figure 2. The crack was represented by reducing in the height of the cross section 10%, as listed in table 1.

Table 1. Crack scenarios of the structural model [3].

Crack Scenario	Element No.	Distance from support (mm)	Crack Size (%)
1	3	75	10
2	8	225	10

The reduction in the height of the cross section reduced the stiffness of the structural model. Then, the vibration properties extracted for the cracked model, as listed in table 2.

Table 2. Modal frequencies of the clamped beam for intact, first and second crack scenario.

Mode	Intact Freq. (Hz)	Cr.Sc.-1 Freq. (Hz)	Cr.Sc.-2 Freq. (Hz)
1	22.721	22.162	22.527
2	142.343	141.52	141.503
3	398.363	398.613	395.586
4	780.087	780.089	779.437
5	1288.396	1283.372	1273.994
6	1922.711	1908.959	1915.23

The proposed new technique using PSO method was written by subroutine in high technical language using MATLAB-software environment [7].

After executing the proposed technique to the adopted structural model with first scenario of crack, the target function history during total iteration of 250 is shown in figure 5. The precise solution was detected of the exact element number 3 and crack size of 0.1 of cross section height, as shown in figure 6.

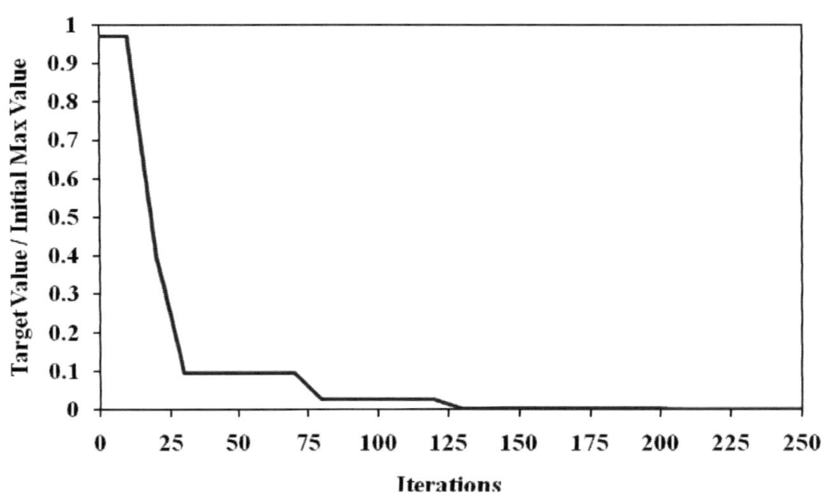

Figure 5. Target function history for first crack scenario during iterations of the proposed new technique.

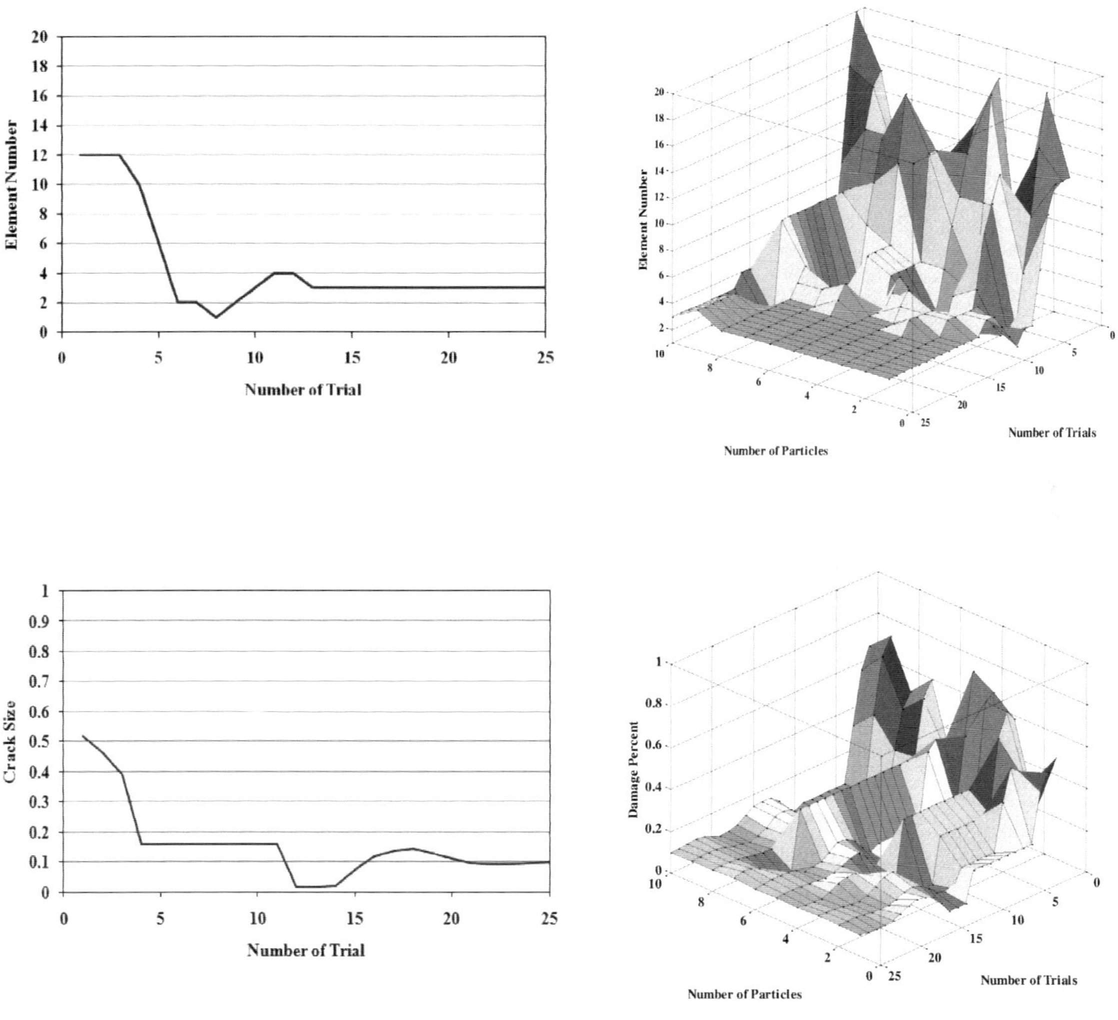

Figure 6. History of crack number and size investigation for first crack scenario of the proposed new technique.

4.2. Second crack scenario

The crack in this scenario was created in the element no.8 at 225 mm from the support, as shown in figure 2. The crack size was the same value of 10%, as listed in table 1. The extracted modal properties for the cracked model are listed in table 2. The results of this case are shown in figures 7 and 8. From figures, the exact element number 8 and crack size of 0.1 of cross section height are explored.

From comparison between the current and Nanda et al [3] study, the better and quick convergence in the target function value was gained. The results show the efficiency and robustness of the proposed new technique.

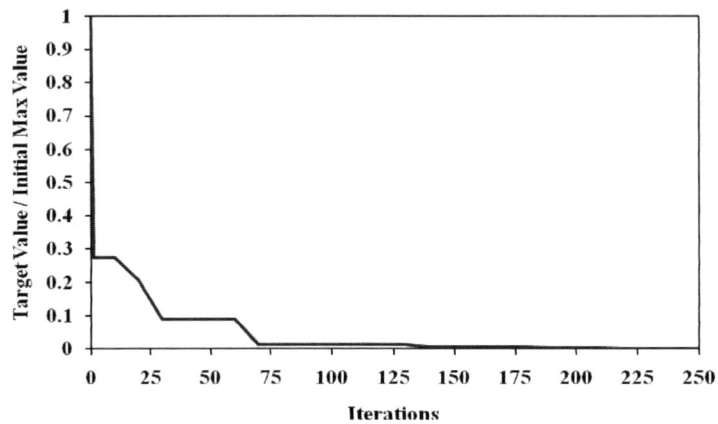

Figure 7. Target function history for second crack scenario during iterations of the proposed new technique.

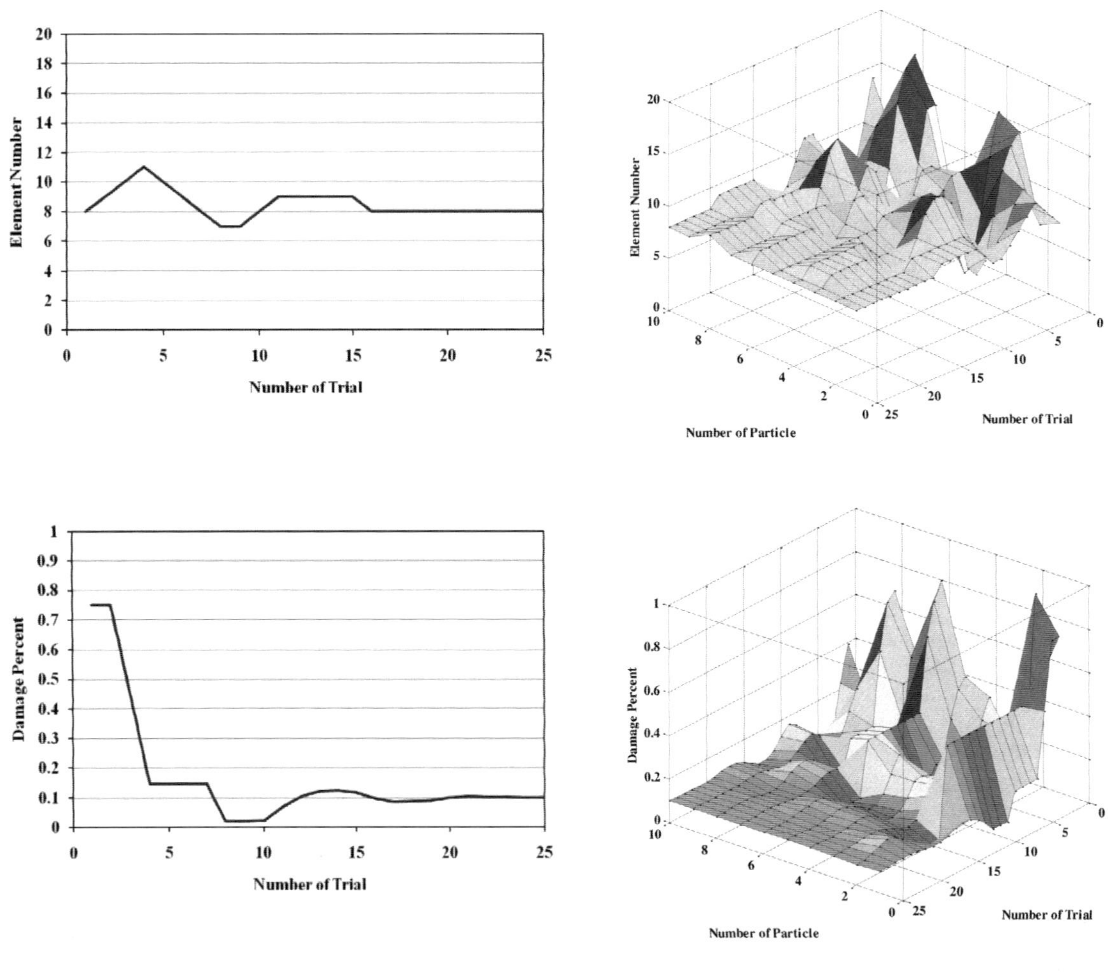

Figure 8. History of crack number and size investigation for second crack scenario of the proposed new technique.

5. Conclusions

A new intelligent technique of non-destructive investigation of crack in a structure using Particle Swarm Optimization (PSO) is proposed. The new technique is based on vibration properties of the structure and included the mode shapes and natural frequencies. The steel clamped beam is adopted in this research study. The target function shows the high sensitivity during assessment of crack in the structure. The exact cracked element number and precise crack size are explored for two crack scenarios, near and far from the support. This new technique exhibits robust and satisfactory performance for investigating the location and size of cracks in the structural model. The experimental testing and other types of structure will study in the future works.

References

[1] Kennedy J and Spears WM 1998Matching algorithm to problems: an experimental test of the Particle Swarm and some Genetic Algorithm on Multi Model Problem Generator*Proc. IEEE Int. Conf. Evolutionary Computation* USA.

[2] Kennedy J and Eberhart RC 1995Particle Swarm Optimization*IEEEProc. Int. Conf.on Neural NetworksIV* (NJ-USA: Picataway).

[3] Bharadwaj N, Damodar M and Dipak K M 2012Vibration based structural damage detection technique using Particle Swarm Optimization with incremental swarm size *Int. J. Aeronautical & Space Science***13** no.(3)323-31.

[4] VaezS R H and Fallah N 2017Damage detection of thin plates using GA-PSO algorithm based on modal data *Arabian J. Science&Engineering*(KSA: Springer) **42** Issue (3)1251-63, doi.org/10.1007/s13369-016-2398-6, KSA.

[5] Wei Z, Liu J and Lu Z 2018Structural damage detection using improved Particle Swarm Optimization *J. Inverse Problem Science&Engineering*(Tylor & Francis) **26** Issue(6)

[6] Academic ANSYS-Software, Inc, Release 14.0, http://www.ansys.com

[7] Academic MATLAB-Software, Release 10.0

Experimental Investigation of Concrete Columns Enhanced by PVC Tubes

Murtaza Allawi Raheemah[1] and Sa'ad Fahad Resan[1,*]

[1] Civil Engineering Department, Engineering College, University of Misan, Amarah, Iraq
[*]Corresponding author; e-mail: sadresan@uomisan.eud.iq

Abstract. An experimental investigation of PVC-concrete columns is presented to assess the efficiency of using PVC tube for enhancing strength and ductility of PVC-concrete columns. Study was conducted to assess provided PVC casing confinement mechanism efficiency for enhancing strength and ductility of considered PVC- concrete columns. To investigate the composite mechanism and confining mechanism of the PVC tube, fabricated specimens are tested under uniaxial compression loading in two modes of load application; composite mode and confining mode. The results showed that the composite mode columns exhibit more strength improving than those of confining mode. The upgrading rate in sustained loading resistance of composite mode columns varied from 2.25 to 1.56, while it was 2.07 to 1.55 for confining mode columns. The same comparison is entirely reversed in scope of ductility, where confining mode columns exhibited more axial and lateral deformations than the corresponding composite mode columns. The results are normalized in term of radial stress (f_r) and concrete compressive strength (f'_c), determined enhanced confined concrete strength (f'_{ce}) could be used to predicate plastic capacity of columns of different mechanism.

Keywords: short column, composite mechanism, confining mechanism, PVC tube, enhanced compressive strength.

1. Introduction

In general, there is no material that sufficiently provide all constructional requirements. The constructional tip therefore consists of an optimization including various materials and techniques of construction, with objective of implementing a structural element at minimum cost to satisfy its requirements. The performance of such element has become a major interested for researchers. Steel hollow sections (SHS), is the most publicly sections used in composite columns. Filling these sections with plain concrete upgrades section strength with a number of advantages as developing greater strength and stiffness due to combine tension properties of steel and compression properties of concrete and ductility enhancing in addition to size reduction [1-6]. On other side, there are many studies deal with confinement mechanism investigation in which retrofitting strips like FRP casing concrete core with or without reinforcement which is mainly aims to upgrading failure strength of concrete by confinement and so upgrading ductility and energy absorbing [7,8].

Developing composite columns beside smart utilizing of materials in composite section intends economic cost benefits and architectural design demands. Polyvinyl chloride (PVC) is a thermoplastic material consists of a polymer resin compounded with varying proportions of stabilizers, lubricants, fillers, pigments, plasticisers and processing aids. Various compounds of these ingredients have been upgraded to get specific properties for different applications. Studies concerned using PVC in composite column are so limited, [9-11] and they were worked to develop composite columns by non-metallic tubes using commercially available plastic pipes filled with concrete. The aims of these studies were to evaluate the upgrading of strength capacity and the effect of slenderness ratio on ultimate strength of PVC-concrete columns. The major purpose of the present study is to assess the efficiency of the PVC tube in enhancing strength and ductility of PVC-concrete columns. To investigate the composite mechanism and confining mechanism of the PVC tube, fabricated specimens are tested under uniaxial compression loading in two modes of load application; in the first mode (composite mode) loading is applied to the entire column, i.e. the PVC tube is utilized as a contributed material in resisting applied loading beside the concrete core. While in the second mode (confining mode), loading is applied to the concrete core, i.e. the PVC tube is utilized a confining material to the concrete core.

2. Experimental Methodology

2.1. Construction Materials

Table 1 depicts material proportion of the adopted batch which developed a cylinder concrete compressive strength of 42.25 MPa and a splitting strength of 3.167 MPa. PVC pipes were used in composite and confining modes and their mechanical properties were considered according to manufactory data sheet. Table 2 shows PVC mechanical properties of used tubes.

2.2. Specimens Description and Test Setting

In the current study, the test specimens are divided into four groups of different sectional characteristic as shown in Table 3 and Figure 1. Each group has five short columns. The group G1 contains columns of PVC tube without concrete filling while G2 contains columns of plain concrete (control columns). Columns for composite mode test are represented in the group G3, while G4 involves columns of confining mode test. Typical configurations of columns of the groups G3 (composite mode) and G4 (confining mode) under different loading fashion are shown in Figure 2. A hydraulic testing machine (MATEST) was used to apply a uniaxial compressive load. Both ends are strengthened with tighten steel rings. Hence, the failure would not occur at specimens' ends [12,13]. Photo 1 shows all column specimens before testing, while Photos 2 shows test setting.

Table 1. Batch proportions

Batch code	Max agg. Size, mm	Cement, kg/m^3	Silica Fume		Binder, kg/m^3	Sand, kg/m^3	Gravel, kg/m^3	Superplastizer		Water/Binder,%
			%	kg/m^3				%	kg/m^3	
I	9.6	824.161	5	43.377	867	1171.176	1474	0.6	5.205	36

Table 2. PVC mechanical properties

Properties	Units	Value
Tensile strength at break	MPa	47.07
Specific gravity	g/cm^3	1.43
Elongation		80%
Modulus of elasticity	MN /m^3	2000

IOP Conf. Series: Materials Science and Engineering **584** (2019) 012047 doi:10.1088/1757-899X/584/1/012047

Table 3. Specimens matrix

No	Groups	Specimen	Section	f'_c, MPa	L, mm	t_p , mm	Dia., mm	λ (L/r)	λ' (D/t)	Description
1	G1	SA1		_	300	2.2	75	8.00	34.09	PVC tube without concrete filling
2		SB1				3.6	75	8.00	20.83	
3		SC1				5.6	75	8.00	13.39	
4		SD1				3.0	63	9.52	21.00	
5		SE1				4.7	63	9.52	13.40	
6	G2	SA3		42.25	300	_	75	8.00	_	concrete columns, without PVC casing
7		SB3					75	8.00		
8		SC3					75	8.00		
9		SD3					63	9.52		
10		SE3					63	9.52		
11	G3	SA5	Sction#1	42.25	300	2.2	75	8.00	34.09	PVC-concrete columns, (composite mode)
12		SB5	Sction#2			3.6	75	8.00	20.83	
13		SC5	Sction#3			5.6	75	8.00	13.39	
14		SD5	Sction#4			3.0	63	9.52	21.00	
15		SE5	Sction#5			4.7	63	9.52	13.40	
16	G4	SA6	Sction#1	42.25	300	2.2	75	8.00	34.09	PVC-concrete columns, (confining mode)
17		SB6	Sction#2			3.6	75	8.00	20.83	
18		SC6	Sction#3			5.6	75	8.00	13.39	
19		SD6	Sction#4			3.0	63	9.52	21.00	
20		SE6	Sction#5			4.7	63	9.52	13.40	

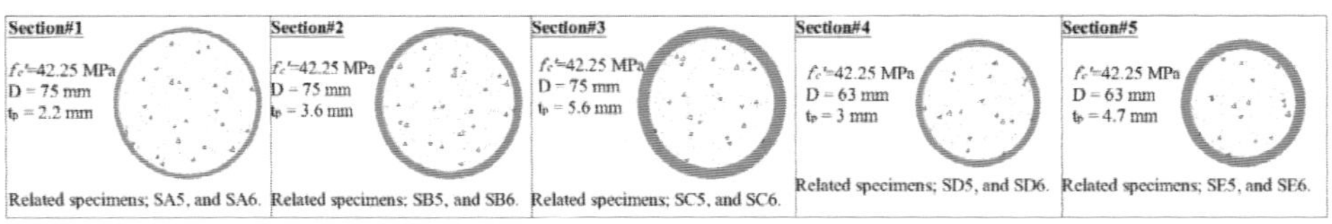

Figure 1. Sectional definitions of used specimens.

a. PVC tubes b. Concrete columns c. Composite mode columns d. Confining mode columns

Photo 1. Columns and their constitutes.

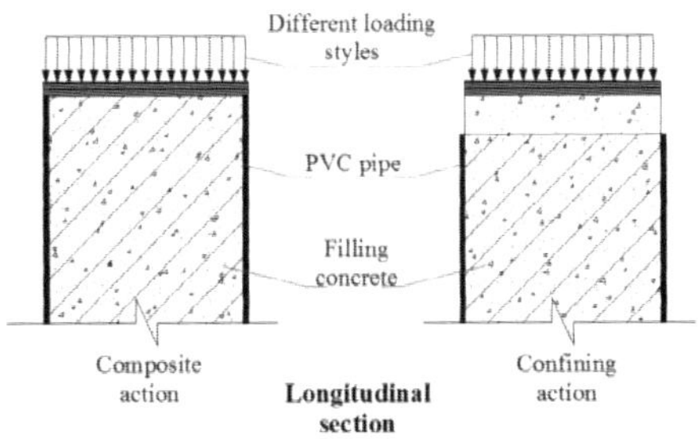

Figure 2. Test loading modes

Photo 2. Test setting

3. Results and Discussion

3.1. Sustained Loading Resistance

Table 4 depicts the obtained test results. A comparative result is also shown in this table. PVC-concrete columns of composite mode exhibited more strength improving (in comparison with their individual constitutes; PVC tube and concrete core) than those of confining mode. The upgrading rate in sustained loading resistance of composite mode columns varied from 2.25 to 1.56, while it was 2.07 to 1.55 for confining mode columns. The same comparison is entirely reversed in scope of ductility, where confining mode columns exhibited more axial and lateral deformations than the corresponding composite mode columns. This reversed variation could be related to the difference in Poisson's ratios between PVC tube and concrete which affected the columns resistance. Poisson's ratio of PVC is about 0.41 in the elastic range [14]. While it is about 0.1 for concrete in the elastic range and can reach values up to 0.75 in case of large strains [15]. Thus in case of loading on the entire section in the initial stage of loading the PVC tube and the concrete act separately due to the faster expansion of the PVC tube in radial direction.

Figures 3 and 4 present load – axial deformation response of specimens of composite and confining mechanism, for different slenderness ratios, the comparison clearly verified the positive enhancement of composite mechanism in scope of strength and stiffness upgrading, as much as verified the positive enhancement of confining mechanism in scope of ductility and energy absorption upgrading.

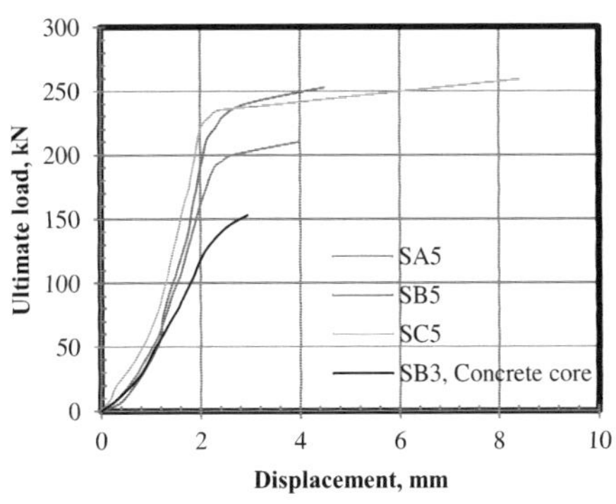

a. $\lambda = 8$

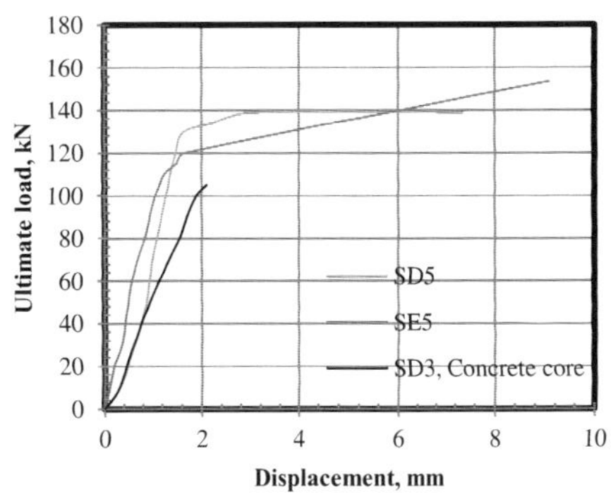

b. $\lambda = 9.52$

Figure 3. Load – axial deformation of composite specimens

IOP Conf. Series: Materials Science and Engineering **584** (2019) 012047 doi:10.1088/1757-899X/584/1/012047

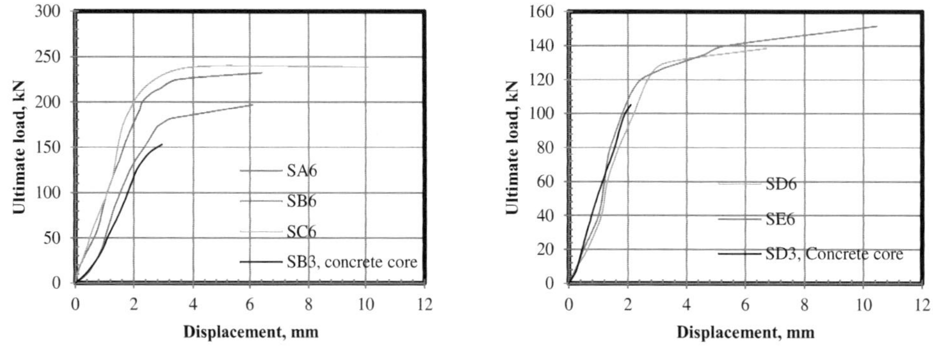

a. λ= 8 b. λ=9.52

Figure 4. Load – axial deformation of confined specimens

3.2. Enhanced Compressive Strength, f'_{ce}

Table 5 exhibits the efficiency of PVC as casing confining materials which it is strongly contributed to enhance concrete compressive strengths. The assigned improving compressive strength decreased as slenderness ratio increased. The increase in strength capacity due to confinement mechanism depends on the tube's D/t ratio, thicker tubes introduce a high increase. Figure 5 demonstrates the PVC role in enhancing compressive strength, while Figure 6 depicts the enhancement in compressive strength of composite and confining mode columns. Table 4 clearly shows that the confinement mechanism of confining mode columns introduces a confinement effectiveness ranged from 2.17 to 1.55 versus a confinement effectiveness ranged from 1.81 to 1.26 for composite mode columns. Enhanced confined concrete strength (f'_{ce}) is experimentally predicated as;

For composite mechanism,

As; $p_{cm}=f_{ce}A_c+f_{yp}A_p$

So;

$$f'_{ce} = (P_{cm}-f_{yp}\,A_p)/A_c \tag{1}$$

For confining mechanism,

$$f'_{ce} = P_{cn}/A_c \tag{2}$$

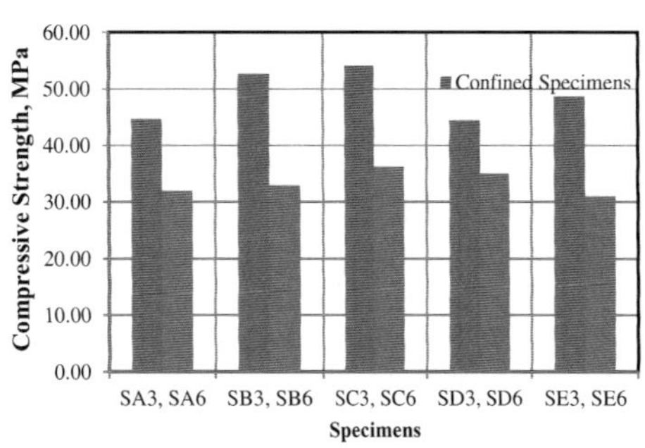

Figure 5. Confined and unconfined compressive strength

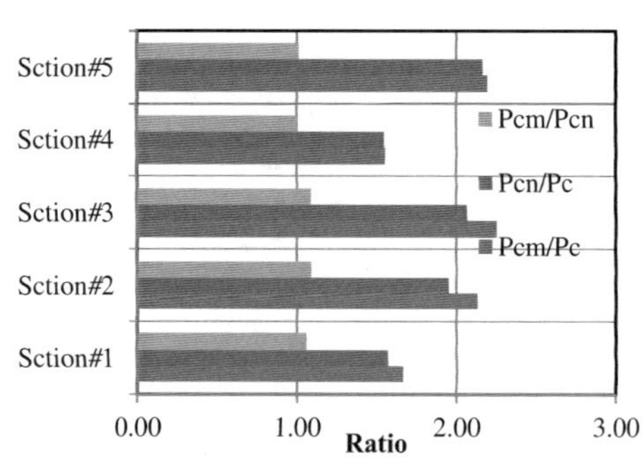

Figure 6. Enhancement in compressive strength

Figures 7 and 8 give a comparison between the ultimate strength of composite and confining mode columns and the ultimate strength of their individual constitutes, respectively. Generally, the obtained load carrying capacity of PVC-concrete is greater than the summation of strength of corresponding PVC tube and concrete core.

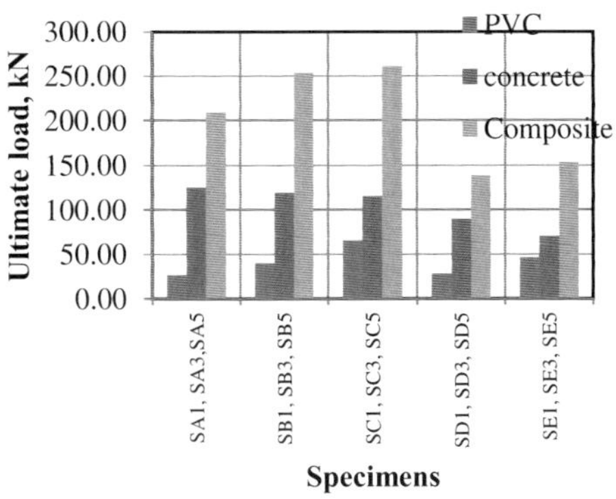

Figure 7. Ultimate strength of PVC tube, concrete core and composite mode columns

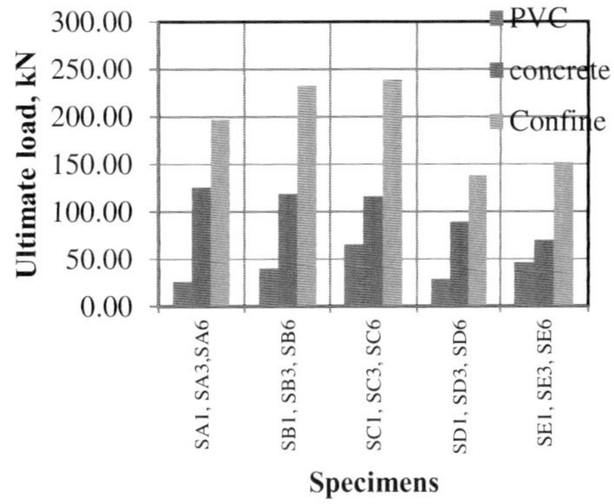

Figure 8. Ultimate strength of PVC tube, concrete core and confining mode columns

Table 4. Test results

No	Groups	Specimen code	Section	Description	λ =L/r	t_p (mm)	D/t_p	f'_c (MPA)	f'_c/f_{yp}	P_p, kN	P_c KN	P_{cm} KN	P_{cn} KN	Axial displacement, mm	Lateral displacement, mm	P_{cm}/P_c	P_{cn}/P_c	P_{cm}/P_{cn}
1		SA1			8.00	2.2	34.09			26.00				8.00	0.80			
2		SB1			8.00	3.6	20.83			40.00				9.33	2.45			
3	G1	SC1		PVC tube without concrete filling	8.00	5.6	13.39			65.10				11.00	2.90			
4		SD1			9.52	3	25.00			28.00				8.25	1.42			
5		SE1			9.52	4.7	15.96			46.00				11.00	1.89			
6		SA3			8.50						125.18			1.53	0.22			
7		SB3			8.85						118.90			2.95	0.43			
8	G2	SC3		concrete columns, without PVC casing	9.40			42.25			115.50			1.01	0.15			
9		SD3			10.53						89.20			2.10	0.32			
10		SE3			11.19						70.00			1.40	0.21			
11		SA5	Sction#1		8.00	2.2	34.09					208.70		3.98	0.62	1.67		
12		SB5	Sction#2		8.00	3.6	20.83					253.20		4.50	0.70	2.13		
13	G3	SC5	Sction#3	PVC-concrete columns, (composite mode)	8.00	5.6	13.39	42.25	0.85			259.90		8.41	1.31	2.25		
14		SD5	Sction#4		9.52	3	25.00					139.00		7.36	1.15	1.56		
15		SE5	Sction#5		9.52	4.7	15.96					153.60		9.08	1.42	2.19		
16		SA6	Sction#1		8.00	2.2	34.09						197.00	7.10	0.85		1.57	1.06
17		SB6	Sction#2		8.00	3.6	20.83						232.30	6.42	0.78		1.95	1.09
18	G4	SC6	Sction#3	PVC-concrete columns, (confining mode)	8.00	5.6	13.39	42.25	0.85				238.70	10.10	1.59		2.07	1.09
19		SD6	Sction#4		9.52	3	25.00						138.40	6.72	1.06		1.55	1.00
20		SE6	Sction#5		9.52	4.7	15.96						151.50	10.45	1.65		2.16	1.01

Table 5. Enhanced Compressive Strength

No	Specimen code	λ= L/r	Ultimate strength, kN	Unconfined concrete compressive strength, $f'c$ MPa	Experimental enhanced concrete compressive strength, $f'ce$	Confinement effectiveness* $f'ce/f'c$
			Concrete columns			
1	SA3	8	125.18	32		
2	SB3	8	118.9	32.95		
3	SC3	8	115.5	36.15		
4	SD3	9.52	89.2	34.98		
5	SE3	9.52	70	31		
			PVC-concrete columns, (composite mode)			
6	SA5	8	208.7		47.29	1.48
7	SB5	8	253.2		59.64	1.81
8	SC5	8	259.9		63.36	1.75
9	SD5	9.52	139		44.07	1.26
10	SE6	9.52	153.6		50.15	1.62
			PVC-concrete columns, (confining mode)			
11	SA6	8	197		50.35	1.57
12	SB6	8	232.3		64.38	1.95
13	SC6	8	238.7		74.70	2.07
14	SD6	9.52	138.4		54.26	1.55
15	SE6	9.52	151.5		67.18	2.17

3.3. Confinement Normalization

Figure 9 shows the normalized confinement effectiveness, the trend of confining approach highlights best enhancing in comparing with composite approach, this difference explains ductility improving of confining mode, while strength improving of composite mode columns are mainly related to the presence of PVC tube as an external reinforcement attribute with concrete to resist applied loading. The results are normalized in terms of radial stress (f_r) and concrete compressive strength (f'_c). The enhanced confined concrete strength (f'_{ce}) could be determined as:

$$f'_{ce}=\alpha 1\, f'_c+ \alpha 2\, f_r \tag{3}$$

Referring to Figure 9;

$\alpha 1=0.905$, $\alpha 2= 10.35$ For confining mechanism

$\alpha 1=0.94$, $\alpha 2= 8.02$ For composite mechanism;

where f_r is determined according to Barlow's formula;

$$f_r=2\ f_{yp}\ t_p / D \tag{4}$$

So, plastic capacity of compact PVC- concrete columns could be predicted as;

$$q_{cm} = f'_{ce}\, Ac + f_{yp}\, A_p \tag{5}$$ For composite mechanism

$$q_{cn} = f'_{ce}\, A_c \tag{6}$$ For confined mechanism

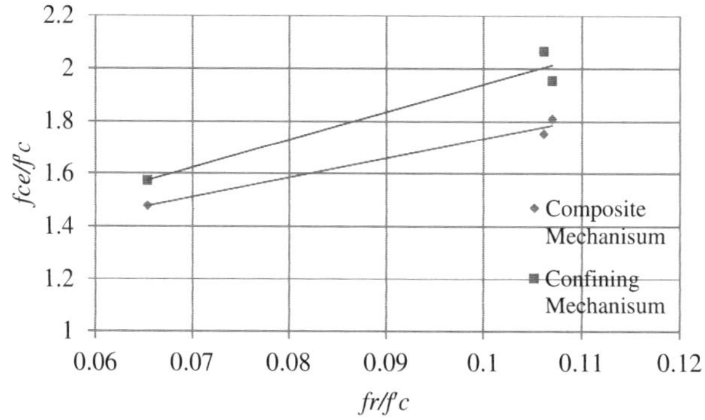

Figure 9. Normalized confined concrete strength for different sections of λ=8

3.4. Failure Modes

The non-composite short concrete cores, suffer from excessive lateral expansion due to unstable propagation of the internal micro-cracks, which cause the strain softening behaviour and eventually the concrete mass loses its integrity and fails in splitting manner. Local buckling occurs in short PVC specimens, as compression strength of the material is not attainted, the column compressed longitudinally until section developed an expansion as hoop deformations within length. The deformations frequently were large and more than pipe thickness. For compacted PVC concrete columns, material failure is dominated in all tested specimens, the assigned failure is shear failure where concrete filling typically initiated failing in a classical shear mode. It was observed, when the PVC tube is cut, interface bonding didn't develop between concrete cores and PVC tube and the confinement exerted by the PVC tube could not fully turn off customary shear failure. Photo .3 depicts assigned typical failure modes of various specimens.

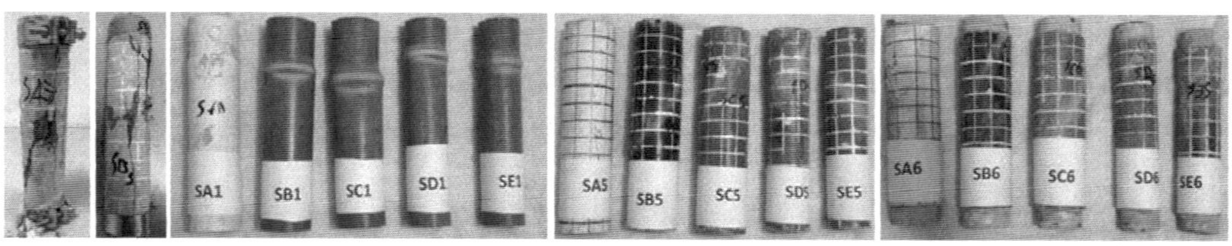

Concrete columns PVC tubes Composite mode columns Confining mode columns

Photo 3. Failure Modes

4. Conclusion

The test results confirmed that the PVC-concrete columns of composite mode exhibited more strength improving than those of confining mode. The upgrading rate in sustained loading resistance of composite mode columns varied from 2.25 to 1.56, while it was 2.07 to 1.55 for confining mode columns. The same comparison is entirely reversed in scope of ductility, where confining mode columns exhibited more axial and lateral deformations than the corresponding composite mode columns. In composite mechanism, PVC tube thickness leads to greater increase in ultimate strength of the PVC-concrete columns. This could be attributed to the effect of PVC tube as an external reinforcement with respect to the overall section area. In confining mechanism, the PVC tube causes the development of a pure triaxial stress field within confined concrete, constraining it during dilation and thereby increasing plastic deformation. Load-axial deformation responses' comparisons clearly verified the positive enhancement of composite mechanism in scope of strength and stiffness upgrading, as much as verified the positive enhancement of confining mechanism in scope of ductility and energy absorption upgrading. For considered PVC concrete columns, material failure is dominated in all tested specimens, the assigned failure is shear failure where concrete filling typically initiated failing in a classical shear mode. It was observed, when the PVC tube is cut, interface bonding didn't develop between concrete cores and PVC tube and the confinement exerted by the PVC tube could not fully turn off customary shear failure. The results are normalized in term of radial stress (f_r) and concrete compressive strength (f'_c), determined enhanced confined concrete strength (f'_{ce}) could be used to predicate plastic capacity of columns of different mechanism.

References

[1] Kvedaras A. K.; Kudzys, A. 2006. The structural safety of hollow concrete-filled circular steel members, *Journal of Constructional Steel Research* 621 (11): 1116–1122.

[2] Benzaid, R.; Chikh, N.; Mesbah, H. 2008. Behavior of square concrete column confined with GFRP composite warp, *Journal of Civil Engineering and Management* 14(2):115–120.

[3] Kuranovas et al. 2007; Kuranovas A.; Kvedaras, A. K. 2007. Behavior of hollow concrete steel tubular composite elements, *Journal of Civil Engineering and Management* 13(2): 131–141

[4] Goode 2007; Goode, C. D. 2008. Composite columns – 1819 tests on concrete-filled steel tube columns compared with Eurocode 4.The Structural Engineer 86(16): 33–38.

[5] Kudzys et al. 2006; Kudzys, A.; Kvedaras, A. K. 2006. Research on structural resistance and safety of tubular composite members, *Mechanika* Issue 5: 5–10.

[6] Han et al. 2004; Han, L-H.; Zhong, T.; Wei, L. 2004. Effects of sustained load on concrete-filled hollow structural steel columns, *Journal of Structural Engineering ASCE* 9: 1392-1404.

[7] Tao, Z., Teng, J.G., Han, L.H. and Lam, L. 2004 "Experimental behavior of FRP-confined slender RC columns under eccentric loading", *Proceedings, Second International Conference on Advanced Polymer Composites for Structural Applications in Construction*, University of Surrey, Guildford, UK, 203-212.

[8] Fitzwilliam, J. and Bisby, L.A. 2010 "Slenderness effects on circular CFRP- confined reinforced concrete columns", *Journal of Composites for Construction*, ASCE, *14*(3), 280-288.

[9] Ahmed Sagban Saadoon, 2010, "Experimental and theoretical investigation of PVC- concrete composite columns", Ph. D thesis, University of Basrah, Engineering college.

[10] Kurt, C. E., (1978). "Concrete filled structural plastic columns."*Journal of the Structural Division*, Vol. 104, No. 1, pp. 55-63

[11] Naftary Kimenju Gathimba, 2015 Performance of UPVC Pipe Confined Concrete Columns in Compression, *M.Sc. thesis*, 2015, University of Agriculture and Technology.

[12] Sa'ad Fahad Resan, 2014, "Experimental investigation of aluminum lightweight concrete composite columns", *Basrah Journal for Engineering Sciences*, Vol.14, No.1.

[13] Feng Zhou and Ben Young, 2009, "Concrete-filled aluminum circular hollow section column tests" *Thin-Walled Structures* Vol. 47, 1272–1280.

[14] Rahman. S., 2007,"PVC pipe & fittings: underground solutions for water and sewer systems in North America" (PDF), 2nd Brazilian PVC Congress, Sao Paulo, Brazil.

[15] Porsch, M., and Hanswille, G., 2006, "Load introduction in composite columns with concrete filled hollow sections", Composite Construction in Steel and Concrete, V, 402-411.

Mechanical Properties of Lightweight Aggregate Moderate Strength Concrete reinforcement with Hybrid Fibers

Haider Araby Ibrahim [1,*], Mohammed Bally Mahdi [2], Basim Jabbar Abbas [3]

[1, 2, 3] Civil Engineering Department, Al-Muthanna University, Samawa, Iraq.
*Corresponding Author: email: arabyhaider@yahoo.com

Abstract. This study aims to detect the mechanical properties of lightweight aggregate moderate strength concrete (LWAMSC) reinforced with single and hybrid fibers in various types and sizes. Moderate strength pumice lightweight aggregate concrete mix with compressive strength 40 MPa at 28 day age with silica fume and superplastizier was used. The fibers used include macro hooked steel fiber with aspect ratio 80, steel fiber with aspect ratio 60, and polypropylene fiber (P.P.F). Two groups of (LWAMSC) mixes were made with reference concrete mix (without fibers). The first group of mixes were substituted three (LWAMSC) mixes with different single fiber including, concrete mix reinforced with Polypropylene (P.P.F) about 0.75 % volume fraction (P1), concrete mix reinforced with macro hooked steel fiber type SF1 with volume fraction 0.75% mix (M1) and concrete mix reinforced with macro hooked steel fiber type SF1 volume fraction 1.5% mix (M2). The second group of fiber reinforcement (0.75% steel fiber type (SF1) + 0.75% steel fiber (SF2) mix (H1), (1% steel fiber (FS1)+ 0. 5% steel fiber type (SF2) mix (H2)),and finaly (1% steel fiber type (SF1)+0.5% PPF mix (H3)). Dry density, strengths of compressive, splitting tensile, and flexural of (LWAMSC) were examined. generally single and hybrid reinforcement fiber of (LWAMSC) show important increase at flexural strength and splitting tensile strength in comparison with reference concrete.The results also show that, the optimum added proportion of fiber steel type (SF1) was (1.0%) and with (0.5%) steel fiber type (SF2) that portion raised the strengths of compressive, flexural, and splitting tensile around (4.34%), (52.93%), and (297%) respectively comparison with control mixture.

Keywords: moderate-strength concretes, lightweight aggregates; dry density, hybrid fibers.

1. Introduction

In a new moderate rise building, the size of columns and the wide of the span between beams concrete structures were increased, it is needed lightweight moderate strength for low stiffness concrete. A diminished concrete density for a similar strength level allows a sparing in dead load for basic foundations, but increase the strength of light weight concrete lead to moderate brittle behavior [1].

Lightweight aggregate moderate strength concrete (LWAMSC) is structural lightweight concrete utilize at concrete building structure with a 28-day compressive strength of (40 MPa) or greater, and has a density less than 1500 kg/m^3 [2], high brittleness low tensile/compressive strength ratio, low flexural strength, larger shrinkage. Therefore, steel fiber is needed for enhanced the mechanical properties of lightweight moderate strength mixture for increasing the tensile/compressive ratio, toughness, and resistance against deformation.[3],[4].To keep on the workability of concrete, a moderate-range water-reducing admixture (superplasticizer) was used within mixing operations with water-cement ratio lower than 0.4 Pumice is a porous, volcanic rock. It is light grey in colour , cheap lightweight aggregates used as lightweight coarse aggregates.The especial application of this type of concrete in the area of ocean engineering structures, long-span bridges and moderate buildings, etc [5].

2. Literature review

Although little knowledge for mechanical properties (LWAMSC) with hybrid fibers, some authors have got adequate results in this field , Jianming Gao, et al. in (1997) [6], found the effect of fiber volume and aspect ratio on the mechanical behaviour is extremely prominent, but strength of compressive is little improved and varied from 70.2 to 85.4 MPa.The ratio of tensile /compressive strength is improved. Strength of flexural varied from 4.94 to 8.7MPa.

While O. Kayali, et al. in (2003) [7] studied the effect of polypropylene and steel fibers on lightweight moderate strength aggregate concrete with Sintered fly ash were used as lightweight aggregates, polypropylene fiber with 0.561% at concrete volume, lead to a 90% addition to the tensile strength and a 20% to the rupture modulus and 1.7% by volume of steel fibers addition at the indirect strength of tensile about 118% and the rupture modulus by about 80%.

In 2011, N.A. Libre et al. [8] observed the increase in mechanical properties and energy absorption capacity when the addition of steel fibers. Polypropylene fibers have a minor effect in mechanical properties of hardened concrete made with both steel and polypropylene fibers. The steel fibers volume fraction at 0.5% ,1% but the PP fibers various 0.2% ,and 0.4%.Furthermore, H.K. Ahmed, et al. in (2017)[9] studied the influence of three types of polypropylene fibers with different volume fraction (Vf) on compressive strength, flexural strength, splitting tensile strength, and static modulus of elasticity. Results show significantly increased in all types of strength with all types of polypropylene fibers and compressive strength it was improved at Vf =0.25% and decrease at Vf =0.75%.

3.Materials and methodology

3.1. Cement

For experimental work, the ordinary Portland cement was used from Al- Douh refectory. Tables 1 and 2 show that the cement test results according to the ASTM standers C150-02 [10].

Table 1. Chemical analysis of cement

Oxides	percentage	Limit of ASTM Specf.C150-02a/2002
CaO	63.2	-
SiO2	18.9	-
Al2O3	3.8	-
Fe2O3	4.6	-
SO3	1.5	$\leq$ 2.3
MgO	1.7	$\leq$ 6.0
L.O.I	1.9	$\leq$ 3.0
L.S.F	0.9	-
I.R.	0.4	$\leq$ 0.75
C3A	2.32	$\leq$ 5.0

Table 2. Physical properties of cement

Physical properties	Test results	Limit of ASTM Specf.C150-02a/2002
Initial setting time (vicat)	65 min.	45min.(Min.)
Final setting time (vicat)	170 min.	375 min. (Max.)
Compressive strength of mortar (MPa) 3-days	19.0	15 (Min.)
Compressive strength of mortar (MPa)7-days	30.5	21 (Min.)

3.2. Aggregate

3.2.1. Fine aggregate

Natural river sand was used as a fine aggregate which is passing through the sieve aggregate of 4.75 mm. Table 3 shows the gradation of the aggregate which is compatible to the required specifications of ASTM C33-01[11].

Table 3. Sieve analysis of fine aggregate

Sieve size	Passing ratio (%)	Limit of ASTMC33-01
9.5mm	100.00	100
4.75mm	94.56	95-100
2.36mm	72.45	80-100
1.18mm	68.40	50-85
600μm	53.32	25-60
300μm	16.60	5-30
150μm	2.12	0-10

3.2.2. Pumice lightweight coarse aggregate

Pumice was used to make preparations for the (LWAMSC) mixtures. The quarry of this stone is found in the north of Iraq in AL- Sulaymania governorate. Pumice stone has a dark colour, moderate permeability, and low density, porous form of vitrified volcanic rock and moderate silicon oxide (SiO2) concrete. Crushed Pumice stone was used as a coarse lightweight aggregate of nominal size (12.5 - 4.75) mm in this work as shown in Figure 1. Pumice pieces were broken manually and graded as per the requirement of ASTM C 330-04 [12]. Table 4 and 5 illustrate the physical properties and the grading for pumice, and the Figure 2 shows the graph of the grading of pumice aggregates.

Figure 1. Pumice pieces

Table 4. Physical properties of lightweight coarse aggregate (pumice)

Tests, ASTM C 127	Test Results	Limits
Loose bulk density (kg/m3)	569	880 (max)
Oven dry bulk specific gravity	1.88	
Absorption % (24 Hours)	2.51	
specific gravity	2.53	

Table 5. Pumice grading of coarse lightweight aggregate

Sieve size (mm)	Percent Passing	Limit required by ASTMStandardC330-04
12.5	100	100
9.5	97	80-100
4.75	12	5-40
2.36	2	0-20
1.18	0	0-10

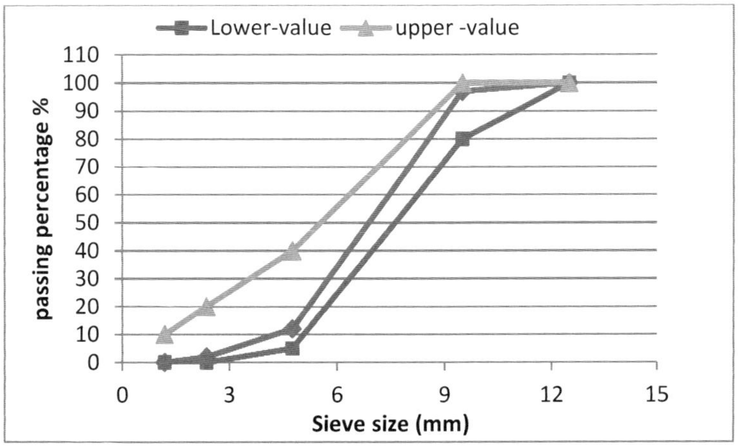

Figure 2. Sieve analysis of Pumice grading of coarse lightweight aggregate.

3.3. High range water reducing admixture (Superplasticizer)

High range water reducing admixture (HRWRA) superplasticizer is a chemical compound used to reduce the amount of water with the same workability for producing moderate strength concrete, according to ASTM-C494 Type G [13]. Because of the fibers addition reducing the workability of mixtures[5]. So (HRWRA) must be used . Table 6 shows the typical properties of superplasticizer.

Table 6. Typical properties of superplasticizer

properties	Specifications
Specific gravity	1.2
Form	Liquid
PH	7-9

3.4. Fibers

Table 7 shows the physical and technical properties of fibers used in this work. Two types of macro hooked steel fibers (SF1and SF2) with an aspect ratio of 80, and 60 and Polypropylene fiber (PPF) with aspect ratio 677 complied with requirements of ASTM C1116-02[14]. Figure 3 shows the two types of fibers that used in this work.

Table 7. Properties of fibers

Property	Steel Fiber, SF1	Steel Fiber, SF2	polypropylene Fibers (P.P.F)
Density (Kg/m3)	7880	7880	7900
Tensile strength (MPa)	1100	1100	360
Diameter (mm)	0.5	0.5	0.018
Length (mm)	40	30	12
Geometry	hooked	hooked	Fibrillated
Aspect ratio	80	60	677

(a) Standard hooked-end steel fibers (b) Polypropylene fibers.

Figure 3. Types of fibers

IOP Conf. Series: Materials Science and Engineering **584** (2019) 012048 doi:10.1088/1757-899X/584/1/012048

3.5. Silica fume

Silica fume utilized in this work was purchased from Sika Company. The chemical composition and physical requirements of the purchased silica fume satisfied the requirements of ASTM C 1240-06 - 06[15].

3.6. Mix Proportions

The volumetric method ACI 211.2 [2] was used for mix design with proportions as 1: 1.33: 0.89 (cement: sand: aggregate) by weight.The reference lightweight aggregate moderate strength concrete mix to give at 28-day age a compressive strength (42 MPa), after many tries, used the mixes with cement content 490 Kg/m3, silica fume 53kg/m3, water-cement ratio 0.31 and slump (80 mm) for good workability was used (HRWRA) 1.2 liter per 100 kg of cement without fiber (0F) as shows in table (8).The high absorption of pumice aggregates directly effects on the quality of the LW concrete. In order to solve this problem, the pre-soaking method was used for pumice aggregates to control the high absorption of pumice aggregates for a fixed time.Generally, 24 hours the time was needed to achieve the 80% of the water absorption [16].Two groups of concrete were made, the first group of mixes were substituted three (LWAMSC) with different single fiber including, concrete mix reinforced with Polypropylene (P.P.F) about 0.75 % volume fraction (P1), concrete mix reinforced with macro hooked steel fiber type SF1 with volume fraction 0.75% mix (M1) and concrete mix reinforced with macro hooked steel fiber type SF1 volume fraction 1.5% mix (M2). In the second group, three (LWAMSC) with hybrid fiber including concrete mix reinforced with a combination of (SF1) steel fiber 0.75% volume fraction + (SF2) steel fiber 0.75% volume fraction (H1). Concrete mix reinforced with a combination of macro (SF1) steel fiber with 1.0 % volume fraction +(SF2) steel fiber and 0. 5% volume fraction (H2). Finally, the concrete mix contains a combination of (SF1) macro steel fiber 1.0% volume fraction + micro PPF with 0.5% volume fraction (H3).

Table 8. Mix Proportions of (LWAMSC)

Mix Design	Steel Fiber % of vol. (SF1)	Steel Fiber % of vol. (SF2)	(P.P) Fiber % of vol.	Pumice coarse aggregates (Kg/m3)	Fine aggregates (Kg/m3)	Cement content (Kg/m^3)	W/C	Silica fume (Kg/m^3)
Ref.- (RF)	0.0	0.0	0.0	435	650	490	0.31	53
0.75%PPF- (P1)	0.0	0.0	0.75	435	650	490	0.31	53
0.75%SF1- (M1)	0.75	0.0	0.0	435	650	490	0.31	53
1.5%SF1- (M2)	1.5	0.0	0.0	435	650	490	0.31	53
0.75%SF1- 0.75%SF2- (H1)	0.75	0.75	0.0	435	650	490	0.31	53
1%SF1- 0.5%SF2- H2)	1.0	0.5	0.0	435	650	490	0.31	53
1%SF1-0.5%PPF- (H3)	1.0	0.0	0.5	435	650	490	0.31	53

3.7. Casting and curing of (LWAMSC)

To produce of lightweight aggregate moderate strength with fiber reinforced concrete, pumice as coarse aggregates and fine aggregates were mixed at the beginning with dry case for one minute. Then, cement and silica fume were added to the mix. Amid the blending activity, the fibers were added and all the materials were mixed for 2 minutes together with the continuous addition of fibers. Hard and continuous mixing was made for the avoidance of fibers to be clumped. Finally, to get a specific mixture containing water,(HRWR) was slowly added to the mixture during the mixing for 3 minutes.

After filling up the moulds, the concrete samples were to rduce the entrapped voids using a vibrating table for a period of 8 to 12 seconds. The samples were maintained in the laboratory for 24 hours,and then stored in the water of a constant 20±2 °C for 7 and 28 days, up to a time the day of the testing.

3.8. Concrete testing

According to ASTM C192-88 [17],the concrete resistance of compression of (LWAMSC) samples were examined during 7 and 28 days of treatment period. For strength of compression at (7) and (28) days in conformance to requirements for ASTM C39-98 [18], utilize six standard steel cubic moulds (150×150×150) mm for casting the samples, while for the strength of splitting tensile emploied two cylindrical moulds of (100×200) mm and the concrete was evaluated according to ASTM C 496-90 [19]. Finally, according to ASTM(C78–98) [20], two prisms of (100×100×400) mm were used for the strength of flexural (Modulus of Rupture) for (28) days.

3.9. Air Dry Density

Air dry density test was performed in accordance with ASTM C567-85[21] by using (100×200) mm cylindrical specimens. The specimens left in moulds for 1 day inside a moisture cabinet, then they were stripped from the moulds and wrapped securely with a polythene bag for 6 days. After that, they were removed from the bag and immersed in water for 1 day. On the second day saturated surface dry and suspended-immersed weight were taken. The specimens were left in the laboratory for 21 days, then the dried specimens were weighted at the 28-day. The air dry density was calculated by using the following equation:-

$$Air\ dry\ density\ (kg/\mathbf{m^3}) = \frac{(A\times 997)}{(B\text{-}C)} \tag{1}$$

where :
A = weight of concrete cylinder, as dried (kg).
B = saturated surface-dry weight of cylinder (kg).
C = suspended-immersed weight of cylinder (kg).

4. Resultant and Discussion

4.1. Bulk density

The hardened concrete densities which were tested at 28-day are shown in Table 9 and Figure 4. The results of the bulk density of the (P.P.F) fibers did not have a significant effect on the concrete density. While, the density of concrete is dirctly influenced at fibers of steel, but the mean density of concrete that not contain steel fibers was 1760 kg/m^3, the average density of hybrid fibers concretes of H1,H2, H3 were 1862, 1856, and 1843 kg/m^3 .This is clearly due to the moderate specific gravity of steel fibers, the density of (LWAMSC) shows the increment of 1.5% steel fiber greater the concrete density at about 116 kg/m^3. The influence fibers steel on growing the concrete density samples very important in the case of lightweight concrete in which the reduction of the density is very significance.

Table 9. Compresive of compression results of (LWAMSC) samples at 7 and 28 days

Mix Design	Average Bulk density (kg/m^3)	Comp. strength 7 days MPa	Comp. strength 28-day MPa
Ref.- (RF)	1768	32.75	41.45
0.75%PPF- (P1)	1769	33.12	42.12
0.75%SF1- (M1)	1840	34.45	44.35
1.5%SF1- (M2)	1860	30.11	37.64
0.75%SF1- 0.75%SF2- (H1)	1862	30.89	38.76
1.0%SF1- 0.5%SF2- (H2)	1856	34.52	43.25
1.0%SF1-0.5%PPF- (H3)	1843	34.35	43.05

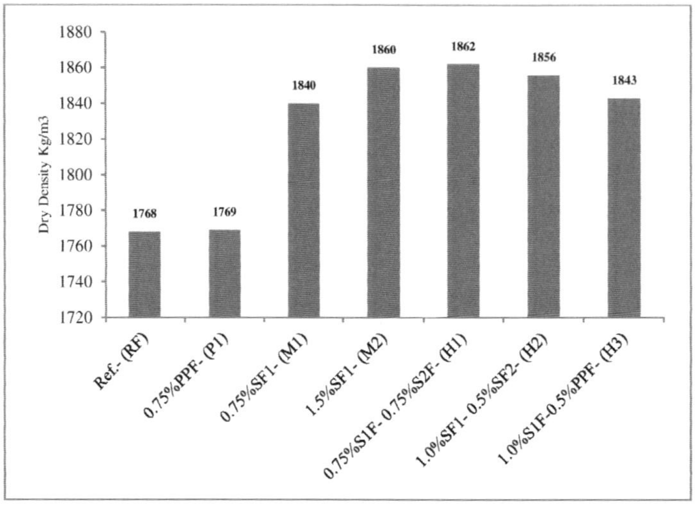

Figure 4. Dry density test resultants of (LWAMSC) at 28-day

4.2. Compressive Strength

Table 9 and Figure 5 explain the increment of the compressive strength results of all (LWAMSC) at 7-day and 28-day within the age. It was noted that the compressive strength of polypropylene fiber type (P1) with volume fraction of 0.75% was slightly better than the refrence concrete (RF) in about 1.13% at 7days and 1.6% at 28-days respectively. Concrete mixes containing single fiber, steel fiber of M1 is showed the higher compressive strength about 5.19%, and 6.97% for 7 and 28 days respectively in comparison with refrence concrete. This is due to the small amount of fibers enhanced the cement paste for micro crack phases and improved the crack strength.

It has been noticed that the addition of fibers in concrete mix of (M2) come to be very hard at 1.5% of steel fiber volume fraction and mixture can not completely compacted, thus the compressive strength of concrete become smaller than to the refrence concrete.The partly exchange to steel fibers type (M2) of 1.5% steel fiber type (SF1) with hybrid fiber (H1, H2, and H3) of low volume fraction of steel fiber type SF2 (0.75% and 0.5%), polypropylene fibers (0.5%) shown that increasing in the compressive strength by about 2.97%, 14.9%, %, and 20.21% respectively at 28-day . This is because of the hybrid fibers with varying types and sizes would suggest different put down conditions. Moreover, this social position can be ascribed to the becoming better in the mechanical strength of bond as the fibers both have the capacity gaudily the microcrack arrangement and to detain their diffusion afterwards up to a certain scale[22].

(LWAMSC) Hybrid fibers mixes with 1.0% of steel fiber type SF1 and 0.5% fraction of volume of (P.P.F) show the moderately advancing in compressive strength. This is because, in comparison with longer and moderate specific gravity fibers, shorter and low specific gravity are more efficient in delaying cracks owing to the moderate specific contact surface of nonmetallic fibers [23].

The using of two types of macro steel fibers in (LWAMSC) with volume fraction 0.75% for each (mix H1), shows decreasing in compressive strength in comparison with reference mix (mix RF).This is due to the pores and honeycombs which can be formed during placing of moderate volume fraction of steel fibers as a result of not completed consolidation [24].

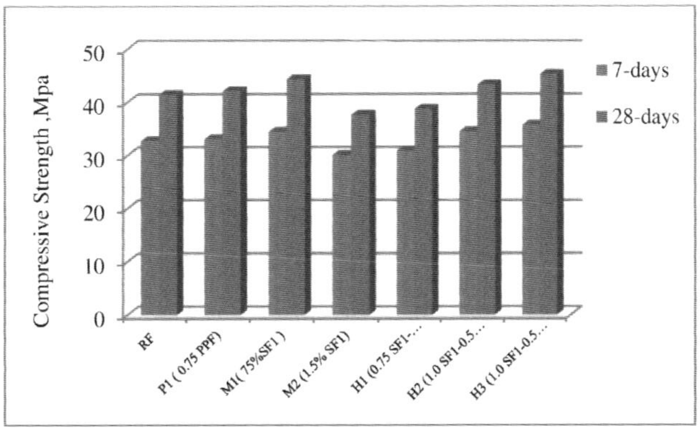

Fig. 5. Compressive Strength test of (LWAMSC) results at 7-day and 28-day

4.3. Splitting Tensile Strength

Table 10 and Figure 6 list the effects of fibers in (LWAMSC) appreciably, increasing the splitting tensile strength for alone type and hybrid fibers concrete mixtures compared to the reference concrete. The amount of increase in splitting tensile strength for (LWAMSC) mixes containing single steel fiber type (M1) with volume fraction 0.75%, and 0.75% polypropylene fiber (P1) is 295.8%, and 221.6% successively comparison to the refrence concrete. This is ascribed to the mechanism of fibers in detention cracking progression and the improvement of the bond between fibers and matrix due to the extra dense calcium silicate hydrate gel obtained from silica fume addition [25].

The comparison between splitting tensile strength values of (LWAMSC) mix having one steel fiber type (M2) with a volume fraction of 1.5% and the hybrid fiber mixes H1, H2, and H3 shows that the increase ratio in strength of splitting tensile of hybrid fiber (LWAMSC) is about 6.85%, 20.25%, and 7.47%, respectively.

Similar to the case of compressive strength, steel-polypropylene fibers mix (H3) shows the moderateest splitting tensile strength relative to all fibrous (LWAMSC) mixes (with mono or hybrid fiber). This is because of the cooperation of hybrid fibers, small fibers may bridge the microcracks more qualification, because they are not thick and their amount in mixture is much paramount than tof the great thick fibers, for the same fiber quantity. As the microcracks grown and join upto very large macrocracks, the long hooked-end fibers become more and more active in crack bridging. In this way, the strength of tensile and ductility can be improved. Long fibers can therefore provide a stable post peak response, but short fibers will be less active because they are being more pulled out, therfore the width of crack increasing [26].

4.4. Flexural Strength

The flexural resistance test proceeds for all (LWAMSC) combined are coincident in Table 10 and Figure 7. The conclusion show that the incorporation of fibers increment the flexural strength in single and multi-fibers heterogeneous suitable to mix together. The percentages of increased in parameter of breach for single steel fibre (LWAMSC) at fragment 0.75%, 1.5% of M1, and M2 are 52.37%, 29.52% and 60.64% for 0.75% of polypropylene fibers (P1) compared with reference concrete. The purpose for this improvement in parameter of fracture is that after table bully, the fibers will move the cargo that the concretion uniform until great by interfacial fetters between them. The principle for this increase in modulus of rupture is that after cement paste cracking, the fibers is carried the load that the concrete without changing up to cracking by means of bond of interfacial between the fibers and the matrix. Therefore, the fibers resist the dissemination of cracks and do not fail without warning, which source an increase in load carrying ability [27].

The comparison between strength of flexural values for (LWAMSC) samples containing 1.5% of steel fiber type (M2) and hybrid fiber samples H1, H2, and H3 shows that the increase ratio of strength of flexural for hybrid fiber (LWAMSC) samples is 20.9%, 35.61%, and 12.02% respectively. It can be observed that (LWAMSC) samples containing a merge of steel and polypropylene fibers (mix H3) show the lowest percentage increase in modulus of rupture relative to mono steel fiber mix (M1). This is because of the lower strength of (P.P.F) tensile and also the not strong bonding between cement matrix and(P.P.F) [28].

Table 10. Results of tensile of splitting and flexural strengths tests for (LWAMSC) mixtures

Mix Design	Splitting tensile resistance 28 day MPa	Flexural strength 28 day MPa
Ref.- (RF)	0.97	3.76
0.75%PPF- (P1)	3.12	3.87
0.75%SF1- (M1)	3.84	5.53
1.5%SF1- (M2)	3.21	4.24
0.75%S1F- 0.75%S2F- (H1)	3.43	5.12
1.0%SF1- 0.5%SF2- (H2)	3.86	5.75
1.0%S1F-0.5%PPF- (H3)	3.45	4.75

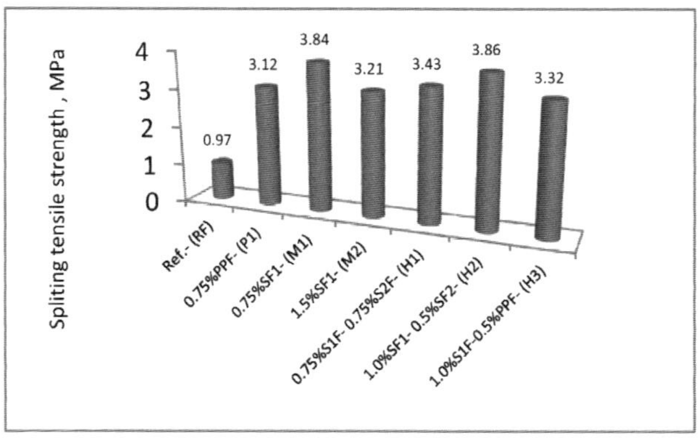

Figure 6. Strength of splitting tensile for overall (LWAMSC) mixtures of 28-day

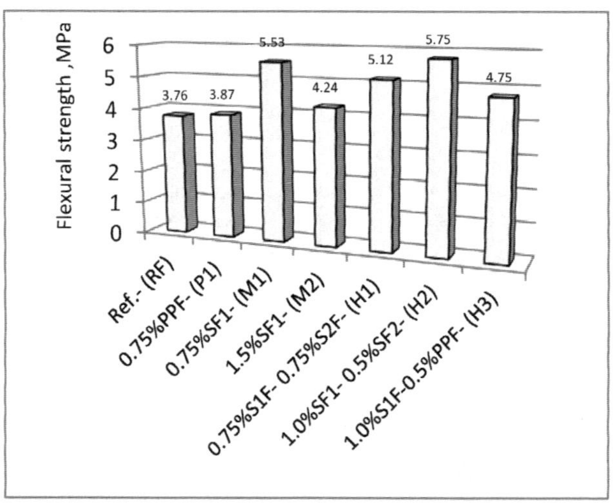

Figure 7. Strength of flexural and ratio of single and hybrid fibers

5. Conclusion

- The addition of steel fibers to (LWAMSC) increases its oven dry density of all mixes. Hybrid (LWAMSC) mixe containing a combination of steel fibers 0.75% steel fiber type (SF1) + 0.5% polypropylene fiber show densities less than that containing 1.5% of mono steel fiber type (SF1).
- The use of 0.75% volume fraction of macro steel fiber type (SF1) and polypropylene fibers increases the cube compressive strength at 28 day by about 5.19%, and 6.97% relative to the refrence concrete, and the partial replacement of macro steel fibers type SF1 with 0.75% of SF2 (H1), 0.5% SF2 (H2), and 0.5% polypropylene fiber (H3) increases the compressive strength at 28 day by about 2.9%,14.9%, and 20.21% compared to the 1.5% SF1- (M2)
- The inclusion of fibers in (LWAMSC) significantly improves the splitting tensile strength for both single and hybrid fiber reinforced samplesrelative to the reference concrete. Hybrid fiber reinforced (LWAMSC) samples show significant increase in splitting tensile strength. The percentage of increase in splitting tensile strength for (LWAMSC) specimens prepared from H1, H2, and H3 mixes is about 6.85%, 20.25%, and 7.47%, respectively attribute to the control mixture.
- The incorporation of fibers increases the flexural strength in single and hybrid fiber concrete compare to the reference concrete. The increase ratio of flexural strength for hybrid (LWAMSC) supplied of mixtures H1, H2, and H3 is 20.9%, 35.61%, and 12.02% respectively relative to the reference concrete.
- The resultant indicated that, the most efficient state addition proportion of hybrid fiber was 1.0% of (SF1) with 0.5% of (SF2) this ratio raised the strengths of compressive, flexural, and split tensile for 4.34%, 52.93%, and 297% successively, comparing to reference concrete.

References

[1] Ries J P, Boyle M J, Bremner T W, Burg R G, Crocker D A, Dodl C L, Robinson G M, Ramme B W, Mccall W C, Mor A A, Rowe S K, Sheetz S R, Snow P G, Speck J F and Vaysburd A M 2010 Guide for Structural Lightweight-Aggregate Concrete Reported by ACI Committee 213 1–38

[2] Prestera J R, Dixon D E and Crocker D A 1990 Standard practice for selecting proportions for structural lightweight concrete (ACI 211.2) *ACI Materials Journal* 87 638–51

[3] Campione G 2005 Mechanical properties of steel fibre reinforced lightweight concrete with pumice stone or expanded clay aggregates *Materials and Structures* 34 201–10

[4] Korolev E V and Inozemtcev A S 2013 Preparation and Research of the High-Strength Lightweight Concrete Based on Hollow Microspheres *Advanced Materials Research* 746 285–8

[5] Libre N A, Shekarchi M, Mahoutian M and Soroushian P 2011 Mechanical properties of hybrid fiber reinforced lightweight aggregate concrete made with natural pumice *Construction and Building Materials* 25 2458–64

[6] Gao J, Sun W and Morino K 1997 Mechanical properties of steel fiber-reinforced, high-strength, lightweight concrete *Cement and Concrete Composites* 19 307–13

[7] Kayali O, Haque M N and Zhu B 2003 Some characteristics of high strength fiber reinforced lightweight aggregate concrete *Cement and Concrete Composites* 25 207–13

[8] Iqbal S, Ali A, Holschemacher K and Bier T A 2015 Mechanical properties of steel fiber reinforced high strength lightweight self-compacting concrete (SHLSCC) *Construction and Building Materials* 98 325–33

[9] Aslam M, Shafigh P, Alizadeh Nomeli M and Zamin Jumaat M 2017 Manufacturing of high-strength lightweight aggregate concrete using blended coarse lightweight aggregates *Journal of Building Engineering* 13 53–62

[10] Yap S P, Chen P Z C, Goh Y, Ibrahim H A, Mo K H and Yuen C W 2018 Characterization of pervious concrete with blended natural aggregate and recycled concrete aggregates *Journal of Cleaner Production* 181 155–65

[11] Shaikh F U A 2016 Mechanical and durability properties of fly ash geopolymer concrete

containing recycled coarse aggregates *International Journal of Sustainable Built Environment* 5 277–87
[12] Test C C, Mortar H, Concrete C, Specimens C C, Concrete L, Freezing R and Concrete S L 2010 Standard Specification for Lightweight Aggregates for Structural Concrete 1 i 1–4
[13] Gorgots V D 2006 About drawdown drilling parameters design procedure at primary formation exposing by a horizontal wellbore *Neftyanoe Khozyaistvo - Oil Industry* 120–1
[14] Gold R 1999 Standard Specification for 95 1–4
[15] PCA 2018 Chemical Admixtures 1–21
[16] Kamal M M, Safan M A, Etman Z A and Kasem B M 2013 Mechanical properties of self-compacted fiber concrete mixes *HBRC Journal* 10 25–34
[17] Test C C, Aggregate C, Aggregate F, Content A, Rooms M and Concrete P 2002 2.2 Astm C192 *Management* 04 1–8
[18] ASTM:C39/C39M-15a 2012 Standard test method for compressive strength of cylindrical concrete specimens *ASTM International* 04 1–7
[19] ASTM C496-96 1996 Standard Test Method for Splitting Tensile Strength of Cylindrical Concrete Specimens *ASTM International, West Conshohocken, PA* 04 1–4
[20] Speci- C C T, Cores T D and Speci- C C T Astm C78 1–3
[21] Smith D R, Lia J M and Ap L 2012 Potential Application of ASTM C 1701 for Evaluating Surface Infiltration of Permeable Interlocking Concrete Pavements *10th International Conference on Concrete Block Paving Shanghai* 1–8
[22] Mozan S A A 2015 Some Properties of Hybrid Fibers High Strength Lightweight Aggregate Concrete 33 815–29
[23] Pereira E B, Fischer G and Barros J A O 2012 Effect of hybrid fiber reinforcement on the cracking process in fiber reinforced cementitious composites *Cement and Concrete Composites* 34 1114–23
[24] Allan M L and Kukacka L E 1995 Strength and durability of polypropylene fibre reinforced grouts *Cement and Concrete Research* 25 511–21
[25] Dawood E T and Ramli M 2011 Contribution of hybrid fibers on the hybrid fibers on the properties of high strength concrete having high workability *Procedia Engineering* 14 814–20
[26] Jitchaiyaphum K, Sinsiri T and Chindaprasirt P 2011 Cellular lightweight concrete containing pozzolan materials *Procedia Engineering* 14 1157–64
[27] Al-Swaidani A M, Aliyan S D and Adarnaly N 2016 Mechanical strength development of mortars containing volcanic scoria-based binders with different fineness *Engineering Science and Technology, an International Journal* 19 970–9
[28] Yasar E, Atis C D, Kilic A and Gulsen H 2003 Strength properties of lightweight concrete made with basaltic pumice and fly ash *Materials Letters* 57 2267–70.

Evaluation of the Performance of MBR-RO Technology for Treatment of Textile Wastewater and Reuse

Zahraa S H[1], Gzar H A[2]

[1] Civil Engineering Department, College of Engineering, Waist University, Iraq. E-mail: zsabah@uowasit.edu.iq
[2] Faculty member, PhD, Civil Engineering Department, College of Engineering, Waist University, Iraq. E-mail: hatemasal@uowasit.edu.iq

Abstract. This study aims to evaluate the performance of membrane bioreactor (MBR) followed by the membrane of reverse osmosis (RO) for textile wastewater treatment and reuse. Firstly, study of the effect of hydraulic retention time (HRT) and the concentration of mixed liquid suspended solids (MLSS) on the removal efficiency of chemical oxygen demand, colour and turbidity were achieved by MBR system. The results showed that the MBR technology became increasingly effective in the treatment of textile wastewater with increasing HRT and MLSS concentrations The highest removal efficiencies were 86.7%, 77.67% and 96.74% for COD, colour and turbidity, respectively when HRT =2days, while the highest removal efficiencies were 88.34%, 83.7% and 97.9% when HRT =2.5days. The second step was studying the effect of operating pressure and RO feed temperature on the removal efficiency of COD and colour of the RO system. MBR effluent has been treated with RO membrane to remove the remaining pollutants. The results showed that, the increasing of the pressure and decreasing of the temperature have a positive effect on removal efficiency. The maximum removal efficiencies obtained for COD and colour were 99.3% and 99.8% respectively at a pressure of 10 bar and 25 °C RO feed temperature.
Keywords: Textile effluent, Treatment, MBR-RO. Permeate water, Reuse

1. Introduction

Textile industries are one of the most water-intensive industries in different stages of textile processing, thus they are the largest producers of wastewater polluted by colours, textile auxiliaries, toxic chemicals and resistant to biological treatment methods [1]. Textile industries consume roughly 0.2 to 0.5 m3 of water per kg of the ultimate product [2].

Water resources have turned into an issues operational and ecological. Wastewater treatment and recycling have become an active tool for sustainable industrial development programs [3]. Reuse and decentralization will be fundamental for addressing human requirements for water and sanitation in developing and developed nations. Rigorous environmental legislation has increased the intensity, potency, and variety of treatment technologies in order to treat and reuse the water in several industrial processes.

MBR technology is one of the most advanced processing techniques used recently for the treatment of wastewater and reuse. MBR is a mix of the traditional biological sludge process and membrane separation process, this technique is characterized by increasing in a suspended biomass, with membranes systems of microfiltration or ultrafiltration. Membrane bioreactors can either be in-series or submerged , where the submerged MBR is better because it requires less energy than the in-series MBR [4]. The membrane includes an important and effective function in the removal of suspended solids and industrial dyes utilized in textile processing [5]. There are four sorts of membranes: microfiltration (MF), ultrafiltration (UF), nanofiltration (NF) and reverse osmosis (RO), respectively. A large portion of these membranes is produced using artificial organic polymers. Membranes (MF and UF) are made of a similar material yet the contrast between them is the variety in pore sizes and techniques of their preparation [6]. While RO membranes are manufactured either from polysulfone coated with aromatic polyamides or cellulose acetate [7]. UF or MF membrane is applied to the treatment of secondary effluents before treatment via the reverse osmosis unit, wherein (UF or MF) membrane is used to eliminate suspended solids and colloids, while RO membrane is used to remove the dissolved solids, organic and ionic materials. MBR-RO technique is used to treat industrial wastewater and recycling [8]. The objective of the present study is to, evaluate the performance of MBR-RO system in the removal of textile wastewater.

2. Materials and Methods

2.1. Wastewater source

The wastewater used in the laboratory experiments was taken from the wastewater collection tank of Al-Kut Textile Factory, Wasit Governorate, Iraq. The collection tank receives the wastewater from all sections of the weaving and knitting factory, and the pumps were designed to raise the wastewater from this tank to a higher altitude and pumping it to tanks of primary settling, neutralization, aeration, secondary settling, etc. The samples of textile effluent were tested in the laboratory to determine the physicochemical characteristics of wastewater as shown in Table (1).

Table 1. Characteristics of textile effluent.

Parameters	Unit	Value
COD	mg/l	280-300
TSS (mg/l)	mg/l	200-250
PH	-	8±0.5
Colour	Pt-Co	350-450
TDS (mg/l)	mg/l	700-800
EC	µS/cm	1455-1602
DO	mg/l	1.6-1.8
Turbidity	NTU	85-100

2.2. Operating conditions

The effluent of wastewater treatment plant was treated by using the MBR technology followed by the RO membrane as shown in Figure 1.

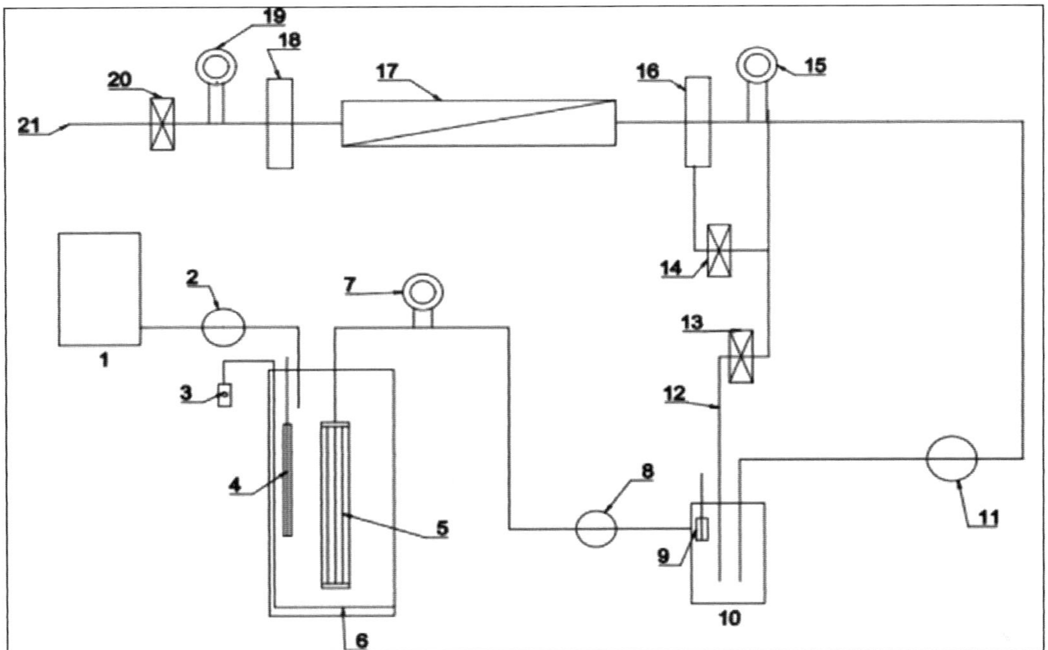

(1) Feed tank, (2) Peristaltic pump, (3) Air pump, (4) Fully automatic submersible water heater, RS electrode (in reactor), (5) UF membrane module, (6) Air bubble diffuser, (7) Vacuum pressure gauge, (8) Peristaltic pump, (9) Fully automatic submersible water heater (in storage tank), RS electrode, (10) Storage tank, (11) Parkside pressure pump, (12) Recycle, (13,14,20) Valves, (15) Influent gauge, (16) Influent flowmeter, (17) Ro membrane module, (18) Effluent flowmeter, (19)Effluent gauge, (21) RO effluent

Figure 1. Schematic view of MBR-RO pilot scale.

2.2.1. MBR pilot unit.

The MBR system consists of a reactor tank made of Plexiglas with a height of 40 cm and a diameter of 16.5 cm, an operating volume of 8L. The hollow fibre ultrafiltration membrane modules were installed inside the bioreactor, the membrane filtration precision is 0.01 μm and the effective surface area is 0.8 m2, the UF membrane characteristics are summarized in Table (2). Air diffuser was installed under the reactor to perform the ventilation required and to maintain the dissolved oxygen concentration in a biological reactor greater than 5 mg / l. The reactor temperature was set at 25 ± 1 ° C. The wastewater was fed inside the reactor by the peristaltic pump with two flow rates of 0.166 and 0.133 l / h. The permeate was withdrawn from (MBR) by using a peristaltic pump and it was stored in the storage tank. The permeable flux ranged from 0.173 to 0.205 L / m2 h when the HRT was 2 days and ranged from 0.14 to 0.164 L / m2 h when the HRT was 2.5 days. The sludge retention time (SRT) was determined within 32 days. The mixed liquid is withdrawn daily to maintain the MLSS concentration between 2000 - 2500 mg /L. The negative pressure gauge was installed on the draw line between the reactor and the peristaltic pressure pump and the manual pressure data were recorded at the end of each filtration process, and the membrane pressure ranged (TMP) from 0.05-0.3 bar. At the end of each experiment, the sludge is withdrawn from the bioreactor, the reactor is washed with distilled water and the membrane was cleaned with an anti-scalant solution to remove the waste layer created on membrane fibres. The MBR operating characteristics are summarized in Table (3) as adopted by [10].

Table 2. Characteristics of UF membrane.

Parameter Process	Unit	Value
Maximum Pressure	MPa/PSI	0.4/58
Maximum Operating Temperature	°C	5-40
Material	-	Hollow fibre
Effective membrane area	m^2	0.8
Filtration precision	μm	0.01

Table 3. The operating conditions of MBR system.

		Average value	
Parameters	Unit	Run1	Run2
Mode of operation		Continuous	Continuous
Inflow	L/day	4	3.2
Permeate flux	L/m2/day	3.94	3.149
(HRT)	day	2	2.5
(SRT)	day	32	32
(MLSS)	mg/L	2200 ± 50	2500±50
MLSS temp.	°C	25 ± 1	25 ± 1
DO	mg/L	5.5 ± 0.2	6.5 ± 0.5
PH		7.4 – 8.4	7.4 – 8.4

2.2.2. RO membrane pilot unit.

The MBR effluent was treated with RO membrane to produce higher quality water. The RO membrane was used as a type of "spiral wound membrane filtration unit". Two pressure gauges and two flow meters were used. Pressure difference was applied across the RO membrane to study the impact of pressure on the removal efficiency. The pressures applied on the reverse osmosis membrane were 2, 4, 6, 8 and 10 bar. Two temperature tests 25±1° C and 37± 1 °C were tested to compare the efficiency of RO membrane treatment in summer and winter. The RO membrane was cleaned at the end of each experiment using a pump with a low pressure (Ritek diaphragm pump USA with workflow 28 LPH). The RO membrane characteristics are summarized in Table (4).

Table 4. Characteristics of RO membrane.

Parameter Process	Unit	Value
Maximum Operating Pressure	Bar	10
Maximum Operating Temperature	°C	45
pH range	-	2-11
Membrane type	-	Spiral-wound element with polyamide thin-film composite membrane
Effective membrane area	m^2	0.37
Minimum salt rejection	-	96%
Manufacturer	-	Film Technology Corp. USA

2.3. Analytical methods

Laboratory analysis was performed to determine the characteristics of MBR influent, MBR and RO effluent. The flux of the MBR effluent is calculated by collecting water in a volumetric flask and analysed by analytical equipment. MLSS concentration was determined using the standard APHA 2540E method. The DO concentration was determined using a dissolved oxygen meter, and the PH meter (WTW – Germany) is used to determine the PH. The turbidity meter (WTW-Germany) is used to determine the turbidity. COD concentration and colour are determined using spectrophotometric (Spectro Direct-Lovibond) tests. Finally, the efficiency of removal (% R) is calculated for all species, as shown in the equation below [11].

$$Removal\ Efficiency\ (\%R) = \left[1 - \frac{Permeate\ Concentration\left(\frac{mg}{L}\right)}{Feed\ Concentration\left(\frac{mg}{L}\right)}\right] \times 100 \quad \ldots\ldots\ldots\ldots\ldots\ldots (1)$$

3. Results and Discussion

Sludge was initially collected from sedimentation tanks in the textile plant and transferred to the MBR tank. The MBR was operated for 30 days prior to the initiation of the experiments in order to achieve the operational conditions specified in Table 2. It was then started to withdrawal the sludge each day in order to maintain MLSS concentration in the aeration tanks. The HRT of the membrane bioreactor system was increased from 2 to 2.5 days to study the effect of HRT on the rate of removal efficiency. The change of HRT from 2 to 2.5 days, increased the removal efficiency of COD from 86.7% to 89.3%, as shown in Figures 2(a) and 2(b), and these values are satisfactory according to previous studies [12]. It has been observed that COD removal efficiencies increase with increasing HRT. The reason for the increased efficiency of COD removal with increasing HRT is that biological reactors have the ability to oxidize the organic material entering the reactor, especially at low inflow, as mentioned in previous studies [13].

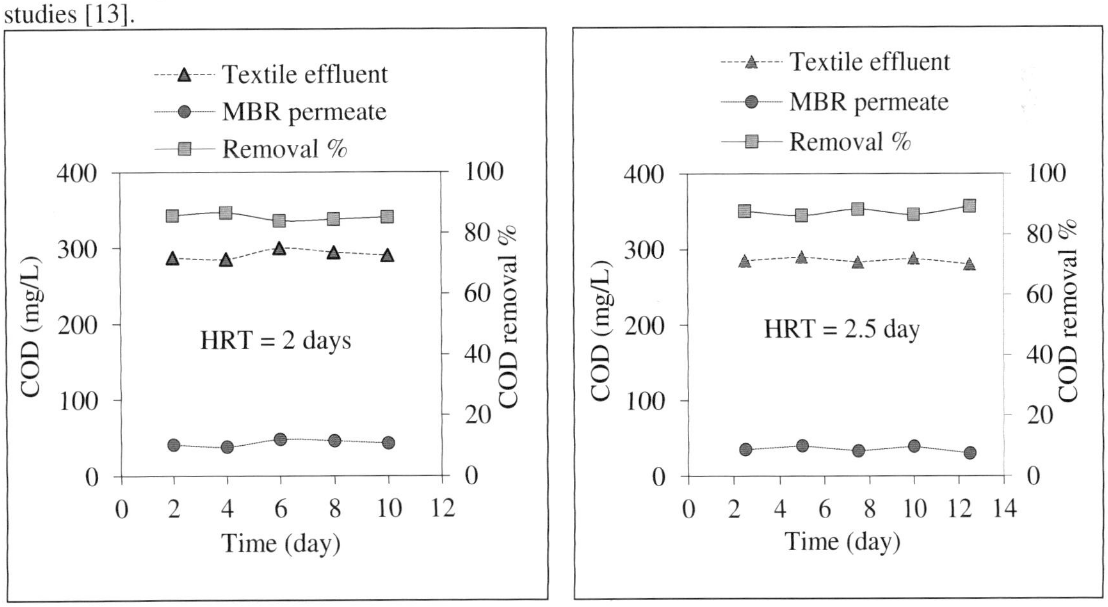

(a) **(b)**

Figure 2. Concentration of COD in textile effluent and the MBR permeate with the time of operation and their removal percentage, at (a) HRT=2 days and (b) HRT= 2.5 days.

Figures 3 (a) and 3 (b) showed the removal efficiency of turbidity achieved in the MBR system is to be 96.74% and 97.9% at HRT = 2 and 2.5 day, which is a good rate to remove as reported in previous studies [1]. This percentage of removal indicates that the turbidity of MBR permeate are very low, and this means that the UF membrane operates properly without any flaws during the experiment period and the produces clean water and free from impurities.

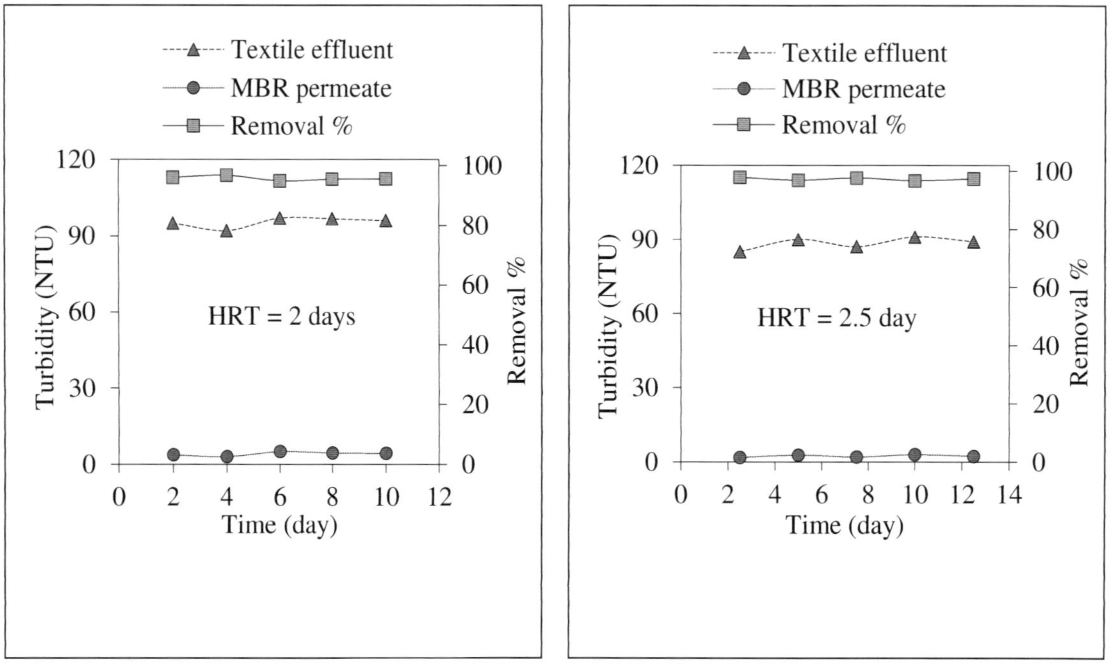

(a) **(b)**

Figure 3. Turbidity in textile effluent and the MBR permeate with the time of operation and their removal percentage, at (a) HRT=2 days and (b) HRT= 2.5 days.

Figures 4 (a) and 4 (b) showed the removal efficiency of colour achieved in the MBR system is to be 77.67% and 83.7% at HRT = 2 and 2.5 day, which is consistent with previous studies [14]. Note that there is an increase in colour removal with increased HRT, because the increased of HRT leads to increased microorganisms, the function of these organisms is to absorb pollutants and colour molecules that enter MBR.

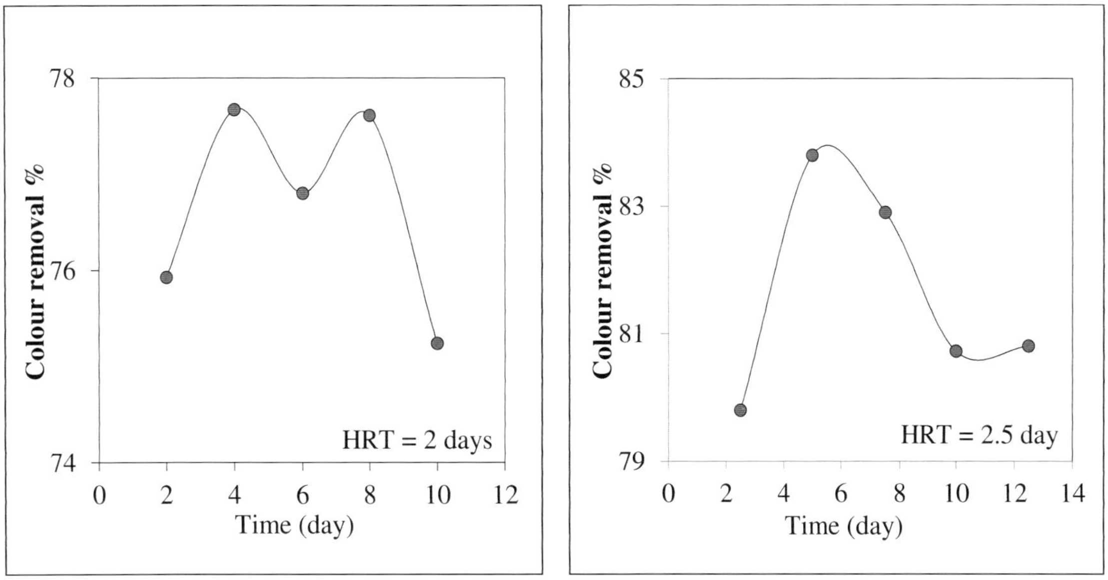

(a) **(b)**

Figure 4. Colour removal with time, at (a) HRT=2 days and (b) HRT= 2.5 days.

MBR effluent was treated with RO membrane unit to treat residual of pollutants to produce high-quality water that conforms to reuse standards. Figures (5) and (6) show that the minimum overall removal efficiencies achieved by RO membrane for COD and colour are 93.4 % and 93%, respectively, at pressure 2 bar and RO feed temperature 37 °C, and the highest overall removal efficiency of COD was 99.3% and colour was 99.8%, at pressure 10 bar and RO feed temperature 25 ° C. We note that the removal efficiencies of COD and colour increases with increasing mechanical pressure applied to the membrane. This can be attributed to the fact that increasing the mechanical pressure applied to the membrane leads to an increase in the free volume available to the active layer of the membrane, which allows components to permeate the created space and the reason for the decrease of the efficiency with high feed temperature is that the high temperature leads to an increase in the pore size of the membrane and thus allows the passage of pollutants through it, as mentioned in previous studies [9].

It is very essential to mix the MBR and RO technologies. this mixture able to achieve the highest levels of removal for COD, colour and salts to provide the production of high quality reusable water for industrial purposes. Although the MBR is extremely effective in removing industrial pollutants (COD, colour and turbidity), but it is ineffective in removing salts. As a result, it is necessary to resort to treatment with RO system to remove the remaining pollutants and achieve reuse standards. MBR and RO showed excellent performance in removing all contaminants with removal rates of more than 93% for COD and colour. And removal rates of total dissolved solids (TDS) ranged from 96-100%. This means that mixing (MBR with RO) successfully processes the textile wastewater and within the reuse specification and this also corresponds to what was mentioned in previous studies [1].

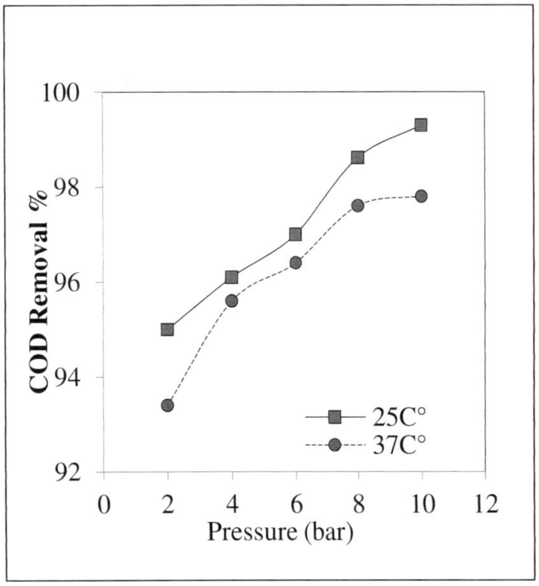

Figure 5. Effect of Feed samples Temperature on COD overall removal at different operating Pressure.

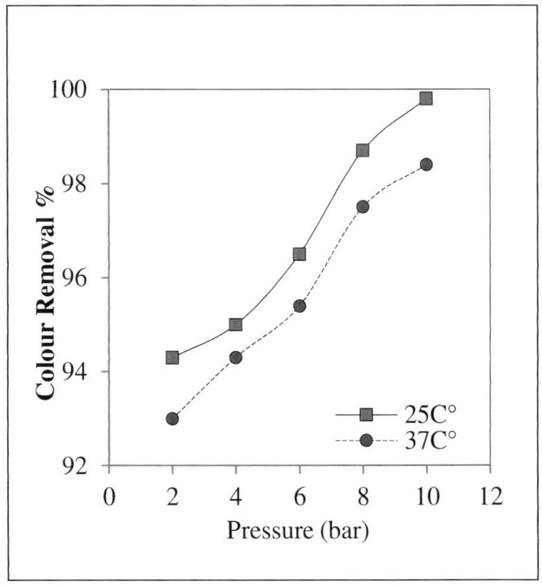

Figure 6. Effect of feed samples temperature on colour overall removal at different operating pressure.

4. Conclusion

Following the integration of the MBR and RO techniques for textile wastewater treatment, it can be concluded that the performance of these technologies are an extremely effective in the wastewater treatment and reuse . It can also be concluded that these techniques are complementary to each other in removing all organic pollutants at rates exceeding 93% for COD and colour, and exceeding 96% for total dissolved solids (TDS). MBR technology proved to be effective in removing organic pollutants, except for salts. It was observed through the results obtained that the HRT has a positive effect on the removal of organic pollutants (COD, colour and turbidity) and it was also observed that MBR technology is an extremely effective way to RO pre-treatment. It has also been demonstrated that the use of RO technology is effective in removing residual organic pollutants and salts within the required limits and the reuse specifications. The results showed that the applied pressure on the RO membrane has a positive effect on the removal efficiency, whereas the feeding temperature has an inverse effect on the removal efficiency.

References

[1] De Jager D, Sheldon M S and Edwards W 2014 Colour removal from textile wastewater using a pilot-scale dual-stage MBR and subsequent RO system *Sep. Puri. Technol.* 135 135-44.

[2] Marcucci M, Nosenzo G, Capannelli G, Ciabatti I, Corrieri D and Ciardelli G 2001 Treatment and reuse of textile effluents based on new ultrafiltration and other membrane technologies *Des.* 138 75-82.

[3] Hoinkis J, Deowan S A, Panten V, Figoli A, Huang RR and Drioli E 2012. Membrane bioreactor (MBR) technology–a promising approach for industrial water reuse. *Proc. Engi.* 33 234-41.

[4] Brauns E, Van Hoof E, Molenberghs B, Dotremont C, Doyen W and Leysen R 2002 A new method of measuring and presenting the membrane fouling potential *Des.* 150 31-43.

[5] Patnaik P 1996 *A comprehensive Guide to The Hazardous Properties of Chemical Substances* (New York: John Wiley)

[6] Baker RW 2012 *Membrane Technology and Applications* (New York: John Wiley)

[7] Lee K P, Arnot T C and Mattia D 2011 A review of reverse osmosis membrane materials for desalination—development to date and future potential *J. Membr. Sci.* 370 1-22

[8] Tam L S, Tang T W, Lau G N, Sharma K R and Chen G H 2007 A pilot study for wastewater reclamation and reuse with MBR/RO and MF/RO systems *Des.* 202 106-13

[9] Gzar H A and shhaieb Z K 2018 The removal ofterasil navy blue dye from Al-Kut textile wastewater using membrance technologies *Inter. J. Engi. Technol.* 7 557-562

[10] Ferdian T A , Reza A and Setiadi T 2010 Evaluation of laboratory and plant scale membrane bioreactors in treating textile wastewater *Evaluation IWA Asp. Conf. e-mail: tjandra@che.itb.ac.id*

[11] Hendricks D 2006 *Water Treatment Unit Processes: physical and chemical* (Boca Raton: CRC press) p 927

[12] Brik M, Schoeberl P, Chamam B, Braun R and Fuchs W 2006 Advanced treatment of textile wastewater towards reuse using a membrane bioreactor *Proc. Bioch.* 41 1751-1757

[13] Metcalf and Eddy 2003 *Wastewater Engineering Treatment and Reuse* (New York: McGaw-Hill)

[14] Badani Z, Ait-Amar H, Si-Salah A, Brik M and Fuchs W 2005 Treatment of textile waste water by membrane bioreactor and reuse *Des.* 185 411-417

IOP Conf. Series: Materials Science and Engineering **584** (2019) 012050 doi:10.1088/1757-899X/584/1/012050

Numerical analysis of thin ring foundations under different loading conditions

Adel A Al-Azzawi[1] and Khalida A Daud[2]

[1]Assistant Professor, College of Engineering, Al-Nahrain University, Baghdad, Iraq. Email: dr_adel_azzawi@yahoo.com
[2]Lecturer, College of Engineering, Al-Nahrain University, Baghdad, Iraq. Email: khalida_dwd@yahoo.com

Abstract. The problem of ring foundation under uniform load or different point loads are a very common one in practice. The resulting independent degrees of freedom for thin depth ring members are the deflection and the twisting angle at any section in the ring. The governing equations for bending of ring members resting on Winkler foundation which derived by the previous researcher are solved in this research using finite differences. Also, finite element methods are used to model the problem by using two-dimensional and three-dimensional elements and spring elements to model the soil. The behaviour under point and distributed loads are investigated and compared with previous analytical results and good agreements were obtained. This study revealed that the maximum deflection decreased by 86% when the loading changed from point to uniformly distributed. Also, when the subgrade reaction was increased from (1.7 to 6 MPa), the maximum deflection decreased by 72% and the maximum moment decreased by 31%.

1. Introduction

For any building or structure, the foundation is must be sufficient to resist applied loads or to support the building. Sufficient and efficient foundations depend on the selection of foundation type and the properties of the soil beneath it. The ring shape footing is the most suitable foundation for heavy structures and it may be considered to be economical for elevated tanks, silos, cooling towers, and chimneys. The foundation with ring shape will enable the construction engineer to reduce the used materials and construction cost. Several researches were implemented on ring members placed on soil by various researchers. In 2008, Mehrjardi [1] made a study to evaluate the bearing capacity and deflection of ring foundations. The response of ring footing and circular footing was compared and presented. The research made a conclusion that the ring foundations are more appropriate and economical than circular ones especially in shoring axisymmetric constructions.

In 2009, Algin [2] studied theoretically the rectangular footing elastic settlement. The surface footing was subjected to eccentric loads and placed on sand deposits. The solution was obtained in closed-form. The resulting influence factors were given in order to calculate the elastic settlement for footing under eccentric load. The expressions were also expanded for the analytical problems of infinitely deep incompressible soil layer. This research is an attempt to understand the elastic and differential settlement of footing under eccentric load.

In 2012, Zhou et. al. [3] used the curved flexural member or beam theory to derive the differential equations for the flexural deformation of ring footings. Different boundary conditions were investigated. The elastic deflection of ring foundation subjected to point load is obtained mathematically through

solving the problem and substituting the boundary conditions. A parameter relating soil stiffness to the flexural member stiffness (non-dimensional) is suggested by the researchers. In that study, the parameter is recognized to govern the ring foundation response results. The mathematical model is verified using finite elements with different variables and the study revealed that such non-dimensional parameter gives a brief indication and guidance in the ring foundation design of special structures.
In 2013, Atalar et.al [4] studied the soil bearing capacity for shallow footing subjected to inclined eccentric load. The researchers conducted small scale laboratory tests in order to obtain the ultimate bearing capacity of a strip footing. The foundation was placed on sand and under inclined eccentric load. The embedment ratio of footing in sand was varying from zero to one. The laboratory outcomes were compared with the empirical relationships and the difference between them was 25% .
In 2015, Naseri and Hosseininia [5] carried out finite difference analysis study on the deflection of ring footing on soil. The aim of the study was to obtain solution in closed-form in order to calculate ring footing elastic deflection. This calculation is made through using the theory of elasticity and defining displacement influence factors. These factors consider the non-homogeneity of soil, the ring dimensions, ring stiffness, and the embedment of ring footing. The obtained results were compared with the previous analytical outcomes and good agreement was obtained.
In 2016, Nayyeri et. al. [6] derived the governing equations of ring on Winkler foundation considering the bending and torsion effects. The soil was simulated as elastic material in the derivation assumptions. The load on foundation was assumed to be a normal concentrated load. Estimation of the deflection and twist has been obtained for a range of practical field applications. The response of rings under several point loads can be calculated using the developed closed-form equations using superposition principle. The finite elements were used to check the validity of the proposed equations in calculating the settlement and angular twist for three ring foundations with different radii and cross sections.
In 2017, Keshavarz A., Kumar J. [7] used the (SCM) stress characteristics method to estimate the footing bearing capacity. The considered problem was ring foundation with smooth and rough bases. Different failure mechanisms were considered for ring footings. Two mechanisms were selected for smooth base case and four mechanisms for rough base case. A curved non-plastic wedge was utilized for a rough base footing. The stress singularities were considered in the analysis at the inner and outer edges of foundations. The bearing capacity coefficients, Nc, Nq and Nc were given in the form of a function of inner to outer radii ratio of footing and internal friction angle of soil. The analysis results were compared with the reported computational and experimental data and good correlation between them were obtained.
Based on previous researches the differential equation derived by Nayyeri et. al. [6] is selected in the present study. The finite difference method is selected to solve the differential equations numerically as the closed-form solution developed by Nayyeri et. al. [6] is limited to simple loading and boundary conditions. The Nayyeri et. al. [6] solution focused on obtaining expressions for deflection and twist only. The 2D and 3D finite elements through ABAQUS software are used in the present study to check the finite difference developed model.

2. Finite difference method

The differential equations for bending of thin ring member on Winkler foundation was derived by Nayyeri et. al. [6]. The derivation was established on basis of small strain theory and linear Hooke's law.

$$\frac{d^2}{ds^2}\left\{-EI\left(\frac{\emptyset}{R}+\frac{d^2w}{ds^2}\right)\right\}-\frac{d}{ds}\left\{\frac{GJ}{R}\left(-\frac{1}{R}\frac{dw}{ds}+\frac{d\emptyset}{ds}\right)\right\}+q-kw=0 \quad (1)$$

$$\frac{d}{ds}\left\{GJ\left(-\frac{1}{R}\frac{dw}{ds}+\frac{d\emptyset}{ds}\right)\right\}-\frac{EI}{R}\left(\frac{\emptyset}{R}+\frac{d^2w}{ds^2}\right)+T-k'\emptyset=0 \quad (2)$$

where s is the distance along the centerline of the ring,
w=w(s) is the centerline ring deflection,
$\emptyset=\emptyset(s)$ is angular twist,

G is the ring shear modulus,
J is the polar moment of inertia of ring section,
k is the vertical subgrade modulus,
k' is the rotational subgrade modulus,
EI is the ring bending rigidity,
q is the distributed load,
M is the bending moment,
V is the shear force,
T is the distributed torque
and R is the radius of ring centerline.

The slab cross section is shown in Figure 1.
The bending and twisting moments can be calculated using the equations following:

$$M = -EI\left(\frac{\emptyset}{R} + \frac{d^2w}{ds^2}\right) \tag{3}$$

$$T = GJ\left(-\frac{1}{R}\frac{dw}{ds} + \frac{d\emptyset}{ds}\right) \tag{4}$$

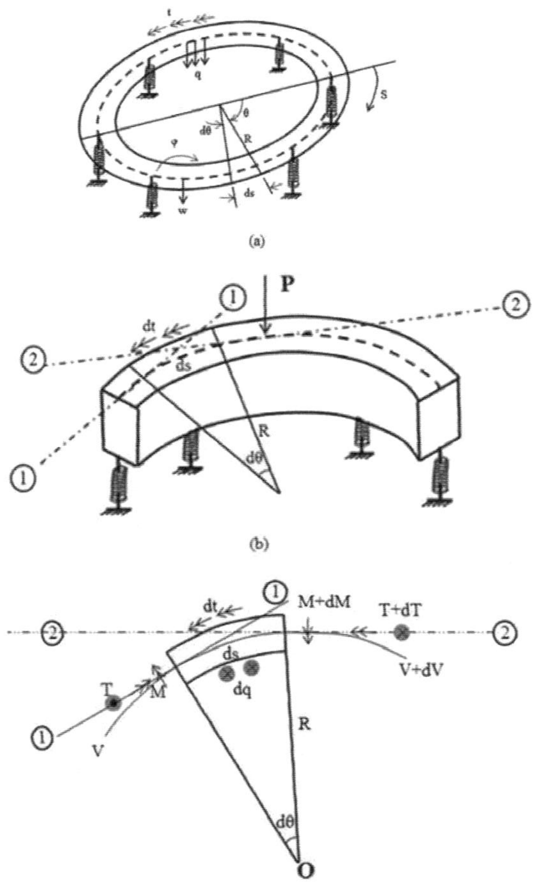

Figure 1. General representation of the prismatic ring member [6]

The finite differences are used to solve the governing differential equation using polar coordinates. At each node, the differential equations of ring foundation are commuted by difference expressions to simulate the ring. The ring peripheral length in the (s) direction is divided into length intervals (Δs or R$\Delta\theta$) (Figure 2).

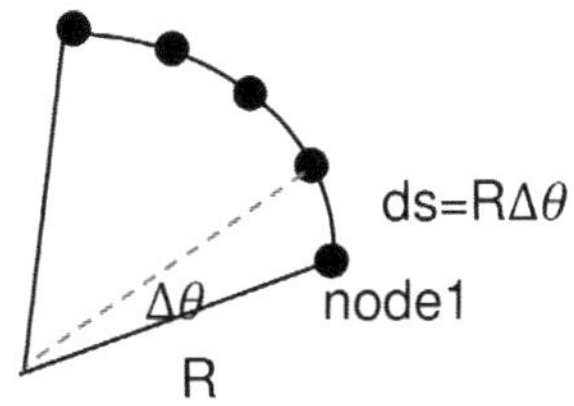

Figure 2. Finite difference mesh

3. Finite element method

The numerical modelling is made using ABAQUS software. The 8 nodes quadrilateral element is used to model the problem in two dimensions and 8 node brick element to model the problem in three dimensions. The soil is modelled using Winkler spring or spring element. Figure 3 shows the numerical model assumed for the ring foundation under point load. The finite element analysis is made carefully through selecting the suitable elements number for modelling the ring problem. A mesh sensitivity analysis is carried out such that more mesh refinement did not affect the solution results.

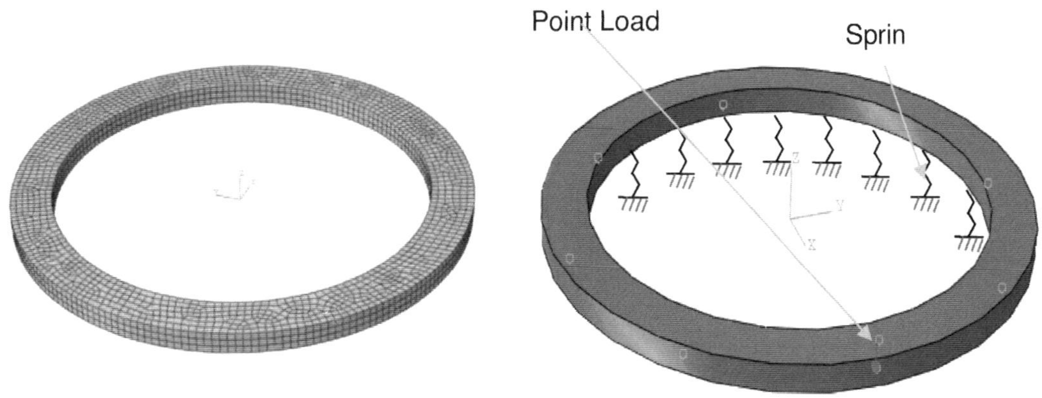

Figure 3. Finite element geometry and loading

4. Verification of the numerical method

The problem of the ring resting on springs was modelled using two numerical methods. Namely, the finite differences and finite elements are used in order to obtain a brief analysis results. Both numerical solutions are based on discretizing the differential equations or the continuum into a member or element of finite size. To study the response of the ring footing under a single point load, three different loading examples solved by Nayyeri et. al. [6] are selected. The finite elements simulation of the ring on springs under a point load was implemented using ABAQUS software. Comparisons of the settlements and angular twists of the ring were made between the analytical Nayyeri et. al. [6] solution and the present study numerical solution. Three different footings examples, case 1, case 2 and case 3 were analysed with different radii of 3050, 6100 and 9150 mm, respectively. The values of width, depths and radii of the ring examples are shown in Figure 4. The ring is under the action of a 10 kN point load with different locations. The ring material elasticity modulus was 20.7 MPa and Poisson's ratio was 0.35 in the selected examples. The subgrade reaction of soil (k) (multiplied by ring width) is taken to be 1.7MPa for the settlement and $k' = k\left(\frac{b^3}{12}\right)$ for twist.

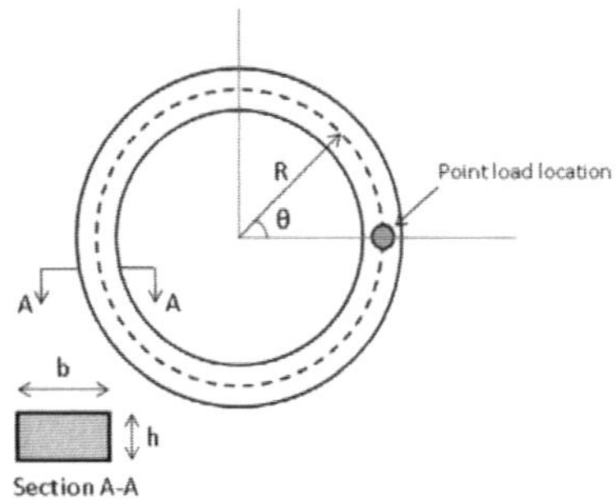

Case number	Radius cm (inch)	Ring width (*b*) cm (inch)	Ring depth (*h*) cm (inch)
1	305 (120)	61 (24)	30.5 (12)
2	610 (240)	91.5 (36)	45.75 (18)
3	915 (360)	122 (48)	61 (24)

Figure 4. Case previously studied by Nayyeri et. al. [6]

Figure 5 shows the deflection or settlement under the ring footing for the first case using finite difference and finite element methods (brick elements) with a radial position (the 10 kN point load at θ=0). Also, Figure 6 shows the other two cases and good agreement is obtained with the previous analytical study of Nayyeri et. al. [6] with a maximum tolerance of 3% and 5% for finite differences and finite elements respectively. The maximum deflection of the ring footing occurs at load application position (θ = 0). The deflection reduces with enlarging the value of θ. The effect of the point load on the deflection for θ= π/2 and beyond can be neglected for the studied cases. Figure 7 show for case 1 the variation of the angle of twist with radial positions. Small differences between the previous analytical and the present study numerical results were obtained.

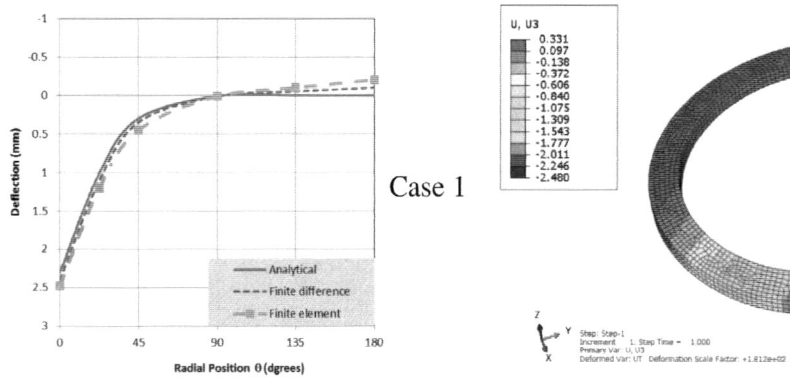

Figure 5. Deflection profile and contour line for case 1

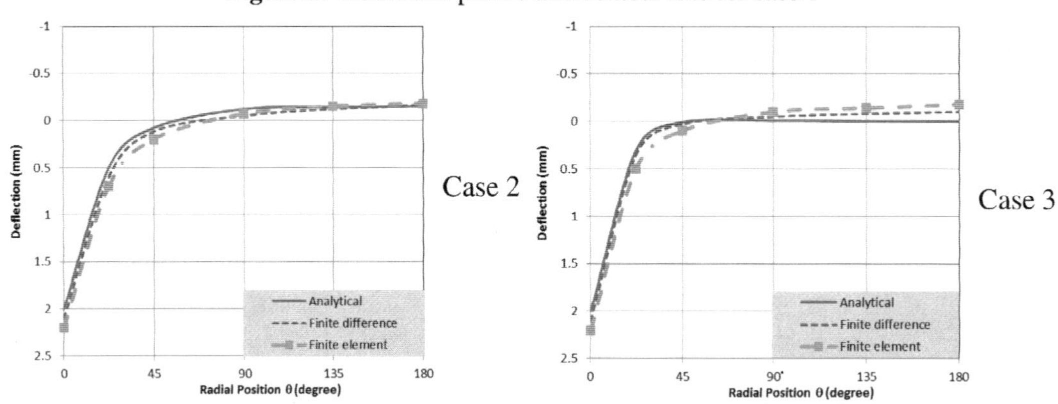

Figure 6. Deflection profile for case 2 and 3

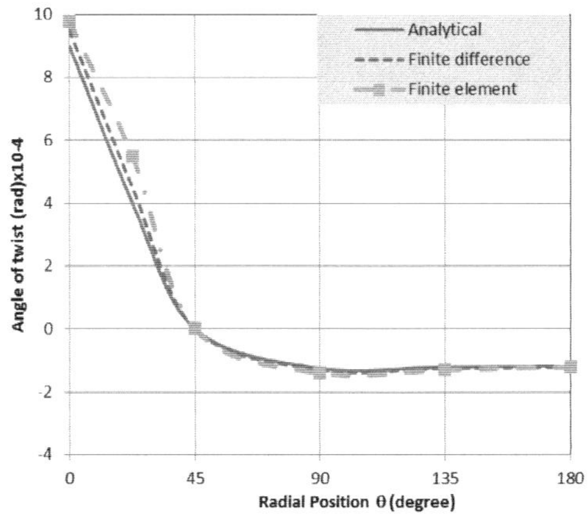

Figure 7. Angle of twist about central axis verses point load position for case 1

5. Parametric study

The influence of subgrade reaction, ring depth and load type on the response of ring foundations were investigated. A case 1 ring foundation under equivalent uniform load (with value equal point load divided by the ring area) is considered in the parametric study.

5.1 Effect of subgrade reaction

The maximum ring settlement is found to be reduced as the soil subgrade reaction was enlarged as shown in Figure 8. When the soil subgrade reaction (multiplied by ring width) was enlarged from (1.7 to 6 MPa), the maximum settlement or deflection reduced by 72%. Figure 9 illustrates the variation of soil subgrade reaction with ring maximum bending moment. It is recognized that the maximum moment reduced with enlarging the subgrade modulus for the same range by 31%.

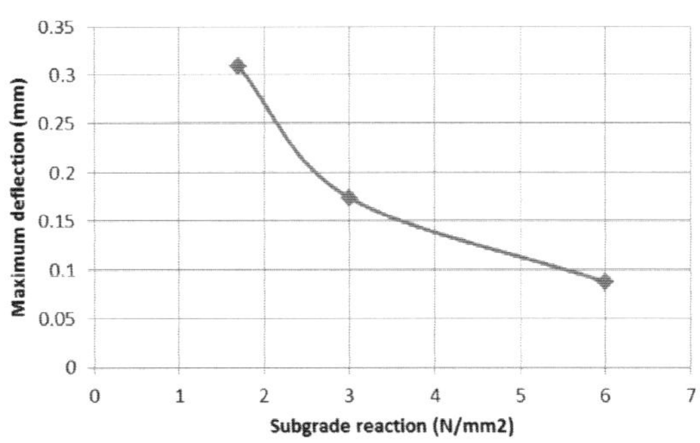

Figure 8. Variation of the maximum deflection with subgrade modulus

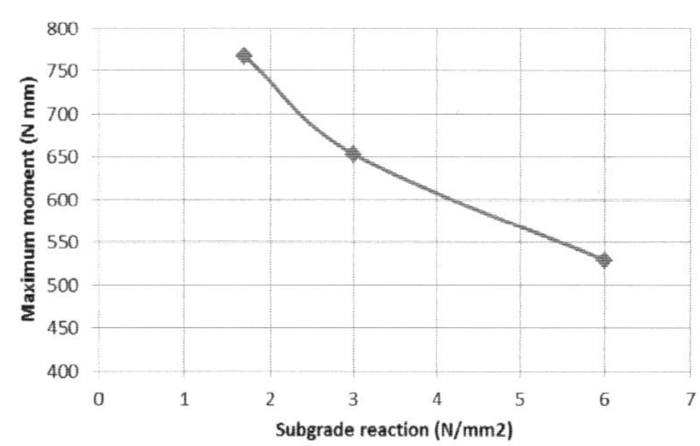

Figure 9. Variation of the maximum moment with subgrade modulus

5.2 Effect of ring depth

Figure 10 shows the variation of ring depth with ring maximum deflection. The maximum settlement or deflection had a lesser value when the ring depth was enlarged. If the ring depth enlarged from (305 to 915 mm), the maximum deflection decreased by 1% because the footing self-weight is neglected in the analysis. Figure 11 shows the variation of ring depth with ring maximum bending moment. It is recognized that the maximum moment increased with increasing ring depth by 52% and this is may be due to the increase in the flexural rigidity of the footing.

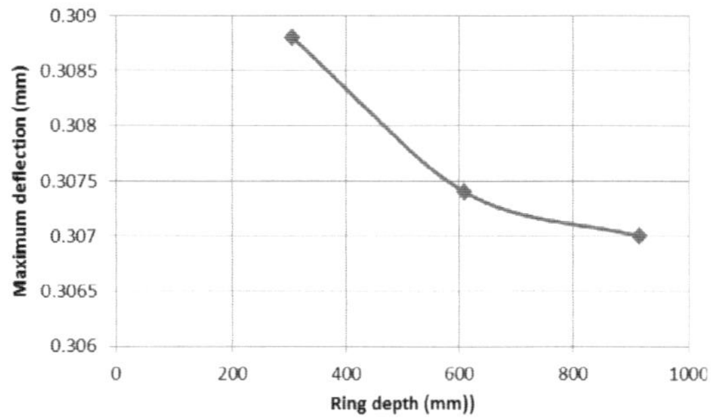

Figure 10. Variation of the maximum deflection with ring depth

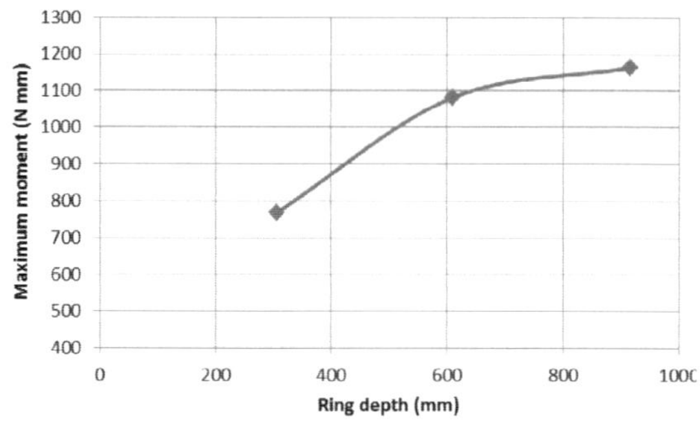

Figure 11. Variation of maximum the moment with ring depth

5.3 Effect of loading type

The effect of load type on ring maximum deflection was investigated. The maximum deflection had a reduction from (2.2 mm to 0.308 mm) or by 86 % when the loading changed from point to uniformly distributed.

6. Conclusions

Numerical simulation for the problem ring footings resting on Winkler foundation was developed herein using finite difference and finite element techniques as the closed-form solutions are limited to very simple applications. These techniques extended the application for analysing ring foundation with complicated geometry and loading.

The results of the previous analytical solution were compared with present numerical solutions and for several foundation sizes and varying positions for verification purposes. The numerical solution results are shown to be comparable with previous analytical solution. The maximum and average relative errors of 3 and 5% were obtained for finite differences and finite elements respectively.

A parametric study was made to study the behaviours of ring foundations under different loading types. The maximum deflection had a reduced value by 86% when the loading changed from point to uniformly distributed with equal resultant force. When the soil stiffness increased or the subgrade reaction increased from (1.7 to 6 MPa), the maximum deflection decreased by 72% and the maximum moment decreased with increasing subgrade by 31%. When the ring depth was increased from (305 to 915 mm), the maximum deflection decreased by 1% because the footing self-weight is neglected in the analysis. While the maximum moment increased with increasing ring depth by 52% and this is may be due to the increase in the flexural rigidity of the footing.

References

[1] Mehrjardi G T 2008 *Bearing Capacity and Settlement of Ring Footings* The14thWorld Conference on Earthquake Engineering.

[2] Algin H M 2009 Elastic Settlement under Eccentrically Loaded Rectangular Surface Footings on Sand Deposits *American Society of Civil Engineers (ASCE)*, 139(10).

[3] Zhou F, Yu T X and Yang L 2012 Elastic behavior of ring-on-foundation *International Journal of Mechanical Sciences* 54: 38–47.

[4] Atalar C, Patra C R, Das B M, and Sivakugan N 2013 *Bearing Capacity of Shallow Foundation under Eccentrically Inclined Load* Proceeding of 18th International Conference on Soil Mechanics and Geotechnical Engineering., 1:10-13.

[5] Naseri M and Hosseininia E 2015 Elastic settlement of ring foundations *Journal of Soils and Foundations*;55(2):284–295

[6] Nayyeri S, Hajali M and Shdid C 2016 Ring foundation on elastic subgrade: an analytical solution for computer modelling using the Lagrangian multiplier method *International Journal for Numerical and Analytical Methods in Geomechanics*

[7] Keshavarz A, and Kumar J 2017 Bearing capacity computation for a ring foundation using the stress characteristics method *Computers and Geotechnics*, 89: 33–42

Properties of different types of concrete containing waste tires rubber- a review

A S Ali[1] and T M Hasan[2,*]

[1,2] Civil Engineering Department, Al-Nahrain University, Baghdad, Iraq.
*Corresponding author, e-mail: eng.tamara1993@gmail.com

Abstract. According to Rubber Manufacturer's Association, about 270 million waste tires are produced in United States each year. Also about 300 million scrap tires are stockpiled at the United States. Significant development in transportation and an enormous increment in the quantity of vehicles create various issues, for example, environmental contamination. Many research studies utilized scrap tires rubber in different types of concrete mixes such as normal concrete and self-compacting concrete to produce rubberized concrete and self-compacting rubberized concrete.
This review paper presents important results and conclusions that obtained from previous studies which used waste tires rubber as aggregates (fine and coarse aggregates). The results showed that the mechanical properties (compressive strength, splitting tensile strength, and flexural strength) decreased by adding rubber particles. Also the fresh properties negatively affected as rubber contain was increased but stay in the acceptable limits. The failure mode of concrete change from brittle to ductile by incorporating scrap tires rubber in concrete.

1. Introduction

The scrap tires rubber non-degradable by nature, so they are left as stockpiles. These stockpiles of tires rubber causing a great health and environmental risk worldwide. Tire fires can happen easily, burning for a long time (months) and creating contamination in the air and ground. Recycling helps to reduce the number of tires in storage. So many researches focused on the use of tires rubber as aggregates to produce sustainable concrete [1]. The use of rubber aggregates in different shapes and sizes saves natural resources, helps to maintain a clean environment, reduces unite weight of concrete, changes behavior of concrete from brittle to ductile, enhances the absorption of energy, and increases resilience [2]. Scrap tires can be used for manufacturing of Portland cement (used as a fuel). In addition, it can be used to produce crumb rubber modifier- recycled asphalt pavement (CRM-RAP) [3, 4]. Fine and coarse aggregate is replaced by waste tires rubber. Waste tires rubber is the main waste materials used to obtain rubberized concrete.

2. Classification of waste tires rubber

The cracker mill process tears apart or reduces the size of waste tires rubber by passing the tires between rotating corrugated steel drums. By using this process, particles in different shapes and sizes are produced. Table 1 and figure 1 show the types of waste tires rubber [5].

Table 1. Waste tires rubber classification

Types	Size	Application in concrete
Shredded/chipped	Shredded Stage (1) Length: 300mm-430 mm Width: 100mm-230mm Stage (2) Length: 100mm-150mm Chip 13mm- 76mm	Gravel replacement
Crumb rubber	0.425mm- 4.75mm	Sand replacement
Ground rubber	0.0075mm- 0.475mm	Cement replacement
Fiber rubber	Length: 8.5mm-21.5mm	Reinforced fiber

Figure 1. Results of water absorption [5]

3. Rubberized concrete (RC)

It is a new structural materials using recycled tires rubber as partial replacement for aggregates (fine and coarse aggregates) in a Portland cement concrete. The percentages of rubber replacement usually ranges from 5% to 25% by weight or volume of aggregates and sometimes reach to 50% and 75%. The percentages more than 25% lead to more negative effect on properties of concrete [6, 7].

3.1. Fresh and hardened properties

Güneyisi et al. (2004), [8] two types of waste tires rubber (chip and crumb rubber) were used instead of coarse and fine aggregates, respectively. The percentages of rubber ranging from 2.5% to 50% by volume of total aggregates. Silica fume was used instead of cement with percentages varying from 5% to 20%. The results showed that the strength decreased as rubber content increased. This reduction in strength can be eliminated by adding silica fume.

Ghaly and Cahill (2005), [9] in their studies fine aggregate (sand) was replaced by crumb rubber. The percentage of crumb rubber was 5%, 10%, and 15% by volume of mixture with water cement ratio 0.47. The test of compressive strength was carried out in 7 and 28 days. The results showed that the compressive strength decreased about 18.9%, 26.1%, and 56.8% in 7 days and about 25.9%, 47.5%, and 59.1% in 28 days when the rubber replacement was 5%, 10%, and 15%, respectively.

Aiello and Leuzzi (2010), [10] used chip tire rubber to replace coarse aggregates with size from 10mm to 25mm and plasticizer about 1% by weight of cement. The percentage of chip rubber was 25.5%, 50.5%, and 75.5% by volume of coarse aggregate. They observed that the workability increased as a chip tire content increased. The results also showed that the compressive strength decreased as rubber percentage increased. The diminution in compressive strength was 47.8%, 54.4, and 62.3% when the rubber replacement was 25.5%, 50.5%, and 75.5%, respectively.

Ganjian et al. (2012), [5] in their studies coarse aggregate was replaced by chip rubber with percentage 5%, 7.5%, and 10% by weight of coarse aggregate. The results displayed that the compressive strength, splitting tensile strength, and modulus of elasticity decreased with increased chip rubber content. The diminution in compressive was about 10% and 23% when the replacement of rubber was 7.5% and 10%. The reduction in modulus of elasticity and splitting tensile strength was about 17%, 24%, and 25% and about 30%, 40%, and 60%, respectively, when the rubber replacement was 5%, 7.5%, and 10%.

Antil (2014), [6] investigated the effects of utilizing crumb rubber as sand in various percentages of replacement (0%, 5%, 10%, 15% and 20%) (by volume). They observed that concrete containing higher proportion of tiers rubber has high toughness. The slump value of concrete increased about 1.08%, with the crumb rubber replacement of 10%. The results also showed the splitting tensile strength, compressive strength, and modulus of rupture of the concrete decreased about 30%, 37%, and 31%, respectively, when 20% fine aggregate was replaced by rubber (crumb rubber).

Ali and Saravanan (2015), [7] in their studies sand (fine aggregate) was replaced by a single gradation of crumb rubber (0.9 mm) varying from 5% to 20% by weight of fine aggregate. The test results demonstrated that the value of slump decreased with the increase in rubber content. The test results also showed that the addition of rubber aggregate led to a decrease in compressive strength and splitting tensile strength of concrete when compared with the reference mix. The reduction in compressive and splitting tensile strength were 29.5% and 60.8% respectively at 28 days of curing (20% replacement by weight of fine aggregate). The increasing in modulus of rupture (flexural strength) was obtained with rubber replaces 10% of the fine aggregate.

Khorami and Ganjian (2015), [11] studied the mechanical properties such as flexural strength, compressive strength, and splitting tensile strength by replacing coarse aggregate with chip rubber (0, 5, 7 and 9%) by weight of gravel. The results displayed that the values of compressive strength increased about 5% with percentage of rubber 5%. The compressive strength values decreased about 10% and 23% with rubber replacement 7% and 9% when compared with the control mix. The reduction in tensile strength was observed up to 7 % replacement. Flexural strength did not significantly affect with the rubber percentage of 5%.

Bhatt et al. (2017), [12] used crumb rubber as a sand replacement with percentage 0%, 2.5%, 5%, 7.5% and 10 % by weight of fine aggregate. They observed that the percentage of crumb rubber about 10% exhibited a linear response between the crumb rubber increasing and the compressive strength. The flexural strength comes optimum when 5% of crumb rubber replaced by fine aggregate.

4. Self-compacting rubberized concrete (SCRC)

Self-compacting concrete (SCC) differs from the ordinary concrete that the fresh properties (flow ability, ability of passing, and resistance to segregation) must be checked to be classified as a self-compacting concrete. The hardened properties are similar. European guideline (EFNARC 2005) [13] has developed several different tests to obtain the three fresh properties of self-compacting concrete. The tests of SCC according to EFNARC 2005 are listed in table 2.

Table 2. Tests of SCC

Characteristic	Test Method	Measured value
Flowability/filling ability	Slump flow	Total spread
	Kajima box	Visual filling
Viscosity/flowability	T_{500}	Flow time
	V-funnel	Flow time
	O-funnel	Flow time
	Orimet	Flow time
Passing ability	L-box	Passing ratio
	U-box	Height difference
	J-ring	Step height, total flow
	Kajima box	Visual passing ability
Segregation resistance	Penetration	Depth
	Sieve segregation	Percent laitance
	Settlement column	Segregation

Self-Compacting Rubberized Concrete (SCRC) is a kind of self-compacting concrete (SCC) with the partial substitution of aggregates (fine and coarse aggregates) by rubber particles made from the scrap tires rubber. The studies about SCRC still limited because adding waste tires rubber to the concrete mixtures exhibits a decrease in strength. On the other hand, SCRC possesses several benefits such as the ability to absorb plastic energy, change failure mode from brittle to ductile, vibration absorption, and acceptable workability. SCRC considered eco-friendly and cheap concrete [14].

4.1. Fresh and hardened properties

Bignozzi and Sandrolini (2006), [15] investigated the properties of self-compacting concrete (SCC) containing ground rubber instead of fine aggregate. The rubber percentages was 0%, 22.2%, and 33.3% by volume of fine aggregate. They observed that the value of slump flow did not affect by adding rubber particles for both percentages. The value of J ring test also did not effected when the rubber percentages was 22.2% but increased with the 33.3%. The results also showed that the compressive strength decreased as rubber contain increased.

Topçu and Bilir (2009), [16] in their studies the aggregates was replaced by rubber particles with rubber contents of 180, 160, and 120 kg/m^3 by weight in self-compacting concrete. The results showed that the slump flow value slightly increased as rubber content increased by adding chemical admixtures. The value of slump flow ranging from 680mm to 800mm. The value of V-funnel falls above or under acceptance criteria of European guideline. The time of V-funnel test ranging from 4.4 to 15.3 sec. The compressive strength decreased by adding rubber particles.

Najim and Hall (2012), [17] investigated the fresh and hardened properties of SCC by utilizing crumb rubber (2mm-6mm size) instead of fine aggregate, coarse aggregate, and both them. The rubber replacement was 15, 10, and 5% by weight. They observed that the best results of compressive strength can be obtained from FCR (fine and coarse aggregates replacement). The flexural strength decreased about 12% by utilizing 5% waste tires rubber (for all types of mixes). The reduction in flexural strength was 40% by using 15% waste tires rubber (for all types of mixes). For splitting tensile strength, the results observed that the splitting tensile strength by using rubber instead of coarse aggregates gives higher splitting tensile strength than other mixes with rubber replacement 5%.

Ismail and Hassan (2015), [18] used crumb rubber as a fine aggregate. The rubber replacement was 0, 5, 10, and 15% by weigh of fine aggregate. The results showed that the slump flow diameter decreased from 700mm for SCC to 675mm for SCC containing 15% rubber particles. L-box ratio decreased from 0.89 to 0.75 and time of V-funnel increased from 6.39sec to 8.75sec. The results also showed that the compressive strength and splitting tensile strength decreased about 30%, 16%, and 14% and about 31%, 24%, and 16% respectively by utilizing 15%, 10%, and 5% rubber replacement.

Güneyisi et al. (2016), [19] utilized crumb rubber and chips rubber instead of fine aggregate and coarse aggregate in self-compacting concrete. The rubber replacement was 5, 10, 15, 20, and 25% by weight of aggregates. The tests results showed that the diameter of slump flow ranging from 680mm to 750mm and from 560mm to 710mm in the mixes containing crumb rubber and chips rubber, respectively. The compressive strength decreased by adding rubber particles but when the rubber replacement was 25% the compressive strength more than 31MPa can be achieved with cement content about 364 kg/m^3.

Padhi and Panda (2016), [20] prepared SCC mixes containing waste tires rubber by weight of fine aggregate. The rubber replacement was 5, 10, and 15% by weight of fine aggregate. The size of rubber particles passing through 4.75 mm sieve. Fresh and hardened tests were performed at 28 days of curing. The results showed that the workability of self-compacting rubberized concrete (SCRC) decreased by using scrap tires rubber. The compressive strength and splitting tensile strength decrease from 45MPa to 27MPa and from 4Mpa to 2.2MPa, respectively, for SCC and SCC containing 15% of rubber particles. The modulus of rupture increased when the percentage of crumb rubber was 5%.

Hilal (2017), [21] in her study two size of rubber particles was used, the first one passing from 1mm sieve (No.18) and the second passing from 4mm sieve (No.5). The rubber percentages was 5, 10, 15, 20, and 25% by weight of fine aggregate. The reduction in compressive strength was 31% and 46% by using No.18 and No.5 crumb rubber, respectively. The splitting tensile strength decreased about 45% and 48% by using No.18 and No.5 crumb rubber, respectively. The results also showed that the flexural strength decrease from 5.6MPa to 3.3MPa by utilizing No.18 crumb rubber, and from 5.6Mpa to 3.1MPa by utilizing No.5 crumb rubber.

5. Water absorption

Bignozzi and Sandrolini (2006), [15] investigated the properties of self-compacting concrete (SCC) containing ground rubber instead of fine aggregate. The rubber percentages was 0%, 22.2%, and 33.3% by volume of fine aggregate. The test results showed that the water absorption increased about 0.8% when the rubber replacement 33.3%.

Gesoğlu and Güneyisi (2011), [22] examined the water absorption capacity of SCC containing crumb rubber instead of fine aggregate. The replacement was 0, 5, 15, and 25% by volume of fine aggregate. The test results observed that the water absorption increased as rubber content increased. The water absorption ranging from 2.8% to 4.3% and from 2.6% to 4.0% at 28 and 90 days, respectively.

Padhi and Panda (2016), [20] used crumb rubber in NC and SCC passing through 4.75mm with rubber replacement 5, 10, and 15% by weight of fine aggregate. The results showed that the water absorption increased from 0.35 to 0.59 for NC and increased from 0.25 to 0.92 for SCC.

Bhatt et al. (2017), [12] used crumb rubber as a sand replacement with percentage 0%, 2.5%, 5%, 7.5% and 10 % by weight of fine aggregate. The results showed that the water absorption increases as crumb rubber proportion increases as shown in figure 2.

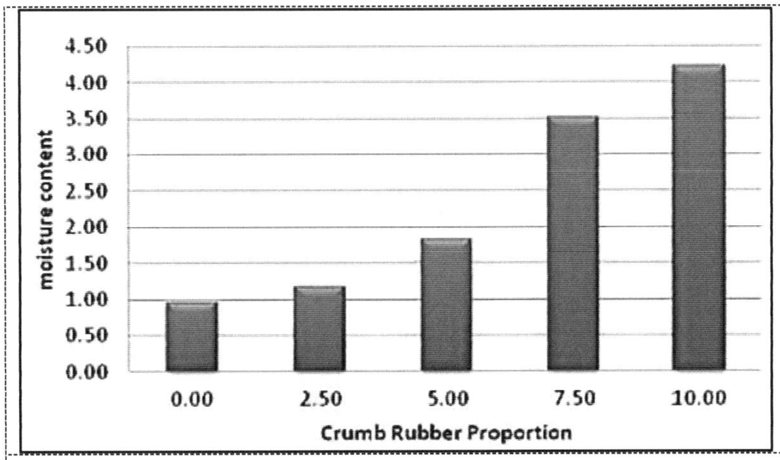

Figure 2. Results of water absorption **[12]**

6. Stress-strain relationship for RC

The addition of waste tire rubbers improved the post-cracking resistance of concrete by absorbing energy. This type of concrete can be used for applications where impact or blast resistance was needed, such as bunkers and jersey barriers, or where vibration damping is required such as foundation pads in railway stations. Due to the positive action of rubber particles [23].

Benazzouk et al. (2003), [24] showed that the ductility in the stress-strain relation increased by adding rubber particles and increasing size of particles. The BI (brittleness index) was computed to evaluate the ductility of concrete as shown in figure 3. The peak value was obtained when the percentage of rubber was 10%. After this percentage, the failure mode change from brittle to ductile. The decline in brittleness index values with rubber percentage more than 10 % lead to increase plastic energy.

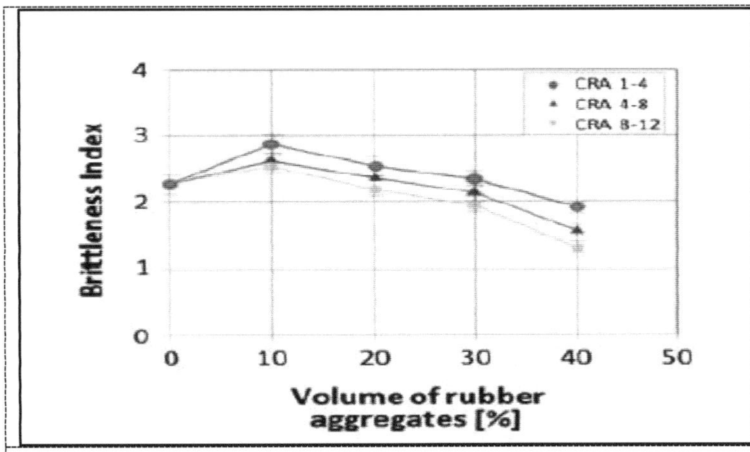

Figure 3. Brittleness index of various types of concrete

Batayneh et al. (2007), [1] studied the behavior of the stress-strain curve for rubberized concrete. Fine aggregate was replaced by rubber (0.075-4.75 mm).The stress–strain behavior of the mix still brittle with rubber percentage 40 % similar to that of the reference specimen, but with rubber percentage 60% and 80% stress-strain curve behavior transition from brittle to ductile failure mode as shown in figure 4.

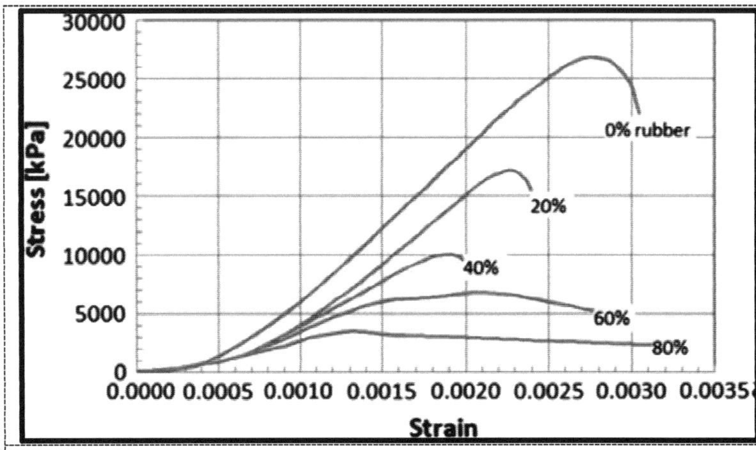

Figure 4. Stress-strain behavior of concrete with various percentages of rubber aggregates

Aiello and Leuzzi (2010), [6] an enhancement in the post-cracking resistance of concrete was achieved by adding waste tires rubber aggregates. Coarse aggregate was replaced by chip rubber with rubber proportion, 50% and 75% by volume of gravel (coarse aggregate). Toughness indices and residual strength factors increased with rubber content increased. The toughness indices and residual strength factors are presented in table 3. Residual strength factors, which are derived directly from toughness indices. It means the level of strength retained after first crack.

Table 3. Toughness indices of rubberized concrete

Amount of rubber content %	Toughness indices			Residual strength factor		Toughness
	I_5	I_{10}	I_{20}	$R_{5,10}$	$R_{10,20}$	
50	4.06	8.72	14.4	93.2	56.8	113
75	4.96	9.92	17.8	99.2	78.8	196

Antil (2014), [7] investigated the behavior of concrete containing 10 %, 15%, and 20% tires rubber. The stress–strain behaviors of the mix with rubber percentage 10% behave in a similar to the reference mix (brittle material). Nonlinear behavior was observed with the other two mixes containing 15% and 20% rubber. When the peak stress was reached, the specimen continued to yield (tough material) as shown in figure 5. The mix with a higher proportion of waste tires rubber has a high toughness.

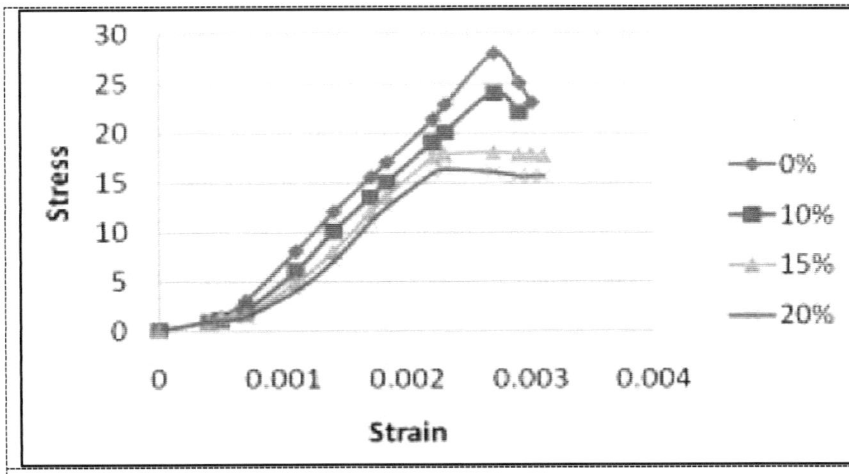

Figure 5. Relationship between stress and strain for different rubber contents

7. Conclusions

From the previous studies that mentioned above, concluded the following points:

- Scrap tires rubber can be used in normal concrete and self-compacting concrete.
- The workability decreased by using waste tires rubber.
- The compressive strength decreased as rubber content increased. Also the splitting tensile strength and flexural strength decreased by utilizing rubber particles.
- In some studies can be noticed the increment in flexural strength when the rubber replacement was 10% by weight of fine aggregate. The flexural strength comes optimum when 5% of crumb rubber replaced by fine aggregate.
- The incorporating rubber particles in concrete lead to convert the failure mode of concrete from brittle to ductile. Also the absorption energy (post-crack resistance) increased by adding scrap tires rubber.

References

[1] Batayneh M, Marie I, and Asi I 2007, *Waste Management* 27, 1870-1876

[2] Md Noor N 2014, *Physical Performance and Durability Evaluation of Rubberized Concrete. PhD. Dissertation* (Japan: University of Kyushu)

[3] Baker T E, Allen T M, Jenkins D V, Mooney T, Pierce L M, Christie R A, and Weston J T 2003, Report to the Legislature as Required by SHB,2308, 1-14

[4] Chatziaras N, Psomopoulos C S, and Themelis N J 2016, *Management of Environmental Quality an International Journal* 27, 178-193

[5] Ganjian E, Khorami M, and Maghsoudi A A 2009, *Construction and Building Materials* 23, 1828-1836

[6] Antil E Y 2014, *International Journal of Emerging Technology and Advanced Engineering* 4, 309-316

[7] Ali A M, and Saravanan A 2015 *Int. Conf. on Engineering Trends and Science and Humanities (ICETSH)*, 60-65

[8] Güneyisi E, Gesoğlu M, and Özturan T 2004, *Cement and concrete research* 34, 2309-2317
[9] Ghaly A M, and Cahill IV J D 2005, *Canadian Journal of Civil Engineering* 32, 1075-1081.
[10] Aiello M A, and Leuzzi F 2010, *Waste Management* 30, 1696-1704
[11] Khorami M, and Ganjian E 2015, *Mechanical Properties of Concrete with Waste Tire Rubber as Coarse Aggregates*, https://www.researchgate.net/publication/266450615
[12] Bhatt B, Khandle P, and Kauswala T 2017, *Global Research and Development Journal for Engineering* 2, 151-155
[13] EFNARC 2005 (European Federation of National Associations Representing for Concrete). The European guidelines for self-compacting concrete specification, production and use
[14] Najim K B, and Hall M R 2010, *Construction and building materials* 24, 2043-2051
[15] Bignozzi M C, and Sandrolini F 2006, *Cement Concrete Research* 36, 735–739
[16] Topçu I B, and Bilir T 2009, *Materials and Design* 30, 3056–3065
[17] Najim K B, and Hall M R 2012, *Construction and Building Materials* 27, 521–530
[18] Ismail M K, and Hassan A A 2016, *ACI Materials Journal* 113, 207
[19] Güneyisi E, Gesoglu M, Naji N, and İpek S 2016, *Archives of civil and mechanical engineering* 16, 9-19
[20] Padhi S, and Panda K C 2016, *Advances in Concrete Construction* 4, 049-069
[21] Hilal N N 2017, *International Journal of Sustainable Built Environment* 6, 191-206
[22] Gesoğlu M, and Güneyisi E 2011, *Construction and building materials 25*, 3319-3326
[23] De Brito J, and Saikia N 2012 *Recycled Aggregate in Concrete: Use of Industrial, Construction and Demolition Waste* (Springer Science & Business Media)
[24] Benazzouk A, Mezreb K, Doyen G, Goullieux A, and Quéneudec 2003, *Cement and Concrete Composites* 25, 711-720

IOP Conf. Series: Materials Science and Engineering **584** (2019) 012052 doi:10.1088/1757-899X/584/1/012052

Experimental study on torsional behavior of fiberious reifoced concrete beams with different concrete strength

Waleed A. Waryosh[1,*], Saad Khalaf mohaisen[1], and Raed Hamed Dkhel[1]

[1] civil engineering department, college of engineering, university of mustansiriyah, Baghdad, Iraq.

*corresponding author, email: waleed.eng73@gmail.com

Abstract. This study is intended to deal with twelve reinforced concrete beams cast into three categories of Concrete type with Normal, Self Compacted and High Strength Concrete. Two beams of each category have steel fibers and the other used as reference one. Test results are presented in graphs also listed in Tables. Test results indicated that, category of High Strength Concrete beams is the strongest category depending onto the ultimate load capacity and the highest stiffness, followed by the Self Compacted Concrete beams category then the Normal Strength Concrete beams category comes later, with an increase of about (3.6-5.3%). Increasing steel fibers contained in concrete beam section will rise the ultimate load capacity of fiberious concrete beam, an increase of about 36% has been obtained. Increasing steel fiber contained with increasing concrete compressive strength of the fiberious concrete section will help increasing the ultimate load capacity of beam, an increase of about 166% has been noted. Due to continuing in rising the steel fibers contained in fiberious concrete section, the rate of development of ultimate load capacity of the fiberious concrete beam will decrease.

1. Introduction

Practically, there are wide range of different structures where torsional loading could be a significant loading condition. The most noticeable are bridges and spandrel beams. In bridges, the torsion could be due to the geometric complexities of horizontally curved bridges and/or to large eccentric vertical loads. The concrete behaves as an elastic material and the reinforcement can be ignored [1]. In RC members under pure torsion, the stiffness of the uncracked member can be predicted by theoretical approaches. After cracking, the member behaves as a composite member and the properties of the concrete, the reinforcing, and their interaction must be considered to accurately predict the member response to torsion. Many buildings and bridge elements are subjected to significant torsional moments that affect the design. Clearly, further research was needed to address this gap in knowledge [2-8].

Self Compacting Concrete is a type of concrete that does not require external or internal compaction, because it becomes leveled and compacted under its self weight. SCC can spread and fill every corner of the formwork, purely by means of its self weight, thus eliminating the need of vibration or any type of compacting effort [9]. The notion behind developing SCC was the concerns regarding the homogeneity and compaction of conventional cast-in-place concrete within intricate (i.e., heavily-reinforced) structures and to improve the overall strength, durability and quality of concrete [10]. An increasing in torsional capacity of concrete beam sections can be obtained using modified technique of strengthening [11,12].

Fiber Reinforced Concrete (FRC) is a composite material in which steel fibers or other tension materials are incorporated to improve the tensile strength and other properties of concrete. The amount of fibers

IOP Conf. Series: Materials Science and Engineering **584** (2019) 012052 doi:10.1088/1757-899X/584/1/012052

added to a concrete mix is measured as a percentage of the total volume of the composite (concrete and fibers) termed (V_f). Typically, the amount of steel fibers varies between (0.0) and (2.0) percent, by volume [13]. The advantage of adding fibers. A matrix include enhancement of compressive strength, tensile strength, flexural toughness, shear strength, durability and resistance to impact [14].

2. Research casting plan

This study aims to investigate different parameters on the torsional behavior of reinforced rectangular concrete section beams. Such as:

1-type of concrete (Normal, high strength and self compacted)
2-concrete strength (20, 40, 60 and 75MPa)
3-percentage of fiberious (0, 0.6 and 1.2%)

It have been cast 12 beams with details as shown in Figure1 and Table 1;

a. Dimensions of (width=100, depth=200 and length=2000mm)
b. Longitudinal reinforcement = 0.0094
c. Stirrups Spacing = 80mm c/c.
d. specimens of normal strength concrete =4
e. specimens of high strength concrete =4
f. specimens of self-compacting concrete =4
g. Steel fiber percentage= (0.6 and 1.2)% .

3. Properties of hardened concrete

Control specimens of cylinders and prisms are cast and tested at age of (28 days) to determine the properties of hardened concrete, as listed in Table1. The properties of these beams consist of cylindrical compressive strength (f'_c), modulus of elasticity (E_c) and modulus of rupture (f_r).

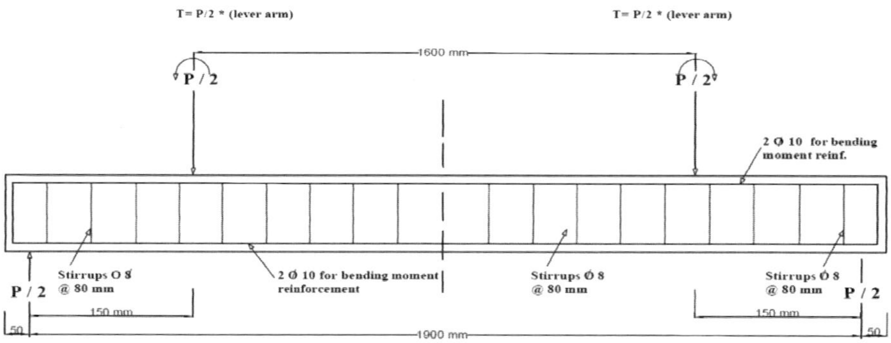

Figure1. Tested beams details

4. Results and discussion

Beams are tested under torsional moments come from eccentricity of the applied load by 750mm from the center of the beam as shown in Figure 2, the test continues up to failure, this failure mode leads to noticeable cracking and this is clear in Figure 3. Two dial gages are poisoned at the points of maximum torsional actions to measure the angle of twist. Another two dial gages are poisoned at the two ends of the tested beam to measure the longitudinal elongation of the beam due to formation and propagation of torsional cracks. The average of angles of twist and the total longitudinal strain are calculated for each tested beam. Test results are shown in Table 2.

IOP Conf. Series: Materials Science and Engineering **584** (2019) 012052 doi:10.1088/1757-899X/584/1/012052

Table1. Hardened concrete properties.

Mix Notation	Beam Symbol	Compressive Strength Nominal (MPa)	Steel Fibers (Vf)	Compressive Strength Tested (MPa)	Modulus of Elasticity (GPa)	Modulus of Rupture (MPa)
NSC	NCA	20	-	21.00	23.20	3.53
	NCB	40	-	39.34	24.33	4.73
	NCA$_{0.6}$	20	0.6%	22.04	23.64	5.76
	NCA$_{1.2}$	40	1.2%	40.22	24.86	7.65
HSC	HSCC	60	-	61.23	33.62	6.84
	HSCD	75	-	73.31	42.53	8.41
	HSCC$_{0.6}$	60	0.6%	65.36	37.87	8.44
	HSCC$_{1.2}$	60	1.2%	68.81	38.82	9.83
SCC	SCCB	40	-	40.7	24.05	4.59
	SCCC	60	-	68.32	39.85	7.76
	SCCB$_{0.6}$	40	0.6%	42.96	24.44	7.69
	SCCB$_{1.2}$	40	1.2%	44.15	24.45	8.08

Figure 2. Eccentricity Detailing at Testing

Figure 3. Deformation of a tested beam

Table 2. Results of tested beams.

Mix Notation	Beam Symbol	Nominal Compressive Strength, (MPa)	Steel Fibers (V_f)	Ultimate Load (P_u) (kN)	Torsional Moment (T), (kN.m)	Angle of Twist (Ø) (rad.)	Total Strain (ε), (%)
NSC	NCA	20	-	25	18.75	0.01165	0.0555
	NCB	40	-	38	28.50	0.01300	0.0440
	NCA$_{0.6}$	20	0.6%	31	23.25	0.01165	0.0471
	NCA$_{1.2}$	20	1.2%	34	25.50	0.01200	0.0471
HSC	HSCC	60	-	57	42.75	0.01570	0.0630
	HSCD	75	-	66.5	49.88	0.02120	0.0770
	HSCC$_{0.6}$	60	0.6%	62	46.50	0.03896	0.0590
	HSCC$_{1.2}$	60	1.2%	66	49.50	0.03320	0.0550
SCC	SCCB	40	-	40	30.00	0.01085	0.0395
	SCCC	60	-	55	41.25	0.01700	0.0630
	SCCB$_{0.6}$	40	0.6%	47	35.25	0.02220	0.0415
	SCCB$_{1.2}$	40	1.2%	51	38.25	0.01220	0.0405

5. Load carrying capacity (Pu)

The ultimate load carrying capacity reflects the maximum torsional moment and represents the ultimate applied load on the tested beam, Figure (4) graphs these values for the tested beams.

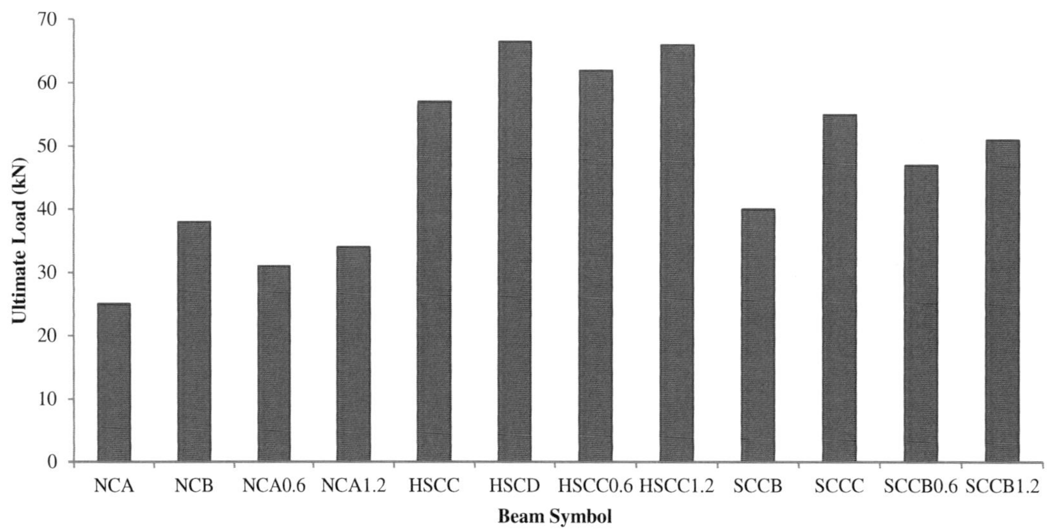

Figure 4. Load carrying capacity of the tested beams

5.1 Concrete type Effect on Ultimate Load

Three samples of concrete NSC, HSC and SCC were cast in recent study, when analysis the three types, it can be seen that, the HSC group is the strongest group depending on the ultimate load capacity and the second group is the SCC group and the NSC group comes later. Also, by contrast between the two beams, the normal strength concrete R-NCB and the self compacted reinforced concrete beam SCCB it can be noted that, the two discussed beams had convergent compression strength equals to about 40MPa, it is found that the carrying capacity of SCCB beams is greater than that of R-NCB by about 5.3%. In same manner, the HSCC and SCCC and the two beams which haves 60MPa nominal compressive strength, the result is the carrying capacity of HSCC is greater than that of SCC by about 3.6%.

5.2 Compressive Strength Effect on Ultimate Load

The results indicated that by augment the concrete compressive strength of the beam, will increases ultimate carrying capacity clearly. However, the rate of increasing in ultimate load capacity of the samples decreases due to continuous increasing compressive strength of the beam.

5.3 Steel fibers Effect on ultimate Load Capacity

The result indicated that augment the steel fibers contained in the sample will increase ultimate load carrying capacity clearly. However, the rate of augment in the ultimate load capacity of the sample will decrease, due to contained in increasing steel fibers contained in the beam sample.

5.4 Improvement of Ultimate Load Carrying Capacity

This work contains three types of concrete with different concrete compressive strength and steel fibers content, the increasing in carrying load capacity will be compared to the reference of each group of concrete type and then will be compared with the weakest beam in this work which it R-NCA as shown in Table 3.

From Table 3, it can be noted that by increasing concrete compressive strength, ultimate load capacity can be improved. the biggest improvement (52%) is in the normal strength concrete samples. It may be noted also that, augment the concrete compressive strength is more effective than utilization of steel fibers in improving ultimate load carrying capacity, for example (52 and 24)% of R-NCB and NCA0.6 in Table 3. Counting rising steel fibers contained, the load carrying capacity may be increased, as shown in Figure 5.

Table3. Improvement of load carrying capacity.

Mix Notation	Beam Symbol	Nominal Compressive Strength (MPa)	Steel Fibers (V_f)	Ultimate Load (P_u) (kN)	Increasing According to the Reference Beam (%)	Increasing According to NCA Beam (%)
NSC	*R-NCA	20	-	25	-	-
	R-NCB	40	-	38	52	52
	$NCA_{0.6}$	20	0.6%	31	24	24
	$NCA_{1.2}$	20	1.2%	34	36	36
HSC	*HSCC	60	-	57	-	128
	HSCD	75	-	66.5	16.7	166
	$HSCC_{0.6}$	60	0.6%	62	8.8	148
	$HSCC_{1.2}$	60	1.2%	66	15.8	164
SCC	*SCCB	40	-	40	-	60
	SCCC	60	-	55	37.5	120
	$SCCB_{0.6}$	40	0.6%	47	17.5	88
	$SCCB_{1.2}$	40	1.2%	51	27.5	104

* this symbols refers to the reference beam in each group

Also, "from Table 3, it can be emphasized that, when comparison adopted according to specific compressive strength. By continuing rising the compressive strength of the section, the load carrying capacity will improve, so that, the HSC beams achieved the highest values (166%). In addition the compressive strength is better than the steel fibers in improvement the load carrying capacity, as shown in Figure 6. "

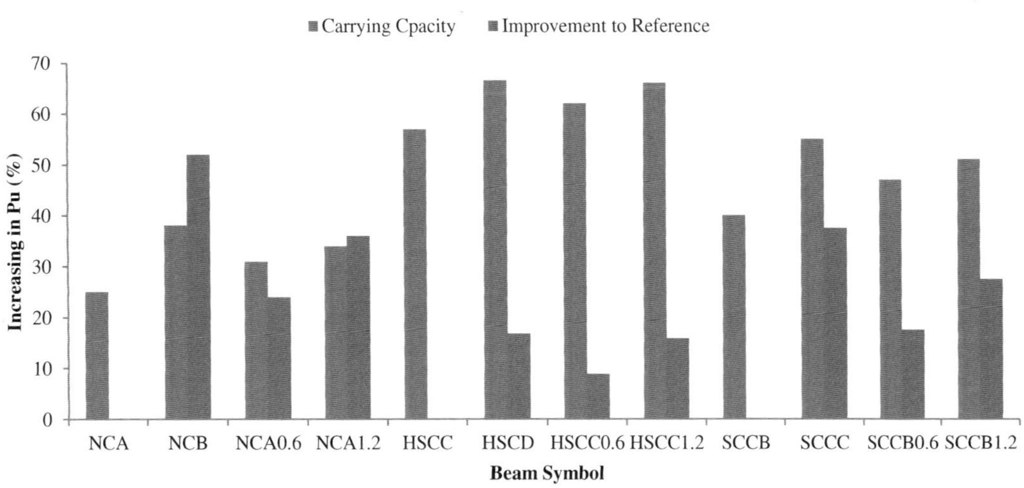

Figure 5. Improving of load carrying capacity in each type of beams.

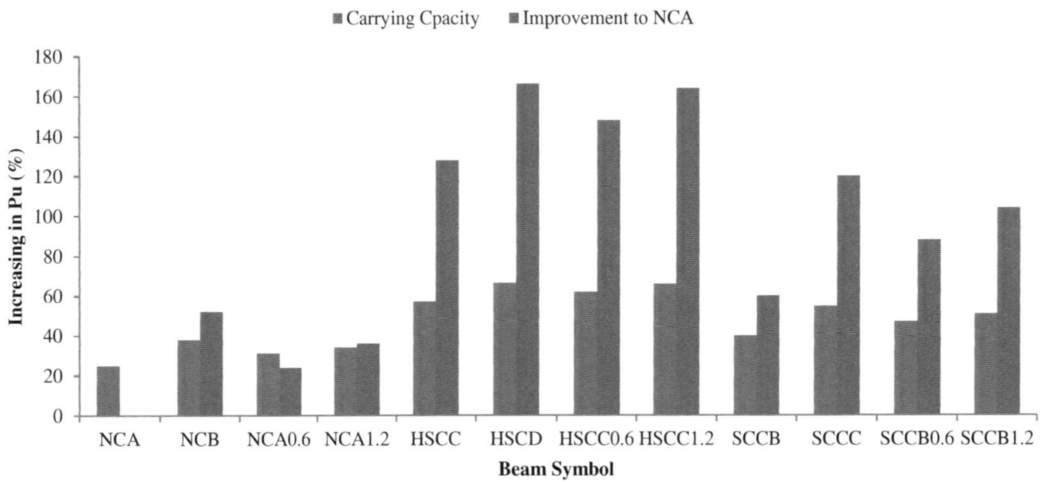

Figure 6. Improving of load carrying capacity according to NCA beam.

6. Twist Angle (Ø)

Average of two twist angles in every tested sample is graphed vs. torsional moment as shown in Figure 7, twist angle is two dimensional deformation with direction of torsional moment path. The comparative of twist angle for all tested samples are at 25kN which was the load of failure of beam sample R-NCA.

6.1. Effect of Concrete Type on Twist Angle

From tests results of three concrete type groups which are normal strength and high strength and self compacted concrete, it may be seen that, the HSC samples of beams is the high stiffness samples according to the twist angle and then the SCC samples and after that the NSC samples. Also, if it has been compared between two samples of beams R-NCB which was concrete beam of normal strength and SCCB which was beams cast of self-compacted concrete where the selected beams samples have an approximate concrete compressive strength near 40MPa, it is found at the applied load of 25kN the twist angle of sample SCCB was 0.0051rad, which is smaller than that of beam sample R-NCB which was 0.0066 rad. Using the same manner, by comparison between beam sample HSCC which was beam with high strength concrete and beam sample SCCC which was beam sample cast with self-compacted concrete where both beams samples haves about 60MPa concrete compressive strength, test result was twist angle of HSC at applied load 25kN is 0.00235 rad. which less than that of SCC which was 0.00295 rad.

6.2. Effect of Concrete Compressive Strength on Twist Angle

In normal strength concrete samples, by comparative analysis between beam sample R-NCA which was about 21.00MPa concrete compressive strength and R-NCB which was about 39.34MPa concrete compressive strength, the test result indicated that the twist angle of beam sample R-NCB is less than that of beam sample R-NCA which they were 0.0066rad. and 0.0116 rad. respectively.

In the HSC beam samples, due comparison between HSCC which was 61.23MPa concrete compressive strength and HSCD which was 73.31MPa concrete compressive strength, test results indicates that, the twist angle of beam sample HSCD is less than that of the beam sample HSCC which they were 0.00216rad. and 0.00235rad. respectively.

For the SCC group, by comparison between SCCB which is 40MPa nominal compressive strength and SCCC which a 60MPa nominal compressive strength, the result is the angle of twist of SCCC is smaller than that of the SCCB which they were 0.00235 rad. and 0.0051 rad. respectively.

Test result indicated that by augment the compressive strength of the beam sample will decrease the twist angle clearly.

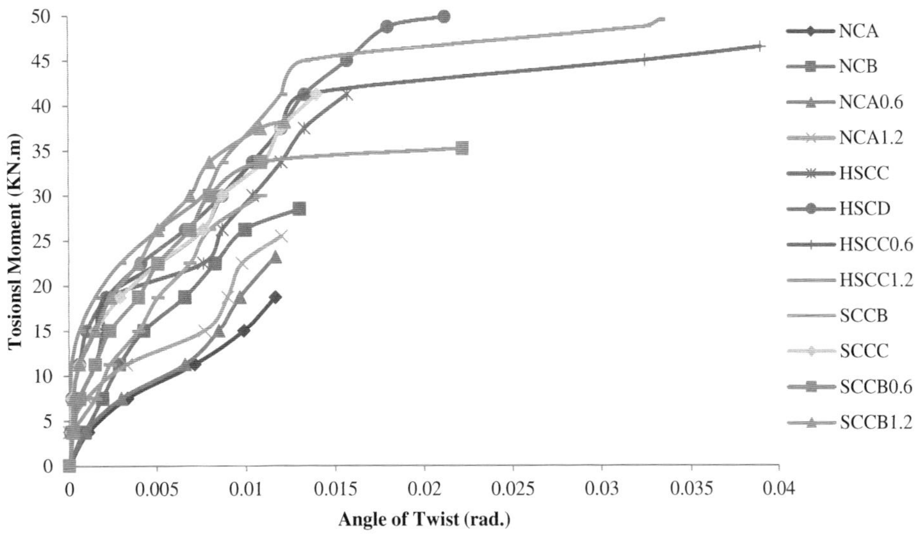

Figure7. Twist angle vs. Torsional moment for all tested beams.

6.3. Effect of Steel fibers on Twist Angle

Tests result indicated that augment of steel fibers contained in concrete beam section will decrease the value of twist angle of concrete beam samples clearly.

6.4. Improvement of Angle of Twist

In the case of the angle of twist, the improvement is decreasing in value which reflects the deformation in the member. Decreasing in angle of twist will be compared according to the angle of twist of the reference beam of each group of concrete type. After then, the decreasing in angle of twist of any beam will be compared with the angle of twist of the weakest beam in this work which it is R-NCA at applied load of 25kN as in Table 4.

Table 4. Improvement of angle of twist.

Mix Notation	Beam Symbol	Nominal Compressive Strength (MPa)	Steel Fibers (V_f)	Angle of Twist (Ø) (rad.)	Decreasing According to the Reference Beam (%)	Decreasing According to NCA Beam (%)
NSC	* NCA	20	-	0.01165	-	-
	NCB	40	-	0.00660	43.3	43.3
	$NCA_{0.6}$	20	0.6%	0.00968	16.9	16.9
	$NCA_{1.2}$	20	1.2%	0.00900	22.7	22.7
HSC	* HSCC	60	-	0.00235	-	79.8
	HSCD	75	-	0	100	100
	$HSCC_{0.6}$	60	0.6%	0	100	100
	$HSCC_{1.2}$	60	1.2%	0	100	100
SCC	* SCCB	40	-	0.00510	-	56.2
	SCCC	60	-	0.00295	42.2	74.7
	$SCCB_{0.6}$	40	0.6%	0.00400	21.6	65.7
	$SCCB_{1.2}$	40	1.2%	0.00240	52.9	79.4

From Table 4, that is clear, the angle of twist can be improved by increasing the compressive strength of the section, therefore; the highest improvement (100%) is in the HSC group. On the other hand, the increase of the compressive strength is better than the use of steel fibers in improving the angle of twist,

for example (43.3 and 16.9)% of NCB and $NCA_{0.6}$ in the Table 4 above. By counting in increasing the steel fibers content, the angle of twist can be decreased but not as increasing the compressive strength of the beam, for example the two beams NCB and $NCA_{1.2}$ when reflected the results (43.3 and 22.7)%, Figure 8 shows these results.
Also, from Table 4 can be emphasized that, when comparison adopted according to specific compressive strength, by continuing in rising the compressive strength of the section the angle of twist will improve, so that, the HSC beams achieved the highest values (100%) and also, the compressive strength is better than the steel fibers fraction in improving the angle of twist, Figure 9 shows these notices.

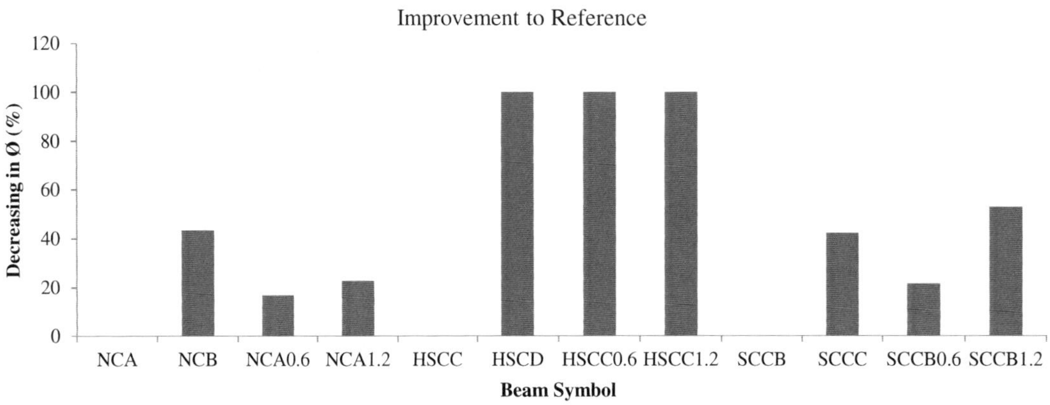

Figure 8. Twist angle improvement in each types of beams.

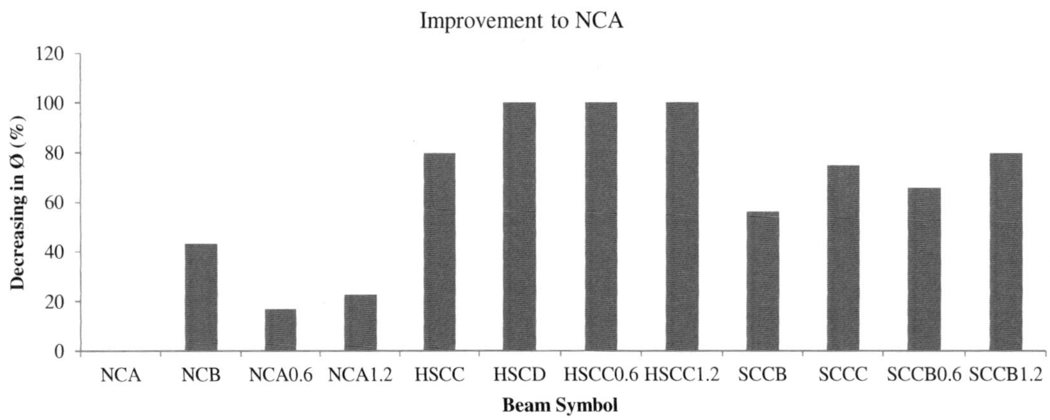

Figure 9. Twist angle Improvement according to R-NCA beam

7. Conclusions

Based on the evidence from the experimental results reported in study, the following conclusions can be drawn. It is emphasized that these conclusions are limited to the variables studied:

- The HSC group of beams is the strongest group according to the ultimate carrying load and the most stiffness and followed by the SCC group and the NSC group comes later, with an increase of about (3.6-5.3%)
- Increasing compressive strength of the concrete section will help increasing the ultimate carrying capacity of beam clearly, an increase of about 166% has been noted.

- By continuing in increasing the compressive strength of the section, the rate of increasing the load carrying capacity of the beam will decrease .
- Increasing the steel fibers content in the section will increase the ultimate carrying capacity of beam, an increase of about 36% has been obtained.
- By continuing in increasing the steel fibers content in the section, the rate of increasing the load carrying capacity of the beam will decrease .
- Increasing the compressive strength of the section will decrease the twist angle of about 100%.
- Increasing steel fibers content in the section will decrease the value of angle of twist of the beam.

References

[1] Collins, M. P., "Stress-Strain Characteristics of Reinforced Concrete in Pure Shear", Inelasticity and Non-Linearity in Structural Concrete, Study No. 8, University of Waterloo Press, 1973, PP. 211-225.

[2] Ghobarah, A., Ghorbel, M. N., and Chidiac, S. E., "Upgrading torsional resistance of reinforced concrete beams using fiber reinforced polymer", J. Compos. Constr., 6(4), 2002, PP. 257–263.

[3] Hii, A. K. Y., and Al-Mahaidi, R., "Torsional strengthening of solid and boxsection RC beams using CFRP composites", Composites in Construction 2005-Third Int. Conf., Lyon, France, Université Lyon I, Lyon, 2005, PP. 59–68.

[4] Hii, A. K. Y., and Al-Mahaidi, R., "Experimental investigation on torsional behaviour of solid and box-section RC beams strengthened with CFRP using photogrammetry", J. Compos. Constr., 10(4), 2006, 321–329.

[5] Salom, P. R., Geirgely, J. and Young, D. T., "Torsional strengthening of spandrel beams with fiber-reinforced polymer laminates", J. Compos. Constr., 8(2), 2004, PP. 157–162

[6] Ban S. A., Ali H. A., & Israa K. M., "Strengthening of reinforced concrete beams under combined torsion and bending using carbon fiber reinforced polymer strips", The Iraqi journal for mechanical and material engineering, Vol. 12, No. 4, (2012), pp. 754-769.

[7] Khalil A., Etman E., Atta A., & Fayed S., "Torsional strengthening of RC box beams using external prestressing technique", IOSR, Journal of mechanical and civil engineering (IOSR-JMCE), Vol. 12,No. 2, (2015), pp. 30-41.

[8] Abeer A. M., Allawi A. A., & Chai H. K., "Theoretical study on torsional strengthening of multi-cell RC box girders", World Acad-emy of Science, Engineering and Technology, International journal of civil science and engineering, Vol. 7, No. 2, (2013), pp. 155-122.

[9] Okamura, H., "Self-Compacting High Performance Concrete", ACI Concrete International, Vol. 19, No. 7, July 1997, PP.50-54.

[10] Campion, J. M., and Jost, P., "Self-Compacting Concrete: Expanding the Possibility of Concrete Design and Placement", ACI Concrete International, Vol. 22, No. 4, April 2000, PP. 31-34.

[11] Ali H. A. & Oday H. H., " Torsional strengthening of prestressed self-compacting concrete box beams using internal transverse con-crete diaphragms technique", Journal of engineering and sustaina-ble development, Vol. 21 , No. 5, (2017), pp.129-145.

[12] Ali H. A. & Oday H. H., "Torsional strength evaluation of rein-forced SCC box beams strengthened internally by opened and closed transverse concrete diaphragms", Proceeding of 3rd international conference on buildings, construction and envi-ronmental engineering-Egypt, 23-25 October, (2017).

[13] Shah, R. H., and Mishra, S. V., "Crack and Deformation Characteristics of SFRC Deep Beams", Journal of Indian Engineering, Vol. (85), May 2004, PP. 44-48.

[14] Gebman, M., "Applications of Steel Fiber Reinforced Concrete in Seismic Beam-Column Joints", M. Sc. Thesis, Faculty of San Diego State University, USA, 2001, (Web Site).

Evaluation of the Dynamic Behavior of Green Concrete under Impact Load by Using CEB-FIP Code

Zainab H. Abdulabbas[1,*], Jumana H. Sahib[2], and Nabaa N. Bashboosh[3]

[1] Civil Engineering Department, Faculty of Engineering, University of Kufa
Najaf- Iraq. Email: zainab.alhasnawi@uokufa.edu.iq
[2] Civil Engineering Department, Faculty of Engineering, University of Kufa
Najaf- Iraq. Email: jumanah.alsabarawi@uokufa.edu.iq
[3] Civil Engineering Department, Faculty of Engineering, University of Kufa
Najaf- Iraq. Email: nabaan.shamsaldeen@uokufa.edu.iq

***corresponding author.**

Abstract. Growing environmental pollution, rapid depletion of natural resources (virgin), need for larger areas of landfill and the increased cost of landfill are the factors that led to focus on the recycling of concrete and the demolition of waste in the new concrete. The principle of recycling for sustainable construction was adopted during the present study by recycling plastic and glass waste as an alternative to aggregates in concrete. Eight concrete mixes were tested, and production of green concrete with waste content of 35% and residual strength about 83% was achieved. The dynamic properties of green concrete under impact load were calculated. Dynamic properties are very useful in the design of civil structures exposed to impact loads such as a gas explosion, runways, etc. CEB-FIP Model code (2010) show the most inclusive forms for expecting the strain rate development of concrete. The dynamic of compressive and flexural calculated by CEB-FIP models under strain rates range 10^{-2} and 10^{0}. This range used to predict dynamic loads in structures of civil engineering within the quasi-static strain level.

1. Introduction

Industrial processes are accompanying with large quantities of non-disintegrated solid waste. It produces hazardous impacts on the environment. Plastic and glass waste are of the most widespread and most pronounced.

The consumption of these substances in other creations is an equitable technique for shrinking their harmful influences. Wastes materials have to be considered as valued supplies just require appropriate management and application [1]. Concrete represents a significant in take advantage of these materials in construction. Obtaining aggregates for concrete and waste disposal from different activities is the current goal. Sustainability is currently a top priority in the construction industry [2]. Utilization of plastic wastes in place of coarse aggregate is characterized with lighter weight in appraisal to normal aggregates, in addition to its elevated crushing resistance [3]. The glass –in concrete- can lead to a reaction and its effectiveness increases by increasing alkali content and particle size to greater than 1 mm. Particles less than 1 mm reduce the occurrence of alkali-silica reaction with the possibility of diminishing their effectiveness by using metakaolin, fly ash, or GGBS [4]. If the glass was crushed to less than 75 μm, the effect of the alkali-silica reaction did not appear, and the durability properties were confirmed. Therefore, the possibility of using glass waste as fine aggregates in concrete and mortar has been identified [5]. The demand to know the impact resistance and dynamic properties of concrete is important in many applications such as: offshore platforms, aircraft impact, transportation facilities, gas explosion, etc. One of important structure is ground storage tank. It is contain many types of liquid such as water, oil, chemicals,…, etc. some of these liquid classified as hazardous materials because any damage in tanks caused the catastrophic humanity and environment. These require the construction of structure nearby the tanks of storage to avoid the leakage in tanks as well as disasters failure. In many countries, a secondary containment as a bund wall was used. It is designed by BS EN 1992-3 [6];this code is dealing with hydrostatic pressure. When failure catastrophic happened, the liquid material was escaped very quickly. In this case, the bund wall falls under dynamic load. The loading impact classified into two major types according to the level of deformation happening in impactor, hard and soft impacts. In the first type, the deformation of the impactor is very unimportant according to that occur in the target. While, in soft impact, both the target and impactor having the same significant deformations as fluid flow beating a bund wall. The various velocities is another property for impact loading. It determined by the performance of structure under impact loading. Othman, [7] was stated that the loading impact more than 40m/s, a local response occurred in structure because the period time of acting load is very short. It is very small to the typical period structure. When the speed was less than 40 m/s, the global responding and damage do not have time to affect. In structures of civil engineering occurring in the low velocity impact during dynamic accidental loading. When the strain rate is high in materials its mechanical characteristics are improved (i.e, increasing in compressive strength, flexural strength, and the Elastic Modulus) [8]. The increasing in strain rate depended on the type of load as shown in the table. 1

Table. 1 Strain rate ranges for various loading types [7].

Load case	Strain rate (s^{-1})
Traffic	$10^{-6} - 10^{-4}$
Gas explosions	$5\times10^{-5} - 5\times10^{-4}$
Earthquake	$5\times10^{-3} - 5\times10^{-1}$
Pile driving	$10^{-2} - 10^{0}$
Aircraft impact	$5\times10^{-2} - 2\times10^{0}$
Hard impact	$10^{0} - 5\times10^{1}$
Hypervelocity impact	$10^{2} - 10^{6}$

There are many existing tests depended on the level of strain rate to estimate dynamic characteristics. The strain rate was range between 10^{-2} to 10^{0} for a hydraulic test machine. While it is up 10 s^{-1} for drop weight impact test machine. The peak strain rate limited by 10 s^{-1} to 10^{4} s^{-1} for Split Hopkinson Pressure Bar due to Ross et al, (1995) cited by [1]. It is feasible to employ codes in which formulae are obtained from investigational tests in the condition of an absence of appropriate experimental devices. CEB-FIP Model Code [9] comprises the most widespread forms for calculating the strain rate development of concrete. The dynamic characteristics for concrete covered up 120MPa and strain rate up 3×10^{2} s^{-1} for compressive as well as flexural [9]. This code is suitable for prediction for the impact loading of fluid flow to strain rate less than 3×10^{2} s^{-1}.

This work estimates mechanical properties of green concrete made with the waste of plastic and glass of juices and soft drink bottles at conditions of quasi static, and goals to predict the dynamic properties under impact load of liquid flow by utilizing of CEB-FIP Model Code (2010).

2. Materials and Methods

2.1. Materials

2.1.1. Cement

Cement Type I (Ordinary Portland cement (O.P.C.)) was used to achieve the experimental program; this cement was in accordance with Iraqi specification (IQS No.5: 1984) [10].

2.1.2. Fine Aggregate

Natural sand was used. The gradation of this sand was found be positioned in the second zone within the Iraqi specification (IQS 45/1984) [11].

IOP Conf. Series: Materials Science and Engineering **584** (2019) 012053 doi:10.1088/1757-899X/584/1/012053

2.1.3. Coarse Aggregate

Coarse aggregate with max. size 20 mm from Badra and Jassan's quarry was employed. Its gradation inside the ranges of Iraqi specification (IQS 45/1984) [10].

2.1.4. Waste Materials

Plastic wastes and glass wastes of bottles of juices and soft drinks were used as aggregate. They were formulated as follows:

- Plastic waste: Bottle caps were distorted approximately to fragments with a dimension of (10×10) mm to use as coarse aggregate, see Figure 1.
- Glass waste: The glass bottles were broken into small pieces that were passed from a 5 mm sieve to be used as an alternative to sand as in a Figure 2.

Figure 1. Plastic pieces.

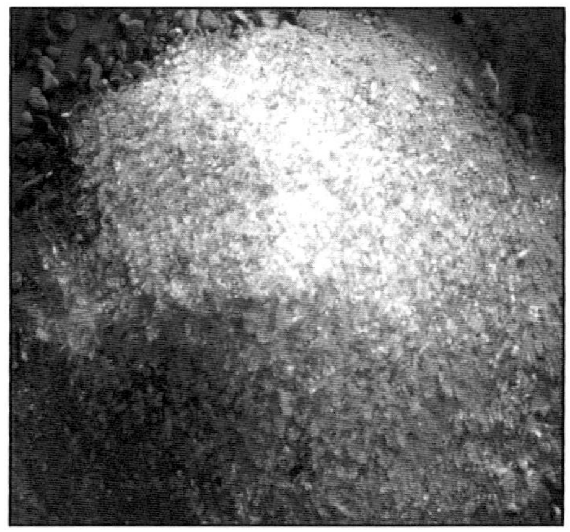

Figure 2. Crashed glass.

2.1.5. Water

Tap water was used in mixing and curing of all specimens.

2.2. Mix design

Concrete mixes were prepared according to British mix design method (B.S. 5328, Part 2:1991) [12]. The target strength was 30MPa at the age of 28 days. Table 2 displays reference mix proportions. Eight mixes were done with the following details:

- C00: the reference mix not including any adding of waste materials.
- CG10: the glass with 10%, was added as a replacement of sand volume.
- CG15: the glass with 15%, was added as a replacement of sand volume.
- CG20: the glass with 20%, was added as a replacement of sand volume.
- CP15: plastic bottle caps with 15% was added as a replacement of gravel volume.
- CP20: plastic bottle caps with 20% was added as a replacement of gravel volume.
- CP25: plastic bottle caps with 25%, was added as a replacement of gravel volume.
- CP20G15: plastic bottle caps with 20% as a replacement of gravel volume and the glass with 15% as a replacement of sand volume were added.

Table 2. Mix proportion.

No.	w/c	Cement, Kg/m^3	Fine aggregate, Kg/m^3	Coarse aggregate, Kg/m^3	Slump, mm
C00	0.5	340	515	1385	100

2.3 Mixing, casting, compaction, and curing

Materials for concrete mixes fitting to mix proportions were prepared, mixed manually, and then cast in the cleaned and oiled mould. The specimens comprise cubes, cylinders, and prisms. The compaction was done by electrical vibration table. After 24±0.5 hrs., moulds were released and specimens were submerged in a curing tank, at laboratory temperature (22±2)°C until testing age.

2.4 Tests

2.4.1. Compressive strength
Test machine with the maximum limit (2000 KN) was used to achieve the test according to (BS.1881: Part 116:1989) [13]. The average of three 100mm cubs for each mix was approved at 7 & 28 days ages.

2.4.2. Flexure strength
ASTM C 293-83 [14] method was followed in this test. (100×100×400) mm prisms were tested at age of 28 days. The average of two tests was adopted for each mix.

2.4.3. Modulus of elasticity
ASTM-C469-06 [15] was adopted to achieve this test using cylinders with 300mm height and 150mm diameter at age of 28 days.

3. CEB-FIP Model Code

Dynamic properties for concrete can be estimated by the Dynamic Increase Factor (DIF). The (DIF) defined by CEB-FIP code [9] as the ratio between dynamic strength and static strength. The mechanical properties for the dynamic state can be calculated from equations below [7]:

- Used for compressive strength

$$DIF_{fc} = \frac{f_{cd}}{f_c} = \left(\frac{\dot{\varepsilon}_c}{3\times10^{-6}}\right)^{0.014} \quad for\ \dot{\varepsilon}_c \le 30\,s^{-1} \tag{1}$$

$$DIF_{fc} = \frac{f_{cd}}{f_c} = 0.012\left(\frac{\dot{\varepsilon}_c}{3\times10^{-6}}\right)^{\frac{1}{3}} \quad for\ 30 \le \dot{\varepsilon}_c \le 300\,s^{-1} \tag{2}$$

- Used for flexural strength

$$DIF_{fc} = \frac{f_{td}}{f_t} = \left(\frac{\dot{\varepsilon}_c}{1\times10^{-6}}\right)^{0.018} \quad for\ \dot{\varepsilon}_c \le 10\,s^{-1} \tag{3}$$

$$DIF_{ft} = \frac{f_{td}}{f_t} = 0.0062\left(\frac{\dot{\varepsilon}_c}{3\times10^{-6}}\right)^{\frac{1}{3}} \quad for\ 10 \le \dot{\varepsilon}_c \le 300\,s^{-1} \tag{4}$$

- Used for modulus of elasticity

$$DIF_{EC} = \frac{E_{cd}}{E_c} = \left(\frac{\dot{\varepsilon}_c}{3\times10^{-6}}\right)^{0.026} \quad for\ 30 \le \dot{\varepsilon}_c \le 300\,s^{-1} \tag{5}$$

Where

f_{cd} = the dynamic compressive strength corresponding to strain rate $\dot{\varepsilon}_c$
f_c = the static compressive strength corresponding to the static strain
f_{td} = the dynamic tensile strength corresponding to strain rate $\dot{\varepsilon}_t$.
f_t = the static tensile strength corresponding to the static strain.
Ecd= the elastic modulus corresponding to strain rate of $\dot{\varepsilon}c$.
Ec =the elastic modulus corresponding to the reference static strain.
f_c,f_t, and Ec depended on quasi static rate as 30×10^{-6}, 1×10^{-6}, and 30×10^{-6} respectively[7]
This paper deal with the impact of liquid flow. The strain, in this case, estimated limit is between 10^{-1} to 10^{0}. Offshore platforms, aircraft impact, transportation facilities, gas explosion… etc are specified in this range of strain rate as shown in table 1.

4. Results and discussions

4.1 Static properties:
In Table 3 and figure 3, many notes can be collected, when plastic waste was used, the compressive strength was reduced. When using glass waste, the decrease in strength was shrunk at 28 days. While combining the two types of waste resulted in clear improvement in compressive strength (i.e, improved than cases of

separate use of waste). In mix CP20G15, the residual compressive strength was 76% and 83% with the entire percentage of wastes 35% at the age of 7 and 28 days, respectively.
The performance of mixes strength is associated with the properties of the wastes that were used. Crushed glass has less abrasion resistance than natural sand, and plastic hardness is less than natural gravel [16].

Table 3. Compressive strength result

No.	C00	CG10	CG15	CG20	CP15	CP20	CP25	CP20G15
Compressive strength MPa. 7 days	24.98	16	19	15	13.79	13.89	13.11	19
Dynamic Compressive strength MPa. 7 days	28.90	18.51	21.98	17.35	15.96	16.07	15.17	21.98
Compressive strength MPa. 28 days	31.01	23.72	28.04	22.08	19.98	20.11	15.93	25.68
Dynamic Compressive strength MPa. 28 days	35.88	27.44	32.44	25.55	23.12	23.27	18.43	29.71

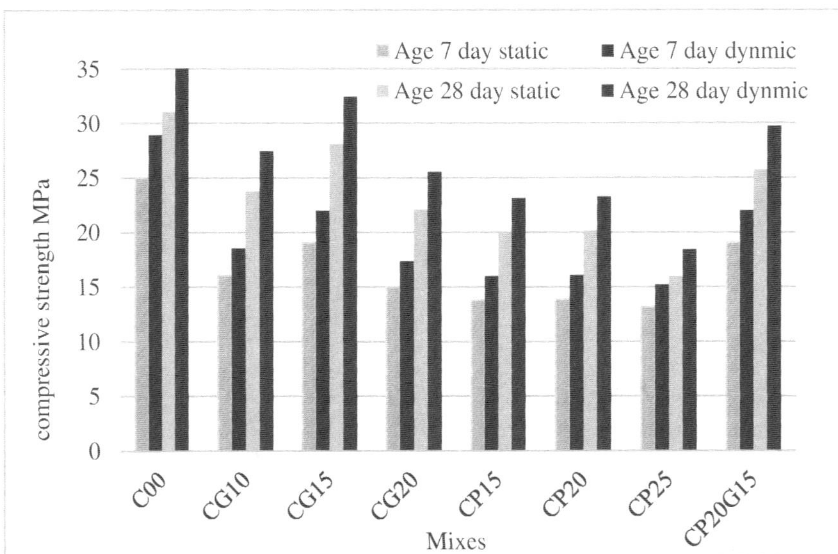

Figure 3. The compressive strength result for static and dynamic strength with age 7 days and 28 days.

The behavior of flexural strengths is denoted in Table 4 and figure 4. The flexural strengths of the mixes containing the glass waste were the closest to the reference mix, while the greatest decrease in the strength was in the mixes containing plastic waste. When the two types of waste were mixed (i.e, CP20G15), the decrease was 50 % of the reference mix. Therefore, It is possible to conclude that the addition of plastic waste has the greatest impact on the flexural strength and this may be due to the nature of the rigidity and limits of the flexibility of the material used [16].

Table 4. Flexural strength results

No.	C00	CG10	CG15	CG20	CP15	CP20	CP25	CP20G15
Flexure strength MPa	2.40	1.68	1.90	1.70	0.70	0.68	0.50	1.20
Dynamic flexure strength MPa	3.08	2.15	2.44	2.18	0.90	0.87	0.64	1.54

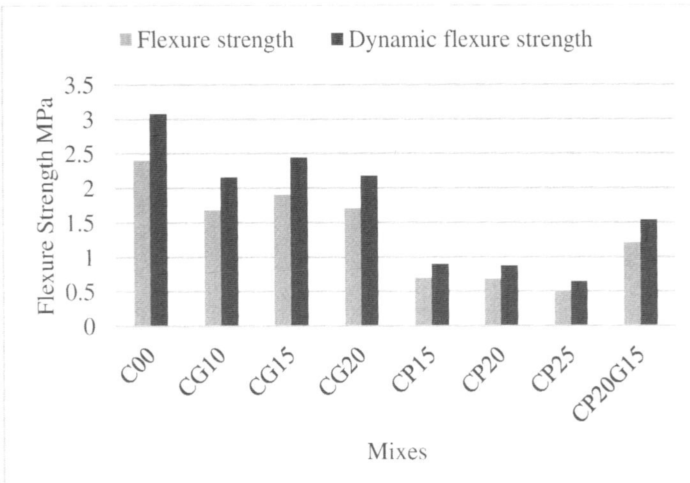

Figure 4. The flexure strength result for static and dynamic strength.

From observing Table 5 and figure 5, can be noticed that; in mixes containing plastic waste the modulus of elasticity values exceeded the mixes containing the glass waste but remained less than that of the reference mix. The modulus of elasticity values was decreased with the increase in the plastic ratio. The value of the elasticity of the mix containing the two types of waste (i.e, CP20G15) lies between the values of the mixes containing the plastic waste and the other containing the glass waste, and less than reference mix by about 3.99 GPa.

Table 5. Modulus of elasticity results.

No.	C00	CG10	CG15	CG20	CP15	CP20	CP25	CP20G15
Modulus of elasticity MPa	28	20.98	23.85	20	26.84	25.92	21.88	24.01
Dynamic modulus of elasticity MPa	40.10	30.05	34.16	28.64	38.44	37.12	31.34	34.39

4.2 Dynamic state

As mentioned earlier, this work deal with the impact of flow liquid. The strain, in this case, estimated to range between 10^{-1} to 10^{0}. The (DIF) is equal to 1.16 for compressive strength. It is 1.282 for flexure strength. The (DIF) for Modules of elasticity is 1.43. The results of compressive strength, flexure strength, and modulus of elasticity in the dynamic state are tabled in table 2, table 3, and table 4 respectively.

The predicted strength attained by CFB-FIP DIF forms for tension and compression is in concurrence with the experimental results of plain concrete depicted in [17] and [8] within the limit of the strain rate studied.

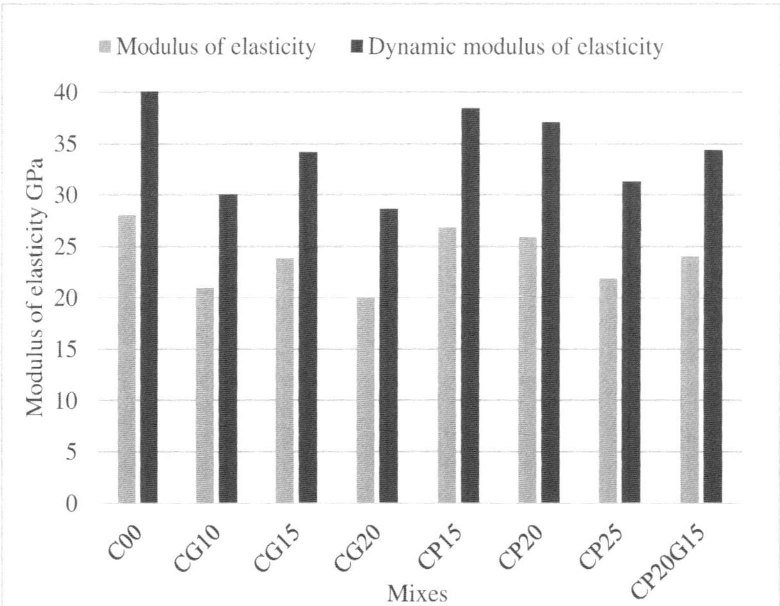

Figure 5. The modulus of elasticity result for static and dynamic strength.

5. Conclusion

Based on the current study, the following points can be inferred:

1. It is possible to produce green concrete with 35% waste content and residual compressive strength of 76% and 83% at the age of 7 and 28 days, respectively.
2. The CEB-FIP Model (2010) DIFs are well suited to this type of green concrete results in both tension and compression as the formulas are close with other experimental studies. It is, therefore, possible to adopt this code to estimate dynamic characteristics within the quasi-static level as an alternative to practical tests, in which testing machines are often expensive, unavailable or specified at a maximum load rate.

References

[1] Ismail Z Z and AL-Hashmi E A 2008 Use of Waste Plastic in Concrete Mix as Aggregate Replacement *Elsevier, Waste Management* pp 2042–6.

[2] Mathew P , Varghese S, Paul T and Varghese E 2013 Recycled Plastics as Coarse Aggregate for Structural Concrete *International Journal of Innovative Research In Science, Engineering And Technology* vol 2 issue 3 pp 687–90.

[3] Subramani T and Pugal V 2015 Experimental Study on Plastic Waste as A Coarse Aggregate For Structural Concrete *International Journal of Application or Innovation In Engineering & Management (IJAIEM)* vol 4 issue 5 pp 144–52.

[4] Byars E, Zhu H and Morales B 2004 ConGlassCrete 1 Project. Final Report *University Of Sheffield, United Kingdom* .

[5] Abbas A H, Al-Deen F M and Abdulah RM 2011 Using of Glass Wastes as a Fine Aggregate in Concrete Mix *Tikrit Journal of Eng Sciences* vol 18 issue 3 pp 81–7.

[6] BS EN 1992-3 2006 Design of concrete structures - Part 3: Liquid retaining and containment structures *UK, British Standards Institution* vol 3 issue 2006 .

[7] Othman H A B 2017 Performance Of Ultra-High Performance Fibre Reinforced Concrete Plates Under Impact Loads *Ph,D* pp 225.

[8] Cusatis G 2011 International Journal of Impact Engineering Strain-rate effects on concrete behavior *International Journal of Impact Engineering* [Internet] vol 38 issue 4 pp 162–70. Available from: http://dx.doi.org/10.1016/j.ijimpeng.2010.10.030

[9] CEB-FIP 2010 fib Model Code for Concrete Structures 2010, design code In: EURO, Comité International du Béton, Lausanne, Switzerland fédération internationale du béton / International Federation for Structural Concrete (fib); 2010. p. 190–350.

[10] IQS No.5 1984 Specification of Portland Cement *Iraqi Specification* .
[11] IQS No.45 1984Specification Of Aggregate From Natural Sources For Concrete And Construction *Iraqi Specification*.
[12] B S 5328 Part 2 1991 Methods for specifying concrete mixes *British Standards Institution* .
[13] B S 1881 Method for Determination of Compressive Strength of Concrete Cubes *British Standards Institution* vol Part 116:1.
[14] ASTM C 293-02 2006 Standard Test Method for flexural strength of concrete
[15] ASTM C 469-02 2006 Standard Test Method for Static Modulus of Elasticity and Poisson's Ratio of Concrete in Compression
[16] Abdulabbas Z H, and Abd-alridha SQ 2016 Evaluation of Using Waste of Bottles in Concrete as Sustainable Construction Abstract : issue 3 pp 782–97.
[17] Bischoff P H and Perry S H 1991 Compressive behaviour of concrete at high strain rates *MateriaLs And Struetures/Matc;Riaux Et Constructions* pp 425–50.

Hydraulic conductivity estimation by using groundwater modelling system program for upper zone of Iraqi aquifers

T S Khayyun[1] and H H Mahdi[2]

[1] Civil Engineering Department, University of Technology, Baghdad, Iraq. Email: thshkhma@yahoo.com
[2] Civil Engineering Department, University of Technology, Baghdad, Iraq. Email: Hasan.hadi3@gmail.com

Abstract. Estimating of soil hydraulic conductivity was so important and essential parameter for designing systems. Hydraulic conductivity measurements need many activities beginning from extract samples. Therefore it is having been found costly and time-consuming. At the present time, geo-statistical ways were used to estimate the hydraulic conductivity on the basis of restricted data. The aim of this research is to characterize hydraulic conductivity for the upper zone of Iraqi aquifers using groundwater modelling system (GMS) software and to build two dimensions geo-statistical analysis model depending on the application of 868 wells.These wells were divided into 778 wells for building the model and 90 wells for the verification. The hydraulic conductivity parameter which was obtained from the geo-statistical analysis model results indicated that the best fitted theoretical model was the inverse distance weighted (IDW) model with a coefficient of determination (R^2) equal to 87.43% and root mean square error (RMSE) equal to 1.853. Also, by conducting many trails, it was found that the optimal distance range for estimating the hydraulic conductivity was 3800 to 5000 meters. A comparison between the kriging and IDW interpolation methods indicated that both methods could estimate the hydraulic conductivity with good accuracy and less error.

1. Introduction

In general, estimating reliable hydraulic conductivity value is one of the fundamental challenges in groundwater modelling applications. Hydraulic conductivity parameter often varies intensely and exhibits spatial variations at an extent range of spatial scales. It is the most important factor for modelling soil hydraulic properties in saturated and unsaturated soils as in [37]. At the level of national scales of tens to hundreds of kilometres and at depths greater than one kilometre, the field information on hydraulic conductivities is relatively scarce as in [11]. Maps of Hydraulic conductivity at such scales of extent areas and depths are usually generated through numerical models that attempt to reflect the geological structure of the region and are simultaneously calibrated on both, natural tracer concentrations and hydraulic heads as in [9].In a popular of hydrologic cycle models, soil hydraulic conductivity can be considered as constant (homogeneous) within an area of 100 to 10,000 km^2 and, this may lead to inappropriate results as in [8]. References [6],[10], [11],[12], [13], [28], [29],[30], [31], [32], [33], [34], [38], [41], [42] and [43] studied hydraulic conductivity variability in heterogeneous aquifers. They were used interpolation estimation procedures such as co-kriging, Inverse Distance Weighted (IDW), Local Polynomial Interpolation (LPI), Global Polynomial Interpolation (GPI), Radial Basis Functions (RBF) and spatial kriging. The results showed that the use of these interpolation methods yields best assessments and a good agree with the results of experimental pumping test, therefore this agreement proves that the distribution of hydraulic conductivity can be estimated economically and efficiently in a difficult and wide area. The aim of this study is to characterize hydraulic conductivity for upper zone of Iraqi aquifers within the thickness of one hundred meters by using GMS software to build two dimensions geo-statistical analysis model depending on available actual field data obtaining from the investigation of drilled wells.

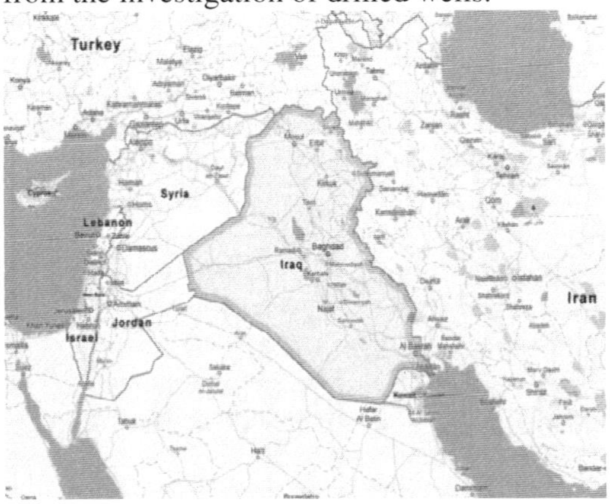

Figure 1. Iraq location map.

Figure 2. Iraq main topographical areas map.

2.Study area

The Republic of Iraq is located in the Middle East at the northernmost extent of the Arabian Gulf, west of Iran, east of Syria, north of Saudi Arabia, and south of Turkey, (figure 1). The total area of Iraq catchment is 437,072 square kilometres, containing 432,162 square kilometres of the land surface as in [48] and [49]. Iraq has four foremost topographical areas: desert region, uplands region, northern highlands, and the alluvial plain (Mesopotamia), (figure 2), as in [48]. The climate in Iraq is mainly of the continental, subtropical semi-arid type, except the north and north-eastern mountainous areas having a Mediterranean climate as in [45], [46], and [47].

3.Material and methods

3.1. Hydrogeology of Iraq

Geologists divide Iraq into six geologic regions: Al-Jazeera area, Mesopotamia plain, Southern Desert, Western Desert, Low Folded Zone, and High Folded Zone, (figure 3). Each of these regions includes a system of formations, aquifers and characterized with specific features. (Table 1) indicates type of formations, transmissivity (T, m^2/d), and hydraulic conductivity (K, m/d)) according to the available data from the performed pumping tests of the drilled wells as in [1, 2, 3, 4, 5, 7, 14, 15, 16, 17, 18, 19, 20, 21, 22, 23, 24, 25, 26, 27, 35, 39, and 40].

Table 1. Hydro-geologic characteristics of study area, [16], [17], [18], [19], [20], and [21].

Region	Formations	T (m^2/d)		K (m/d)	
		from	to	from	to
Jazira area	Fat'ha	2	246	0.1	6.3
	Injana	2	1274	0.1	20
	Quaternary Sediments	5	257	0.2	24.7
Mesoptampia plain	Euphrates	-	-	-	-
	Northern part (Injana, Mukdadiya, Bai Hassan formations, in addition to Quaternary sediments)	4	503	0.1	47
	Northern part (Mukdadiya aquifer within the vicinity of Samara)	3	774	0.1	55.4
	Northern part (Bai Hassan aquifer within the vicinity of Samara)	4	829	0.1	43
	Northern part (Quaternary sediments aquifer within the vicinity of Samara)	21	555	1.5	29.7
	Northern part (Quaternary sediments aquifer within the vicinity of Baghdad)	6	1089	0.5	126
	Northern part (Quaternary sediments aquifer within Mandali vicinity)	4	1382	0.1	374.8
	Central Part (within the vicinity of Karbala city)	10	165	1	27
	Central Part (within the vicinities of Kut and Ali Al-Gharbi)	13	742	0.8	72
	Southern Part	36	439	1.1	36.6
Southern Desert	Hartha	-	163.3	-	-
	Tayarat	0.5	1800	Very low	20.3
	Umm ErRadhuma	3	2100	0.1	21.1
	Dammam	3.1	4752	0.1	100
	Ghar – Euphrates	21	246	1.3	14
	Dibdibba	15	265	0.3	25.1
	Quaternary Sediments	14	964	2.9	74
Western Desert	Suffi	36	350	-	-
	Ga`ara (unconfined)	0.5	620	0.1	13.9
	Ga`ara (confined)	10	250	0.1	19
	Mulussa	0.8	397	0.03	4.4
	Ubaid	7	200	0.1	4
	Muhaiwir	9	1243	0.1	9.3
	Rutbah	2	7000	0.2	19.8
	Ms`ad	15	104	0.2	1.1
	Hartha	2	588	0.1	9.2
	Tayarat	202	15529	0.2	25.6

	Umm ErRadhuma	2	1550	0.1	21.1
	Akashat	2	620	0.1	5.2
	Dammam	6	5000	0.1	16
	Euphrates	3	1750	0.2	29.5
	Injana	21	927	0.8	40.2
Low Folded Zone	Injana	2	1274	0.1	18.5
	Mukdadiya and Bai Hassan	3	2800	0.1	24
	Quaternary sediments	5	539	0.1	47.3
	carbonate (within Sinjar area)	4	337	0.1	6.8
	Injana (within Sinjar area)	2	1274	0.1	11.4
	Quaternary sediments(within Sinjar area)	15	91	1.4	6.7
	Fatha (within Tigris River Sub Province)	8	360	0.1	8
	Injana (within Tigris River Sub Province)	3	626	0.1	18.5
	Khazir – Gomel Sub Province	100	7000	0.001	0.21
	Injana (within Dohuk – Alqosh Sub Province)	6.5	622	0.1	16.8
	carbonate (mainly PilaSpi) (within Dohuk – Alqosh Sub Province)	4	375	0.1	5.1
	Mukdadiya and Bai Hassan (within Erbil Sub Province)	3	1296	0.1	30.4
	Quaternary sediments (within Erbil Sub Province)	10	1225	0.4	35.2
	AltunKupri Sub Province	4	1738	0.1	47.9
	Mukdadiya and Bai Hassan (within Dibiga Sub Province)	5	1650	0.1	28.5
	Mukdadiya and Bai Hassan (within Makhmour Sub Province)	6	1350	0.1	25.6
	Mukdadiya and Bai Hassan (within Kirkuk – Hawija – TuzKhurmatu Sub Province)	3	2775	0.1	57.8
	Mukdadiya and Bai Hassan (within Cham Chamal – QadirKaram – Qara Too Sub Province)	3	1950	0.1	42.6
	Mukdadiya and Bai Hassan (within Kalar – Khanaqeen Sub Province)	15	1170	0.1	24
	Quaternary sediments (within Kalar – Khanaqeen Sub Province)	20	540	0.4	35.5
	Mukdadiya and Bai Hassan (within QaraTappa – Al-Sa'adiyah Sub Province)	10	1120	0.1	23
	Quaternary sediments (within QaraTappa – Al-Sa'adiyah Sub Province)	12	510	0.3	25.8
	Mukdadiya and Bai Hassan (within Mandili town and its vicinities - Mandili – Zurbatiya – Teeb Sub Province)	3	445	0.1	12.2

	Mukdadiya and Bai Hassan (within Zurbatiya – Badra vicinity - Mandili – Zurbatiya – Teeb Sub Province)	7	533	0.1	14.1
	Mukdadiya and Bai Hassan (within Teeb and Fakka areas and their vicinities- Mandili – Zurbatiya – Teeb Sub Province)	12	290	0.5	15.5
High Folded Zone	Bekhme Karst Aquifer	3	864	0.1	18
	PilaSpi Fractured Karst Aquifer	3	2800	0.1	23.9
	Zakho Basin	3	654	0.1	2.5
	NW Barzany Basin	3	134	0.1	5
	Atrush–Dinarta Sub-Basin	3	2800	0.1	23.9
	Central Harir Basin (Diana –MergaSoor Sub-Basin)	6	662	0.1	14.6
	Central Harir Basin (Harir–Shaqlawa Sub-Basin)	3	611	0.1	19.1
	Central Dokan Basin (Sangasar Sub-Basin)	3	540	0.1	14.6
	Central Dokan Basin (Rania –Choarqurna Sub-Basin)	15	864	0.2	12.8
	Central Dokan Basin (Bangard–Surdash Sub-Basin)	14	597	0.2	4.4
	Sharazoor Basin (Bazian–Qaradagh Sub-Basin)	4	749	0.2	20.6
	Sharazoor Basin (Sarchinar–Arbat Sub-Basin)	5	409	0.2	4.9

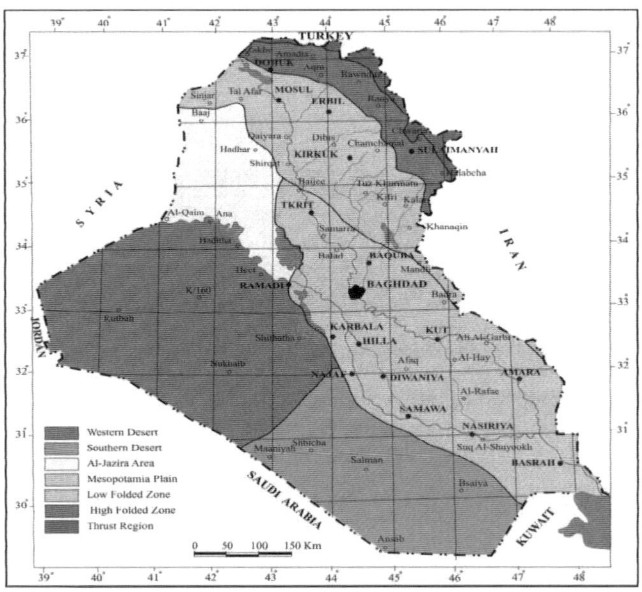

Figure 3. Iraq main geologic regions map, [44].

3.2. Division of data

In this study, 868 well information was used to build geo-statistical model including (x, y, and z) coordinates after processing in order to exchange it to UTM coordinate system, as well as the hydraulic conductivity (k) in (meter per day) unit. The data are randomly separated into two sets (778 for building the geo-statistical model) and (90 for verification) depending on available data within the areas and as possible as covering all of the catchment areas of Iraq. The selection of two data sets was completed by using statistical criteria after many trials in order to meet optimum statistical distribution. (Table 2) summarizes ranges and statistical criteria for both sets. (figures 4 and 5) illustrate the hydraulic conductivity values for both data sets.

Table 2. Ranges and statistical criteria for data sets of the hydraulic conductivity.

Hydraulic conductivity sets	Min. m/d	Max. m/d	Range m/d	Mean m/d	Median	Standard deviation
All data (868)	0	24.7	24.7	3.1637	0.9	5.0180
Building model data (778)	0	24.7	24.7	3.1829	0.9	5.00081
Verification data (90)	0.01	24.7	24.69	2.9973	0.9	5.19052

3.3. GMS software

In this research, the 2D Scatter Point Module integrated with The Groundwater Modeling System (GMS) software was used in order to build the geo-statistical model. GMS represent a comprehensive graphical user environment for conducting groundwater simulations. The entire GMS system contains of a graphical user interface and a sum of analysis codes. The two dimensional (2D) Scatter Point module is implemented to interpolate from sets of 2D scattered data to (Triangulated Irregular Network (TIN) , meshes and grids). Several interpolation techniques are supported, including IDW and kriging as in [50].

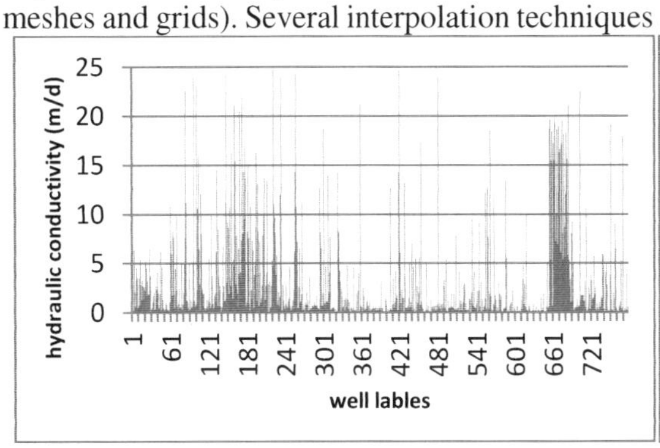

Figure 4. Building geo-statistical model.

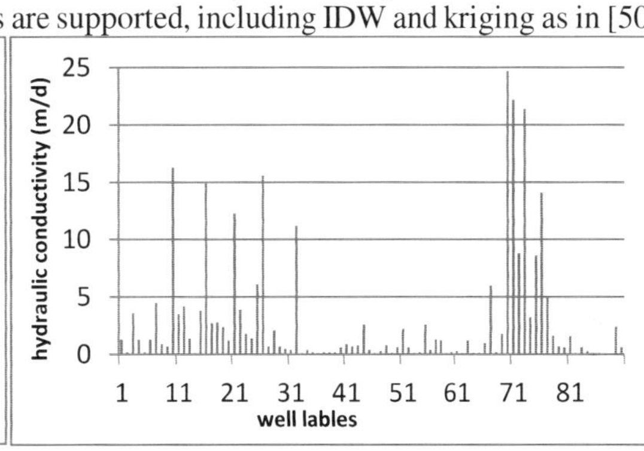

Figure 5. Verification data.

3.4. Interpolation techniques

One of the best commonly used interpolation techniques for point data is inverse distance weighted (IDW). Inverse distance weighted methods are depended on the assumption that the interpolating surface should be most influenced by the nearby points and less influenced by the more distant points. The result of the interpolated surface is a weighted average of the point data; at the interpolated surface, the weight assigned to each point reduces as the distance to the interpolation position increases. In this research, the simplest form of inverse distance weighted technique (constant nodal function) or occasionally called the "Shepard's method" was used as in [51].

Over the past several decades, the kriging technique has become an essential tool in the field of geo-statistics. Kriging is depended on the assumption of treated the parameter as a regionalized variable. A regionalized variable defined as variable intermediate between a completely deterministic variable and a truly random variable that it varies in a continuous way from one location to the next and thus, points that are neighboring each other have a certain degree of spatial correlation, but points that are generally separated are statistically independent. Kriging is a set of linear regression procedures which minimize assessment variance from a predefined covariance model.

GMS program offers two kriging methods (simple and ordinary kriging). In this research, ordinary method was used as an efficient method which recommended by previous researchers work in the same field. The first step in ordinary kriging is to create a variogram from the scatter point set to be interpolated. A variogram involves two parts: an experimental variogram and a model variogram, Figure (6). If the value to be interpolated (f) is assumed, the experimental variogram is constructed by calculating the variance (g) of every point in the set concerning every one of the other points and plotting distance (h) between the points versus the variances. There are many formulas can be used to calculate the variance, but it is typically calculated as one half the difference in (f) squared.

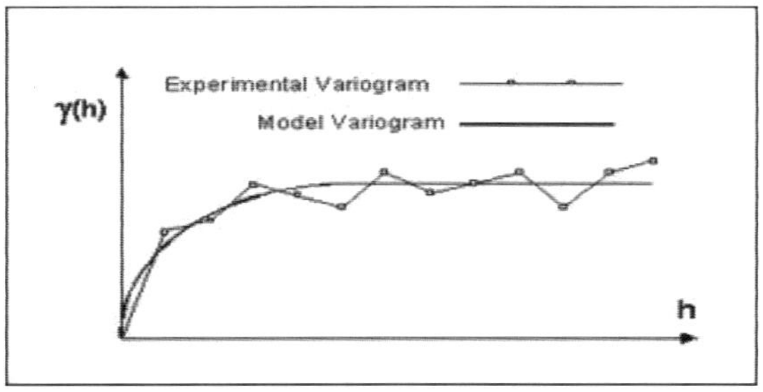

Figure 6.Experimental and model variogram.

A model variogram represents a simple mathematical function that models the style in the experimental variogram. Once the model variogram is created, it was used to calculate the weights which were used in kriging as in [43].

In order to conclude the accuracy of any given method of spatial interpolation, Root Mean Square Error value (RMSE) was used as in [32]. The techniques of spatial interpolation estimate values of the variables at unobserved places in geographic space depended on the values at observed places. RMSE is a commonly used measure of the differences between model predicted values, and the actually values observed from what is being estimated or modeled. The technique which gave the lowest RMSE value was considered to be the most appropriate technique as in [32].

3.5. Triangulated Irregular Network (TIN)

The TIN technique represents a surface as a set of connecting, non-overlapping triangles, where within each triangle the surface created as a plane. In order to construct a TIN in GMS program, it should be available a set of points called TIN vertices. In this research, GMS program was used to construct TIN, (figure 7) and (figure 8).

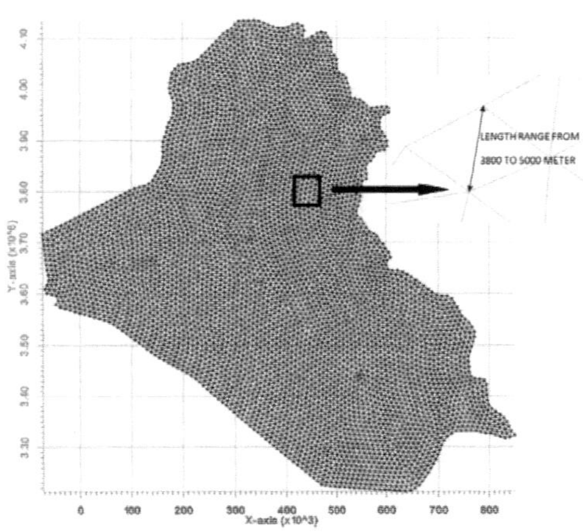

Figure 7.TIN was used for the geo-statistics model.

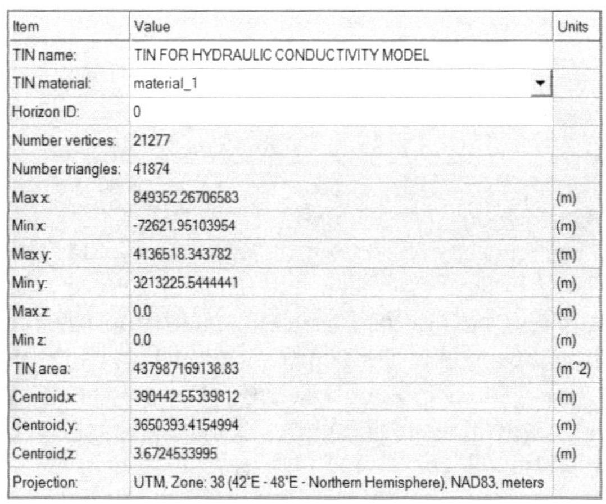

Item	Value	Units
TIN name:	TIN FOR HYDRAULIC CONDUCTIVITY MODEL	
TIN material:	material_1	
Horizon ID:	0	
Number vertices:	21277	
Number triangles:	41874	
Max x:	849352.26706583	(m)
Min x:	-72621.95103954	(m)
Max y:	4136518.343782	(m)
Min y:	3213225.5444441	(m)
Max z:	0.0	(m)
Min z:	0.0	(m)
TIN area:	437987169138.83	(m^2)
Centroid.x:	390442.55339812	(m)
Centroid.y:	3650393.4154994	(m)
Centroid.z:	3.6724533995	(m)
Projection:	UTM, Zone: 38 (42°E - 48°E - Northern Hemisphere), NAD83, meters	

Figure 8.TIN properties interface.

4.Results and discussion

The geo-statistical interpolation methods (IDW and Kriging) were applied based on hydraulic conductivity data of observed spaces in order to estimate this parameter for unobserved spaces. By using each of these methods, a total of 21,277 hydraulic conductivity points were estimated in wholly areas of Iraq, (figure 9) and (Table 3).

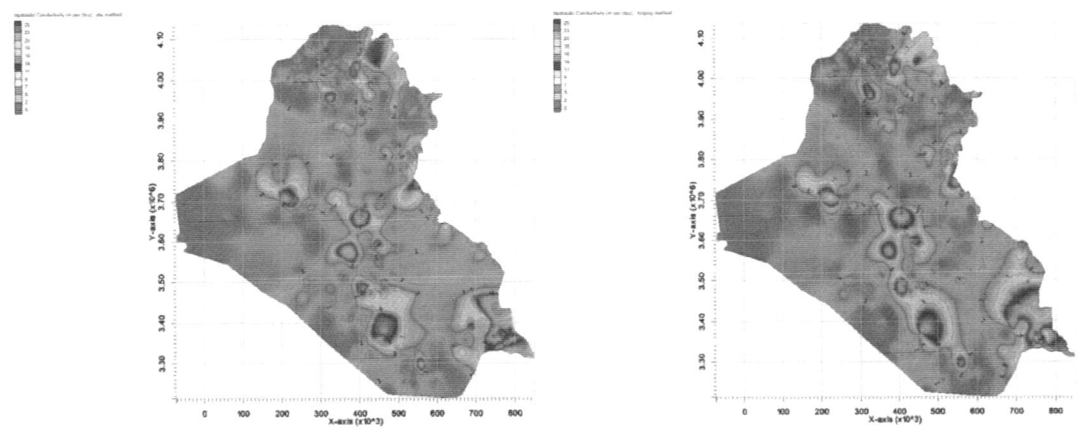

(IDW) method (Ordinary Kriging) method

Figure 9. Hydraulic conductivity distribution.

As can be seen, the two methods indicated well accuracy distribution for hydraulic conductivity values, (figure10). The convergence in results of both methods reflects the appropriate selection for model data building and verification data. For the available used data, the values of hydraulic conductivity over 24.7 m/d, were excluded because it was very little values and to obtain reliable results. There is high convergence in results of two methods in the west desert region, particularly in Al-Anbar governorate while a clear difference in results was shown in the north part of the high folded region particularly in Erbil governorate. Also, results indicated that the high values were found in south desert particularly in al Basra governorate and in many parts of Mesopotamia plain particularly in Baghdad, Babylon, and east of Thi-Qar governorates while the low values were found in west desert region, al- Jazeera region, parts of high folded region, and parts of low folded region.

Table 3. Ranges and statistical criteria for estimated Hydraulic conductivity (m/d).

STATISTICAL DETAILS	IDW	Kriging
Min.	0.02574564	0
Max.	23.6914	22.99957
range	23.66566	22.99957
Mean	3.768914	3.672453
Median	2.75598	2.399603
Standard deviation	3.557749	3.647328
RMSE	1.852755	1.993618
MAE	0.879478	0.827022
MBE	-0.12692	-0.33804
R^2	87.43%	85.97%

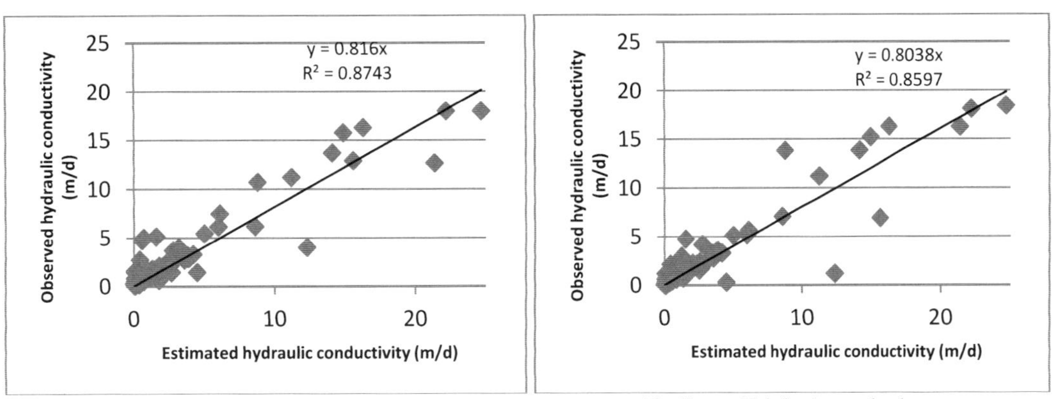

(IDW) method (Ordinary Kriging) method

Figure 10. Comparison between actual and calculated values.

5. Conclusions and Recommendations

Hydraulic conductivity parameter (K) is one of the parameters governing both the direction and the magnitude of groundwater velocity, and therefore, is one of the most key parameters affecting groundwater flow and pollutants transport. Geo-statistics interpolation methods proved a very interesting tool modeling for estimating hydraulic conductivity. The purpose of this research was to confirm the possibility to use IDW and kriging techniques as reliable tools for estimating the hydraulic conductivity parameter. The scares of field data influence the results and restricted the confirmation of the accuracy in the verification process. The accuracy of the results increases with minimizing study area and as possible as available more information and data. The hydraulic conductivity parameter which obtained from geo-statistical analysis model results indicated that the best fitted theoretical model was the inverse distance weighted (IDW) model with R^2 equal to 87.43%. Also, by conducting many trails, it was found that the optimal distance for estimating the hydraulic conductivity was 4800 meters. A comparison between the kriging and IDW interpolation methods indicated that both methods could estimate hydraulic conductivity with good accuracy and less error. The methods were used in this study can still be significantly improved, mainly by refining the kriging and IDW strategy of hydraulic conductivity parameter and distribution of samples. By improving the methods, the obtained assessment will be more precise and more reliable, which will allow for extend estimation uses.

References

[1] Al-dabaj A A A H and Al-Khashab S N A 2001 Hydrogeological and Hydrochemical study of Al-Maania quadrangle, Sheet NI –38 – 5, Scale 1:250000, *State Company of Geological Survey and Mining (GEOSURV)*, Baghdad.

[2] Al-dabaj A A A H and Al-Khashab S N A 2001 Hydrogeological and Hydrochemical study of Ansab quadrangle, Sheet NI –38 – 10, Scale 1:250000, *State Company of Geological Survey and Mining (GEOSURV)*, Baghdad.

[3] Al-dabaj A A A H and Al-Khashab S N A 2001 Hydrogeological and Hydrochemical study of Haditha quadrangle, Sheet NI –38 – 5, Scale 1:250000, *State Company of Geological Survey and Mining (GEOSURV)*, Baghdad.

[4] Al-dabaj A A A H and Al-Khashab S N A 2001 Hydrogeological and Hydrochemical study of Al-Salman quadrangle, Sheet NI –38 – 6, Scale 1:250000, *State Company of Geological Survey and Mining (GEOSURV)*, Baghdad.

[5] Al-dabaj A A A H and Al-Khashab S N A 2002 Hydrogeological and Hydrochemical study of Al-Ramadi quadrangle, Sheet NI –38 – 9, Scale 1:250000, *State Company of Geological Survey and Mining (GEOSURV)*, Baghdad.

[6] Sefelnasr A M 2007 Development of groundwater flow model for water resources management in the development areas of the Western Desert, Egypt. *DSc, Martin Luther University, Germany*.

[7] Al-Fatlawi A N 2010 Geological and Hydrogeological Characteristics of Umm Er Radhuma aquifer West of Iraq. *Euphrates Journal of Agriculture Science*, 2(4), 263-275.

[8] Braud I, Haverkamp R, Arrúe J L and López M V 2003 Spatial variability of soil surface properties and consequences for the annual and monthly water balance of a semiarid environment (EFEDA Experiment). *Journal of hydrometeorology*, *4*(1), 121-137.

[9] Castro M C and Goblet P 2003 Calibration of regional groundwater flow models: Working toward a better understanding of site-specific systems. *Water Resources Research*, *39*(6).

[10] Biondo C S 1995 *Geostatistical Analysis of Hydraulic Conductivity in Heterogeneous Aquifers* (No. AFIT/GEE/ENP/95-D-01). AIR FORCE INST OF TECH WRIGHT-PATTERSON AFB OH.

[11] Patriarche D, Castro M C and Goovaerts P 2005 Estimating regional hydraulic conductivity fields-A comparative study of geostatistical methods. *Mathematical Geology*, *37*(6), 587-613.

[12] Bissell E G and Aichele S S 2004 Geostatistical analysis of effective vertical hydraulic conductivity and presence of confining layers in the shallow glacial drift aquifer, *Oakland County, Michigan. US Department of the Interior, US Geological Survey.*

[13] Ghorbani D S, Homaei M, and Mahdian M H 2010 Effect of Land Use Change on Spatial Variability of Infiltration Parameters. *Iranian Journal of Irrigation and Drainage, 4*(2), 206-221.

[14] Al-Basrawi N H and Al-Jiburi H K 2001 Hydrogeological and Hydrochemical study of Al-Najaf quadrangle, Sheet NI –38 – 2, Scale 1. *State Company of Geological Survey and Mining (GEOSURV), Baghdad.*

[15] Al-Basrawi N H and Al-Jiburi H K 2002 Hydrogeological and Hydrochemical study of Wadi Houran quadrangle, Sheet NI –38 – 14, and H4 Sheet NI –38 – 15 Scale 1:250000. *State Company of Geological Survey and Mining (GEOSURV), Baghdad.*

[16] Al-Basrawi N H and Al-Jiburi H K 2007 Hydrogeology. *Iraqi Bulletin of Geology and Mining, Geology of Iraqi Western Desert*, 125-144.

[17] Al-Basrawi N H and Al-Jiburi H K 2009 Hydrogeology of al-Jazira area. *Iraqi Bulletin of Geology and Mining*, (3), 71-84.

[18] Al-Basrawi N H and Al-Jiburi H K 2009 Hydrogeology. *Iraqi Bulletin of Geology and Mining*, (3), 77-91.

[19] Al-Basrawi N H and Al-Jiburi H K 2011 Hydrogeology of the Mesopotamia plain. *Iraqi Bulletin of Geology and Mining*, (4), 83-103.

[20] Al-Basrawi N H and Al-Jiburi H K 2012 Hydrogeology of the low folded zone. *Iraqi Bulletin of Geology and Mining*, (5), 133-157.

[21] Al-Basrawi N H and Al-Jiburi H K 2014 Hydrogeology of the high folded zone. *Iraqi Bulletin of Geology and Mining*, (6), 163-183.

[22] Al-Jiburi H K 2004 Hydrogeological and Hydrochemical study of Al-Baghdad quadrangle, Sheet NI –38 – 10, Scale 1:250000. *State Company of Geological Survey and Mining (GEOSURV), Baghdad.*

[23] Al-Jiburi H K 2005 Hydrogeological and Hydrochemical study of Al-Amara quadrangle, Sheet NI –38 – 4, Scale 1:250000. *State Company of Geological Survey and Mining (GEOSURV), Baghdad.*

[24] Al-Jiburi H K 2005 Hydrogeological and Hydrochemical study of Al-Qaiyara quadrangle, Sheet NI –38 – 1, Scale 1:250000. *State Company of Geological Survey and Mining (GEOSURV), Baghdad.*

[25] Al-Jiburi H K 2005 Hydrogeological and Hydrochemical study of Al-Kut quadrangle, Sheet NI –38 – 15, Scale 1:250000. *State Company of Geological Survey and Mining (GEOSURV), Baghdad.*

[26] Al-Jiburi H K 2005 Hydrogeological and Hydrochemical study of Al-Nasriya quadrangle, Sheet NI –38 – 3, Scale 1:250000. *State Company of Geological Survey and Mining (GEOSURV), Baghdad.*

[27] Al-Jiburi H K 2007 Hydrogeological and Hydrochemical study of Al-Mosul quadrangle, Sheet NI –38 – 13, Scale 1:250000. *State Company of Geological Survey and Mining (GEOSURV), Baghdad.*

[28] Kareem H 2018 Study of Water Resources by Using 3D Groundwater Modeling in Al-Najaf Region, Iraq. *Doctoral dissertation, Cardiff University.*

[29] Hébert É, Côté J, and Smith M 2016 A geostatistical analysis of variations of permeability within a compacted dam core. In *Proceedings of the 17th Nordic Geotechnical Meeting, Reykjavik, Islande*,845-854.

[30] Gundogdu K S, Ali Osman Demir D and Demirtas C 2018 Determination of suitable spatial interpolation method for mapping on some soil physical properties.

[31] Issazadeh L, Umar M I, Al-Sulaivany S I and Hassanpour J 2018 Geostatistical Analysis of the Permeability Coefficient in Different Soil Textures. *Contemporary Agriculture*, *67*(2), 119-124.

[32] Mohanty B P, Ankeny M D, Horton R and Kanwar R S 1994 Spatial analysis of hydraulic conductivity measured using disc infiltrometers. *Water Resources Research*, *30*(9), 2489-2498.

[33] Moneer H S, Hadi A A and Saad S M 2014 A Comparative Study of Interpolation for Mapping Soil Physical Properties. *Journal of Kerbala University ,12*(2), 221-232.
[34] Moosavi A A and Sepaskhah A R 2012 Spatial variability of physico-chemical properties and hydraulic characteristics of a gravelly calcareous soil. *Archives of Agronomy and Soil Science*, *58*(6), 631-656.
[35] Al-Basrawi N H 2006 Hydrogeological and Hydrochemical study of Al-Najaf quadrangle, Sheet NI –38 – 8, and Abadan Sheet NI –38 – 5 Scale 1:250000. *State Company of Geological Survey and Mining (GEOSURV), Baghdad.*
[36] Oosterbaan R J and Nijland H J 1994 Determining the Saturated Hydraulic Conductivity, Chapter 12, In *Ritzema, H.P., Ed., Drainage Principles and Applications, International Institute for Land Reclamation and Improvement (ILRI), Wageningen, 1125.*
[37] Goovaerts P 1999 Geostatistics in soil science. state-of-the-art and perspectives. *Geoderma*, *89*(1-2), 1-45.
[38] Saud Q J and Ridha Muhammad A A 2010 Hydrogeological and Hydrochemical study of Sulaimaniya quadrangle, Sheet NI –38 – 3 – 10, Scale 1:250000. *State Company of Geological Survey and Mining (GEOSURV), Baghdad.*
[39] Al-Muqdadi S W 2012 Groundwater investigation and modeling-western desert of Iraq. *Geo-Engineering and Mining Technische Universität Bergakademie Freiberg Chair of Hydrogeology institute.*
[40] Ali S M 2012 Hydrogeological environmental assessment of Baghdad area. *Department of Geology, College of Science, University of Baghdad, Baghdad, Iraq*. p, 6.
[41] Painter S L, Woodbury, A D and Jiang Y 2007 Transmissivity estimation for highly heterogeneous aquifers: comparison of three methods applied to the Edwards Aquifer, Texas, USA. *Hydrogeology Journal*, *15*(2), 315-331.
[42] Al-Mussawi W H 2014 Optimum Management Models for Groundwater Use in Karbala Desert Area, *College of Engineering, Al-Mustansiriyah University*.
[43] Xu Z, Wang X, Chai J, Qin Y and Li Y 2017 Simulation of the Spatial Distribution of Hydraulic Conductivity in Porous Media through Different Methods. *Mathematical Problems in Engineering*, *2017*.
[44] Yacoub S Y 2011 Stratigraphy of the Mesopotamia plain. *Iraqi Bulletin of Geology and Mining*, (4), 1-36.
[45] https://www.weatheronline.co.uk/reports/climate/Iraq.htm.
[46] http://www.fao.org/nr/water/aquastat/data/query/index.html.
[47] http://agromet.gov.iq/index.php.
[48] https://www.loc.gov/rr/frd/cs/profiles/Iraq.
[49] https://www.nationmaster.com/country-info/stats/Geography/Land-boundaries/Border-countries.
[50] https://www.aquaveo.com/software/gms-groundwater-modeling-system-introduction
[51] https://www.xmswiki.com/index.php?title=GMS:GMS

Acknowledgments
We thank the anonymous referees for their useful suggestions. I would like to give a sincere thanks to Dr.Naseer Al-Basrawi/ Iraq geological survey (GEOSURV), for providing necessary information. I would like to give special thanks to the dean, head of the department of civil engineering/university of technology without whom, this research paper would not have been possible.

Experimental investigation for the behavior of concrete slab rests on sandy soil contains cavities

Muhammad Abed Attiya[1], Laith Jawad Aziz[2], Ola Mazen Makki[3]

[1]Assistant professor, College of engineering, civil engineering department, University of Kufa, Najaf, Iraq. Email: mohammedw.alfatlawi@uokufa.edu.iq, [2]laith.aljarrah@uokufa.edu.iq
[2]Assistant professor, College of engineering, civil engineering department, University of Kufa, Najaf, Iraq. Email: laith.aljarrah@uokufa.edu.iq
[3] Graduate student, College of engineering, civil engineering department, University of Kufa, Najaf, Iraq. Email:aulam.alibraheemi@uokufa.edu.iq

Abstract. Cavities formation into soil media is a very common problem that occurs in the gypseous soils (in which the gypsum content equals 2% and more). The gypsum dissolves by water flow leaching and leaves cavities in different shapes and sizes. In slab on grade system, the American concrete institute (ACI 360) recommends, the concrete slab transforms the load to the ground by about less than 50% of the allowable bearing capacity thereof, but when the soil contains cavities and excavation, the concrete slab behavior will change. In order to investigate the changing in behavior, twelve concrete slabs were prepared and tested on a soil media contains cavities. The investigated parameters were: cavity depth, steel wire reinforcement existence, bearing load plate shapes, and the load distributed parallel or perpendicular to the cavity. From the experimental results, it has been concluded that, the plain and reinforced concrete slab capacity decreases by (30.7%) and (7.4%) when a cylindrical continuous cavity forms at (30% mm) depth from the soil surface. However, when the cavity forms at a distance equal to (15%) from the soil surface, the plain and reinforced concrete slab capacity reduced by (70%) and (61%) respectively.
Keywords: Slab on grade, Cylindrical cavity, Slab on voided soil, Soil supported slab, slightly gypseous soil.

1. Introduction

Concrete slabs, which are upheld by a soil medium, is a widely used type of constructions because of its durability, mastery to overcome subgrade weakness, and ability to resist hard climate conditions. It is difficult to get an integral relation between concrete slabs and soil action because of the complexity of soil itself. The soil is a non-homogeneous and anisotropic medium, it behaves as a non-linear state, and moreover, its properties may change and vary with climate and time. In contrast to concrete which can be analysed as an isotropic, linear behavior material [1]. Gypsum ($CaSO_4$) is a salt diffuses in many of the world soil lands. It covers 6.7% of the world gypsum's land. [2] Gypsum soils (which are the soils that contain more than 2% of gypsum content [3]) are a strong kind of soil and have good properties when it is dry. The worrisome matter is when constructing a heavy building on it then the soil gets wet. In that case, a gigantic loss of strength will happen, and the worst is when the water flows inside the soil causing reduce of mass by gypsum leaching [1]. When gypsum leaching occurs, cavities in different shapes and sizes will form inside the soil [4] (see Figure 1.). Cavities existence weakens the bearing capacity of the soil [5]. As a result, and in the first level, cavities may cause differential settlement for the building, and next, the passive effect may aggravate to a construction collapse. It is worth to mention that, the ACI-360 committee defines slabs on grade as the member which transfer the load to the soil by less than 40%. [6] In this paper, the effect of cavities caused by gypsum dissolution on the concrete slabs on grade was discussed.

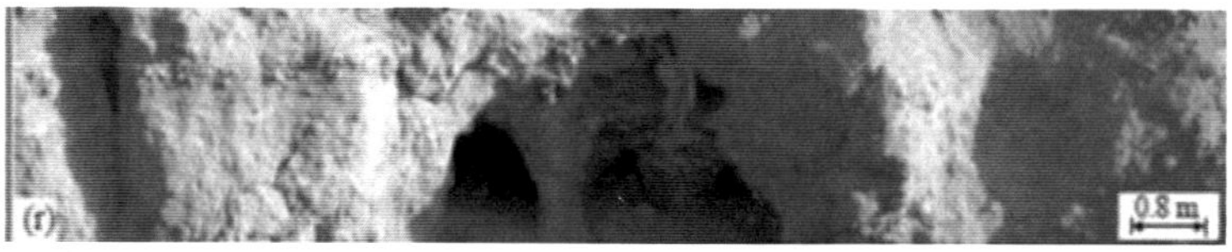

Figure 1. The cavity in different shapes and sizes [4].

2. Experimental procedure

To simulate the slab on a cavitied soil system as a real matter, a concrete slab and soil container box were made in the laboratory. The system was provided by a hydraulic press gaged jack of (20 Ton) capacity. All the details are explained below.

2.1 Soil preparation work

A wooden box of (1500x1500x1300 mm) was made as a container for the sandy soil. These dimensions were detected from the elastic theory curves method [7]. According to this theory, the load diffused laterally in the soil by 2.5B in each direction and 2B for the depth (where B is the slab width). The wooden box contains two doors that used for formation cavities. The cavities are supposed to be in continuous cylindrical shape. It was made by putting a PVC plastic pipe into the soil media in specified depth, then the soil above it will be compacted. After compacting finishing, the PVC pipe pulls out from the box doors, as shown in figure 2. The amount of soil which is filled the container was detected from the standard Procter test. The soil was well graded sand with a gypsum content equal 4.8 %. The soil was chosen as gypous to simulate the nature reality case. The test showed that the optimum density of the soil was 1844 Kg/m^3 and the optimum water content was 12%.

Figure 2. Cavities work formation.

2.2 Press system configuration

A steel frame of I-sections was built around the wooden container. The frame was used to hold a gaged hydraulic jack press. The jack was hanging using its adjustable rotary ring, which is used to tweak the hunger arm, where welding is Forbidden to avoid the thermal damage of the oil inside the jack as shown in figure 3.

IOP Conf. Series: Materials Science and Engineering **584** (2019) 012055 doi:10.1088/1757-899X/584/1/012055

Figure 3. Method of hanging jack.

2.3 Test specimens

In this research, twelve concrete slabs were prepared. Ten specimens are of (600*600*40 mm) dimensions, three of them were reinforced by (9 ϕ 4 mm) at (65 mm) spacing, and the others were plain. The compressive strength of the concrete (f ′c) equals (30 MPa). The maximum aggregate size in the mixture was (9.5 mm). The last two specimens were of (600*400*40 mm) and (600*300*40 mm). One plain slab and one reinforced slab were used as a reference and tested on soil without a cavity. Cavities in two positions were made, (350 and 180 mm) below the soil surface (which are equal (30 and 15%) of the soil continuum total height). The steel bearing plate used were: square of (200*200*20 mm), rectangular of (300*200*20 mm) parallel to the cavity direction, and rectangular of (300*200*20 mm) perpendicular to the cavity direction. Table 1 shows the specimens details.

Table 1. Tested specimen details (all dimensions in mm)

Group	Slab	Dimension	Reinf. details	Steel plate dim.
Reference	SP.S.	600*600*40	———	200*200*20
	SR.S.	600*600*40	4@65mesh	200*200*20
Group 1: 15% cavity depth	SP.S.30	600*600*40	———	200*200*20
	SR.S.30	600*600*40	4@65mesh	200*200*20
	SP.RP.30	600*600*40	———	300*200*20 (parallel to cavity direction)
	SP.RV.30	600*600*40	———	300*200*20 (Vertical to cavity direction)
Group 2: 15 % cavity depth	SP.S.15	600*600*40	———	200*200*20
	SR.S.15	600*600*40	4@65mesh	200*200*20
	SP.RP.15	600*600*40	———	300*200*20(parallel to cavity direction)
	SP.RV.15	600*600*40	———	300*200*20 (vertical to cavity direction)
	RPP.S.15	600*٣00*40 Parallel to cavity	———	200*200*20
	RPV.RV.15	600*٤00*40 vertical to cavity	———	300*200*20 (Vertical to cavity direction)

Three dial gages were fixed at the concrete slab centre and two opposite diagonal corners to read the deflection at these points. The gages were fixed at the top surface of the concrete slab specimen as shown in figure 4.

Figure 4. Dial- gages installation

3. Tests results and discussions

The load deflection curves and comparing studies were discussed in the following section.

3.1. References models results

As shown in figure (5), when a slab tested sandy soil without cavity, the ultimate load of the plain slab was (17KN) and the reinforced was (32KN). The increment in the ultimate load equals (70.58%) when the slab is reinforced. The figure below shows the behavior of concrete slabs under loading.

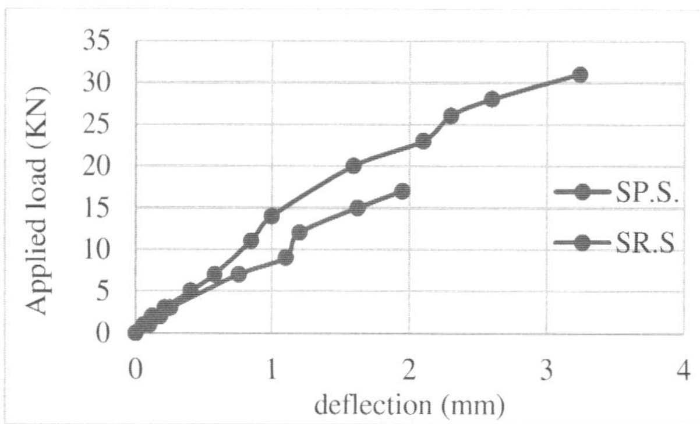

Figure 5. References results.

3.2. Cavity depth effect on reinforced concrete slabs

The failure load of the reinforced concrete slab reduced by (7.4%) and (61%) when there is a cavity located at depth (30%) or (15%) respectively, below the surface of the slab as in figure 6.

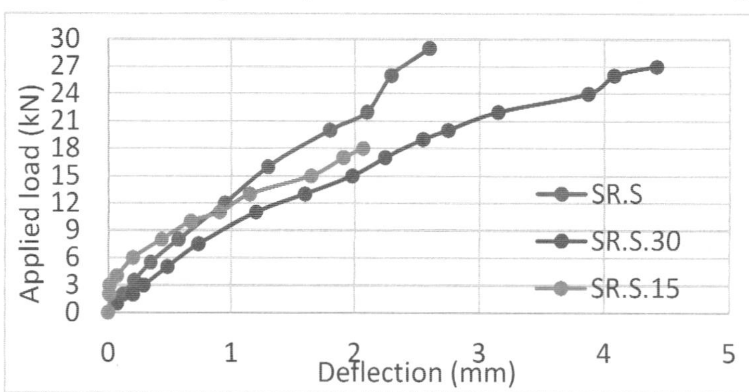

Figure 6. Effect of cavity depth on reinforced concrete slabs of group one.

3.3. Cavity depth effect on plain concrete slabs

The difference in the ultimate load of the slabs in the case of no cavity, cavity existence at depth (30%), and cavity existence at depth (15%) were illustrated in figure 7. The results show that the slab loss about (30.7 %) of its capacity when there is a cavity located at depth equals (30%) and about (70%) when there is a cavity located at depth equals (15%).

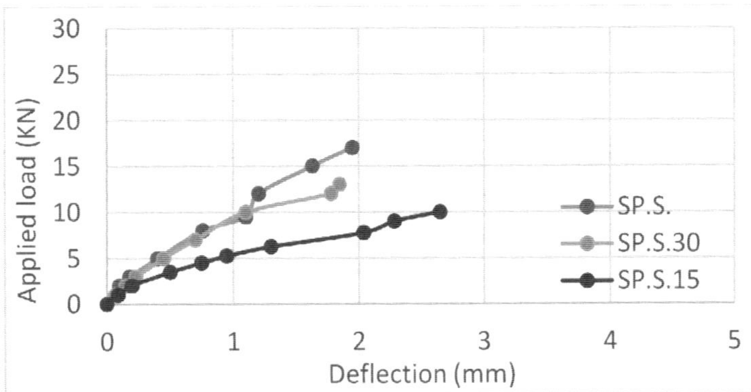

Figure 7. Comparison between the cavity depth difference using a square plate.

3.4. Comparison between plain and reinforced slab at two cavity depths

The effect of steel wire reinforcement on the ultimate load of the slab is illustrated in figure 8. The ultimate slab load enhanced by about (107 %) compared with the case of plain concrete. Figure 9 explained a comparison study between the specimens SP.S.15 and SR.S.15, to maintain the effect of steel wires existence. The results showed that the concrete slab capacity increased to (80%) when the slab is reinforced and on cavity at depth equals (15%).

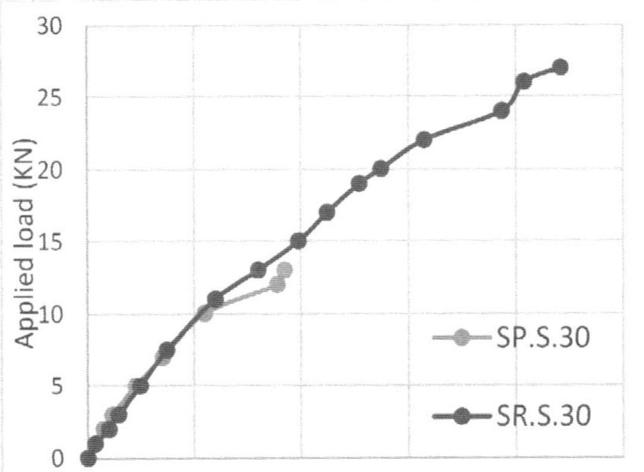

Figure 8. Comparison between plain and reinforced slab for Group one.

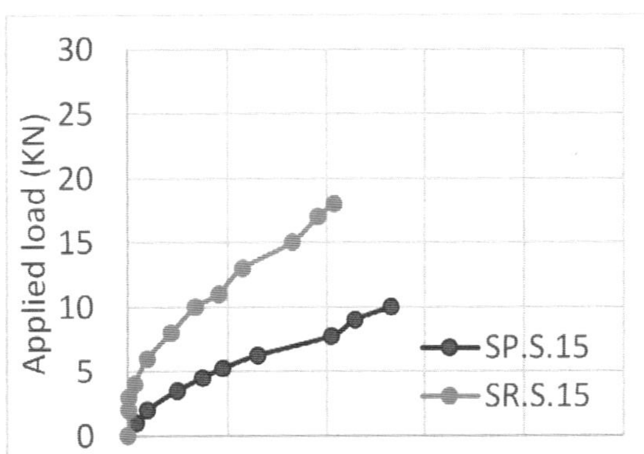

Figure 9. Comparison between plain and reinforced slab for Group two

3.5. Bearing plate shape effect on slabs with 30% cavity depth

As concluded from the experimental work, when a cavity forms at any depth, the soil will divide into two regions. The first region is the soil which lies on the sides of cavity (shoulders), and the second region is the soil lies in between slab bottom and the upper surface of the cavity (which is sometimes be a very thin shell and called by cover). Every group will share in load resistance but the shoulder zone is stiffer than the cover zone. As shown in figure 10, the slab SP.RV.30 shows a failure load greater than the others, because the rectangular perpendicular plate transforms the load into the shoulder zone. Also, the slab SP.RP.30 shows the min. Failure load and max. deflection, since that the rectangular parallel bearing plate distributes the load towards the cavity (i.e.) the cover zone. From SP.RP.30 and SP.RV.30 results, the slab loss (50 %) of its total capacity when the load was distributed towards the cavity direction. Whereas, when using a square plate in the specimen (SP.S.30), the ultimate load capacity of the slab reduces by about (15.38 %) from the case of using a perpendicular bearing plate (SP.RV.30). The slab reduces (30 %) of its capacity when using a parallel rectangular plate instead of a square bearing plate. From the results of (SP.S.30) test, the concrete slab capacity reduces (30.77 %) of its total capacity in case of cavity existence at depth (30%).

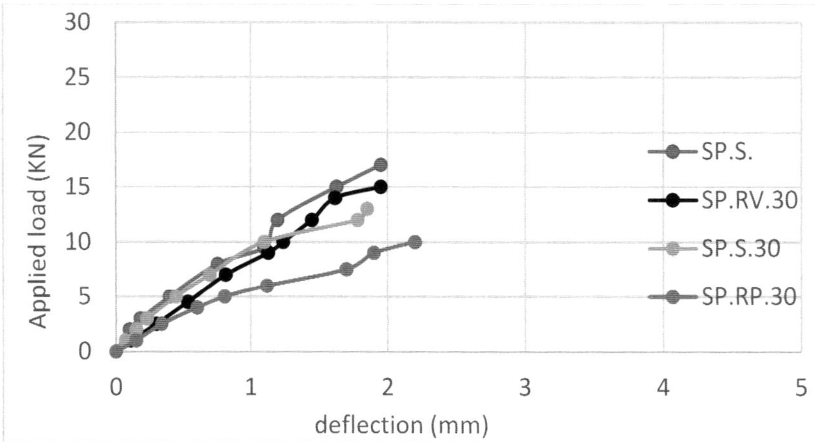

Figure 10. Effect of bearing plate shape on tested slabs for group one.

3.6. Effect of bearing plate shape on slabs on cavity 15% depth

The slab SP.RV.15 shows a failure load greater than the others as shown in figure 11, because of that the rectangular perpendicular plate transforms the load into the shoulder zone. Also, the slab SP.RP.15 shows the minimum failure load and maximum deflection, due to that the rectangular parallel to cavity direction plate distributes the load towards the cavity ((i.e.) the cover zone). The slab loses (69.3 %) of its ultimate load capacity when cavity exists at depth equals (15%), Also the slab capacity reduces by (20 %) when using steel plate parallel to the cavity direction instead of a square plate. Form curve SP.RP.15 and SP.RV.15, the slab capacity reduces to (62.5 %) when the load distributed towards the cavity direction by the rectangular parallel steel plate. While when using a square bearing plate (SP.S.15), the ultimate load capacity of the slab reduces a (30 %) of its total capacity from the case of using a perpendicular rectangular bearing plate.

3.7. Effect of minimizing slab dimensions

When using a square steel plate, and the slab area reduces from (600*600*40 mm) to (600*300*40 mm), the ultimate load capacity reduced by (5.26 %) from the slab SP.S.15 (total area) and (78.9 %) from (SP.S.) the case of without cavity (as explained in figure 12).

When changing the area of slab from (600*600*40 mm) to (600*400*40 mm) the ultimate load of concrete slab increases by (23.07 %) from the case of total slab area (SP.RV.15). Figure 13 illustrated the effect of decreasing slab dimensions.
This is a because a combination effect of two parameters, the first effect is that, the smaller slab area leads to little soil bearing pressure (i.e. the soil will not collaborate in resist the load such as the past

cases) and that will hasten slab failure, the second effect is that, when slab area is reduced, the deflection in the z- direction will reduced too, and that may late the slab failure.

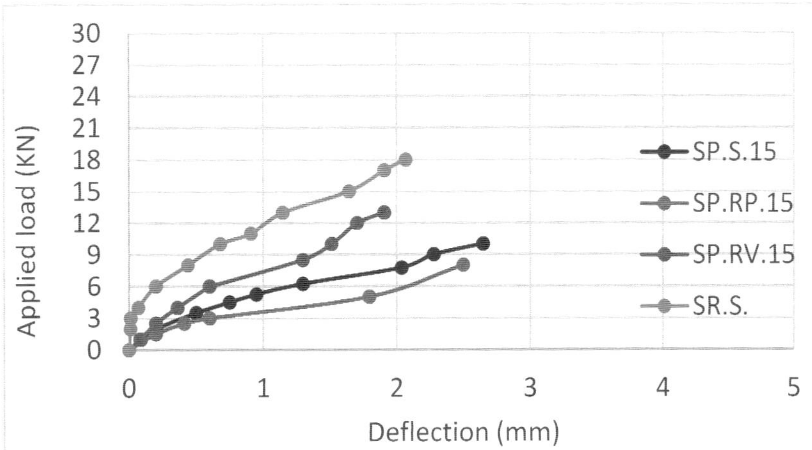

Figure 11. Effect of bearing plate shape of slabs of Group two.

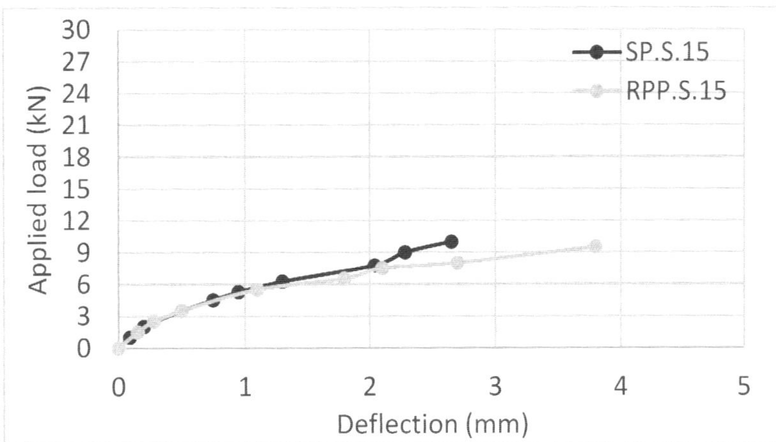

Figure 12. Effect of reducing the slab dimensions.

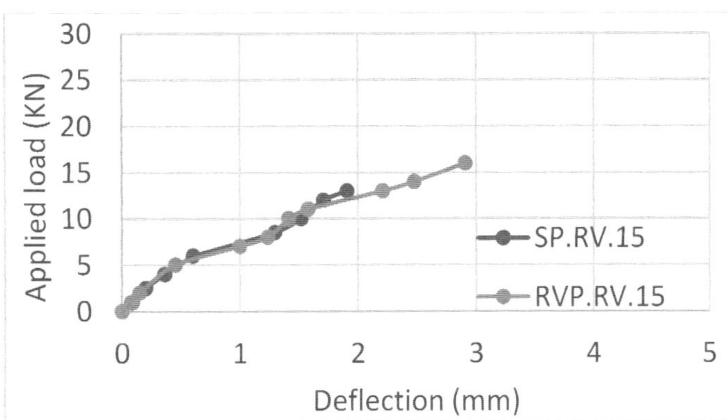

Figure 13. Effect of reducing slab dimensions.

3.8. Effect of cavity depth difference and using rectangular parallel steel bearing plate

The difference in the ultimate load of the slabs in the case of no cavity, cavity existence at depth (30%), and cavity located at depth (15%) were shown in figure 14. The results show the slab losses about (20 %) of its total ultimate load when the cavity location change from (30 to 15%).

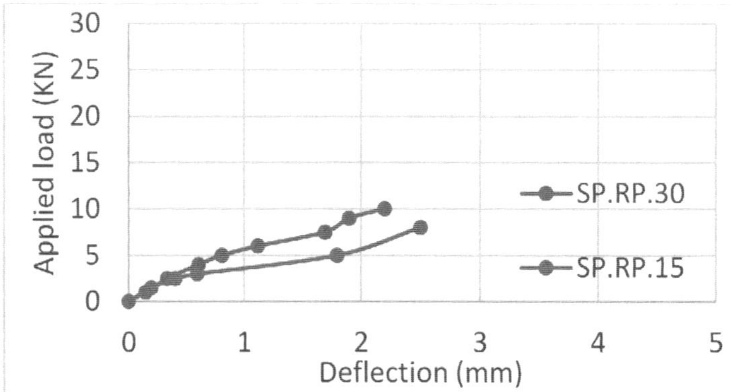

Figure 14. Comparison between cavity depth difference using rectangular plate.

3.9. Effect of cavity depth difference and using rectangular perpendicular steel bearing plate

The difference in the ultimate load of the slabs in the case of no cavity, cavity located at depth (30%), and cavity located at depth (15% mm) are shown in figure15. The results show that the slab loss about (15.38 %) of its total ultimate load when the cavity converge to the surface by (170 mm). Noting that, placing the bearing plate perpendicular to the cavity leads to increase in the overall slab resistance because the load has been distributed towards the stronger soil bath.
When minimizing the slab dimensions from (600*600*40 mm) to (600*400*40 mm), and using a rectangular perpendicular plate, the ultimate load of the slab increases (23.07 %) because the lesser slab span leads to minimizing cracks in the other smaller direction.

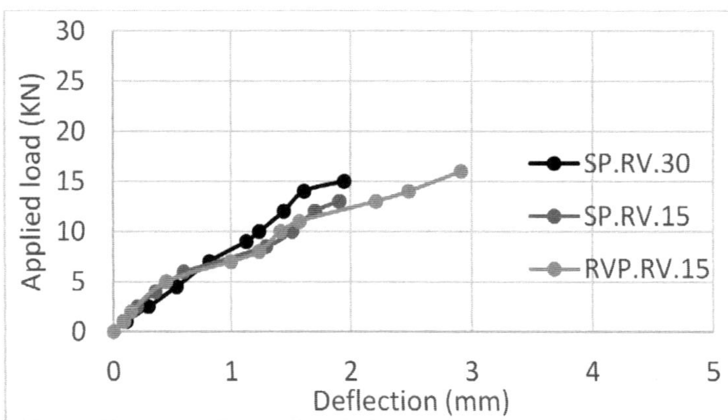

Figure 15. Rectangular vertical plate with cavity depths comparison.

4. Failure modes

All the specimens failed by flexural. For specimen SP.S. and SR.S. (Figure 16. A and B), the failure line occurs in the three paths. The direction of the paths depends on the gravel's particles in concrete and soil.

For specimens SP.S.30 (Figure 16.C), SP.RP.30 (Figure16.D), SP.S.15 (Figure 16.E), RPP.S.16 (Figure 16.G) and RVP.RV.15 (Figure 16.F), the failure path occurs parallel to cavity direction, which is the weak path.

For specimens SP.RV.30 (Figure 17.H.), SP.RP.15 (Figure 17.I), and SP.RV.15 (Figure 17.J), the specimens crashed for three part, because the rectangular steel plate transfers the load in a path vertical to the cavity direction. For specimens SR.S.30 the cracks separated in a direction parallel to the cavity (Figure 17.K)), and the transversal cracks was little when comparing with specimen SR.S.15 (Figure 17.L)), due to the shallower cavity.

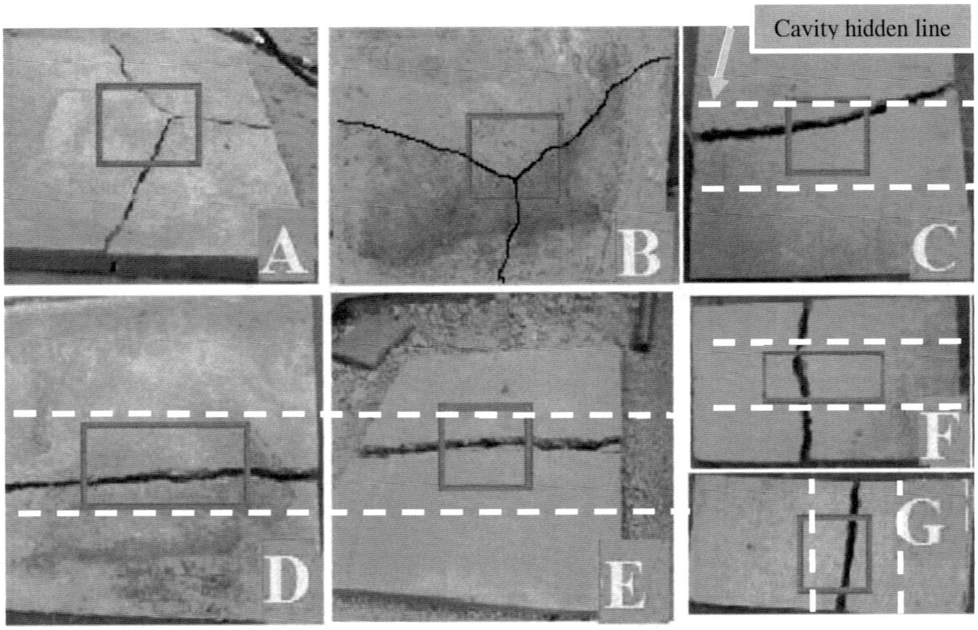

Figure 16. Failure mode of specimens.

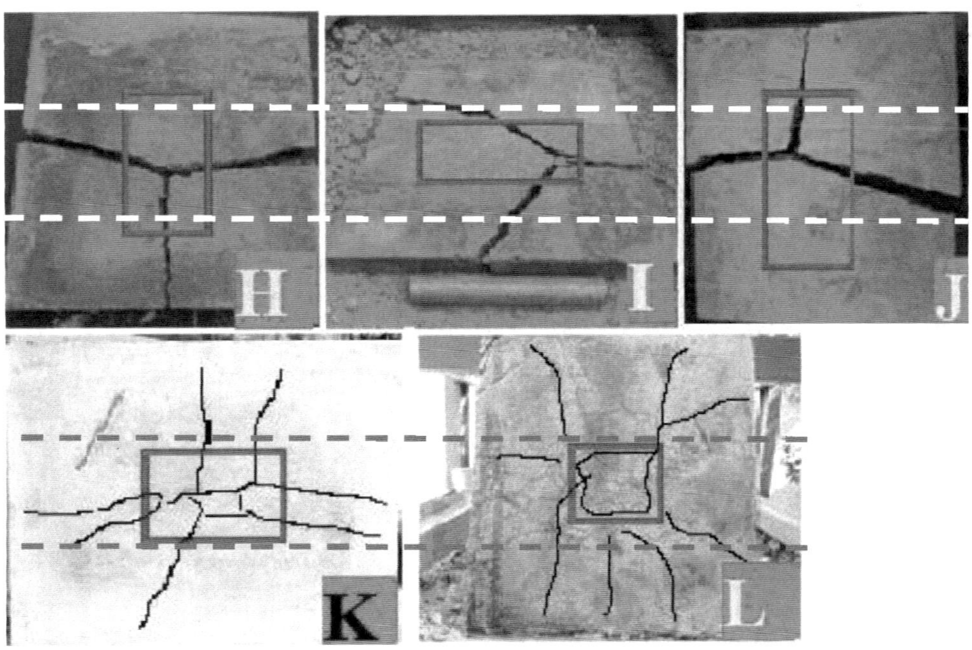

Figure 17. Failure mode of specimens.

5. Conclusions

In accordance with the results, the following points have been found:
1. The existence of steel wire reinforcement leads to increase the slab on grade capacity by (70.58 %) in the case of without cavity.
2. The ultimate bearing capacity of the reinforced concrete slab decreases by (7.4 %) and (61 %) in the case of cavity existence at depth (30%) and (15% mm), respectively.
3. The plain concrete slab capacity decreases (30.77%) of its capacity when a cavity exists at depth (30 %), and (70 %) when cavity is found at depth (15%) in case of square loading plate.
5. In the case of cavity existence at depth equals (30% and 15%), the ultimate capacity of reinforced concrete slab rises to (115 and 80 %), respectively, when compared with the plain slab.
6. The slab loses about (50 %) of its ultimate capacity when the load distributed towards the cavity direction by the rectangular plate (in the case of cavity existence at depth 30%), and when using a square plate for the same cavity depth, the ultimate load reduces by (15.38 %) when comparing with using perpendicular rectangular plate.
7. The slab loss about (62.5 %) of its ultimate capacity when the load distributed (by the rectangular plate) towards the cavity which is of (15%) depth, and when using a square plate, the ultimate load reduces by (30 %) when compared with the perpendicular rectangular plate.
8. When the slab dimension reduces to (600*300*40 mm), the ultimate bearing capacity of the slab reduced by (5.26 %) at (15%) cavity depth.
10. When the slab dimension reduces from (600*600*40 mm) to (600*400*40 mm), the ultimate bearing capacity of the slab increased by (23.07 %) at cavity depth equals (15%).
11. When the cavity depth reduces from 30% to 15%, the ultimate load of slab reduces (30.7 %) when using square bearing plate, (25 %) when using the rectangular parallel to cavity direction, and (6.6 %) in the case of the load distributed by rectangular plate oriented perpendicular to cavity direction.

References

[1] Ahmed F, Said M, and Najah L 2012 *International Journal of Scientific and Research Publications* 2(9) 1-5.
[2] Buringh P 1960 *Soils and soil conditions in Iraq* (Directorate General of Agricultural Research and Projects) p 13.
[3] Seleam S 2006 *Journal of engineering* 3 (13) 712 726.
[4] Aziz L J 2002 *Influence of water on the behavior of strip footing above a cavity* (Master dissertation in Civil Engineering Department Technology University/Iraq).
[5] Aziz L J 2008 *Lateral resistance of single pile embedded in sand with cavities* (PHD thesis in Department of Civil Engineering Technology University/Iraq).
[6] ACI Committee 360 1992 *Design of Slabs on Grade*. American Concrete Institute, Michigan, pp. 57.
[7] Budhu M 2011 *Soil mechanics and foundations* (New York John: Wiley & Sons, INC).

Bearing capacity and settlement of shallow footings on liquefiable soil subjected to seismic loading

Ahmed M. Sadeq Al-Salakh[1] and Bushra Suhale Albusoda[2]

[1] Graduate Student, Civil Engineering Department, University of Baghdad, Baghdad, Iraq; Faculty member at Civil Engineering Department, University of Kufa, Najaf, Iraq. Email: ahmedm.alsalakh@uokufa.edu.iq
[2] Professor, Civil Engineering Department, University of Baghdad, Baghdad, Iraq. Email: dr.bushra_albusoda@coeng.uobaghdad.edu.iq

Abstract. In liquefying soils, Shallow foundations may experience a reduction in bearing capacity and increase in settlement and tilt due to seismic loading. Therefore, the buildings on shallow footings may settle and tilt excessively. In this paper, the bearing capacity and settlement of shallow foundation on liquefiable deposits has studied. The study includes a shaking table test, reviewing of theoretical equations (available in literature), which estimating settlement of shallow footings due to seismic loading, calibration, and verification of these equations with data from the shaking table test for improved soil by grouting and unimproved soil. It is important to note that the grouting materials used in this study are the CKD and Bentonite slurries. A modification to the seismic bearing capacity and settlement equations had been done to account for the liquefaction state. Good convergence has been gotten between predicted and measured settlement.

Keywords: Theoretical evaluation, CKD, Grouting, Liquefaction, bearing capacity, Settlement, Bentonite, granular piles.

1. Introduction

Problem of the seismic bearing capacity of shallow foundations is far from being considered solved. Many researchers stated that the response of a footing to dynamic loads is affected by:
(1) Nature and magnitude of dynamic loads,
(2) Number of pulses and
(3) The strain rate response of soil.

Shallow foundations for seismic loads are usually designed by the equivalent static approach. The foundations are considered as eccentrically loaded and the ultimate bearing capacity is accordingly estimated. To account for the effect of dynamic nature of the load, the bearing capacity factors are determined by using dynamic angle of internal friction which is taken as 2-degrees less than its static value [1].

International Building Code permits an increase of 33 % in allowable bearing capacity when earthquake loads in addition to static loads are used in design of the foundation. This recommendation may be reasonable for dense granular soils, stiff to very stiff clays or hard bedrocks but is not applicable for friable rock, loose soils susceptible to liquefaction or pore water pressure increase [2].

2. Review of related previous works

The seismic effects on the bearing capacity of foundation were studied by Meyerhof during the middle of last century and many other researchers, who adopted a pseudo-static approach. By applying horizontal and vertical accelerations to the center of gravity of the structure, the problem is reduced to a static case of bearing capacity with inclined and eccentric loads. Nevertheless, in these methods the inertial effects on the soil mass beneath the foundation are not considered and this could have important effects on the overall system response.

Isolated column footings, strip footings, mat footings, and even pile foundations all may fail during seismic events. Such failures are generally attributed to liquefaction. However, a number of failures have occurred where field conditions indicate there was only partial saturation or a dense soil and therefore liquefaction alone is a very unlikely explanation. Rather, the reasons for the seismic settlements of these foundations seem to be that the bearing capacity was reduced [3]. Richards et al [3] proposed a simplified approach to estimate the dynamic bearing capacity que and seismic settlement S_{Eq} of a strip footing for assumed failure surfaces as in equation (1). There is hardly any experimental verification of these theoretical solutions. There is a need for such validation [1].

$$S_{eq}(\mathrm{m}) = 0.174\frac{V^2}{A.g}\left(\frac{k_h}{A}\right)^{-4}.tan\alpha AE \quad (1)$$

Where: S_{Eq} is the settlement due to earthquake, V = peak velocity for the design earthquake (m/sec), A=acceleration coefficient for the design earthquake, g = acceleration due to gravity (9.81 m/sec2), the values of kh^* and $\tan\alpha_{AE}$ can be obtained from some well-known figures which related the required factor with angle of internal friction and bearing capacity as mentioned in [4] as well as in table 1 [4].

Generally, the behavior of any structure during earthquake depend on many items such as earthquake magnitude, acceleration, duration, height of the structure/width of the foundation (aspect ratio), foundation boundaries, soil properties and condition, and probability of liquefaction triggering [1].

In present study the specified structure has an aspect ratio < 0.8 according to this ratio and as cited in many researches the structure will suffer from vertical settlement only without any visible tilting [5], while in dry soil the foundation will suffer from excessive settlement and sliding in the wards of shaking.

The propagation of the acceleration amplitude through soil depth will cause additional stress and rearrangement to the soil particles to produce more dense soil, which causes a settlement and sliding failure to the shallow structures and infrastructures.

Table 1. Variation of $\tan\alpha_{AE}$ with k_h and soil friction angle φ (Compiled from [3])

	$\tan\alpha_{AE}$				
k_h^*	$\phi' = 20°$	$\phi' = 25°$	$\phi' = 30°$	$\phi' = 35°$	$\phi' = 40°$
0.05	1.10	1.24	1.39	1.57	1.75
0.10	0.97	1.13	1.26	1.44	1.63
0.15	0.82	1.00	1.15	1.32	1.48
0.20	0.71	0.87	1.02	1.18	1.35
0.25	0.56	0.74	0.92	1.06	1.23
0.30		0.61	0.77	0.94	1.10
0.35		0.47	0.66	0.84	0.98
0.40		0.32	0.55	0.73	0.88
0.45			0.42	0.63	0.79
0.50			0.27	0.50	0.68
0.55				0.44	0.60
0.60				0.32	0.50

In the settlement equation, it is clear that it neglects the effect of saturation or PWP generation, changes in soil density due to shaking, mitigation, and changes in effective stress or bearing capacity failure. To understand the effect of these factors, it is very important to understand the effect of each item on soil behavior individually and together.

As well as the settlement formula [3] suggested a simplified approach to evaluate the dynamic bearing capacity q_{ue} and seismic settlement S_{Eq} of a strip footing for assumed failure surfaces. The proposed seismic bearing capacity (q_{uE}) is given by equation (2) as follow:

$$q_{uE} = cN_{cE} + qN_{qE} + \tfrac{1}{2}\gamma BN_{\gamma E} \quad (2)$$

Where, γ = is the soil unit weight, q= γD_f, D_f is the depth of the foundation,

N_{cE}, N_{qE}, *and* $N_{\gamma E}$ are the factors of seismic bearing capacity.
These factors are related to the magnitude of dynamic angle of internal friction of the soil (φ_{dyn}) and $\tan\psi = k_h / (1-k_v)$
k_h and k_v are the horizontal and vertical coefficients of acceleration due to earthquake.
For static case, $k_h = k_v = 0$ and equation (1) becomes:

$$q_u = cN_c + qN_q + \frac{1}{2}\gamma B N_\gamma \tag{3}$$

In which N_c, N_q *and* N_γ are the static bearing capacity factors.
In practice and according to many investigations that carried out and recorded in many works of literature, the shallow foundation may or may not suffer from a reduction in bearing capacity during dynamic and/or earthquake loading.
At the same time of bearing capacity changes, there is an increase in total and differential settlement under shallow foundation during earthquake loading, this behavior is due to the densification of the soil strata and destroys the soil structure during shaking.
The previous explanation is for the non-liquefaction condition of the soil, in case of liquefaction is occurs both behaviors of bearing capacity and settlement are changed as follow:

- Bearing capacity of shallow foundation during earthquake loading and liquefaction of the underneath soil will decrease sharply after a few time period or loading cycles.
- The amount of total and differential settlement will be an increase in an excessive manner when the liquefaction has happened.
- Thus, the bearing capacity at earthquake loading will not increase as suggested in some researches if liquefaction triggering may occur in loose soils, friable rocks, and sensitive clay [4].

In dynamic and seismic loading, the angle of internal friction will change with the progress of loading cycles (decrease until the failure state will occur due to bearing capacity failure or due to excessive deformations and settlements).
In this paper, the bearing capacity and settlement of shallow foundation on liquefiable deposits has studied. The study includes a shaking table test, reviewing of theoretical equations (available in literature), which estimating settlement of shallow footings due to seismic loading, calibration, and verification of these equations with data from the shaking table test for improved soil by grouting and unimproved soil. A modification to the seismic bearing capacity and settlement equations had been done to account for the liquefaction state. Find out an adequate formula for estimating footings settlement so as to introducing the effect of liquefaction process which accompany with generation of PWP and increasing in the acceleration amplitude within soil column.

3. Suggested modifications on dynamic Bearing Capacity and Settlements of Shallow Footing on Liquefiable Soil formulas.

In liquefaction state, the reduction in the value of the angle of internal friction becomes noticeable during a few seconds of the earthquake which produce a rapid bearing capacity failure.
After checking the soil susceptibility to liquefaction, the selection of the suitable angle of internal friction must be achieved to avoid sudden bearing capacity failure.
In present work, the calculation of bearing capacity during an earthquake executed (simulated by shaking table test for the soil layering and sensors distribution shown in figure 1). The liquefiable soil considered in this study is poorly graded sand with relative density of 33%. The soil was improved by permeation grout using cement kiln dust (CKD) and Bentonite slurries. The grouting process was started from bottom to top by gradual steps with uniform injection process and very low injection pressure (less than 0.05 bars), using grouting machine connected to air pressure compressor so as to permit to the grouting material to pumping through pumping hose, depending on the depth of the injection and overburden pressure of the soil.
The magnitude of design acceleration must be taken at foundation level, that producing from the transfer of acceleration amplitude from the deep origin. The correct selection of the acceleration magnitude will prevent any undesirable or unexpected failure.

The comparison between acceleration values at different levels in the shaking table test shows the effect of foundation level on its behavior during shaking.

The acceleration at the foundation level will become larger than its value at the source due to the following:

- Effect of inertia mass motion.
- Effect of loose soil strata behavior during shaking.
- Effect of generation of PWP and variation of stress states.

Figure 2 presents the variation of acceleration with the depth of soil. It is clear that when the acceleration at hard strata is 0.2g the resulting surface acceleration at foundation level become more than 0.35g.

When saturated soil (the water table is close to ground level) exposed to an earthquake, the pore water pressure will increase. If pore water pressure continues to increase, liquefaction will take place in a few seconds as shown in figure 3. This effect will result in a decrease in bearing capacity of the soil. Therefore, it is proposed to include the effect of pore water pressure in the determination of the dynamic bearing capacity of the soil.

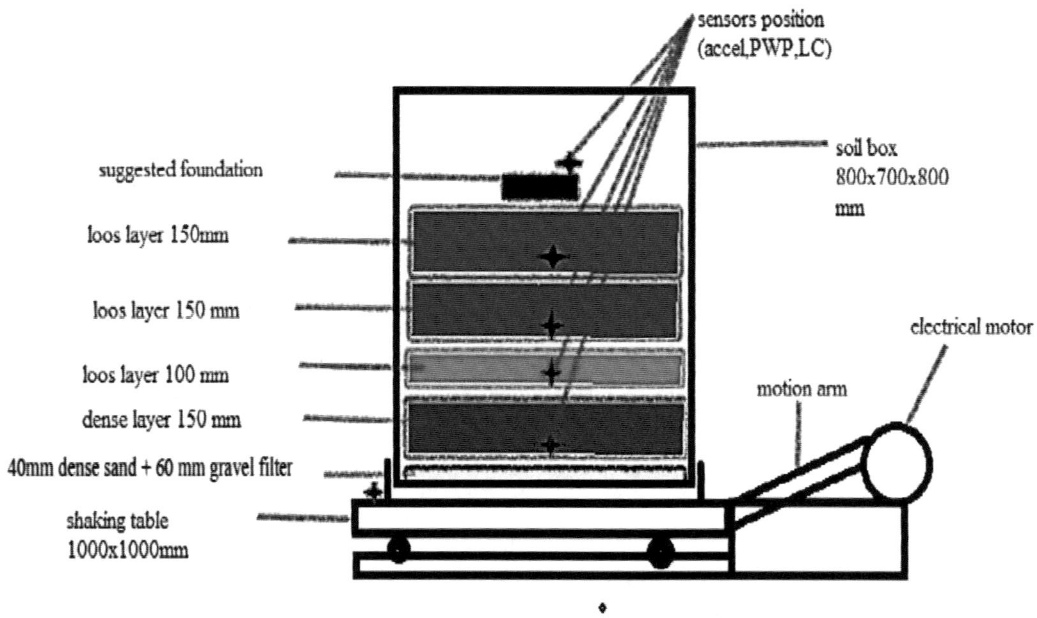

Figure 1. Shaking table schematic diagram explain soil layers and distribution of transducers.

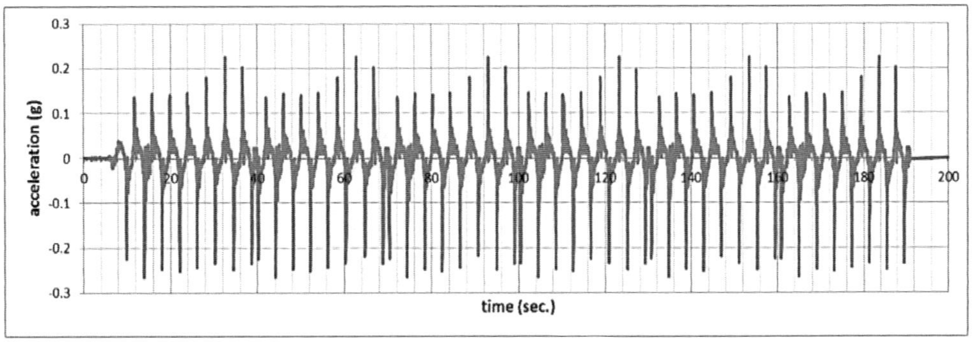

(a)

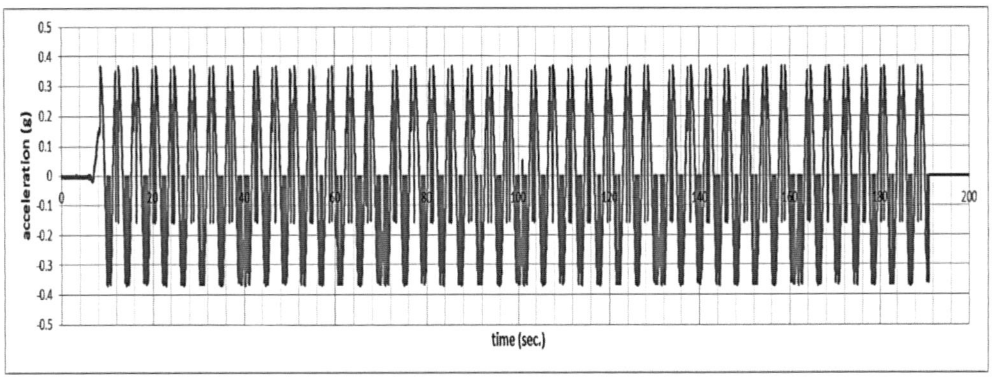

(b)

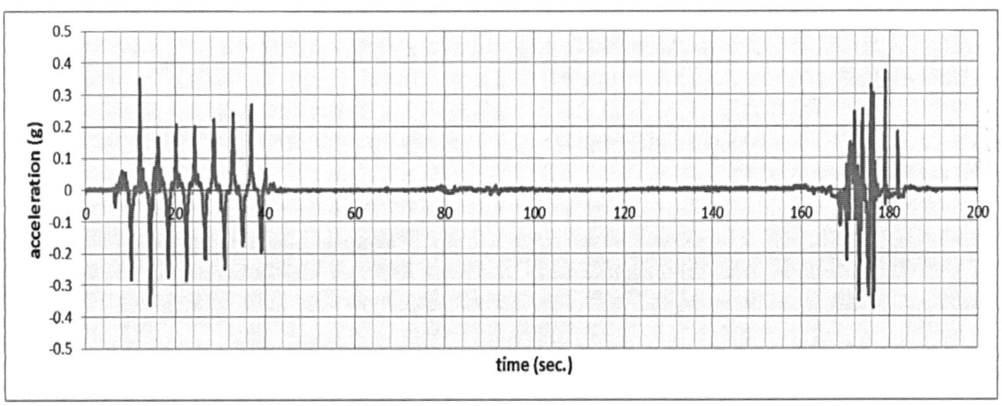

(c)

Figure 2. (a) Acceleration (g) Vs time (sec.) at bottom saturated sand layer for 1Hz.; (b) Acceleration (g) vs time (sec.) at 1st saturated sand layer for 1Hz.; (c) Acceleration (g) Vs time (sec.) at foundation for 1Hz.

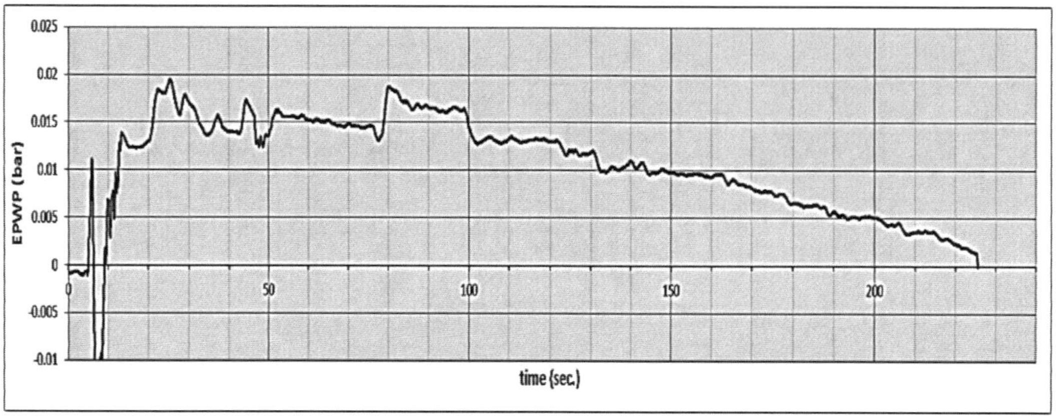

Figure 3. Generation of PWP (bar) Vs time (sec.) at top layer of saturated soil model.

The more suitable and meaningful expression which relate the effects of PWP effective stress changes is the pore water pressure ratio (r_u), this factor can be introduced in the bearing capacity equation as a control parameter for all cases as follow:

$$qu = (1 - ru) * (C.Nceq. + q.Nqe.. + \frac{1}{2}\gamma B.N^{\gamma}e) \quad (4)$$

When the soil is dry r_u value become equal to zero thus there is no effect on the original equation, at r_u equal to 0.5 which mean that the resulting bearing capacity will reduce to the half magnitude due to the generation of PWP and reduction in effective stress, and when r_u become equal or near to 1 which mean that the liquefaction has occurred in this case the bearing capacity will has zero value which compatible with actual or field situation. When liquefaction mitigation methods applied on the liquefiable soils the following changes and improvement recognized:

- The soil strength increases to resist the shaking.
- There are specific reductions in the generation of pore water pressure during shaking.
- There is an increase in soil density due to filling voids as shown in table 2.

These effects will lead to maintain the propagation of acceleration along the soil column during shaking which leads to fixing the acceleration value from bottom to top layers.

- The above changes in mitigated soil produce a noticeable increase in soil stability and the bearing capacity of the mitigated soils.

Table 2. Effect of mitigation method (mixing percent and dilution fraction) on soil unit weight.

Sat. sand unit weight kN/m^3	Sand + Bentonite slurry grout unit weight, kN/m^3						Sand + CKD slurry grout unit weight,kN/m^3						Granular pile equivalent unit weight kN/m^3
	3%		5%		10%		3%		5%		10%		
	3/1	6/1	3/1	6/1	3/1	6/1	3/1	6/1	3/1	6/1	3/1	6/1	
18.5	19.3	18.92	19.5	19.5	19.24	19.2	19.3	19.4	19.25	19	18.65	19	18.65

The presence of water within soil strata will make the soil settle due to lubrication of the particles which tend to become closer to each other. This phenomenon recognized in the experiment when the saturated soil left for one day leading to settlement. When the shaking or earthquakes are applied there are noticeable generations of PWP which cause soil liquefaction.

The change in soil unit weight due to shaking in the dry state is differing from a change in soil density due to liquefaction mitigation process. In the first case, the densification causes a specific settlement due to soil particles rearrangement while in soil mitigation the change in density is due to replacement of the water inside the voids by grouting materials without and rearrangement or densification of soil particles. This behavior exactly recognized during shaking tests at all applied acceleration. The main difference between the non-mitigated soil and the mitigated one is the magnitude of the settlement.

It is important to note that the simplified approach proposed by [3] is derived to estimate settlement for non-liquefied soils. This may explain the divergence between experimental and the simplified approach, since the settlement measured experimentally for liquefied soil post liquefaction.

Also, at mitigated soil the estimation of settlement according to the same formula required a modification so as to adjust the settlement magnitude.

Settlement formula didn't take into account the soil density, degree of saturation, generation of PWP, and reduction in effective stress during shaking. A modified formula for settlement estimation suggested.

This modification makes Richards formula takes the effect of soil density and pore water pressure generation during an earthquake in consideration, See equations (5) and (6).

For frequency amplitude ≤ 0.75 Hz.

$$S_{Eqm} = S_{Eq} * \left(\frac{\gamma soil}{\gamma new}\right)^{a} * \left(\frac{b}{(c-r_u{}^{d})}\right) \ (r_u \leq 0.90) \quad (5)$$

a, b, c, and d: are an empirical coefficient from statistical analysis by SPSS software and it equal to (58.8, -1.85, 1, and -0.133 respectively) the regression coefficient ($R^2 = 0.95$).

For frequency amplitude > 0.75 Hz.

The empirical coefficients become as follow: (-35.1, -321.6, -468.8 and -3.83 respectively) the regression coefficient ($R^2 = 0.86$).

Where γ_{new} is the new or improved soil unit weight after mitigation or improvement process. In case of r_u (0.9-1) there are additional formula can be introduce which cover the effect of pore water pressure generation and the reduction in effective stress. This formula can be written as follow:

$$S_{Eqm} = S_{Eq} * \left(\frac{\gamma soil}{\gamma new}\right)^{e} * \left(\frac{1}{r_u}\right)^{g} \quad (6)$$

Where e, is an empirical coefficient from statistical analysis which equal to 1.5 and g = $(\gamma_{soil} / (1- (\gamma_{new} - \gamma_{soil})) /100$, the regression coefficient ($R^2 = 0.89$).

The above-modified formula has been obtained using statistical analysis by using SPSS software. A good convergence between modified formula and experimental measurement noticed in table 3 and Figures 4, 5, 6 and 7.

The settlement of the foundation during shaking before and after soil mitigation were recorded for some cases and calculated using original and modified formula.

It is observed that the original formula is suitable for high acceleration levels without probability of liquefaction triggering.

Table 3. Settlements of saturated pure, mitigated soils at different acceleration levels depending [3], modified, and experimental measurements.

Settlement formula	**Saturated sand**			**Sand +5% Bentonite+6/1**			**Sand +CKD +6/1**			**Granular piles**		
	0.05g	0.1g	0.2g	0.05g	0.1g	0.2g	0.05g	0.1g	0.2g	0.05g	0.1g	0.2g
Formula of [3]	0.05	0.83	11.6	0.05	0.83	11.6	0.05	0.83	11.6	0.05	0.83	11.6
Modified Ritchard	2.0	17.6	25.3	0.064	1.56	22.2	0.065	1.03	11.6	0.05	1.4	15.15
Experimental results	2.5	19.1	26.0	1.2	2.8	31.0	0.1	0.75	10.2	0.3	3.5	15.35

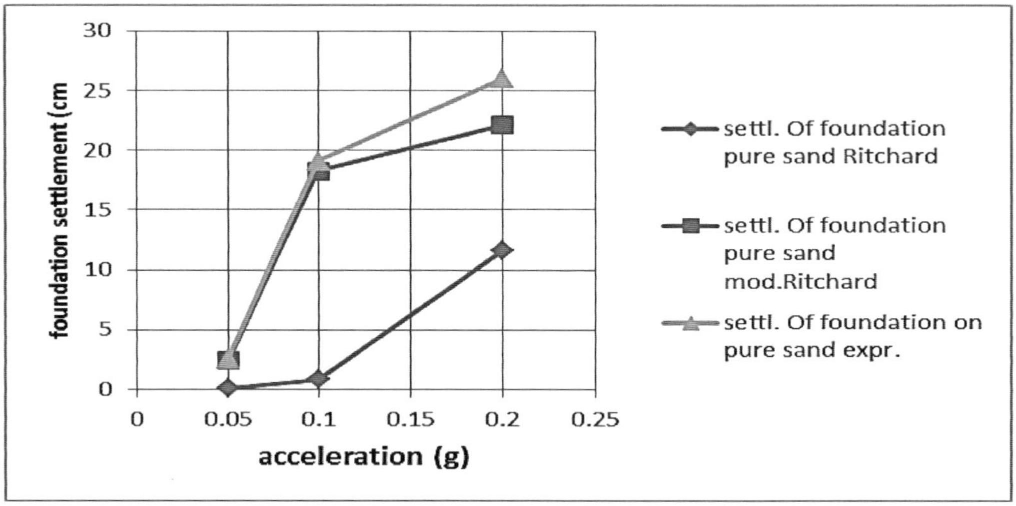

Figure 4. Relationship between settlement of foundation and acceleration on saturated pure sand.

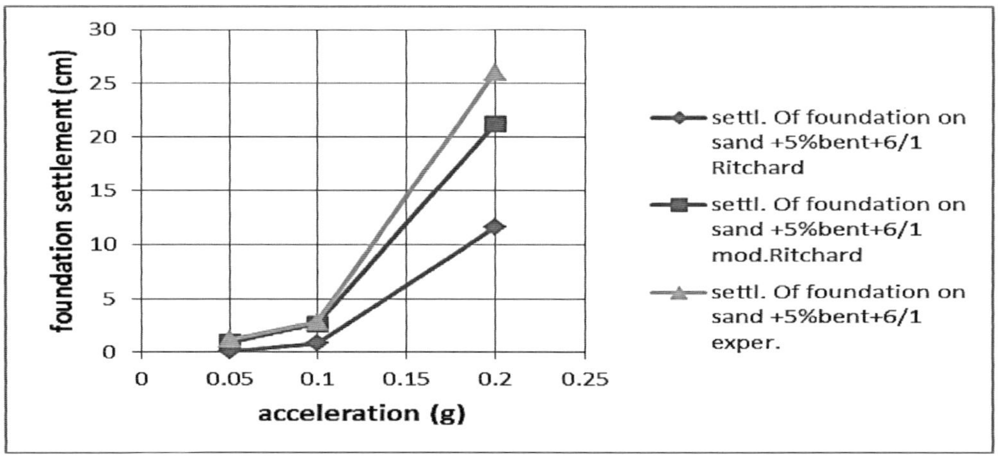

Figure 5. Relationship between settlement of foundation and acceleration on saturated sand + 5% Bentonite. + 6/1 dilution percent.

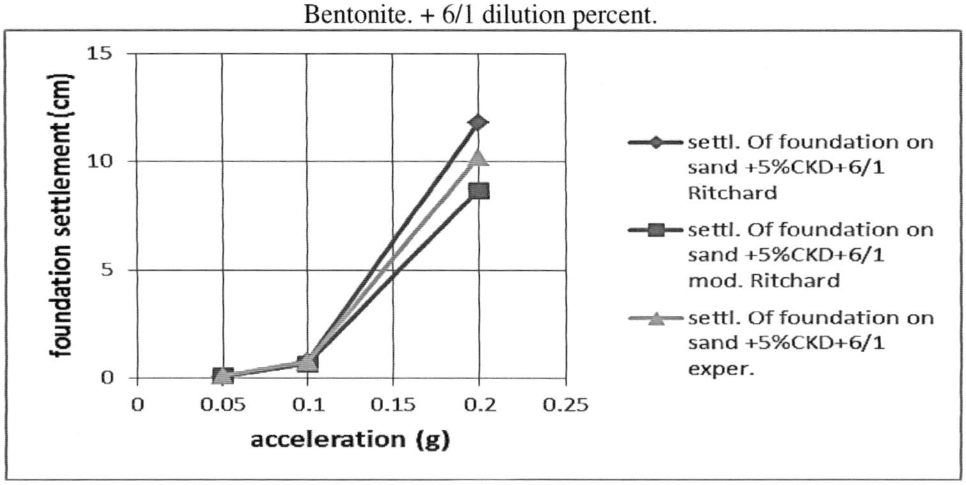

Figure 6. Relationship between settlement of foundation and acceleration on saturated sand + 5% CKD + 6/1 dilution percent.

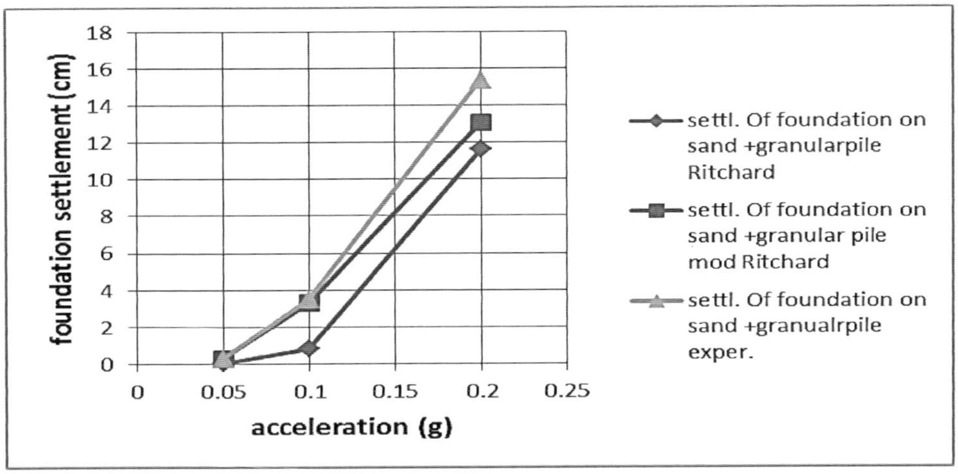

Figure 7. Relationship between settlement of foundation and acceleration on saturated sand+ granular pile.

4. Conclusions

A shallow footing resting on liquefiable poorly graded saturated sand was subjected to horizontal acceleration by a manufactured controlled and instrumented shaking table. The acceleration of the foundation and of the sand deposit is quite different from the input table motion. A modification on bearing capacity equation was introduced for seismic design taking into account the buildup in pore water pressure. This modification makes the equation applicable on liquefiable soil. Another modification was prepared for [3] equation of settlement to take into account the liquefaction triggering due to the buildup in pore water pressure. Good convergence was observed between the predicted and measured settlement.

References

[1] Vijay K. Puri and Shamsher Prakash 2007 ON Foundations under Seismic Loads 4th International Conference on Earthquake Geotechnical Engineering June 25-28, 2007, Paper No. 1118.

[2] Robert W. Day 2006 Foundation Engineering Handbook I edition, McGraw-Hill Companies Inc.

[3] R. Richards Jr., D. G. Elms, and M. Budhu 1993 Seismic Bearing Capacity and Settlements of Foundations (Journal of geotechnical engineering 1993 Vol.119) pp. 627-662.

[4] Braja M. Das and G. V. Ramana 2011 Principles of Soil Dynamics Second Edition (Stamford, CT 06902, Cengage Learning press.).

[5] Gazetas,G. M. Apostou and J.Anasta- sopoular 2004 Seismic Bearing Capacity Failure and Overtiming of Terveler Building in Adapazari 1999 Proc.5th. International conference on case histories in Geo. Engg. (New York) SOAP11(1-51).

[6] Gopal S. P. Madabhushi and Stuart K. Haigh 2010 Liquefaction Induced Settlement of Structures Fifth International Conference on Recent Advances in Geotechnical Earthquake Engineering and Soil Dynamics, 2010.

[7] H.H. Schwab 1981 Settlement and Pore Pressure Due to Cyclic Loading (Technical University of Darmstadt, Federal Republic of Germany).

[8] Indrajit Chowdhury and Shambhu P. Dasgupta 2016 Dynamic Bearing Capacity of Shallow Foundation Under Earthquake Force (Indian Geotechnical Jouranal, 2016).

[9] Jonathan D. Bray and Jorge Macedo 2017 Simplified Procedure for Estimating Liquefaction-Induced Building Settlement Proceedings of the 19th International Conference on Soil Mechanics and Geotechnical Engineering, Seoul 2017 pp 101-118.

[10] Juan Carlos Tiznado A. and Danilo Paillao V. 2014 Analysis of the Seismic Bearing Capacity of Shallow Foundations (Journal of Construction, 2014, Vol. 13(2)), pp. 40 – 48.

[11] C. W. Lu 2017 A Simplified Calculation Method for Liquefaction-Induced Settlement of Shallow Foundation (Journal of Earthquake Engineering) Vol. 21 pp1385- 1405.

[12] Riccardo Conti 2018 Simplified Formulas for the Seismic Bearing Capacity of Shallow Strip Foundations (Soil Dynamics and Earthquake Engineering Vol. 104, 2018) pp. 64–74.

[13] D.K. Karamitros, G.D.Bouckovalas, and Y.K.Chaloulos 2013 Seismic Settlements of Shallow Foundations on Liquefiable Soil with a Clay Crust (Soil Dynamics and Earthquake Engineering, Vol. 46-2013), pp 64–76.

[14] Shideh Dashti, Jonathan D. Bray, Juan M. Pestana, Michael Riemer, and Dan Wilson 2010 Mechanisms of Seismically Induced Settlement of Buildings with Shallow Foundations on Liquefiable Soil (Journal of Geotechnical and Geoenvironmental Engineering, ASCE, January 2010-151).

[15] S. Prakash, V.K. Puri, and S. Kumar 2014 Seismic Settlement of Shallow Foundations Tenth U.S. National Conference on Earthquake Engineering Frontiers of Earthquake Engineering, July 21-25, 2014, Anchorage, Alaska.

[16] A. Stamatpolous, Lydia N. Balla, and Aris C. Stamatpolous 2004. Earthquake-Induced Settlement as a Result of Densification, Measured in Laboratory Tests 13th World Conference on Earthquake Engineering Vancouver, B.C., Canada August 1- 6, 2004, Paper No. 3291.

[17] Vijay K. Puri and Shamsher Prakash 2013 Shallow Foundations for Seismic Loads: Design Considerations Seventh International Conference on Case Histories in Geotechnical Engineering, 2013, paper No. OSP6.

[18] Xiaoming Yuan, Shangjiu Meng, Zhaoji Shi, and Rui Sun 2000 A Procedure for Evaluation of Differential Settlements of buildings during earthquake 12th world conference on earthquake engineering, Auckland, New Zealand, Sunday,30 January-Friday, 4 february,2000, paper No. 1465.

[19] Zana Karimi, Zach Bullock, Shideh Dashti, Abbie Liel, and Keith Porter 2017 Seismic Settlement of Shallow-Founded Structures on Liquefiable Ground University of Colorado Boulder, 2017.

IOP Conf. Series: Materials Science and Engineering **584** (2019) 012057 doi:10.1088/1757-899X/584/1/012057

Assessment of Soil Tunnel Interaction in Sand Soil

Omar Yaseen Almashhadany[1, *] **and Bushra Suhale Albusoda** [1]

[1] College of Engineering, University of Baghdad, Baghdad, Iraq.
*Corresponding author, email: omar.yasin66@yahoo.com

Abstract. In this study, series of shaking table tests were carried out to investigate the effect of tunnel model depth and direction of ground acceleration on the ground surface settlement and the dynamic earth pressure on the tunnel wall. Box tunnel model is embedded in dry and saturated soil (Karbala Sand). It subjected to three input sinusoidal motions 0.05g, 0.1g, and 0.2g. Two relative densities were used, 30% for the upper layer and 70% for the lower layer. The results obtained indicate that the settlement of soil surface increases with an increase in ground acceleration. The increase in tunnel model depth leads to an increase in settlement especially at 0.2g earthquake loading (for saturated and dry soil). The change of direction of seismic loading leads to a small change in the settlement for dry soil. The results appear that dynamic earth pressure (DEP) was positive in dry soil and negative in saturated soil in all test. In saturated soil, the effect of direction of ground acceleration and depth of tunnel model on DEP in 0.2g is quite little.

1. Introduction

The large underground structures and tunnels subjected to large damage during/after the earthquake leads to cause large deformation or even collapse of these structures. In seismic zones, tunnel structures must be designed to withstand significant seismic forces and static overburden loads. The response of tunnel subjected to seismic loading has been connected to the response of surrounding soil due to inertia response and deformation of soil that controls the response of the tunnel. Therefore, main focus when studying the response of tunnel is on the soil- tunnel interaction effect of uplift force on a tunnel in liquefied soil, transverse raking deformation, mechanisms of transferring the load between the surrounding soil and tunnel, and failure mechanisms of the tunnel.
The seismic response of the underground structures has been extensively investigated through a series of numerical [9, 10], and experimental studies [11, 13]. However, some seismic response researches of box-shaped tunnels being still studying, including seismic earth pressure on the side wall of the tunnel, seismic displacement above and around the structure and a complex mechanism of dynamic deforming of the tunnel during vibration.
In previous studies, all researchers have examined the response of tunnel structure under seismic loading. Their studies focused on the racking and the uplift displacement of the tunnel structure. In this study, the focus will be on studying the effect of soil saturation on the tunnel response at different depths and the horizontal earth pressure on the tunnel side walls.

2. Experimental Programs

2.1. Physical properties of soil

In the present study, Karbala sand has been used. Standard tests were carried out according to ASTM standards to determine the physical and mechanical properties of sand. Figure 1 shows the grain size distribution of the soil [4], while the physical properties of sandy soil have been presented in table 1.

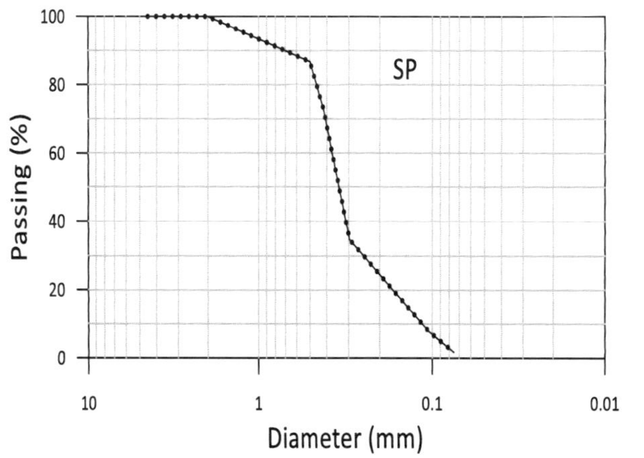

Figure 1. Grain size distribution for sand.

Table 1. Soil properties.

Index properties	Specific gravity	Maximum void ratio	Minimum void ratio	Angle of internal fraction (ϕ)
	ASTM standard [8]	ASTM standard [5]	ASTM standard [6]	ASTM standard[3]
Values	2.63	0.86	0.65	310

2.2. Tunnel model

A square plastic tunnel has been used with dimensions (25×25) mm and thickness of 1.5 mm. The length of the tunnel is 760 mm, (less than the width of the steel box (800 mm)) to avoid any interaction between the tunnel and the steel box that affect the behavior of the tunnel. The two ends of the tunnel are closed by paste and silicone to prevent the entrance of soil and water inside the tunnel. Also, the tunnel ends are fixed by two-rod bolts that connected with two sides of the steel box to prevent the movement in any direction. Also, the pressure sensor has fixed on the right wall of the tunnel, see figure 2.

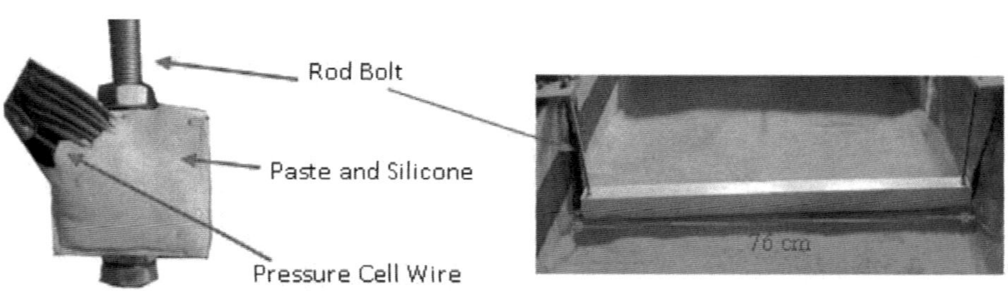

Figure 2. Tunnel model.

The mechanical properties of the tunnel model have determined by the tensile test according to ASTM specification [7]. The results of the test have shown table 2.

Table 2. Mechanical properties of the tunnel section.

Properties	Moment of Inertia (mm^4)	Poisson's Ratio v	Modulus of Elasticity E (N/mm^2)
Value	1953.13	0.4	106.5

2.3. Shaking table

A shaking table was used to study the response of tunnel structure subjected to seismic loading. The shaking table is manufactured by [2] and developed in this study. It includes three parts: 1) Shaking table base and Electrical motor, 2) Steel box; and 3) Damping system, see figure 3.

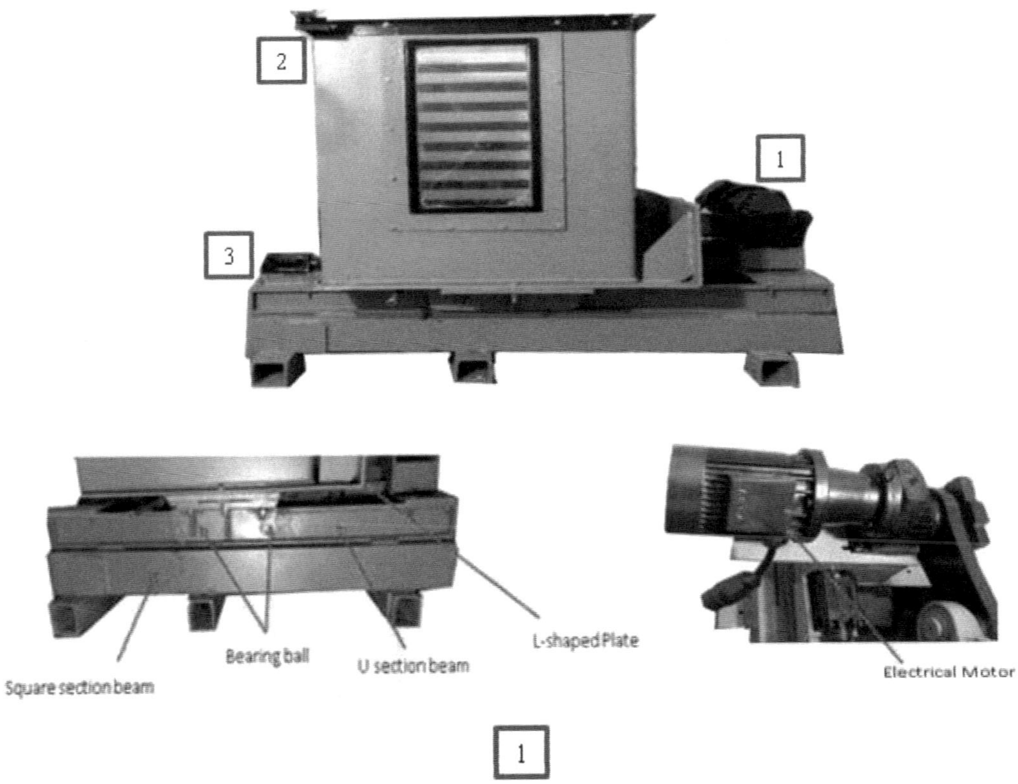

Figure 3. Shaking table: 1) Shaking table base and electrical motor 2) Steel box; and 3) Damping system.

The soil-tunnel models were subjected to three different types of input motion 0.05g, 0.1g and 0.2g during all tests using shaking table for 30 seconds. The input motion time histories have presented in figure 4.

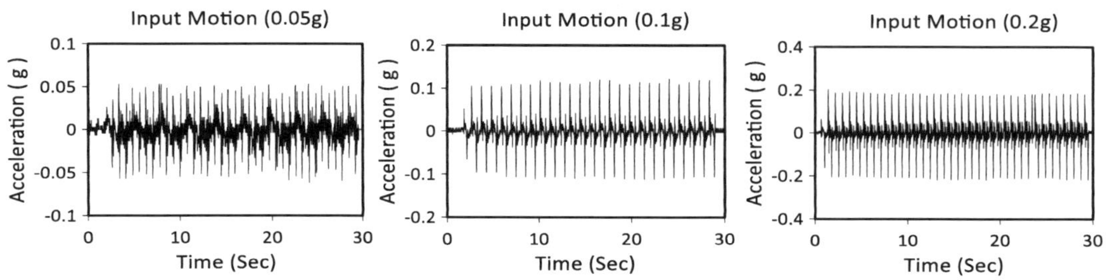

Figure 4. Input motions.

2.4. Seismic loading test of tunnel model

The tamping method was used for soil preparation to get the required relative density. Two relative densities are used (30% for the first layer and 70%for the bottom layer). When the foundation level of the tunnel has reached, the soil surface has leveled and the tunnel placed at the middle of the container. Rod bolts (10mm) have been used to fix the tunnel ends. Then fix the pressure cell on the tunnel. After the tunnel installation has finished, the filling of soil has continued until reaching the final layer. Then the LVDT sensor is placed above the middle of the tunnel to record the settlement of the soil surface, see figures 5.

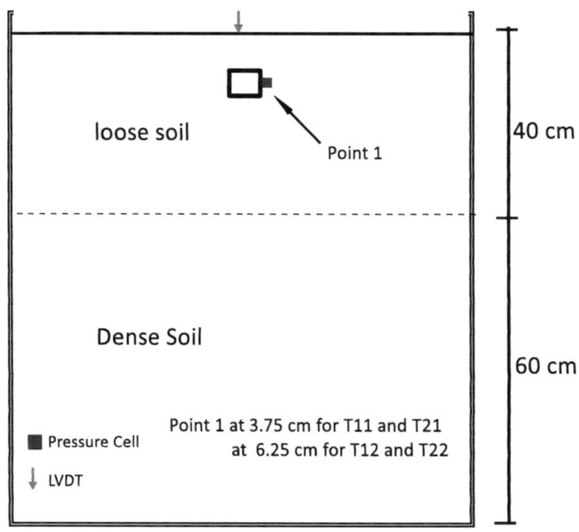

Figure 5. Sensors distribution.

3. Results and Discussions

The model tests include studying the effect of the following parameters: i) soil saturation, ii) tunnel depth (H/W=1, 2) where H is the thickness of the soil layer above tunnel and W is tunnel width. , iii) input motion (0.05g, 0.1g, and 0.2g), and iv) direction of dynamic loading with respect to tunnel direction. The model tests divided into four groups (T11, T12, T21, and T22). Each group consist of 6 tests, see figures 6 and figure 7.

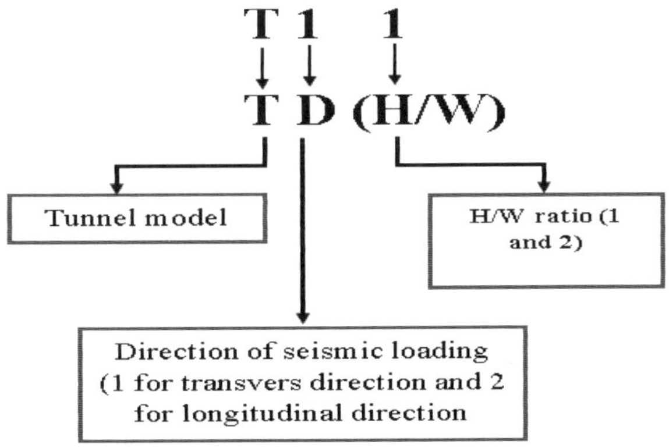

Figure 6. Group definition.

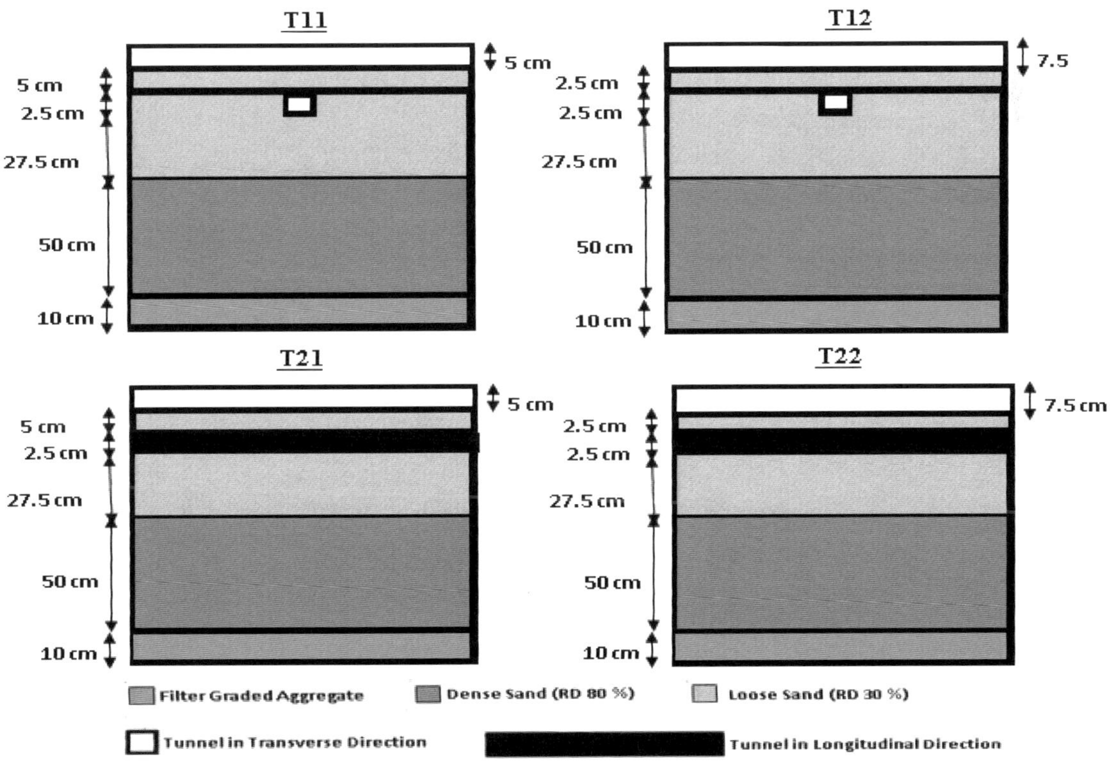

Figure 7. Schematic diagram for groups.

3.1. Soil surface settlement

The settlements of the ground surface have recorded above the center of the tunnel model using LVDT. In dry cases, the soil surface settlement in all groups increases as the ground acceleration increases due to more densification of loose sandy soil and the full slip of soil around the tunnel. The seismic load 0.05g has little effect on the settlement of the soil surface. The settlement of soil is quite similar in all groups. In other words, the effect of tunnel model depth and direction of ground acceleration is seemed low. Also, the effect of direction on the soil surface settlement is small for ground acceleration 0.1g and 0.2g. While the effect of tunnel model depth in 0.1g is noticeable where the settlement of T12 and T22 is less than T11 and T21 respectively. It is observed a contrary behavior at ground acceleration 0.2g where the settlement of T11 and T21 is less than T12 and T22 respectively. This behavior can be attributed to the thickness of the loose soil layer above the tunnel structure in T12 and T22 is greater than its thickness in T11 and T21 due to the large slip of soil around the tunnel because of the highly seismic load. This results associate the findings by [1].

For saturated soil, the settlement is greater with the increase in ground acceleration if compared with the dry soil test results due to the increases in pore water pressure. The increase of ground acceleration leads to reduce the shear strength of soil, especially for input motion 0.1g and 0.2g. Because the increase of pore water pressure in voids between particles lead to increase the spacing between these particles as result decrease the contact pressure between soil particles and the friction between soil and tunnel. In cases of ground acceleration 0.05g, the soil surface settlement in T11 and T12 is greater than the settlement T21 and T22 because the transverse direction of seismic loading on the tunnel section in T11 and T12 lead to decrease in shear strength of soil around tunnel more than the longitudinal direction in seismic loading in T21 and T22. The effect of tunnel model depth in this case of loading is quite little. The effect of ground acceleration direction on the soil surface settlement is small for all groups in 0.1g. The increase in depth of the tunnel model leads to an increase in the settlement of the soil surface, while, in the cases of 0.2g ground acceleration (highly ground shaking) the settlement are largely increased if compared with 0.05g and 0.1g because the high buildup of pore water pressure leads to largely decrease in soil resistance and increasing the slip of soil around the tunnel. For ground acceleration 0.2g, the liquefaction occurs in all cases lead to an increase in the settlement. The settlement in T12 and T22 is greater than T11 and T21. This behavior may be due to the greater thickness of the loose soil layer.

Figure 8 presents the comparison between the final soil surface settlement in dry and saturated soil for all cases. For cases of dry soil, the settlement at the end of seismic loading 0.2g is about (39 to 47) times that of 0.05g, while, it is about (2.4 to 3.2) times that of 0.1g. In saturated soil, the settlement for seismic loading 0.2g is about (8.3 to 23.6) times that of 0.05g, while, it is about (2.05 to 2.51) times of 0.1g.

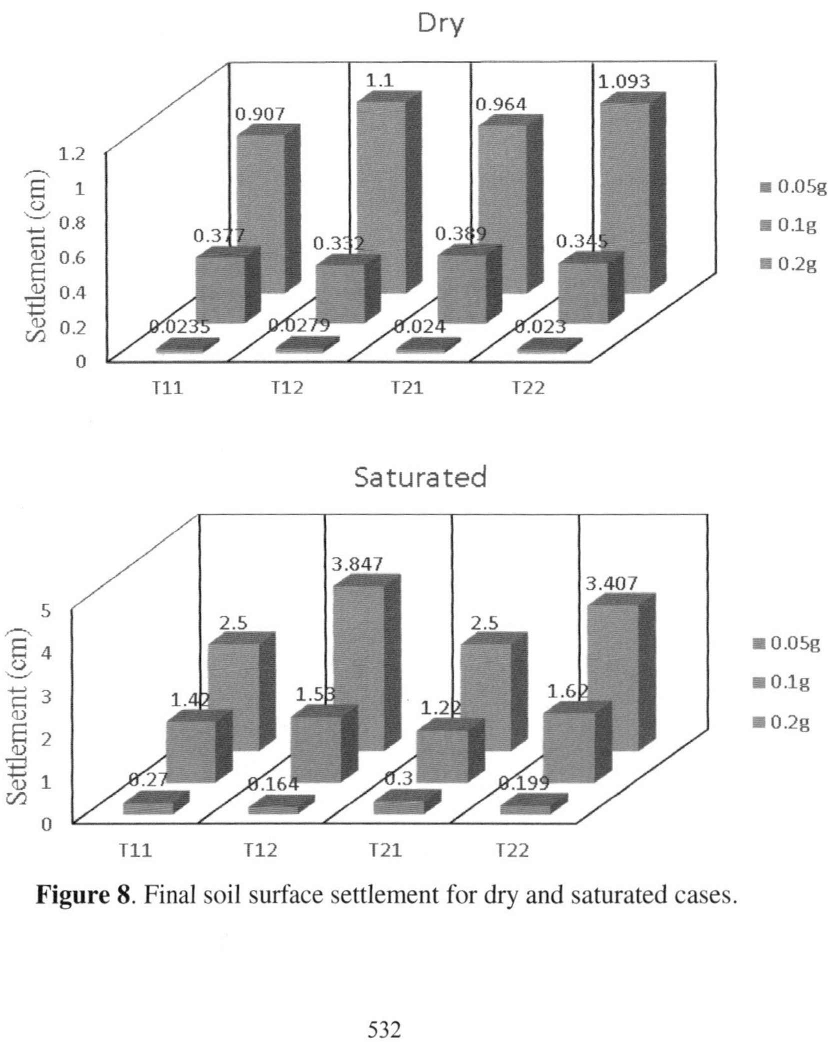

Figure 8. Final soil surface settlement for dry and saturated cases.

3.2. Horizontal earth pressures on the tunnel side-wall

The dynamic earth pressure (DEP) has been measured on the right wall of the tunnel structure in all cases. In dry cases, generally, the DEP on the wall of the tunnel structure increases with increases of ground acceleration. This increment in DEP may be due to the densification of loose soil and yielding phenomena around the tunnel during shaking. During the ground acceleration 0.05g, the DEP in all groups is very low. The effect of depth of the tunnel on the response of DEP is clear. The DEP for T12 and T22 is greater than T11 and T21, while, the effect of the direction of ground acceleration on DEP is little, see figure 9a.

For ground acceleration 0.1g and 0.2g, it is clear that the DEP increase with increasing the magnitude of seismic loading. In 0.1g, when increasing the depth of tunnel the DEP increases, hence when H/W equals 2 the DEP in T12 and T22 greater than T11 and T21.

The effect of ground acceleration direction can clearly be shown between T12 and T22 more than that between T11 and T21. The effect of ground acceleration can be attributed to the transmission of seismic wave loading from soil to tunnel structure, in T11 and T12 the wave of loading is perpendicular on the tunnel section that is lead to increase the DEP, while, in T21 and T22 the wave of loading is parallel to the wall of tunnel structure. In T11 and T12, the presence of tunnel structure in this direction causing non-uniform distribution of wave loading in soil; therefore, the intensity of wave loading is large on the tunnel section, while, in T21 and T22 the wave loading distributed uniformly in the soil because it is parallel to the wall of the tunnel structure. This may cause a decrease in DEP. Besides, the stresses that result from the transverse shear wave on the wall of the tunnel in T11 and T12 greater than that result from the longitudinal shear wave on the wall of the tunnel in T21 and T22.

The ground acceleration 0.2g is considerably very large if compared with 0.05g and 0.1g. The response of DEP in all groups is similar because of more densification of loose soil around the tunnel, therefore, the DEP increased. The DEP in T12 and T22 is greater than T11 and T21 respectively. The effect of direction of ground acceleration on DEP is low, see figure 9b and figure 9c. This response of the DEP in dry soil approximately like that findings by [12].

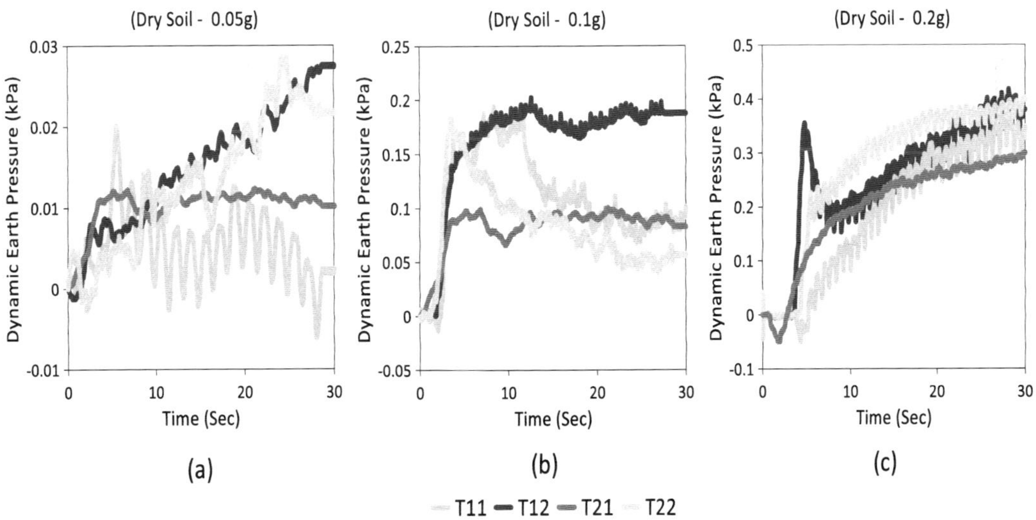

Figure 9. Dynamic earth pressure for dry cases.

The behavior of saturated soil around the tunnel is quite different from dry soil. The effect of pore water pressure highly appears in this case. In all saturated cases, the DEP is negative, while, it's positive in dry cases. The negative DEP increases with increases in ground acceleration because the pore water pressure increased when ground acceleration increased. The negative DEP increases as the ground acceleration increases due to pore water pressure buildup. The buildup in pore water pressure leads to reduce the total thrust on the wall of the tunnel and cause liquefaction of soil around the tunnel (especially in seismic loading 0.2g). Besides, in saturated soil, the slip of soil around the tunnel is more than that in dry soil. In figure 10, it can see the increment of negative DEP with the increase in seismic loading. Also, it can observe the effect of the depth of tunnel structure in all cases because the DEP in T12 is greater than T11 and DEP in T22 is greater than T21, while, the effect of direction of ground acceleration on the DEP in all cases is quite little for the three input motions.

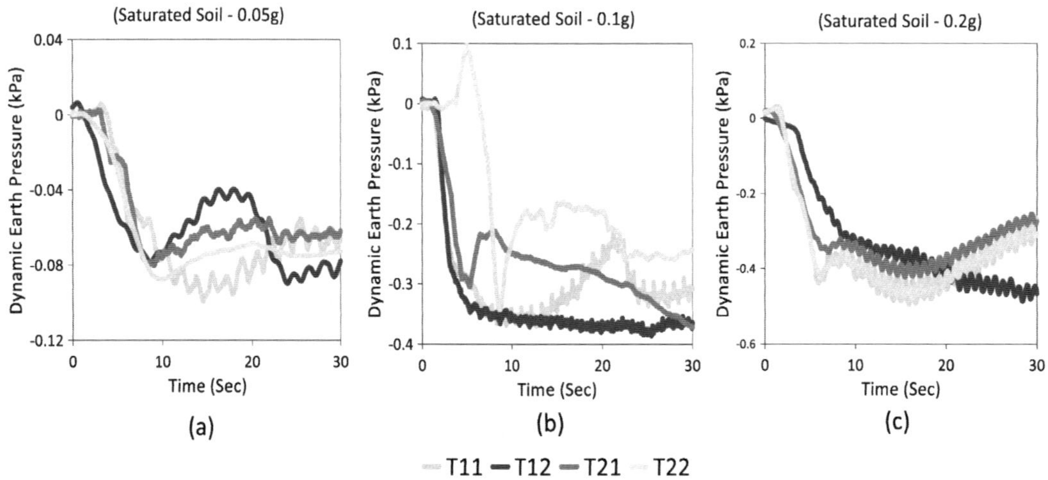

Figure 10. Dynamic earth pressure for saturated cases.

Figure 11 and figure 12 show a comparison between the peak value of DEP for all groups in dry and saturated soil. For dry soil, the maximum DEP observed in T12 that's reached to (+0.027 kPa), (+0.2 kPa), and (+0.417 kPa) for the seismic loading 0.05g, 0.1g, and 0.2g respectively. Also, in saturated soil, the maximum DEP observed in T12. It was equal to (-0.099 kPa), (-0.38 kPa), and (-0.51 kPa) for the seismic loading 0.05g, 0.1g, and 0.2g respectively.

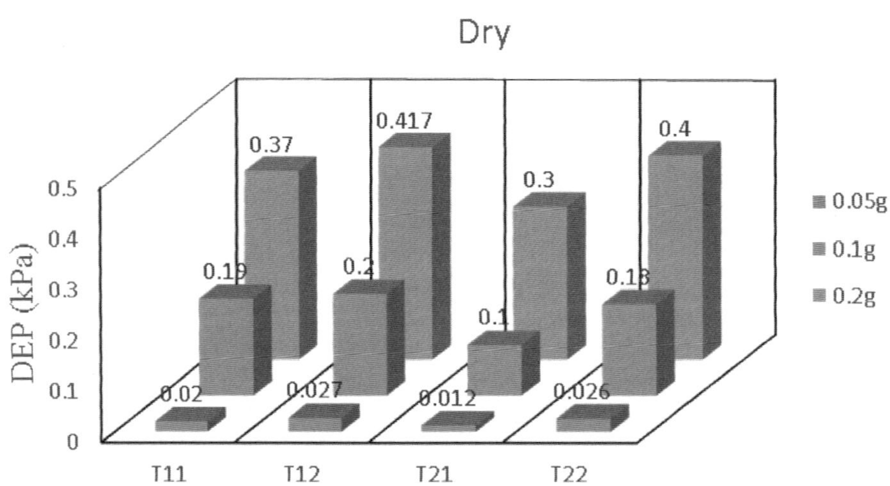

Figure 11. Peak of dynamic earth pressure for dry cases.

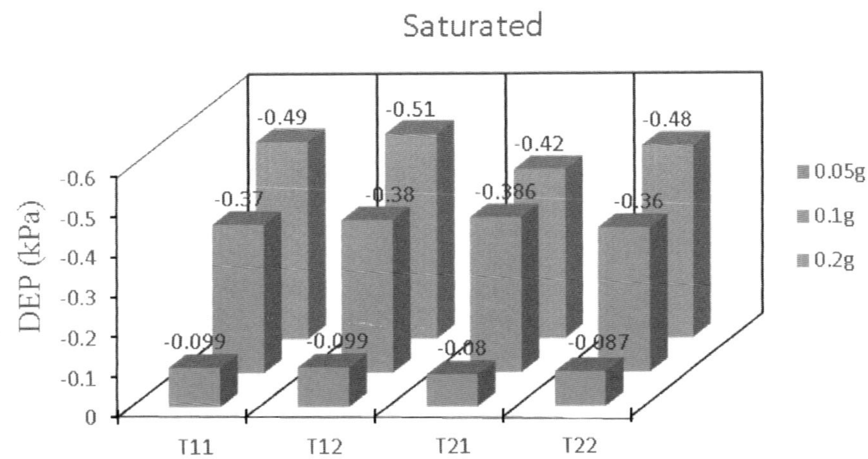

Figure 12. Peak of dynamic earth pressure for saturated cases.

4. Conclusions

1. In dry cases, the soil surface settlement increases with an increase in ground acceleration in all groups. The increase in the settlement is due to more densification of loose sandy soil around the tunnel. The full slip of soil around the tunnel leads to increase settlement of soil. The effect of depth of tunnel model and direction of ground acceleration for 0.05g is quiet little, while, for ground acceleration 0.1g and 0.2g the effect of direction on the soil surface settlement is small. The effect of tunnel model depth is noticeable.

2.For saturated cases, the soil settlement increases with increase in ground acceleration due to the increases of pore water pressure with increases of ground acceleration that is lead to reduce the shear strength of soil, especially for input motion 0.1g and 0.2g. The effect of tunnel model depth and direction of ground acceleration for 0.05g and 0.1g seemed clear. For ground acceleration 0.2g, because of occurrences of liquefaction, these factors have a small effect.

3. In dry cases, the increment in positive DEP is due to the densification of loose soil and yielding phenomena around the tunnel during shaking, while in saturated cases, the negative DEP, due to increment in pore water pressure, leads to reduce the total thrust on wall of tunnel and causing liquefaction of soil around the tunnel especially in seismic loading 0.2g. Further, in saturated soil, the soil slip around the tunnel is more than that in dry soil.

4. The effect of direction of ground acceleration and tunnel model depth on DEP are tangible in dry cases for 0.05g and 0.1g, while, this factor has little effect in 0.2g. In saturated cases, the effect of tunnel model depth on the results of DEP is greater than the effect of direction of ground acceleration.

References

[1] Abuhajar O El Naggar H and Newson T 2015 Seismic soil-culvert interaction Can. Geotech J (52) 1–19

[2] Al-Recaby M K Fattah M Y and Karim H H 2016 Dynamic Behavior of Pile Group Model in Two-Layer Sandy Soil Subjected to Lateral Earthquake Excitation Global Journal of Engineering Science and Research Management vol 3 no 8 pp 57- 80

[3] ASTM D3080-1998 Standard Test Method for Direct Shear Test of Soils under Consolidated Drained Conditions American Society for Testing and Materials

[4] ASTM D422-2001 Standard Test Method for Particle Size-Analysis of Soils American Society for Testing and Materials

[5] ASTM D4253-2000 Standard Test Method for Maximum Index Density and Unit Weight of Soils Using a Vibratory Table American Society for Testing and Materials

[6] ASTM D4254-2000 Standard Test Method for Minimum Index Density and Unit Weight of Soils and Calculation of Relative Density American Society for Testing and Materials

[7] ASTM D638-2003 Test Method for Tensile Properties of Plastics American Society for Testing and Materials
[8] ASTM D854-2005 Standard Test Method for Specific Gravity of Soil Solids by Water Pycnometer American Society for Testing and Materials
[9] Baziar M H Moghadam M R Kim D S and Choo Y W 2014 Effect of underground tunnel on the ground surface acceleration Tunnelling and Underground Space Technology (44) 10–22
[10] Lanzano G Bilotta E Russo G and Silvestri F 2015 Experimental and numerical study on circular tunnels under seismic loading European Journal of Environmental and Civil Engineering, 19(5) 539-563
[11] Tsinidis G Heron C Pitilakis K and Madabhushi S P G 2015b Centrifuge modelling of the dynamic behavior of square tunnels in sand In Taucer F. and Apostolska R. (eds.) Experimental research in earthquake engineering - EU-SERIES concluding workshop, Geotechnical Geological and Earthquake Engineering, (35) 509-523, Springer International Publishing, Switzerland
[12] Tsinidis G Rovithis E Pitilakis K and Chazelas J L 2016b Seismic response of boxtype tunnels in soft soil: experimental and numerical investigation Tunn.Undergr. Space Technol. (59) 199–214
[13] Yu Haitao Yuan Yong Li Chong Yan Xiao and Yuan Juyun 2016 Multi-point shaking table test for long tunnels subjected to non-uniform seismic loadings – Part I Theory and validation Soil Dynamics and Earthquake Engineering. (108) 10.1016/j.soil dyn.

IOP Conf. Series: Materials Science and Engineering **584** (2019) 012058 doi:10.1088/1757-899X/584/1/012058

Seismic Behavior of Composite Simply Supported Bridge Decks supported on Elastomeric Bearings

Sara k AL Rubaee[1] and Ammar A Abdul Rahman[2]

[1] Civil Engineering Department, AL-Nahrain University, Baghdad, Iraq. Email: saraka94@yahoo.com
[2] Civil Engineering Department, AL-Nahrain University, Baghdad, Iraq. Email: ammar.arahman@yahoo.com

Abstract. This paper studies the behavior of bridges' superstructure under seismic action. The bridge under investigation is one of the typical bridges usually used in overpasses in Iraq, which is made of simply supported steel built up girders and reinforced concrete deck slab isolated from the substructure by elastomeric bearing pads. A three-dimensional finite element model was developed using CSI Bridge Software. Time history analysis was carried out using true-recorded earthquake took place in Iraq near the city of Halabja in 2017. The results of the analysis show that the displacements of the elastomeric bearing pads under transverse seismic loading are much higher than the permissible displacements' limit which will produce failure of these bearings and potential failure of the bridge superstructure.

1. Introduction

Simply supported composite bridge decks is usually used for overpasses inside Iraqi cities. This type of deck consists of reinforced concrete deck slab resting over longitudinal girders (built-up steel girders) and connected with the concrete deck slab by shear connecters to insure the composite action between steel and concrete. Most researches on bridges under seismic loading concentrates on the seismic behavior of the bridge substructure such as the piers including the cross heads and foundation since these elements are in contact with soil and transfer the loads from the seismic activity to the whole bridge. It is necessary to study the behavior of the bridge superstructure under seismic loading and to develop retrofit strategies against it. Saadeghvaziri in 2008 [2] studied the seismic behavior and capacity/ demand analysis in order to evaluate the seismic vulnerability of three multi-span simply supported bridge and to develop retrofit strategies and investigate the effect of a parameter on the seismic performance on bridges such as steel bearing, and the impact between girders. Zasiah and Nazmus in 2010 [3] made a seismic analysis of three span deck girder bridge. The analysis carried out was by using the software package ANSYS in a two-step procedure. The first step was by using equivalent static earthquake loading and the second step by using response spectrum analysis and made a comparative study between the two analyses. Cardone, D., Perrone and S. Sofia in 2011 [4] proposed multi-performance seismic device based on super-elastic SMA "Shape memory alloy". They used this device in seismic retrofit of simply supported and continuous bridge decks to investigate the behavior of bridge deck before and after retrofitting using nonlinear time history analysis through SAP2000 software.

2. Bridge Description

AL-Batha Bridge is one of the typical common simply supported bridges used in Iraq, which located in AL-Nasiriya city southern Iraq. It consists of two spans with total length of 53m designed by the supervisor of this work, as illustrated in figure 1.

Each span is 26.5m in length and a carriageway width of 12.75 m with two sidewalks of 1.775 m each, figure 2. The bridge deck consists of reinforced concrete slab 220mm thick and seven built-up steel girders. Figure 3 shows the dimensions of the steel girder. The steel girders rest on elastomeric pads without any other lateral stiffness except end steel diaphragms to connect the steel girders together. Shear connectors made of steel channels are used to connect the deck slab with the girders which gives complete connection between the steel girders and the deck slab.

3. Finite Element Modeling

The three dimensional modelling of the bridge superstructure was developed using finite element software CSI Bridge, as shown in figure 4. Steel built-up girders were modelled using homogenous shell elements. Concrete deck slab with its reinforcements were modeled using 4-node layered shell element, which has six degrees of freedom per node. The reinforcement in the concrete deck was modeled as a smeared layer of steel "equivalent area of each reinforcement bar divided by reinforcement bar spacing", as illustrated in figure 5. Complete bond between the top flange of the built up girder and the R.C. deck slab was assumed to represent shear stud connector. Elastomeric bearing pads were modeled as hysteretic isolator link element and the 8-node solid element was used to model the supporting concrete box below the bearing. Frame element was used to model the v-type steel bracing for the end and intermediate diaphragms. Table 1 shows the properties of the materials used in modelling the deck.

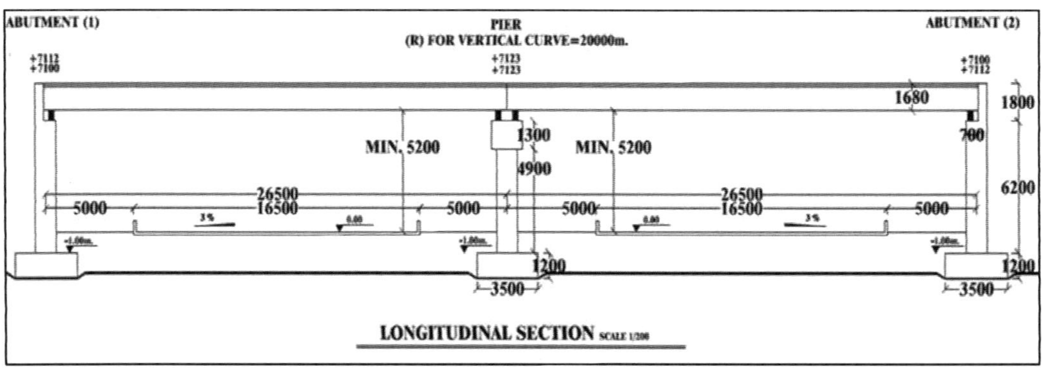

Figure 1. Longitudinal section of AL-Batha Bridge.

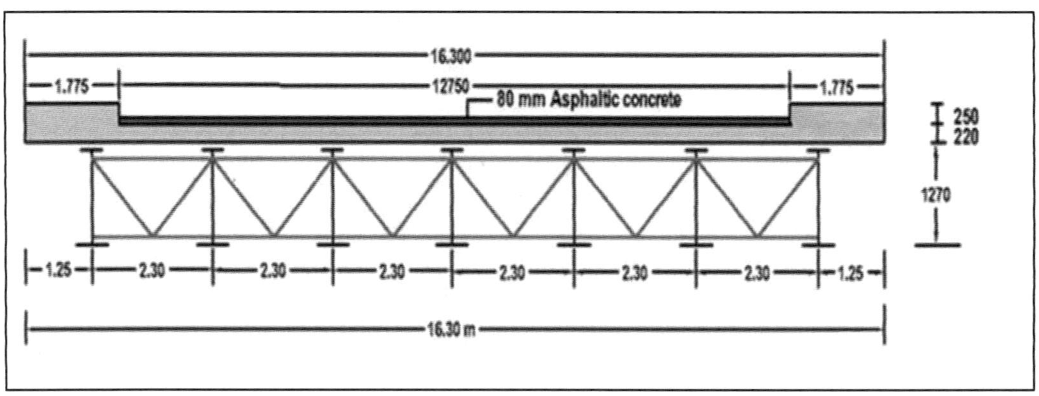

Figure 2. Components of composite simply supported bridge deck.

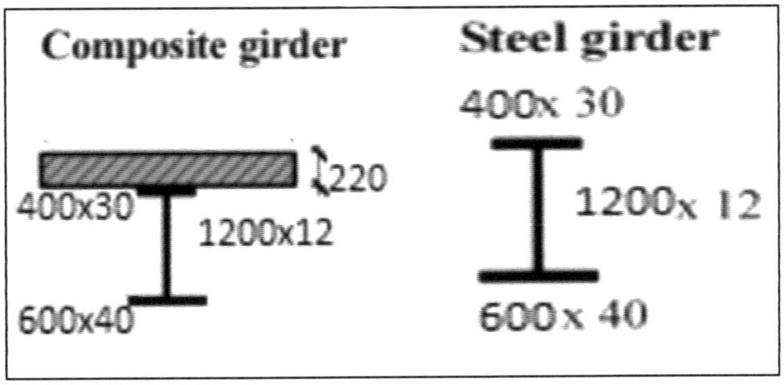

Figure 3. Detail and dimensions of built-up steel girder.

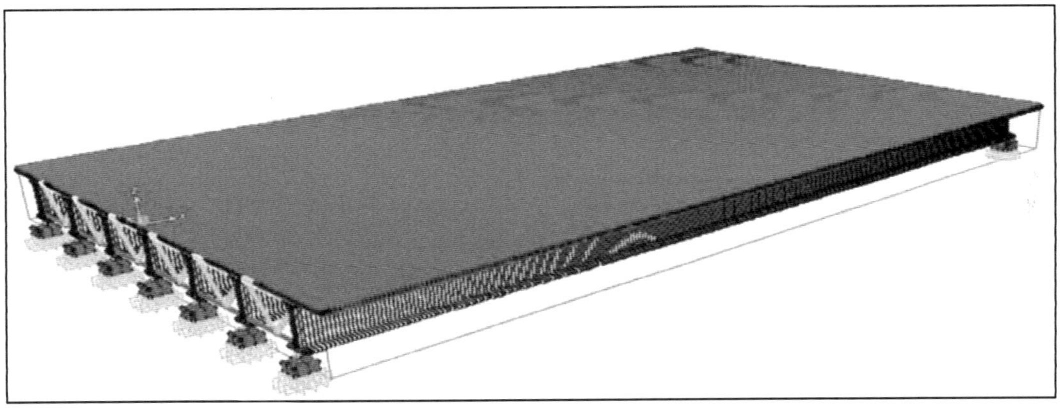

Figure 4. Finite element modeling of composite simply supported bridge deck.

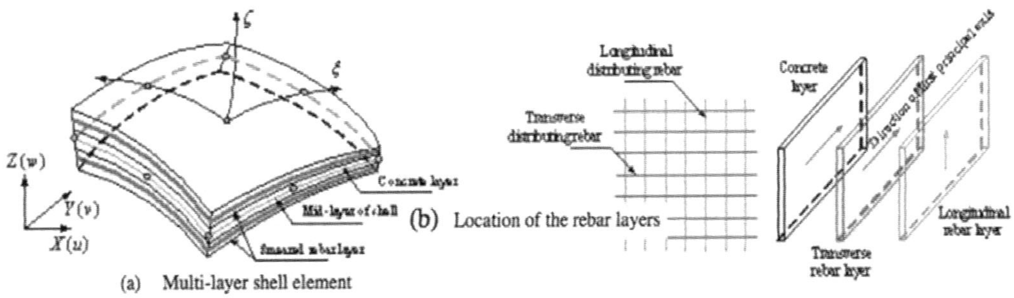

Figure 5. multi-layered shell elemnet that used to model the reinforcement concrete deck slab of bridge.

Table 1. The material properties of composite deck.

Concrete		A50 Steel	
E_C	27805.57 MPa	E_S	200000 MPa
f'_c	35 MPa	f_s	345 MPa
υ	0.19	υ	0.3
γ_C	24.5 KN/m^3	γ_S	78.6 KN/m^3

3.1 Elastomeric Bearing Pads

The elastomeric bearing pads are designed to transmit the forces from the superstructure to substructure and receive translations and rotational displacements by elastic deformation. Main action is like a hyper-elastic body. It consists of natural rubber with steel plates inserted inside the rubber making it as layers of rubber, as shown in figure 6. Two nodes hysteretic rubber isolator link element was used to model the elastomeric pads with stiffness in three directions which composite of six springs (axial, shear, pure bending, and torsion), as shown in figure7. This element provides non-linear properties for two horizontal directions and linear properties for remaining axial and three rotational directions.

The top and bottom faces of the elastomeric bearing pads are actually covered by epoxy resin with no bolt to fix the pad in place, which is modelled as a rigid spring element with high stiffness.

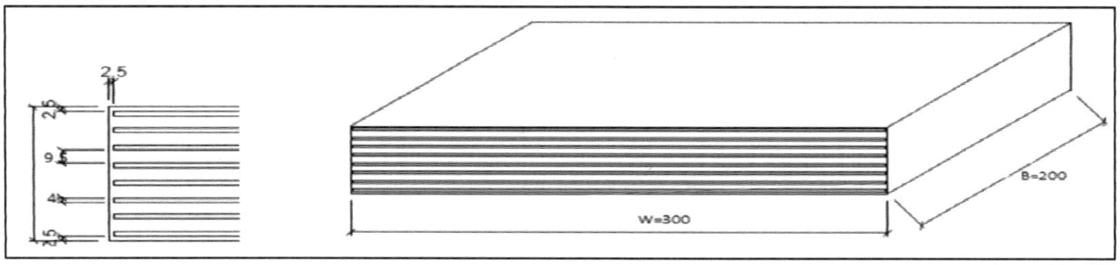

Figure 6. elastomeric bearing pads (all dimension in mm).

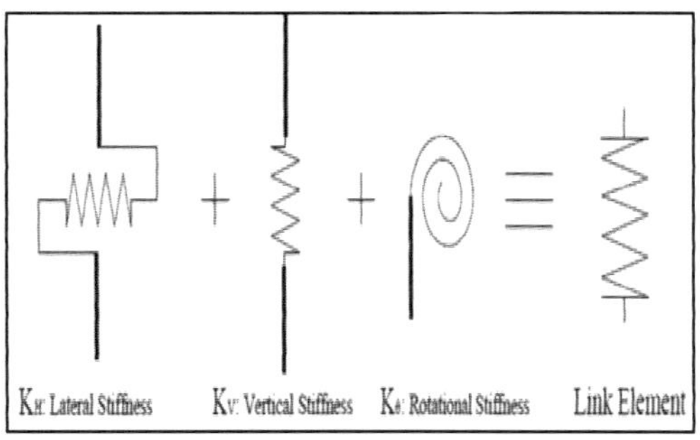

Figure 7. Rubber isolator link element of elastomeric bearing pad.

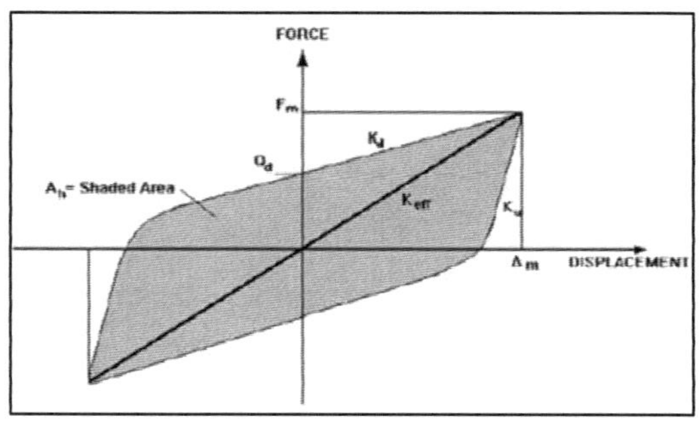

Figure 8. Force-displacement hysteresis curve of isolator.

The hysteresis curve produced under actual lateral displacement of an isolator is a combination of linear-elastic and elastic-perfectly plastic force-displacement relationship. Rectangular hysteresis curve will not be produced by an isolator, which has a curve translation on unloading and reloading. Equivalent bi-linear approximation was used for the analysis to define the isolator properties such that damping is defined equal to the area under hysteresis curve in figure 8, as calculated below:

$$\beta = \frac{1}{2\pi}\left(\frac{4\,Q_d}{K_{Heff}} * \frac{\Delta_m - \Delta}{\Delta^2}\right) \tag{1}$$

Lateral stiffness of bearing pads was determined from the following equation depending on hysteresis curve based on empirical data [1] with a given properties shown in table 2.

$$K_{Heff} = K_d + \frac{Q_d}{\Delta} \tag{2}$$

Where: K_d post-elastic stiffness, Q_d characteristic strength, Δ specific displacement, Δ_m maximum displacement.

The vertical stiffness of elastomeric bearing pads depend on reducing the area of rubber, which is calculated based on the overlapping area of bearing between the top and the bottom portion at a specific displacement.

$$K_v = \frac{Ec\,Ar}{Tr} \tag{3}$$

Where: Ar reduce area of bearing, Ec compressive modulus is a function of material constant and shape factor of isolator, Tr total thickness of elastomer.

$$E_c = E[1 + 2KS^2] \tag{4}$$

$$Ar = A_b\left[1 - \left(\frac{\Delta}{B_b} + \frac{\Delta}{w_b}\right)\right] \tag{5}$$

Where:E elastic modulus, K material constant, S shape factor, A_b bounded area, B_b w_b bounded dimension.

Table 2. Properties of isolator.

Elastomer Properties	kN, mm	Isolator Dimension	mm
Shear strain%	50	Rubber cover thickness	2.5
Shear modulus, G	0.000707	No. of rubber layer	4
Material constant, k	0.7	Rubber layer thickness, t_r	9.5
Elastic modulus, E	0.00263	No. of steel plate	5
Bulk modulus, $\boldsymbol{E_\infty}$	1.5	Steel plate thickness, t_s	4
Damping	0.05	Total height, H	63
Gravity	9810	Total elastomer thickness, T_r	38
Vertical Stiffness	**kN, m**	**Hysteresis Loop Properties**	**kN, m**
Bounded area, $\boldsymbol{A_b}$	0.06	Characteristic strength, Qd	305.1
Bounded perimeter	0.0769	Post-elastic stiffness,Kd	457.05
Shape factor, Si	3.05	Elastic stiffness, ke	25351.27
Reduce area of rubber	0.05016	Effective stiffness, K_{Heff}	7898.50
Vertical stiffness of isolator, $\boldsymbol{K_V}$	48683.976	Area of hysteresis loop, Ah	33453
Moment of inertia ,I	3.255E0-4	Equivalent viscous damping, β_{eff}	0.4

4. Seismic Analysis of Bridge Superstructure.

Seismic analysis of the typical simply supported bridge superstructure in Iraq was carried out using nonlinear time history analysis and the equation of motion in this analysis was performed, using direct integration method in order to study the behavior of superstructure under seismic loading. In this study, the response of the structure was evaluated at each step and at the end of any step, the equation of motion was investigated.

The input data of loading was a function of acceleration - time history figure 9. In this study, the earthquake hit Iraq-Iran border 30 km southwest of Iraqi city of Halabja on 12 November 2017 with a magnitude 7.3 Mw as recorded by International stations and with 4.9 Mw as recorded at the station of Baghdad was used in this study as the input seismic activity. This earthquake was chosen for the following reasons:

- To investigate the behavior of typical bridge superstructure under true recorded earthquake happened in Iraq.
- It is the largest earthquake recorded in Iraq until the date of preparing this study that felt clearly and strongly in Baghdad.

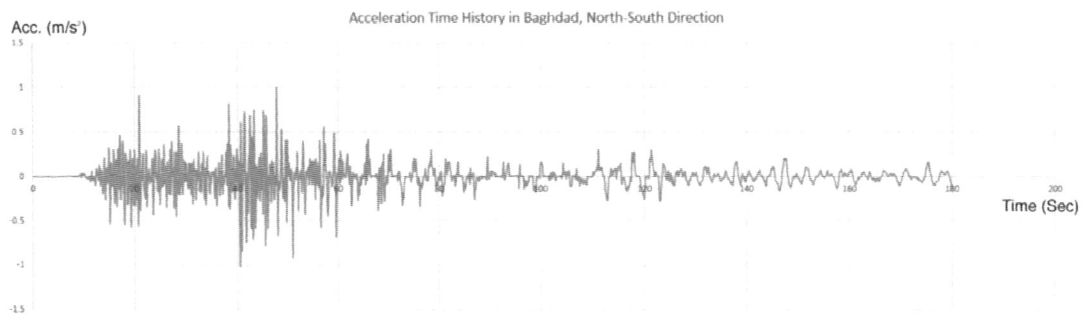

Figure 9. Iraq-Iran recorded acceleration time hiistory in Baghdad, north-south (Y) direction.

5. Analysis Results

The response obtained from the bridge superstructure analysis was defined in terms of displacements limits under the effect of self-weight load and earthquake ground motion, which was applied in the transverse direction of the bridge deck, because the stiffness of the superstructure under lateral load is not enough to produce the required lateral resistance and makes it the point of research and inspection. Elastomeric bearing pads were the weakest component of the bridge superstructure and the failure of the bearings governed the behavior of superstructure under the effect of the earthquake. When the displacement of such bearing exceeds the permissible limit, steel girder slip from its place and the superstructure susceptible to toppling. The maximum lateral displacement of elastomeric bearing pads approaches 261.3mm occurred in the bearings, consequently instability of these bearings in transverse direction due to excessive displacements. The maximum transverse displacement of steel girder due to seismic loading approached to 388.9mm occurred in the bottom web of the left exterior girder. Figure 10 showed the maximum and minimum displacements of superstructure component A) elastomeric pads, B) in steel girder. Maximum longitudinal stresses due to the same loading in the bottom flange of steel girder is relatively higher than that in the top web with a magnitude of 381.0 mm due to the connection with bearings. Finally, maximum stresses in the deck slab when subjected to the earthquake was presented. Figure 11 shows the distribution of stresses in a) reinforcement concrete deck slab, b) 7-steel girders.

Figure 12 shows how the steel girders slipped from their positions during the effect of applying the earthquake to the superstructure of the bridge.

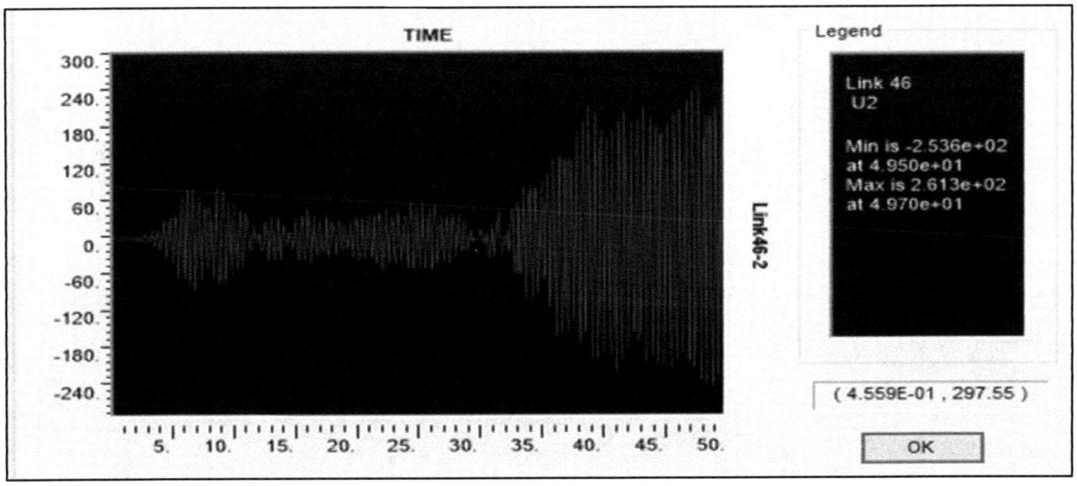

(a)

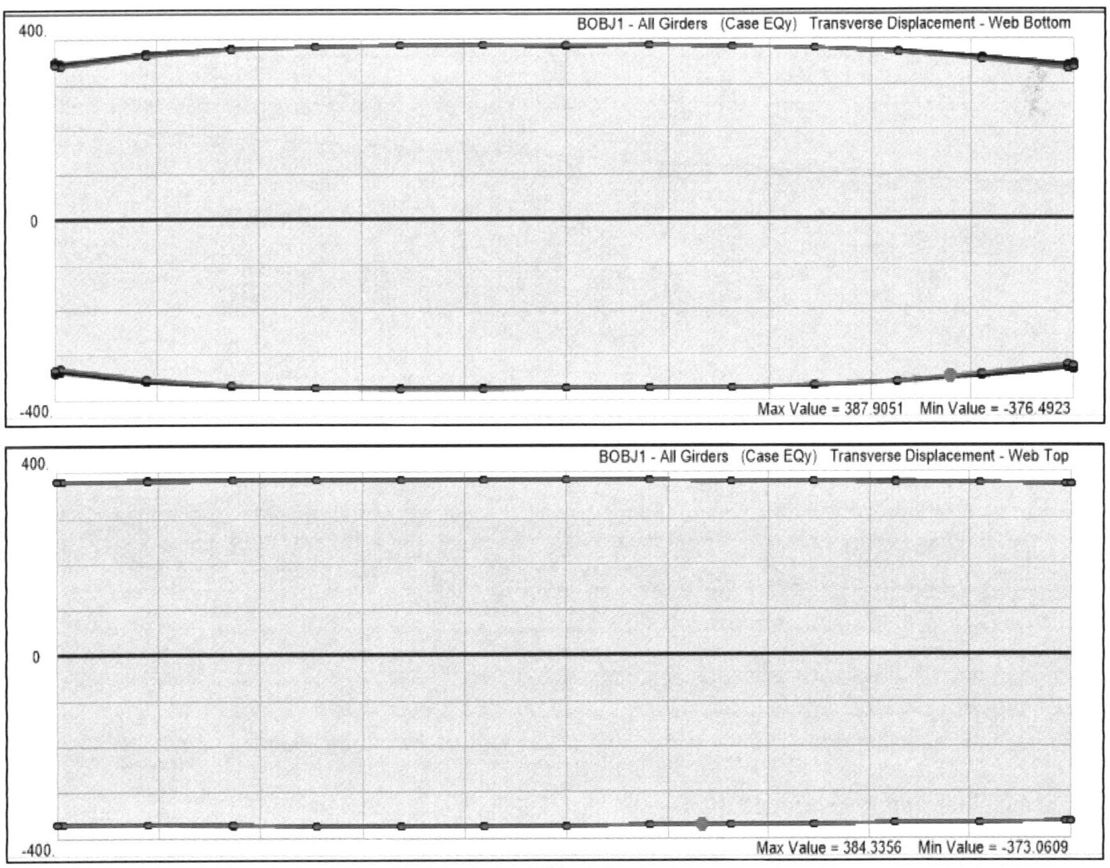

(b)

Figure 10. Maximum and minimum displacement of superstructure component (a) elastomeric pads, (b) top and bottom web of steel girder

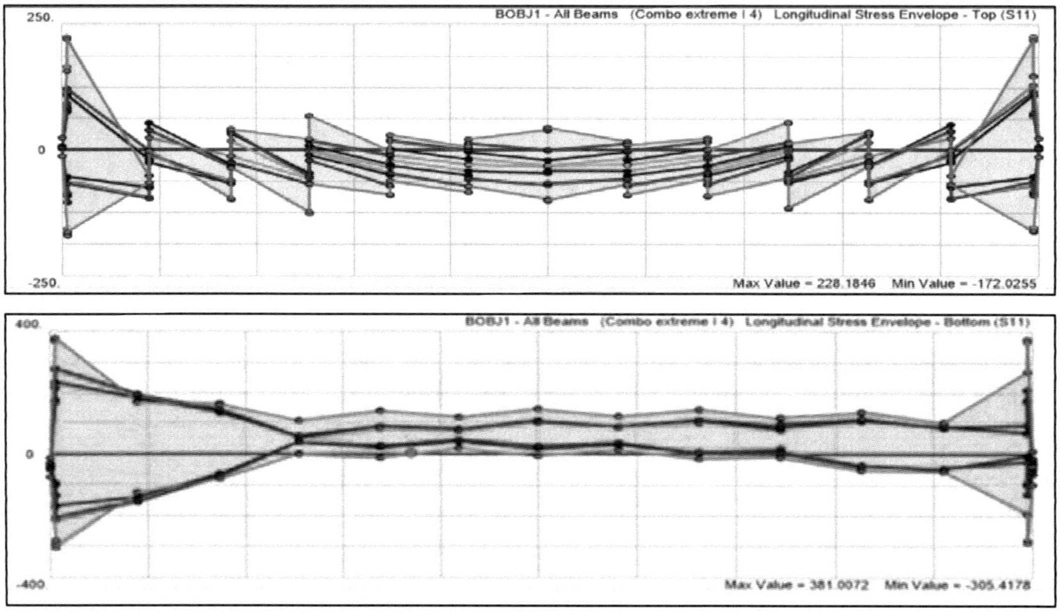

(a)

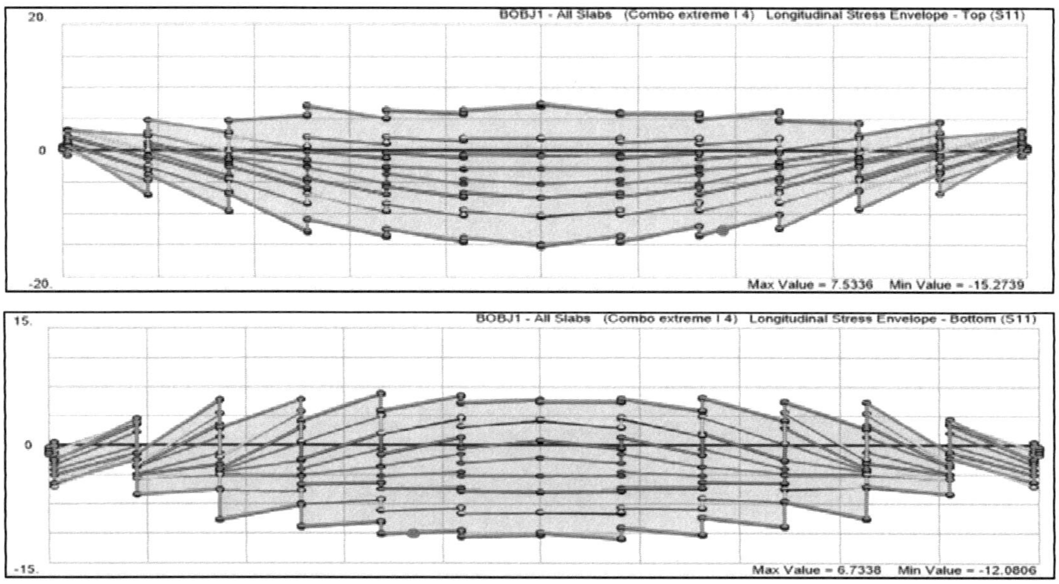

(b)

Figure 11. The distribution of stress in: (a) seven-steel girder, (b) reinforcement concrete deck slab.

6. Conclusions and Recommendations

From this analysis, the bearings were under deformations exceed the acceptable limit (30.5mm) according to the specification of bearing pads manufacturer. This conclusion demonstrate that under such a magnitude of earthquake not exceeding 5.0 Mw the typical simply supported bridge superstructure will fail due to the failure of elastomeric pads which is the main connecting part of the superstructure with the substructure which governs the whole behavior. Consequently, a retrofit strategy such as shape memory alloy, bombers, and restrained component or pads should be used to prevent superstructure from toppling for this type of bridge decks.

The bumpers are a restrainer block installed between the steel girders in order to prevent them from dropping from the elastomeric pads and piers during seismic loading. It remains in elastic range and produces reaction force for restraining the displacement of a steel girder to reduce seismic response. This type of proposed section is recommended to use, which provides energy dissipations capacity. It is not difficult to build on a crosshead of a pier and not expensive compared to other dissipations devices. FEA of this type of restrainer will be investigated in the future through complete analysis of such new deck configuration.

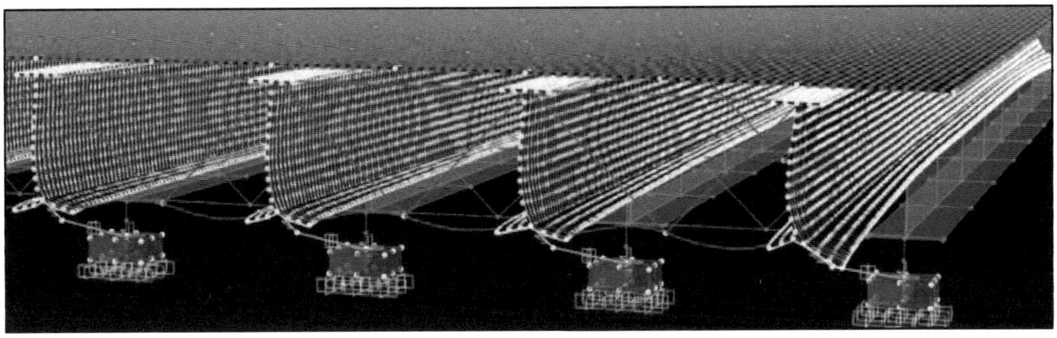

Figure 12. Steel Girders out of their positions after the Earthquake.

References

[1] Ronald L M 2001 *The Seismic Design Handbook* vol. 1, ed F Naeim (Boston: Springer) chapter 6 pp. 723-756

[2] Saadeghvaziri, M A and Motlagh, A R 2008 Seismic Behavior and Capacity/Demand Analyses of Three Multi-Span Simply Support Structures *J. Eng. Struct.* **30** pp. 54-66.

[3] Zasiah T, Nazmus S and K M Amanat 2010 *IABSE-JSCE 2nd Int. joint Conf. on Advances in Bridge Engineering – II* (Bangladesh: University of Science and Technology, Dhaka) pp. 285-295

[4] D cardone, G perrone, and S Sofia 2011 Seismic Response of Simply Supported Deck Bridges with Auxiliary Superelastic Devices *J. Proc. Eng.* **14** pp. 2315-2322.

[5] CSI Analysis Reference Manual, CSI computers and structures, Inc. 2013.

Effect of Partial Replacement with Crushed Concrete on Settlement and Bearing Capacity of Soft clay

Maki Jafar Mohammed Al-Waily

Al Furat Al-Awsat Technical University, Al Musaib Technical College - Iraq. Email: maki_jafar@yahoo.com

Abstract. This paper aims to examine the effect of partial replacement with crushed concrete as a recycled material to enhance the bearing capacity and to decrease the settlement of soft clay. To clarify the effect of scale factor, four cubic steel soil tank with four dimensions of (250 x 250 x 250) mm; (300 x 300 x 300) mm;(400 x 400 x 400) mm; and (500 x 500 x 500) mm are used to carry out the experimental works, by using a foursizes for each square and strip footings with varying the depth of replacement (0.2 , 0.4 and 0.6 width of square footing) and (0.33, 0.67 and 1 width of strip footing).The effect of replacement wasevaluated with regard to settlement control and bearing capacity. It is found from results of experimental works that a significant decreasing in settlement and increasing in bearing capacity of treated soil with increasing the depth of the partial replacement layer for both square and strip footing compared to untreated soil. The study also shows that the use of partial replacement layer with a depth of 0.4 B for strip footings results in reducing the settlements by an averaged percentage of 72 % and increasing the bearing capacity with an average of 96%. Also the results demonstrates that clear increasing in bearing capacity for all depth of replacement under square footing with indistinct in settlement

1. Introduction

These guidelines, Because of low bearing capacity and high settlement of cohesive or soft clay soils, the improving of its properties in the site is one of the main challenges faced by researchers, civil and geotechnical engineers. The need for rapid and economic improvement of the natural clay soils is widely apparent in many countries of the world, especially emerging countries with significant infrastructure development. The conventional methods of soft clay soil treatments include the following techniques: (a) Sand drain, (b) Preloading (c) Stone columns, (d) Heat treatment, (e) Chemical mixing, (f) electro-osmosis, and (f) deep dynamic compaction.
The bearing capacity of the foundations built on soft soils can be enhanced by a layer of compact sand or gravel. [1] examined the final bearing capacity of the foundations on a clay-based foundation in the two cases, in the first case of a solid layer existed on a soft layer, and the second being the presence of a soft layer covering a stiff layer. The studies were based on a model test using circular footing and a bar using a combination of layer thickness and clay strength. The results of the study were presented in the graphs that used to assess the bearing capacity of the footings constructed on the clayey soil. [2] conducted a field load test and compared these results with the corresponding numerical results using a column model more than the circular footings based on the granular soil under the soft clay and claimed that there is a good match between them.The classic method to improve soft soils is by replacing those soft soils with stronger, better soil or material to improve its properties such as a bearing capacity or reduce the anticipated settlement. The depth of the replaced sand is depended on the required bearing capacity and the allowable settlement. Sometimes the using of this method leads to great heights of soil replacement and hence the excessive cost and effort, therefore the feasible method is by substituting the

soft soil by a granular material in a specific area under the footing in similar to the trench, as many have suggested practically, theoretically and experimentally such as [3-12].

2. Laboratory model tests

2.1. Formatting Loading frame and accessories

Figure (1) shows a schematic diagram of the devices that used to conduct the model tests of this study. Four steel containers (soil tank) in the shape of a cube with four sizes as follows:(A), (250 x 250 x 250) mm; (B),(300x 300x 300) mm;(C),(400x 400x 400) mm; and (D),(500x500x500) mm, all soil tanks with thickness of 4 mm. The soil tanks are manufactured to be rigid enough to keep the plain strain as low as possible in all directions. The interior walls of the soil tank are polished smooth to minimize contact with the soil as much as possible using a coating layer. The loading system consists of a hydraulic jack operated by a hydraulic pump (compressor) to apply the load on the soil bed. The ram of hydraulic jack is a attached with a load cell to identify the applied pressure with assistance of data logger to record the load. The transducer is put up on the footing model to read settlement with aid of the same data logger.

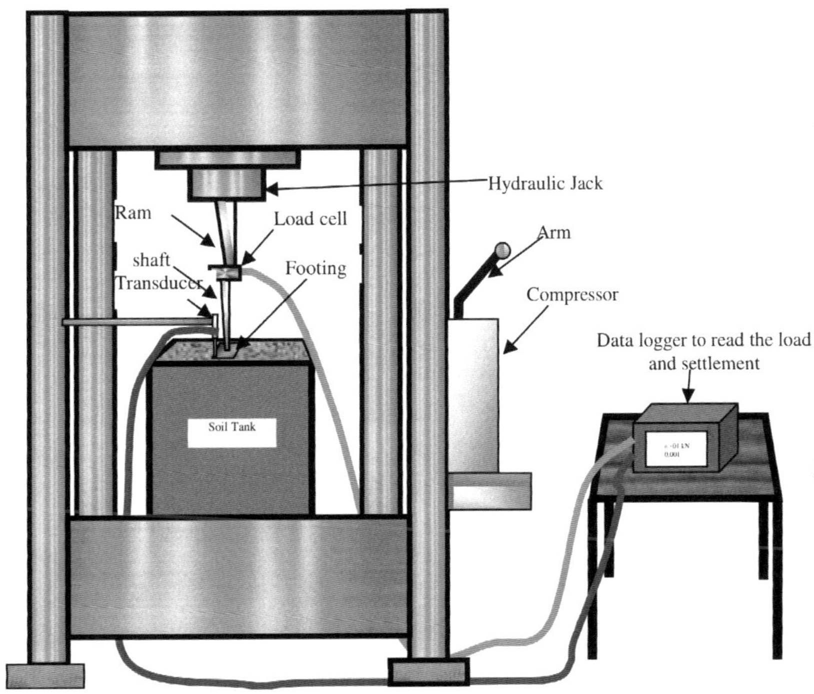

Figure 1. Loading frame and accessories.

Models of footing

The models of footing were made of a steel plate of thickness 8 mm with a two shape and six sizes as follows:

1- Square footing with dimensions of 100 mm x 100 mm, tested in the soil tank (A).
3- Strip footing with dimensions of 30 mm x240 mm, tested in the soil tank (A)
1- Square footing with dimensions of 120 mm x 120mm, tested in the soil tank (B)
3- Strip footing with dimensions of 35 mmx280 mm, tested in the soil tank (B)
3- Square footing with dimensions of 130mm x 130 mm, tested in the soil tank (C)
4- Strip footing with dimensions of 40 mmx320mm, tested in the soil tank (C)
5- Square footing with dimensions of 150 x150 mm, tested in the soil tank (D)

6- Strip footing with dimensions of 50x400 mm, tested in the soil tank (D) The load is transferred to the footing through the shaft attached with load cell which was placed between the footing and the load cell.

2.2. Test materials

2.2.1 Soft clay .The natural soft clayey soil was subjected to laboratory tests to determine the physical and mechanical properties. The soft clayey soil is classified as a low plasticity clay or (CL) according to the unified classification system. The soft clay bed is prepared by placing it in layers, with a certain water content. The unconfined compression test was performed to obtain the shear strength of the soil. The soft clay bed has been mixed and placed handling in the soil tank according to the testing program. A heavy plastic hammer was used to obtain the required density of the soil bed. The characteristics of the clay bed are shown in Table 1.

2.2.2 ***Crushed concrete*.** Standard cubes of concrete dimensions of 150mm x 150mm x 150mm were used as replacement material underneath the footing after crushing with a special steel hammer. The crushed concrete are sieved and Table 2 shows the results of sieving analysis

Table 1. Properties of Soil.

Properties	Value
Liquid limit, L.L. (%)	40
Plastic limit, P.L. (%)	23
Plasticity index, P.I. (%)	17
Specific Gravity	2.65
Maximum dry unit weight (kN/m^3)	18.65
Degree of Saturation	95 %
Optimum moisture content	10%
Sand content	15
Silt content	37
Clay content	48
Classification (Unified Soil Classification System)	CL

Table 2. Soil Used Properties.

Properties	**Value**
D10	2.65
D30	6.95
D60	11.15
C_c	1.63
C_u	4.21
Specific Gravity	2.66
Dry unit weight (kN/m^3)	17.97

2.3. Preparation and procedure

The first step of preparation of soil sample used in model tests was placing the layers of soft clay at bottom of the soil tank. Each layer has a thickness of 50 mm and it compacted with the heavy plastic hammer to reach the required density. This technique was continued to get to the final thickness of 25 cm in the soil tank (A), or 30 cm in the soil tank (B), or 40 cm in the soil tank (C), or 50 cm in the soil tank (D) according to height of the soil tank. After preparing the clay bed, the models of footing were placed in a centre position of the top surface of clay bed in the cases of untreated model tests.

In the case of treated model test with replacement of crushed concrete, the holes with the same plane area of footing was made under the model footings with the various depths as follows:

1- 0.2B, 0.4 B and 0.6 B in all models of square footing, (B= width of footing)

2- 0.33B, 0.67 B and 1 B in all models of strip footing.
The holes were then filled with crushed concrete as a replacement material as shown in Figure 2.The incremental load was applied by a hydraulic jack until reaching failure and more. The settlement of the footing was measured using displacement transducer placed on the side of the footing. Both, load and settlement were recorded with aids of digital unit or data logger.

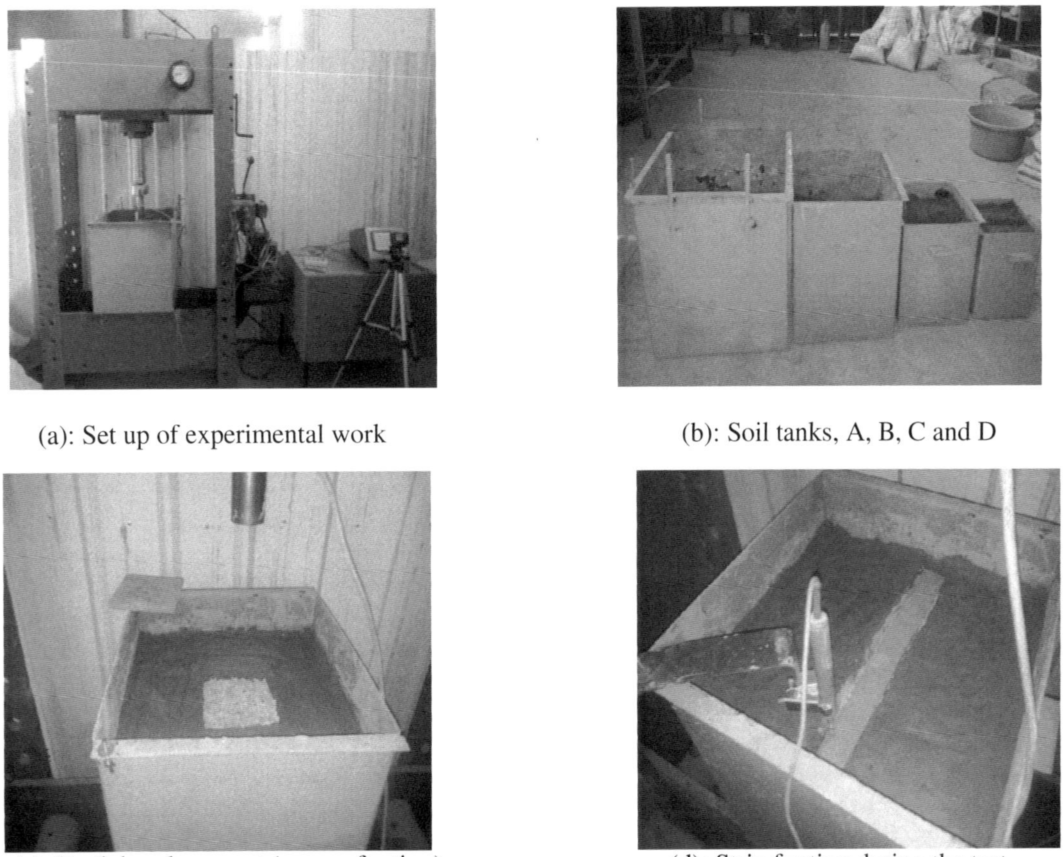

(a): Set up of experimental work

(b): Soil tanks, A, B, C and D

(c): Partial replacement (square footing)

(d): Strip footing during the test

Figure 2. Experimental Works.

2.4. Testing program
32 model tests were carried out to investigate the effect of scaling factor and replacement material on settlement and of square and strip footing.

3. Results and discussions
3.1. Settlement analysis
The Figures from (3 to 10) showed the relation between the applied pressure and settlement for 32 model tests for different sizes of square and strip footings on untreated and treated of a samples of soft clay bed tested in the different steel soil tanks. The replacement technique of soft clay by using crushed concrete as alayer under footing was used to improve the settlement of soil sample.
Four sizes of square footing models with sides of (100 mm, 130mm, 140 mm and 150 mm) respectively, were tested in four cubic soil tanks with sides of (A), 250 mm; (B), 300 mm; (C), 400 mm and ; (D), 500 mm, respectively. Three depths of replacement layer of crushed concrete 0.2B, 0.4B and 0.6B were adopted in this study.

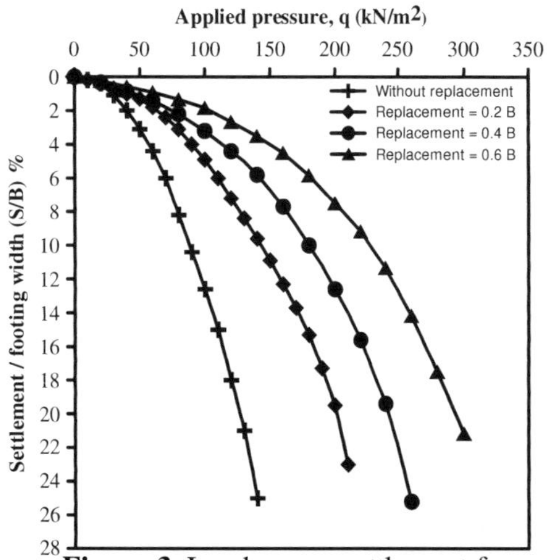

Figure 3. Load versus settlement for square footing in soil tank A

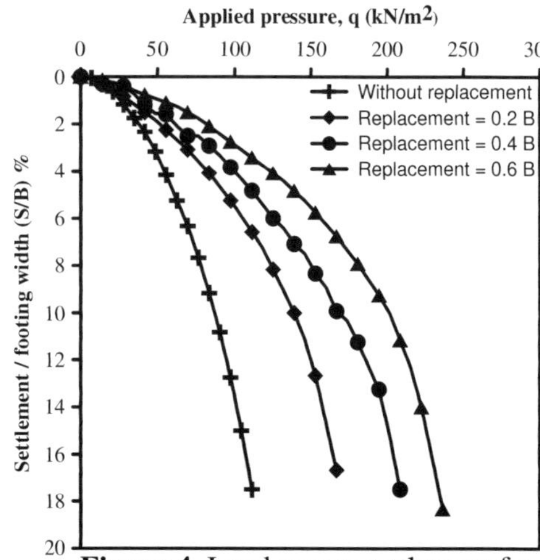

Figure 4. Load versus settlement for square footing in soil tank B

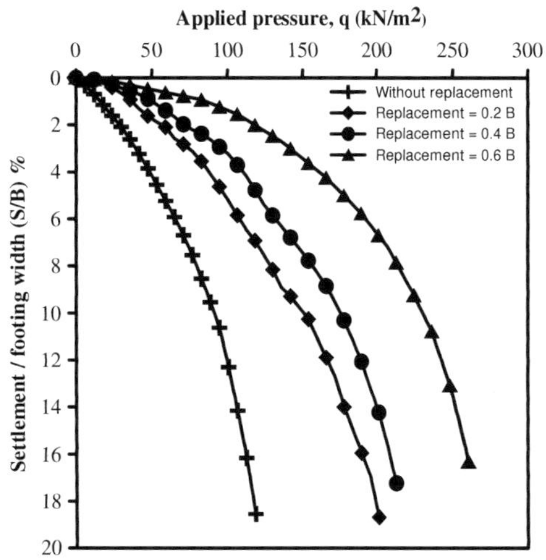

Figure 5. Load versus settlement for square footing in soil tank C

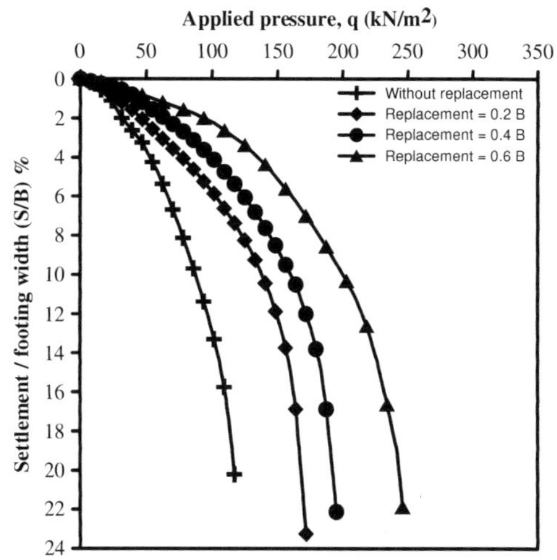

Figure 6. Load versus settlement for square footing in soil tank D

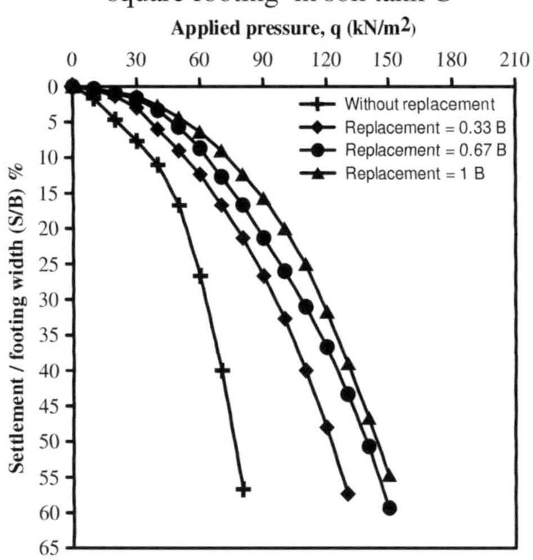

Figure 7. Load versus settlement for

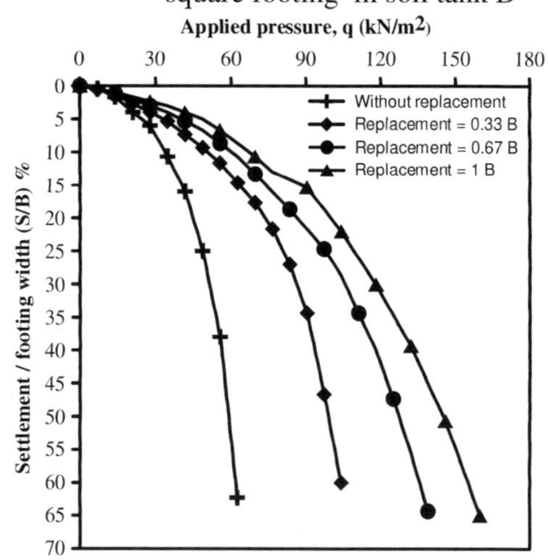

Figure 8. Load versus settlement for

strip footing in soil tank A

strip footing in soil tank B

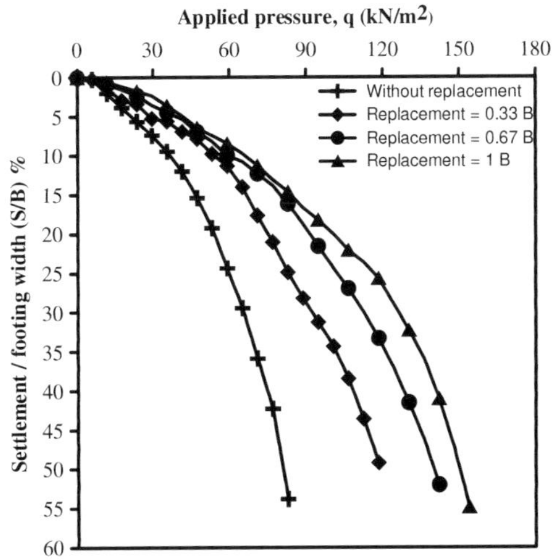

Figure 9. Load versus settlement for strip footing in soil tank C

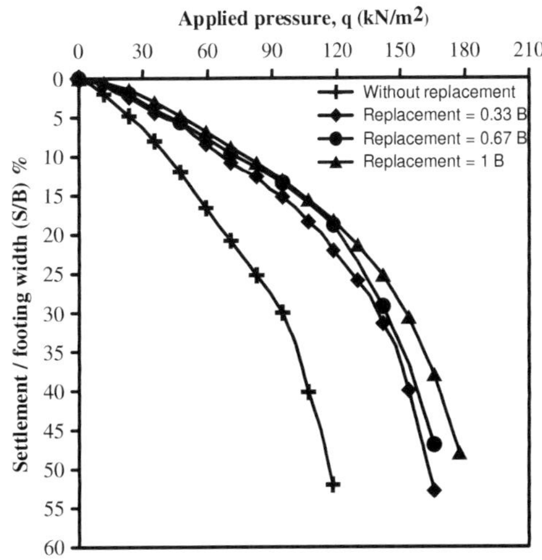

Figure 10. Load versus settlement for strip footing in soil tank D

To demonstrate the effect of the replacement layer of crushed concrete, on the settlement of the treated soils, the ratio called is S_r or settlement ratio which is defined as follows:

$$S_r = \frac{S_{treated}}{S_{untreated}} \tag{1}$$

where $S_{treated}$=Settlement of treated soil with crushed concrete corresponding to the same applied the load at failure of untreated soil.

$S_{untreated}$= Settlement of untreated soil with crushed concrete that equal to 10% of the width of footing according to criteria of Terzaghi to define the load failure.

The results of settlement reduction ratio (S_r) of 16 model tests of square footings are summarized in Table 3. It can be noticed from data of Table 3 that the S_r for replacement layer depth of 0.2 B is ranged from 0.48 to 0.51 for 4 sizes of square footing (100, 120, 130 and 150) mm, i.e. the settlement of treated soil is decreased to approximately 50% percent or equal to half of the untreated soil. Also, the S_r for soil treated with thecrushedconcretedepth of 0.4 B were (0.28, 0.39, 0.41 and 0.36) for the same previous of footing sizes, respectively with an average equal to 0.36. i.e. the settlement was decreasing by percentage of 64%. While the average the settlement of soil treated with 0.6 B depth of crushed concrete is reduced with the averaged percentage of 71%. These results showed that no clear difference between settlement reduction of soil treated with 0.4B and 0.6B because the stress bulb is decreased with depth. Four sizes of strip footing models with sizes of (30 x 240) mm^2, (35 x 280) mm^2, (40x320) mm^2 and (45 x 360) mm^2 are tested in the four same previous soil tanks:(A), 250 mm; (B), 300 mm; (C), 350 mm and ; (D), 500 mm, respectively. Three depths of replacement layer of crushed concrete were 0.33B, 0.67B and 1B are adopted in present work. Figures from (7 to 10) demonstrate the relationship between the applied pressure and deformation ratio (Settlement/footing width) for four sizes of strip footing. Table4 presents a summary of the results of the values of the settlement reduction ratio (S_r) due to the use of the crushed concrete layer under the strip footing in three different depths, (0.33B, 0.67B and 1B). The following points can be drawn about S_r from tabulated data of partial replacement of strip footings with three depths of partial replacement layers.

1. The relation between the settlement ratio (S_r) and size of strip footing is not obvious.
2. The averaged value of (S_r) for soil treated with crushed concrete for four sizes of the strip footing at a depth of a replacement layer of 0.33Bwas 0.42. The minimum value of S_r was in strip footing of (50 x 400) mm and the maximum value was in strip footing of (30 x 240) mm.

4. The values of S_r obtained from models of soil treated with replacement layer of 0.67B in the four sizes of the strip footing were (0.26), (0.31), (0.27) and (0.32) respectively.
5. The S_r was ranged from (0.11) to (0.23) with an average value of (0.17) for replacement layer depth of 1B.
6. The settlement reduction is more visible in strip footing than square footing at three depths of replacement layer.

Table 3. Settlement reduction ratio (S_r) of square footing models

Soil tank size	Size of square footing	Depth of replacement layer		
		0.2 B	0.4B	0.6B
(250 x 250 x 250) mm	(100 x 100) mm	0.51	0.28	0.23
(300 x 300 x 300) mm	(120 x 120) mm	0.52	0.39	0.29
(400 x 400 x 400) mm	(130 x 130) mm	0.59	0.41	0.33
(500 x 500 x 500) mm	(150 x 150) mm	0.48	0.36	0.32
Average		0.52	0.36	0.29

Table 4. Settlement ratio (S_r) of strip footing models

Soil tank size	Size of strip footing	Depth of replacement layer		
		0.33B	0.67B	1B
(250 x 250 x 250) mm	(30 x 240) mm	0.38	0.26	0.18
(300 x 300 x 300) mm	(35 x 280) mm	0.44	0.31	0.23
(400 x 400 x 400) mm	(40 x 320) mm	0.43	0.27	0.11
(500 x 500 x 500) mm	(50 x 400) mm	0.47	0.32	0.18
Average		0.42	0.28	0.17

3.2. Bearing capacity analysis

The bearing capacity ratio (BCR) was used in the present work to demonstrate the effectiveness of using the crushed concrete under footings in both types of square and strip footing. The BCR was defined as the ratio between the bearing capacity of treated soil with crushed concrete (B.C treated) and bearing capacity of untreated soil (B.C untreated) at the settlement level equal to 0.1 widths of footing (0.1 B).

$$BCR = \frac{B.C_{treated}}{B.C_{untreated}} \quad (2)$$

It can see clearly from Figs. from 3 to 10that the use of the partial replacement technique is affected the bearing capacity. To make the comparison between untreated and treated soil with crushed concrete, the values of (BCR) are collected in Tables 5 and 6.
The following main points can be inferred from the data in Tables 5 and 6 when using the partial replacement method for both the square and strip footings and the three replacement depths for the bearing capacity ratio BCR:
1. The maximum value of BCR was in a model of square footing with side 100 mm for all depths of partial replacement layer (0.2B, 0.4B and 0.67).
2. No clear differences between the BCR values in the model footing of sizes (120 x 120) mm^2 and (130 x 130) mm^2.
3. The bearing capacity ratio increases significantly when the replacement layer depth is increased for both the types of footing.
4. In general, the effect of using a layer of crushed concrete under the square footing region was more effective than strip footing especially in the depth of 0.6B.
5. The BCR for treating soil when using square footing of side130 mm are (1.62, 1.92 and 2.33) for three depths of replacement layer (0.2B, 0.4B and 0.6B) respectively. These values of BCR were closed to average values for all treatment.

6. The average value of BCR in treating soil with partial replacement layer of depths (0.33B, 0.67B, and 1B) were 1.48, 1.69 and 1.88 respectively on a strip footing.

Table 5. Bearing capacity ratio (BCR) of square footing models

Soil tank size	Size of square footing	Depth of replacement layer		
		0.2 B	0.4B	0.6B
(250 x 250 x 250) mm	(100 x 100) mm	1.85	2.16	2.39
(300 x 300 x 300) mm	(120 x 120) mm	1.60	1.93	2.31
(400 x 400 x 400) mm	(130 x 130) mm	1.62	1.92	2.33
(500 x 500 x 500) mm	(150 x 150) mm	1.58	1.83	2.29
Average		1.66	1.96	2.33

Table 6. Bearing capacity ratio (BCR) of strip footing models

Soil tank size	Size of strip footing	Depth of replacement layer		
		0.33B	0.67B	1B
(250 x 250 x 250) mm	(30 x 240) mm	1.43	1.70	1.97
(300 x 300 x 300) mm	(35 x 280) mm	1.47	1.71	1.88
(400 x 400 x 400) mm	(40 x 320) mm	1.45	1.61	1.77
(500 x 500 x 500) mm	(50 x 400) mm	1.59	1.73	1.88
Average		1.48	1.69	1.88

4. Conclusions

The following main points can be drawn from the results of present work

4.1. Settlement ratio SR

1. The relation between the settlement ratio (Sr) and size of strip footing is not obvious.
2. The settlement ratio (Sr) is increased clearly with increasing the depth of partial replacement.
3. The settlement reduction is more visible in strip footing than square footing at any depths of partial replacement layer.

4.2. Bearing capacity ratio BCR

1. The maximum value of BCR was in a model of square footing of less dimensions.
2. The bearing capacity ratio increases significantly when the partial replacement layer depth is increased for both the types of footing (square and strip).
3. In general, the effect of using a layer of crushed concrete under the square footing region was more effective than strip footing.

References

[1] Brown, J. D., and Meyerhof, G. G.1969 An experimental study of the ultimate bearing capacity of layered clay foundations. *Proc., 7th International Conference on Soil Mechanics and Foundations Engineering*, pp. 45-51.

[2] Ibrahim, K. M., 2014 Bearing capacity of circular footing resting on granular soil overlying soft clay." HBRC Journal, volume12, pp. 71-77,DOI: 10.1016/j.hbrcj.2014.07.004

[3] Madhav, M. R., and Vitkar, P. P. 1978 Strip Footing on Weak Clay Stabilized with a Granular Trench or Pile." *Canadian Geotechnical Journal*, Vol. 15, pp. 605–609,DOI: 10.1139/t03-075

[4] Madhav, M. R.; Kurapati, R. and Sakleshpur, V. A. 1986 Bearing capacity of strip footing in reinforced granular bed over soft non-homogeneous ground stabilized with granular trench, *Proc. 15th Asian Regional Conference on Soil Mechanics and Geotechnical Engineering*, Japanese Geotechnical Society Special Publication pp. 1348-1353. DOI: 10.3208/jgssp.SEA-06.

[5] Das, B. M. 1988 Bearing Capacity of Shallow Foundation on a Granular Trench in Clay, *Proc. 5th Australia-New Zealand Conference on Geomechanics*: Barton, ACT: Institution of Engineers, Australia. pp. 278-282. ISBN: 0858254085

[6] Sabry, M. A., 1995 Ultimate Capacity of Strip Footing on the Granular trench in Saturated Medium Stiff clay. *Journal of Egyptian Society of Engineering*, Volume 34(1), pp. 9-13.

[7] Fattah, M. Y., Al-Baghdadi, W., Omar, M., Shanableh, A. 2010 Analysis of Strip Footings Resting on Reinforced Granular Trench by the Finite Element Method, *International Journal for Geotechnical Engineering*, Vol. 4, Issue 4, pp. 471-482, J. Ross Publishing, Inc, DOI: 10.3328/IJGE.2010.04.04.471-482

[8] Unnikrishnan, N. Johnson, A.S. Rajan, S. 2010 Response of Strip Footings Supported on Granular Trench, *Proc. Indian Geotechnical Conference* , GEOtrendz, IGS Mumbai Chapter & IIT Bombay.

[9] El Sawwaf, M., and Nazir, A., 2012 Behavior of Eccentrically Loaded Small-Scale Ring Footings Resting on Reinforced Layered Soil, *Geotechnical and Geoenvironmental Engineering*, ASCE, Vol. 138, No. 3, pp. 376-384,DOI: 10.1061/(ASCE)GT.1943-5606.0000593.

[10] Ornek, M., Laman, M., Demir, A., and Yildiz, A. 2012 Prediction of bearing capacity of circular footings on soft clay stabilized with granular soil, *Soils and Foundations*, Volume 52(1), pp 69-80,DOI: 10.1016/j.sandf.2012.01.002

[11] Fattah, M. Y., Al-Waily, M. J., 2015 Bearing Capacity of Foundations Resting on a Trench of Local Reclaimed Asphalt Pavement Material, *Global Journal of Engineering Science and Research Management*, ISSN 2349-4506

[12] Bhattacharya, P. and Kumar, J. 2017 Bearing capacity of foundations on soft clays with granular column and trench, *Soils and Foundations*, Vol. 57, pp 488-495. DOI: 10.1016/j.sandf.2017.05.013

An experimental study of flow over V- shape crump weir crest

Hasan Mahdi Mohammed Al-Khateeb[1,*], Jumana Hadi Sahib[1], Huda Hadi Hasan Al-Yasisri[1]

[1] Civil Engineering Dep., Faculty of Engineering, University of Kufa, Najaf, Iraq

* E-mail: hasan.alkhateeb@uokufa.edu.iq

Abstract. Weirs are considered important structures controlling flow conditions and measuring flow rates in open channels. Crump weir represent good solution tool to control flow depth in streams, especially where controlling fish zones is required. However, installing crump weirs in rivers and relatively wide streams may be accompanied with difficulty of controlling same water level over the entire weir crest length. This study aimed to modify the crest of crump weir to be V-shaped rather than normal crest. Four wooden crump weir models were adopted. Each of the four models had longitudinal equilateral triangle section with the same horizontal base length, L (78.5cm), width, W (29.5cm) and height, P (20cm), but they differ by the middle crest height, P' as (20, 17.5, 15 and 12 cm) according to models 1, 2, 3 and 4, respectively.. The three v-shaped crested crumps and the normal crump were subjected to eight flow rates and five flow depths at steady state modular flow conditions. Data gathered were used to conclude a nonlinear multiple regression formula of discharge coefficient for the modified models tested. The formula concluded was calibrated and verified with very good determination factor (0.958). Effect of crest modification on water level was clear at relatively low flow rates and vice versa.

1- Introduction

Weirs are considered as hydraulic structures usually used for indirect flow measurement and/or controlling water elevation in basins and channels [1],[2]. When installing a weir in an open channel, a relationship between flow rate and water flow height above weir may be derived according to weir geometry on the basis of the Bernoulli equation [3]. Water viscosity and surface tension as well as weir surface roughness and geometry are all parameters included in that relationship. However, surface tension effect is very small compared to the other parameters and can be neglected [4]. The relationship "flow rate equation" is, principally, consist of dimensions related to weir geometry and upstream water depth, whereas the other parameters are included in a term called "discharge coefficient, Cd" in the equation. The Cd is assumed to cover effects of vertical curvature of the upstream flow streamlines in the vicinity of weir which cause head differences at both upstream and over weir. It also covers the assumption of considering pressure distribution upstream weir is hydrostatic [5]. According to downstream flow conditions, flow over weir may be modular or non-modular. In modular flow, the weir operates undrowned, and the upstream head is not affected by the downstream head, whereas in non-modular, the weir operates drowned, and the upstream head is affected by changes in the downstream head. Therefore, it is possible to determine the flow rate depending on only upstream head measurement in modular flow in contrast to non-modular, which has to include downstream head by means of "submergence factor" [6],[7], [8].

Weirs installed in open channels can take a verity of applications, such as in rivers, irrigation channels, water treatment plants, industries…etc. However, the most popular type of weirs adopted in rivers and relatively wide streams are long base weirs for their structural stability and simplicity, and also comparative good accuracy in flow measurement. Crump weir is considered one of the Broad crested weirs (long base weirs)[9],[10]. Its hydraulic behaviour approaches to the stepped weirs that considerably dissipate hydraulic jump energy in downstream side that may endanger stability of the weir

[11],[12]. Standard crump weir has a triangular section along flow direction with an upstream slope of 1 (vertical) to 2 (horizontal) and a downstream slope of 1 to 5, and the intersection of upstream and downstream surfaces forms a straight line crest, horizontal and at right angles to the direction of flow in the approach channel [10]. However, some modifications for the geometry of the weir have been investigated for certain purposes. Servais carried out a study through which several physical models of low-cost modifications to the Crump weir investigated in order to improve fish passage in England and Wales. Her study included the hydraulic investigation for installing baffles with different numbers, geometry and arrangement [13]. AL-Naely et. al. studied the effect of longitudinal flow openings (holes) penetrating upstream and downstream faces of crump weir, horizontally, to do as energy dissipaters, and as an improver for the discharge coefficient [14].

Triangular profile flat-V weir is a modification for the crump weir to measure a wider range of discharges by adopting a transverse symmetrical V-shaped crest, having small side-slopes (typically 1:10, 1:20 or 1:40) in the planes normal to the flow direction (upstream and downstream faces)[15],[16],[17],[18],[19]. Its longitudinal section is triangular same as that of crump weir but with possibility of equal (1:2) longitudinal slopes rather than the standard slopes (1:2) and (1:5) of upstream and downstream faces of standard crump weir [20]. Keller studied a standing inclined crest Crump weir. He concluded that at relatively large heads, the weir behaves as one half of a flat-V Crump weir for the same transversal crest slope. At lower heads, the flow cross section becomes strongly asymmetrical with a significant decrease in discharge coefficient value, [21].

Through the present study, a modified crump weir is adopted. The modified weir have triangular section along flow direction with equal longitudinal slopes as (1 vertical: 2 horizontal) for the upstream and downstream plane faces, with a transverse symmetrical V-shaped crest of three slopes (1:8, 1:4, 1:2.5). So, the invert of the V-shaped crest of the adopted modified weir represent horizontal line parallel to flow direction rather than the common triangular profile flat-V weir which has a point invert not as line, Figure 1. According to available literature, such modified weir may be investigated for the first time through present study. Justification for such study is that it investigate a weir that provide solution for wider flow range in open channels. This modified weir if replicated to form a structure consisting of successive attached weir segments, transversally, can provide a solution for the problem of uneven distribution of crest water depth along wide section channels. Test experiments of four weir models for different water flow rates and depths (upstream and downstream) have been evaluated. The results used to conclude a multiple regression equation for the discharge coefficient of the modified weir for variables driven from dimensional analysis.

2. Methodology

2.1. Hypothesis and constrains

This study assumes constant values as constrains for some variables related to the models tested and conditions of experiments. The variables assumed to have constant values are: weir length, width, crest height of sides and longitudinal slopes (upstream and downstream slopes), whereas middle crest height is variable. Hydraulic variables considered are flow rate and flow depth in upstream and downstream sides. Water temperature was nearly constant throughout experiments (around 25 ±1 °C). Channel bottom slope is zero (horizontal).

2-2- Dimensional Analysis

Variables effecting flow over weirs may be categorized into three groups; weir geometry, flow conditions and flowing fluid (water) properties. Table (1) lists the variables related to the testing conditions:-

Table 1. Principal variables related to the testing conditions.

Variable	Meaning	Dimensions
	Weir geometry	
L	Weir length	L
P	Weir height at crest terminals	L
P'	Weir height at weir centerline	L
W	Weir width	L
	Flow conditions	
C_s	Submergence ratio	unit less
g	Gravitational acceleration	LT^{-2}
H	Total upstream head	L
h_d	Total downstream head	L
Q	Flow rate	$L^3\,T^{-1}$
S	Channel bed slope	unit less
	Fluid properties	
ρ	Water mass density	ML^{-3}
μ	Water dynamic viscosity	$ML^{-1}T^{-1}$
σ	Water surface tension	MT^{-1}

Discharge over the weir according to modular flow may be expressed as:-

$$Q = f\,(L, P, P', W, g, H, S, \rho, \mu, \sigma) \tag{1}$$

or

$$f\,(Q, L, P, P', W, g, H, S, \rho, \mu, \sigma) = 0 \tag{2}$$

According to the π-theorem:-
m = 11 (number of variables)
n = 3 (number of the primary units involved in the equations above, MLT)
π-terms = 11 – 3 = 8 (number of dimensionless terms)

Considering g, H and ρ as repeating variables, eq. (2) can be expressed as:-

$$f_1\,(\pi_1, \pi_2, \pi_3, \pi_4, \pi_5, \pi_6, \pi_7, \pi_8)=0 \tag{3}$$

where:-

$\pi_1 = g^{a1}\ H^{b1}\ \rho^{c1}\ Q$ $\qquad \pi_5 = g^{a1}\ H^{b1}\ \rho^{c1}\ W$

$\pi 2 = g^{a1}\ H^{b1}\ \rho^{c1}\ \mu$ $\qquad \pi 6 = g^{a1}\ H^{b1}\ \rho^{c1}\ S$

$\pi 3 = g^{a1}\ H^{b1}\ \rho^{c1}\ \sigma$ $\qquad \pi 7 = g^{a1}\ H^{b1}\ \rho^{c1}\ L$

$\pi_4 = g^{a1}\ H^{b1}\ \rho^{c1}\ P$ $\qquad \pi_8 = g^{a1}\ H^{b1}\ \rho^{c1}\ P'$

According to dimensional form, π1 may be written as:-

$\pi_1 = \rho^{a1}\ g^{b1}\ H^{c1}\ Q \ \rightarrow\ M^0\,L^0\,T^0 = (ML^{-3})^{a1}\ (LT^{-2})^{b1}\ (L)^{c1}\ L^3T^{-1}$

hence, for *M*: $a1 = 0$

for *T*: $-2b1-1=0$ *thus* $b= -0.5$

for *L*: $-3a_1+b_1+c_1+3=0$, thus $c_1 = -2.5$

therefore, $\pi_1 = \rho^0 g^{-0.5} H^{-2.5} Q$ ➔ $\pi 1 = \frac{Q}{\sqrt{gH^5}}$

following the same procedure, the other π- terms can be written as:-

$\pi 2 = \frac{\mu}{\rho\sqrt{gH^3}}, \pi 3 = \frac{\sigma}{\rho g H^2}, \pi 4 = \frac{P}{H}, \pi 5 = \frac{W}{H}, \pi 6 = \frac{S}{H}, \pi 7 = \frac{L}{H}, \pi 8 = \frac{P\prime}{H}$

But, $\pi 2 = \frac{\mu}{\rho\sqrt{gH^3}} = \frac{1}{Re}$, and $\pi 3 = \frac{\sigma}{\rho g H^2} = \frac{1}{We}$

Where, Re and We are Reynolds number and Weber number, respectively.

Both of Re and We can be ignored in present study. This due to that flow over weirs with a depth exceeding 3 cm involves ignorable We [22] and Re [23]. For horizontal channel bed of present study, π6 of bed slope can be ignored. Simplifying and combining some of π- terms, such as π4 and π8, eq. (3) can be written as:-

$$f_1\left(\frac{Q}{\sqrt{gH^5}}, \frac{P}{P\prime}, \frac{W}{H}, \frac{L}{H}\right) = 0 \tag{4}$$

or

$$\frac{Q}{\sqrt{gH^5}} = f\left(\frac{P}{P\prime}, \frac{W}{H}, \frac{L}{H}\right) \tag{5}$$

Rewriting eq. (5) in the form of coefficient of discharge Cd as:-

$$C_d = f\left(\frac{P}{P\prime}, \frac{W}{H}, \frac{L}{H}\right) \tag{6}$$

2.3. Experimental Work

2.3.1. Modified Crump Weir Models

Four wooden models have been adopted in present study. Wood sheets of 5mm thick cut into desired dimensions and assembled through fixing with nails and pasted together by silicon to produce the models shown in Fig. 1. Each of the four models had longitudinal equilateral triangle section with the same base length, L (78.5cm), width, W (29.5cm) and height, P (20cm), but they differ by the middle height, P' (20, 17.5, 15 and 12 cm) according to models 1, 2, 3 and 4, respectively. Wooden stiffeners fixed inside the models to strengthen them against flow stresses. Through experiment runs, each model fixed in the experimental flume by means of two terminal screws and adequate silicon rubber to prevent movement and provide water tightness, Fig. 2. The silicon rubber margins fixing the model in the flume have thickness of 2.5mm to get total weir width of 30cm. All models painted with red color to be clearly distinguished in the flume. According to ratio of P'/P the models 1, 2, 3 and 4 denoted as P'/P=1, P'/P=0.875, P'/P=0.75, P'/P=0.6, respectively.

2.3.2. Flume and accessories

All experiments of the present study carried out using a rectangular flume of 15m length, 0.3m width and 0.45m depth. Walls of the flume are made of acrylic glass to enable visual observation. Flume bed is made of stainless steel. Flume operation is controlled by an electrical board. Sluice gate was situated at the entrance and downstream gate at the end to control water depth in the flume. The flume provided with point gauges to measure water depth. Water is circulating from main downstream tank into upstream tank by means of variable speed pump forcing water through steel pipe at which float-type flow meter is installed. Wave suppressor made of vertical steel piers is installed close to the entrance to distribute the flow and break waves, Fig.3.

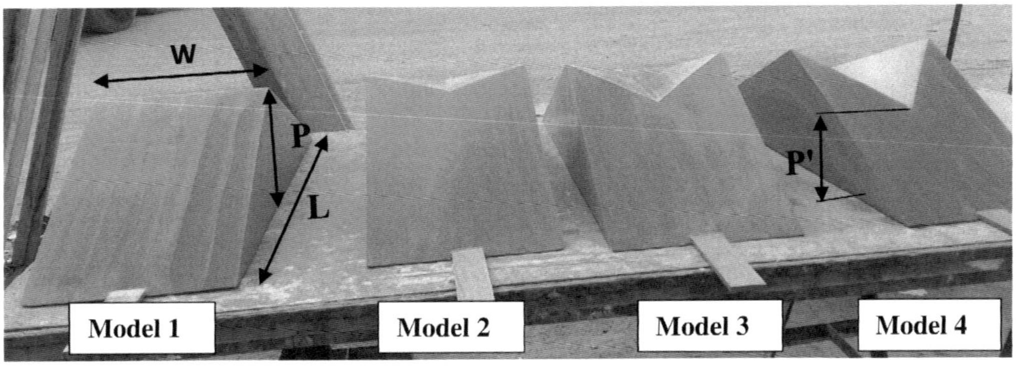

Fig. 1: Modified Crump Weir synthesized for present study.

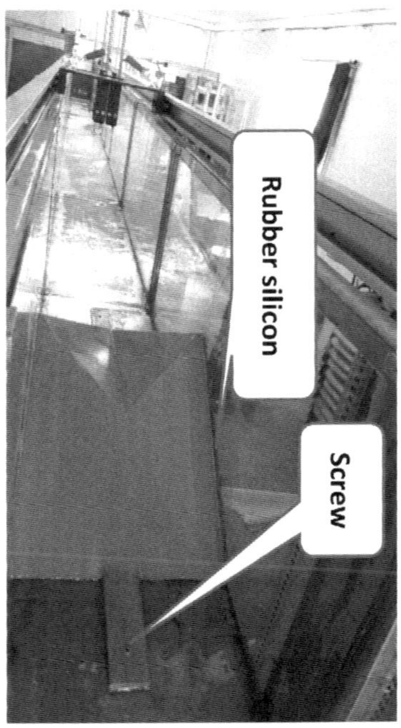

Fig. 2: Fixing the model in the experimental flume.

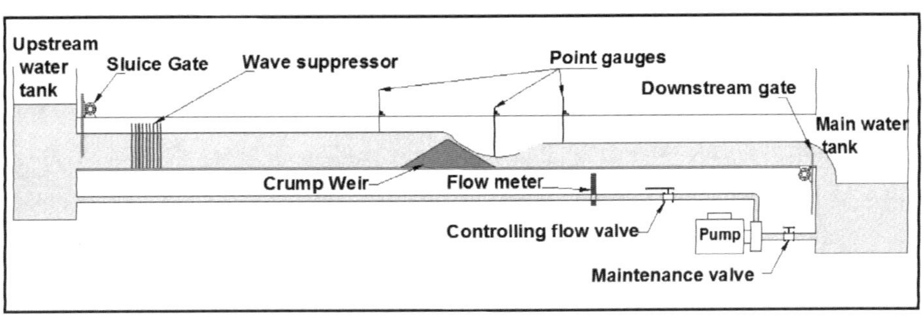

Fig. 3: Sketch of the experimental flume

2.3.3. Testing procedure

First, flow meter readings checked and calibrated with quantitative method (flow volume with time watch). The bed of the flume was set horizontal for all runs. Models installed at the middle of flume length. For each model, eight flow rates (8, 23, 40, 70, 95, 110, 117, 123 m3/hr) were passed. For each of the eight flow rates, downstream gate rise (hs) was maintained for five depths (20, 17.5, 15, 12 and 0.0 cm). Net of 168 experiment runs carried out throughout the present study. In each run, flow depths at four locations were recorded after 5 minutes passed of continuous operation. The locations agreed with WMO [10]. They were at the weir crest (from water surface to weir crest center hcr), at 0.48m upstream weir center (from water surface to channel bottom, H), at 0.8m downstream weir center (from water surface to channel bottom, hd1) and at 1.14m downstream weir center (from water surface to channel bottom, hd2). At the same location where (H) measured, water depth from water surface to weir crest upper level (y) was calculated by subtracting weir crest height (P=20cm) from (H) reading. Also, length of hydraulic jump (Lj) was measured when occurred downstream the weir, Fig.4 and Fig.5.

3. Results and discussion

3.1. Flow over weir evaluation

3.1.1. Approach water depth at weir upstream, y.

The transversal V- modification of weir crest facilitates passage of water over the crest lowering approach water surface level for a given flow rate. Fig.6 illustrates results of water depth at approach (y) verses flow rate (Q) range for the four models. It is obvious that increasing (Q) lead to increase (y), and this is logical as higher flow rate requires higher upstream water depth for a given weir. But, for a given flow (Q) when comparing (y) for different models (P'/P), it is clear that decreasing weir crest middle height (P') lead to enlarge crest triangular open flow area that facilitating flow over weir and resulted in lower water depths (y). The overall average reduction in (y) can be calculated as ratio of average decrease in (y) values to average of (y) of model 1 (P'/P=1). The overall average reduction is 7% for model 2 (P'/P=0.875), 17.3% for model 3 (P'/P=0.75) and 32% for model 4 (P'/P=0.6). This reduction may effect downstream flow condition including strength (turbulence) and length of hydraulic jump as well as energy dissipation as illustrated in following sections below.

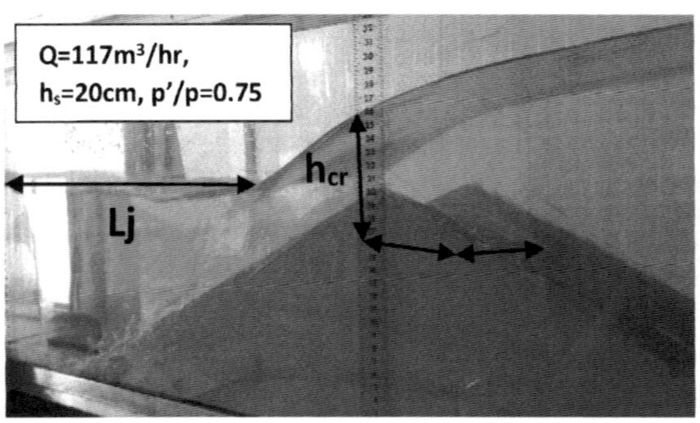

Fig. 4: Measurement of h_{cr} and L_j

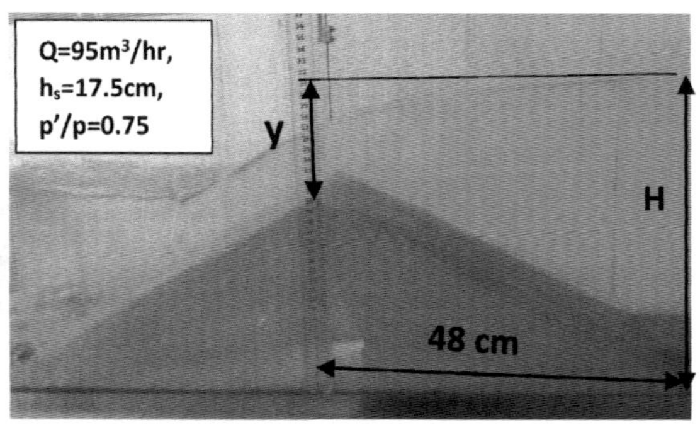

Fig. 5: Measurement of y and H

3.1.2. Hydraulic Parameter H/P versus Froude Number Fr.

The parameter (H/P) effects the discharge coefficient of any weir, directly. As this parameter increases, discharge coefficient increases too. In other words, minimizing weir height (P) to ignorable value leading to ignoring effect of the weir as retarding to flow, and vice versa. On the other hand, Froude number (Fr) indicates local ratio between gravitational and inertial forces of flow. So, as (Fr) of approaching flow upstream the weir increases, that mean water flow depth (hcr) decreases and/or flow velocity increases. Logically, for a given weir height (p) in modular flow, increasing upstream water flow depth (H) mean increasing total upstream head, so, it lead to increase flow rate and hence velocity rather than increase flow depth at weir crest section which is mean increase (Fr). But, as continuing with increasing (H) reaching certain values, the rate of increase in (Fr) decreases due to flow depth increase at weir crest section. That is why trend of the relationship between (H/P) verse (Fr) is approximately linear for certain limits then changes to curvilinear as illustrated in a through d of Fig. 7 for all models. However, comparing the four curves for similar (H/P) values indicates that the modified models resulted in decreasing (Fr) for a given (H/P) with maximum decrease in Model 3 of (P'/P=0.75). Values of (Fr) was calculated from the well-known formula (Fr= v/√(g*D)), in which, (v) is flow velocity at weir crest section, and (D) the hydraulic depth which was calculated by dividing flow cross sectional area at the weir crest centre over the top width of flow. The eight (Fr) values of each of the four curves belong to the same flow rates.

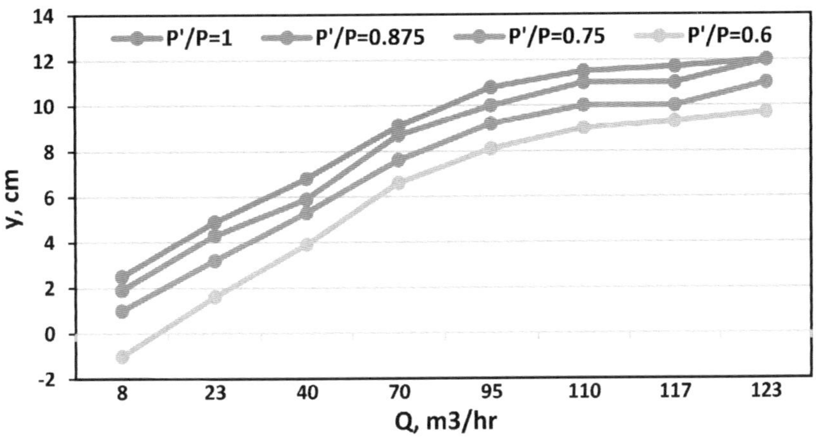

Fig. 6: Approach water depth (y) verses flow rate (Q)

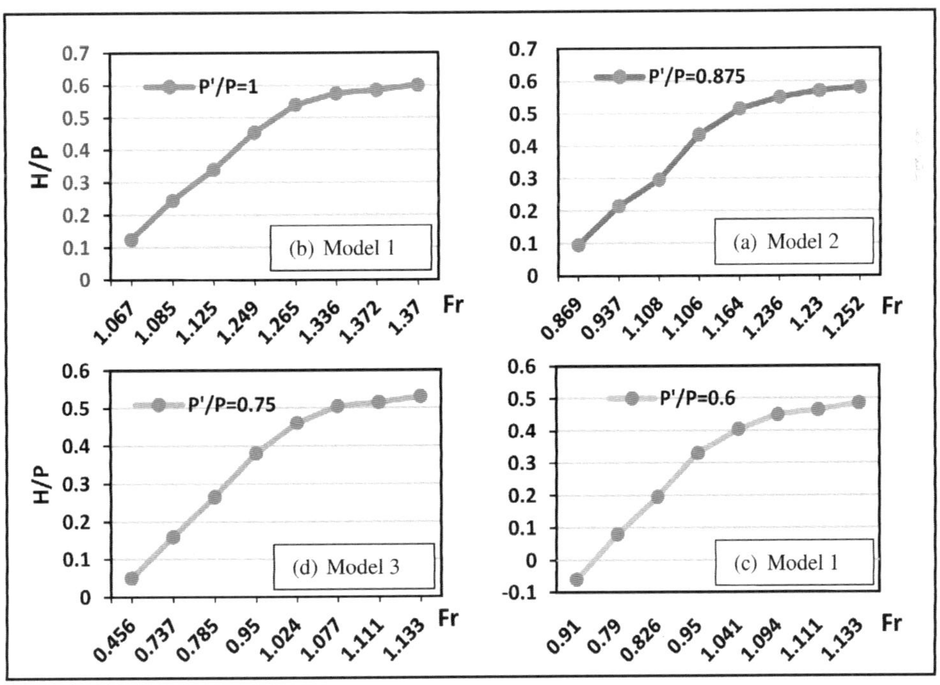

Fig. 7: H/P vs (Fr) for (a) model 1, (b) model 2, (c) model 3 and (d) model 4.

3.1.3. Length of Hydraulic Jump (Lj).

Fig. 8 (a throguh d) illustrates the relationship between the length of hydraulic jump (Lj) and the flow rate for downstream water depth ratios hs/P as 1, 0.875, 0.75 and 0.6, respectively. For (hs/P= 1 and 0.875), model 4 (of P'/P=0.6) resulted in the lowest length, because downstream water depth (hs) equals and exceeds (P') that lead to reduce jump length. For (hs/P= 0.75 and 0.6) all modified models (2,3 and 4) resulted in jump length longer than that of model 1 of (P'/P=1) with highest values for model 3 of (P'/P=0.75).

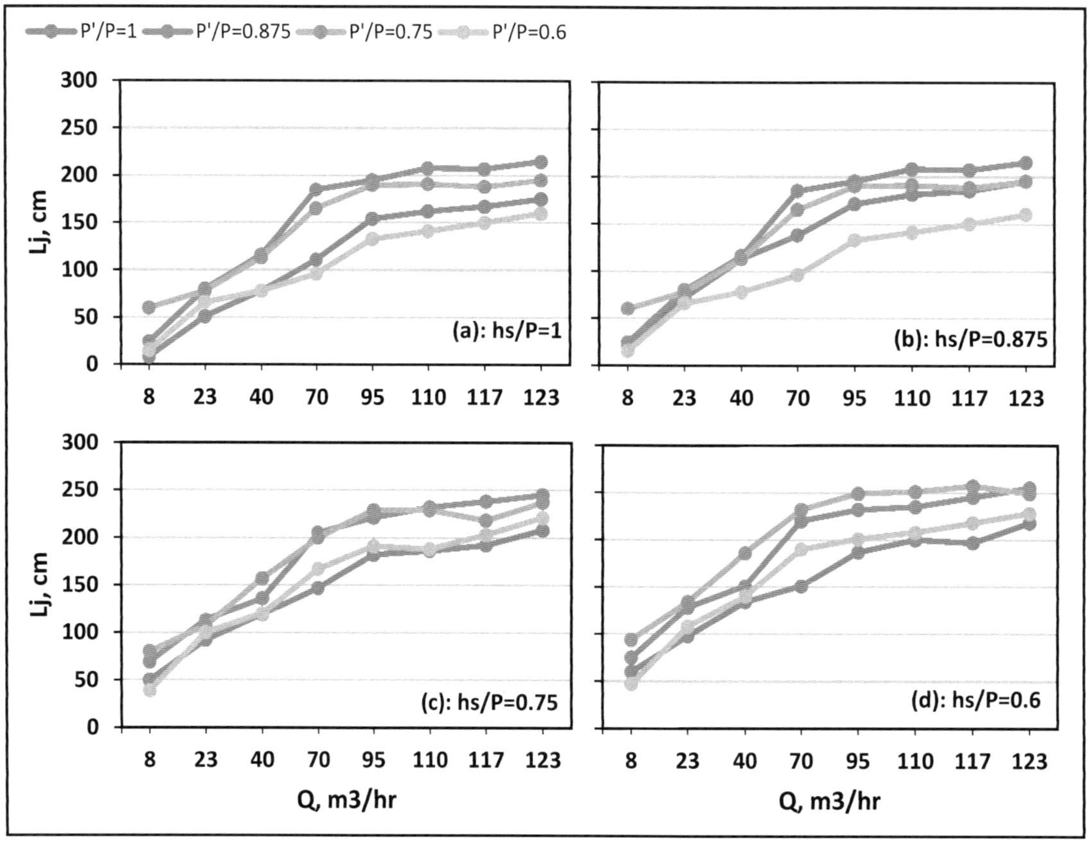

Fig. 8: (Lj) vs (Q) for a: hs/P=1, b: hs/P=0.875, c: hs/P=0.75 and d: hs/P=0.6

3.1.4. Energy Dissipation (ΔE)

Energy dissipation (ΔE) of flow within weir zone is important to reduce downstream turbulence, hence, reduce scour and erosion. Usually, inertial forces (kinetic energy) is proposed to be reduced more than gravitational (potential energy). The dissipation is calculated as a ratio of energy difference between upstream and downstream sides to the upstream one as [$\Delta E = (Eu-Ed)/Eu$]. The upstream energy is ($Eu= H +3yc /2$), where (yc) is the critical flow depth as [$yc=(Q^2/gW^2)(1/3)$]. The downstream energy is [$Ed=y1+ 1.1\ Q^2/(A^2\ g)$], where (y1) is downstream flow depth before hydraulic jump, and (A) is the flow cross sectional area at (y1) location [24]. In present study, focusing was on energy dissipation when the downstream gate was lowered down, completely, (hs=0). Values of (hd1) were considered as the needed (y1) values. Fig.9 (a through d) shows the results of energy dissipation calculation for the four models. The eight points of each of the four fcurves belong to the same flow rates. However, values of upstream approach (Fr) decreased with decreasing (p') according to models 2, 3 and 4, respectively when compared with that of model 1 of (P'/P=1). This is because of decreasing flow velocity and increasing flow depth with decreasing (p'). At relatively low (Fr), effect of the weir in dissipating flow energy decreases. This is clear in models 3 and 4, and consequently, less energy dissipation. Model 2 with (P'/P=0.875) showed similar energy dissipation as Model 1 with no opening (P'/P=1).

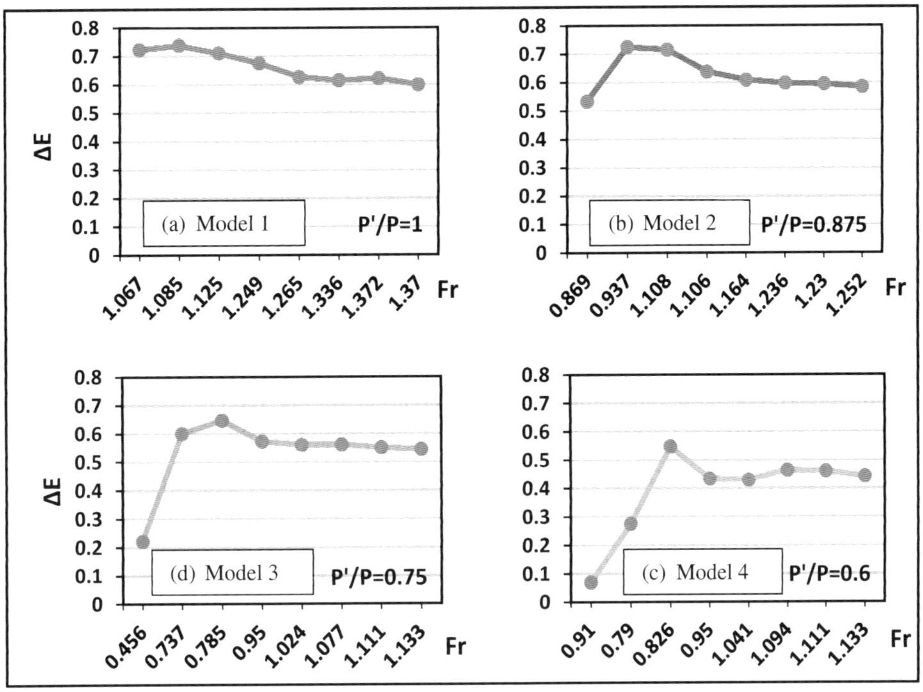

Fig. 9: (ΔE) vs (Fr) for (a) model 1, (b) model 2, (c) model 3 and (d) model 4.

3.2. Regression formula for discharge coefficient

On the basis of dimensional analysis results explained in section (2-2) above, and in particular eq. (5), multiple regression analysis was carried out for about 80% of the experimental results obtained, whereas the rest 20% were used for verification. The selection of the 80% and 20% was random process. All results adopted were belong to modular flow conditions. Statistical computer software SPSS was used to perform the non-linear multiple regression. The resulted basic equation is:-

$$Cd = \left(\frac{\bar{p}}{p}\right)^{a} \times \left(\frac{w}{H}\right)^{s} \times \left(\frac{L}{H}\right)^{z} + b \tag{7}$$

The factors a, s, z and b produced by the software to have the developed formula as:-

$$Cd = \left(\frac{\bar{p}}{p}\right)^{6.09} \times \left(\frac{w}{H}\right)^{0.606} \times \left(\frac{L}{H}\right)^{-0.003} + 0.712 \tag{8}$$

The coefficient of determination R^2 of this formula is (0.958). To test (verify) this formula, the rest 20% of the observed experimental results were compared with the corresponding results of the formula. Fig. 10 show a comparison between results predicted by the formula (theoretical) and those observed (actual). Good agreement between the predicted and observed results that justify using this formula to estimate the discharge coefficient Cd of the weirs modified in this study.

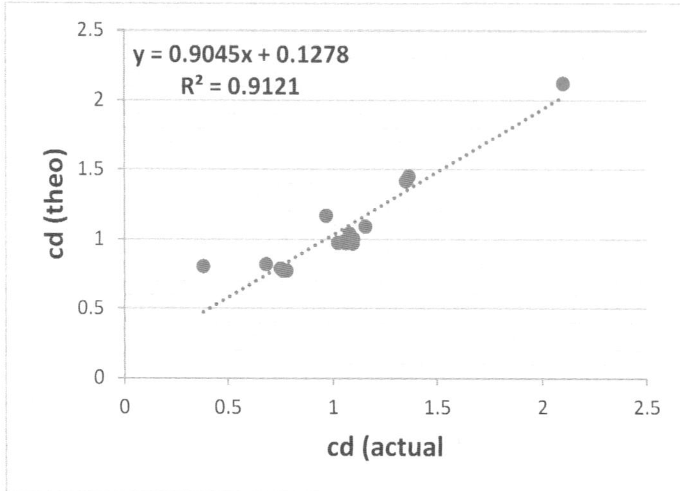

Fig. 10: Comparison between experimental results and predicted by formula (8)

4. Conclusions

The following conclusions may derived from the present study:-

a- The overall average reduction in approach water depth upstream the modified crump weir models is 7% for model 2 (P'/P=0.875), 17.3% for model 3 (P'/P=0.75) and 32% for model 4 (P'/P=0.6).
b- The modified models resulted in increasing (Fr) for a given (H/P) with maximum increase in Model 3 of (P'/P=0.75).
c- For relatively high downstream water depth (hs/P= 1 and 0.875), model 4 (of P'/P=0.6) resulted in lowering the length of hydraulic jump, while for (hs/P= 0.75 and 0.6) all modified models (2,3 and 4) resulted in jump length longer than that of model 1 of (P'/P=1) with highest values for model3 of (P'/P=0.75).
d- Model 2 with (P'/P=0.875) showed similar energy dissipation as Model 1 with no opening (P'/P=1).
e- The non-linear multiple regression formula developed in this study to estimate coefficient of discharge Cd of the modified models 2, 3 and 4 has good coefficient of determination R^2 as (0.958) and good agreement with observed data in verification process.

Acknowledgments
Authors wishing to acknowledge technical staff of the Hydraulic Laboratory at the Faculty of Engineering, Kufa University for assistance throughout the experimental work of the present study.

References

[1] Chin D A 2006 *Water Resources Engineering*, 2nd edition, (Upper Saddle River: Prentice Hall).
[2] Rady R M H 2011, 2D-3D Modeling of Flow Over Sharp-Crested Weirs *Journal of Applied Science Research* pp. 2495-2505.
[3] Chadwick A & Morfett J 1993 *Hydraulics in Civil and Environmental Engineering* (London; New York: E & FN Spon), pp.397-411.
[4] Chadwick A & Morfett J 1986 *Hydraulics in Civil Engineering* (London; Allen and Unwin) p.492.
[5] El Alfy K S 2005 Effect of Vertical Curvature of Flow at Weir Crest on Discharge Coefficient," *Ninth Int. Water Technology Conf (Egypt)* pp:249-262.
[6] Skogerboe Gaylord V, Hyatt M Leon and Austin Lloyd H 1967 *Design and Calibration of Submerged Open Channel Flow Measurement Structures: Part 4 – Weirs* Reports, Paper 80.

[7] Mays Larry 1999 *Hydraulic Design Handbook* (New York: McGraw Hill).
[8] Dabral P P, Pandey P K, Tushar Kumar and Sourav Chakraborty 2014 Determination of Discharge Coefficient and Head-Discharge Relationships Of Different Hydraulic Structures *Journal of Indian Water Resources Society* **34** (1) pp:40-52.
[9] Crump E S and Burkitt F H 1952 A new method of gauging stream flow with little afflux by means of a submerged weir of triangular profile *Proc. of the Institution of Civil Engineers* **1** (2) pp: 223–242.
[10] Wodd Meteorological Organization 1971*Use of Weirs and Flumes in Stream Gauging* Technical Note No. 117, Report of the Commission for Hydrology, WMO-No. 26. TP, Geneva: Switzerland p 27.
[11] Chafi C, Hazzab A and Seddini A Study of Flow and Energy Dissipation in Stepped Spillways *Jordan Journal of Civil Engineering* **4** (1) pp: 1–11.
[12] Chanson H 2001 Hydraulic Design of Stepped Spillways and Downstream Energy Dissipaters *Dam Engineering* **11** (4) pp: 205–242.
[13] Servais S A 2006 *Physical modelling of low-cost modifications to the Crump weir in order to improve fish passage: Development of favorable swimming conditions and investigation of The hydrometric effect* Ph.D. Thesis, Engineering Systems Department, Cranfield University, Shrivenham, England.
[14] AL-Naely H, Al-Khafaji Z, Khassaf S 2018 Effect of Opening Holes on the Hydraulic Performance for Crump Weir *International Journal of Engineering* (IJE), IJE TRANSACTIONS C: Aspects **31** (12) :pp:2022-2027.
[15] White W R 1966 The Flat-Vee Weir *HR Wallingford Report INT 56*, cited by [18].
[16] White W R 1968 The Flat-Vee Triangular Profile Weir with 1:2 Upstream and Downstream Slopes *HR Wallingford Report INT 66* cited by [18].
[17] White W R 1970 The Crump Profile Flat-Vee Weir *HR Wallingford Report EX 473* cited by [18].
[18] Herschey R W 1985 *Stream flow Management* (Elsevier Applied Science Publishers).
[19] Bos M G 1989 *Discharge Measurement Structures* (ILRI Publication 20 Netherlands).
[20] White W R 1968 The Flat-Vee Triangular Profile Weir with 1:2 Upstream and Downstream Slopes *HR Wallingford Report INT 66* cited by [18].
[21] Keller R J 1989 Sloping Crest Crump Weir *Journal of Irrigation and Drainage Engineering* **115** (2) pp: 231–238.
[22] Novak P Guinot, Jeffery V A and Reeve D E 2010 *Hydraulic Modelling – an Introduction* 1st edition (London: Spon Press).
[23] Chanson H 2004 *The Hydraulics of Open Channel Flows: An Introduction* 2nd edition, (Oxford, UK: Butterworth-Heinemann).
[24] Boes R M and Hager W H Hydraulic Design of Stepped Spillways *Journal of Hydraulic Engineering* **129** (9) pp:671–679.

IOP Conf. Series: Materials Science and Engineering **584** (2019) 012061 doi:10.1088/1757-899X/584/1/012061

The effect on the interfacial strength of the length/width ratio of FRP sheet bonded to concrete

W. Al-Juboori[1], and L. Weekesb[2]

[1] Dep. Of Civil Engineering, Faculty of Engineering, University of Kufa
Email: Wisams.aljuburi@uokufa.edu.iq
[2] Address: School of Computing, Science and Engineering, University of Salford, Salford M5 4WT, Email: L.Weekes@salford.ac.uk

Abstract. Fibre reinforced polymer laminates are commonly used for the retrofit of ailing reinforced concrete structures for both shear and flexure. Recent studies have been undertaken regarding bond behaviour, but these have not advanced significantly to apply in a general manner to FRP sheets. This paper presents the effect of the length/width ratio on the bond stress and strength between FRP laminates and concrete. The experimental and the theoretical study give the same effect of the length/width ratio. Experimental bond models were tested of 7 specimens with 3 bonding techniques and compared with predictions from finite element models. Mesh sensitivity studies were carried out for the FE modelling, and compatible crack patterns were observed between FE models and the experimental samples. The FE shows clearly the distribution of principal stress in the concrete and explains the failures observed in the experimental tests. This study shows that the width of CFRP sheet has more influence than the length. Also it can be concluded that increasing the CFRP sheet width leads to a propensity for concrete de-bonding. This also illustrates how existing expressions for bond strength match with the observed results. In general the results show that an increase in width provides more load capacity than a relative increase in length even some design methods do not interest to the effective of width/length sheet ratio.

Keywords: Shear test; Concrete; FRP; Strengthening; Bond strength; Bond stress; Theoretical derivation.

1. Introduction

External bonding of fibre reinforced polymer (FRP) plates or sheets have emerged as a popular method for the retrofit strengthening of reinforced concrete (RC) structures [1-2]. This technology offers unique advantages with respect to the traditional strengthening techniques, including good immunity to corrosion, low weight and excellent mechanical properties. Furthermore, the hand lay-up allows this reinforcement to be adapted to the shape of any structural element [3-5]. The performance of the FRP-to-concrete interface in providing an effective stress transfer is of crucial importance [6-8]. Indeed, a number of failure modes in FRP-strengthened RC members are directly caused by de-bonding the FRP from the concrete. Therefore, for the safe and economic design of externally bonded FRP systems, a sound understanding of the behaviour of FRP-to-concrete interfaces needs to be developed and a reliable bond-slip model established.

The anchorage between concrete and FRP sheets plays an important role in reinforcing design. As a brittle material, FRP shows a lack of ductility, and the failure mode occurs suddenly without preceding yielding. Recently, many studies have been undertaken to understand the bond behaviour between concrete and FRP sheets, but current understanding is not sufficiently advanced for FRP sheets to be applied in general, on actual structures. For these reasons, in terms of adhesion, it is necessary to confirm the safety of FRP sheets, or to seek a bond length and width considered to provide an adequately safe bond stress, so as to prevent the appearance and progress of bond rupture. In cases where the bond rupture or anchorage rupture is unavoidable, it is necessary to obtain the rupture strength experimentally, based on reliable data that adequately consider safety factors. This paper presents a new expression for the prediction of the bond stress and strength of FRP sheets with concrete that compares favourably with experimental results.

1.1 Research significance and objective
The broad objective of this paper is to understand the influence of the length and the width of the FRP on the ultimate stress/strain at de-bonding. The specific objectives of this paper are:

- To study the influence of the length and width of FRP sheet on the interfacial shear stress distribution in the FRP during de-bonding;
- To propose interfacial shear stress and bond strength models that can provide effective bond length and effective bond width for externally bonded FRP sheet; and
- To increase the database on bond behaviour between FRP sheet and concrete.

2. Experimental Programme

2.1 Test set-up
FRP–concrete bonding has been investigated by testing a number of specimens with CFRP reinforcement bonded to concrete prisms. The nominal dimensions of the prismatic concrete blocks are as follows: length (L = 280 mm), width (b = 200 mm) and height (h = 90 mm). The CFRP sheet was bonded in the centre on both sides of the concrete block. The concrete blocks had been sawn into two separate parts as shown in Figure 1, and then re-jointed by two threaded rods.

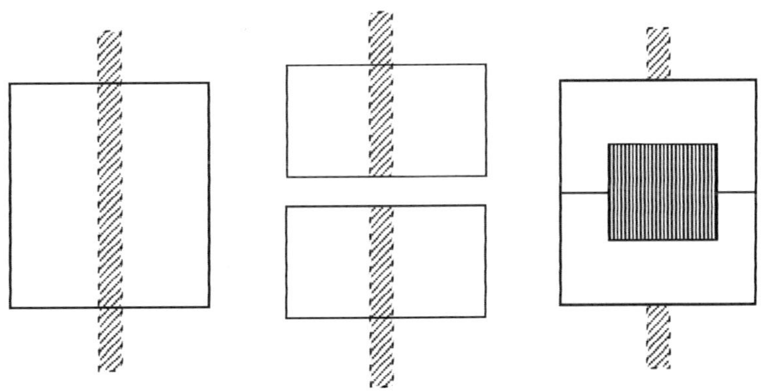

Figure 1. Process of preparing two concrete blocks for bond testing.

In a typical test, the tensile load is applied to the ends of a steel bar, which has been cast inside the concrete block. Figure 2 shows the steel bars inside the timber moulds, which have also been cut into two parts with concrete block. In addition, there are two longitudinal holes provided to facilitate clamping of the sample 'halves' prior to testing whilst loading into the tensile test machine. These were released before testing, as shown in Figure 3(c). Also a plastic membrane was placed between the two parts of concrete block to prevent any additional bond from the adhesive.

The samples were to be tested in a 50kN tensile test machine. To negate the effects of any eccentricity causing moment on the sample during testing, a 'ball and socket' connection was employed, as shown in Figure 3(b). Global slip was measured using two LVDTs that were set up on both sides of concrete

IOP Conf. Series: Materials Science and Engineering **584** (2019) 012061 doi:10.1088/1757-899X/584/1/012061

block close to the edge of the boundary surface of two concrete parts as shown in Figure 3(a). The surfaces of the concrete blocks, where the CFRP sheet would be glued, were ground to a fine finish with a stone wheel to remove the top layer of mortar, just until the aggregate was visible (approximately 2-3 mm); due to the very small dimension of the marble powder glued to the wheel, the concrete surface was very smooth (see Figure 2).

Figure 2. Casting processes of specimens.

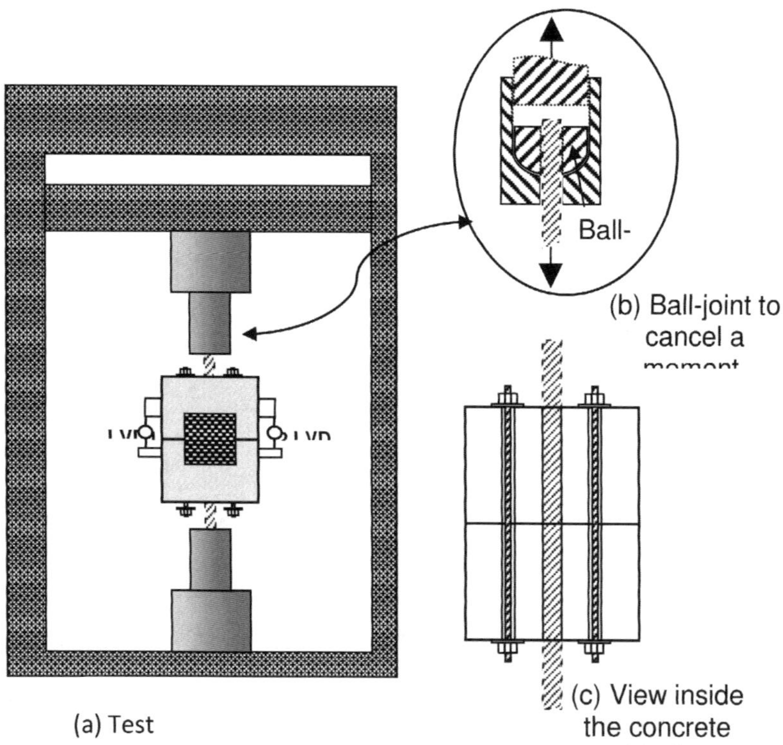

Figure 3. Detail of bond-slip test.

2.2 Material properties

- **Concrete:** The average compressive strength ($f_c{}' = 51\ MPa$) at 28 days and at the date of specimen testing was evaluated from uniaxial compression tests.
- **CFRP system:** Material properties for the unidirectional CFRP sheets were based upon information supplied by [Weber]. The nominal thickness, t_f, of the carbon fibres contained in the FRP sheet was equal to 0.1178 mm. The tensile strength and the Young's modulus of the composite sheet based on the sheet thickness were equal to 3900 MPa and 240 GPa, respectively.

2.3 Test programme

The experimental programme comprised of 7 specimens with 3 bonding techniques as shown in Figure 4. Each specimen was made up of two separate concrete sections without any bond between them other than the CFRP sheets. All these specimens had the same bond area. The purpose of this test was to understand the effects of changing the dimensions of the CFRP sheet (length and width) and the difference in strength contribution of CFRP for sheets and strips. To examine the consistency in results, three specimens were provided for arrangement B1-1, while two arrangements were provided for B2-3 and B2-4.

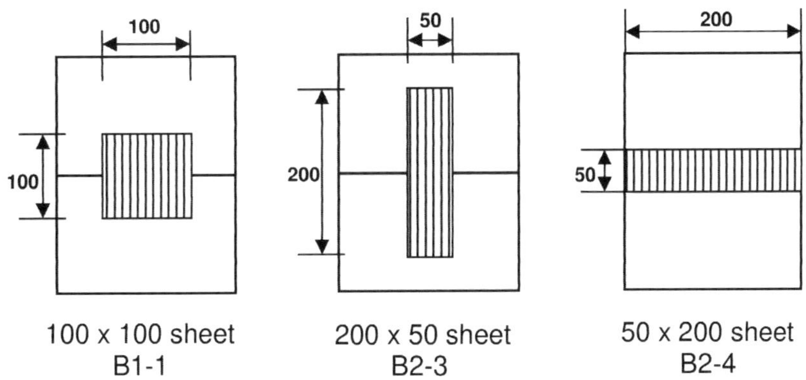

Figure 4. Detail and techniques of specimens tested.

3. Test results and discussion

All the figures in this section show the tensile load at the bar ends versus the average reading of the two gauges which were mounted on both sides of concrete block. To understand the shear stress at the FRP-to-concrete interface, the bond area is constant in almost all specimens. Therefore all results are compared with B1-1, which has CFRP sheet on both sides (100 mm x 100 mm). Figure 5 illustrates the experimental bond-slip results of the specimens.

First of all, as some of these graphs show, there is a difference in behaviour of specimens which have the same strengthening technique. Some of the reasons for this may be as follows:

- The eccentricity between the top and bottom grips
- Differences in thickness of adhesive
- Difference in concrete strength which affects the bond-slip for the CFRP-concrete interface
- On occasions, adhesive can enter between the contacting surfaces of the concrete blocks. This may lead to changes in bond-slip behaviour.

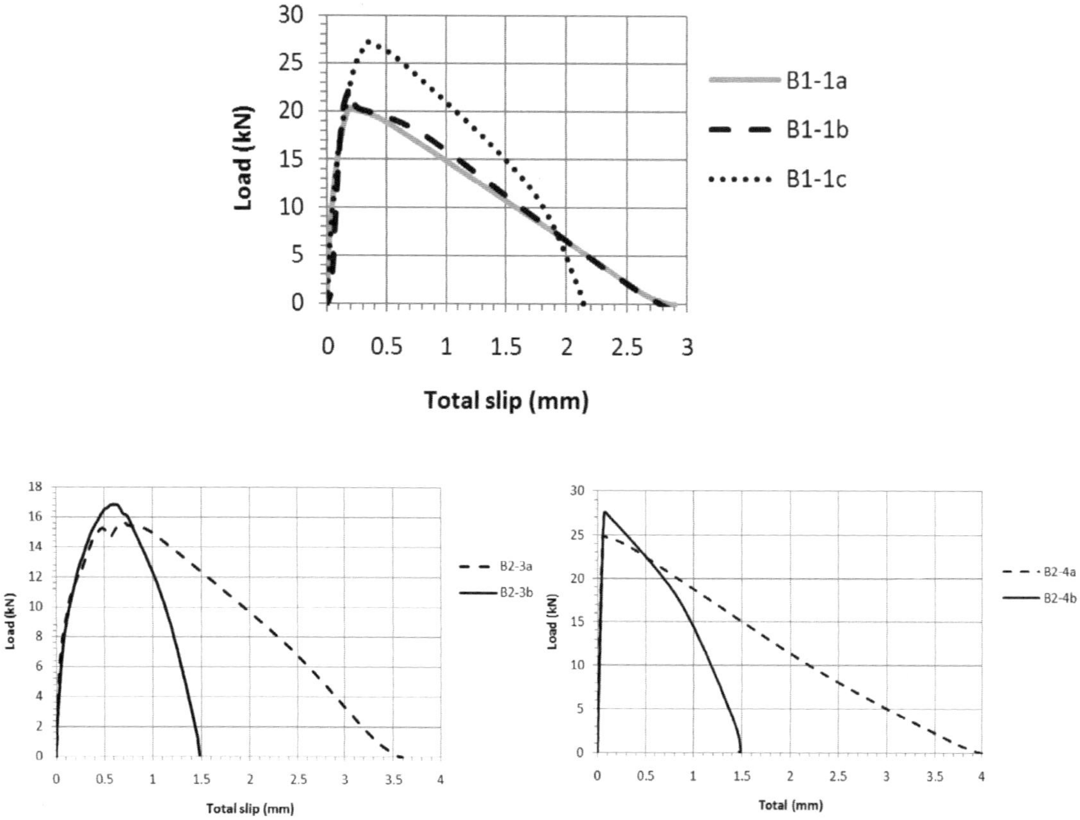

Figure 5. Load vs. total slip experimental bond-slip results of specimens.

Figure 6. Typical failures of specimens

Figure 6 shows selected photos after failure of the specimens. All of the samples failed by de-bonding (there were no failures by rupture). Some of these photos show that there is laminar failure (a type of de-bonding) happening in the concrete. The concrete used for the samples was normal concrete but its strength was high (51MPa) if it is compared with most commonly used concrete strengths. It can be seen that there are two types of de-bonding failure in the specimens of this study (adhesive de-bonding and concrete de-bonding). Table 1 illustrates that an increase in CFRP sheet width leads to a propensity for concrete de-bonding rather than adhesive de-bonding. Almost all of the specimens which failed with concrete de-bonding gave high load capacity. Because there is more than one specimen for each case, all the curves present the average behaviour of each case (the details are shown in Figure 5).

Table 1. Specimen data.

Specimen	Maximum load	Failure mode
B1-1a	20.3	adhesive de-bonding
B1-1b	22.4	adhesive de-bonding
B1-1c	27.5	concrete de-bonding
B2-3a	15.6	adhesive de-bonding
B2-3b	16.9	adhesive de-bonding
B2-4a	24.93	concrete de-bonding
B2-4b	27.5	concrete de-bonding

Figure 7 focuses on the effect of the total length to width ratio of CFRP sheet. The three cases (B1-1, B2-3, and B2-4) in this figure have the same area (10000 mm2) and the same thickness (t=0.1178 mm). This appears to indicate that the width of CFRP sheet has more influence than the length. Also it can be concluded that increasing the CFRP sheet width leads to a propensity for concrete de-bonding as shown in Figure 6. The theoretical implications of this are explored in the following section.

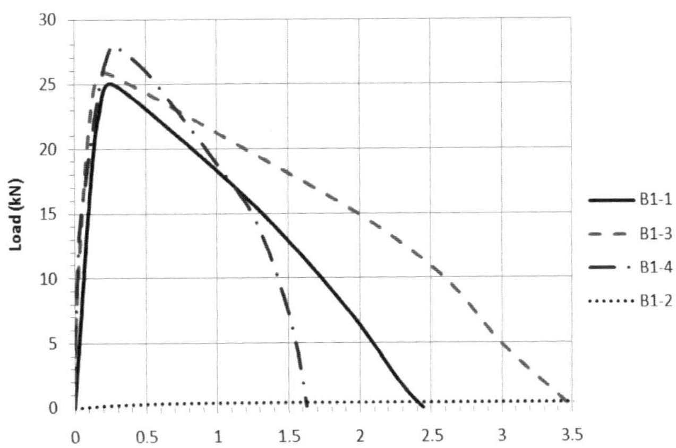

Figure 7. Influence length and width of the CFRP sheet.

4. Numerical analysis

The Finite Element method is a numerical method which can approximate and solve complex structural problems to within acceptable boundaries. Finite element analysis was first developed by the aircraft industry to predict the behaviour of metals forming for wings. The ANSYS finite element program has been comprehensively developed to the extent that it has applications across the whole engineering spectrum [14]. In particular, civil engineers are frequently interested in modelling materials such as steel and concrete, the latter requiring complex methodology in its representation. As concrete is an orthotropic material that exhibits nonlinear behaviour during loading, this behavior is numerically implemented in ANSYS [15]. A number of previous researchers have used the finite element method to provide insight into the behaviour of the FRP-concrete bonded joints, and CFRP-strengthened RC beams. Hemmaty et al. [16] considered a nonlinear adherence-shear law based on the experimental studies between concrete and reinforcement in the modelling of reinforced concrete elements. While modelling the adherence-shear relationship, they used a nonlinear spring/damper element COMBIN39 (element in ANSYS) for their main modelling. Also, X.Z. Lu et al. in 2009 [17] used COMBIN39 to model the interface between the FRP elements and the supports. The study of X.Z. Lu et al. [17]

IOP Conf. Series: Materials Science and Engineering **584** (2019) 012061 doi:10.1088/1757-899X/584/1/012061

presented a numerical study of the FRP stress distribution at de-bonding failure in U-jacketed or side-bonded beams using a rigorous FRP-to-concrete bond–slip model and assuming several different crack width distributions. This element type COMBIN39 was used in the present study. Huyse et. al [18] presented a paper concerning analysis of reinforced concrete structures using the ANSYS nonlinear concrete model. This paper considers the practical application of nonlinear models in the analysis of reinforced concrete structures. The results of some analyses performed using the reinforced concrete model of ANSYS are presented and discussed. The differences observed in the response of the same reinforced concrete beam, caused by variations in a material model that is always basically the same, are emphasized. The consequences of small changes in modelling are discussed and it is shown that satisfactory results may be obtained from relatively simple and limited models.

Santhakumar et. al. in 2004 [19] presented a numerical study to simulate the behaviour of retrofitted reinforced concrete beams strengthened with CFRP laminates using ANSYS. The effect of retrofitting on un-cracked and pre-cracked reinforced concrete beams was studied, and the behaviour of beams obtained from the numerical study showed good agreement with the experimental data. There was no significant difference in behaviour between the un-cracked and pre-cracked retrofitted beams. Al-Mahaidi et al. [20] studied the behaviour of three shear deficient T-beams strengthened using web-bonded CFRP plate. The experimental results have shown that repairing the beams with CFRP strips enhances their shear capacity. The increase in strength ranged between 68% and 87%. Nonlinear finite element modelling and analysis with DIANA was used to investigate the behaviour of these beams assuming plane stress conditions and perfect bond between the concrete surface and the web bonded CFRP strips. Finite element analysis was shown to be capable of predicting the ultimate strength, stiffness of the beams and strain levels in CFRP plates with reasonable accuracy. The cracking patterns and crack inclinations produced by the finite element model were also comparable to the patterns observed from testing. Fanning [21] presented nonlinear models for reinforced and post-tensioned concrete beams. The finite element software used (ANSYS) included dedicated numerical models for the nonlinear response of concrete under loading. These models usually included a smeared crack analogy to account for the relatively poor tensile strength of concrete, a plasticity algorithm to facilitate concrete crushing in compression regions, and a method of specifying the amount, the distribution and the orientation of any internal reinforcement. The numerical model adopted by ANSYS is discussed in this paper. Appropriate numerical modelling strategies are recommended and comparisons with experimental load-deflection responses are discussed for ordinary reinforced concrete beams and for post-tensioned concrete T-beams.

The finite element modelling of experimental specimens using ANSYS Ver.12 is presented here. The ultimate purpose of a finite element analysis is to recreate mathematically the behaviour of an actual engineering system. Three and two dimensional nonlinear finite element analysis is used here to simulate the performance of the experimental model and thereby to employ these methods for further investigation.

Initially, the FEA requires meshing of the structure, and, after loading, stresses and strains are calculated at integration points within these elements. An important step in the FE modelling is the selection of an appropriate mesh density. A convergence of results is obtained when an adequate number of elements are used in a structure (i.e. a mesh sensitivity study). This is practically achieved when an increase in the mesh density has a negligible effect on the results. Therefore, in this FE modelling, a convergence study was carried out initially to determine an appropriate mesh density.

At first ANSYS models for implicit analysis were created within the ANSYS pre-processor. They comprised SOLID65 elements to represent the concrete, and SOLID46 elements were used to simulate the CFRP sheets and adhesive layer, with LINK8 elements representing the steel bar. The properties of CFRP sheets and adhesive are illustrated in Table 2. For a convergence study, a quarter of the full specimen was used by taking advantage of symmetry of the specimen and loading, as shown in Figure 8.

Table 2. Summary of material properties of CFRP and adhesive.

Material	Elastic modulus	Poisson's ratio
CFRP	E_x=240 GPa E_y=31 GPa E_z=31 GPa	$\nu_{xy} = 0.22$ $\nu_{yx} = 0.22$ $\nu_{yz} = 0.3$
adhesive	E=10 GPa	$\nu = 0.3$

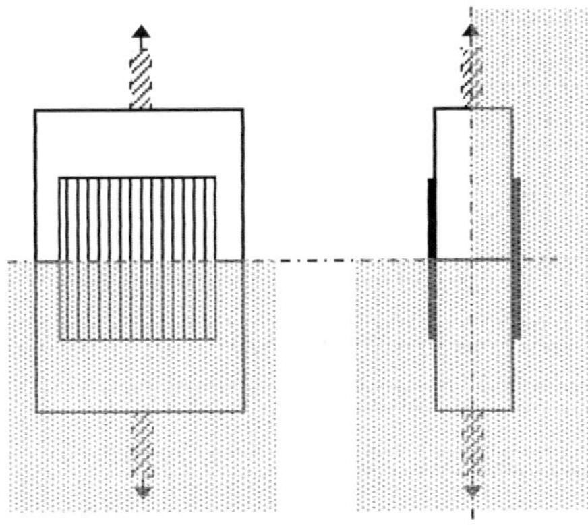

Figure 8. Use of a quarter model.

Figure 9 illustrates three different mesh densities for the same case. The difference was in the width of elements (6.25 mm, 12.5 mm, and 25.0 mm). The comparisons between the experimental results and the FEA results of the three different element widths (25 mm, 12.5 mm and 6.25 mm) are shown in figures 10 to 12 respectively. These figures represent total slip in the CFRP sheet and the total tensile load. It is clear that FE predictions are in close agreement with corresponding test results. It should be noted that the use of a 12.5 mm element width gives the best compatibility between experimental and FE results. Experimental behaviour was close to theoretical behaviour in the elastic case. In contrast, there was disparity with the inelastic case as shown in figures 9 to 11. This is because of the difficulties of the ANSYS program in representing the actual solution when the stiffness value is small or when it is zero. Also in this stage, many types of elements were used to simulate the CFRP sheet with adhesive. The use of layered element SOLID46 provides the best solution for the behaviour and de-bonding process of the CFRP sheet. This element provides full connection between the CFRP sheet and adhesive. This is compatible with the actual physical case. Also it provides a simple approach for simulating the CFRP sheet in single or double layers.

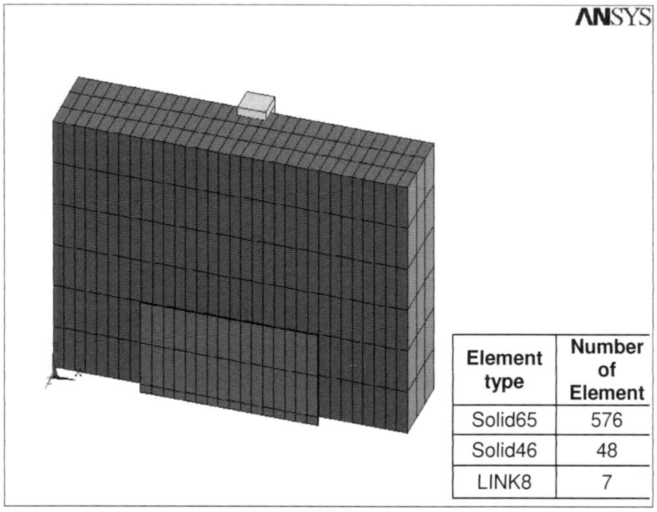

Element type	Number of Element
Solid65	576
Solid46	48
LINK8	7

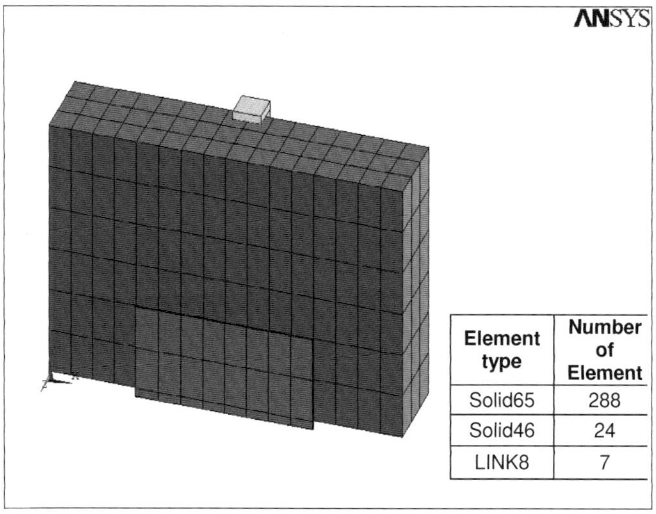

Element type	Number of Element
Solid65	288
Solid46	24
LINK8	7

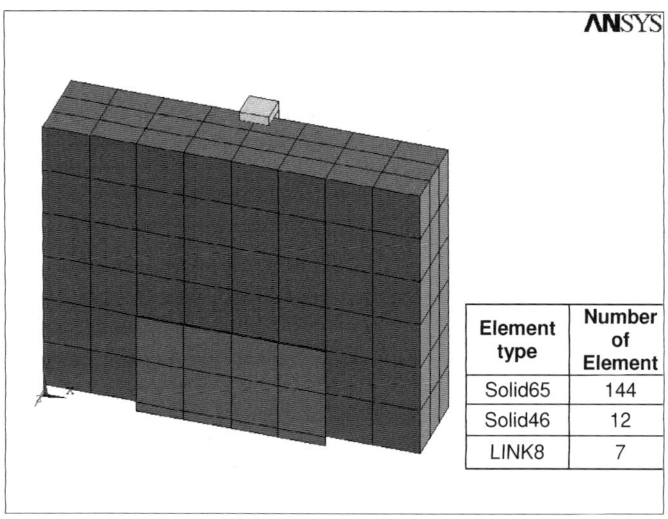

Element type	Number of Element
Solid65	144
Solid46	12
LINK8	7

Figure 9. Mesh for half of model in ANSYS (for example: B1-1).

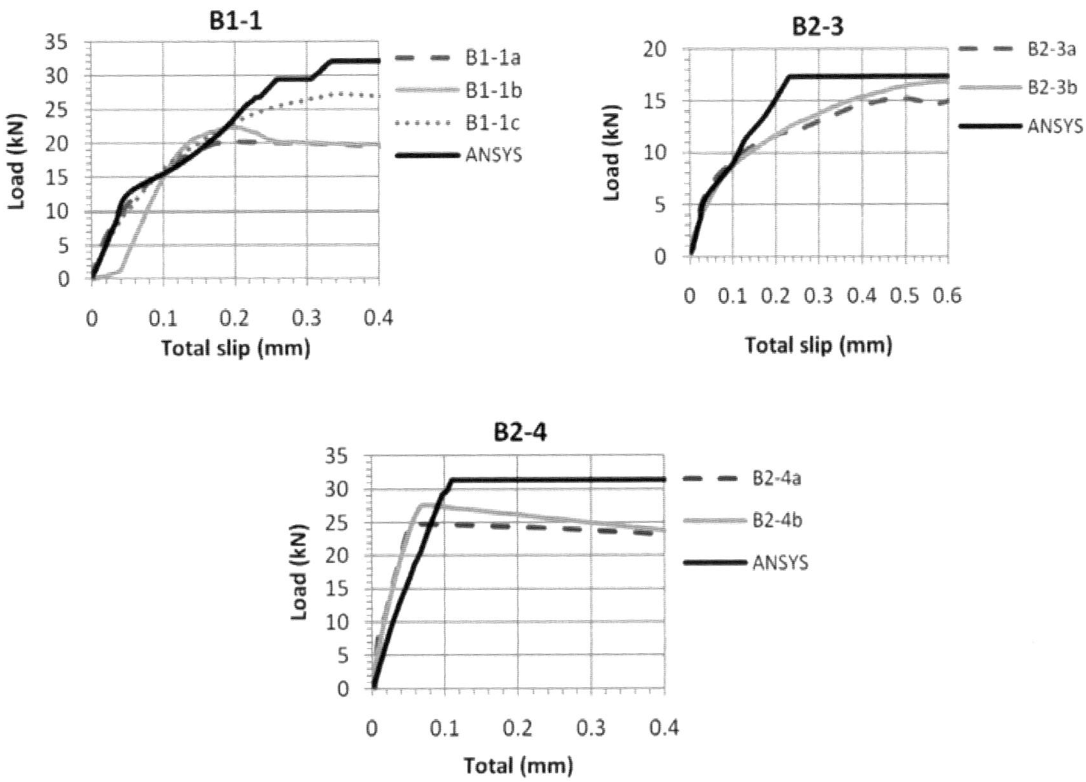

Figure 10. Total slip versus tensile load for pure tension samples (width elements = 25mm).

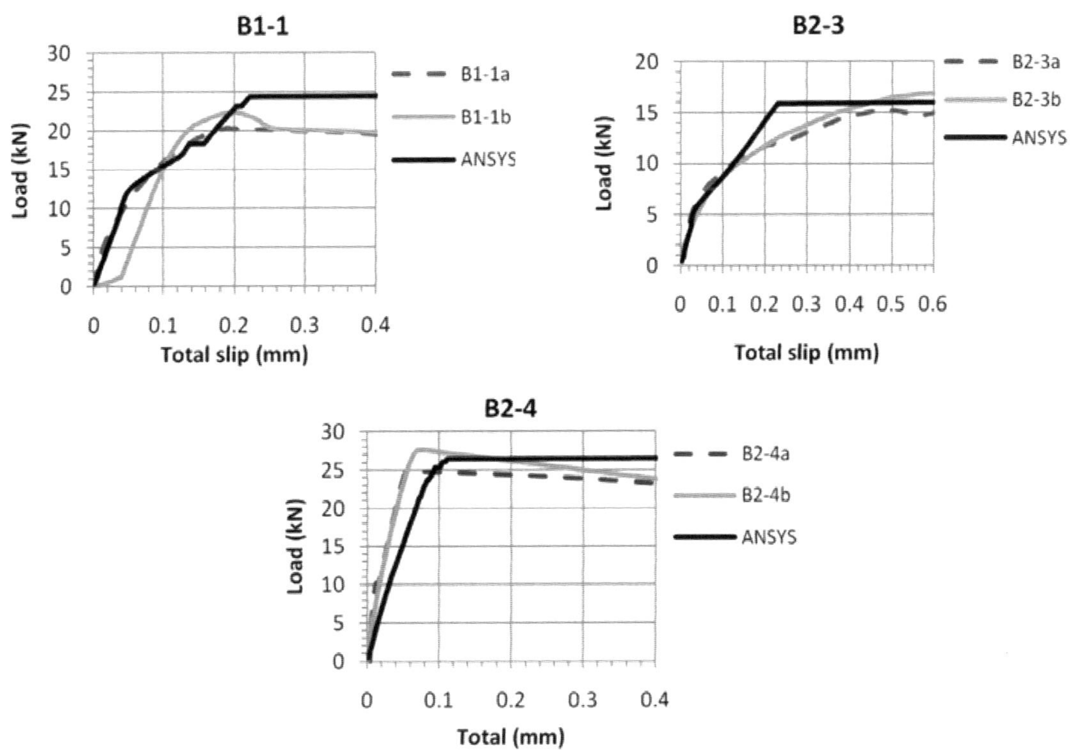

Figure 11. Total slip versus tensile load for pure tension samples (width elements = 12.5mm)

IOP Conf. Series: Materials Science and Engineering **584** (2019) 012061 doi:10.1088/1757-899X/584/1/012061

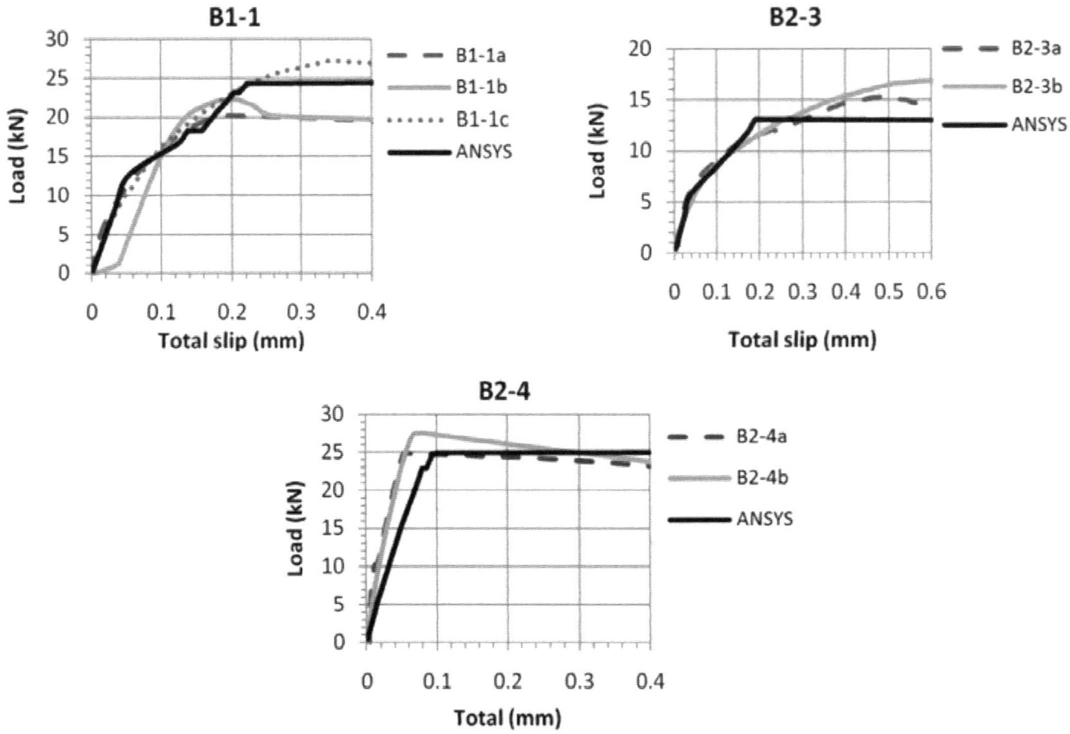

Figure 12. Total slip versus tensile load for pure tension samples (width elements = 6.25mm)

Symbols shown at the element centre are based on the status of all of the element's integration points. If any integration point in the element has crushed, the crushed (octahedron) symbol is shown at the centre. If any integration point has cracked or cracked and then closed, the cracked symbol is shown at the element integration point or at the centre. If at least five integration points have cracked and then closed, the cracked and closed symbol is shown at the element centre. Finally, if more than one integration point has cracked, the circle outline at the element centre shows the average orientation of all cracked planes for that element. Figures 13 and 14 show the crack patterns of concrete sections. The similarity between the theoretical and experimental results provides additional evidence to show that the FE model of this study is an acceptable simulation for CFRP and for the bond relationship between CFRP and the concrete section.

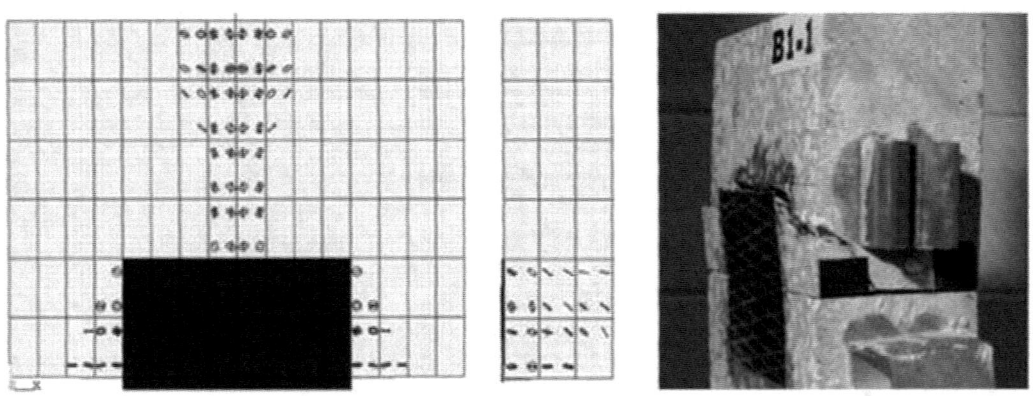

Figure 13. The theoretical and experimental crack for B1-1.

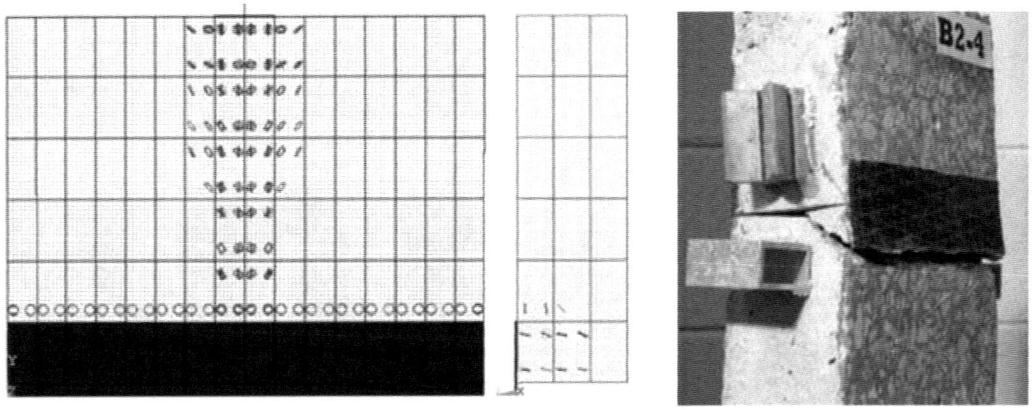

Figure 14. The theoretical and experimental crack for B2-4.

5. Influence of width to length ratio of sheet

In the experimental work of this study, three samples (B1-1, B2-3, and B2-4) have the same area (10000 mm2). However, they give broad range of load capacity because they have different width to length ratios of sheet, as illustrated in section 3. To understand this influence, we will concentrate on the contour stress and make comparisons between the existing design methods to understand how they address this effect. Slices of the half FE model are taken to show the contour stress distribution on the outer surface of the concrete section and at the same depth, as shown in Figure 15.

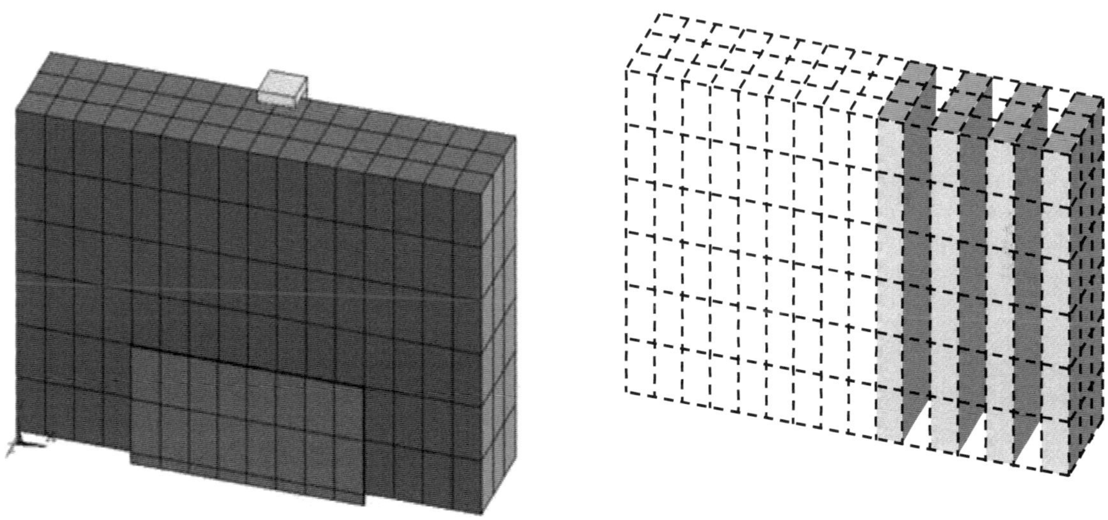

Figure 15. The slices of FE model.

Figure 16 shows the contour stress of FE models (width of the elements = 12.5mm). There are many points that can be noted in this figure. The value of the stress ranged between (-4 MPa to 40 MPa). The maximum stress (at an applied load = 15 kN) in B2-3, B1-1, and B2-4 was38.19 MPa, 11.37 MPa, and 9.45 MPa respectively. Even with all the specimens at the same applied load and having the same area of CFRP sheet, the distribution of stress and maximum stress are different. This provides an initial understanding for the effect of the ratio of the width/ length of the sheet. In addition to this, the stress in the case of sample B1-1 was concentrated in small area and reached deeper into the model than for other cases. This led to failure of the B1-1 samples with a low load capacity. However, we can conclude that the width of CFRP sheet has more influence than the length. Figure 17 compares the existing design methods of bond strength. The Sato and Lzumo studies do not account for the effect in width/length sheet ratio. Therefore their prediction for bond strength differed from experimental observations. When compared with other studies, it is the predictions of Iso which provide the closest agreement.

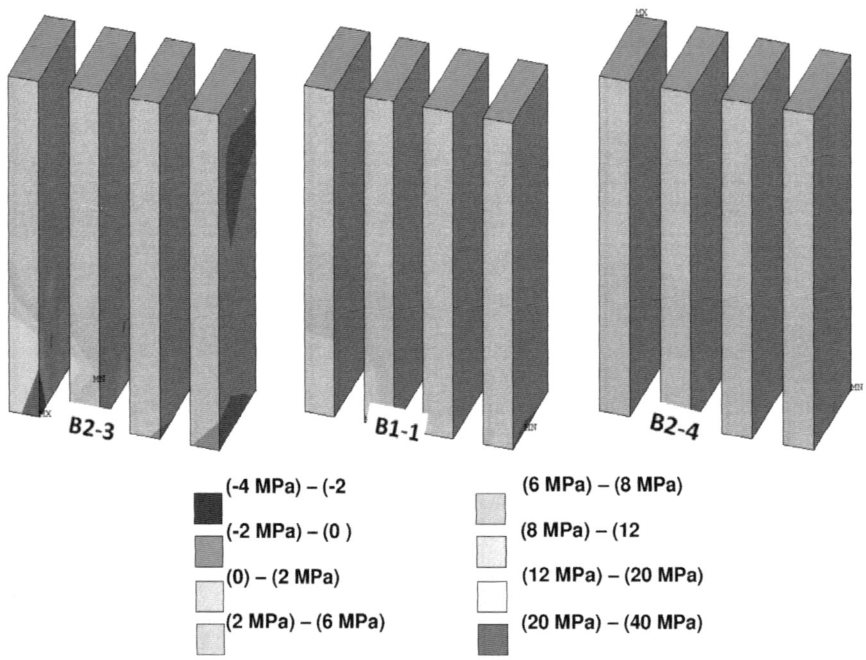

Figure 16. Contours of principal stress

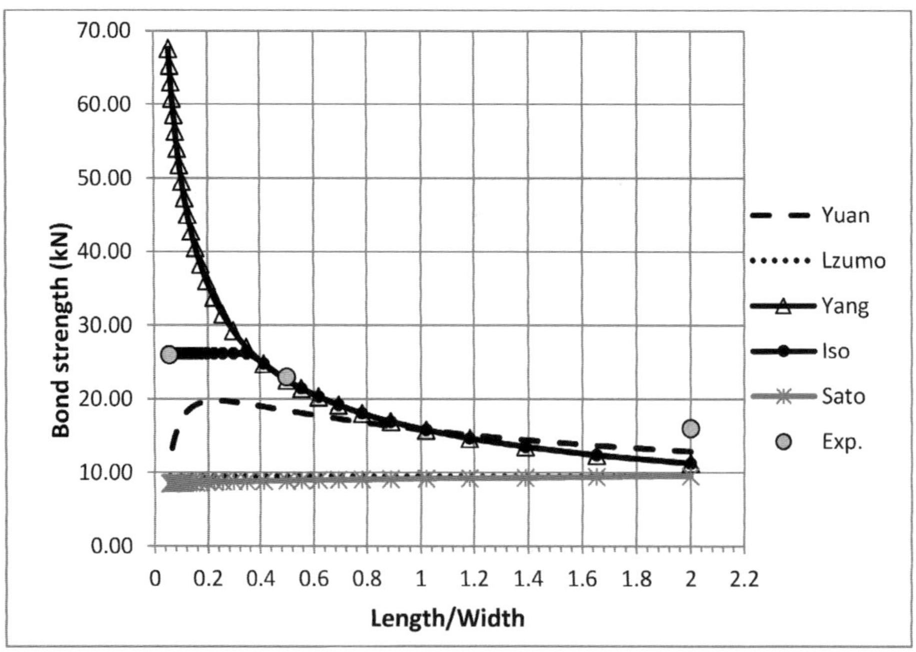

Figure 17. Comparison between existing predictions.

6. Conclusion

A modified expression for the prediction of bond strength has been presented which demonstrates a high level of accuracy when compared with experimental tests, and it has been corroborated with finite element analysis. It is proposed that this model should form the basis of improved design rules for concrete members strengthened with FRP.

The work has also led to the following conclusions:

i. The width of CFRP sheet has more influence than the length.
ii. Increasing the CFRP sheet width leads to concrete de-bonding.

References

[1] Al-Rousan R., Issa M., Fatigue performance of reinforced concrete beams strengthened with CFRP sheets, Construction and Building Materials 25 (2011) 3520–3529.

[2] Lu XZ, Chen JF, Teng JG, Ye LP, Jiang JJ Rotter JM. RC beams shear-strengthened with FRP: Stress distributions in the FRP reinforcement. Construction and Building Materials 23 (2009) 1544–1554.

[3] Cottone A., Giambanco G., Minimum bond length and size effects in FRP-ubstrate bonded joints, Engineering Fracture Mechanics 76 (2009) 1957–1976.

[4] Yuan H, Teng JG, Seracino R, Wu ZS, Yao J. Full-range behaviour of FRP-to-concrete bonded joints. Engineering Structures 26 (2004) 553–64.

[5] Yao J., Teng J.G., Chen J.F., Experimental study on FRP-to-concrete bonded joints, Composites: Part B 36 (2005) 99–113

[6] Pellegrino C., Tinazzi D., and Modena C., Experimental Study on Bond Behavior between Concrete and FRP Reinforcement, Journal of Composites for Construction, 12 (2008).

[7] Bonacci J. F., and Maalej M., Externally Bond FRP for Service-Life Extension of R Infrastructure, Journal of Infrastructure Systems, Vol.6, No. 1, March 1, 2000.

[8] Lu XZ, Teng JG, Ye LP, Jiang JJ. Bond–slip models for FRP sheets/plates externally bonded to concrete. Eng Struct 27(2005) 938–950.

[9] Yang YX, Yue QR, Hu YC. Experimental study on bond performance between carbon fiber sheets and concrete. Journal of Building Structures 22 (2001) 36–42

Evaluation Al-Safeena and Sayed Jawda Intersections in Kerbala City

Huda M. Al-Murshidy[1], Hussein A. Ewadh[2] and Raid R. A. Almuhanna[3,*]

[1] MSc Student, College of Engineering, University of Kerbala, Kerbala, Iraq.
[2] College of Engineering, University of Babylon, Hilla, Iraq.
[3] College of Engineering, University of Kerbala, Kerbala, Iraq.

*Corresponding author, email: raidalmuhanna@yahoo.com

Abstract: Increase of traffic volumes at intersections is one of the significant reasons that cause congestion. The current study aims to analyze, assess and improve traffic performance of (Al-Safeena roundabout) and (Sayed Jawda intersection) in Kerbala city. In order to implement these goals, data of the traffic volume and geometric layout for (Al-Safeena roundabout) and (Sayed Jawda intersection) are manually collected. To achieve the traffic analysis process, (SIDRA version 7) traffic program was used. Al-Safeena intersection improved by adding a new lane for each approach, this reduce the delay times, fuel consumption, and stops. This process will reduce ratio of volume to capacity (v/c) to value 0.7 and the delay rate value to 15.3 sec and LOS to C. On other hand, traffic operation at Sayed Jawda intersection is relatively enhanced by dividing the width of approach to four lanes, however, level of service is not upgraded satisfactorily. Change Sayed Jawda intersection to roundabout reduces the value of v/c to 0.8, and the value of average delay to 24 sec and LOS C.

Keywords: Traffic congestion, Roundabout, Intersection, Peak Hour Factor, LOS.

1. Introduction

Traffic congestion is of great concern in many big cities. The intersection is characterized by slow speed, long journey times and an increase in the number of vehicles in the queue [2]. The delay is a significant problems happen in any mean of traffic, the delay can be defined "the additional travel time experienced by a driver, passenger, or pedestrian" Two factors that affect delay are lane group volume and lane group capacity [3]. A roundabout is defined as "a circular traffic intersection featuring yield control on all entering legs, one-way continuous flow within the circulatory roadway, channelization of approaches, and appropriate geometric curvature to keep circulating speeds low" [4]. When analyzing intersections, the capacity, level of service and delay are considered as fundamental factors, therefore, the capacity and level of service should be taken fully into account when evaluating traffic performance at intersections [5]. Al-Safeena roundabout and Sayed Jawda intersection in Karbala city are an important intersection and are heavily congested as they connecting major streets. Al-Saffena and Sayed Jawda intersections have the following characteristics:

1. They have a very high traffic volume on each approaches.
2. Many activities are located around these intersections.

The SIDRA software, up on methodology extension in Australia, also employ a gap acceptance approach to model roundabout operations [1].

2. Objective of the Study:

The current traffic study aims at several basic objectives

1. Find the exist LOS at both intersections to evaluate the traffic performance.
2. Improve the current LOS for Al-Safeena and Sayed Jawda intersections.

3. Description of Site:

Al-Safeena intersection is a four-leg roundabout intersection located in Karbala city, this roundabout very important due to connection two major street Fatima Al-Zahraa street with Al-Iskan street as shown in Figure (1).

Sayed Jawda intersection is a four leg intersection located in Karbala city as shown in Figure (2).

Figure 1. Satellite Image of Al-Safeena roundabout (satellite photographs for Kerbala, 2016).

Figure 2. Satellite Image of Sayed Jawda Intersection (satellite photographs for Kerbala, 2016).

4. Methodology
The steps required to complete the analysis of intersections shown in Figure (3).

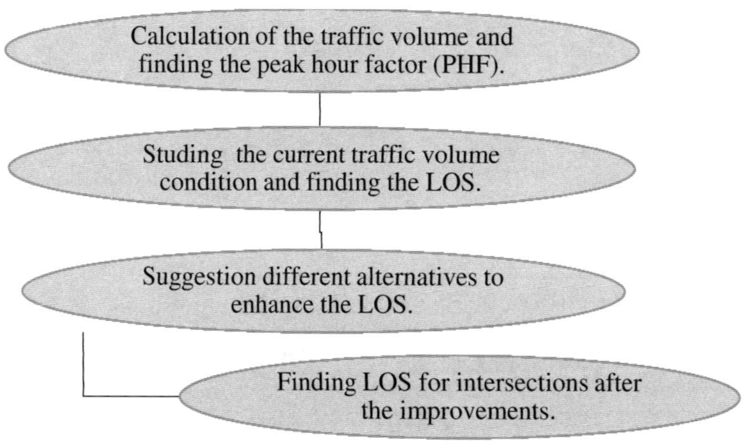

Figure 3. Steps Required to Complete the Analysis of Intersections.

5. Data Collection
The data collection for Al- Safeena roundabout and Sayed Jawda include traffic volume and geometric data.

5.1 Traffic Volume
The calculation of traffic volume distribution by movements was done manually for each approach in four days namely (Sunday, Tuesday, Thursday and Friday) starting on (3/4/ 2018 for five hours (730-830,830-930,1330-1430,1430-1530,1530-1630) for each day, the reason for choosing these hour is the highest traffic volumes falling within those periods. For calculating PHF, traffic volumes were recorded for every 15 minutes at each approach. Tables (1) and (2) show the traffic volume at each approach across Al-Safeena roundabout and Sayed Jawda intersection.

IOP Conf. Series: Materials Science and Engineering **584** (2019) 012062 doi:10.1088/1757-899X/584/1/012062

5.2 Geometrical Data

The current geometric design that includes the number of lane and width of lane for Al-Safeena roundabout and Sayed Jawda intersection as shown in Tables (3 and 4).

Table 1. Traffic Volume for Al-Safeena Roundabout.

Time	E			W			N			S.		
	R	TH	L	R	TH	L	R	TH	L	R	TH	L
730-745	60	336	45	93	143	89	82	219	46	67	224	100
745-800	83	156	42	92	118	92	94	226	38	68	227	108
800-815	84	155	43	92	119	93	92	228	41	67	230	109
815-830	80	163	43	81	263	83	97	220	42	78	174	103
830-845	86	203	40	82	168	81	92	184	35	93	148	91
845-900	62	168	38	64	238	90	85	189	54	122	153	40
900-915	84	165	38	34	291	88	82	238	36	36	140	91
915-930	79	166	33	37	261	84	78	223	42	35	133	80
1330-1345	99	361	18	58	34	3	36	291	120	173	287	60
1345-1400	110	199	27	43	86	63	34	149	112	128	182	72
1400-1415	107	203	27	49	162	90	38	263	97	72	260	74
1415-1430	123	186	25	57	72	26	39	270	82	139	308	75
1430-1445	117	172	20	39	41	34	36	239	83	134	297	72
1445-1500	80	202	12	53	48	27	42	199	84	73	256	98
1500-1515	53	137	13	61	12	70	44	215	215	74	285	102
1515-1530	60	152	11	56	2	14	43	193	69	105	224	106
1530-1545	54	126	12	51	37	35	37	178	71	116	262	104
1545-1600	67	155	17	32	48	35	202	38	60	288	109	113
1600-1615	77	175	20	55	41	40	37	191	83	137	272	101
1615-1630	74	86	23	40	179	85	38	71	88	159	200	116

IOP Conf. Series: Materials Science and Engineering **584** (2019) 012062 doi:10.1088/1757-899X/584/1/012062

Table 2. Traffic Volume for Al-Sayed Jawda Intersection.

Time	E			W			N			S		
	R	TH	L	R	TH	L	R	TH	L	R	TH	L
730-745	88	293	74	55	110	33	211	116	43	46	224	118
745-800	84	229	71	53	138	49	196	68	50	53	148	49
800-815	82	215	67	47	136	50	180	115	57	46	144	95
815-830	85	245	65	49	141	61	179	113	62	50	142	104
830-845	99	140	29	39	96	19	134	98	275	41	212	140
845-900	80	246	39	37	80	29	112	127	62	50	185	94
900-915	84	199	38	25	92	22	101	133	62	37	177	96
915-930	83	190	29	33	88	22	100	120	58	46	170	91
1330-1345	67	242	68	38	99	33	185	72	36	38	202	93
1345-1400	78	230	56	39	78	36	201	93	35	39	191	36
1400-1415	69	243	69	41	78	47	183	81	47	44	192	66
1415-1430	62	231	68	44	82	48	184	88	47	36	183	78
1430-1445	48	207	76	36	70	38	165	74	78	36	168	52
1445-1500	59	197	57	32	85	35	173	76	79	35	169	52
1500-1515	60	193	64	31	79	31	167	68	69	34	170	57
1515-1530	67	200	72	35	67	34	172	85	68	36	177	55
1530-1545	33	196	65	34	67	34	163	81	71	38	177	49
1545-1600	47	41	68	37	63	25	167	82	53	40	187	50
1600-1615	41	53	75	46	70	47	174	72	63	44	195	59
1615-1630	36	51	77	54	68	58	174	74	77	145	204	72

Table 3. Geometric Characteristics for Al- Safeena Roundabout

Approach direction	No. lane	Width of lane(m)
E	3	4
W	3	3.5
N	3	3.5
S	3	2.9

Table 4. Geometric Characteristics for Sayed Jawda intersection

Approach direction	No.lane	Width of lane(m)
E	4	3.2
W	3	3.5
N	3	3.5
S	3	3.5

6. Analysis and Results.

To evaluate and analyze the performance of intersections, the SIDRA program version 7 was used and through this program the following information was calculated:

- The degree of saturation(v/c)
- Average delay
- LOS

6.1 Determine Peak Hour Volume:

The traffic account every 15 min. is carried out from (7:30-9:30) a.m. and (13:30-16:30) p.m. for Al-Safeena roundabout and Sayed Jawda intersection. The peak hour for Al-Safeena and for Sayed Jawda located in (7:30-830) p.m. The PHF for Al-Safeena roundabout and Sayed Jawda intersection shown in Tables (5 and 6), where the number of vehicles accounted is equal to 5628 veh./hr for Al-Safeena roundabout .the number of vehicles for Sayed Jawda intersection is 5282 veh./hr.

6.2 Calculation of the PHF (Peak Hour Factor):

The peak hour factor defined as "the ratio of total hourly volume to the maximum 15- min rate of flow within the hour" as following equations (1) and (2) (HCM, 2000).

$$PHF = \frac{Hourly\ volume}{peak\ rate\ of\ flow\ (within\ our)} \quad (1)$$

$$PHF = \frac{Hourly\ volume}{4*V15min} \quad (2)$$

V15= The volume that is through peak 15 minutes of peak hour, on veh/15min

Variation of Traffic volume for Al-Safeena and Sayed Jawda intersections during peak period shown in Figure (4 and 5).

Table 5. PHF for Each Direction for Al-Safeena Roundabout.

Direction	PHF
E	0.93
W	0.97
N	0.98
S.	0.956

Table 6. PHF for Each Direction for Al-Sayed Jawda Intersection.

Direction	PHF
E	0.93
W	`0.94
N	0.932
S	0.92

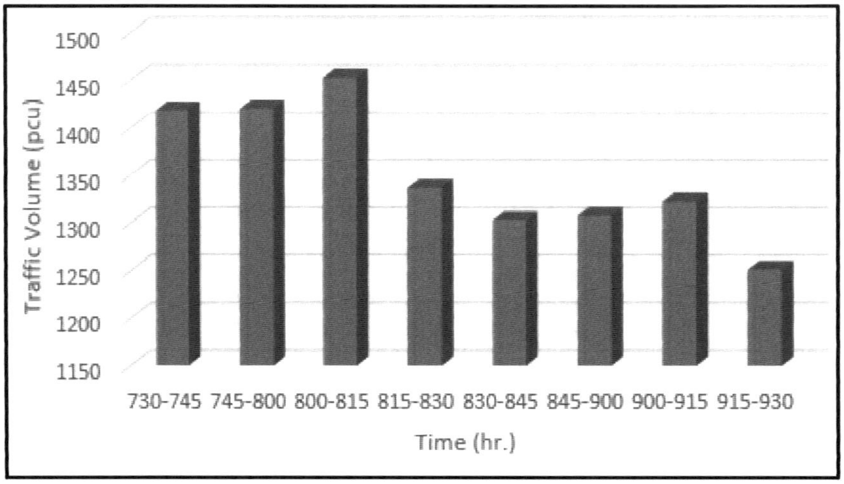

Figure 4. Variation of Traffic Volume at Al-Safeena Roundabout During (730-930).

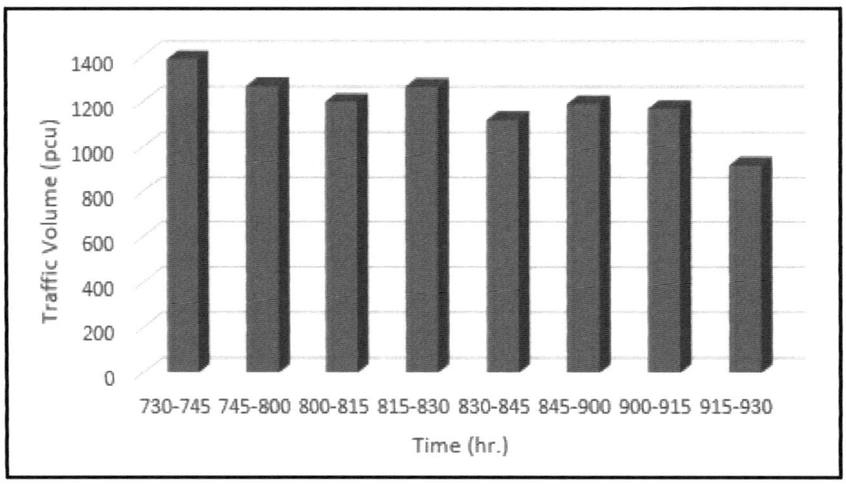

Figure 5. Variation of Traffic Volume During (730-930) at Sayed Jawda Intersection.

6.3. Existing LOS

The program SIDRA version 7 was invented to analyze the traffic conditions and to obtain the current capacity as well as to obtain the ratio of speed to volume as well as calculating the delay for each approach of each traffic movement. After setting the peak hour that represents the design hour volume, it is very important to guess the level of service in the current flow and geometric properties at Al-Safeena roundabout and Sayed Jawda intersection. When the service level is set to the current case, the delay rate must be calculated at Al-Safeena and Sayed Jawda because for LOS account, the average delay perform the major parameter. By using SIDRA 7 program, the average delay for AL-Safeena intersection at its existing geometric is (71) sec./veh. this mean,Al- Safeena will operate in LOS (F). Table (7) show the average delay

and LOS. The average delay for Al-Sayed Jawda intersection at existing geometric is (123) sec/veh so Sayed Jawda intersection will operate in LOS (F). Table (8) shows the average delay and LOS for Sayed Jawda intersection for each approach that connected with this Intersection.
The Geometric layout for Al-Safeena roundabout and Sayed Jawda intersection shown in Figure (6 and 7).

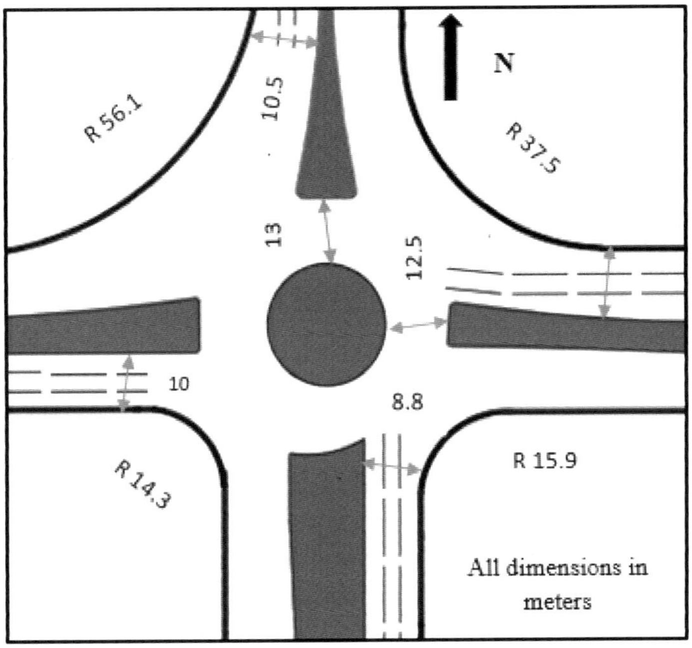

Figure 6. Geometric Layout of Al-Safeena Roundabout.

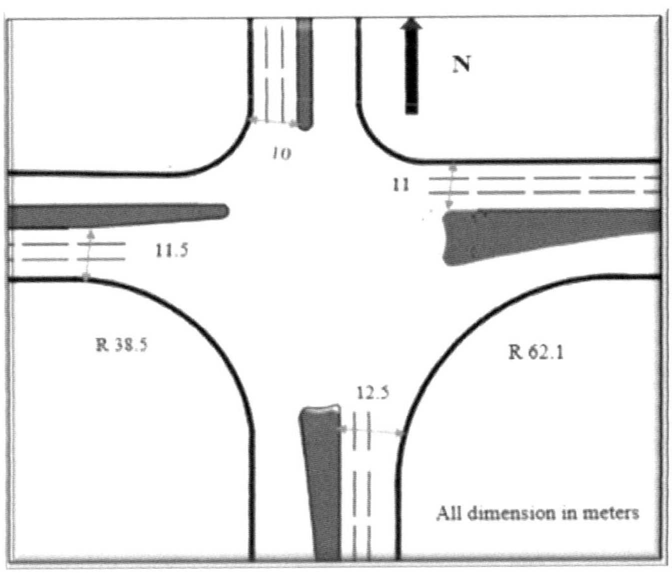

Figure 7. Geometric Layout of Sayed Jawda Intersection.

Table 7. LOS at Al-Safeena Roundabout.

Direction	Approach		Intersection	
	Delay (sec)	LOS	Delay(sec)	LOS
Northbound	178	F	71	F
Eastbound	78	F		
Southbound	154	F		
Westbound	60	F		

Table 8. LOS at Sayed Jawda Intersection.

Direction	Approach		Intersection	
	Delay (sec)	LOS	Delay(sec)	LOS
Northbound	110	F	123.5	F
Eastbound	147	F		
Southbound	92	F		
Westbound	147	F		

7. Design of Proposals

7.1 design of proposal for Al-Safeena roundabout.

The level of service of al Safeena intersection improve by decrase the width of middle island for east approach from 9.8 to 4.5 so this increase the no of lane for the east approach (when the no of lane increase the approach capacity increase so the LOS enhanced . The no of lanes for the north and west aaproach increase in same method for east approach as shown in Figure (8) .The north approach increase throught the increase of entry radius this lead to ncrease flare length and entry width and consequently the capacity of approach increase The improvement of AL Safeena shown in Figure (8).In these improvement v/c rduce to 0.7 and the LOS for this intersection enhanced to (C),as shown in Table (9) and Figure (9).

IOP Conf. Series: Materials Science and Engineering **584** (2019) 012062 doi:10.1088/1757-899X/584/1/012062

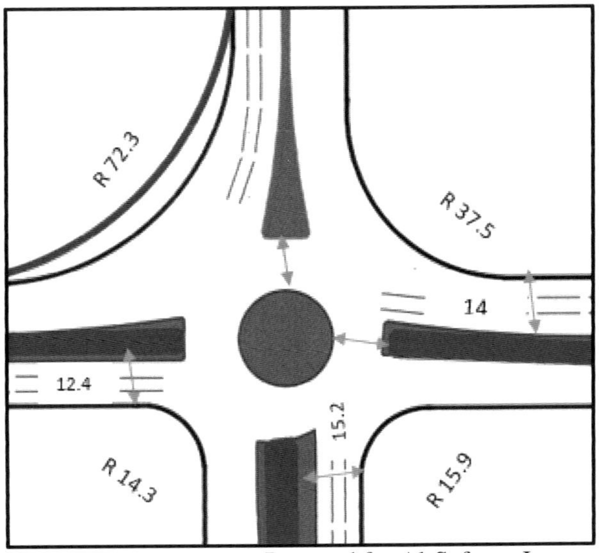

Figure 8. Improvement Proposal for Al-Safeena Intersection.

Movement Performance - Vehicles

Mov ID	OD Mov	Demand Flows Total veh/h	Demand Flows HV %	Deg. Satn v/c	Average Delay sec	Level of Service	95% Back of Queue Vehicles veh	95% Back of Queue Distance m	Prop. Queued	Effective Stop Rate per veh	Average Speed km/h
South: RoadName											
3	L2	355	0.0	0.679	16.7	LOS C	5.0	38.0	0.81	1.04	46.1
8	T1	872	0.0	0.679	15.2	LOS C	5.1	39.0	0.79	0.99	48.4
18	R2	386	0.0	0.668	21.2	LOS C	4.3	32.7	0.84	1.06	43.5
Approach		1613	0.0	0.679	17.0	LOS C	5.1	39.0	0.80	1.02	46.6
East: RoadName											
1	L2	190	0.0	0.286	9.0	LOS A	1.0	7.7	0.71	0.73	49.9
6	T1	852	0.0	0.641	17.8	LOS C	3.7	28.3	0.83	1.02	47.2
16	R2	326	0.0	0.499	13.4	LOS B	2.4	18.0	0.78	0.90	47.0
Approach		1368	0.0	0.641	15.5	LOS C	3.7	28.3	0.80	0.95	47.5
North: RoadName											
7	L2	186	0.0	0.242	7.4	LOS A	0.8	6.3	0.64	0.64	50.9
4	T1	968	0.0	0.591	13.5	LOS B	3.6	27.2	0.76	0.93	50.0
14	R2	391	0.0	0.697	23.3	LOS C	5.0	37.9	0.88	1.13	42.0
Approach		1545	0.0	0.697	15.2	LOS C	5.0	37.9	0.78	0.95	47.8
West: RoadName											
5	L2	402	0.0	0.456	9.7	LOS A	2.2	16.9	0.69	0.78	49.4
2	T1	717	0.0	0.424	9.5	LOS A	2.0	15.2	0.69	0.77	52.8
12	R2	381	0.0	0.722	26.2	LOS D	4.9	37.2	0.87	1.13	40.8
Approach		1500	0.0	0.722	13.8	LOS B	4.9	37.2	0.73	0.87	48.3
All Vehicles		6026	0.0	0.722	15.4	LOS C	5.1	39.0	0.78	0.95	47.6

Figure 9 . Shown the Movement Intersection for Al- Safeena Roundabout after the Improvement .

Table 9. LOS at Al-Safeena Roundabout by SIDRA 7.

Direction	Approach		Intersection	
	Delay (sec)	LOS	Delay(sec)	LOS
Northbound	17	C	15.4	C
Eastbound	13.3	B		
Southbound	15.1	C		
Westbound	15.5	C		

7.2 Design of Proposals for Sayed Jawda

The LOS at Sayed Jawda Intersection was enhanced by remarking approaches pavement into four lanes instead of three. The traffic lanes that have low traffic volume are assigned to be shared. In this proposal the LOS enhanced to E, which is not enhanced to accepted level as shown in Figure (10), the LOS and delay for this intersection are shown in Table (10).

Movement Performance - Vehicles

Mov ID	OD Mov	Demand Flows Total veh/h	Demand Flows HV %	Deg. Satn v/c	Average Delay sec	Level of Service	95% Back of Queue Vehicles veh	95% Back of Queue Distance m	Prop. Queued	Effective Stop Rate per veh	Average Speed km/h
South: RoadName											
3u	U	43	3.0	0.808	67.6	LOS E	16.5	125.8	1.00	0.90	27.6
3	L2	252	0.0	0.808	67.3	LOS E	17.9	136.2	1.00	0.90	28.2
8	T1	513	0.0	0.808	64.0	LOS E	19.8	150.8	0.99	0.89	30.4
Approach		809	0.2	0.808	65.2	LOS E	19.8	150.8	0.99	0.90	29.5
East: RoadName											
1u	U	57	3.0	0.890	81.3	LOS F	18.2	141.9	1.00	0.95	25.0
1	L2	212	3.0	0.890	81.0	LOS F	19.3	150.3	1.00	0.95	25.4
6	T1	572	3.0	0.890	75.3	LOS E	22.8	177.5	0.99	0.96	27.9
Approach		840	3.0	0.890	77.1	LOS E	22.8	177.5	0.99	0.96	27.0
North: RoadName											
7u	U	48	3.0	0.986	91.0	LOS F	26.0	202.3	0.98	1.01	23.5
7	L2	459	3.0	0.986	90.5	LOS F	26.7	208.0	0.97	1.03	23.9
4	T1	717	3.0	0.986	80.7	LOS F	40.0	311.5	0.97	1.06	26.7
Approach		1224	3.0	0.986	84.8	LOS F	40.0	311.5	0.97	1.05	25.5
West: RoadName											
5u	U	57	3.0	0.970	86.7	LOS F	25.2	196.5	0.97	0.99	24.1
5	L2	303	3.0	0.970	86.5	LOS F	26.2	204.4	0.97	1.00	24.5
2	T1	852	3.0	0.970	78.4	LOS E	38.6	300.6	0.97	1.05	27.3
Approach		1212	3.0	0.970	80.8	LOS F	38.6	300.6	0.97	1.04	26.4
All Vehicles		4085	2.4	0.986	78.2	LOS E	40.0	311.5	0.98	1.00	26.8

Figure 10. The Intersection Movements and LOS for Sayed Jawda Intersection after Pavement Remarking.

Table 10. LOS for Sayed Jawda intersection after Pavement Remarking by SIDRA 7

Direction	Approach		Intersection	
	Delay (sec)	LOS	Delay(sec)	LOS
Northbound	65	E	78	E
Eastbound	78	E		
Southbound	84	F		
Westbound	77	E		

Changing Sayed Jawda intersection to roundabout reduced the v/c to 0.8 and LOS to C as shown in Figure 11 and Table 11. This roundabout with circulatory width 14m and island diameter of 30m, the improvement for Sayed Jawda is shown in Figure 12.

Movement Performance - Vehicles

Mov ID	OD Mov	Demand Flows Total veh/h	Demand Flows HV %	Deg. Satn v/c	Average Delay sec	Level of Service	95% Back of Queue Vehicles veh	95% Back of Queue Distance m	Prop. Queued	Effective Stop Rate per veh	Average Speed km/h
South: RoadName											
3u	U	25	0.0	0.650	22.2	LOS C	4.4	33.1	0.80	1.02	47.1
3	L2	464	0.0	0.650	16.2	LOS C	4.4	33.1	0.80	1.02	46.1
8	T1	928	0.0	0.619	15.4	LOS C	3.9	29.5	0.79	0.98	49.8
18	R2	304	0.0	0.698	28.7	LOS D	4.2	32.0	0.88	1.11	40.6
Approach		1722	0.0	0.698	18.1	LOS C	4.4	33.1	0.81	1.01	46.9
East: RoadName											
1u	U	25	0.0	0.575	32.2	LOS D	2.6	19.5	0.88	1.04	42.7
1	L2	188	0.0	0.575	23.9	LOS C	2.6	19.5	0.88	1.04	41.9
6	T1	883	0.0	0.753	26.2	LOS D	5.1	38.5	0.90	1.20	43.6
16	R2	334	0.0	0.852	48.2	LOS E	6.2	47.0	0.95	1.37	33.4
Approach		1429	0.0	0.852	31.2	LOS D	6.2	47.0	0.91	1.21	40.6
North: RoadName											
7u	U	39	0.0	0.535	29.2	LOS D	2.4	18.2	0.84	0.98	44.6
7	L2	182	0.0	0.535	19.3	LOS C	2.4	18.2	0.84	0.98	43.7
4	T1	973	0.0	0.707	20.4	LOS C	4.8	36.7	0.86	1.11	46.7
14	R2	397	0.0	0.848	41.8	LOS E	7.0	52.8	0.94	1.37	35.5
Approach		1590	0.0	0.848	25.8	LOS D	7.0	52.8	0.88	1.16	43.0
West: RoadName											
5u	U	36	0.0	0.603	24.4	LOS C	3.2	24.4	0.83	1.01	45.0
5	L2	389	0.0	0.603	19.2	LOS C	3.5	26.4	0.82	1.00	45.2
2	T1	701	0.0	0.603	15.7	LOS C	3.6	27.2	0.80	0.97	49.1
12	R2	349	0.0	0.813	40.0	LOS E	6.1	46.4	0.92	1.29	36.1
Approach		1475	0.0	0.813	22.6	LOS C	6.1	46.4	0.84	1.06	44.3
All Vehicles		6216	0.0	0.852	24.1	LOS C	7.0	52.8	0.85	1.11	43.7

Figure 11. LOS at Sayed Jawda Intersection after Pavement Remarking and Changing to Roundabout.

Table 11. LOS for Sayed Jawda after Pavement Remarking and Changing to Roundabout by SIDRA.

Direction	Approach		Intersection	
	Delay (sec)	LOS	Delay(sec)	LOS
Northbound	18	C	24.1	C
Eastbound	22.6	C		
Southbound	25	D		
Westbound	31	D		

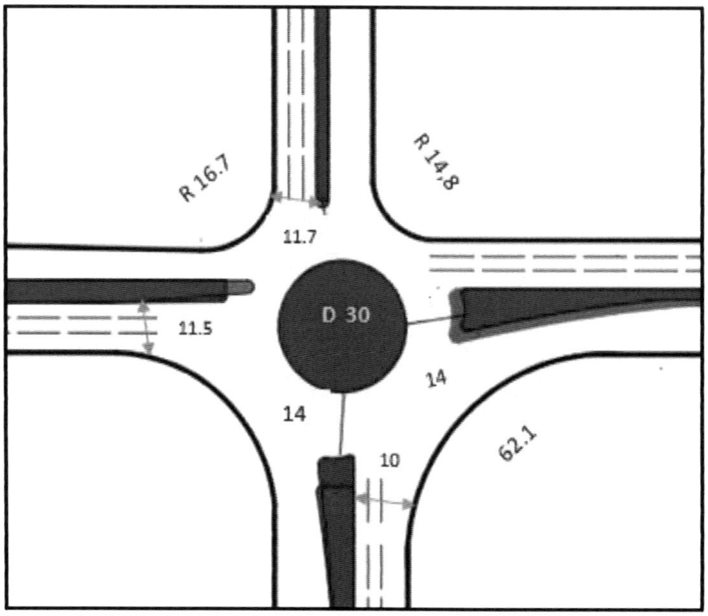

Figure 12. Improvement Proposal for Sayed Jawda Intersection.

8. Conclusions

The results indicated that the LOS for Al-Safeena roundabout was unacceptable. The level of service of al Safeena intersection was improved by decrasing the width of the middle island for east approach from 9.8m to 4.5m which increases the no of lane for the east approach (When the no of lanes increases, the approach capacity also increases and the LOS enhanced). The no of lanes for the south and west aaproach increases in the same manner as of the east approach. The north approach width increases through the increase of entry radius this lead to increase flare length, and entry width and consequently the capacity of approach capacity increases. Following these improvements, the LOS for this intersection enhanced to (C). Also the unacceptable LOS for Sayed Jawda was enhanced by dividing the width of the approaches to four lanes so each movement in each lane is isolated and the low traffic lanes are shared, the level of service of Sayed Jawda intersection was not enhanced and still work in LOS E. Changing Sayed Jawda intersection to roundabout will enhance the LOS to (C).

References

[1] Akcelik R 1986 SIDRA version 2.2 input and output (Austrian road research board) Technical manual, ATM No. **19.**

[2] Federal Highway Administration 1995 Improving Traffic Signal Operations (U.S. Department of Transportation), p 1.

[3] Khisty C J and Lall B k 1998 Transportation Engineering Handbook Second edition (Prentice –hall International) Inc.

[4] Kittelson & Associates Inc Roundabouts an Informational Guide Report FHWSRD-00-067 FHWA, U.S. Department of Transportation, 2000.I am at, Federal Highway Administration, Publication No. FHWA-RD-00-067, June 2000.

[5] Zegeer D.C 1986 Field Validation of Intersection Capacity Factors (Transportation Research record 1091, Transportation research Board Washoington, DC).

IOP Conf. Series: Materials Science and Engineering **584** (2019) 012063 doi:10.1088/1757-899X/584/1/012063

Experimental study for strengthening octagonal castellated steel beams using circular and octagonal ring stiffeners

H W Al-Thabhawee[1]* , A Mohammed[1]
[1]Department of Civil Engineering , College of Engineering, Al-Kufa University

E-mail:* hayder.althabhawi@uokufa.edu.iq

Abstract: Castellated beams made by cutting web of hot rolled steel I-section in particular zigzag pattern and recollecting the two parts by welding, to form a castellated beam with hexagonal openings, or more increasing of section depth by placing expansion plates between these parts to fabricate castellated beam with octagonal openings. The main advantage of fabrication this type of beam is increasing section depth, which leads to improve its behavior compared with original beam. So as to use the castellated beams with expansion plates into its best advantages, the failure modes due to web opening need to be avoided. This paper aims to provide added strength of castellated beams with expansion plates by using two types of stiffeners (circular and octagonal ring) around octagonal web opening. This strengthening technique was suggested to reinforce web portion and avoid failure due to web-post buckling. Standard (IPE140) were chosen as a parent section for fabricating four specimens as well as original beam. The important observation through this study is that using ring stiffener leads to have significant effect on ultimate strength. In case of using circular ring stiffeners, the ultimate strength was increased up to (188%) compare with original beam. Whereas the increment in ultimate load at using octagonal ring stiffeners was reach to about (77.6%). Consequent, using circular ring stiffeners to reinforced this type of beams was more efficient than using octagonal ring stiffeners. It is worthwhile to mention that it can be used (37%) as additional steel material (Expansion plates and ring stiffeners) for increasing the load capacity of castellated beams up to (288%) relative to original beam.

1- Introduction:

In recent years, with the technology of steel structures progress, more types of steel sections were developed to improve mechanical properties of this type of structural and also to obtain sections that allow usage for more aesthetic applications by satisfying the architectural needs. The perforated section, such as, castellated and cellular members are one of the best example of newly developed section. During the second world war, castellated beams with hexagonal opening that fabricated by cutting the web portion of I-section along its centerline and shafted then re-jointed by welding as shown in Figure 1 were one such improvement occurred in built-up structural members to modify hot rolled I-section steel beams by increasing their resistance against bending due to the increased depth of section[1]. Some design cases made its advantageous to expanse the depth of section more by placing expansion plates between the two halves of the tee sections which are called "Expansion Plates" as shown in Figure 2.

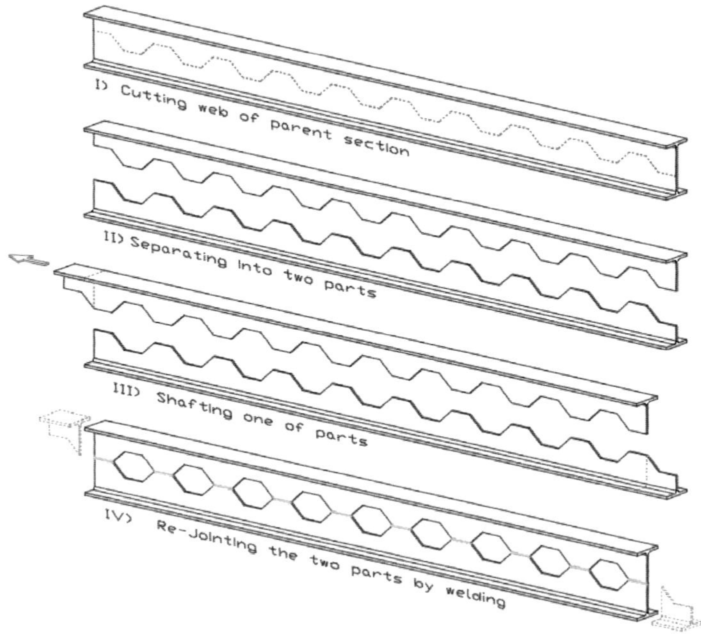

Figure 1. Fabrication of Castellated Beam with Hexagonal opening (CB-H).

In additional to increase depth of castellated beams, these processes enhance area of web opening which turns into octagonal shape and can be used to pass the accommodate services. Also, the elastic stiffness and bending strength of castellated beam about its principal axis are heightened. The structural properties allow castellated steel beam with expansion plate (CB-O) to be used in long span applications with moderate loading conditions for supporting floors and roofs.

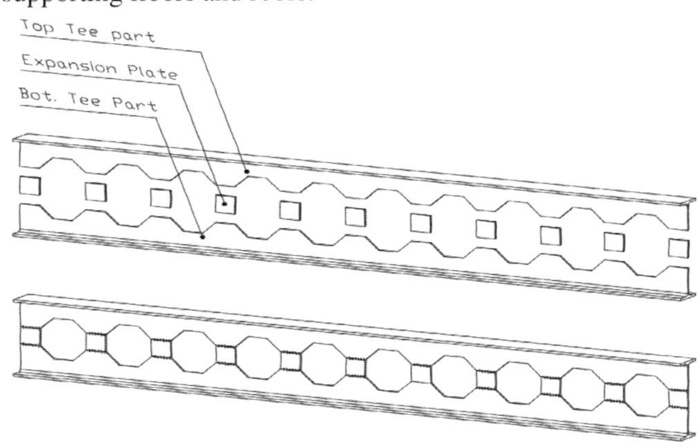

Figure 2. Fabrication of Castellated Beam with expansion plate (Octagonal opening) (CB-O). [2]

In order to use the castellated beams with expansion plates (CB-O) into its best advantages, the failure modes due to large web opening need to be avoided. The structural behavior of the castellated beams are essentially differs from solid web beam as with the existence of openings in the web portion which may cause various failure modes such as Vierendeel mechanism and web-post buckling due to shear. These models of failure do not occur with plain webbed beams since they are a direct result of the different way in which shear is transferred through the perforated web but failure by the formation of a flexural mechanism or by lateral-torsional instability are entirely similar to the equivalent modes of solid web beams. All these failure modes happen due to the instability of the web portion. A proper strengthening process will avoid these types of failures and improves the performance of (CB-O).

The present work aims to improve structural behavior of castellated steel beams with expansion plates (CB-O) by welding two types of stiffener around octagonal web opening (Circular and octagonal Ring) as shown in Figure 3 which can be used to strengthening web portion and avoid failure due to Vierendeel mechanism or web-post buckling. In order to understand performance of cases, five experimental specimens are tested in Engineering Collage Laboratories of Kufa University.

a) Circular Ring stiffeners. b) Octagonal Ring stiffeners.

Figure 3. Reinforced Octagonal web opening of castellated beam with expansion plate by : (Circular ring stiffeners. & Octagonal ring stiffeners).

2- Previous Studies and Research Gap :

During second war castellated beams with hexagonal opening were one of the important modification happened in steel members as built up section from hot rolled I-section by increasing the depth of it which lead to increase the stiffness and the ultimate strength. In (2011), Tsavdaridis and D'Mello, studied the structural behavior of castellated beams with webs of closely spaced openings by testing seven experimental specimens. The purpose of their work was to examine the potential impacts of various shapes and sizes of openings on the failure mode and ultimate strength of castellated beams. Finite element models of castellated beams with hexagonal, circular and elongated web openings were developed and analyzed by using ANSYS v11.0 software; The results were compared with seven experiments test specimens. They concluded that the maximum shear stresses spread from the mid-height of the web-post nearer to the flanges in the direction of the axial forces [3]. Soltani et al. in (2012), developed a numerical model to investigate the structural behavior of castellated beams with hexagonal and octagonal openings up to failure. Their principal objective was to study the web post buckling. Their work showed that the use of expansion plates could make castellated steel beams more vulnerable to occur web-post buckling. Also, they concluded that depth of these expansion plates must not be more than (15%) of the overall hexagonal opening depth [4]. In (2015), Tsavdaridis and Galiatsatos investigated closely spaced cellular beams with double concentric transverse stiffeners by using finite element analysis. From the results, they noted that Vierendeel failure occurred more frequently for very closely spaced sections. However, as the spacing increased, the influence of the transverse stiffener to strength of the section was decreased and web post buckling failure happened more often. Moreover, they suggested a maximum spacing between the web openings [5].

In the previous literatures, it can be noted that there are a lack in experimental study to investigate the structure behavior of (CB-O) as shown in Figure 2. Predominantly, the castellated beams with hexagonal openings and cellular beams with circular openings are widely used in the constructions due to ease of the manufacturing. In addition, most of numerical researches indicated that the use of expansion plates could make castellated steel beams more vulnerable to occur web-post buckling in web segment. Few studies have been carried out to strength the octagonal opening of castellated beams with expansion plates in order to avoid web post buckling. Also, It can be noted that there is no regulated knowledge of how (CB-O) would behave if a stiffeners are placed around the octagonal openings. This research aims to examine the effectiveness of reinforcing (CB-O) by using stiffeners around edge of octagonal openings to improve its structural behavior and avoid web post buckling failure in web segment at high load.

3- Experimental Program:

In this study , hot rolled I-sections (IPE 140) were chosen as a parent section to fabricating four specimens of castellated beams as well as a control beam. For all specimens, flange thickness, web thickness and span length were (t_f=6.9 mm), (t_w=4.7mm) and (L=1700mm) respectively. A plasma CNC machine was used to cut the parent I-section along its web for making the zigzag pattern, then rejoining the two halves together using electrode welding to make a castellated beam with hexagonal opening or adding the square plate (as expansion plate) and welding all parts together as shown in Figure 4. The material properties were obtained from tension tests on flat tensile specimens according to ASTM A370[7].

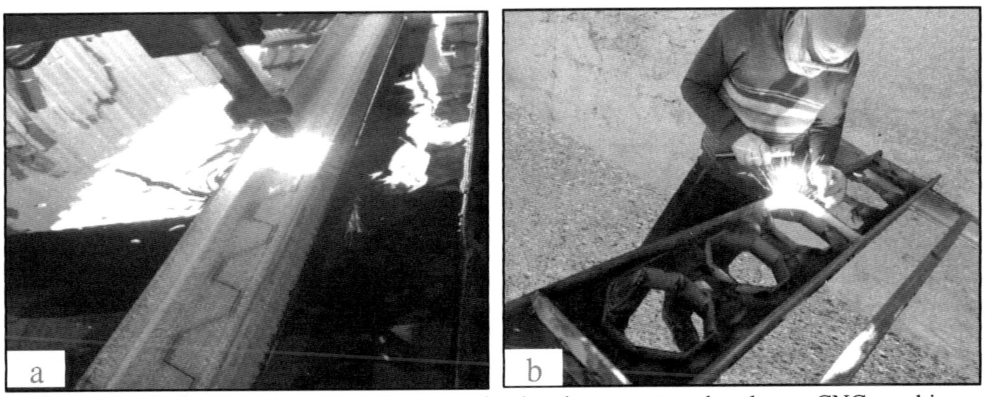

Figure 4 . a) Cutting parent I-section to make the zigzag pattern by plasma CNC machine
b) Rejoining the two halves together and placing the ring stiffeners

As aforementioned, this work focused on using two types of stiffener (circular and octagonal rings) that it will placed around octagonal web opening of (CB-O) to prevent failures due to Vierendeel failure and web-post-buckling [8]. Circular and octagonal rings are used as stiffeners around edge of octagonal web opening of (CB-O). The circular ring stiffeners insulated by welding at tangent point of each sides of octagonal web opening edges as shown in Figure 3(a). But the octagonal ring stiffeners placed by welding at along length of each sides of octagonal web opening edges as shown in Figure 3(b). In order to obtain castellated beam with hexagonal opening (CB-H) that have equal side, cutting angle which used to cut the web of parent I-section was (45°) to achieve equal edge sides of hexagonal openings. The purpose of fabricating the castellated beam with equal edges hexagonal opening was to obtain octagonal castellated beam with equal edge sides opening after adding the expansion plate. The dimensions and details of experimental cases are shown in Figure 5 and Figure 6. The first specimen (CB-00) was parent beams (IPE140) which used as control beam as shown in Figure 5(a). Whilst, Figure 5(b) shows dimension and details of hexagonal castellated beam (CBC-01) which was fabricated by cutting angle of web (45°) in order to obtain equal side of hexagonal web opening.

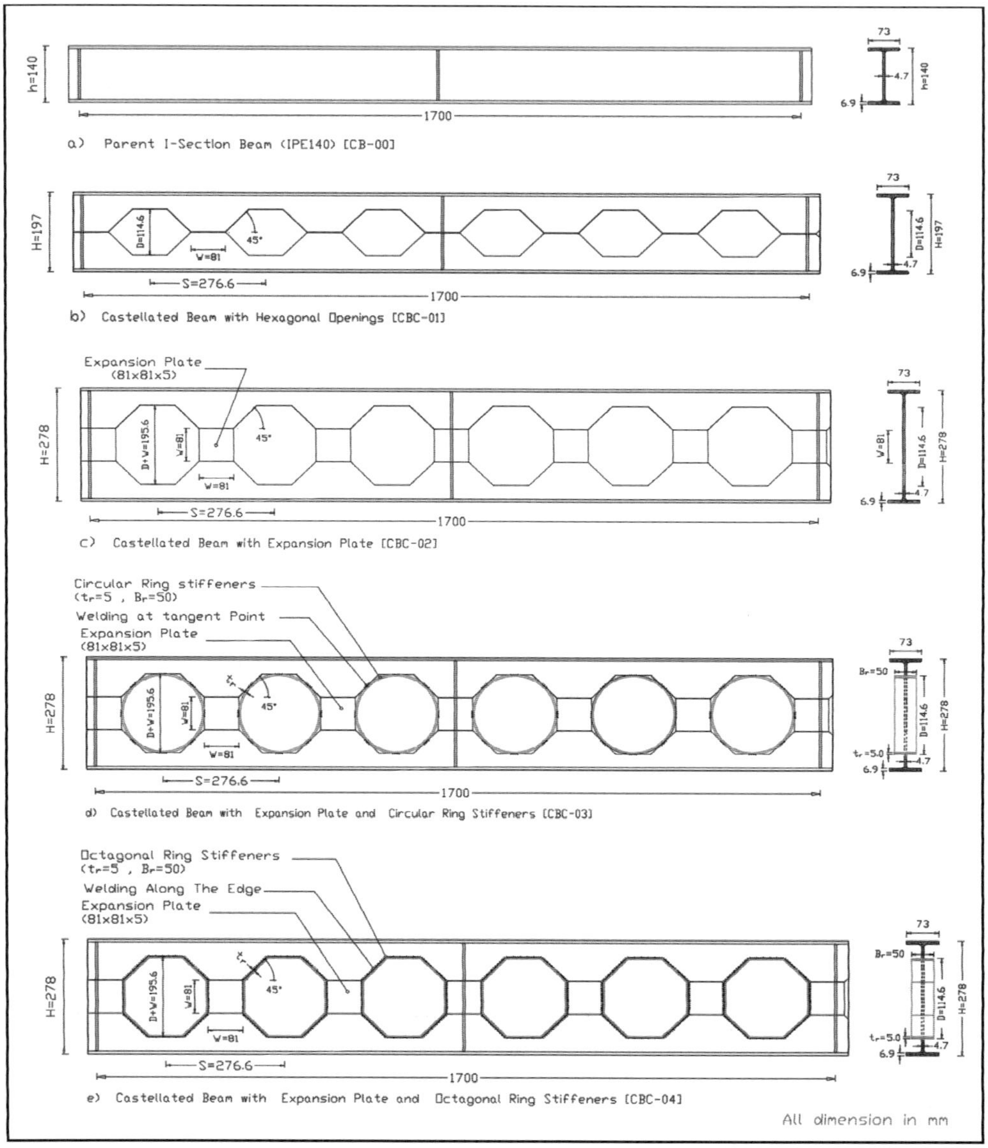

Figure 5. Dimensions and notations of all specimens.

This specimen have total depth (H=197mm) and expansion depth ratio (H/h=1.4). The purpose of fabricating the castellated beam with equal edges hexagonal opening was to obtain castellated beam with equal sides octagonal web opening after placing square expansion plates as shown in Figure 5(c). Consequently, specimen (CBC-02) was octagonal castellated beam with square expansion plates (81mm x 81mm x 5mm) which placed between upper and lower tee section to increase its depth up to (H=278mm) and expansion depth ratio (H/h =1.985 ≈2.0). In order to test the effectiveness of reinforcing (CB-O) using stiffeners around edge of octagonal web openings, specimen (CBC-03) and specimen (CBC-04) that have same dimensions of specimen (CBC-02) are fabricated as shown in Figure 4(d) and Figure 5(e). In specimen (CBC-03), the shape and geometry of equal sides octagonal web openings allowed to use circular ring as stiffeners to reinforce (CB-O). The circular ring stiffeners that have thickness (t_r=5mm) and width (B_r=50mm) are welded at tangent point with each side of octagonal web opening as shown in Fig.(3-a). So as to strengthening specimen (CBC-04),octagonal ring stiffeners that have thickness (t_r=5mm) and width (B_r=50) are placed by welded all sides of the stiffeners with octagonal web opening as shown in Fig.(3-b). Dimensional and Material Properties of Specimens are shown in Table 1.

Figure 6. a) Hexagonal castellated beam (CBC-01), **b)** Octagonal castellated beam (CBC-02) **c)** Octagonal castellated beam with circular ring stiffeners (CBC-03), **d)** Octagonal castellated beam with octagonal ring stiffeners (CBC-04)

Table 1. Dimensional and Material Properties of Specimens

Specimens	Open Shape	H	$\frac{H}{h}$	$\frac{L}{S}$	$\frac{D}{h}$	Ring Stiffeners			Material Properties (MPa)		
						Shape	B_r	t_r	f_y	f_u	E_s
CB-00	Plain	D=140	1.00	–	–	–	–	–			
CBC-01	Hexagonal	197.3	≈ 1.40	6	0.82	–	–	–			
CBC-02	Octagonal	278.3	≈ 2.00	6	0.82	–	–	–	279	432	2.01×10^5
CBC-03	Octagonal	278.3	≈ 2.00	6	0.82	Circular					
CBC-04	Octagonal	278.3	≈ 2.00	6	0.82	Octagonal					

Where : *H: overall depth of specimen.* ,
h: parent I- section height.
D: opening height (114.6mm).
w: clear distance between two successive openings (81.0mm).
S: spacing between two openings center to center (276.6mm).
t_r: thickness of ring stiffeners.
B_r: width ring stiffeners.

In order to ensure load distribution from concentrated force and end supports and to avoid failure due to web buckling by compression force, transverse stiffeners were utilized at supports and mid-span. All specimens were simply supported and lateral supports are used to prevent lateral torsional buckling. The specimens were set in a test machine with adequate care taken to ensure that beams were correctly placed in the test machine and the mid-span of the beam was in line with the hydraulic jack centerline as shown in Figure 7. The applied load produced using load machine of (1000 kN) capacity at Engineering Collage Laboratories of Kufa University

4. Results Discussion:

Control specimen (CB-00): Hot rolled I-sections (IPE 140) which was chosen as a parent section was tested to compare the experimental results of this beam (as control beam) with castellated beam with and without ring stiffeners. The results obtained from experimental test demonstrate that control beam failed by flexural mechanism which mean the sections above and below the neutral axis yielded in compression on upper section and tension on lower section. The testing of control beam and its failure mode presented in Figure 8. When the applied load was reach to about (62.0kN), first sign of yielding was seen at upper and lower flange with vertical deflection (9.2mm). Whilst, the maximum vertical deflection recorded was (47.0mm) at ultimate load (78.8 kN).

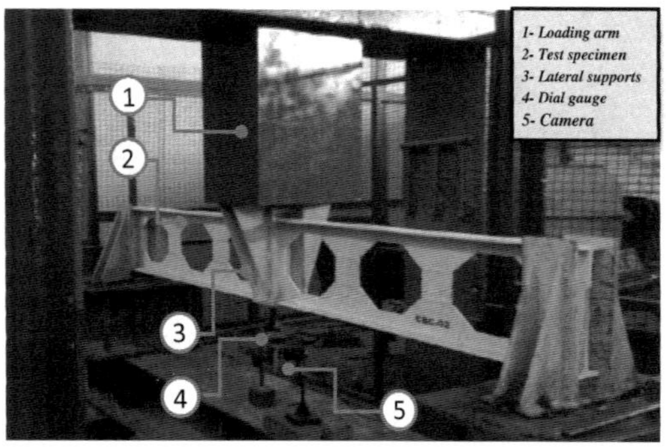

Figure 7. Details of Testing Setup.

Figure 8. Experimental setup of control beam (CB-00) and failure mode

Specimen (CBC-01): This specimen (Hexagonal castellated beam) failed by plastic hinges formation at four corners of the panel close to applied point load and in the region of the beam where both moment and shear are present. The failure mode of this specimen (castellated beam with hexagonal opening) is observed in Figure 9 and Figure 10 shows the sequence of yielding point at failure zone. In Figure 11 the load-deflection curve for this specimen is presented and compared with results of control specimen (CB-00). As the applied load was reach to about (77.0kN) at point (I), first sign of yielding was seen along the line marked (1) in Figure 10.

Figure 9. Failure model of specimen (CBC-01)

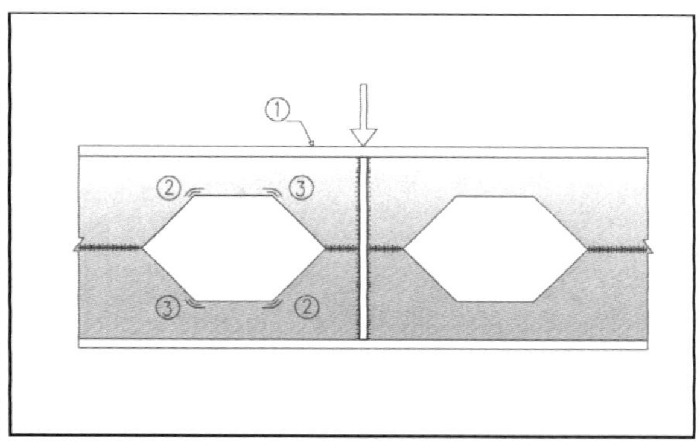

Figure 10. Sequence of yielding in specimen (CBC-01)

While the applied load was being increased to (81.0 kN) at point (II), the yielding started firstly at the corners of hexagonal opening marked (2) and then at opposite corners marked (3). Yielding at four corners of hexagonal opening of this specimen became prominent when the applied load was gradually raised and it can be shown that a well-defined of Vierendeel mechanism had occurred when the applied load reached to maximum value at (87.6kN) at point (III). The castellated beam sustained the maximum load only briefly before web-post buckling occurred in the first panel on the lift hand side of the applied load. Consequently, the beam started to unload shortly thereafter as shown in region between point (III) to failure load (86.5kN) at point (IV) in Fig.(6). It is worth to mention that the ultimate load of specimen (CBC-01) which is equal (87.6 kN) increases by (11.1%) when comparing with the ultimate load of control beam without adding any material.

***Specimen (CBC-02)*:** In this castellated beam, its depth was increased by adding expansion plate. Figure 12 shows the load-deflection curve of castellated beam with expansion plate (octagonal web opening). It can be noted that first sign of yielding was observed, on the inside face of the compression flange near the applied load, when the load was (89.0kN). The applied load was gradually raised to ultimate load. After the applied load reach to value (120kN), it was observed that unloading started. Unloading rate was gradual at the beginning because the deformation in web but at (96.0kN) web-post buckling failure occurred at the panel close to the point load as shown in Figure 13.

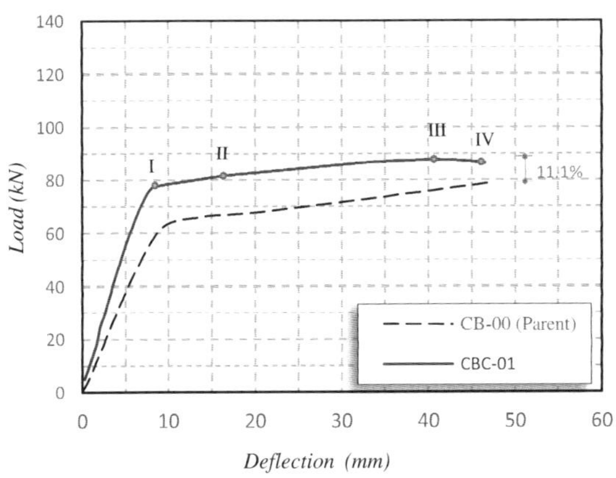

Figure 11. Load-deflection curve of (CBC-01) and (CB-00).

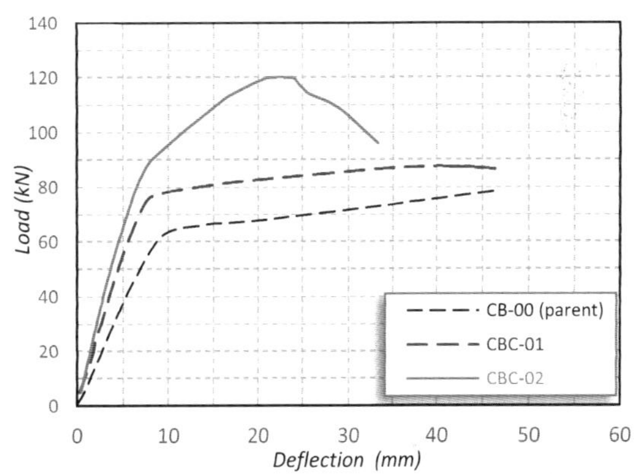

Figure 12. Load-deflection curve of (CBC-02), (CBC-01) and (CB-00).

Figure 13. Failure mode (Web-Post Buckling) in specimen (CBC-02)

The ultimate load of specimen (CBC-02) which was equal (120.0 kN) increased by (52.2%) and (37%) when comparing with the ultimate load of control specimen (CB-00) and hexagonal castellated beam (CBC-01) respectively. But, the ductility index of this beam which was equal (4.0) decreased to (38%) as comparing to ductility index of (CBC-01) which was (5.5). It is worth to mention that the ductility index is the ratio of deflection at failure load of beam to the deflection at yielding point. In general, a high ductility index indicates that a beam tested are capable of undergoing large deformations prior to failure. Figure 14 is explained the web-post buckling due to shear force. It can be noted that the horizontal shear force (V_{wp}) which was acting along the welded joint stress between expansion plate and web-post in bending, caused tensile stress along parts (AB and GH) and compression along parts (CD and EF). In the web-post buckling, the compressed parts tend to move away from the longitudinal plane of the web while the tensioned parts tend to remain in the starting position. This mode of failure occurs usually in inelastic regime with a significant yielding of the section. In general, increasing the height of castellation beam causes the web-post buckling failure which as a results lead to rapid failure [9][10].

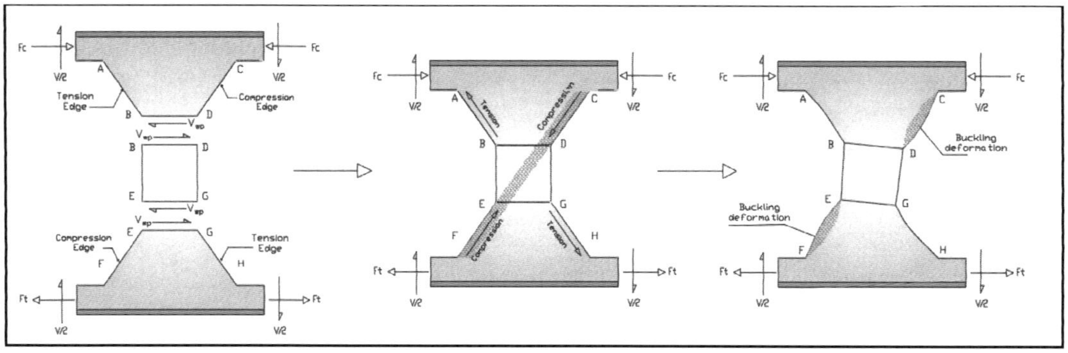

Figure 14. Web-Post buckling due to shear force.

Specimen (CBC-03): In this test, investigating the efficiency of using circular ring stiffeners for strengthening (CB-O) was searched. In previous literature, different type of stiffeners like transvers stiffeners are tested to improve structure behavior of octagonal castellated beam but ring stiffeners are not used in experimental test. Consequently, the circular ring stiffeners were placed around the octagonal web opening to avoid web-post buckling which was occurred due to web opening as shown Figure 3 (a).

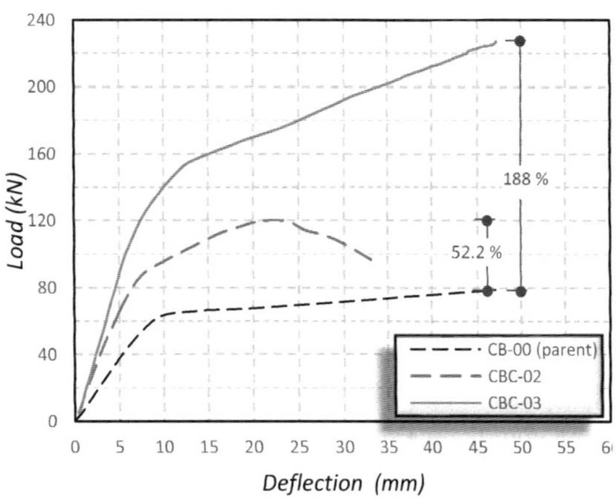

Figure 15. Load-deflection curve of (CBC-03), (CBC-02) and (CB-00)

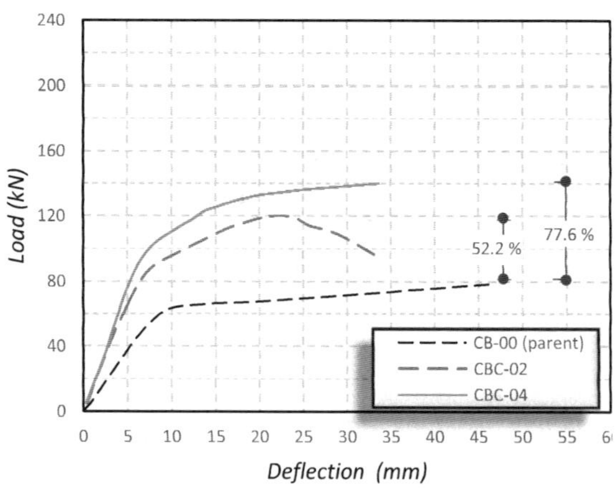

Figure 16. Load-deflection curve of (CBC-03), (CBC-02) and (CB-00).

It can therefore be observed that using circular ring stiffeners seem to have drastic effect on the elastic stiffness and the ultimate strength of octagonal castellated beams as well as considerably increases their ductility. It appears from Figure 15 that there is significant increase in ultimate strength of specimen (CBC-03) which was reach to (227.0 kN) increased by (188 %) and (89%) as comparing with the ultimate load of control beam and octagonal castellated beam without stiffeners (CBC-02) respectively. Moreover, the ductility index of (CBC-03) which was equal (5.9) increased to (44.6%) when comparing to ductility index of (CBC-02) which was (4.08). The important notation through this test is that the proposed strengthening technique (Circular ring stiffener) prevented the web-post buckling failure due to shear force and improved behavior of octagonal castellated beams as shown in Figure 15 and Figure 17.

(a): (CBC-03) at Failure load (227 kN) (b): (CBC-02) at Failure load (120 kN)

Figure 17. Modes of failure of specimen (CBC-02) and (CBC-03) at ultimate load.

Specimen (CBC-04): In this specimen , other strengthening technique was suggested to avoid web-post buckling in castellated beam with expansion plate. In specimen (CBC-4), octagonal ring stiffeners are provided to reinforce the octagonal web opening see Figure 3(b), instead of using circular ring stiffeners, that were used in previous specimen (CBC-03). The Load-deflection curve of (CBC-04) and (CBC-02) as well as control beam is shown in Figure 16 to observe effect of using octagonal ring stiffeners. It can be noted that using octagonal stiffener leads to have significant effect on the elastic stiffness and the ultimate strength. Moreover, it increases ductility of beams and prevents unloading which occurred due to web post buckling in specimen (CBC-02). In case of using octagonal ring stiffeners, the increment in ultimate load and ductility index were about (77.6%) and (9.8%) respectively. Figure 18 shows model failure of octagonal castellated beam with octagonal ring stiffeners at ultimate load (140kN) and comparison with model failure of octagonal castellated beam without ring stiffeners.

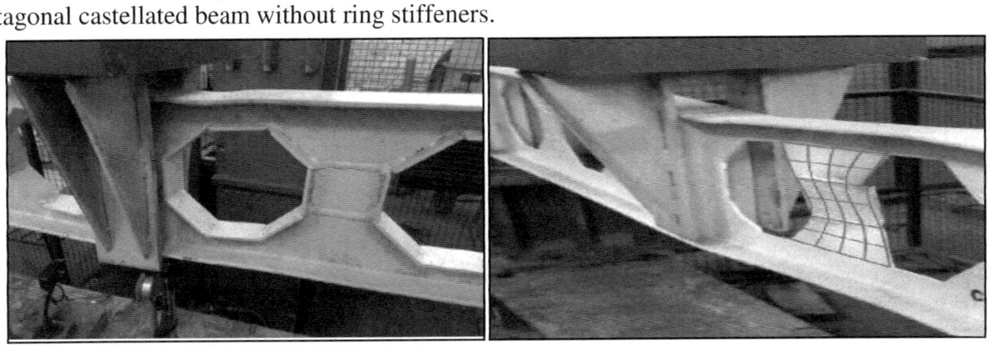

(a): (CBC-04) at Failure load (140 kN) (b): (CBC-02) at Failure load (120 kN)

Figure 18. Failure Modes of specimen (CBC-02) and (CBC-04) at ultimate load.

5- Economic Feasibility for Using Ring Stiffeners in Octagonal Castellated Beam

One of the most important advantages of castellated beams is their economy, the economy depends on many factors such as loading, depth, and span length requirements. The researchers continuously attempt to find new ways to decrease the cost of structures, circular ring stiffeners around octagonal web openings could provide a great cost saving. In this research, hexagonal castellated beam doesn't consume any additional steel material but the increasing in strength compared with parent section was limited. While (CB-O) consumes amount of steel about (7.5%) with increase in strength about (11.1%) compared with parent section. In spite of octagonal castellated beam which was reinforced by adding circular and octagonal ring stiffeners consume a considerable amount of steel about (37%) and (39%) of original weight respectively but with great increasing in strength reach up to (188%) and (77.6%) compared with ultimate load capacity of original beam. Figure 19 shows relationship between ratio of weight and ultimate load of all specimens to original beam. The important observation through this study is that the proposed strengthening techniques prevents web-post buckling due to shear force occurred in octagonal castellated beam and improves its structural behavior. In this figure, the maximum ultimate load was noted in using circular ring stiffeners which reach to (227) about (288%) respectively to original beam with adding amount of steel (square expansion plates and circular ring stiffeners) about (37%) compared with total weight of original beam.

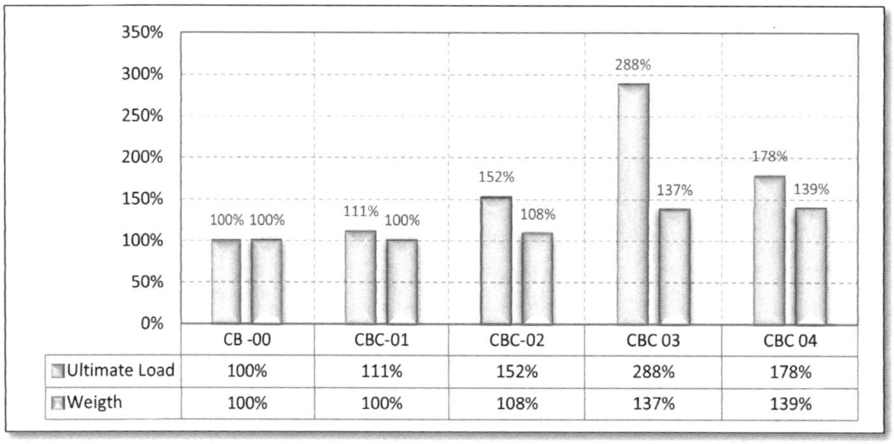

Figure 19. Economic feasibility of using ring stiffeners with castellated beams

6. CONCLUSION

From this study, it could be concluded:

1- The ultimate load of octagonal castellated beam with expansion plates (CBC-02) which was equal (120.0 kN) increased by (52.2%) and (37%) when comparing with the ultimate load of control specimen (CB-00) and hexagonal castellated beam (CBC-01) respectively. But, the ductility index of this beam which was equal (4.0) decreased to (38%) as comparing with ductility index of (CBC-01) which was (5.5).

2- It can be found that using ring stiffener leads to have significant effect on the stiffness and the ultimate strength. Moreover, it increases ductility of beams and prevents unloading which occurred due to web post buckling in (CB-O). In case of using circular ring stiffeners, the increment in ultimate load and ductility index were about (188%) and (44.6%) respectively. But, using octagonal ring stiffeners, the increment in ultimate load and ductility index were about (77.6%) and (9.8%) respectively.

3- It can be seen that using circular ring stiffeners to reinforced octagonal castellated beam with expansion plates was more efficient than using octagonal ring stiffeners. In case of using circular ring instead of octagonal ring as stiffeners, the differential of increment for the ultimate load reach to (110.4%) relative to original beam.

4- Providing circular ring around octagonal web opening of castellated beam with expansion plate (CB-O) can be improved the stiffness response and ultimate load capacity of these beams up to (288 %) relative to ultimate load of the parent I-section with using only (37%) of origin weight as additional material (Expansion plates and circular ring stiffeners).

References:

[1] Ellobody E (2011). Interaction of buckling modes in castellated steel beams. *Journal of Constructional Steel Research*, **67** (5), 814–825.

[2] Al-thabhawee H W and AL-kannoon M A (2018). Improving Behavior of Castellated Beam by Adding Spacer Plat and Steel Rings. *Journal of University of Babylon, Engineering Sciences* , **26** (4) , 331 – 344.

[3] Tsavdaridis K D and D'Mello C (2011). Web Buckling Study of The Behavior And Strength of Perforated Steel Beam With Different Novel Web Opening Shapes. *Journal of Constructional Steel Research*, **67** , 1605-1620.

[4] Soltani et al. (2012). Nonlinear FE analysis of the ultimate behavior of steel castellated beams. *Journal of Constructional Steel Research* **70**, 101–114.

[5] Tsavdaridis K D and Galiatsatos G (2015). Assessment of Cellular Beams with Transverse Stiffeners and Closely Spaced Web Openings. *Thin-Walled Structures*, **94** , 636-650.

[6] Patil S A and Kumbhar P D (2016). Comparative Study of Transverse Stiffeners and Stiffeners along the Opening Edge used for Castellated Beam. *International Journal of Innovative Research in Science, Engineering and Technology*, **5**, 8516- 8522.

[7] ASTM E8M-99. Standard Test Methods for Tension Testing of Metallic Materials. [Metric], ASTM.

[8] Kerdal D and Nethercott D A (1984). Failure Modes for Castellated Beams. *Journal of Constructional Steel Research*, **4**, 295–315.

[9] Liu T and Chung K (2003). Steel beams with large web openings of various shapes and sizes: finite element investigation. *Journal of Constructional Steel Research*, **59** (9), 1159–1176.

[10] Jamadar A M and Kumbhar P D (2015). Parametric Study of Castellated Beam with Circular and Diamond Shaped Openings. *International Research Journal of Engineering and Technology,* **2**, 715–722.